Current Cancer Research

For other titles published in this series, go to
www.springer.com/series/7892

Marcelo G. Kazanietz
Editor

Protein Kinase C in Cancer Signaling and Therapy

Springer

Editor
Marcelo G. Kazanietz, Ph.D.
Department of Pharmacology
University of Pennsylvania School of Medicine
1256 Biomedical Research Building II/III
421 Curie Blvd.
Philadelphia, PA 19104-6160
USA
marcelog@upenn.edu

ISBN 978-1-60761-542-2 e-ISBN 978-1-60761-543-9
DOI 10.1007/978-1-60761-543-9
Springer New York Dordrecht Heidelberg London

Library of Congress Control Number: 2010929851

Printed on acid-free paper

Humana Press is a part of Springer Science+Business Media (www.springer.com)

Acknowledgements

The field of PKC and cancer expanded dramatically in the last decades, and there are many people that greatly contributed to our understanding of the basic regulation of these kinases and their implication in cancer progression. I owe a lot to those pioneers in the field. There are many, but my special recognition goes to the late Dr. Yasutomi Nishizuka, Dr. Peter M. Blumberg, Dr. Peter J. Parker, Dr. Alexandra Newton, Dr. Susan Jaken, and Dr. Daria Mochly-Rosen. As a post-doctoral researcher in the early '90s I avidly read their papers, which were a source of inspiration for my own career.

I am heartily thankful to Dr. Peter M. Blumberg, an amazing and generous mentor. Peter, you are right: "Mentoring is for life".

A special thanks to the authors of the different chapters in this book. They are a group of extraordinary scientists who made seminal contributions to the field of PKC and cancer.

I am grateful to Dr. Wafik El-Deiry for his invitation to participate in this outstanding book series.

I owe a great deal to my current and past laboratory members at the University of Pennsylvania. Your insights have challenged and strengthened me.

I want to acknowledge the editors at Springer for advice and excellent editorial work and support.

Above all, I want to thank my family for their love, support, and patience. Nati, my wife and love of my life, and my sons Julian and Diego, are my source of inspiration.

Acknowledgments

Contents

Contributors

Antonia Avila-Flores
Department of Immunology and Oncology, Centro Nacional de Biotecnología/CSIC, Madrid, Spain

Alakananda Basu
Department of Molecular Biology & Immunology, University of North Texas Health Science Center, Fort Worth, TX, USA

Jennifer D. Black
Department of Pharmacology and Therapeutics, Roswell Park Cancer Institute, Buffalo, NY, USA

Peter M. Blumberg
Laboratory of Cancer Biology and Genetics, Center for Cancer Research, National Cancer Institute, National Institutes of Health, Bethesda, MD, USA

Andrew P. Bradford
Department of Obstetrics and Gynecology, School of Medicine, Anschutz Medical Campus, University of Colorado Denver, CO, USA

Chaya Brodie
William and Karen Davidson Laboratory of Cell Signaling and Tumorigenesis, Department of Neurosurgery, Hermelin Brain Tumor Center, Henry Ford Hospital, Detroit, MI, USA
Mina and Everard Goodman Faculty of Life-Sciences, Bar-Ilan University, Ramat-Gan, Israel

Silvia Carrasco
Department of Immunology and Oncology, Centro Nacional de Biotecnología/CSIC, Madrid, Spain

Mitchell F. Denning
Department of Pathology, Cardinal Bernardin Cancer Center, Loyola University Chicago, Maywood, IL, USA

Maria T. Diaz-Meco
Department of Cancer and Cell Biology, University of Cincinnati College of Medicine, Cincinnati, OH, USA

Alan P. Fields
Department of Cancer Biology, Mayo Clinic College of Medicine, Jacksonville, FL, USA

Tamas Geczy
Laboratory of Cancer Biology and Genetics, Center for Cancer Research, National Cancer Institute, National Institutes of Health, Bethesda, MD, USA

Verline Justilien
Department of Cancer Biology, Mayo Clinic College of Medicine, Jacksonville, FL, USA

Marcelo G. Kazanietz
Department of Pharmacology, University of Pennsylvania School of Medicine, Philadelphia, PA, USA

Noemi Kedei
Laboratory of Cancer Biology and Genetics, Center for Cancer Research, National Cancer Institute, National Institutes of Health, Bethesda, MD, USA

Jeewon Kim
Department of Chemical and Systems Biology, Stanford University, School of Medicine, Stanford, CA, USA
Stanford Comprehensive Cancer Center, Stanford University, School of Medicine, Stanford, CA, USA

Christer Larsson
Center for Molecular Pathology, Malmö University Hospital, Lund University, Malmö, Sweden

Nancy E. Lewin
Laboratory of Cancer Biology and Genetics, Center for Cancer Research, National Cancer Institute, National Institutes of Health, Bethesda, MD, USA

Stephanie L. Lomonaco
William and Karen Davidson Laboratory of Cell Signaling and Tumorigenesis, Department of Neurosurgery, Hermelin Brain Tumor Center, Henry Ford Hospital, Detroit, MI, USA

Gry Kalstad Lønne
Center for Molecular Pathology, Malmö University Hospital, Lund University, Malmö, Sweden

Patricia S. Lorenzo
Cancer Research Center of Hawaii, Natural Products & Cancer Biology Program, University of Hawaii at Manoa, Honolulu, HI, USA

Sofia D Merajver
Division of Hematology and Oncology, Department of Internal Medicine, University of Michigan, Ann Arbor, MI, USA

Isabel Mérida
Department of Immunology and Oncology, Centro Nacional de Biotecnología/CSIC, Madrid, Spain

Daria Mochly-Rosen
Department of Chemical and Systems Biology, School of Medicine, Stanford University, Stanford, CA, USA

Jorge Moscat
Department of Cancer and Cell Biology, University of Cincinnati College of Medicine, Cincinnati, OH, USA

Alexandra C. Newton
Department of Pharmacology, University of California at San Diego, La Jolla, CA, USA

Mary E. Reyland
Department of Craniofacial Biology, School of Dental Medicine, Anschutz Medical Campus, University of Colorado Denver, CO, USA

Natalia A Riobo
Department of Biochemistry and Molecular Biology, Thomas Jefferson University, Philadelphia, PA, USA

Devin T Rosenthal
Division of Hematology and Oncology, Department of Internal Medicine, University of Michigan, Ann Arbor, MI, USA

Enrique Rozengurt
Division of Digestive Diseases and CURE, Department of Medicine, Digestive Diseases Research Center, UCLA School of Medicine, David Geffen School of Medicine, University of California, Los Angeles, CA, USA

Juan Tao
Laboratory of Cancer Biology and Genetics, Center for Cancer Research, National Cancer Institute, National Institutes of Health, Bethesda, MD, USA

Andrea Telek
Laboratory of Cancer Biology and Genetics, Center for Cancer Research, National Cancer Institute, National Institutes of Health, Bethesda, MD, USA

Alex Toker
Department of Pathology, Beth Israel Deaconess Medical Center, Harvard Medical School, Boston, MA, USA

Lauren Van Wassenhove
Division of Hematology and Oncology, Department of Internal Medicine, University of Michigan, Ann Arbor, MI USA

Ajit K. Verma
Department of Human Oncology, School of Medicine and Public Health, University of Wisconsin, Madison, WI, USA

Q. Jane Wang
Department of Pharmacology and Chemical Biology, University of Pittsburgh School of Medicine, Pittsburgh, PA 15261, USA

Lei Xiao
Department of Anatomy and Cell Biology, University of Florida, Gainesville, FL, USA

Dazhi Yang
Laboratory of Cancer Biology and Genetics, Center for Cancer Research, National Cancer Institute, National Institutes of Health, Bethesda, MD, USA

Kiyotsugu Yoshida
Medical Research Institute, Tokyo Medical and Dental University, Tokyo, Japan

Part I
Regulation of PKC Isozyme Function: From Genes to Biochemistry

Chapter 1
Protein Kinase C in Cancer Signaling and Therapy: Introduction and Historical Perspective

Alex Toker

Keywords Protein Kinase C • Diacylglycerol • Phorbol ester • Signal transduction

The Protein Kinase C (PKC) signal relay pathway represents one of the best understood mechanisms by which extracellular signals elicit cellular responses through the generation of lipid second messengers. Thirty years have elapsed since the late Yasutomi Nishizuka discovered an enzymatic activity that was dependent on calcium and phosphatidylserine and activated by diacylglycerol (DG) (Takai et al. 1979a). During the 1980s, biochemical studies focused primarily on elucidating the mechanisms by which DG, calcium and cofactors such as phosphatidylserine control PKC activity. During this decade, it also became evident that multiple isoforms of PKCs exist in mammals and other organisms. The 1990s saw a flurry of research which identified the mechanisms of PKC phosphorylation by autophosphorylation and also by upstream kinases. Studies also revealed substrates of PKC that transduce the lipid signal, leading to the realization that PKCs control a multitude of cellular responses and phenotypes in response to virtually all cellular agonists. Toward the end of the millennium genetic studies using homologs recombination to knock out PKC isozymes in various organisms began to unravel the specific functions of PKCs in physiology. It was not until the third decade of PKC research that all of the information that had been collected over these years could be translated into therapeutic benefit, whereby small-molecule inhibitors were developed for specific therapeutic interventions in human pathophysiologies. Much of this work is ongoing and clinical trials using PKC antagonists are yielding exciting and potentially fruitful results.

This chapter is devoted to a review of the key mechanisms that control PKC activity through lipid second messengers, phorbol esters, phosphorylation, and protein interactions. When considering the key findings in the history of PKC, it is essential

A. Toker (✉)
Department of Pathology, Beth Israel Deaconess Medical Center, Harvard Medical School, Boston, MA, USA
e-mail: atoker@bidmc.harvard.edu

M.G. Kazanietz (ed.), *Protein Kinase C in Cancer Signaling and Therapy*, Current Cancer Research, DOI 10.1007/978-1-60761-543-9_1,

to recognize the landmark studies by Hokin and Hokin who in the mid-1950s were the first to recognize that extracellular agonists in the form of acetylcholine could stimulate the incorporation of radiolabeled phosphate into phospholipids (Hokin and Hokin 1953, 1954). These studies gave birth to an entirely new field of lipid signaling, and subsequent studies by Michell showed that the phospholipase-mediated hydrolysis of minor membrane phosphoinositides such as $PI45P_2$ is responsible for the release of two key second messengers, DG and inositol trisphosphate (IP_3) (Lapetina and Michell 1973a, b). At the same time, Berridge, Irvine, and Schulz revealed that one byproduct of this reaction, soluble inositol phosphates such as IP_3, elicit the release of calcium from intracellular stores (Streb et al. 1983). In the late 1970s, Nishizuka and colleagues had identified a highly active protein kinase activity from rat brains that was sensitive to magnesium, and so termed it PKM. Realizing that they were probably dealing with a proteolytic fragment of a protein kinase, they subsequently purified and characterized the holoenzyme and found that it was activated by phospholipids such as phosphatidylserine. Because they found that this activity was also enhanced by the calcium-activated protease calpain, they termed it PKC (Takai et al. 1979a, b). However, they also recognized that crude preparations of phospholipids were more effective at activating this new enzymatic activity than phosphatidylserine alone, and their quest to identify the molecular species within these crude preparations informed them that diacylglycerol serves to activate PKC. The field had at this point come full circle with the elegant mechanism we are so familiar with today, whereby extracellular signals stimulate phosphoinositide-specific phospholipases inducing hydrolysis of $PI45P_2$, leading to DG and IP_3 and calcium production, which represent the rate-limiting signals required for maximal PKC activation.

An equally important landmark finding in the field was the discovery that tumor-promoting substances collectively known as phorbol esters are potent activators of PKC. Phorbol esters such as PMA (phorbol 12-myristate 13-acetate) are potent tumor promoters that are the biologically active components of the phytotoxins in the sap of the *Euphorbiaceae* family of tropical plants. Nishizuka and colleagues reported that phorbol esters bind to and activate PKC (Castagna et al. 1982), and this was made possible by the synthesis of hydrophilic versions of these compounds such as phorbol dibutyrate by Blumberg and colleagues. This also led to the identification of high-affinity binding sites on PKC that bind to both DG and PMA, suggesting that despite an altogether very different structure, phorbol esters activate PKC by molecular mimicry of DG. Immediately thereafter, Sando and Anderson found that exposure of cells to phorbol esters resulted in the rapid redistribution of PKC to the plasma membrane, leading to enzymatic activation (Kraft et al. 1982). For decades, this membrane translocation mechanism served as an effective molecular readout for the activation state of PKC. Although still in use, this biochemical readout has given way to genetically encoded FRET-based biosensors that report on the activation of PKC in intact cells (Violin et al. 2003).

The next major landmark finding in the field was the sequencing and cloning of the first PKC isoform, termed the major phorbol ester receptor, made by Parker, Waterfield and colleagues in the mid-1980s (Parker et al. 1986). This coincided with the identification and cloning of the PKCα, PKCβ, and PKCγ isoforms (Coussens et al. 1986), eventually leading to the realization that there exist 10 distinct mammalian PKC isoforms expressed with varied distribution and abundance in cells and tissues. The notion that multiple PKC isoforms exist was actually first realized by Huang and colleagues who detected multiple isozymes of the calcium- and phospholipid-activated PKC in rat brains (Huang et al. 1986). The cloning of PKC revealed a conserved catalytic kinase domain most similar to that of the protein kinase A, but a unique regulatory domain comprising two copies of the C1 domain (C1a and C1b) which represents the ligand (DG)-binding site, a C2 domain which coordinates calcium binding, and an amino-terminal region called the pseudosubstrate because it comprises the optimal PKC phosphorylation consensus sequence except that the phospho-acceptor is replaced by an alanine. The PKC family is now classified into: conventional isoforms PKCα, PKCβI and the alternative splice variant PKCβII, and PKCγ that are activated by DG, calcium and phosphatidylserine; novel PKC isoforms PKCδ, PKCε, PKCη, and PKCθ that are activated by DG and phosphatidylserine but are calcium-insensitive; and atypical PKC isoforms PKCζ and PKCι/λ that are both DG- and calcium-insensitive. Although an additional family member was identified and termed PKCμ, it was subsequently reclassified into the PKD family of kinases because the catalytic kinase core of PKD is more similar to calcium- and calmodulin-dependent protein kinases that it is to PKCs. A series of elegant biochemical and biophysical studies in the 1980s and into the early 1990s by Newton, Parker, Nishizuka, Blumberg and other laboratories provided exquisite detail into the mechanisms of the molecular activation of PKC isoforms by DG, calcium, and phosphatidylserine. These studies revealed that the two membrane targeting modules of PKCs, the C1 and C2 domains, serve to position PKC at the inner leaflet of the plasma membrane, whereby the C1 and DG interaction leads to an increase of the enzyme for anionic phospholipids such as phosphatidylserine, and in turn calcium facilitates the interaction of the C2 domain with the same anionic phospholipids. The chapters by Newton and also by Kazanietz and Merida review in detail the role of calcium, DG, and anionic phospholipids in PKC activation. Equally important was the finding that C1 domains also bind phorbol esters such as PMA and PDBU, but an important distinction exists between phorbol ester and DG binding to C1 domains. Phorbol esters bind to C1 domains with several orders higher affinity than does DG, and thus in cells exposed to PMA, PKC is recruited and retained on membranes with different temporal kinetics than is observed with DG as the physiological ligand that is transiently released following $PI45P_2$ hydrolysis. Blumberg reviews the mechanisms of phorbol ester activation of PKCs in his chapter.

In 1989, the Fabbro laboratory was the first to report the posttranslational modification of PKC in the form of phosphorylation (Borner et al. 1989). Subsequent studies by Newton and other laboratories revealed three key phosphorylation sites in the catalytic domain of PKCs that are highly conserved (Keranen et al. 1995). It was later found that phosphorylation of these three key sites, known as the activation loop residue, turn motif and hydrophobic sites, is required for maximal PKC activity in cells. Biochemical and cell-based assays showed that the two carboxyl-terminal sites, the turn motif and hydrophobic sites, are regulated by an intermolecular autophosphorylation reaction in conventional PKCs, but it also became clear that the activation loop, whose phosphorylation is absolutely required for maximal PKC activity in cells, was catalyzed by an upstream kinase. It was not until Alessi and Cohen identified the PDK-1 (Phosphoinositide-Dependent Kinase-1) enzyme as the upstream kinase that phosphorylates the equivalent motif on the related AGC kinase family member Akt/PKB (Alessi et al. 1997), that the Toker, Parker, and Newton laboratories went on to show that PDK-1 also phosphorylates the activation loop residue in all PKC family members (Chou et al. 1998; Dutil et al. 1998; Le Good et al. 1998). Very recent studies by Sabatini and colleagues have added further complexity to the model of PKC phosphorylation, suggesting that in addition to autophosphorylation, the mTor protein kinase in the TORC2 complex can also catalyze phosphorylation of the carboxyl-terminal residues in PKCs (Sarbassov et al. 2004). The regulation of PKC phosphorylation is discussed in detail in the chapter by Newton.

Finally, the scaffolding of PKC isoforms in proximity to both substrates as well as upstream activators such as DG and phosphatidylserine is critical for efficient signal relay. In the early 1990s, Mochly-Rosen and colleagues were the first to identify a family of proteins that interact with the active conformation of PKC (Mochly-Rosen et al. 1991). For this reason, they termed them RACKS (Receptors for Activated C KinaseS) and, in a series of studies, revealed the mechanism by which activated PKCs directly bind with RACKS and position them to discrete locations, thus facilitating downstream signaling. Identification of the PKC:RACK binding sites also permitted the generation of specific peptides which could be introduced into cells, thus uncoupling the interaction and terminating PKC signaling. Subsequently, other PKC scaffolding proteins such as STICKS (Substrates That Interact with C Kinase) as well as 14-3-3 proteins were found to directly bind PKC isozymes, thus providing additional regulation in downstream signal relay. Similarly, Scott and colleagues found a separate family of proteins terms AKAPs (A Kinase Anchoring Proteins) that act as true scaffolds because they directly assemble signaling complexes comprising kinases such as PKA and PKC, as well as phosphatases such as calcineurin (Klauck et al. 1996). A number of AKAPs are now known to spatially and temporally coordinate the assembly of PKC isoforms in discrete cellular locations, thus ensuring both efficiency and specificity in signal transmission. The regulation of PKC activation and downstream signaling by adapter proteins is covered in the chapter by Mochly-Rosen.

In summary, a number of landmark findings in the 30 years since the discovery of PKC by Nishizuka have propelled this field into the signaling limelight time and time again. Ultimately, this made it possible to begin to investigate the relevance and importance of PKC isoforms in pathophysiology, which in turn facilitated the development of chemical inhibitors designed to attenuate PKC activity in clinical settings. The following chapters are authored by investigators who made seminal contributions in the field of PKC and lipid signaling and they discuss in detail the major mechanisms of PKC activation by lipid second messengers, phorbol esters, phosphorylation, and scaffolding proteins.

References

Alessi, D. R., James, S. R., Downes, C. P., Holmes, A. B., Gaffney, P. R., Reese, C. B., et al. (1997). Characterization of a 3-phosphoinositide-dependent protein kinase which phosphorylates and activates protein kinase Balpha. *Current Biology, 7*, 261–269.

Borner, C., Filipuzzi, I., Wartmann, M., Eppenberger, U., & Fabbro, D. (1989). Biosynthesis and posttranslational modifications of protein kinase C in human breast cancer cells. *The Journal of Biological Chemistry, 264*, 13902–13909.

Castagna, M., Takai, Y., Kaibuchi, K., Sano, K., Kikkawa, U., & Nishizuka, Y. (1982). Direct activation of calcium-activated, phospholipid-dependent protein kinase by tumor-promoting phorbol esters. *The Journal of Biological Chemistry, 257*, 7847–7851.

Chou, M. M., Hou, W., Johnson, J., Graham, L. K., Lee, M. H., Chen, C. S., et al. (1998). Regulation of protein kinase C zeta by PI 3-kinase and PDK-1. *Current Biology, 8*, 1069–1077.

Coussens, L., Parker, P. J., Rhee, L., Yang-Feng, T. L., Chen, E., Waterfield, M. D., et al. (1986). Multiple, distinct forms of bovine and human protein kinase C suggest diversity in cellular signaling pathways. *Science, 233*, 859–866.

Dutil, E. M., Toker, A., & Newton, A. C. (1998). Regulation of conventional protein kinase C isozymes by phosphoinositide-dependent kinase 1 (PDK-1). *Current Biology, 8*, 1366–1375.

Hokin, L. E., & Hokin, M. R. (1953). The incorporation of 32P into the nucleotides of ribonucleic acid in pigeon pancreas slices. *Biochimica et biophysica acta, 11*, 591–592.

Hokin, M. R., & Hokin, L. E. (1954). Effects of acetylcholine on phospholipides in the pancreas. *The Journal of Biological Chemistry, 209*, 549–558.

Huang, K. P., Nakabayashi, H., & Huang, F. L. (1986). Isozymic forms of rat brain $Ca2^{+}$-activated and phospholipid-dependent protein kinase. *Proceedings of the National Academy of Sciences of the United States of America, 83*, 8535–8539.

Keranen, L. M., Dutil, E. M., & Newton, A. C. (1995). Protein kinase C is regulated in vivo by three functionally distinct phosphorylations. *Current Biology, 5*, 1394–1403.

Klauck, T. M., Faux, M. C., Labudda, K., Langeberg, L. K., Jaken, S., & Scott, J. D. (1996). Coordination of three signaling enzymes by AKAP79, a mammalian scaffold protein. *Science, 271*, 1589–1592.

Kraft, A. S., Anderson, W. B., Cooper, H. L., & Sando, J. J. (1982). Decrease in cytosolic calcium/phospholipid-dependent protein kinase activity following phorbol ester treatment of EL4 thymoma cells. *The Journal of Biological Chemistry, 257*, 13193–13196.

Lapetina, E. G., & Michell, R. H. (1973a). Phosphatidylinositol metabolism in cells receiving extracellular stimulation. *The FEBS Letters, 31*, 1–10.

Lapetina, E. G., & Michell, R. H. (1973b). A membrane-bound activity catalysing phosphatidylinositol breakdown to 1, 2-diacylglycerol, D-myoinositol 1:2-cyclic phosphate an D-myoinositol 1-phosphate. Properties and subcellular distribution in rat cerebral cortex. *The Biochemical Journal, 131*, 433–442.

Le Good, J. A., Ziegler, W. H., Parekh, D. B., Alessi, D. R., Cohen, P., & Parker, P. J. (1998). Protein kinase C isotypes controlled by phosphoinositide 3-kinase through the protein kinase PDK1. *Science, 281*, 2042–2045.

Mochly-Rosen, D., Khaner, H., & Lopez, J. (1991). Identification of intracellular receptor proteins for activated protein kinase C. *Proceedings of the National Academy of Sciences of the United States of America., 88*, 3997–4000.

Parker, P. J., Coussens, L., Totty, N., Rhee, L., Young, S., Chen, E., et al. (1986). The complete primary structure of protein kinase C – the major phorbol ester receptor. *Science, 233*, 853–859.

Sarbassov, D. D., Ali, S. M., Kim, D. H., Guertin, D. A., Latek, R. R., Erdjument-Bromage, H., et al. (2004). Rictor, a novel binding partner of mTOR, defines a rapamycin-insensitive and raptor-independent pathway that regulates the cytoskeleton. *Current Biology, 14*, 1296–1302.

Streb, H., Irvine, R. F., Berridge, M. J., & Schulz, I. (1983). Release of $Ca2^{+}$ from a nonmitochondrial intracellular store in pancreatic acinar cells by inositol-1, 4, 5-trisphosphate. *Nature, 306*, 67–69.

Takai, Y., Kishimoto, A., Iwasa, Y., Kawahara, Y., Mori, T., & Nishizuka, Y. (1979a). Calcium-dependent activation of a multifunctional protein kinase by membrane phospholipids. *The Journal of Biological Chemistry, 254*, 3692–3695.

Takai, Y., Kishimoto, A., Kikkawa, U., Mori, T., & Nishizuka, Y. (1979b). Unsaturated diacylglycerol as a possible messenger for the activation of calcium-activated, phospholipid-dependent protein kinase system. *Biochemical and Biophysical Research Communications, 91*, 1218–1224.

Violin, J. D., Zhang, J., Tsien, R. Y., & Newton, A. C. (2003). A genetically encoded fluorescent reporter reveals oscillatory phosphorylation by protein kinase C. *The Journal of Cell Biology, 161*, 899–909.

Chapter 2
Regulation of Conventional and Novel Protein Kinase C Isozymes by Phosphorylation and Lipids

Alexandra C. Newton

Abstract The amplitude of protein kinase C signaling is precisely controlled by mechanisms that regulate the amount of protein kinase C in the cell that is available to become activated with appropriate stimuli. Two mechanisms critically control the amount and activity of protein kinase C in cells. First, a series of phosphorylation events prime conventional and novel protein kinase C isozymes into stable, signaling-competent species. Second, signals that cause phospholipid hydrolysis cause protein kinase C to bind to membrane lipids, an interaction that allosterically activates the kinase. Deregulation of either step alters the amplitude of protein kinase C signaling in the cell, resulting in pathophysiological states. This chapter focuses on the molecular mechanisms by which phosphorylation and lipid binding control protein kinase C.

Keywords Protein kinase C • Diacylglycerol • Phosphorylation • C1 domain • C2 domain

Abbreviations

AGC kinases	Protein kinases A, G and C
DG	Diacylglycerol
PI3 kinase	Phosphatidylinositol 3 kinase
PDK-1	Phosphoinositide-dependent kinase-1
PH	Pleckstrin homology
PHLPP	PH domain Leucine-rich repeat Protein Phosphatase
PS	Phosphatidylserine
PIP_2	Phosphatidylinositol-4,5-bisphosphate
PIP_3	Phosphatidylinositol-3,4,5-trisphosphate
RACK	Receptor for activated C kinase
TORC2	Target of rapamycin complex 2

A.C. Newton (✉)
Department of Pharmacology, University of California at San Diego,
La Jolla, CA 92093-0721, USA
e-mail: anewton@ucsd.edu

M.G. Kazanietz (ed.), *Protein Kinase C in Cancer Signaling and Therapy*,
Current Cancer Research, DOI 10.1007/978-1-60761-543-9_2,

2.1 Introduction

The ten members of the protein kinase C family are grouped into three classes that are defined by the composition of their regulatory modules (Nishizuka 1995; Newton 2001) (Fig. 2.1, left). These, in turn, dictate the cofactor requirements for activity (Fig. 2.1, right). Conventional isozymes (protein kinase C α, the alternatively spliced βI and βII, and γ) are composed of two tandem C1 domains (Hurley et al. 1997), allowing them to respond to diacylglycerol, and a C2 domain, which binds anionic membranes in a Ca^{2+}-dependent manner. The C1 domain also stereospecifically binds the anionic phospholipid, phosphatidylserine (Johnson et al. 1998, 2000). It is also the binding site for the potent tumor promoting phorbol esters, which bind the same site as diacylglycerol (Sharkey et al. 1984). The C2 domain of conventional protein kinase C isozymes binds anionic phospholipids, with a modest (but not stereospecific) preference for phosphatidylserine

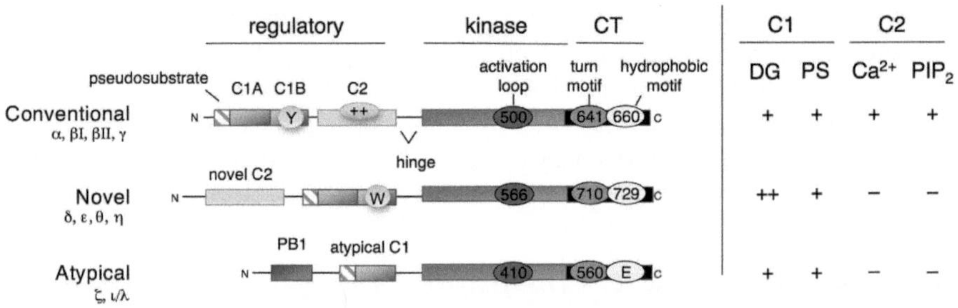

Fig. 2.1 Schematic of protein kinase C family members showing membrane-targeting modules in amino-terminal regulatory moiety and phosphorylation sites in carboxyl-terminal kinase moiety. Conventional isozymes (α, alternatively spliced βI and βII, γ) have a tandem C1 domain that confers specificity for diacylglycerol and phosphatidylserine and a C2 domain that binds anionic phospholipids via a Ca^{2+}-occupied ligand binding site and via a basic patch distal to the Ca^{2+} site (*oval* with ++), with selectivity for PIP_2. Novel isozymes (δ, ε, θ, η) also have tandem C1 domains, but Trp at position 22 (*circle* with W) in the C1B domain confers an order of magnitude higher affinity for diacylglycerol than the C1B domain of conventional protein kinase C isozymes, which have a Tyr at position 22 of domain (*circle* with Y; numbering of (Hurley et al. 1997)). Atypical isozymes have a single C1 domain whose highly basic ligand-binding pocket is unable to bind diacylglycerol but retains binding to phosphatidylserine. In addition, atypical protein kinase C isozymes have a protein–protein interaction PB1 domain. All isozymes contain an autoinhibitory pseudosubstrate sequence directly preceding the C1 domain (*stippled area*) and have a proteolytically labile hinge segment that separates the regulatory moiety from the kinase moiety. The kinase domain contains three conserved phosphorylation sites modified during the maturation of the enzyme into a catalytically competent species: the activation loop in the kinase domain (*dark gray circle*; Thr 500 in protein kinase C βII; Thr 566 in protein kinase C ε; Thr 410 in protein kinase C ζ), and the turn motif (*medium gray circle*; Thr 641 in protein kinase C βII; Thr 710 in protein kinase C ε; Thr 560 in protein kinase C ζ) and the hydrophobic motif (*light gray circle*; Ser 660 in protein kinase C βII; Ser 729 in protein kinase C ε; and the phosphomimetic Glu 579 in protein kinase C ζ) in the carboxyl tail (CT). Table on *right* summarizes second messenger (diacylglycerol (DG) or Ca^{2+}) and phospholipid (phosphatidylserine (PS) or PIP_2) binding to the two membrane-targeting modules, the C1 and C2 domains, in the three subclasses of isozymes

(Medkova and Cho 1999; Johnson et al. 2000; Conesa-Zamora et al. 2001), and a significant preference for phosphatidylinositol-4,5-bis phosphate (PIP_2) mediated by a basic patch distal to the Ca^{2+}-binding site (Fig. 2.1, ++ in C2 domain of conventional protein kinase C isozymes) (Corbalan-Garcia et al. 2007). Novel isozymes (protein kinase C δ, ε, η, and θ) also contain two tandem C1 domains, conferring diacylglycerol sensitivity, but they contain a "novel" C2 domain that does not bind Ca^{2+} and does not serve as a membrane-binding module. Atypical isozymes (ζ and ι/λ) possess an "atypical" C1 domain whose highly basic ligand-binding pocket does not allow ligand binding, so these isozymes respond to neither diacylglycerol nor Ca^{2+} (Kazanietz et al. 1994; Pu et al. 2006). These isozymes contain a PB1 protein-binding domain which poises this class of protein kinase C isozymes at discrete intracellular locations (Lamark et al. 2003). All protein kinase C isozymes have an autoinhibitory segment, the pseudosubstrate (Fig. 2.1, stippled box), that occupies the substrate-binding cavity in the absence of lipid binding, thus maintaining the kinase in an autoinhibited state. Engagement of the membrane-binding modules provides the energy to release the pseudosubstrate from the substrate-binding cavity, allowing downstream signaling. All isozymes are also regulated by a conserved segment at the carboxyl-terminal tail (Fig. 2.1, CT) that controls the stability of the kinases, serves as a docking site for key regulatory molecules, and provides a phosphorylation switch for kinase function (phosphorylation sites indicated by ovals).

2.2 Regulation of Protein Kinase C by Priming Phosphorylation

Before protein kinase C is competent to respond to lipid second messengers, it must first be processed by a series of ordered and tightly coupled phosphorylation events at three conserved positions in the carboxyl-terminal half of the protein (Fig. 2.2) (Newton 2003; Parker and Murray-Rust 2004). These phosphorylation events are required to structure protein kinase C into a catalytically competent and stable species. It is this phosphorylated species that transduces signals. Constructs of protein kinase C that cannot be phosphorylated are shunted to the detergent-insoluble fraction of cells and degraded. Thus, it is important to note that phosphorylation is not only required for the catalytic competence of protein kinase C, but also to protect the mature (but inactive) enzyme from degradation.

The first phosphorylation is catalyzed by the upstream kinase phosphoinositide-dependent kinase-1 (PDK-1) and occurs on a conserved Thr on a loop near the entrance to the active site of the kinase core (Chou et al. 1998; Dutil et al. 1998; Le Good et al. 1998). This phosphorylation triggers two ordered phosphorylations on the carboxyl-terminal tail of protein kinase C. These sites, identified by mass spectrometry in the mid-1990s, are referred to as the turn motif and the hydrophobic motif (Keranen et al. 1995). The species of conventional protein kinase C found in the detergent-soluble fraction of mammalian cells is quantitatively phosphorylated

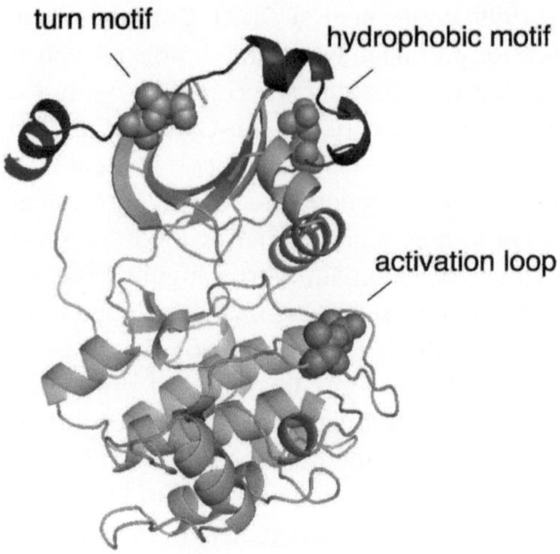

Fig. 2.2 Structure of kinase domain of conventional isozyme, protein kinase C βII, showing the position of the three priming phosphorylations. Phosphorylated residues are shown in space filling rendition. These phosphorylations are the activation loop site on a segment near the entrance to the active site and the turn motif and hydrophobic motif on the carboxyl tail (CT) (dark segment of structure). Note how the turn motif and hydrophobic motif clamp the CT around the upper lobe of the kinase core

at the two carboxyl-terminal sites but may have incomplete occupancy of the activation loop site. Species that are not phosphorylated at the carboxyl-terminal sites are targeted for degradation. Note that protein kinase C isozymes are controlled by additional phosphorylations on Ser, Thr and Tyr that fine-tune the function of specific isozymes (reviewed in (Gould and Newton 2008)); this chapter focuses on the priming phosphorylations.

2.2.1 Activation Loop Phosphorylation and PDK-1

PDK-1 serves as the upstream kinase for many members of the AGC superfamily of kinases, catalyzing the phosphorylation of a Thr on a segment near the entrance to the active site referred to as the activation loop (Taylor and Radzio-Andzelm 1994). Phosphorylation on this Thr correctly aligns residues for catalysis. Phosphorylation of PDK-1 substrates is controlled by the conformation of the substrate (Toker and Newton 2000; Mora et al. 2004); conformational changes that unmask the activation loop site promote the phosphorylation by PDK-1. In the case of protein kinase C family members, newly synthesized protein kinase C is membrane-associated in a conformation in which the pseudosubstrate sequence is expelled from the substrate-binding cavity, thus unmasking the activation loop site. Thus, newly synthesized protein kinase C is constitutively phosphorylated by PDK-1.

In striking contrast, the activation loop site of Akt (protein kinase B) is masked until agonist stimulation recruits the kinase to the membranes. Engaging its membrane-targeting module, a PH domain, to phosphatidylinositol-3,4,5,-tris phosphate (PIP_3) unmasks the PDK-1 site. Thus, the phosphoinositide-dependence of Akt derives from the phosphoinositide-dependence of unmasking the activation loop site.

PDK-1 docks on the carboxyl-terminal tail of the newly synthesized protein kinase C (Fig. 2.3, species of protein kinase C on upper left), specifically recognizing the hydrophobic phosphorylation motif, to phosphorylate the activation loop Thr (Thr500 on protein kinase C βII). This is generally considered to be the first step in the processing of protein kinase C. Phosphorylation at this site is required to continue the processing of protein kinase C by phosphorylation at the two carboxyl-terminal sites: mutation of the activation loop Thr to a nonphosphorylatable neutral residue results in an inactive protein kinase C (Cazaubon et al. 1994; Orr and Newton 1994). Because unphosphorylated protein kinase C is not stable, constructs that cannot be phosphorylated at the activation loop are degraded. Consistent with this, embryonic stem cells lacking PDK-1 have reduced levels of conventional and novel protein kinase C isozymes (Balendran et al. 2000).

2.2.2 *Carboxyl-Terminal Phosphorylations and TORC2*

The immediate consequence of phosphorylation at the activation loop is the phosphorylation of two conserved sites on the carboxyl-terminus: first on the turn motif, so named because the analogous position in protein kinase A is at the apex of a turn; and, secondly, on the hydrophobic motif, so named because it is flanked by hydrophobic residues (Keranen et al. 1995; Newton 2001). Phosphorylation at both sites depends on the intrinsic catalytic activity of protein kinase C, suggesting that they are catalyzed by autophosphorylation. Enzymological studies with pure conventional protein kinase C βII have shown that this is the case for the hydrophobic motif: under conditions where the enzyme is a monomer, it incorporates phosphate at this position in a concentration-independent manner, revealing intramolecular autophosphorylation (Behn-Krappa and Newton 1999). The mechanism of phosphorylation of the turn motif is not clear. However, recent reports have established that phosphorylation of the turn motif depends on the mTORC2 complex, a structure comprising the kinase mTOR, sin1, rictor, and mLST8 (Facchinetti et al. 2008; Ikenoue et al. 2008; Jacinto and Lorberg 2008). Specifically, protein kinase C cannot be processed by phosphorylation in cells lacking this complex and, because the unphosphorylated species is unstable, it is degraded (Guertin et al. 2006; Ikenoue et al. 2008). Whether this complex assists by noncatalytic mechanisms, for example chaperoning or positioning newly synthesized protein kinase C for processing by phosphorylation, or whether it directly phosphorylates protein kinase C is unclear. It is noteworthy, however, that mTORC2 is not able to phosphorylate protein kinase C in vitro (Ikenoue et al. 2008).

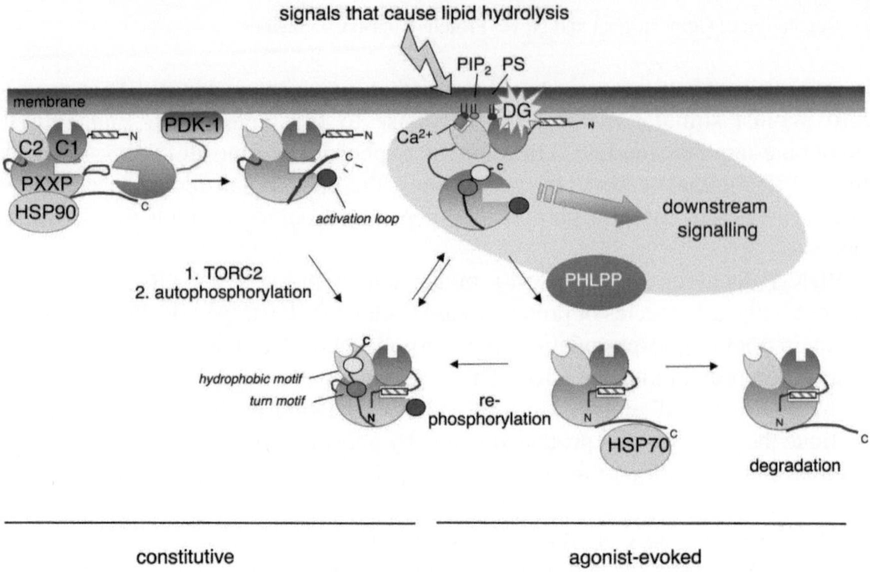

Fig. 2.3 Schematic illustrating the life cycle of conventional protein kinase C. Newly synthesized protein kinase C (species in *top left*) associates with a membrane fraction in a conformation in which the autoinhibitory pseudosubstrate (*stippled rectangle*) is removed from the substrate-binding cavity (*rectangular indentation in large circle*), thus exposing the activation loop phosphorylation site to phosphorylation by PDK-1 which is docked to carboxyl tail. Hsp90 binds this newly synthesized species on a surface that depends on an intramolecular clamp formed between a conserved PXXP motif on the carboxyl tail and a helix in the kinase domain. The structural integrity of the clamp and the interaction with Hsp90 are essential for the processing of protein kinase C PDK-1, constitutively docked to the carboxyl-terminal tail, phosphorylates the activation loop, correctly aligning residues in the active site for catalysis. One of the immediate consequences of this phosphorylation is the tightly coupled phosphorylation of the turn motif and then the hydrophobic motif. Phosphorylation of the turn motif is rate-limiting, depends on a functional mTORC2 complex, and is required to complete the processing of protein kinase C. The mechanism of this phosphorylation has yet to be elucidated. The third and last phosphorylation on the hydrophobic motif occurs by intramolecular autophosphorylation. The fully phosphorylated enzyme localizes primarily to the cytosol or is scaffolded to specific intracellular locations by protein–protein interactions (Schechtman and Mochly-Rosen 2001). This phosphorylated "mature" species adopts a conformation in which the pseudosubstrate occupies the substrate-binding cavity, thus autoinhibiting the enzyme (*bottom left species*). This processing by phosphorylation is constitutive. In response to signals that cause phospholipid hydrolysis, protein kinase C associates with cellular membranes and becomes activated. For conventional protein kinase C isozymes, activation typically requires hydrolysis of PIP$_2$ to generate two second messengers: Ca^{2+} and diacylglycerol (DG). Ca^{2+} binds the C2 domain, promoting the association of protein kinase C with the plasma membrane via interaction with anionic lipids, and importantly PIP$_2$. The enzyme then diffuses in two-dimensional space until the C1 domain finds its membrane-embedded ligand, diacylglycerol. This interaction is strengthened by stereospecific binding to phosphatidylserine (PS). The binding energy provided by engaging both the C1 and C2 domains on membranes releases the pseudosubstrate, allowing substrate phosphorylation and downstream signaling (*top, species on right*). This "open" conformation of protein kinase C (pseudosubstrate exposed and hinge between regulatory moiety and kinase domain unmasked) is sensitive to dephosphorylation; PHLPP initiates the dephosphorylation of the hydrophobic motif, with PP2A-type phosphatases contributing to the complete dephosphorylation of the priming sites. This dephosphorylated species associates with a detergent-insoluble fraction and is degraded. Hsp70 sustains the signaling lifetime of dephosphorylated protein kinase C by binding the dephosphorylated turn motif and promoting the rephosphorylation of protein kinase C

However, the turn motif may very well be modified by another kinase: while lack of PDK-1 prevents phosphorylation of the activation loop and the hydrophobic motif, the turn motif has recently been reported to be efficiently phosphorylated in PDK-$1^{-/-}$ cells (Ikenoue et al. 2008). Consistent with another kinase catalyzing the phosphorylation of the turn motif, but not the hydrophobic motif, a GST-fusion construct of the carboxyl-terminal tail is efficiently phosphorylated at the turn motif, but not the hydrophobic motif (Ikenoue et al. 2008). These priming phosphorylations are constitutive for conventional protein kinase C isozymes. For novel isozymes, basal phosphorylation of the priming sites is high but can increase modestly upon agonist stimulation. For atypical protein kinase C isozymes, the PDK-1 step displays the highest agonist sensitivity. Note that atypical isozymes contain a constitutive negative charge (Glu) at the phospho-acceptor position of the hydrophobic motif.

The processing of protein kinase C by phosphorylation depends on the binding of Hsp90 to a carboxyl-terminal motif conserved amongst the AGC kinases that comprises the sequence PXXP. This PXXP motif forms an intramolecular clamp with residues in the kinase domain that provides a surface for binding Hsp90. Disruption of the clamp, or inhibition of Hsp90, prevents the processing of protein kinase C by phosphorylation (Gould et al. 2009).

Once protein kinase C is phosphorylated at the two carboxyl-terminal sites, phosphorylation of the activation loop becomes dispensable. In fact, mass spectrometric analysis has revealed that about half the protein kinase C in brain extracts or in mammalian cultured cells is not phosphorylated on the activation loop despite quantitative phosphorylation on the carboxyl-terminal sites (Keranen et al. 1995). Thus, phosphorylation of the activation loop site is required to process protein kinase C, but once the mature conformation is achieved, the phosphorylation state of the activation loop site does not impact the activity of protein kinase C.

2.3 Regulation of Protein Kinase C by Lipid Second Messengers

2.3.1 Conventional Protein Kinase C

The processing of conventional protein kinase C by phosphorylation occurs at the membrane (Sonnenburg et al. 2001), but once phosphorylated, the "mature" enzyme is released to the cytosol where it adopts an autoinhibited conformation with the pseudosubstrate occupying the substrate-binding cavity (Dutil and Newton 2000). This species of protein kinase C is also poised at specific intracellular locations by protein scaffolds (Schechtman and Mochly-Rosen 2001). Phospholipase C-catalyzed hydrolysis of PIP_2 results in elevated levels of two second messengers required for agonist-evoked activation of conventional protein kinase C isozymes: diacylglycerol and Ca^{2+} (Nishizuka 1995). Biophysical and molecular imaging studies suggest that in the absence of these ligands, conventional protein kinase C bounces on and off the membrane by diffusion-controlled mechanisms, with membrane interactions of far too low an affinity to retain protein kinase C there (Schaefer et al. 2001). However, elevation of intracellular Ca^{2+} recruits protein kinase C to the plasma membrane via a low-affinity interaction of the Ca^{2+}-bound C2 domain (Oancea and Meyer 1998; Nalefski and Newton 2001) (Fig. 2.3). The affinity of this interaction is too low to activate protein kinase C, but it has the important biological function of poising protein kinase C on the plasma membrane, where it can effectively search for its membrane-embedded ligand, diacylglycerol, in two-dimensional space (Nalefski and Newton 2001). The interaction of the C1 domain with diacylglycerol-containing membranes is strengthened by the stereo-specific interaction with phosphatidylserine (Johnson et al. 1998, 2000). The enrichment of phosphatidylserine at the plasma membrane thus likely contributes to the translocation of PKC to the plasma membrane (Yeung et al. 2008). Once the C1 domain is engaged, the binding energy of protein kinase C to membranes is sufficiently high to release the autoinhibitory pseudosubstrate from the substrate-binding cavity, allowing substrate binding and phosphorylation. Note that the affinity of each membrane-targeting module of conventional protein kinase C isozymes, the C1 and C2 domains, is too low to allow pseudosubstrate release in response to physiological levels of diacylglycerol or Ca^{2+} alone. However, the C1 domain of conventional protein kinase C isozymes binds membranes containing phorbol esters (functional analogs of diacylglycerol) with two orders of magnitude higher affinity than membranes containing diacylglycerol (Mosior and Newton 1996). Thus, phorbol ester treatment of cells recruits protein kinase C to membranes with sufficiently high affinity to promote pseudosubstrate release in the absence of Ca^{2+} binding to the C2 domain. Importantly, under physiological conditions, activation of conventional protein kinase C requires the coordinated binding of the C1 and C2 domains to membranes, with Ca^{2+} binding to the C2 domain pretargeting protein kinase C to membranes where it can efficiently engage diacylglycerol.

Conventional protein kinase C isozymes are primarily recruited to the plasma membrane, despite the relatively high levels of diacylglycerol at the Golgi, where novel isozymes are primarily recruited (Carrasco and Merida 2004; Gallegos et al. 2006; Dries et al. 2007). The molecular basis for the selective translocation of conventional protein kinase C isozymes to the plasma membrane is likely accounted for by their affinity for PIP_2, a lipid found primarily on the plasma membrane (Yeung et al. 2006). This lipid binds a basic surface on conventional C2 domains (Corbalan-Garcia et al. 2007; Marin-Vicente et al. 2008; Evans et al. 2006; Landgraf et al. 2008). Phosphatidylserine is also enriched at the plasma membrane relative to Golgi membranes (Yeung et al. 2008), an enrichment that may also contribute to the targeting of conventional protein kinase C isozymes, which are tightly regulated by phosphatidylserine, to the plasma membrane. Thus, the unique phospholipid composition of the plasma membrane, which results in the most negatively charged membrane surface in cells (Yeung et al. 2006), favors the recruitment of conventional protein kinase C isozymes.

2.3.2 Novel Protein Kinase C

The affinity of the C1B domain of novel protein kinase C isozymes for diacylglycerol-containing membranes is two orders higher than that of conventional protein kinase C isozymes (Giorgione et al. 2006). This low vs. high affinity binding depends on the nature of the hydrophobic residue at position 22 of the C1B domain: when present as a Trp, as it is in novel isozymes, the domain binds diacylgycerol membranes with high affinity and when present as a Tyr, as it is in conventional protein kinase C isozymes, the domain binds with low affinity (Dries et al. 2007). Thus, the isolated C1B domain of novel enzymes, but not conventional isozymes, is recruited to membranes following agonist-evoked increases in diacylglycerol. This enhanced affinity for diacylglycerol allows novel protein kinase C isozymes to translocate to membranes in response to physiological increases in diacylglycerol. Because basal levels of diacylglycerol are relatively high at Golgi, significant levels of novel isozymes are localized at this membrane (Carrasco and Merida 2004). Agonist-evoked increases in diacylglycerol increase the association of novel protein kinase C isozymes with Golgi and, to a lesser extent, plasma membrane (Gallegos et al. 2006). There are also differences in the cellular locations of individual members of the novel protein kinase C family driven by differences in lipid interactions (Stahelin et al. 2005). For example, the C2 domain of protein kinase C ε also binds phosphatidic acid, an interaction that also tunes the membrane interaction of this isozyme (Pepio and Sossin 1998).

2.4 Termination of Protein Kinase C Signaling

Signaling by protein kinase C is terminated by the removal of the activating second messengers. However, prolonged activation of protein kinase C, as occurs with phorbol esters, results in the "down-regulation" of protein kinase C (Parker et al. 1995; Leontieva and Black 2004; Gould and Newton 2008). Membrane-bound protein kinase C adopts an open (pseudosubstrate-exposed) conformation that exposes the phosphorylation sites to cellular phosphatases (Fig. 2.3). The first dephosphorylation event appears to be catalyzed by the recently discovered hydrophobic motif phosphatase, the PH domain Leucine-rich repeat Protein Phosphatase (PHLPP) (Gao et al. 2005, 2008). Its selective dephosphorylation of the hydrophobic motif shunts protein kinase C to the detergent-insoluble fraction of cells, where it is further dephosphorylated at the turn motif and activation loop by additional phosphatases, including PP2A-type phosphatases (Hansra et al. 1996; Gao et al. 2008). The dephosphorylated species is targeted for degradation. Interestingly, nature has devised a mechanism to "rescue" protein kinase C from degradation: the molecular chaperone Hsp70 specifically binds the dephosphorylated turn motif, an event that stabilizes protein kinase C. This binding is proposed to promote the rephosphorylation of the enzyme at the priming sites, thus sustaining the signaling lifetime of the enzyme (Gao and Newton 2002, 2006). In addition to agonist-stimulated degradation, the total cellular levels of protein kinase C, independent of activation or phosphorylation state, have recently been shown to be controlled by a protein kinase C-interacting E3 ligase, RINCK, which ubiquitinates protein kinase C and targets it for proteasomal degradation (Chen et al. 2007). There are likely to be additional ligases that control the degradation of specific species of protein kinase C. Of particular interest will be the identification of ligases that control the phorbol ester-dependent down-regulation.

Protein scaffolds are essential for coordinating components of signaling pathways (Smith et al. 2006), and they play key roles in poising specific protein kinase C isozymes near regulatory molecules and substrates (Mochly-Rosen 1995; Jaken and Parker 2000; Schechtman and Mochly-Rosen 2001). It is protein scaffolds, rather than subtle changes in second messenger affinities, that confer specificity in signaling by the structurally similar protein kinase C isozymes within each subclass. Thus, for example, specific scaffolds for the conventional isozymes protein kinase C α and protein kinase C βII promote isozyme-specific signaling. One class of scaffolds termed Receptors for Activated C Kinase (RACKs) specifically recognizes sequences that are exposed in the active conformation of protein kinase C. The first RACK was in fact identified as a protein kinase C βII-specific adaptor (Ron et al. 1994). These scaffolds finely tune the location of protein kinase C isozymes within the cell via protein–protein interactions, stabilizing the active conformation. Mochly-Rosen and coworkers have taken advantage of sequences on specific isozymes that either directly bind the RACK scaffolds or sequences within the protein kinase C isozyme that intramolecularly bind and mask the RACK-binding sequence in the inactive conformation (Souroujon and Mochly-Rosen 1998) to generate peptide inhibitors and activators, respectively (Schechtman and Mochly-Rosen 2001). Additionally, the last three amino acids of protein kinase C α encode a PDZ ligand which has been shown to bind the PDZ-domain containing protein PICK1 (Staudinger et al. 1997). Furthermore, a recent proteomics approach identified several other potential partners for this PDZ ligand on protein kinase C α (Stiffler et al. 2007). Thus, whereas lipids acutely control the activation state of protein kinase C by releasing the pseudosubstrate, protein partners poise protein kinase C isozymes at precise intracellular locations to control substrate access and interactions with regulatory molecules (phosphatases, E3 ligases, chaperones, etc).

2.5 Spatiotemporal Dynamics of Protein Kinase C Signaling

The advent of genetically encoded reporters revolutionized the study of the spatiotemporal dynamics of protein kinase C signaling (Sakai et al. 1997; Oancea and Meyer 1998; Oancea et al. 1998; Violin and Newton 2003; Violin et al. 2003). The ability to simultaneously visualize protein kinase C translocation, protein kinase C activity, and the second messengers, diacylglycerol and Ca^{2+}, has revealed that protein kinase C isozymes have a unique signature of activation depending on the cellular location (Gallegos et al. 2006; Gallegos and Newton 2008). In response to

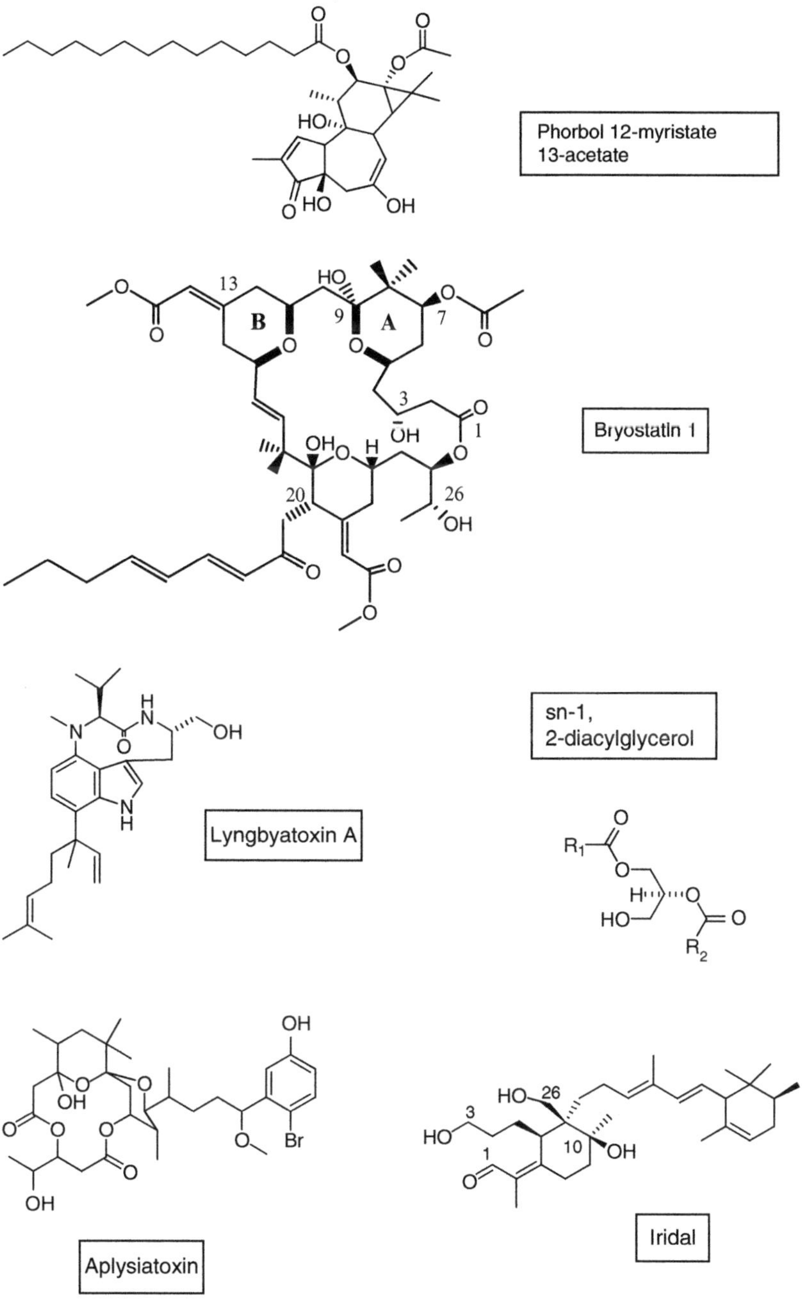

Fig. 3.2 Structural diversity of ligands with high affinity for typical C1 domains

3.4 DAG-Lactones as Manipulable Ligands for C1 Domains

The exogenous natural products identified as high affinity ligands for C1 domains all possess synthetically complex, constrained backbones that maintain the appropriate orientation of hydrogen-bonding substituents and hydrophobic regions. The tigliane backbone of the phorbol esters has eight chiral centers; the macrocyclic ring of bryostatin 1 has 11. The endogenous ligand, DAG, in contrast, has only a single chiral center but its simplicity is counterbalanced by its relatively low affinity, reflecting the flexibility of its conformation. Using the DAG structure as a starting point, the Marquez group sought to constrain the structure in order to eliminate this flexibility. The guiding concept was that the constraint imposed upon a flexible ligand when it binds to its binding site is associated with a loss of entropy, which translates thermodynamically into a less favorable free energy of binding. If the ligand in the unbound state can already be constrained into the binding conformation, then this loss of entropy upon binding will not occur and the free energy of binding will be correspondingly enhanced. Comparison of different approaches for constraining DAG revealed that the DAG-lactone structure successfully accomplished this objective (Marquez et al. 1999). Optimization of the pattern of substitution on the side chains of the DAG-lactone has then provided ligands with binding affinities approaching those of the phorbol ester (Nacro et al. 2001), and a convenient stereosynthesis for the single chiral center has been developed (Kang et al. 2004). Such DAG-lactones have provided powerful probes for approaching a variety of issues related to ligand – C1 domain interactions (Marquez and Blumberg 2003).

3.5 Nature of the Interactions of Ligands with the C1 Domain

The three-dimensional structures of multiple DAG-responsive C1 domains have been solved by NMR (Hommel et al. 1994; Xu et al. 1997), and we were able to determine the crystal structure of the C1b domain of PKC δ in complex with phorbol ester (Zhang et al. 1995). This structure revealed that the C1 domain functions as a hydrophobic switch. The phorbol ester inserts into a hydrophilic cleft formed by the pulled apart strands of a β-sheet at the top of the C1 domain. This upper surface of the C1 domain surrounding the cleft is hydrophobic and the phorbol ester, upon insertion, completes the hydrophobic surface, favoring the penetration of the C1 domain – phorbol ester complex into the lipid membrane. The C1 domain does not appreciably change conformation upon phorbol ester binding, eliminating possible allosteric models for its mechanism of action. Rather, the binding changes the association preference of the hydrophobic face of the C1 domain. Two different structural features of the ligand contribute to the hydrophobicity of the ligand – C1 domain complex. While the first feature is the coverage of the hydrophilic cleft of the C1 domain, which may be similar between ligands, the second is the contribution made by the hydrophobic side chains projecting from the ligand, for which great diversity is possible. As described above, different acyl substituents on the phorbol ester can lead to very different biological responses.

While models typically imply an ordered process in which the ligand first binds to the C1 domain and this binding then induces translocation and membrane association of the C1 domain containing protein, the opposite order is more plausible, especially for highly lipophilic ligands like physiological DAGs. Their high octanol–water partition coefficient (log P) dictates that they will not be free in aqueous solution at an appreciable concentration. Rather, they will essentially be present exclusively in the lipid bilayer. The plausible sequence of events is therefore that the C1 domain is in equilibrium with the membrane, rapidly associating and dissociating. When associated with the membrane, the binding of ligand by the C1 domain would stabilize its association, slowing its rate of release and thus shifting the equilibrium distribution of receptor in favor of membrane association. This expectation was confirmed in elegant studies using stopped flow spectroscopy to determine the on- and off-rates of the C1b domain of PKCβ to phospholipid vesicles (Dries and Newton 2008). The extent of negative charge on the vesicles had marked effect on the rate of association, whereas diacylglycerol or phorbol ester predominantly regulated the rate of dissociation. The situation might be different for highly potent, relatively hydrophilic ligands. In this case, the ligands would have significant aqueous solubility and would thus have the potential to bind to the free C1 domain.

Molecular modeling has provided additional insights into the binding of the phorbol ester and other ligands to the C1 domain. Different structural classes of ligands in fact do not bind in exactly the same fashion. Rather, different ligands utilize overlapping combinations of similar and distinct elements to form hydrogen bonds with the C1 domain (Pak et al. 2001). In addition, different ligand templates when bound to the C1 domain project their hydrophobic side chains at different orientations relative to the C1 domain. These different orientations must necessarily be reflected in the preferred orientation of the C1 domain relative to the lipid bilayer surface. Although the orientations of the hydrophobic groups may differ, for most of the ligands there is an element of overall similarity in that the hydrophobic groups project into the lipid bilayer. Bryostatin, which has unique biology, contrasts markedly in that the large upper portion (encompassing rings A and B) of this macrocyclic lactone appears to cap the top of the C1 domain (Kimura et al. 1999). Intriguingly, this portion of the molecule is a major contributor to the unique biology of the compounds (Keck et al. 2008, 2009). Finally, in the case of DAG and the DAG-lactones, potent constrained synthetic analogs of diacylglycerol, modeling indicates that these compounds have two alternate binding orientations, designated sn-1 and sn-2. The two orientations differ in whether it is the carbonyl of the sn-1 or the sn-2 position that, along with the hydroxyl of the sn-3 position, hydrogen bonds to the C1 domain (Sigano et al. 2003; Marquez and Blumberg 2003). DAG prefers the sn-1 binding orientation. The DAG-lactones prefer the sn-2 orientation. Once again, the binding orientation controls which substituent projects into the lipid bilayer, leading to different preferences for the pattern of substitution on these two closely related families of molecules.

3.6 Role of the Lipid Environment in Ligand Binding to the C1 Domain

The actual binding of ligand with the C1 domain is measured in the presence of phospholipid, leading to formation of a ternary complex of ligand – C1 domain – lipid. In contrast, limitations in methodology largely restrict the quantitative modeling to the binary complex of ligand and C1 domain. If the C1 domain thus represents only a half-site for binding, what are the contributions of the other half-site, *viz.* the phospholipid? First, it is clear that the phospholipid, albeit important, is not essential. Thus, binding of phorbol ester to the C1 domain could be measured in the absence of phospholipid albeit with substantially reduced affinity (Kazanietz et al. 1995a). This observation has suggested the strategy of antagonizing PKC with ligands that would bind to the C1 domain but would destabilize membrane insertion, thereby maintaining PKC in the wrong cellular location for interaction with its substrates (see discussion in Blumberg et al. 2008). Second, interactions of ligand with the phospholipid headgroups provide an explanation for the striking inconsistency between the predicted pharmacophore derived from comparison of the various classes of ligands for PKC and the X-ray crystallographic structure of the binding complex. The 9-OH in the phorbol ester structure constitutes the third element of the postulated pharmacophore but has no interaction with the C1 domain. Similarly, in the DAG-lactones, although only one carbonyl (either sn-1 or sn-2) is involved in interaction with the C1 domain, both carbonyls are critical for its binding as determined by biochemical measurements in the presence of phospholipid (Kang et al. 2003, 2005). The role of the C9–OH of the phorbol ester in lipid interaction is supported by a recent, elegant modeling study (Hritz et al. 2004). Likewise, appropriate substitutions on the hydrophobic domains of the DAG-lactones demonstrate marked effects on binding potencies, consistent with the suitability or lack thereof of the substituent to interact with the phospholipid bilayer (Kang et al. 2006).

3.7 Influence on Activity of the Pattern of Side Chain Substitution on the Ligands

As described above, an early observation was that different phorbol esters could induce different patterns of biological response, where the only variable between ligands was the nature of the substituents on the phorbol ester. Thus, among 12-deoxyphorbol 13-monoesters, the short-chain substituted 13-acetate or 13-phenylacetate were inflammatory but not tumor-promoting (Fürstenberger and Hecker 1972), whereas the more lipophilic 13-tetradecanoate was not only inflammatory but also a potent tumor promoter (Zayed et al. 1984). Similarly, the symmetrically substituted phorbol 12,13-diesters with unsaturated side chains were likewise either not tumor-promoting or weakly tumor-promoting (Fürstenberger and Hecker 1972), unlike the corresponding derivatives with saturated chains that were tumor-promoting.

Similar differences were seen in mechanistic analysis. The paradigmatic tumor-promoting phorbol ester PMA induced a complicated pattern of translocation of the PKCδ isoform, visualized with the GFP fusion construct (Wang et al. 1999). PMA first induced GFP-PKCδ translocation to the plasma membrane of Chinese hamster ovary cells followed by redistribution of the GFP-PKCδ to the nuclear membrane and internal membranes. The tumor-promoting 12-deoxyphorbol 13-tetradecanoate induced a similar pattern of translocation, whereas the nonpromoting 12-deoxyphorbol 13-acetate or 13-phenylacetate induced translocation directly to the internal and nuclear membranes.

The relative paucity and lack of availability of derivatives of the natural products that interact with the C1 domains has limited detailed exploration of the influence of the pattern of hydrophobic domain substitution on biological activity. This situation is changing with the development of the DAG-lactone as a potent, synthetically facile template for affording C1 domain ligands. Using combinatorial chemistry, the Marquez group has begun to generate libraries of DAG-lactones varying only in their hydrophobic domains (Duan et al. 2004, 2008). These initial libraries were evaluated in a battery of different biological assays (Duan et al. 2008). These assays included binding to PKCα versus RasGRP3, induction of AP-1 transcriptional activity versus induction of transformation in JB6 mouse epidermal cells, induction of α-secretase activity in rat neuroblastoma cells, enhancement of cellular motility in MCF-10A human breast cancer cells, sensitization of LNCaP human prostate cancer cells to radiation induced apoptosis, or costimulation together with IL-12 of interferon-γ production in human NK cells. The striking conclusion was that different biological responses reflected different structure activity relations for the hydrophobic substituents. The plausible model is that important contributors to this diversity in structure activity relations are the local lipid microdomains with which the DAG receptors interact and which form the half-sites for binding, along with the C1 domains. The different hydrophobic domains on the ligand could thus be thought of as providing selective "zip codes" contributing to specificity.

3.8 Localization of Ligand and Its Relation to Localization of DAG Receptor

As an approach to determine the degree to which the kinetics and localization of ligand determined the kinetics and localization of the PKC isoforms, living cells expressing PKC isoforms fused to appropriate GFP variants were treated with a series of fluorescent phorbol ester derivatives that could be imaged in real time and were compatible with simultaneous imaging of the GFP (Braun et al. 2005). In the case of PKCα, this isoform localized to the plasma membrane regardless of the pattern of distribution of the phorbol ester, indicating that the ligand was necessary to drive localization but that other factors, dictating plasma membrane localization, were dominant in determining the final isoform distribution. In contrast, for PKCδ and RasGRP3, the pattern of DAG receptor localization mirrored that of the ligands.

Interestingly, the fluorescent ligands revealed that the rates of uptake depended very much on the lipophilicities of the compounds, with the derivatives having similar lipophilicity to that of PMA requiring 30–60 min for equilibration, whereas the more hydrophilic derivatives penetrated quickly. In addition, the lipophilic derivatives did not distribute uniformly within the cell as they penetrated. Rather, they first accumulated in the plasma membrane before transferring to internal membranes, explaining the pattern of translocation observed for PKCδ in response to PMA. This pattern presumably reflects the delay occasioned by the plasma membrane-dissolved PMA needing to transfer from the plasma membrane into the aqueous phase of the cytoplasm before it transfers from the internal aqueous phase into the internal membranes. Finally, the presence of overexpressed PKC delayed the transfer of the fluorescent phorbol ester into internal membranes, reflecting the sequestration of free ligand in the membrane as a result of receptor binding. This observation is a reminder that receptor can act as a sink for ligand. Under conditions of limiting ligand, multiple receptors at a single location within the cell will compete for the available DAG, potentially leading to antagonism of the receptors of weaker affinity. A further intriguing concept from these studies is that different ligands can activate different receptors in a different temporal sequence, depending on the sites of these receptors within the cell. If a consequence of activation of one receptor is the inhibition of another – the model of "the first one out closes the gate behind him/her" – then different sequences of activation could lead to different responses.

3.9 Role of Cellular Context in Determining Ligand Structure Activity Relations

Lipid microdomains provide only one element in the cellular environment that will differentially regulate ligand recognition and structure activity relations, but the profound influence of overall cellular environment is unambiguous. Comparison of the relative potencies of PMA and of bryostatin 1 to induce translocation of PKCα and PKCδ in two different cell types, mouse epidermal cells (Szallasi et al. 1994a) and mouse 3T3 fibroblasts (Szallasi et al. 1994b) provides a clear example. Here, the same ligands acting on the same PKC isoforms showed markedly different selectivities in the two different cell types.

One obvious candidate contributing to differential regulation by cellular context is the internal calcium level. Calcium interaction with the C2 domains of the classic PKC isoforms α, β, and γ will enhance the association of these isoforms with the membrane, whereas the association with the membranes of the novel PKC isoforms δ, ε, η, and θ will not be affected.

Another factor of course will be the different lipid compositions, whether of different cellular membranes or different compositions in the corresponding membranes of different cells. Different PKC isoforms show different dependence on the types of phospholipids required for activity. For example, PKCδ and PKCα show strong dependence on phosphatidylserine, whereas PKCε and PKCγ do not (Medkova and Cho 1998; Ananthanarayanan et al. 2003; Stahelin et al. 2004). PKCη is uniquely responsive to cholesterol sulfate (Ikuta et al. 1994). Likewise, different phospholipid compositions cause different structure activity relations for different DAG receptors, shifting the relative potencies of ligands. Thus, higher negative phospholipid compositions favored PDBu binding to PKCα, whereas a low negative phospholipid composition favored the binding to RasGRP (Lorenzo et al. 2000).

The concept of the hydrophobic switch makes the strong prediction that ligand binding and receptor activation are coupled to the energy constraints of the overall conformational change induced by the switching mechanism. All mechanisms that hold the receptor in either an open or closed conformation will thus facilitate or impede the binding and the coupled conformational change associated with the hydrophobic switching. A powerful illustration of this principle is provided by chimeras prepared between different PKC isoforms (Acs et al. 1997). Phorbol ester induced translocation of the (C1 domain containing) regulatory domain of PKCα with 30-fold greater potency when this domain was coupled to the catalytic domain of PKCε in lieu of the normal PKCα catalytic domain. In addition, since the different classes of DAG receptors presumably have differential conformational consequences linked to this switching mechanism, it would thus be expected that they would have different dependence on the different components of the cellular environment.

3.10 Different Functional Roles for the C1 Domain as a Hydrophobic Switch

The C1 domain functioning as a hydrophobic switch responsive to ligand binding may have two distinct consequences. The more general consequence is that it stabilizes association of the DAG receptor at hydrophobic surfaces, typically at cellular membranes. In the case of the PKCs, this membrane association is also associated with stabilization of an unfolded conformation of the enzyme, leading to extraction of the pseudosubstrate region from the catalytic site and enzyme activation. In the case of PKD, in contrast, enzymatic activation is a consequence of phosphorylation, typically by PKC, and the role of the C1 domain is for membrane association of the activated enzyme (Wang 2006). Since it is known that PKC isoforms such as PKCδ can be activated by tyrosine phosphorylation downstream of oxidative stress (Kikkawa et al. 2002; Steinberg 2008), it is plausible in such cases that C1 ligands under such conditions might again be influential for localization of the enzyme but no longer for its enzymatic activity.

Nonetheless, absolute enzymatic activity is not the most relevant parameter for determining biological consequences. For PKC or PKD, the ability to phosphorylate downstream targets will depend not only on the intrinsic activity of the enzyme but also on the concentration of its substrate, which in turn will be determined by the proximity of the enzyme and its substrates in cases where both are positionally restricted. Position will also determine proximity of the enzyme to other modifying factors, such as phosphatases that can antagonize substrate phosphorylation or remove phosphates from the regulatory sites on the enzymes themselves (Gallegos et al. 2006; Gould and Newton 2008).

3.11 Access of the C1 Domains to Its Ligands

The C1 domain is found in the context of the intact receptor. Because the structure of the C1 domain is one of a hydrophobic face with a hydrophilic cleft, it would be surprising if that hydrophobic surface were normally exposed rather than occluded in the absence of ligand. In the case of PKCγ, in fact, the kinetics of membrane translocation indicated that ligand only induced translocation after a latency period, whereas a fragment of PKCγ with the C1 domain was translocated immediately (Oancea and Meyer 1998). In contrast, translocation of PKCγ was immediate in response to elevated calcium. These results suggested that the C1 domain was inaccessible in the native enzyme and required time for the enzyme to unfold.

Elegant confirmation of this model is found for β2-chimaerin (Canagarajah et al. 2004). X-ray crystallography revealed that the C1 domain was in fact held in the occluded state by interactions with multiple domains – the N terminus, the SH2 domain, the RacGAP domain, and the linker domain between the C1 and the SH2 regions. Mutations that destabilized these contacts with the C1 domain enhanced the potency of phorbol esters for inducing translocation. A similar situation was shown to prevail for α2-chimaerin (Colon-Gonzalez et al. 2008).

To the degree that the C1 domain is occluded in the unstimulated receptor, all those elements of cellular context which influence the ability to expose the C1 domain will in parallel influence the potency of ligands to bind to the C1 domain.

3.12 Potential of the C1 Domain as a Therapeutic Target: Opportunities for Selectivity

A potential obstacle to the use of the C1 domains as targets for drug development is the high degree of conservation of the domain. The basis for the interaction of phorbol ester with the C1 domain is hydrogen bonding between the ligand and the peptide backbone of the residues forming the binding cleft. It is thus less sensitive to the specific residues constituting the cleft than would have been the case if the hydrogen bonding were to the head groups of the amino acids. On the other hand, the actual formation of the ternary binding complex integrates the interactions of both the ligand and the C1 domain with the lipid bilayer. For example, the C1b domain of the novel PKC isoforms contains tryptophan (W) at position 22, whereas the classical isoforms contain tyrosine in that position. Dries et al. (2007) showed that W22 in place of Y22 in the C1 domain markedly enhanced its affinity for lipid vesicles containing DAG. More generally, the ability of the C1 domains to interact with and to insert into the bilayer will depend on the nature of the many residues on the outer face of the C1 domain. These surfaces show marked variety, as revealed by modeling (Blumberg et al. 2008). Finally, the formation of the ternary complex will reflect the energetics of the conformational change in the receptor and the interactions of the receptor as a whole with the membrane. This multiplicity of contributing elements discussed above provides great opportunities for diversity.

3.13 Potential of the C1 Domain as a Therapeutic Target: Antagonistic Functions of the Diverse DAG Receptors

Just as the complex C1 domain–ligand–membrane interactions provide a strong basis for selective drug design, so the diversity of DAG signaling pathways provides a strong underlying rationale for the targeting of C1 domains. It has become clear that different isoforms of PKC can have antagonistic functions. The strongest example is perhaps the contrast between PKCδ, which in most systems is proapoptotic and growth inhibitory, and PKCε, which is antiapoptotic and growth promoting (Griner and Kazanietz 2007). Thus, for example, in the mouse skin system, overexpression of PKCε enhances the development of carcinomas (Aziz et al. 2007) whereas overexpression of PKCδ inhibits phorbol ester induced tumor promotion (Reddig et al. 1999). The implication is that a selective activator of PKCδ might accomplish the same overall result as a selective inhibitor of PKCε.

Not only may one isoform of PKC be antagonistic of another PKC isoform but one family of DAG receptors may be antagonistic of another. The chimaerins, for example, function as inhibitors of Rac and are predicted to be tumor suppressors (Griner and Kazanietz 2007; Yang and Kazanietz 2007). They thus stand in contrast to the typical PKC. Likewise, activation of DAG kinase, of which several isoforms recognize DAG at their C1 domains, terminates DAG signaling and should thereby suppress response through all of the DAG receptor families.

A more complicated possibility is afforded by RasGRP (Topham and Prescott 2001). RasGRP binds to DAG kinase zeta. This binding is enhanced in the presence of phorbol ester and the binding drives localization of DAG kinase to the membrane where it can suppress DAG signaling. Since RasGRP requires phosphorylation by PKC for its function as an exchange factor for Ras (Teixeira et al. 2003; Brodie et al. 2004; Zheng et al. 2005), a selective C1 domain targeted ligand for RasGRP which does not lead to PKC activation would be predicted to induce translocation of DAG kinase zeta through the adapter protein function of RasGRP without leading to the downstream consequences of signaling through PKC or active RasGRP. Such a ligand would thus lead to antagonism of physiologically activated DAG receptor functions.

3.14 C1 Domain Ligands as Clinical Candidates: Bryostatin 1

Although there are strong mechanistic arguments for why C1 domains represent attractive targets for development of drugs directed at DAG receptors such as PKC, the most powerful argument is the reality that several such drugs are already in clinical trials. The most extensively studied of these is bryostatin 1. Bryostatin 1 was identified as part of the very productive natural products screening program of the Pettit group, evaluating marine sources for antiproliferative activity against the P388 leukemia cell line (Pettit 1991). Currently, bryostatin 1 either as a single agent or in combination is the subject of 38 clinical trials which have been completed, are in progress, or are being instituted.

Bryostatin 1 is a macrocyclic lactone, possessing 11 chiral centers. After initial reports that bryostatin 1 functioned as a potent PKC ligand able to induce several similar effects as did the phorbol esters (Berkow and Kraft 1985; Smith et al. 1985), a critical observation by Kraft and coworkers was that in some other instances bryostatin 1 failed to induce a typical phorbol ester response. Importantly, in such instances, bryostatin 1 was paradoxically able to antagonize the response to phorbol ester if both agents were co-applied (Kraft et al. 1986). This antagonism proved to be the rule rather than the exception. Phorbol esters block differentiation of Friend erythroleukemia cells in response to hexamethylene bisacetamide and bryostatin 1 reverses this block (Dell'Aquila et al. 1987). Phorbol esters induce differentiation of HL-60 promyelocytic leukemia cells and bryostatin 1 inhibits this differentiation (Kraft et al. 1986). Phorbol esters block cell–cell communication in primary mouse epidermal cells and bryostatin 1 leads to restoration of cell–cell communication (Pasti et al. 1988). Phorbol esters induce arachidonic acid release in mouse C3H 10T1/2 cells and bryostatin 1 inhibits this release (Dell'Aquila et al. 1988). Phorbol esters induce attachment and block proliferation in U937 leukemia cells whereas bryostatin 1 blocks attachment and restores proliferation (Ng and Guy 1992; Asiedu et al. 1995; Grant et al. 1996; Vrana et al. 1998; Keck et al. 2009). Finally and of particular significance, we showed that bryostatin 1 failed to function as a tumor promoter in mouse skin and indeed inhibited tumor promotion by phorbol ester (Hennings et al. 1987). These findings provided strong motivation for the evaluation of bryostatin 1 treatment in those cancers for which PKC was implicated.

While bryostatin 1 provides an example of the potential of C1 domain ligands to act as antagonists of PKC action, unfortunately the mechanism(s) responsible for the functional antagonism exerted by bryostatin 1 remains unresolved. It is clear, for example, that in some systems bryostatin 1 treatment induces a response similar to that of PMA but of transient duration. This is the case, for example, for inhibition of cell–cell communication in mouse epidermal cells (Pasti et al. 1988). Here, bryostatin 1 induces a response at 1 h but the response is largely lost by 4 h, whereas that by PMA is persistent. Inhibition of epidermal growth factor binding behaves similarly in these cells (Sako et al. 1987). On the other hand, the failure of bryostatin 1 to induce arachidonic acid release in C3H10T1/2 cells was evident at 30 min, the earliest time at which response to PMA could be observed, suggesting an absolute lack of stimulation of PKC in this system for this response (Dell'Aquila et al. 1988). The C3H10T1/2 system further illustrates that bryostatin 1 does not induce a single pattern of response in a given cell. Thus, whereas bryostatin 1 failed to induce arachidonic acid release in the C3H10T1/2 cells, in this same system, it led to persistent inhibition of EGF binding, similar to the response to PMA.

A second mechanistic difference between bryostatin 1 and PMA lies in its effect on downregulation of PKC isoforms. In multiple cell types, whereas PMA causes dose-dependent down regulation of PKC isoforms such as PKCα or PKCδ (Gould and Newton 2008), bryostatin 1 causes dose-dependent downregulation of PKCα but, for PKCδ, it induces a biphasic pattern of downregulation, with maximal downregulation at low doses and protection of PKCδ from downregulation at higher doses (Szallasi et al. 1994a, b). In the case of HOP92 cells, bryostatin 1 likewise affords a biphasic dose response curve for stimulation of proliferation and we were able to show that this pattern of biological response was inversely mirrored by the level of downregulation of PKCδ (Choi et al. 2006). This effect on PKCδ appeared to be responsible for the effect of bryostatin 1 on cell proliferation in this instance, since suppression of PKCδ expression with siRNA rendered the proliferative response insensitive to bryostatin 1 and overexpression of PKCδ inhibited cell growth. These results are consistent with the typical antiproliferative effect of PKCδ.

A third mechanistic difference between bryostatin 1 and PMA, already discussed, is the different pattern of translocation of PKC delta that it induces (Wang et al. 1999). While PMA induces translocation initially to the plasma membrane and subsequently to the internal membranes and the nuclear membrane, bryostatin 1 induces translocation directly to these internal sites, with little or no translocation to the plasma membrane. This pattern of translocation is similar to that induced by the nontumor-promoting phorbol esters and contrasts with that by 12-deoxyphorbol 13-tetradecanoate, for example.

Mutational analysis of the individual C1 domains of PKCδ indicated that both C1 domains contributed to both the descending and the ascending phases of the dose response curve for downregulation of PKCδ, albeit with a greater contribution from the C1b domain (Lorenzo et al. 1999). These findings are consistent with other studies on the relative importance of these two domains on ligand responsiveness by PKCδ (Bögi et al. 1998). Importantly, they do not provide support for the model that interaction at one C1 domain could drive downregulation while interaction at the second domain could be protective. Further insight is provided by chimeras between the PKCα and PKCδ isoforms. The protection from downregulation at high bryostatin concentrations is observed in chimeras with the PKCδ catalytic domain, and not with the PKCδ regulatory domain which contains the C1 domains responsible for bryostatin 1 binding (Lorenzo et al. 1997).

The great challenges of *de novo* synthesis of bryostatins, arising from their complex structures, and the difficulties of isolation of bryostatins from natural sources have been major impediments to investigation of their structure activity relations. A breakthrough was the design of simplified bryostatin derivatives, termed bryologues, by the Wender group (Wender et al. 1999). This group showed that potent activity for binding to PKC was preserved in derivatives simplified in the A and B rings, comprising carbons 1–14. This group concluded that this portion of the molecule functioned simply as a linker region for the molecule.

While this conclusion may remain valid in terms of the ability of bryostatins to bind to PKC, it ignored the biological question of whether such derivatives in fact possessed the unique biological pattern of response of bryostatin or whether these derivatives rather were phorbol-like molecules built on a bryostatin skeleton. This issue was addressed by the Keck group. They showed that a series of bryologue derivatives lacking functionalization on the A and B rings maintained nM potency for PKC but, when assayed on the U937 leukemia cells, acted essentially like PMA rather than like bryostatin 1 (Keck et al. 2008). In this system, as described above, PMA inhibits cell growth and induces cell attachment. Bryostatin 1 shows a much reduced effect on these parameters and inhibits the PMA response when both compounds are applied together (Ng and Guy 1992; Asiedu et al. 1995; Grant et al. 1996; Vrana et al. 1998). The Keck group proceeded to show that another bryologue, now incorporating the pattern of substitution of bryostatin 1 in the A-ring, functioned in this system like bryostatin 1, neither inhibiting U937 cell proliferation nor inducing cell attachment, while blocking the effect of PMA on these parameters (Keck et al. 2009). These studies establish that this "linker region" is a major contributor to the unique pattern of biological response to the bryostatins. According to computer modeling (Kimura et al. 1999), this region of bryostatin overlays the hydrophobic face of the C1 domain, providing a cap unique to this class of molecules.

It should now be possible to identify the specific groups responsible for this activity and potentially to open the way to the design of much simplified molecules incorporating these structural features. Further, understanding of the mechanisms by which these structural features translate into inhibition of many PKC responses may provide generalizable strategies for the development of C1 targeted antagonists of PKC. Of course, PKC is only one of the families of DAG receptors. It remains to be clarified whether bryostatin differentially affects the function of these other classes of receptors.

3.15 C1 Domain Ligands as Clinical Candidates: Ingenol 3-Angelate (PEP005)

PEP005 (ingenol 3-angelate) is a second C1 domain-directed ligand that is well advanced in drug studies, with 18 clinical trials completed, in progress, or recruiting for treatment of actinic keratosis and nonmelanotic skin cancer. Ingenol 3-angelate is a constituent of traditional medicines derived from *Euphorbia peplus* and *Euphorbia antiquorum* (Adolf et al. 1983), which were reputed to have activity against skin cancer (Ogbourne et al. 2007). Conceptually, ingenol 3-angelate seems analogous to the short-chain substituted 12-deoxyphorbol derivatives, which are inhibitors of tumor promotion whereas the more hydrophobic congeners are potent tumor promoters. Similarly, the long chain substituted ingenol 3-esters are potent tumor promoters (Opferkuch and Hecker 1982), whereas ingenol 3-angelate was reported not to be tumor-promoting (Adolf et al. 1983). Of further note, both ingenol 3-angelate (Adolf et al. 1983) and its analog ingenol 3-angelate 20-acetate (Zayed et al. 1998) proved to have high local toxicity in the animal experiments, again in parallel with the short-chain substituted 12-deoxyphorbol 13-monoesters.

Analysis of the mechanism of action of ingenol 3-angelate indicated that, while it bound to and activated PKC isoforms, it displayed several differences relative to PMA (Kedei et al. 2004). First, consistent with its more hydrophilic nature, ingenol 3-angelate was less able than PMA to stabilize association of the C1 domain with lipid surfaces, as measured by surface plasmon resonance; and, at higher concentrations, ingenol 3-angelate antagonized the association driven by PMA. Second, ingenol 3-angelate induced IL-6 secretion in WEHI-231 cells with a biphasic dose response curve, contrasting with a more monophasic curve for PMA, and the absolute level of induction was higher. In addition, ingenol 3-angelate showed substantial differences in downregulation of PKC isoforms compared with PMA and these differences were markedly dependent on the specific cell type. For example, whereas PMA was twofold more potent than ingenol 3-angelate for downregulation of PKCδ in the WEHI-231 cells, in the Colo-205 cells PMA was 25-fold less potent, for a relative shift in potencies of 50-fold. Similarly, in the Colo-205 cells ingenol 3-angelate was 125-fold more potent for downregulation of PKCδ than for down regulation of PKCα, whereas in the WEHI-231 cells ingenol 3-angelate was threefold less potent for downregulation of PKCδ than for downregulation of PKCα.

At the whole animal level, ingenol 3-angelate showed toxicity for the LK-2 squamous cell carcinoma line grown *sc* and, in this system, a critical contributor to this toxicity was acute inflammation associated with neutrophil infiltration. Likewise, ingenol 3-angelate was reported to induce, both in vitro and in vivo, induction of MIP-2, TNF-α, and IL-1β, all involved in neutrophil attraction and activation (Challacombe et al. 2006).

3.16 C1 Domain Ligands Selective for Subsets of DAG Receptors or C1 Domains

A goal of the combinatorial chemistry strategy with the DAG-lactones discussed above was to probe the potential of variation in the hydrophobic domain of the ligand for generating selective ligands. Indeed, among the early compounds to emerge from this effort were DAG-lactones with appreciable selectivity for RasGRP1/3 relative to PKCs (Pu et al. 2005). The DAG-lactone 130C037 bound RasGRP1/3 with affinities of 3–4 nM; the affinity for PKCε was 30 nM and that of PKCα was 340 nM. Similar selectivity was observed in intact cells for induction of translocation. Likewise, 130C037 was able to stimulate ERK phosphorylation only in HEK-293 cells overexpressing RasGRP3, whereas PMA stimulated ERK phosphorylation in the control cells as well as the overexpressing cells.

The above results were with intact C1 domain containing proteins. It is also clear that compounds can possess marked selectivity among isolated C1 domains. For example, the RasGRP selective DAG-lactone 130C037 bound with high affinity (1.8 nM) to the C1b domain of PKC delta, whereas it had 1,400-fold less affinity for the C1a domain as well as 1,500-fold less affinity for the C1a domain of PKC alpha (and even less affinity for the C1b domain of PKC alpha). In extensive work, Irie and coworkers (Nakagawa et al. 2006; Irie et al. 2005) similarly described benzolactams which displayed orders of magnitude differences in their affinities for different individual C1 domains.

A limitation in the analysis of selectivity for isolated C1 domains, as distinct from selectivity for the intact receptors, is that this selectivity may not carry over to the intact proteins. The DAG-lactone 130C037, cited above, provides a good example (Pu et al. 2005). Whereas its measured affinity in vitro for either C1 domain of PKCα was at least 1,500-fold weaker than its affinity for the C1b domain of PKCδ, for the intact PKC isoforms it had only fourfold weaker binding affinity for PKCα than for PKCδ . It is thus clear that other elements in the intact PKC may make a major contribution to the formation of the ternary complex with a ligand.

3.17 A Widening Window of Opportunities for C1 Domain Directed Ligands

C1 domains have been subclassified into two groups – DAG-responsive and DAG-unresponsive or "atypical" (Hurley et al. 1997). As with virtually every other aspect of DAG receptor function, the emerging picture is more complicated. "Atypical" C1 domains can be further subdivided into two groups, those that are sufficiently divergent so that they have lost the overall geometry of the binding cleft and those where the geometry is retained but where other factors impair binding. The C1 domains of Raf and of KSR (Kinase Suppressor of Ras) provide examples of divergent binding clefts. One of the two loops of the binding cleft has undergone the deletion of several residues and the geometry is correspondingly distorted (Mott et al. 1996; Zhou et al. 2002). On the other hand, modeling of the C1 domains of the atypical PKC isoforms zeta and iota indicate that these "atypical" C1 domains retain a binding cleft geometry similar to that of the C1b domain of PKC delta (Pu et al. 2006). They differ, however, in the presence of multiple arginine residues lining the rim of the binding cleft. Since this portion of the C1 domain inserts into the lipid bilayer in the presence of ligand, these charged residues could interfere with the formation of the ternary ligand – C1 domain – lipid complex. Furthermore, the modeling indicates that the arginine residues can swing into and occlude the binding cleft, thereby competing with ligand. In support of this explanation for the lack of responsiveness of the C1 domains of the atypical PKC isoforms to phorbol esters, we showed that mutation of the corresponding residues in the C1b domain of PKC delta to arginine progressively led to loss of ligand-binding activity. Conversely, mutation of the arginine residues in the isolated C1 domains of PKC zeta and iota to the corresponding residues in the C1b domain of PKC delta restored phorbol ester responsiveness. Whether one can design ligands that will exploit the structural peculiarities of such atypical C1 domains that retain the binding cleft geometry remains to be determined.

The DAG/phorbol ester responsiveness of the C1 domains of the Vav isoforms is unclear. The Vav family members function as guanyl exchange factors for the small GTPase Rho (Hornstein et al. 2004; Swat and Fujikawa 2005). We showed that Vav1 bound neither [^{3}H]PDBu nor [^{3}H]bryostatin under conditions in which very weak binding affinity should still have been detectable (Kazanietz et al. 1994). On the other hand, modeling suggests that the C1 domain of Vav2 is very similar to that for PKC (Heo et al. 2005), and crystallographic analysis of a Vav1 fragment including the C1 domain together with the DH and PH domains likewise indicates that the binding cleft in the C1 domain is preserved (Rapley et al. 2008). This cleft is located in apposition to the DH domain, which might both prevent access by ligand and prevent association with the lipid bilayer, which provides one of the important elements of the overall pharmacophore. Nonetheless, the retention of the binding site geometry raises the exciting possibility that appropriately modified ligands could access this binding site and disrupt the activating function of Vav proteins for Rho family members.

The C1 domain of RasGRP2 has a seemingly homologous sequence to that of the other members of the RasGRP family. Although it was able to bind to anionic phospholipid vesicles as did the C1 domains of the other family members, it did not respond to either the addition of exogenous DAG or of phorbol ester with translocation when expressed in cells or with enhanced association with phospholipid vesicles in vitro (Johnson et al. 2007). In light of its close sequence homology, the basis for its lack of ligand recognition should be of considerable interest.

3.18 C1 Domains with Reduced Affinity

Intermediate between C1 domains with high affinity for DAG/phorbol ester and those without measureable affinity (as yet) are those C1 domains with reduced activity, but where the reduced affinity raises the question of whether these C1 domains are still capable of recognizing physiological levels of endogenous DAG. For example, we have shown that the C1 domains of MRCK alpha and beta indeed bind PDBu but with 60–90-fold weaker affinities than the C1 domain of PKC delta (Choi et al. 2008). Since it is not likely that there would be such differences in the concentration of endogenous DAG, a probable explanation is that other coregulators in the case of MRCK contribute to the membrane association, complementing the contributions from the liganded C1 domain itself.

3.19 Role of the Individual C1 Domains in the Responsiveness of PKC to Ligands

The different subfamilies of C1 domain containing receptors for DAG/phorbol ester fall into two categories. PKC and PKD possess twin C1 domains. The members of the other subfamilies contain only a single C1 domain responsive to DAG/phorbol ester. Just as there may be different affinities and selectivities for the individual C1 domains of these latter receptor subfamilies, so there are differences between the individual C1 domains, C1a and C1b, in the PKC isoforms. These differences provide further opportunities for potential drug design, since not only might one exploit the combination of selectivities of the two C1 domains in a single isoform, but one might also be able to take advantage of the different spacing between the C1 domains of the classic and novel PKCs.

Unfortunately, the individual roles and specificities of the C1a and C1b domains are still not clearly understood, with the underlying complication being that different technical approaches have yielded partially inconsistent conclusions. The initial strategy taken by this laboratory was to mutate the C1a domain, the C1b domain, or both domains in PKCα and PKCδ, introducing a P11G mutation into the domain structure. For the isolated domain, this mutation had been shown to lead to a 125-fold loss of binding affinity for PDBu (Kazanietz et al. 1995b). The mutated PKC isoforms were introduced into cells and their responsiveness to various phorbol esters and other PKC ligands were quantitated for membrane translocation.

For PKCδ expressed in mouse 3T3 cells, mutation in the C1a domain caused little shift in the dose response curve for translocation by PMA; mutation in the C1b domain caused a 20-fold shift; and mutation in both C1a and C1b caused a 140-fold shift (Szallasi et al. 1996). These results argued that both domains bound PMA but that the C1b domain played the predominant role in PMA binding. This pattern of selectivity proved to be a function of the specific ligand (Bögi et al. 1998). Whereas PMA showed selectivity for the C1b domain, analysis of these same mutants with a series of ligands revealed that indolactam V and octylindolactam V behaved like PMA, with selectivity for the C1b domain, whereas mezerein, 12-deoxyporbol 13-phenylacetate, and bryostatin 1 were affected to the same degree by mutation either in the C1a or the C1b domain. Interestingly, these latter compounds are all not tumor-promoting, whereas PMA and the octylindolactam V are tumor-promoting. The conclusion is that different ligands showed different relative dependence on the C1a and C1b domains.

Analysis of the relative roles of the C1a and C1b domains of PKCα, using the same approach, yielded a different picture (Bögi et al. 1999). Here, of the three ligands examined, not only mezerein but also PMA and octylindolactam V showed comparable dependence on the C1a and the C1b domains. Interestingly, the double mutant of PKCα showed no additional loss of potency for PMA compared to the single mutants. The conclusion is that different PKC isoforms behave differently in the relative contributions to ligand interaction of their C1a and C1b domains.

A difficulty in the above approach is that the effect of the P11G mutation in the C1 domain was determined for the isolated C1 domain. If the C1 domain is stabilized in the context of the intact protein, then perhaps the mutation in the intact protein is not causing the same magnitude of loss of activity as was demonstrated in the isolated C1 domain. This concern is highlighted by further mutational studies, which indicated that the mutation P11R in the C1b domain caused only a 3.6-fold reduction in binding affinity (Pu et al. 2006). For PKCδ, an alternative approach was provided from studies of the interaction of dioxolanone derivatives with the C1 domains (Choi et al. 2007). Dioxolanones are DAG-lactone derivatives. These compounds have an affinity similar to that for the DAG-lactones. However, modeling revealed that they possessed an additional point of interaction with the C1b domain, *viz.* Q27, which was not involved in the binding of the phorbol esters or DAG-lactones. The mutation Q27E correspondingly led to a several thousand fold loss of binding affinity for the dioxolanones whereas it had a more modest 20–60-fold effect for binding of the corresponding DAG-lactones or PDBu. Introduction of this mutation into the C1b domain of the intact PKCδ blocked its translocation in response to dioxolanone whereas response to phorbol ester or DAG-lactone was preserved. The retained response for these other ligands provides a powerful positive control for the effect of the mutation on the PKC itself. In the case of the C1a domain, the mutation caused marked loss of binding activity to the isolated C1a domain but did not inhibit translocation of the intact PKCδ mutated in the C1a domain. These findings demonstrate the predominant role of the C1b domain for translocation of PKCδ in response to this class of DAG analogs.

As discussed above, compounds such as 130C037 have been shown to have marked selectivity for individual C1 domains which may not directly translate into differences in activity on the intact PKC. Contributing factors may be the additional elements in the intact protein which contribute to membrane binding or C1 domain stability. Support for this suggestion comes from the analysis of chemically synthesized C1 domains (Irie et al. 2004a). These authors showed that the inclusion of additional basic residues at the C terminus of some C1 domains, *e.g.* RasGRP3, substantially enhanced the measured binding affinities. Likewise, for some of the C1 domains, a reduction in the assay temperature from 37 to 4°C yielded a much higher level of binding activity (B_{max}; Shindo et al. 2001).

To further explore the issue of the relative contributions of C1 domain structure versus positional context of the C1 domain within PKC, we prepared a series of mutants of PKCδ containing all combinations of one or two C1a or C1b domains in their normal positions or in the respective positions for one another (Pu et al. 2009). This approach again showed that the identity of the C1 domain was the primary determinant of binding and translocation, with the C1b domain making the major contribution and the C1a domain making only a minor contribution. This result was true not only for phorbol ester but also for mezerein and DAG.

3.20 Conclusion

Reflecting their central role in cellular control, the families of signaling proteins that integrate the information from the varying levels of DAG with that of other second messengers and signaling pathways provide attractive opportunities for drug development. Although cancer represents a major therapeutic target for these proteins, their impact is as diverse as the underlying biology that they mediate (DiazGranados and Zarate 2008; Chen and LaCasce 2008; Lee et al. 2008; Sun and Alkon 2006; Churchill et al. 2008; Dempsey et al. 2007; Farhadi et al. 2006). Because of the limited factors influencing specificity, largely encompassed by the geometry of the catalytic site, kinase inhibitors represent one productive approach for the PKCs and PKDs. The appreciably greater complexity of the C1 domain in the context of the cellular environment poses a correspondingly greater challenge for rational drug design. On the other hand, this complexity potentially provides the basis for a level of specificity beyond that achievable with catalytic site inhibitors. Moreover, the reality that only two of the classes of DAG receptors are protein kinases means that standard strategies of enzymatic inhibitor design are not even available for most of these other classes of targets. A critical theme in drug design is that of whether a target is "druggable." Here, the power of natural products asserts itself. We know that the C1 domain is a druggable target because natural products, designed by nature and acting through C1 domains, are in clinical trials. The challenge is to build on this opportunity.

Acknowledgments This contribution was supported by the Intramural Research Program of the NIH, National Cancer Institute, Center for Cancer Research.

References

Acs, P., Bögi, K., Lorenzo, P. S., Marquez, V. E., Biro, T., Szallasi, Z., et al. (1997). The catalytic domain of protein kinase C chimeras modulates the affinity and targeting of phorbol ester-induced translocation. *The Journal of Biological Chemistry, 272*, 22148–22153.

Adolf, W., Chanai, S., & Hecker, E. (1983). 3-O-angeloylingenol, the toxic and skin irritant factor from latex of *Euphorbia antiquorum* L. (Euphorbiaceae) and from a derived Thai purgative and anthelimintic (vermifuge) drug. *Journal of the Science Society of Thailand, 9*, 81–88.

Ananthanarayanan, B., Stahelin, R. V., Digman, M. A., & Cho, W. (2003). Activation mechanisms of conventional protein kinaseC isoforms are determined by the ligand affinity and conformational flexibility of their C1 domains. *The Journal of Biological Chemistry, 278*, 46886–468894.

Asiedu, C., Biggs, J., Lilly, M., & Kraft, A. S. (1995). Inhibition of leukemia cell growth by the protein kinase C activator bryostatin correlates with the dephosphorylation of cyclin-dependent kinase 2. *Cancer Research, 55*, 3716–3720.

Aziz, M. H., Manoharan, H. T., Sand, J. M., & Verma, A. K. (2007). Protein kinase Cepsilon interacts with Stat3 and regulates its activation that is essential for the development of skin cancer. *Molecular Carcinogenesis, 46*, 646–653.

Battaini, F., & Mochly-Rosen, D. (2007). Happy birthday protein kinase C: Past, present and future of a superfamily. *Pharmacological Research, 56*, 461–466.

Berenblum, I., & Shubik, P. (1947). The role of croton oil applications, associated with a single painting of a carcinogen, in tumour induction of the mouse's skin. *British Journal of Cancer, 1*, 379–382.

Berkow, R. L., & Kraft, A. S. (1985). Bryostatin, a non-phorbol macrocyclic lactone, activates intact human polymorphonuclear leukocytes and binds to the phorbol ester receptor. *Biochemical and Biophysical Research Communications, 131*, 1109–1116.

Blumberg, P. M. (1980). In vitro studies on the mode of action of the phorbol esters, potent tumor promoters: Part 1. *Critical Reviews in Toxicology, 8*, 153–197.

Blumberg, P. M. (1981). In vitro studies on the mode of action of the phorbol esters, potent tumor promoters: Part 2. *Critical Reviews in Toxicology, 8*, 199–234.

Blumberg, P. M., Acs, P., Bhattacharyya, D. K., & Lorenzo, P. S. (2000). Inhibitors of protein kinase C and related receptors for the lipophilic messenger sn-1, 2-diacylglycerol. In J. S. Gutkind (Ed.), *Signal Transduction and Cell Cycle Inhibitors* (pp. 347–364). Totowa, NJ: Humana Press.

Blumberg, P. M., Kedei, N., Lewin, N. E., Yang, D., Czifra, G., Pu, Y., et al. (2008). Wealth of opportunity – the C1 domain as a target for drug development. *Current Drug Targets, 9*, 641–652.

Bögi, K., Lorenzo, P. S., Acs, P., Szallasi, Z., Wagner, G. S., & Blumberg, P. M. (1999). Comparison of the roles of the C1a and C1b domains of protein kinase C alpha in ligand induced translocation in NIH 3T3 cells. *FEBS Letters, 456*, 27–30.

Bögi, K., Lorenzo, P. S., Szallasi, Z., Acs, P., Wagner, G. S., & Blumberg, P. M. (1998). Differential selectivity of ligands for the C1a and C1b phorbol ester binding domains of protein kinase C delta: Possible correlation with tumor-promoting activity. *Cancer Research, 58*, 1423–1428.

Braun, D. C., Cao, Y., Wang, S., Garfield, S. H., Hur, G. M., & Blumberg, P. M. (2005). Role of phorbol ester localization in determining protein kinase C or RasGRP3 translocation: Real-time analysis using fluorescent ligands and proteins. *Molecular Cancer Therapeutics, 4*, 141–150.

Brodie, C., Steinhart, R., Kazimirsky, G., Rubinfeld, H., Hyman, T., Ayres, J. N., et al. (2004). PKC delta associates with and is involved in the phosphorylation of RasGRP3 in response to phorbol esters. *Molecular Pharmacology, 66*, 76–84.

Canagarajah, B., Leskow, F. C., Ho, J. Y., Mischak, H., Saidi, L. F., Kazanietz, M. G., et al. (2004). Structural mechanism for lipid activation of the Rac-specific GAP, beta2-chimaerin. *Cell, 119*, 407–418.

Castagna, M., Takai, Y., Kaibuchi, K., Sano, K., Kikkawa, U., & Nishizuka, Y. (1982). Direct activation of calcium-activated, phospholipid-dependent protein kinase by tumor-promoting phorbol esters. *The Journal of Biological Chemistry, 257*, 7847–7851.

Challacombe, J. M., Suhrbier, A., Parsons, P. G., Jones, B., Hampson, P., Kavanagh, D., et al. (2006). Neutrophils are a key component of the antitumor efficacy of topical chemotherapy with ingenol-3-angelate. *The Journal of Immunology, 177*, 8123–8132.

Chen, Y. B., & LaCasce, A. S. (2008). Enzastaurin. *Expert Opinion on Investigational Drugs, 17*, 939–944.

Cho, W., & Stahelin, R. V. (2005). Membrane-protein interactions in cell signaling and membrane trafficking. *Annual Review of Biophysics and Biomolecular Structure, 34*, 119–151.

Choi, S. H., Czifra, G., Kedei, N., Lewin, N. E., Lazar, J., Pu, Y., et al. (2008). Characterization of the interaction of phorbol esters with the C1 domain of MRCK (Myotonic dystrophy kinase related Cdc42 binding kinase) alpha/beta. *The Journal of Biological Chemistry, 283*, 10543–10549.

Choi, S. H., Hyman, T., & Blumberg, P. M. (2006). Differential effect of bryostatin 1 and phorbol 12-myristate 13-acetate on HOP-92 cell proliferation is mediated by down-regulation of protein kinase C delta. *Cancer Research, 66*, 7261–7269.

Choi, Y., Pu, Y., Peach, M. L., Kang, J. H., Lewin, N. E., Sigano, D. M., et al. (2007). Conformationally constrained analogues of diacylglycerol (DAG). 28. DAG-dioxolanones reveal a new additional interaction site in the C1b domain of PKC delta. *Journal of Medicinal Chemistry, 50*, 3465–3481.

Churchill, E., Budas, G., Vallentin, A., Koyanagi, T., & Mochly-Rosen, D. (2008). PKC isozymes in chronic cardiac disease: Possible therapeutic targets? *Annual Review of Pharmacology and Toxicology, 48*, 569–599.

Colon-Gonzalez, F., & Kazanietz, M. G. (2006). C1 domains exposed: From diacylglycerol binding to protein-protein interactions. *Biochimica et Biophysica Acta, 1761*, 827–837.

Colon-Gonzalez, F., Leskow, F. C., & Kazanietz, M. G. (2008). Identification of an autoinhibitory mechanism that restricts C1 domain-mediated activation of the Rac-GAP alpha2-chimaerin. *The Journal of Biological Chemistry, 283*, 35247–35257.

Corbalan-Garcia, S., & Gomez-Fernandez, J. C. (2006). Protein kinase C regulatory domains: The art of decoding many different signals in membranes. *Biochimica et Biophysica Acta, 1761*, 633–654.

Dell'Aquila, M. L., Herald, C. L., Kamano, Y., Pettit, G. R., & Blumberg, P. M. (1988). Differential effects of byrostatins and phorbol esters on arachidonic acid metabolite release and epidermal growth factor binding in C3H10T1/2 cells. *Cancer Research, 48*, 3702–3708.

Dell'Aquila, M. L., Nyugen, H. T., Herald, C. L., Pettit, G. R., & Blumberg, P. M. (1987). Inhibition by bryostatin 1 of the phorbol ester-induced blockage of differentiation in hexamethylene bisacetamide-treated Friend erythroleukemia cells. *Cancer Research, 47*, 6006–6009.

Dempsey, E. C., Cool, C. D., & Littler, C. M. (2007). Lung disease and PKCs. *Pharmacological Research, 55*, 545–559.

Diamond, L., O'Brien, T. G., & Baird, W. M. (1980). Tumor promoters and the mechanism of tumor promotion. *Advances in Cancer Research, 32*, 1–74.

DiazGranados, N., & Zarate, C. A., Jr. (2008). A review of the preclinical and clinical evidence for protein kinase C as a target for drug development for bipolar disorder. *Current Psychiatry Reports, 10*, 510–519.

Driedger, P. E., & Blumberg, P. M. (1980). Specific binding of phorbol ester tumor promoters. *Proceedings of the National Academy of Sciences of the United States of America, 77*, 567–571.

Dries, D. R., Gallegos, L. L., & Newton, A. C. (2007). A single residue in the C1 domain sensitizes novel protein kinase C isoforms to cellular diacylglycerol production. *The Journal of Biological Chemistry, 282*, 826–830.

Dries, D. R., & Newton, A. C. (2008). Kinetic analysis of the interaction of the C1 domain of protein kinase C with lipid membranes by stopped-flow spectroscopy. *The Journal of Biological Chemistry, 283*, 7885–7893.

Duan, D., Lewin, N. E., Sigano, D. M., Blumberg, P. M., & Marquez, V. E. (2004). Conformationally constrained analogues of diacylglycerol. 21. A solid-phse method of synthesis of diacylglycerol lactones as a prelude to a combinatorial approach for the synthesis of protein kinase C isozyme-specific ligands. *Journal of Medicinal Chemistry, 47*, 3248–3254.

Duan, D., Sigano, D. M., Kelley, J. A., Lai, C. C., Lewin, N. E., Kedei, N., et al. (2008). Conformationally constrained analogues of diacylglycerol. 29. Cells sort diacylglycerol-lactone chemical zip codes to produce diverse and selective biological activities. *Journal of Medicinal Chemistry, 51*, 5198–5220.

Farhadi, A., Keshavarzian, A., Ranjbaran, Z., Fields, J. Z., & Banan, A. (2006). The role of protein kinase C isoforms in modulating injury and repair of the intestinal barrier. *Journal of Pharmacology and Experimental Therapeutics, 316*, 1–7.

Fujiki, H., & Sugimura, T. (1987). New classes of tumor promoters: Teleocidin, aplysiatoxin, and palytoxin. *Advances in Cancer Research, 49*, 223–264.

Fürstenberger, G., & Hecker, E. (1972). Zum Wirkungsmechanismus cocarcinogener Pflanzeninhaltsstoffe. *Planta Medica, 22*, 241–266.

Gallegos, L. L., Kunkel, M. T., & Newton, A. C. (2006). Targeting protein kinase C activity reporter to discrete intracellular regions reveals spatiotemporal differences in agonist-dependent signaling. *The Journal of Biological Chemistry, 281*, 30947–30956.

Gomez-Fernandez, J. C., Torrecillas, A., & Corbalan-Garcia, S. (2004). Diacylglycerols as activators of protein kinase C (review). *Molecular Membrane Biology, 21*, 339–349.

Gould, C. M., & Newton, A. C. (2008). The life and death of protein kinase C. *Current Drug Targets, 9*, 614–625.
Grant, S., Turner, A. J., Freemerman, A. J., Wang, Z., Kramer, L., & Jarvis, W. D. (1996). Modulation of protein kinase C activity and calcium-sensitive isoform expression in human myeloid leukemia cells by bryostatin 1: Relationship to differentiation and Ara-C induced apoptosis. *Experimental Cell Research, 228*, 65–75.
Griner, E. M., & Kazanietz, M. G. (2007). Protein kinase C and other diacylglycerol effectors in cancer. *Nature Reviews. Cancer, 7*, 281–294.
Hecker, E. (1968). Cocarcinogenic principles from the seed oil of Croton tiglium and from other Euphorbiaceae. *Cancer Research, 28*, 2338–2349.
Hecker, E. (1978). Structure-activity relationships in diterpene esters irritant and cocarcinogenic to mouse skin. In T. J. Slaga, A. Sivak, & R. K. Boutwell (Eds.), *Carcinogenesis* (Vol. 2, pp. 11–48). NY: Mechanisms of Tumor Promotion and Cocarcinogenesis Raven Press.
Hennings, H., Blumberg, P. M., Pettit, G. R., Herald, C. L., Shores, R., & Yuspa, S. H. (1987). Bryostatin 1, an activator of protein kinase C, inhibits tumor promotion by phorbol esters in SENCAR mouse skin. *Carcinogenesis, 8*, 1343–1346.
Hennings, H., & Boutwell, R. K. (1970). Studies on the mechanism of skin tumor promotion. *Cancer Research, 30*, 312–320.
Heo, J., Thapar, R., & Campbell, S. L. (2005). Recognition and activation of Rho GTPases by Vav1 and Vav2 guanine nucleotide exchange factors. *Biochemistry, 44*, 6573–6585.
Hommel, U., Zurini, M., & Luyten, M. (1994). Solution structure of a cysteine rich domain of rat protein kinase C. *Nature Structural Biology, 1*, 383–387.
Hornstein, I., Alcover, A., & Katzav, S. (2004). Vav proteins, masters of the world of cytoskeleton organization. *Cellular Signalling, 16*, 1–11.
Hritz, J., Ulicny, J., Laaksonen, A., Jancura, D., & Miskovsky, P. (2004). Molecular interaction model for the C1b domain of protein kinase C-γ in the complex with its activator phorbol-12-myristate-13-acetate in water solution and lipid bilayer. *Journal of Medicinal Chemistry, 47*, 6547–6555.
Hurley, J. H., Newton, A. C., Parker, P. J., Blumberg, P. M., & Nishizuka, Y. (1997). Taxonomy and function of C1 protein kinase C homology domains. *Protein Science, 6*, 477–480.
Ikuta, T., Chida, K., Tajima, O., Matsuura, Y., Iwamori, M., Ueda, Y., et al. (1994). Cholesterol sulfate, a novel activator for the eta isoform of protein kinase C. *Cell Growth Diff, 9*, 943–947.
Irie, K., Masuda, A., Shindo, M., Nakagawa, Y., & Ohigashi, H. (2004a). Tumor promoter binding of the protein kinase C C1 homology domain peptides of RasGRPs, chimaerins, and Unc13s. *Bioorganic & Medicinal Chemistry, 12*, 4575–4583.
Irie, K., Nakagawa, Y., & Ohigashi, H. (2004b). Indolactam and benzolactams compounds as new medicinal leads with binding selectivity for C1 domains of protein kinaseC isozymes. *Current Pharmaceutical Design, 10*, 1371–1385.
Irie, K., Nakagawa, Y., & Ohigashi, H. (2005). Toward the development of new medicinal leads with selectivity for protein kinase C isozymes. *Chemical Record, 5*, 185–195.
Itai, A., Kato, Y., Tomioka, N., Iitaka, Y., Endo, Y., Hasegawa, M., et al. (1988). A receptor model for tumor promoters: Rational superposition of teleocidins and phorbol esters. *Proceedings of the National Academy of Sciences of the United States of America, 85*, 3688–3692.
Johnson, J. E., Goulding, R. E., Ding, Z., Partovi, A., Anthony, K. V., Beaulieu, N., et al. (2007). Differential membrane binding and diacylglycerol recognition by C1 domains of RasGRPs. *Biochemical Journal, 406*, 223–236.
Kang, J. H., Benzaria, S., Sigano, D. M., Lewin, N. E., Pu, Y., Peach, M. L., et al. (2006). Conformationally constrained analogues of diacylglycerol. 26. Exploring the chemical space surrounding the C1 domain of protein kinase C with DAG-lactones containing aryl groups at the sn-1 and sn-2 positions. *Journal of Medicinal Chemistry, 49*, 3185–3203.
Kang, J. H., Chung, H. E., Kim, S. Y., Kim, Y., Lee, J., Lewin, N. E., et al. (2003). Conformationally constrained analogues of diacylglycerol (DAG). Effect on protein kinase C (PK-C) binding by

the isosteric replacement of sn-1 and sn-2 esters in DAG-lactones. *Bioorganic & Medicinal Chemistry, 11*, 2529–2539.

Kang, J. H., Peach, M. L., Pu, Y., Lewin, N. E., Nicklaus, M. C., Blumberg, P. M., et al. (2005). Conformationally constrained analogues of diacylglycerol (DAG). 25. Exploration of the sn-1 and sn-2 carbonyl functionality reveals the essential role of the sn-1 carbonyl at the lipid interface in the binding of DAG-lactones to protein kinase C. *Journal of Medicinal Chemistry, 48*, 5738–5748.

Kang, J. H., Siddiqui, M. A., Sigano, D. M., Krajewski, K., Lewin, N. E., Pu, Y., et al. (2004). Conformationally constrained analogues of diacylglycerol. 24. Asymmetric synthesis of a chiral ®-DAG-lactone template as a versatile precursor for highly functionalized DAG-lactones. *Organic Letters, 6*, 2413–2416.

Kazanietz, M. G. (2005). Targeting protein kinase C and "non-kinase" phorbol ester receptors: Emerging concepts and therapeutic implications. *Biochimica et Biophysica Acta, 1754*, 296–304.

Kazanietz, M. G., Barchi, J. J., Jr., Omichinski, J. G., & Blumberg, P. M. (1995a). Low affinity binding of phorbol esters to protein kinase C and its recombinant cysteine-rich region in the absence of phospholipids. *The Journal of Biological Chemistry, 270*, 14679–14684.

Kazanietz, M. G., Bustelo, X. R., Barbacid, M., Kolch, W., Mischak, H., Wong, G., et al. (1994). Zinc finger domains and phorbol ester pharmacophore. Analysis of binding to mutated form of protein kinase C zeta and the *vav* and *v-raf* proto-oncogene products. *The Journal of Biological Chemistry, 269*, 11590–11594.

Kazanietz, M. G., Wang, S., Milne, G. W., Lewin, N. E., Liu, H. L., & Blumberg, P. M. (1995b). Residues in the second cysteine-rich region of protein kinase C delta relevant to phorbol ester binding as revealed by site-directed mutagenesis. *The Journal of Biological Chemistry, 270*, 21852–21859.

Keck, G. E., Kraft, M. B., Truong, A. P., Li, W., Sanchez, C. C., Kedei, N., et al. (2008). Convergent assembly of highly potent analogues of bryostatin 1 via pyran annulation: Bryostatin look-alikes that mimic phorbol ester function. *Journal of the American Chemical Society, 130*, 6660–6661.

Keck, G. E., Poudel, Y. B., Welch, D. S., Kraft, M. B., Truong, A. P., Stephens, J., et al. (2009). Substitution on the A ring confers to bryologues the unique biological activity characteristic of bryostatins and distinct from that of the phorbol esters. *Organic Letters, 284*, 1302–1312.

Kedei, N., Lundberg, D. J., Toth, A., Welburn, P., Garfield, S. H., & Blumberg, P. M. (2004). Characterization of the interaction of ingenol 3-angelate with protein kinase C. *Cancer Research, 64*, 3243–3255.

Kikkawa, U., Matsuzaki, H., & Yamamoto, T. (2002). Protein kinase C delta (PKC delta): Activation mechanisms and functions. *Journal of Biochemistry, 132*, 831–839.

Kimura, K., Mizutani, M. Y., Tomioka, N., Endo, Y., Shudo, K., & Itai, A. (1999). Docking study of bryostatins to protein kinase Cδ Cys2 domain. *Chemical and Pharmaceutical Bulletin, 47*, 1134–1137.

Kishimoto, A., Takai, Y., Mori, T., Kikkawa, U., & Nishizuka, Y. (1980). Activation of calcium and phospholipid-dependent protein kinase by diacylglycerol, its possible relation to phosphatidylinositol turnover. *The Journal of Biological Chemistry, 255*, 2273–2276.

Koehler FE (1887) Köhler's Medizinal-Pflanzen in naturgetreuen Abbildungen mit kurz erläuterndem Texte (ed. Pabst G) Gera-Untermhaus, Germany

Kraft, A. S., Smith, J. B., & Berkow, R. L. (1986). Bryostatin, an activator of the calcium phospholipid-dependent protein kinase, blocks phorbol ester-induced differentiation of human promyelocytic leukemia cells HL-60. *Proceedings of the National Academy of Sciences of the United States of America, 83*, 1334–1338.

Lee, M. R., Duan, W., & Tan, S. L. (2008). Protein kinase C isozymes as potential therapeutic targets in immune disorders. *Expert Opinion on Therapeutic Targets, 12*, 535–552.

Leung, T., Chen, X. Q., Tan, I., Manser, E., & Lim, L. (1998). Myotonic dystrophy kinase-related Cdc42-binding kinase acts as a Cdc42 effector in promoting cytoskeletal reorganization. *Molecular Biology of the Cell, 18*, 130–140.

Lorenzo, P. S., Beheshti, M., Pettit, G. R., Stone, J. C., & Blumberg, P. M. (2000). The guanine nucleotide exchange factor RasGRP is a high-affinity target for diacylglycerol and phorbol esters. *Molecular Pharmacology, 57*, 840–846.

Lorenzo, P. S., Bögi, K., Acs, P., Pettit, G. R., & Blumberg, P. M. (1997). The catalytic domain of protein kinase C delta confers protection from down regulation induced by bryostatin 1. *The Journal of Biological Chemistry, 272*, 33338–33343.

Lorenzo, P. S., Bögi, K., Hughes, K. M., Beheshti, M., Bhattacharyya, D., Garfield, S. H., et al. (1999). Differential roles of the tandem C1 domains of protein kinase C delta in the biphasic down-regulation induced by bryostatin 1. *Cancer Research, 59*, 6137–6144.

Marquez, V. E., & Blumberg, P. M. (2003). Diacylglycerol (DAG) and DAG-lactones as selective activators of protein kinase C (PK-C). *Accounts of Chemical Research, 36*, 434–443.

Marquez, V. E., Nacro, K., Benzaria, S., Lee, J., Sharma, R., Teng, K., et al. (1999). The transition from a pharmacophore-guided approach to a receptor-guided approach in the design of potent protein kinase C ligands. *Pharmacology and Therapeutics, 82*, 251–261.

Medkova, M., & Cho, W. (1998). Differential membrane-binding and activation mechanisms of protein kinase C-alpha and -epsilon. *Biochemistry, 37*, 4892–4900.

Mott, H. R., Carpenter, J. W., Zhong, S., Ghosh, S., Bell, R. M., & Campbell, S. L. (1996). The solution structure of the Raf-1 cysteine-rich domain: A novel Ras and phospholipid binding site. *Proceedings of the National Academy of Sciences of the United States of America, 93*, 8312–8317.

Nacro, K., Sigano, D. M., Yan, S., Nicklaus, M. C., Pearce, L. L., Lewin, N. E., et al. (2001). An optimized protein kinase C activating diacylglycerol combining high binding affinity (Ki) with reduced lipophilicity (log P). *Journal of Medicinal Chemistry, 44*, 1892–1904.

Nakagawa, Y., Irie, K., Yanagita, R. C., Ohigashi, H., Tsuda, K. I., Kashiwagi, K., et al. (2006). Design and synthesis of 8-octyl-benzolactam-V9, a selective activator for protein kinase Cε and η. *Journal of Medicinal Chemistry, 49*, 2681–2688.

Nakamura, H., Kishi, Y., Pajares, M. A., & Rando, R. R. (1989). Structural basis of protein kinase C activation by tumor promoters. *Proceedings of the National Academy of Sciences of the United States of America, 86*, 9672–9676.

Newton, A. C. (2004). Diacylglycerol's affair with protein kinase C turns 25. *Trends in Pharmacological Sciences, 25*, 175–177.

Newton, A. C., & Johnson, J. E. (1998). Protein kinase C: A paradigm for regulation of protein function by two membrane-targeting modules. *Biochimica et Biophysica Acta, 1376*, 155–172.

Ng, S. B., & Guy, G. R. (1992). Two protein kinase C activators, bryostatin-1 and phorbol-12-myristate-13-acetate, have different effects on haemopoietic cell proliferation and differentiation. *Cellular Signalling, 4*, 405–416.

Oancea, E., & Meyer, T. (1998). Protein kinase C as a molecular machine for decoding calcium and diacylglycerol signals. *Cell, 96*, 307–318.

Ogbourne, S. M., Hampson, P., Lord, J. M., Parsons, P., De Witte, P. A., & Suhrbier, A. (2007). Proceedings of the First International Conference on PEP005. *Anti-Cancer Drugs, 18*, 357–362.

Opferkuch, H. J., & Hecker, E. (1982). On the active principles of the spurge family (Euphorbiaceae) IV. Skin irritant and tumor promoting diterpene esters from *Euphorbia ingens E. Mey*. *Journal of Cancer Research and Clinical Oncology, 103*, 255–268.

Pak, Y., Enyedy, I. J., Varady, J., Kung, J. W., Lorenzo, P. S., Blumberg, P. M., et al. (2001). Structural basis of binding of high-affinity ligands to protein kinase C: Prediction of the binding modes through a new molecular dynamics method and evaluation by site-directed mutagenesis. *Journal of Medicinal Chemistry, 44*, 1690–1701.

Pasti, G., Rivedal, E., Yuspa, S. H., Herald, C. L., Pettit, G. R., & Blumberg, P. M. (1988). Contrasting duration of inhibition of cell-cell communication in primary mouse epidermal cells by phorbol 12, 13-dibutyrate and by bryostatin 1. *Cancer Research, 48*, 447–451.

Pettit, G. R. (1991). The bryostatins. *Fortschritte der Chemie Organischer Naturstoffe, 57*, 153–195.

Pu, Y., Garfield, S. H., Kedei, N., & Blumberg, P. M. (2009). Characterization of the differential roles of the twin C1a and C1b domains of protein kinase C-delta. *The Journal of Biological Chemistry, 284*, 1302–1312.

Pu, Y., Peach, M. L., Garfield, S. H., Wincovitch, S., Marquez, V. E., & Blumberg, P. M. (2006). Effects on ligand interaction and membrane translocation of the positively charged arginine residues situated along the C1 domain binding cleft in the atypical protein kinase C isoforms. *The Journal of Biological Chemistry, 281*, 33773–33788.

Pu, Y., Perry, N. A., Yang, D., Lewin, N. E., Kedei, N., Braun, D. C., et al. (2005). A novel diacylglycerol-lactone shows marked selectivity in vitro among C1 domains of protein kinase C (PKC) isoforms alpha and delta as well as selectivity for RasGRP compared with PKC alpha. *The Journal of Biological Chemistry, 280*, 27329–27338.

Rapley, J., Tybulewicz, V. L. J., & Rittinger, K. (2008). Crucial structural role for the PH and C1 domains of the Vav1 exchange factor. *EMBO Reports, 9*, 655–661.

Reddig, P. J., Dreckschmidt, N. E., Ahrens, H., Simsiman, R., Tseng, C. P., Zou, J., et al. (1999). Transgenic mice overexpressing protein kinase C delta in the epidermis are resistant to skin tumor promotion by 12-O-tetradecanoylphorbol-13-acetate. *Cancer Research, 59*, 5710–5718.

Sako, T., Yuspa, S. H., Herald, C. L., Pettit, G. R., & Blumberg, P. M. (1987). Partial parallelism and partial blockade by bryostatin 1 of effects of phorbol ester tumor promoters on primary mouse epidermal cells. *Cancer Research, 47*, 5445–5450.

Shao, L., Lewin, N. E., Lorenzo, P. S., Hu, Z., Enyedy, I. J., Garfield, S. H., et al. (2001). Iridals are a novel class of ligands for phorbol ester receptors with modest selectivity for the RasGRP receptor family. *Journal of Medicinal Chemistry, 44*, 3872–3880.

Shindo, M., Irie, K., Nakahara, A., Ohigashi, H., Konishi, H., Kikkawa, U., et al. (2001). Toward the identification of selective modulators of protein kinase C (PKC) isozymes: Establishment of a binding assay for PKC isozymes using synthetic C1 peptide receptors and identification of the critical residues involved in the phorbol ester binding. *Bioorganic & Medicinal Chemistry, 9*, 2073–2081.

Sigano, D. M., Peach, M. L., Nacro, K., Choi, Y., Lewin, N. E., Nicklaus, M. C., et al. (2003). Differential binding mnodes of diacylglycerol (DAG) and DAG-lactones to protein kinase C (PK-C). *Journal of Medicinal Chemistry, 46*, 1571–1579.

Silinsky, E. M., & Searl, T. J. (2003). Phorbol esters and neurotransmitter release: More than just protein kinase C? *British Journal of Pharmacology, 138*, 1191–1201.

Slaga, T. J., Fischer, S. M., Nelson, K., & Gleason, G. L. (1980). Studies on the mechanism of skin tumor promotion: Evidence for several stages in promotion. *Proceedings of the National Academy of Sciences of the United States of America, 77*, 3659–3663.

Smith, J. B., Smith, L., & Pettit, G. R. (1985). Bryostatins: Potent, new mitogens that mimic phorbol ester tumor promoters. *Biochemical and Biophysical Research Communications, 132*, 939–945.

Stahelin, R. V., Digman, M. A., Medkova, M., Ananthanarayanan, B., Rafter, J. D., Melowic, H. R., et al. (2004). Mechanism of diacylglycerol-induced membrane targeting and activation of protein kinase C delta. *The Journal of Biological Chemistry, 279*, 29501–29512.

Steinberg, S. F. (2008). Structural basis of protein kinase C isoform function. *Physiological Reviews, 88*, 1341–1378.

Stone, J. C. (2006). Regulation of Ras in lymphocytes: Get a GRP. *Biochemical Society Transactions, 34*, 858–861.

Sun, M. D., & Alkon, D. L. (2006). Protein kinase C pharmacology: Perspectives on therapeutic potentials as antidementic and cognitive agents. *Recent Patents on CNS Drug Discovery, 1*, 147–156.

Swat, W., & Fujikawa, K. (2005). The Vav family. At the crossroads of signaling pathways. *Immunologic Research, 32*, 259–265.

Szallasi, Z., Bögi, K., Gohari, S., Biro, T., Acs, P., & Blumberg, P. M. (1996). Non-equivalent roles for the first and second zinc fingers of protein kinase δ. *The Journal of Biological Chemistry, 271*, 18299–18301.

Szallasi, Z., Denning, M. F., Smith, C. B., Dlugosz, A. A., Yuspa, S. H., Pettit, G. R., et al. (1994a). Bryostatin 1 protects protein kinase C-delta from down-regulation in mouse keratinocytes in parallel with its inhibition of phorbol ester-induced differentiation. *Molecular Pharmacology, 46*, 840–850.

Szallasi, Z., Krsmanovic, L., & Blumberg, P. M. (1993). Nonpromoting 12-deoxyphorbol 13-esters inhibit phorbol 12-myristate 13-acetate induced tumor promotion in CD-1 mouse skin. *Cancer Research, 53*, 2507–2512.

Szallasi, Z., Smith, C. B., Pettit, G. R., & Blumberg, P. M. (1994b). Differential regulation of protein kinase C isozymes by bryostatin 1 and phorbol 12-myristate 13-acetate in NIH 3T3 fibroblasts. *The Journal of Biological Chemistry, 269*, 2118–2124.

Teixeira, C., Stang, S. L., Zheng, Y., Beswick, N. S., & Stone, J. C. (2003). Integration of DAG signaling systems mediated by PKC-dependent phosphorylation of RasGRP3. *Blood, 102*, 1414–1420.

Topham, M. K. (2006). Signaling roles of diacylglycerol kinases. *Journal of Cellular Biochemistry, 97*, 474–484.

Topham, M. K., & Prescott, S. M. (2001). Diacylglycerol kinase zeta regulates Ras activation by a novel mechanism. *The Journal of Cell Biology, 152*, 1135–1143.

Vrana, J. A., Saunders, A. M., Chellappan, S. P., & Grant, S. (1998). Divergent effects of bryostatin 1 and phorbol myristate acetate on cell cycle arrest and maturation in human myelomonocytic leukemia cells (U937). *Differentiation, 63*, 33–42.

Wang, Q. J. (2006). PKD at the crossroads of DAG and PKC signaling. *Trends in Pharmacological Sciences, 27*, 317–323.

Wang, Q. J., Bhattacharyya, D., Garfield, S., Nacro, K., Marquez, V. E., & Blumberg, P. M. (1999). Differential localization of protein kinase C delta by phorbol esters and related compounds using a fusion protein with green fluorescent protein. *The Journal of Biological Chemistry, 274*, 37233–37239.

Wender, P. A., Cribbs, C. M., Koehler, K. F., Sharkey, N. A., Herald, C. L., Kamano, Y., et al. (1988). Modeling of the bryostatins to the phorbol ester pharmacophore on protein kinase C. *Proceedings of the National Academy of Sciences of the United States of America, 85*, 7197–7201.

Wender, P. A., Hinkle, K. W., Koehler, M. F., & Lippa, B. (1999). The rational design of potential chemotherapeutic agents: Synthesis of bryostatin analogues. *Medicinal Research Reviews, 19*, 388–407.

Xu, R. X., Pawelczyk, T., Xia, T. H., & Brown, S. C. (1997). NMR structure of a protein kinase C-γ phorbol-binding domain and study of protein-lipid micelle interactions. *Biochemistry, 36*, 10709–10717.

Yang, C., & Kazanietz, M. G. (2003). Divergence and complexities in DAG signaling: Looking beyond PKC. *Trends in Pharmacological Sciences, 24*, 602–608.

Yang, C., & Kazanietz, M. G. (2007). (2007) Chimaerins: GAPS that bridge diacylglycerol signaling and the small G-protein Rac. *Biochemical Journal, 403*, 1–12.

Zayed, S. M. A. D., Farghaly, M., Taha, H., Gotta, H., & Hecker, E. (1998). Dietary cancer risk conditional cancerogens in produce of livestock fed on species of spurge (*Euphorbiaceae*) I. Skin irritant and tumor-promoting ingenane-type diterpene esters in *E. peplus*, one of several herbaceous *Euphorbia* species contaminating fodder of livestock. *Journal of Cancer Research and Clinical Oncology, 124*, 131–140.

Zayed, S., Sorg, B., & Hecker, E. (1984). Structure activity relations of polyfunctional diterpenes of the tigliane type, V1. *Planta Medica, 34*, 65–69.

Zhang, G., Kazanietz, M. G., Blumberg, P. M., & Hurley, J. H. (1995). Crystal structure of the cys2 activator-binding domain of protein kinase C delta in complex with phorbol ester. *Cell, 81*, 917–924.

Zheng, Y., Liu, H., Coughlin, J., Zheng, J., Li, L., & Stone, J. C. (2005). Phosphorylation of RasGRP3 on threonine 133 provides a mechanistic link between PKC and Ras signaling systems in B cells. *Blood, 105*, 3648–3654.

Zhou, M., Horita, D. A., Waugh, D. S., Byrd, R. A., & Morrison, D. K. (2002). Solution structure and functional analysis of the cysteine-rich C1 domain of kinase suppressor of Ras. *Journal of Molecular Biology, 315*, 435–446.

Chapter 4
Diacylglycerol Signaling: The C1 Domain, Generation of DAG, and Termination of Signals

Isabel Mérida, Silvia Carrasco, and Antonia Avila-Flores

Abstract Diacylglycerol (DAG) is a simple lipid consisting of a glycerol molecule linked through ester bonds to two fatty acids in positions 1 and 2. In spite of, or may be thanks to, its small size and simple composition, DAG exerts multiple functions as a key intermediate in lipid metabolism, as a critical component of biological membranes and as a relevant second messenger. DAG-dependent functions are important not only in the transduction of signals from activated receptors, but also in the regulation of cell metabolism. The correct control of these two processes guarantees the adequate maintenance of homeostasis during cell growth and development, and its deregulation has been related to malignant transformation.

DAG exerts its function through direct binding to its target proteins, characterized by the presence in their sequences of at least one conserved 1 (C1) domain, with different specificities and affinities for this lipid. This interconnection probably fostered the appearance of numerous mechanisms that control DAG production, and clearance is necessary to allow the correct function of its target proteins, which have also increased in diversity and number throughout evolution. The DAG signaling network holds much promise as a target for the treatment of conditions such as cancer. A better understanding of the mechanisms that regulate DAG generation and clearance as well as the exact role of this lipid in the activation of C1-containing proteins is indispensable to identify new approaches for the better and more effective manipulation of DAG-regulated functions.

Keywords Diacylglycerol • Protein kinase C • C1 domain • Cancer • RasGRP • Signal transduction • Phosphatidic acid • Diacylglycerol kinase

I. Mérida (✉), S. Carrasco, and A. Avila-Flores
Department of Immunology and Oncology, Centro Nacional de Biotecnología/CSIC, Madrid, E-28049, Spain
e-mail: imerida@cnb.csic.es

M.G. Kazanietz (ed.), *Protein Kinase C in Cancer Signaling and Therapy*,
Current Cancer Research, DOI 10.1007/978-1-60761-543-9_4,

4.1 DAG Metabolism

Diacylglycerol (DAG) occupies a central role in lipid metabolism; it is a perfect module to which new components can be added for the synthesis of more complex lipids, acting at the same time as a source of free fatty acids. Bacteria, yeast, plants and animals all have the ability to metabolize DAG, a critical function that makes DAG essential for cell growth and development. In recent years, great advances have been made in the understanding of DAG metabolism at the molecular level. One of the most striking discoveries has been the characterization of multiple enzyme isoforms catalyzing the same chemical reactions, as phospholipases C and D (PLC and PLD) or diacylglycerol kinases (DGK), suggesting distinct and new functional roles in the metabolic pathways.

4.1.1 De Novo DAG Synthesis

There are two main pathways for DAG synthesis in yeast and mammals (Athenstaedt and Daum 1999): one is from glycerol-3-phosphate (G3P; as a result of triacylglycerol mobilization) and the second from dihydroxyacetone-3-phosphate (DHAP; glycolysis intermediate). These two precursors undergo several modifications, including two acylation steps that give rise first to lysophosphatidic acid (LPA) and then to phosphatidic acid (PA); PA is then transformed into DAG through the action of phosphatidic acid phosphohydrolases (PAP) (Nanjundan and Possmayer 2003) (Fig. 4.1).

In these two pathways, acylation in the first position of DAG chain takes place in different subcellular localizations (Athenstaedt and Daum 1999), with addition of only saturated fatty acids in mitochondria and peroxisome, and both saturated and unsaturated fatty acids in the endoplasmic reticulum (ER). The second acylation is performed principally in the ER membrane, where two proteins with acyltransferase activity are located (Athenstaedt and Daum 1999; Tan et al. 2001). Although both catalyze the same reaction, their specificity differs and they incorporate distinct fatty acids into LPA. The subcellular site of the first acylation, together with the specificity of the LPA acyltransferases, allows generation of PA with different fatty acid compositions.

Once PA is generated, the action of PAP will metabolize it to DAG. Two PAP activities, PAP1 and PAP2, have been described (Athenstaedt and Daum 1999; Brindley et al. 2002), based on differences in enzymatic activity and subcellular localization. PAP1 requires Mg^{2+} for catalysis and its only substrate is PA. It is located in the cytosol, from which it translocates to internal membranes such as that of the ER. PAP2, which localizes at membranes, does not require Mg^{2+} for catalysis and it is substrate-promiscuous.

Recent detailed studies concluded that the enzymatic activity corresponding to PAP2 is exerted by a family of enzymes with broad substrate specificity (between others, PA, LPA, sphingosine-1-phosphate and choline-1-phosphate) (Brindley et al. 2002).

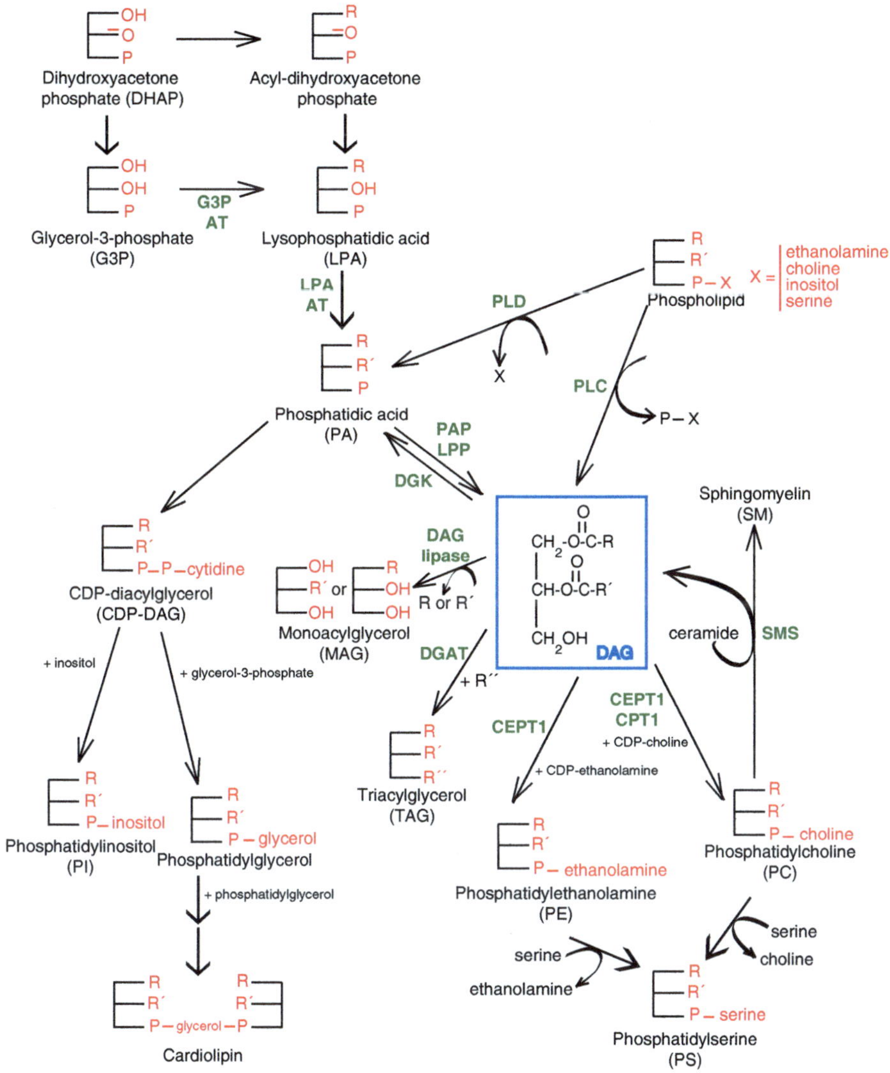

Fig. 4.1 Pathways that regulate DAG metabolism. The figure illustrates the pathways leading to diacylglycerol (DAG, highlighted by the blue box) generation and consumption together with the metabolites generated from this lipid. The main enzymes involved in DAG production and degradation (encircled ones are directly implicated in signaling) are shown in green (other enzymes have been omitted for simplicity), groups that change during the reactions are shown in red (OH = hydroxyl group, R and R′ = fatty acids, P = phospho group, X = choline, ethanolamine, inositol or serine) and the three-carbon invariable chain is in black. AT = acyltransferase, PAP = phosphatidic acid phosphohydrolases, LPP = lipid phosphate phosphatases, DGK = diacylglycerol kinase, PLD = phospholipase D, PLC = phospholipase C, DGAT = diacylglycerol acyltransferase, CDP = cytidine diphosphate, CEPT1 = choline/ethanolamine phosphotranspherase 1, CPT1 = choline phosphotranspherase 1, SMS = sphingomyelin synthase

These enzymes are now known as lipid phosphate phosphatases (LPP), and some of them have been characterized, including LPP1, LPP2, LPP3, SPP1 (sphingosine-1-phosphate phosphatase) and LPAP (LPA phosphatase). Studies of their subcellular localization are controversial and LPP enzymes have been found both in internal and plasma membranes. PAP activity (originally known as PAP1) has been recently cloned in yeast. Sequence information has been used to search for mammalian orthologs and these studies have revealed that the previously characterized, Lipin1, is a mammalian enzyme with PAP activity.

4.1.2 Alternative Pathways for DAG Synthesis

In addition to de novo synthesis, three alternative pathways can generate DAG, through the action of sphingomyelin synthase (SMS), PLC, and PLD. In the last two cases, DAG generation is highly dependent on extracellular stimulation, and DAG generated by these mechanisms is not usually consumed with a metabolic purpose.

SMS activity is responsible for sphingomyelin (SM) synthesis from phosphatidylcholine (PC) by catalyzing the replacement of a glycerol molecule by ceramide, resulting in a reaction that releases DAG (Fig. 4.1). Two SMS were recently cloned: SMS1 and SMS2, which catalyze SM synthesis from PC in the Golgi lumen and in the plasma membrane, respectively (Huitema et al. 2004).

4.1.3 The Role of DAG as a Lipid Precursor

DAG can act as a precursor of phosphatidylethanolamine (PE) and PC (Fig. 4.1). Two mammalian enzymes, choline/ethanolamine phosphotranspherase (CEPT1) and choline phosphotranspherase (CPT1), catalyze the incorporation of activated alcohols to DAG (Henneberry et al. 2002). CEPT1 is located in the ER and the external nuclear membrane, whereas CPT1 is a Golgi enzyme. Phosphatidylserine (PS) is synthesized, from PE and PC, by the action of two transferases (Kuge and Nishijima 1997) that catalyze the exchange of the ethanolamine or choline group for a serine in a reaction that takes place in the ER or in the Golgi apparatus (Fig. 4.1).

DAG can also be metabolized into triacylglycerol (TAG) by esterification of a new fatty acid in the free position of the glycerol moiety (Fig. 4.1). This activity is catalyzed by diacylglycerol acyltransferases (DGAT) in the ER or the plasma membrane (Cases et al. 2001). TAG is the main energy store and through a lipase-catalyzed reaction, it can be reconverted to DAG as a precursor for complex lipid synthesis.

DAG can also serve as a substrate for diacylglycerol lipases that hydrolyze the fatty acid in position 1 or 2, generating monoacylglycerols (MAG) (Fig. 4.1). DAG lipases are also strongly linked to signaling functions; in platelets, in response to thrombin, its combined action with PLC allows the release of arachidonic acid, an intermediate in thromboxane and prostaglandin synthesis (Smith et al. 1991). In neurons, this activity is necessary during retrograde synaptic transmission for the generation of 2-arachidonoyl-glycerol, an endocannabinoid (Yoshida et al. 2006).

In addition, DGK activity phosphorylates DAG, transforming it into PA, which is essential for phosphatidylinositol (PI) and cardiolipin production (Fig. 4.1). In bacteria, DAG phosphorylation has a metabolic function recycling DAG into the cytidine diphosphate-DAG pathway for phospholipid synthesis and preventing the lethal accumulation of DAG in bacterial membranes. Bacterial DGKs belong to one of two protein families. DgkA is an integral membrane protein, with three membrane-spanning domains (Loomis et al. 1985). A second prokaryotic DGK isoform, DgkB, has been recently identified (Jerga et al. 2007). This family represents soluble enzymes that share a common catalytic core signature sequence with the mammalian DGKs (Miller et al. 2008).

In multicellular organisms that originated as early as *Dictyostelium discoideum* and *Caenorhabditis elegans*, a family of cytosolic proteins is responsible for DAG phosphorylation (Kanoh et al. 2002). This indicates that higher organisms evolved a highly conserved function to include additional mechanisms that permit enzymes to reach the membrane. This specific regulation positions the DGK family enzymes as a perfect link between signaling and metabolism.

4.1.4 The Regulation of DAG Levels

The oldest role of DAG, as a basic membrane component and metabolic intermediate, is highly conserved throughout evolution. Considering the numerous metabolic pathways in which DAG is implicated, cells must rigorously control its production and clearance to guarantee a permanent reservoir of this lipid. Indeed, many mechanisms have developed throughout evolution to maintain correct levels during cell growth.

In *Saccharomyces cerevisiae*, the PI carrier Sec14p controls the PC synthesis rate (Bankaitis et al. 2005). When its expression is disrupted, the CDP-choline pathway leads to increased PC synthesis and, as a consequence, increased DAG use. The subsequent reduction in DAG levels alters Golgi secretory functions and affects cell viability (Kearns et al. 1997). This function is conserved in more complex organisms, as it has been described for Nir2, a functional equivalent of Sec14p in mammals (Litvak et al. 2005). Indeed, a *nir2* knockout mouse model shows embryonic lethality (Lu et al. 2001), and ablation of the *nir2* ortholog (*Dm-rdgB*) in *Drosophila melanogaster* induces retinal degeneration (Hardie 2003; Milligan et al. 1997). Enzymes with PAP activity are also implicated in mechanisms related to DAG control. In *S. cerevisiae*, ScPAP1 is necessary for correct cell growth and cytokinesis (Katagiri and Shinozaki 1998); in *D. melanogaster*, the mammalian LPP orthologs Wunen and Wunen2 are essential for correct germinal cell migration through the mesoderm (Santos and Lehmann 2004), and another recently described LPP, *lazaro*, is also linked to *Drosophila* phototransduction (Garcia-Murillas et al. 2006).

Dysregulation of DAG metabolism has been linked to the pathophysiology of several human diseases such as diabetes or malignant transformation. Enhanced lipid biosíntesis, particularly DAG, PA and PLA is a characteristic feature of cancer. Accordingly, changes on the levels of lipases and phospholipases have been described as prognostic markers of malignancies. For instance, LPA acyltransferase-β (LPA-ATβ) has been identified as a prognostic marker in ovarian cancer. Its inhibition reduces tumor formation in mouse ovary, an effect that it is not due to reduced LPA levels, but to the blockade of DAG synthesis, which is translated into reduced activity of its effectors (Springett et al. 2005). LPP-1 mRNA levels are decreased in the majority of ovarian cancers, contributing to the elevated levels of LPA observed in the ascites of ovarian cancer patients (Tanyi et al. 2003). These and other observations suggest that enzymes that participate in the DAG synthetic and degradative pathways would represent rational therapeutic targets for cancer.

4.2 DAG Response: The C1 Domain

In eukaryotes, a host of proteins have evolved the ability to bind to DAG and are thus activated by DAG-dependent signaling, creating additional levels of control to meet the complex needs of multicellular organisms. Alterations in the mechanisms that govern DAG generation and consumption are translated into aberrant localization/activation of DAG-regulated proteins, ultimately resulting in pathological conditions. All proteins that bind DAG directly, and thus respond to its presence, have at least one C1 domain, consisting of a conserved 50-amino-acid sequence bearing the $HX_{11-12}CX_2CX_{12-14}CX_2CX_4HX_2CX_{6-7}C$ motif (Hubbard et al. 1991). C1 domains were initially described as domains that bind phorbol esters (Ono et al. 1989); their capacity to bind DAG and other related compounds such as bryostatins, indolactanes or merezeins was confirmed later (Kazanietz et al. 2000).

Study of the residues necessary for interaction with phorbol esters led to the description of two types of C1 domain, typical and atypical (Hurley et al. 1997). The function of the atypical C1 domains has not been fully established, although some studies point to a role of protein and/or membrane interaction (Colon-Gonzalez and Kazanietz 2006). DGK is the largest family of proteins with two atypical C1 domains; other proteins with atypical C1 domains are Vav, Raf, ROCK, CRIK, C1-TEN, NORE or Lfc.

Proteins with typical C1 domain are candidates for DAG modulation. Since the 1980s, when PKC family members were described as the main effectors of cellular DAG, six additional families have been reported, increasing the number of proteins modulated by direct interaction with this lipid (Fig. 4.2) (Brose et al. 2004; Hall et al. 2005; Spitaler and Cantrell 2004; Yang and Kazanietz 2003) chimaerins, DGK (β and γ), PKD, Munc13, RasGRP and MRCK. All are characterized by the presence in their sequences of at least one conserved 1 (C1) domain, with different specificities and affinities for DAG. This range of specificities and affinities augments the complexity of DAG-dependent responses and facilitates discrimination by

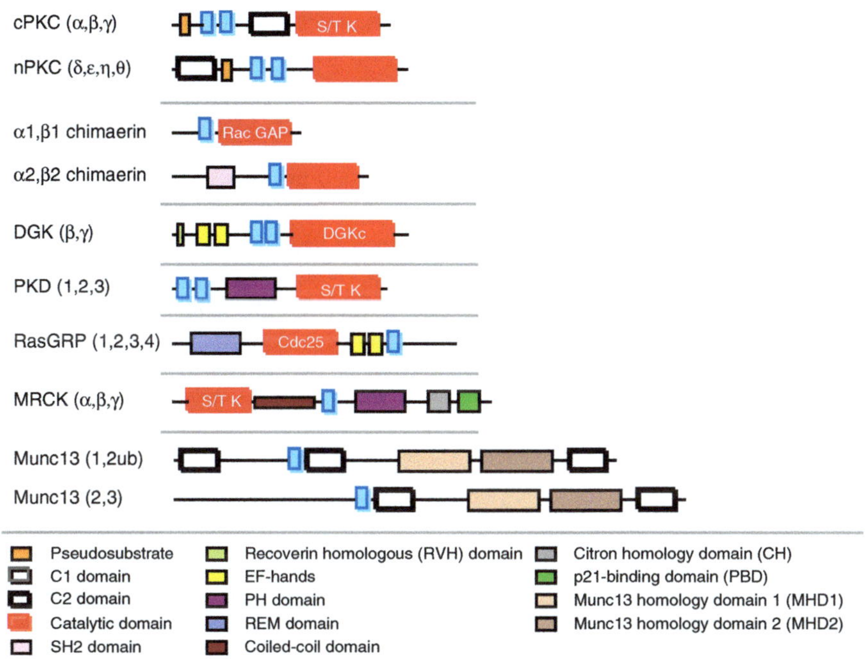

Fig. 4.2 C1 domain-containing proteins. The primary structures of C1 domain-containing proteins. PKC=Protein kinase C, DGK=diacylglycerol kinase, PKD=protein kinase D, RasGRP=Ras guanine-releasing protein, MRCK=myotonic dystrophy kinase-related Cdc42-binding kinase, Munc13=mammalian unc13, S/T K=Ser/Thr kinase, Rac GAP=Rac GTPase-activating protein, DGKc=DGK catalytic region, Cdc25=Cdc25 homology domain

the target proteins of the appropriate DAG pool among the numerous reservoirs of this lipid in the cell. These DAG-modulated proteins have distinct catalytic activities (Fig. 4.2), except the Munc13 family members, which are exclusively scaffolding proteins.

The principal DAG effect on responsive proteins is considered to be protein translocation to membranes, mediated by its direct binding to the C1 domain (Johnson et al. 1998). The latest discoveries have nonetheless broadened this concept to include modulation of protein activity and localization in specific cell membrane subdomains as C1-mediated DAG functions (Hall et al. 2005).

The recent description of the β2-chimaerin crystal structure (Canagarajah et al. 2004), the first for a complete C1 domain-containing protein, has helped to clarify C1 domain function. This study showed that, in the protein inactive conformation, both the catalytic and the C1 domains are buried. This specific folding is stabilized by intramolecular interactions that cover C1 domain hydrophobic residues and the DAG binding groove, yielding a protein unable to sense DAG. According to these data, β2-chimaerin would require previous signals that would expose the catalytic and C1 domains and subsequently allow DAG binding. This would promote closer contact with DAG-enriched membranes, favoring protein activity.

This model concurs with the earliest reports, later forgotten, on the PKC C1 domain, which proposed that DAG binding to the C1 domain served as an activator of protein activity (Newton and Koshland 1989).

These results reconcile all previous data showing that PKC (Parekh et al. 2000) and PKD (Auer et al. 2005) C1 domains are not exposed in the absence of stimulus. Nonetheless, it cannot explain how DAG-regulated proteins reach the membrane. Numerous PKC-interacting proteins (Poole et al. 2004) have been described as scaffolding proteins, important for translocation of PKC family members. For other DAG-regulated proteins, similar functions could be ascribed to the few interacting proteins reported to date, Tmp21 for β2chimaerin (Wang and Kazanietz 2002), actin for RasGRP1/2 (Caloca et al. 2003, 2004), heterotrimeric GTPase for PKD (Diaz Anel and Malhotra 2005; Oancea et al. 2003), synaptobrevin for Munc13 (Betz et al. 2001) or Nck for α2chimaerin (Wegmeyer et al. 2007).

Membrane specificity is important for defining downstream effectors of some DAG-regulated proteins, as demonstrated for PKD (Marklund et al. 2003) and RasGRP (Perez de Castro et al. 2004; Sanjuan et al. 2003). Due to their C1 domain specificities, these proteins can localize to internal membranes (Bivona et al. 2003; Liljedahl et al. 2001), where DAG fatty acids are mostly saturated (Carrasco and Merida 2004; Henneberry et al. 2002). This implies that certain C1 domains recognize and bind to DAG species earlier considered part of the metabolic pool, revealing that they are competent for signaling. Localization to distinct cell membranes is thus probably achieved as a result of the combination of binding to scaffolding proteins, C1 domain recognition of different DAG species, and the presence of other regulatory domains in the protein sequence (Colon-Gonzalez and Kazanietz 2006); in addition, DAG specificity and phorbol ester binding can be modified by the number of C1 domains in the sequence (Stahelin et al. 2005) or by phosphorylation of nearby residues, respectively (Thuille et al. 2005). Other recently described C1 domain functions, including binding to proteins (Oancea et al. 2003; Pang and Bitar 2005; Prekeris et al. 1998) or other lipids such as PS (Bittova et al. 2001), contribute also to membrane specificity.

The C1 domain thus emerges as a DAG-dependent regulatory module that controls protein activation and determines specific subcellular sites at which the protein must remain activated until DAG returns to basal levels. For more precise control of DAG-dependent signaling, C1 domain-containing proteins can also coordinate their actions within the same pathway, as is the case of PKD (Wang 2006) and RasGRP (Zheng et al. 2005), both PKC phosphorylation targets.

C1 domain function in defining specific protein localization is extremely important, as seen in unc13/Munc13, a protein family with no known catalytic domain (Fig. 4.2). Here, C1 domain function appears to be the assembly of the exocytosis machinery (Basu et al. 2005; Betz et al. 2001) at a membrane site at which DAG enrichment promotes membrane instability and fusion; this facilitates the secretion of neurotransmitters in neurons (Betz et al. 1998) or insulin in pancreatic cells (Kang et al. 2006; Kwan et al. 2006). As a consequence, *C. elegans* and mouse unc13/Munc13 knockout models show defective neurotransmitter release, which provokes severe alterations in motor coordination (Maruyama et al. 2001; Varoqueaux et al. 2005).

4.3 DAG Receptors and Oncogenesis

Since the early identification of the PKCs as intracellular receptor for the tumor promoting phorbol esters (Kikkawa et al. 1983), the role of DAG in the context of oncogenesis has been extensively investigated. After several years of intensive research, it is now generally accepted that each of the ten existing PKC isoforms contribute differently to cancer development and progression. Among the multiple PKC family members, several isoforms have tissue specific and even opposite role in tumor initiation. An example is PKCδ, proposed as an antiproliferative molecule in animal models of skin cancer (Reddig et al. 1999), whereas other authors report a role for this isoform in the survival of breast and lung cancer (Clark et al. 2003; McCracken et al. 2003; Grossoni et al. 2007). PKCα has been reported as a mediator of cell proliferation in head and neck cancer cell lines and also a predictive biomarker for disease free survival in head and neck cancer patients (Cohen et al. 2009).

From the initial studies, PKCε emerged as a protein with clear oncogenic properties (Cacace et al. 1993). PKCε overexpression is associated with oncogenesis from multiple organ sites, including breast, lung, prostate and head and neck (Bae et al. 2007; Cornford et al. 1999; Martinez-Gimeno et al. 1995; Pan et al. 2005). Albeit the exact etiology of PKCε overexpression remains to be fully elucidated, it is clear that this isoform is emerging through the literature as an important biomarker and potential drug target for many cancer types (Gorin and Pan 2009).

The characterization of nonkinase receptors for DAG has further extended the possible mechanisms by which this lipid second messenger can exert its functions. The characterization of a family of exchange factors for Ras family GTPases containing a conserved C1 domain directly couples elevation of DAG membrane levels with Ras activation. These proteins are designated RasGRP (Ras guanine releasing proteins) or CalDAG-GEF (calcium and DAG regulated guanine nucleotide exchange factors). The high expression of RasGRP family members in hematopoietic and nervous system suggest the existence of tissue specificity for DAG-mediated Ras activation. RasGRP2/CalDAG-GEF 1, which primarily targets Rap1 and Rap2 and is related to the regulation of integrin-mediated adhesion, has also been identified as a leukemogenic protooncogene in a murine model (Dupuy et al. 2001). A role for CalDAG-GEF1 as an oncogene in human hematologic malignancies has not been demonstrated, but the human RasGRP2 locus has been found to be differentially expressed in lymphoma cells of patients, where the disease progressed from low-grade follicular lymphoma to aggressive diffuse large cell lymphoma (Martinez-Climent et al. 2003).

A third family of DAG receptors is represented by the chimaerins, a family of GTPase-activating protein (GAPs) also regulated by DAG. Mammalian genomes contain two chimaerin loci, each of which produces at least two splice variants: a full length transcript (α2 and β2-chimaerin respectively) and a truncated transcript (α1 and β1-) that lacks the N-terminal SH2 domain (Hall et al. 2001). Recent studies have implicated β2-chimaerin as a tumor suppressor. Levels of this protein are reduced in multiple types of cancer, including breast tumors and malignant gliomas

(Yang et al. 2005; Yuan et al. 1995). Accordingly, overexpression of β-2chimaerin GAP domain in mouse mammary cancer cell lines reduces the growth rate and metastatic potential of tumors in vivo (Menna et al. 2005). There are no data showing that attenuation of β2-chimaerin levels in healthy epithelial tissue predisposes or is related to tumorigenesis. Nevertheless, experiments in *D. melanogaster* have demonstrated that reduction of the single chimaerin gene, *RhoGAP5A*, in the fly eye results in an increase in cell number and aberrant cell–cell adhesion, consistent with a progression to a more "tumor-like" phenotype.

The data in fruit fly and cancer cell models suggest that the role of β2-chimaerin is directly related to the inactivation of Rac activity downstream of epidermal growth factor receptor (EGFR) (Bruinsma et al. 2007; Wang et al. 2006). Interestingly, experiments in the fly eye reveal a role for chimaerin in shutting down ERK activation at the plasma membrane, a signal that has been directly linked to the regulation of adherens junctions. Downregulation of adherent junctions appears to be critical for metastatic transformation of epithelial tumors, a model has been thus proposed where downregulation of β2chimaerin would be related to increased ERK activation at the plasma membrane where it would disrupt adherent junctions (Bruinsma et al. 2007).

4.4 Termination of DAG Signaling. Diacylglycerol kinases

The diacylglycerol kinases (DGK) are a family of signaling proteins that modulate DAG levels by catalyzing its conversion to phosphatidic acid (PA) (Merida et al. 2008). DGKs consume DAG providing a mechanism for termination of DAG signaling pathway but, at the same time, they generate PA, which is also an important modulator of signaling molecules. Mammalian DGK comprise an extended family, currently with ten members classified into five different subtypes based on the presence of different regulatory domains in their primary sequences (Fig. 4.3). DGK diversity is further increased by alternative splicing, which produces several isoforms with distinct domain structures (Caricasole et al. 2002; Ding et al. 1997; Ito et al. 2004; Kai et al. 1994; Murakami et al. 2003; Sakane et al. 2002). The three mammalian type I DGK have characteristic Ca^{2+}-binding EF hands and a recoverin motif in the N-terminus, while the two type II isoforms have PH domains. Members of the type IV group contain C-terminal ankyrin repeats and a PDZ binding sequence together with a MARCKS homology region upstream of the catalytic site. The single type V member has a Rho-binding domain, and the only type III member has the simplest structure, with no regulatory region.

Proteins of this family are conserved in multicellular organisms, including *D. discoideum*, which has a single gene (*dgkA*) that encodes an enzyme related to mammalian DGKθ (Ostroski et al. 2005). Disruption of the *dgkA* locus alters myosin II assembly, raising the intriguing possibility that DG and/or PA may have a role in controlling cytoskeletal organization in this organism. Analysis of *D. melanogaster* or *C. elegans* genomes reveals the presence of members for each of the different DGK subtypes, suggesting nonredundant functions.

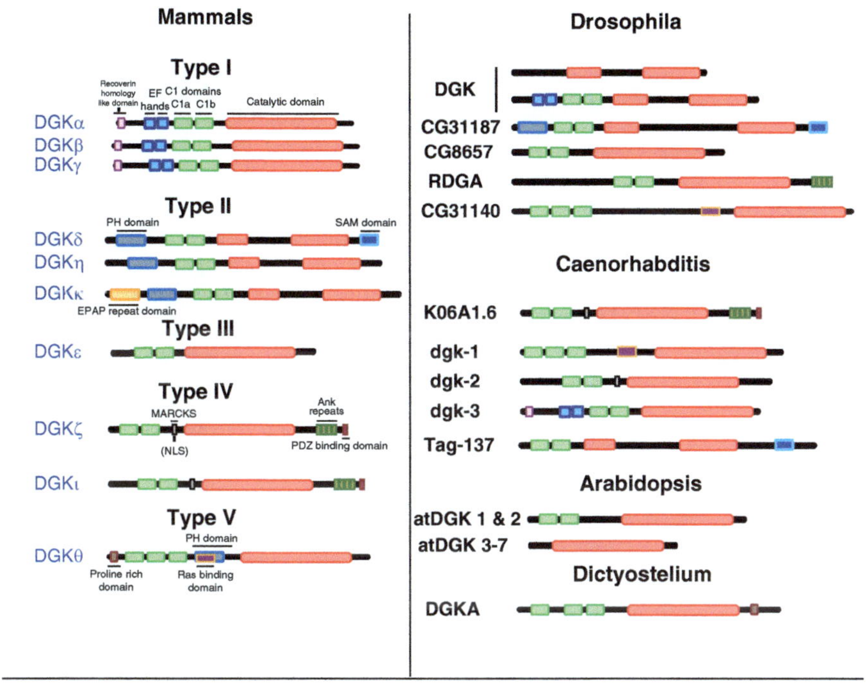

Fig. 4.3 The DGK family. The different DGK isoforms present in multicellular organisms are represented, indicating the distinct regulatory domains. In mammals, ten DGK have been cloned; they are characterized by a common catalytic region and have been grouped into five subtypes, depending on the presence of different regulatory motifs in their primary sequences. For the other organisms, the sequence number of the identified DGK is provided

DGK activity has also been reported in several plant species. Plant DGKs fall into three distinct clusters, simpler in organization than mammalian DGK since none contains a regulatory region. Cluster I DGK contains two cysteine-rich domains, while clusters II and III have only the characteristic catalytic region (Wanga et al. 2006). The major role of DGK in plants is related to PA generation in response to biotic challenges such as microbial elicitation and abiotic stress, including chilling, salts, drought and dehydration (den Hartog et al. 2003; Meijer and Munnik 2003; Ruelland et al. 2002). DGK-derived PA levels also accumulate during various developmental processes, including root elongation (Gomez-Merino et al. 2005).

4.4.1 DGK Structure

Analysis of the mammalian DGK primary sequence reveals a C-terminal conserved catalytic domain that is common to the DGK superfamily that also includes the recently identified bacterial DAGKB as well as the sphingosine kinase (SPK) and ceramide kinase (CEK) families. This domain is subdivided into a conserved motif called DGKc in SMART, which contains the sequence ϕϕϕGGDGT (ϕ represents any hydrophobic residue), and an accessory domain (SMART ID DGKa). DGKc has sequence similarity with the catalytic site of SK, another lipid kinase family that phosphorylates sphingosine to form sphingosine-1-phosphate (Pitson et al. 2002). The SK catalytic site contains a highly conserved ϕϕϕGGDGT motif reminiscent of the DGK signature. Mutation of the second G in both the DGK and SK motifs abolishes their kinase activity. Multiple sequence alignment has shown conservation in the catalytic regions of these two lipid kinase families and the catalytic domains of phosphofructokinase (PFK) and polyphosphate/NAD-ATP kinase (PPNK) (Labesse et al. 2002). This signature encompasses both the ATP- and substrate-binding sites in the crystal structure of PFK, suggesting that DGK and SK may have a similar ATP-binding site to catalyze phosphorylation of their substrates using a shared specific mechanism. An extensive review recently summarized the properties of these two structurally related lipid kinases (Wattenberg et al. 2006). The recent resolution of the bacterial DAGKB structure has allowed the characterization of the common catalytic core that extends to the two regions previously identified in SMART as the common and accessory DAGK domains (Miller et al. 2008).

DGK family members share another conserved structure, found at least twice in all DGK, which is homologous to the PKC phorbol ester/DG-binding, C1-type motifs (Cho 2001). The presence of C1 domains in the DGK sequence originally led to consideration of these motifs as responsible for DG binding. Nevertheless, sequence analysis indicated that, with the exception of the first C1 in DGKβ and DGKγ (Shindo et al. 2003), the C1 regions lack the key residues that define a canonical C1-like, phorbol ester-binding domain (Hurley and Misra 2000). The participation of DGK C1 domains in the enzymatic activity of this family thus remains a matter of debate. Some studies have suggested that these conserved motifs are required for activity (Houssa et al. 1997), while others have found them dispensable for DG phosphorylation in vitro (Sakane et al. 1996). Some well-characterized plant DGKs lack C1 domains, suggesting that these domains are not necessary for activity (Snedden and Blumwald 2000). Whereas the DGK C1 domains may not be needed for catalytic DG binding, they do appear to be critical for membrane targeting. Mutations that disrupt one of the C1 domains impair receptor-dependent translocation of GFP-DGKζ chimaeras to the plasma membrane in live Jurkat T cells (Santos et al. 2002); this is also the case for DGKθ in response to G protein-coupled receptors (van Baal et al. 2005) and for DGKγ (Shirai et al. 2000). Targeting to the membrane may be fostered by C1 domain interaction with lipids and/or proteins. Accordingly, DGK C1 domains are proposed to bind different lipids including PS and cholesterol, as well as PI3 kinase derivatives, and proteins including β arrestin and Rho (Cipres et al. 2001; Fanani et al. 2004; McMullan et al. 2006; Nelson et al. 2007).

Whereas all DGK catalyze the same reaction, the presence of diverse regulatory regions confers specificity to the distinct DGK isoforms by restricting their site of action and/or their activation mechanisms. Most DGK are cytosolic in unstimulated cells and translocation to membranes appears as a general mechanism that modulates the spatio/temporal activation of this family. The nonconserved regulatory domains appear to govern subtype-specific DGK translocation, providing specific mechanisms based on protein–protein and/or protein–lipid interactions. Several recent reviews have explored these mechanisms in depth (Merida et al. 2008; Sakane et al. 2008; Topham and Epand 2009).

4.4.2 DGKs and DAG Signal Termination

From early on, the main function attributed to the DGK family has been that of negative regulation of DAG receptors. The DGK were initially recognized as modulators of classical and novel PKC family members. The identification of several additional families of DAG-regulated proteins with distinct functions provides new insight into the complex, strategic role of DGKs in the regulation of biochemical networks. Some examples include regulation of RasGRP1 by DGKα and DGKζ (Jones et al. 2002; Topham and Prescott 2001), of RasGRP3 by DGKι (Regier et al. 2005), and of β2-chimaerin by DGKγ in mammals (Yasuda et al. 2007). One of the more interesting examples of the negative role exerted by DGK in the regulation of DAG effectors is that exerted by DGK-1 (DGKθ ortholog) in *C. elegans* through negative regulation of UNC-13 (Nurrish et al. 1999). The recent generation of animal models deficient in different DGK isoforms has further highlighted the important role of these proteins as negative regulators of DAG-mediated functions. Thus, DGKδ haploinsufficiency results in increased diacylglycerol content and reduced peripheral insulin sensitivity, signaling and glucose transport. This contribution of DGKδ to hyperglycemia-induced peripheral insulin resistance was further confirmed by the identification of reduced DGK delta expression and DGK activity in skeletal muscle from type 2 diabetic patients (Chibalin et al. 2008).

4.4.3 DGKs and Cancer

The negative regulatory function of DGKs in DAG-mediated effects would suggest a suppressor role in malignant transformation for this family. However, and probably reflecting the complex roles of DAG-regulated molecules in cancer (Griner and Kazanietz 2007), DGK are reported to act both as tumor suppressors and as positive regulators of survival and proliferation in transformed cells (Filigheddu et al. 2007). Of particular interest is the case of DGKα that several studies link with the maintenance of viability of tumor cells. For instance, it is proposed that DGKα-dependent signals contribute to the maintenance of viability of several human melanoma cell lines that express higher levels of this isoform than normal human epidermal melanocytes (Yanagisawa et al. 2007). Attenuation of DGKα expression significantly enhanced tumor necrosis factor (TNF)-α-induced apoptosis suggesting a specific effect of this isoform as a suppressor of TNFα-induced apoptosis.

Different lines of evidence suggest that prevention of apoptosis is due to the participation of DGKα in NF-kB activation. The exact mechanisms linking DGKα activity and NFkB activation remain undefined, but the lack of effect of a kinase dead enzyme suggests a role depending either on DAG generation and/or lack of PA.

The specific role of DGKα and not other type I enzymes suggest a role for this particular isoform in the prevention of apoptosis. The recent characterization of the regulation of DGKα by Src-dependent phosphorylation (Baldanzi et al. 2008) suggests that the regulation of DGKα by Src kinases could be important for the NF-KB activation and prevention of apoptosis. A role for c-Src as positive regulator of TNF-α-mediated NF-κB activation was recently reported in endothelial cells (Itoh et al. 2005). Thus, the association/activation of DGK with c-Src may be critical for the activation of DGKα leading to the prevention of apoptosis. The blockade of NF-kB activity in several cancer cells is related to the suppression of carcinogenesis and metastasis, suggesting that the manipulation of DGKα activity could be of interest in the treatment of other malignancies.

Studies in the anaplastic large cell lymphoma (ALCL) Karpas also reveal high constitutive DGKα activity (Bacchiocchi et al. 2005). Pharmacological inhibition of DGKα impairs the growth rate of NPM/ALK cells as well as the EGF-dependent growth of cells expressing a chimeric EGFR/ALK receptor, identifying DGKα as a possible therapeutic target in the treatment of ALCL lymphomas.

The function of DGKα as a positive regulator of cell proliferation was first described in T cell lines showing that the low levels of DGKα activity in activated T cells appear to be required for IL-2-mediated G1-S transition (Flores et al. 1999). The inhibitory properties of the DGKα inhibitor R59459 in IL-2-dependent proliferation of the lymphocyte cell line CTLL-2 were similar to those of rapamycin, implying that both drugs act on the same pathway. Accordingly, rapamycin has been proposed to inhibit mTOR by blocking PA-dependent activation of this protein (Foster 2009). A role for DGKα as a positive modulator of cell cycle progression and migration are not specific of T lymphocytes and has also been described for other tyrosine kinase receptors. Experiments in endothelial cells demonstrated that activation of DGKα in response to activation of tyrosine kinase receptors vascular endothelial growth factor (VEGF) receptor-2 is required for ligand-induced chemotaxis, proliferation and angiogenesis (Baldanzi et al. 2004). A similar role for DGKα is required for HGF-induced cell motility and proliferation in endothelial and epithelial cells (Cutrupi et al. 2000). The requirement for DGKα in the correct transduction of VEGF and HGF receptor-dependent signals appears to be a direct consequence of the interaction between DGKα and Src family kinases. The DGKα carboxy-terminus can bind Src kinases, and in this case, there are two nonmutually exclusive options. DGKα protein can act as a scaffold, maintaining the tyrosine kinases in an appropriate conformation for activation, or DGKα might either generate or metabolize a lipid needed to inhibit a phosphatase activity. Attenuation of DGKα expression and/or function by different mechanisms has been shown to impair both HGF and v-Src-induced cell scatter and migration, further demonstrating a connection between Src kinases, DGKα and HGF-mediated signals. DGKα inhibition results in uncoupling the downregulation of E-cadherin-mediated intercellular adhesions from cell migration, suggesting a role for this particular isoform in the signals required for HGF-and v-Src-stimulated epithelial cell motility. Although the exact role of DGKα in the regulation of cell motility is still undefined, experiments strongly suggest that DGKα is recruited to membrane ruffles where probably participates in the regulation of small GTPases like Rac. In fact, different DGK isoforms have been proposed to act as modulators of Rac membrane targeting and activation through multiple mechanisms. DGKγ has been shown to negatively regulate platelet-derived growth factor (PDGF) and epidermal growth factor (EGF)-induced Rac activation and membrane ruffling by enhancing the activity of β2-chimerin (Tsushima et al. 2004; Yasuda et al. 2007). In neurons and skeletal myoblasts, DGKζ interaction with syntrophins regulates Rac activation by favoring RhoGDI dissociation (Abramovici et al. 2009). The exact mechanism by which DGKα modulates Rac activation is not fully elucidated, although it could be related to the role of PA as activator of PI (4) P5 Kinase activity and the role of both lipids, PA and PIP2, impairing RhoGDI affinity for Rac. In this regard, DGKα has been shown to associate and activate PI (4) P5 kinase in vitro (Jones et al. 2000).

Contradictory effects on proliferation are also reported for DGKζ. The nuclear localization of this isozyme in some cell types correlates with accumulation on the G1 phase of the cell cycle (Evangelisti et al. 2007; Topham et al. 1998). The DGKζ negative effect on cell cycle progression is presumably related to the reduction of DG nuclear levels, although the exact mechanism remains unknown. In vitro studies demonstrate that DGKζ interacts with the hypophosphorylated Retinoblastoma protein (pRb), a key tumor suppressor controlling S phase entry, and such interaction leads to an increase in DGKζ activity (Los et al. 2006). A reciprocal regulation between these two proteins may exist, since overexpression of DGKζ in C2C12 myoblasts leads to pRb hypophosphorylation (Evangelisti et al. 2007).

While DGKζ exerts a negative regulation on cell proliferation, as a result of attenuation of nuclear DG levels, DGKζ-produced PA positively regulates mTOR activity hinting for a positive role of this isozyme in cell growth and survival (Avila-Flores et al. 2005). Although PLD has been proposed as the main source of the PA that regulates mTOR activity, DGKs and LPA acyltransferases represent either alternative or PLD interconnected sources of PA (Tang et al. 2006); (Hornberger et al. 2006); (Foster 2007). This suggests that mTOR is mainly activated by PA-derived of biosynthetic pathways suggesting a direct connection between mTOR activation and phospholipid synthesis. mTOR is a master regulator, which integrates different signaling pathways that sense the availability of nutrients, and oxygen and PA-dependent regulation would provide an expected connection with lipid metabolism.

As is also the case with the different DAG receptors, DGK function in cancer appears to be highly dependent expression levels and on cellular context. Several expression profile studies for instance demonstrate differential DGKα levels in normal *vs.* transformed cells, pointing to DGKα expression as a potential biomarker. Validation of this distinct DGK isoform expression, together with a careful assessment of their precise function in normal and transformed cells, represents an important challenge to the full evaluation of the potential of DGK as a therapeutic target in cancer.

4.5 Concluding Remarks and Perspectives

A strict control of the synthesis, metabolism and compartmentalization of cellular DAG membrane levels enables DAG to perform its dual role as a key player in the biosynthesis and degradation of glycerolipids and as modulator of C1-containing proteins. We are still far away from understanding exactly how DAG fulfills these two functions, how the distinct DAG pools are maintained, the interrelationships between the multiple pathways that regulate DAG levels and the mechanisms by which DAG regulates the spatio/temporal activation of its multiple effectors. Experimental evidence suggests that a tight control of DAG membrane levels guarantees correct transition from quiescence to proliferative states and/or regulation of apoptosis in untransformed cells. Defects in the activity and/or expression of enzymes responsible for DAG metabolism would result in deregulation of DAG membrane levels. This could lead to the sustained activation and/or activation at the wrong compartment of DAG effectors that in turn would contribute to cell transformation. The DAG signaling network holds high promises as a target for the treatment of malignant diseases. The recent progress in our understanding of DAG-regulated processes only emphasizes the need for additional studies to evaluate the use of metabolic enzymes as prognostic and/or diagnostic marker as well as the potential therapeutic manipulation of DAG generation and clearance in the design of novel and more personalized cancer therapies.

Acknowledgments We are grateful to the members of the Mérida lab for contributions and discussions. This work was supported in part by grants RD067002071035 from the Carlos III Institute (Spanish Ministry of Health), BFU2007-62639 (Spanish Ministry of Education) and S-SAL-0311 from Comunidad de Madrid.

References

Abramovici, H., Mojtabaie, P., Parks, R. J., Zhong, X. P., Koretzky, G. A., Topham, M. K., et al. (2009). Diacylglycerol kinase zeta regulates actin cytoskeleton reorganization through dissociation of Rac1 from RhoGDI. *Molecular Biology of the Cell, 20*, 2049–2059.

Athenstaedt, K., & Daum, G. (1999). Phosphatidic acid, a key intermediate in lipid metabolism. *European Journal of Biochemistry, 266*, 1–16.

Auer, A., von Blume, J., Sturany, S., von Wichert, G., Van Lint, J., Vandenheede, J., et al. (2005). Role of the regulatory domain of protein kinase D2 in phorbol ester binding, catalytic activity, and nucleocytoplasmic shuttling. *Molecular Biology of the Cell, 16*, 4375–4385.

Avila-Flores, A., Santos, T., Rincon, E., & Merida, I. (2005). Modulation of the mammalian target of rapamycin pathway by diacylglycerol kinase-produced phosphatidic acid. *The Journal of Biological Chemistry, 280*, 10091–10099.

Bacchiocchi, R., Baldanzi, G., Carbonari, D., Capomagi, C., Colombo, E., van Blitterswijk, W. J., et al. (2005). Activation of alpha-diacylglycerol kinase is critical for the mitogenic properties of anaplastic lymphoma kinase. *Blood, 106*, 2175–2182.

Bae, K. M., Wang, H., Jiang, G., Chen, M. G., Lu, L., & Xiao, L. (2007). Protein kinase C epsilon is overexpressed in primary human non-small cell lung cancers and functionally required for

proliferation of non-small cell lung cancer cells in a p21/Cip1-dependent manner. *Cancer Research, 67*, 6053–6063.
Baldanzi, G., Cutrupi, S., Chianale, F., Gnocchi, V., Rainero, E., Porporato, P., et al. (2008). Diacylglycerol kinase-alpha phosphorylation by Src on Y335 is required for activation, membrane recruitment and Hgf-induced cell motility. *Oncogene, 27*, 942–956.
Baldanzi, G., Mitola, S., Cutrupi, S., Filigheddu, N., van Blitterswijk, W. J., Sinigaglia, F., et al. (2004). Activation of diacylglycerol kinase alpha is required for VEGF-induced angiogenic signaling in vitro. *Oncogene, 23*, 4828–4838.
Bankaitis, V. A., Phillips, S., Yanagisawa, L., Li, X., Routt, S., & Xie, Z. (2005). Phosphatidylinositol transfer protein function in the yeast Saccharomyces cerevisiae. *Advances in Enzyme Regulation, 45*, 155–170.
Basu, J., Shen, N., Dulubova, I., Lu, J., Guan, R., Guryev, O., et al. (2005). A minimal domain responsible for Munc13 activity. *Nature Structural and Molecular Biology*, *12*(11), 1017–1018.
Betz, A., Ashery, U., Rickmann, M., Augustin, I., Neher, E., Sudhof, T. C., et al. (1998). Munc13-1 is a presynaptic phorbol ester receptor that enhances neurotransmitter release. *Neuron, 21*, 123–136.
Betz, A., Thakur, P., Junge, H. J., Ashery, U., Rhee, J. S., Scheuss, V., et al. (2001). Functional interaction of the active zone proteins Munc13-1 and RIM1 in synaptic vesicle priming. *Neuron, 30*, 183–196.
Bittova, L., Stahelin, R. V., & Cho, W. (2001). Roles of ionic residues of the C1 domain in protein kinase C-alpha activation and the origin of phosphatidylserine specificity. *The Journal of Biological Chemistry, 276*, 4218–4226.
Bivona, T. G., Perez De Castro, I., Ahearn, I. M., Grana, T. M., Chiu, V. K., Lockyer, P. J., et al. (2003). Phospholipase Cgamma activates Ras on the Golgi apparatus by means of RasGRP1. *Nature, 424*, 694–698.
Brindley, D. N., English, D., Pilquil, C., Buri, K., & Ling, Z. C. (2002). Lipid phosphate phosphatases regulate signal transduction through glycerolipids and sphingolipids. *Biochimica et Biophysica Acta, 1582*, 33–44.
Brose, N., Betz, A., & Wegmeyer, H. (2004). Divergent and convergent signaling by the diacylglycerol second messenger pathway in mammals. *Current Opinion in Neurobiology, 14*, 328–340.
Bruinsma, S. P., Cagan, R. L., & Baranski, T. J. (2007). Chimaerin and Rac regulate cell number, adherens junctions, and ERK MAP kinase signaling in the Drosophila eye. *Proceedings of the National Academy of Sciences of the United States of America, 104*, 7098–7103.
Cacace, A. M., Guadagno, S. N., Krauss, R. S., Fabbro, D., & Weinstein, I. B. (1993). The epsilon isoform of protein kinase C is an oncogene when overexpressed in rat fibroblasts. *Oncogene, 8*, 2095–2104.
Caloca, M. J., Wang, H., & Kazanietz, M. G. (2003). Characterization of the Rac-GAP (Rac-GTPase-activating protein) activity of beta2-chimaerin, a 'non-protein kinase C' phorbol ester receptor. *The Biochemical Journal, 375*, 313–321.
Caloca, M. J., Zugaza, J. L., Vicente-Manzanares, M., Sanchez-Madrid, F., & Bustelo, X. R. (2004). F-actin-dependent translocation of the Rap1 GDP/GTP exchange factor RasGRP2. *The Journal of Biological Chemistry, 279*, 20435–20446.
Canagarajah, B., Leskow, F. C., Ho, J. Y., Mischak, H., Saidi, L. F., Kazanietz, M. G., et al. (2004). Structural mechanism for lipid activation of the Rac-specific GAP, beta2-chimaerin. *Cell, 119*, 407–418.
Caricasole, A., Bettini, E., Sala, C., Roncarati, R., Kobayashi, N., Caldara, F., et al. (2002). Molecular cloning and characterization of the human diacylglycerol kinase beta (DGKbeta) gene: Alternative splicing generates DGKbeta isotypes with different properties. *The Journal of Biological Chemistry, 277*, 4790–4796.
Carrasco, S., & Merida, I. (2004). Diacylglycerol-dependent binding recruits PKCtheta and RasGRP1 C1 domains to specific subcellular localizations in living T lymphocytes. *Molecular Biology of the Cell, 15*, 2932–2942.

Cases, S., Stone, S. J., Zhou, P., Yen, E., Tow, B., Lardizabal, K. D., et al. (2001). Cloning of DGAT2, a second mammalian diacylglycerol acyltransferase, and related family members. *The Journal of Biological Chemistry, 276*, 38870–38876.

Chibalin, A. V., Leng, Y., Vieira, E., Krook, A., Bjornholm, M., Long, Y. C., et al. (2008). Downregulation of diacylglycerol kinase delta contributes to hyperglycemia-induced insulin resistance. *Cell, 132*, 375–386.

Cho, W. (2001). Membrane targeting by C1 and C2 domains. *The Journal of Biological Chemistry, 276*, 32407–32410.

Cipres, A., Carrasco, S., & Merida, I. (2001). Deletion of the acidic-rich domain of the IL-2Rbeta chain increases receptor-associated PI3K activity. *FEBS Letters, 500*, 99–104.

Clark, A. S., West, K. A., Blumberg, P. M., & Dennis, P. A. (2003). Altered protein kinase C (PKC) isoforms in non-small cell lung cancer cells: PKCdelta promotes cellular survival and chemotherapeutic resistance. *Cancer Research, 63*, 780–786.

Cohen, E. E., Zhu, H., Lingen, M. W., Martin, L. E., Kuo, W. L., Choi, E. A., et al. (2009). A feed-forward loop involving protein kinase Calpha and microRNAs regulates tumor cell cycle. *Cancer Research, 69*, 65–74.

Colon-Gonzalez, F., & Kazanietz, M. G. (2006). C1 domains exposed: From diacylglycerol binding to protein-protein interactions. *Biochimica et Biophysica Acta, 1761*, 827–837.

Cornford, P., Evans, J., Dodson, A., Parsons, K., Woolfenden, A., Neoptolemos, J., et al. (1999). Protein kinase C isoenzyme patterns characteristically modulated in early prostate cancer. *The American Journal of Pathology, 154*, 137–144.

Cutrupi, S., Baldanzi, G., Gramaglia, D., Maffe, A., Schaap, D., Giraudo, E., et al. (2000). Src-mediated activation of α-diacylglycerol kinase is required for hepatocyte growth factor-induced cell motility. *The EMBO Journal, 19*, 4614–4622.

den Hartog, M., Verhoef, N., & Munnik, T. (2003). Nod factor and elicitors activate different phospholipid signaling pathways in suspension-cultured alfalfa cells. *Plant Physiology, 132*, 311–317.

Diaz Anel, A. M., & Malhotra, V. (2005). PKCeta is required for beta1gamma2/beta3gamma2- and PKD-mediated transport to the cell surface and the organization of the Golgi apparatus. *The Journal of Cell Biology, 169*, 83–91.

Ding, L., Bunting, M., Topham, M. K., McIntyre, T. M., Zimmerman, G. A., & Prescott, S. M. (1997). Alternative splicing of the human diacylglycerol kinase zeta gene in muscle. *Proceedings of the National Academy of Sciences of the United States of America, 94*, 5519–5524.

Dupuy, A. J., Morgan, K., von Lintig, F. C., Shen, H., Acar, H., Hasz, D. E., et al. (2001). Activation of the Rap1 guanine nucleotide exchange gene, CalDAG-GEF I, in BXH-2 murine myeloid leukemia. *The Journal of Biological Chemistry, 276*, 11804–11811.

Evangelisti, C., Tazzari, P. L., Riccio, M., Fiume, R., Hozumi, Y., Fala, F., et al. (2007). Nuclear diacylglycerol kinase-{zeta} is a negative regulator of cell cycle progression in C2C12 mouse myoblasts. *The FASEB Journal, 12*, 3297–3307.

Fanani, M. L., Topham, M. K., Walsh, J. P., & Epand, R. M. (2004). Lipid modulation of the activity of diacylglycerol kinase alpha- and zeta-isoforms: Activation by phosphatidylethanolamine and cholesterol. *Biochemistry, 43*, 14767–14777.

Filigheddu, N., Cutrupi, S., Porporato, P. E., Riboni, F., Baldanzi, G., Chianale, F., et al. (2007). Diacylglycerol kinase is required for HGF-induced invasiveness and anchorage-independent growth of MDA-MB-231 breast cancer cells. *Anticancer Research, 27*, 1489–1492.

Flores, I., Jones, D. R., Cipres, A., Diaz-Flores, E., Sanjuan, M. A., & Merida, I. (1999). Diacylglycerol kinase inhibition prevents IL-2-induced G1 to S transition through a phosphatidylinositol-3 kinase-independent mechanism. *Journal of Immunology, 163*, 708–714.

Foster, D. A. (2007). Regulation of mTOR by phosphatidic acid? *Cancer Research, 67*, 1–4.

Foster, D. A. (2009). Phosphatidic acid signaling to mTOR: Signals for the survival of human cancer cells. *Biochimica et Biophysica Acta, 1791*(9), 949-955.

Garcia-Murillas, I., Pettitt, T., Macdonald, E., Okkenhaug, H., Georgiev, P., Trivedi, D., et al. (2006). lazaro encodes a lipid phosphate phosphohydrolase that regulates phosphatidylinositol turnover during Drosophila phototransduction. *Neuron, 49*, 533–546.

Gomez-Merino, F. C., Arana-Ceballos, F. A., Trejo-Tellez, L. I., Skirycz, A., Brearley, C. A., & Dormann, P. (2005). Arabidopsis AtDGK7, the smallest member of plant diacylglycerol kinases (DGKs), displays unique biochemical features and saturates at low substrate concentration: The DGK inhibitor R59022 differentially affects AtDGK2 and AtDGK7 activity in vitro and alters plant growth and development. *The Journal of Biological Chemistry, 280*, 34888–34899.

Gorin, M. A., & Pan, Q. (2009). Protein kinase C epsilon: An oncogene and emerging tumor biomarker. *Molecular Cancer, 8*, 9.

Griner, E. M., & Kazanietz, M. G. (2007). Protein kinase C and other diacylglycerol effectors in cancer. *Nature Reviews. Cancer, 7*, 281–294.

Grossoni, V., Falbo, K., Kazanietz, M., de Kier Joffe, E., & Urtreger, A. (2007). protein kinase C delta enhances proliferation and survival of murine mammary cells. *Molecular Carcinogenesis, 46*, 381–390.

Hall, C., Lim, L., & Leung, T. (2005). C1, see them all. *Trends in Biochemical Sciences, 30*, 169–171.

Hall, C., Michael, G. J., Cann, N., Ferrari, G., Teo, M., Jacobs, T., et al. (2001). alpha2-chimaerin, a Cdc42/Rac1 regulator, is selectively expressed in the rat embryonic nervous system and is involved in neuritogenesis in N1E-115 neuroblastoma cells. *The Journal of Neuroscience, 21*, 5191–5202.

Hardie, R. C. (2003). TRP channels in Drosophila photoreceptors: The lipid connection. *Cell Calcium, 33*, 385–393.

Henneberry, A. L., Wright, M. M., & McMaster, C. R. (2002). The major sites of cellular phospholipid synthesis and molecular determinants of Fatty Acid and lipid head group specificity. *Molecular Biology of the Cell, 13*, 3148–3161.

Hornberger, T. A., Chu, W. K., Mak, Y. W., Hsiung, J. W., Huang, S. A., & Chien, S. (2006). The role of phospholipase D and phosphatidic acid in the mechanical activation of mTOR signaling in skeletal muscle. *Proceedings of the National Academy of Sciences of the United States of America, 103*, 4741–4746.

Houssa, B., Schaap, D., van der Wal, J., Goto, K., Kondo, H., Yamakawa, A., et al. (1997). Cloning of a novel human diacylglycerol kinase (DGKtheta) containing three cysteine-rich domains, a proline-rich region, and a pleckstrin homology domain with an overlapping Ras-associating domain. *The Journal of Biological Chemistry, 272*, 10422–10428.

Hubbard, S. R., Bishop, W. R., Kirschmeier, P., George, S. J., Cramer, S. P., & Hendrickson, W. A. (1991). Identification and characterization of zinc binding sites in protein kinase C. *Science, 254*, 1776–1779.

Huitema, K., van den Dikkenberg, J., Brouwers, J. F., & Holthuis, J. C. (2004). Identification of a family of animal sphingomyelin synthases. *The EMBO Journal, 23*, 33–44.

Hurley, J., & Misra, S. (2000). Signaling and subcellular targeting by membrane-binding domains. *Annual Review of Biophysics and Biomolecular Structure, 29*, 49–79.

Hurley, J. H., Newton, A. C., Parker, P. J., Blumberg, P. M., & Nishizuka, Y. (1997). Taxonomy and function of C1 protein kinase C homology domains. *Protein Science, 6*, 477–480.

Ito, T., Hozumi, Y., Sakane, F., Saino-Saito, S., Kanoh, H., Aoyagi, M., et al. (2004). Cloning and characterization of diacylglycerol kinase iota splice variants in rat brain. *The Journal of Biological Chemistry, 279*, 23317–23326.

Itoh, S., Lemay, S., Osawa, M., Che, W., Duan, Y., Tompkins, A., et al. (2005). Mitochondrial Dok-4 recruits Src kinase and regulates NF-kappaB activation in endothelial cells. *The Journal Biological Chemistry, 280*, 26383–26396.

Jerga, A., Lu, Y. J., Schujman, G. E., de Mendoza, D., & Rock, C. O. (2007). Identification of a soluble diacylglycerol kinase required for lipoteichoic acid production in Bacillus subtilis. *The Journal of Biological Chemistry, 282*, 21738–21745.

Johnson, J. E., Zimmerman, M. L., Daleke, D. L., & Newton, A. C. (1998). Lipid structure and not membrane structure is the major determinant in the regulation of protein kinase C by phosphatidylserine. *Biochemistry, 37*, 12020–12025.

Jones, D. R., Sanjuan, M. A., & Merida, I. (2000). Type Ialpha phosphatidylinositol 4-phosphate 5-kinase is a putative target for increased intracellular phosphatidic acid. *FEBS Letters, 476*, 160–165.

Jones, D. R., Sanjuan, M. A., Stone, J. C., & Merida, I. (2002). Expression of a catalytically inactive form of diacylglycerol kinase alpha induces sustained signaling through RasGRP. *The FASEB Journal, 16*, 595–597.

Kai, M., Sakane, F., Imai, S., Wada, I., & Kanoh, H. (1994). Molecular cloning of a diacylglycerol kinase isozyme predominantly expressed in human retina with a truncated and inactive enzyme expression in most other human cells. *The Journal of Biological Chemistry, 269*, 18492–18498.

Kang, L., He, Z., Xu, P., Fan, J., Betz, A., Brose, N., et al. (2006). Munc13-1 is required for the sustained release of insulin from pancreatic beta cells. *Cell Metabolism, 3*, 463–468.

Kanoh, H., Yamada, K., & Sakane, F. (2002). Diacylglycerol kinases: Emerging downstream regulators in cell signaling systems. *Journal of Biochemistry (Tokyo), 131*, 629–633.

Katagiri, T., & Shinozaki, K. (1998). Disruption of a gene encoding phosphatidic acid phosphatase causes abnormal phenotypes in cell growth and abnormal cytokinesis in Saccharomyces cerevisiae. *Biochemical and Biophysical Research Communications, 248*, 87–92.

Kazanietz, M. G., Caloca, M. J., Eroles, P., Fujii, T., Garcia-Bermejo, M. L., Reilly, M., et al. (2000). Pharmacology of the receptors for the phorbol ester tumor promoters: Multiple receptors with different biochemical properties. *Biochemical Pharmacology, 60*, 1417–1424.

Kearns, B. G., McGee, T. P., Mayinger, P., Gedvilaite, A., Phillips, S. E., Kagiwada, S., et al. (1997). Essential role for diacylglycerol in protein transport from the yeast Golgi complex. *Nature, 387*, 101–105.

Kikkawa, U., Takai, Y., Tanaka, Y., Miyake, R., & Nishizuka, Y. (1983). Protein kinase C as a possible receptor protein of tumor-promoting phorbol esters. *The Journal of Biological Chemistry, 258*, 11442–11445.

Kuge, O., & Nishijima, M. (1997). Phosphatidylserine synthase I and II of mammalian cells. *Biochimica et Biophysica Acta, 1348*, 151–156.

Kwan, E. P., Xie, L., Sheu, L., Nolan, C. J., Prentki, M., Betz, A., et al. (2006). Munc13-1 deficiency reduces insulin secretion and causes abnormal glucose tolerance. *Diabetes, 55*, 1421–1429.

Labesse, G., Douguet, D., Assairi, L., & Gilles, A. M. (2002). Diacylglyceride kinases, sphingosine kinases and NAD kinases: Distant relatives of 6-phosphofructokinases. *Trends in Biochemical Sciences, 27*, 273–275.

Liljedahl, M., Maeda, Y., Colanzi, A., Ayala, I., Van Lint, J., & Malhotra, V. (2001). Protein kinase D regulates the fission of cell surface destined transport carriers from the trans-Golgi network. *Cell, 104*, 409–420.

Litvak, V., Dahan, N., Ramachandran, S., Sabanay, H., & Lev, S. (2005). Maintenance of the diacylglycerol level in the Golgi apparatus by the Nir2 protein is critical for Golgi secretory function. *Nature Cell Biology, 7*, 225–234.

Loomis, C. R., Walsh, J. P., & Bell, R. M. (1985). sn-1,2-Diacylglycerol kinase of Escherichia coli. Purification, reconstitution, and partial amino- and carboxyl-terminal analysis. *The Journal of Biological Chemistry, 260*, 4091–4097.

Los, A. P., Vinke, F. P., de Widt, J., Topham, M. K., van Blitterswijk, W. J., & Divecha, N. (2006). The retinoblastoma family proteins bind to and activate diacylglycerol kinase zeta. *The Journal of Biological Chemistry, 281*, 858–866.

Lu, C., Peng, Y. W., Shang, J., Pawlyk, B. S., Yu, F., & Li, T. (2001). The mammalian retinal degeneration B2 gene is not required for photoreceptor function and survival. *Neuroscience, 107*, 35–41.

Marklund, U., Lightfoot, K., & Cantrell, D. (2003). Intracellular location and cell context-dependent function of protein kinase D. *Immunity, 19*, 491–501.

Martinez-Climent, J. A., Alizadeh, A. A., Segraves, R., Blesa, D., Rubio-Moscardo, F., Albertson, D. G., et al. (2003). Transformation of follicular lymphoma to diffuse large cell lymphoma is associated with a heterogeneous set of DNA copy number and gene expression alterations. *Blood, 101*, 3109–3117.

Martinez-Gimeno, C., Diaz-Meco, M. T., Dominguez, I., & Moscat, J. (1995). Alterations in levels of different protein kinase C isotypes and their influence on behavior of squamous cell carcinoma of the oral cavity: Epsilon PKC, a novel prognostic factor for relapse and survival. *Head and Neck, 17*, 516–525.

Maruyama, H., Rakow, T. L., & Maruyama, I. N. (2001). Synaptic exocytosis and nervous system development impaired in Caenorhabditis elegans unc-13 mutants. *Neuroscience, 104*, 287–297.

McCracken, M. A., Miraglia, L. J., McKay, R. A., & Strobl, J. S. (2003). Protein kinase C delta is a prosurvival factor in human breast tumor cell lines. *Molecular Cancer Therapeutics, 2*, 273–281.

McMullan, R., Hiley, E., Morrison, P., & Nurrish, S. J. (2006). Rho is a presynaptic activator of neurotransmitter release at pre-existing synapses in C. elegans. *Genes and Development, 20*, 65–76.

Meijer, H. J., & Munnik, T. (2003). Phospholipid-based signaling in plants. *Annual Review of Plant Biology, 54*, 265–306.

Menna, P. L., Peche, L. Y., Alonso, D. F., & Gomez, D. E. (2005). Role of beta2-chimaerin in the behaviour of murine mammary carcinoma cells in response to extracellular matrix components. *International Journal of Molecular Medicine, 15*, 91–95.

Merida, I., Avila-Flores, A., & Merino, E. (2008). Diacylglycerol kinases: At the hub of cell signalling. *The Biochemical Journal, 409*, 1–18.

Miller, D. J., Jerga, A., Rock, C. O., & White, S. W. (2008). Analysis of the Staphylococcus aureus DgkB structure reveals a common catalytic mechanism for the soluble diacylglycerol kinases. *Structure, 16*, 1036–1046.

Milligan, S. C., Alb, J. G., Jr., Elagina, R. B., Bankaitis, V. A., & Hyde, D. R. (1997). The phosphatidylinositol transfer protein domain of Drosophila retinal degeneration B protein is essential for photoreceptor cell survival and recovery from light stimulation. *The Journal of Cell Biology, 139*, 351–363.

Murakami, T., Sakane, F., Imai, S., Houkin, K., & Kanoh, H. (2003). Identification and characterization of two splice variants of human diacylglycerol kinase eta. *The Journal of Biological Chemistry, 278*, 34364–34372.

Nanjundan, M., & Possmayer, F. (2003). Pulmonary phosphatidic acid phosphatase and lipid phosphate phosphohydrolase. *American Journal of Physiology. Lung Cellular and Molecular Physiology, 284*, L1–L23.

Nelson, C. D., Perry, S. J., Regier, D. S., Prescott, S. M., Topham, M. K., & Lefkowitz, R. J. (2007). Targeting of diacylglycerol degradation to M1 muscarinic receptors by beta-arrestins. *Science, 315*, 663–666.

Newton, A. C., & Koshland, D. E., Jr. (1989). High cooperativity, specificity, and multiplicity in the protein kinase C-lipid interaction. *The Journal of Biological Chemistry, 264*, 14909–14915.

Nurrish, S., Segalat, L., & Kaplan, J. M. (1999). Serotonin inhibition of synaptic transmission: Galpha(0) decreases the abundance of UNC-13 at release sites. *Neuron, 24*, 231–242.

Oancea, E., Bezzerides, V. J., Greka, A., & Clapham, D. E. (2003). Mechanism of persistent protein kinase D1 translocation and activation. *Developmental Cell, 4*, 561–574.

Ono, Y., Fujii, T., Igarashi, K., Kuno, T., Tanaka, C., Kikkawa, U., et al. (1989). Phorbol ester binding to protein kinase C requires a cysteine-rich zinc-finger-like sequence. *Proceedings of the National Academy of Sciences of the United States of America, 86*, 4868–4871.

Ostroski, M., Tu-Sekine, B., & Raben, D. M. (2005). Analysis of a novel diacylglycerol kinase from Dictyostelium discoideum: DGKA. *Biochemistry, 44*, 10199–10207.

Pan, Q., Bao, L. W., Kleer, C. G., Sabel, M. S., Griffith, K. A., Teknos, T. N., et al. (2005). Protein kinase C epsilon is a predictive biomarker of aggressive breast cancer and a validated target for RNA interference anticancer therapy. *Cancer Research, 65*, 8366–8371.

Pang, H., & Bitar, K. N. (2005). Direct association of RhoA with specific domains of PKC-alpha. *American Journal of Physiology. Cell Physiology, 289*, C982–C993.

Parekh, D. B., Ziegler, W., & Parker, P. J. (2000). Multiple pathways control protein kinase C phosphorylation. *The EMBO Journal, 19*, 496–503.

Perez de Castro, I., Bivona, T. G., Philips, M. R., & Pellicer, A. (2004). Ras activation in Jurkat T cells following low-grade stimulation of the T-cell receptor is specific to N-Ras and occurs only on the Golgi apparatus. *Molecular and Cellular Biology, 24*, 3485–3496.

Pitson, S. M., Moretti, P. A. B., Zebol, J. R., Zareie, R., Derian, C. K., Darrow, A. L., et al. (2002). The nucleotide-binding site of human sphingosine kinase 1. *The Journal of Biological Chemistry, 277*, 49545–49553.

Poole, A. W., Pula, G., Hers, I., Crosby, D., & Jones, M. L. (2004). PKC-interacting proteins: From function to pharmacology. *Trends in Pharmacological Sciences, 25*, 528–535.

Prekeris, R., Hernandez, R. M., Mayhew, M. W., White, M. K., & Terrian, D. M. (1998). Molecular analysis of the interactions between protein kinase C-epsilon and filamentous actin. *The Journal of Biological Chemistry, 273*, 26790–26798.

Reddig, P. J., Dreckschmidt, N. E., Ahrens, H., Simsiman, R., Tseng, C. P., Zou, J., et al. (1999). Transgenic mice overexpressing protein kinase Cdelta in the epidermis are resistant to skin tumor promotion by 12-O-tetradecanoylphorbol-13-acetate. *Cancer Research, 59*, 5710–5718.

Regier, D. S., Higbee, J., Lund, K. M., Sakane, F., Prescott, S. M., & Topham, M. K. (2005). Diacylglycerol kinase iota regulates Ras guanyl-releasing protein 3 and inhibits Rap1 signaling. *Proceedings of the National Academy of Sciences of the United States of America, 102*, 7595–7600.

Ruelland, E., Cantrel, C., Gawer, M., Kader, J. C., & Zachowski, A. (2002). Activation of phospholipases C and D is an early response to a cold exposure in Arabidopsis suspension cells. *Plant Physiology, 130*, 999–1007.

Sakane, F., Imai, S., Kai, M., Yasuda, S., & Kanoh, H. (2008). Diacylglycerol kinases as emerging potential drug targets for a variety of diseases. *Current Drug Targets, 9*, 626–640.

Sakane, F., Imai, S., Yamada, K., Murakami, T., Tsushima, S., & Kanoh, H. (2002). Alternative splicing of the human diacylglycerol kinase delta gene generates two isoforms differing in their expression patterns and in regulatory functions. *The Journal of Biological Chemistry, 277*, 43519–43526.

Sakane, F., Kai, M., Wada, I., Imai, S., & Kanoh, H. (1996). The C-terminal part of diacylglycerol kinase alpha lacking zinc fingers serves as a catalytic domain. *The Biochemical Journal, 318*(Pt 2), 583–590.

Sanjuan, M. A., Pradet-Balade, B., Jones, D. R., Martinez, A. C., Stone, J. C., Garcia-Sanz, J. A., et al. (2003). T cell activation in vivo targets diacylglycerol kinase alpha to the membrane: A novel mechanism for Ras attenuation. *Journal of Immunology, 170*, 2877–2883.

Santos, A. C., & Lehmann, R. (2004). Germ cell specification and migration in Drosophila and beyond. *Current Biology, 14*, R578–R589.

Santos, T., Carrasco, S., Jones, D. R., Merida, I., & Eguinoa, A. (2002). Dynamics of diacylglycerol kinase zeta translocation in living T-cells. Study of the structural domain requirements for translocation and activity. *The Journal of Biological Chemistry, 277*, 30300–30309.

Shindo, M., Irie, K., Masuda, A., Ohigashi, H., Shirai, Y., Miyasaka, K., et al. (2003). Synthesis and phorbol ester binding of the cysteine-rich domains of diacylglycerol kinase (DGK) isozymes. DGKgamma and DGKbeta are new targets of tumor-promoting phorbol esters. *The Journal of Biological Chemistry, 278*, 18448–18454.

Shirai, Y., Segawa, S., Kuriyama, M., Goto, K., Sakai, N., & Saito, N. (2000). Subtype-specific translocation of diacylglycerol kinase alpha and gamma and its correlation with protein kinase C. *The Journal of Biological Chemistry, 275*, 24760–24766.

Smith, W. L., Marnett, L. J., & DeWitt, D. L. (1991). Prostaglandin and thromboxane biosynthesis. *Pharmacology and Therapeutics, 49*, 153–179.

Snedden, W. A., & Blumwald, E. (2000). Alternative splicing of a novel diacylglycerol kinase in tomato leads to a calmodulin-binding isoform. *Plant Journal, 24*, 317–326.

Spitaler, M., & Cantrell, D. A. (2004). Protein kinase C and beyond. *Nat Immunology, 5*, 785–790.

Springett, G. M., Bonham, L., Hummer, A., Linkov, I., Misra, D., Ma, C., et al. (2005). Lysophosphatidic acid acyltransferase-beta is a prognostic marker and therapeutic target in gynecologic malignancies. *Cancer Research, 65*, 9415–9425.

Stahelin, R. V., Digman, M. A., Medkova, M., Ananthanarayanan, B., Melowic, H. R., Rafter, J. D., et al. (2005). Diacylglycerol-induced membrane targeting and activation of protein kinase

Cepsilon: Mechanistic differences between protein kinases Cdelta and Cepsilon. *The Journal of Biological Chemistry, 280*, 19784–19793.
Tan, I., Seow, K. T., Lim, L., & Leung, T. (2001). Intermolecular and intramolecular interactions regulate catalytic activity of myotonic dystrophy kinase-related Cdc42-binding kinase alpha. *Molecular and Cellular Biology, 21*, 2767–2778.
Tang, W., Yuan, J., Chen, X., Gu, X., Luo, K., Li, J., et al. (2006). Identification of a novel human lysophosphatidic acid acyltransferase, LPAAT-theta, which activates mTOR pathway. *Journal of Biochemistry and Molecular Biology, 39*, 626–635.
Tanyi, J. L., Hasegawa, Y., Lapushin, R., Morris, A. J., Wolf, J. K., Berchuck, A., et al. (2003). Role of decreased levels of lipid phosphate phosphatase-1 in accumulation of lysophosphatidic acid in ovarian cancer. *Clinical Cancer Research, 9*, 3534–3545.
Thuille, N., Heit, I., Fresser, F., Krumbock, N., Bauer, B., Leuthaeusser, S., et al. (2005). Critical role of novel Thr-219 autophosphorylation for the cellular function of PKCtheta in T lymphocytes. *The EMBO Journal, 24*, 3869–3880.
Topham, M. K., Bunting, M., Zimmerman, G. A., McIntyre, T. M., Blackshear, P. J., & Prescott, S. M. (1998). Protein kinase C regulates the nuclear localization of diacylglycerol kinase-zeta. *Nature, 394*, 697–700.
Topham, M. K., & Epand, R. M. (2009). Mammalian diacylglycerol kinases: Molecular interactions and biological functions of selected isoforms. *Biochimica et Biophysica Acta, 1790*(1), 416–424.
Topham, M. K., & Prescott, S. M. (2001). Diacylglycerol kinase zeta regulates Ras activation by a novel mechanism. *The Journal of Cell Biology, 152*, 1135–1143.
Tsushima, S., Kai, M., Yamada, K., Imai, S., Houkin, K., Kanoh, H., et al. (2004). Diacylglycerol kinase gamma serves as an upstream suppressor of Rac1 and lamellipodium formation. *The Journal of Biological Chemistry, 279*, 28603–28613.
van Baal, J., de Widt, J., Divecha, N., & van Blitterswijk, W. J. (2005). Translocation of diacylglycerol kinase theta from cytosol to plasma membrane in response to activation of G protein-coupled receptors and protein kinase C. *The Journal of Biological Chemistry, 280*, 9870–9878.
Varoqueaux, F., Sons, M. S., Plomp, J. J., & Brose, N. (2005). Aberrant morphology and residual transmitter release at the Munc13-deficient mouse neuromuscular synapse. *Molecular and Cellular Biology, 25*, 5973–5984.
Wang, H., & Kazanietz, M. G. (2002). Chimaerins, novel non-protein kinase C phorbol ester receptors, associate with Tmp21-I (p23): Evidence for a novel anchoring mechanism involving the chimaerin C1 domain. *The Journal of Biological Chemistry, 277*, 4541–4550.
Wang, H., Yang, C., Leskow, F. C., Sun, J., Canagarajah, B., Hurley, J. H., et al. (2006). Phospholipase Cgamma/diacylglycerol-dependent activation of beta2-chimaerin restricts EGF-induced Rac signaling. *The EMBO Journal, 25*, 2062–2074.
Wang, Q. J. (2006). PKD at the crossroads of DAG and PKC signaling. *Trends in Pharmacological Sciences, 27*, 317–323.
Wanga, X., Devaiaha, S. P., Zhang, W., & Welti, R. (2006). Signaling functions of phosphatidic acid. *Progress in Lipid Research, 45*, 250–278.
Wattenberg, B. W., Pitson, S. M., & Raben, D. M. (2006). The sphingosine and diacylglycerol kinase superfamily of signaling kinases: Localization as a key to signaling function. *Journal of Lipid Research, 47*, 1128–1139.
Wegmeyer, H., Egea, J., Rabe, N., Gezelius, H., Filosa, A., Enjin, A., et al. (2007). EphA4-dependent axon guidance is mediated by the RacGAP alpha2-chimaerin. *Neuron, 55*, 756–767.
Yanagisawa, K., Yasuda, S., Kai, M., Imai, S., Yamada, K., Yamashita, T., et al. (2007). Diacylglycerol kinase alpha suppresses tumor necrosis factor-alpha-induced apoptosis of human melanoma cells through NF-kappaB activation. *Biochimica et Biophysica Acta, 1771*, 462–474.
Yang, C., & Kazanietz, M. G. (2003). Divergence and complexities in DAG signaling: Looking beyond PKC. *Trends in Pharmacological Sciences, 24*, 602–608.

Yang, C., Liu, Y., Leskow, F. C., Weaver, V. M., & Kazanietz, M. G. (2005). Rac-GAP-dependent inhibition of breast cancer cell proliferation by {beta}2-chimerin. *The Journal of Biological Chemistry, 280*, 24363–24370.

Yasuda, S., Kai, M., Imai, S., Kanoh, H., & Sakane, F. (2007). Diacylglycerol kinase gamma interacts with and activates beta2-chimaerin, a Rac-specific GAP, in response to epidermal growth factor. *FEBS Letters, 581*, 551–557.

Yoshida, T., Fukaya, M., Uchigashima, M., Miura, E., Kamiya, H., Kano, M., et al. (2006). Localization of diacylglycerol lipase-alpha around postsynaptic spine suggests close proximity between production site of an endocannabinoid, 2-arachidonoyl-glycerol, and presynaptic cannabinoid CB1 receptor. *The Journal of Neuroscience, 26*, 4740–4751.

Yuan, S., Miller, D. W., Barnett, G. H., Hahn, J. F., & Williams, B. R. (1995). Identification and characterization of human beta 2-chimaerin: Association with malignant transformation in astrocytoma. *Cancer Research, 55*, 3456–3461.

Zheng, Y., Liu, H., Coughlin, J., Zheng, J., Li, L., & Stone, J. C. (2005). Phosphorylation of RasGRP3 on threonine 133 provides a mechanistic link between PKC and Ras signaling systems in B cells. *Blood, 105*, 3648–3654.

Chapter 5
Regulation of PKC by Protein–Protein Interactions in Cancer

Jeewon Kim and Daria Mochly-Rosen*

Abstract Protein kinase C (PKC) was first identified in 1977 by Nishizuka's group as a proteolytically activated protein kinase. It was subsequently found that the enzyme is activated by calcium and anionic phospholipids. Unsaturated diacylglycerol (DAG) was then found to be an essential activator of PKC, linking PKC activation to tyrosine kinase- or G-protein-coupled receptor-mediated inositol phospholipid hydrolysis. In addition to phospholipids, DAG, and calcium (depending on the isozyme), PKC isozymes are also regulated by protein–protein interactions. In a variety of experimental models, PKC isozymes have been found to mediate different and often opposing roles in tumor growth, and their activities are further regulated by these multiple intra- and intermolecular protein–protein interactions.

In clinical samples, the levels of protein or activities of PKCs are dysregulated when compared to normal tissue of the same origin and correlate with poor prognosis. PKC regulates multiple aspects of tumorigenesis, including cell proliferation, angiogenesis, metastasis, and apoptosis, making it a major regulator in the transformation to malignant phenotype. This review focuses on the current understanding of PKC regulation by protein–protein interactions as it relates to cancer. We summarize known roles for each domain of PKC and discuss intramolecular interactions that regulate the activation state of the enzyme, as well as intermolecular interactions that determine the specificity of the signaling of each PKC isozyme. We also demonstrate how identification of the molecular sites of specific protein–protein interactions within PKC and between PKC and other proteins has led to the

---

*DM-R is the founder and share holder of KAI Pharmaceuticals, Inc., a company that plans to bring PKC regulators to the clinic. However, none of the work described in this study is based on or supported by the company. Other authors have no disclosure.

J. Kim and D. Mochly-Rosen (✉)
Department of Chemical and Systems Biology, Stanford University, School of Medicine, Stanford, CA 94305, USA
e-mail: jwonkim@stanford.edu; mochly@stanford.edu

M.G. Kazanietz (ed.), *Protein Kinase C in Cancer Signaling and Therapy*, Current Cancer Research, DOI 10.1007/978-1-60761-543-9_5,

design of effective isozyme-selective activators and inhibitors of PKC and discuss how these pharmacological tools can assist in determining the role of specific PKC isozymes in tumorigenesis.

Keywords Angiogenesis • Cancer • Metastasis • Protein kinase C • Protein–protein interaction

5.1 Protein Kinase C

Protein kinase C (PKC), a family of related isozymes, is described in detail in section 1.2, "PKC isozymes; genes and structure (see table of contents)" Part 2. Briefly, there are ten members in this family that can be subdivided into the conventional cPKC isozymes (PKC α, βI, βII and γ), the novel nPKC isozymes (PKC δ, ε, η and θ), and the atypical aPKC isozymes (PKC ζ and λ/ι; Fig. 5.1). These families differ in the composition and order of the domains; all share substantial homology in the C3/C4 catalytic domain. However, while cPKCs and nPKCs have a C1 domain with two repeats of the Cys-rich diacylglycerol (DAG)-binding domain, the atypical PKCs have only one Cys-rich domain and do not bind DAG. The C2 domain is not present in the aPKC and these PKCs can be activated, in part, by interaction with the Cdc42-GTP-Par6 complex through the PB1 domain.

PKC is activated by multiple steps in a process involving calcium, phosphatidylserine (PS), and DAG binding, release of pseudosubstrate interaction with the catalytic

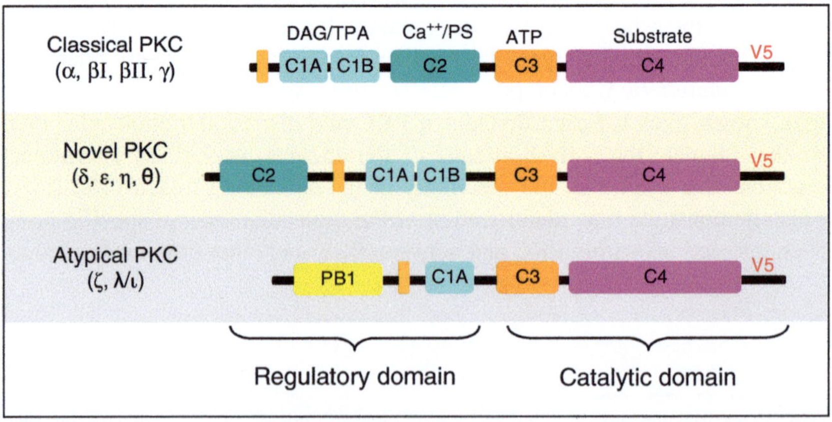

Fig. 5.1 The architecture of the domains of PKC isozymes. PKC isozymes are classified based on their dependence on specific second messengers for their activation. The subfamilies differ mainly in the composition of their regulatory domain. Conventional PKCs are sensitive to calcium and DAG, which bind to their C2 and C1 domains, respectively. Novel PKCs are not sensitive to calcium but are more sensitive to DAG through its binding to the C2. Atypical PKCs lack a C2 domain and a functional C1 domain and are therefore sensitive neither to calcium nor to DAG. Atypical PKCs have a PB1 domain that provides unique interactions with PB1 domain of Par6. The location of the pseudosubstrate sequence within the regulatory domains is also indicated; this site keeps PKC in an inactive state in the absence of stimuli (*TPA* 12-*O*-tetradecanoylphorbol 13-acetate, *PS* phosphatidyl serine)

site (as well as other intramolecular interactions), and binding of the enzyme with its isozyme-selective receptor for active C-kinases (RACK) and its substrates (Takai et al. 1979a; Nishizuka 1986; Mochly-Rosen et al. 1991). With activation of G-protein-coupled receptors or tyrosine receptor kinases, phsopholipase C is activated (PLC-β or PLC-γ) and hydrolyzes $PtnIns_{4,5}P_2$ (PIP_2) into DAG and IP_3, and leads to an increase in intracellular calcium concentration. For classical PKCs, calcium binding to the C2 domain increases the affinity of enzyme for PS at the cell membrane (Medkova and Cho 1998; Kohout et al. 2002). PKC also binds DAG through its C1 domain and through conformational changes, the autoinhibitory pseudosubstrate site is released from the catalytic domain. With these conformational changes, PKC becomes an "open" form and the catalytic and RACK-binding domains are available to access RACK and substrates (Mochly-Rosen et al. 1991; Mochly-Rosen 1995). Activation leads to translocation of PKCs from the cytosol to the cell membranes where they are localized at specific sites in the cells and are anchored by RACK. Therefore, translocation places each PKC isozyme near its specific substrate and away from other substrates (Mochly-Rosen 1995). PKC is also activated by phosphorylation in serine/threonine residues by transphosphorylation and autophosphorylation or by proteolysis (Parekh et al. 2000). Further, both the regulatory and catalytic domains of the enzyme participate in intra- and intermolecular protein–protein interactions, fine-tuning the multistep events that lead to PKC activation and localization in close proximity to its substrates (Kheifets and Mochly-Rosen 2007).

5.2 PKC Isozymes Are Regulated by Multiple Protein–Protein Interactions

PKC isozymes are activated when the levels of the lipid-derived second messenger, DAG, are elevated in the cell (Takai et al. 1979b). Whereas activation of the cPKCs requires elevation of intracellular calcium, the nPKC isozymes are independent of calcium. As mentioned above, activation of the c and nPKC isozymes is associated with movement or translocation of the enzyme from the cell soluble to the cell particulate fraction, an effect that can be determined readily after cell fractionation (Kraft and Anderson 1983). Because the second messenger that triggers this translocation is derived from lipids, it was expected that the activated PKCs bind at the plasma membrane. However, work using a variety of cellular models demonstrated that the location of individual PKC isozymes after activation is not restricted to the plasma membrane; activated PKC isozymes can be found inside the nucleus (Disatnik et al. 1994), on contractile elements and cell–cell contacts (Vallentin et al. 2001), in the mitochondria (Churchill et al. 2005), on the Golgi apparatus (Lehel et al. 1995), in the endoplasmic reticulum (Qi and Mochly-Rosen 2008), and on other fibrillar structures in the cell (Vattemi et al. 2004). These observations suggested to us that localization of activated individual PKC isozymes is mediated, in part, by their binding to anchoring proteins termed RACKs (for receptors for active C kinases)

(Mochly-Rosen et al. 1991; Mochly-Rosen 1995). Compartmentalization of individual isozyme near a subset of substrates and away from others provides the means for isozyme-selective functions. Further, a set of substrate proteins collectively termed STICKs (substrates that interact with C-kinases; Jaken and Parker 2000) further provide anchoring to select subcellular sites. Translocation of individual PKC isozymes from one location to another may be facilitated by their bindings to other proteins, such as annexins. For example, we showed that annexin I selectively binds PKCβII (Ron and Mochly-Rosen 1994) and annexin V selectively binds PKCδ (Kheifets et al. 2006). PKCδ binding to annexin V is required for its translocation upon activation (Kheifets et al. 2006). Furthermore, a number of intramolecular interactions stabilize the enzyme in the inactive state and in the active state. These intramolecular interactions can be between the regulatory and the catalytic halves of the enzyme and between domains within each of them. Therefore, PKC must contain many protein–protein interaction sites, and inhibition of these interactions should change the activity state of the enzymes. The following is a review of some of these protein–protein interactions and how identification of these sites can lead to new types of pharmacological agents that regulate PKC function *in vivo*.

5.2.1 The Pseudo-Substrate Site

Although the inhibitory role of the regulatory domain was demonstrated already by the early finding that PKC is activated by proteolysis (Inoue et al. 1977), the mapping of the first intramolecular protein–protein interaction site in PKC was identified by House and Kemp (1987). These authors found a sequence that precedes the C1 domain in cPKC that mimics a substrate consensus sequence for PKC, except instead of having a Ser/Thr in the position that is to be phosphorylated, that PKC sequence has an Ala. This site, termed pseudosubstrate site (ψ-substrate), binds to the catalytic site in the C4 domain and keeps the enzyme in the inactive state. Activation leads to dissociation of the ψ-substrate sequence from the catalytic site and frees the catalytic domain to interact with substrates. The evidence supporting this conclusion includes the findings that PKC enzyme lacking this sequence is catalytically active (Pears et al. 1990), that a peptide corresponding to the catalytic domain of PKC is an activator of PKC (House et al. 1989) and that a mutation of the PKC ψ-substrate sequence from Ala to Glu to mimic the negative charge of the phosphorylated enzyme also leads to active PKC (Pears et al. 1990). Interestingly, the ψ-substrate site also participates in protein–lipid interaction; the positive Arg residues in that site interact with the negative head groups of phospholipids (Mosior and McLaughlin 1991).

5.2.2 C1 Domain

The C1 domain contains two repeats of a Cys-rich domain. In most isozymes except PKCγ, only one of these two domains actively binds the second messenger DAG (Ono et al. 1989). This domain alone can mediate anchoring to the plasma membrane, as shown by studies using the GFP-C1 domain (Hurley and Meyer 2001). In addition to binding DAG, the C1 domain contains sequences that participate in protein–protein interactions. These include potential interactions with small G proteins (Ghosh et al. 1994), interaction with actin (Prekeris et al. 1996), and subcellular targeting to Golgi (Schultz et al. 2003).

5.2.3 *The C2 Domain*

The C2 domain is a β sandwich, made of two sheets of four β strands each (Rizo and Sudhof 1998). This domain binds the negatively charged phospholipid, PS, and in cPKCs, the C2 domain coordinates calcium binding (Takai et al. 1979a), which increases the affinity of the domain to membranes (Nalefski and Newton 2001). Studies by Newton and collaborators also demonstrated that the C2 domain interacts with the V5 domain at the end of the catalytic half of the enzyme (Edwards and Newton 1997); PKCβI and PKCβII, which differ only in their V5 domain, have different sensitivities for calcium (Edwards and Newton 1997; Keranen and Newton 1997). Many studies from our laboratory demonstrate that the C2 domain is also involved in multiple intramolecular interactions as well as interaction between PKCs and their RACK (Ron et al. 1995; Chen et al. 2001; Inagaki et al. 2003a, b).

5.2.3.1 The C2 Domain Mediates Intermolecular Interactions

In 1992, we showed that a recombinant fragment containing the C2 domain of synaptotagmin binds the PKC-binding protein, RACK, at about a 100-fold lower affinity relative to PKC binding (Mochly-Rosen et al. 1992). These data suggested that regions within the PKC C2 domain also contain a RACK-binding site. We identified short peptides derived from the C2 domain of different PKC isozymes as selective inhibitors of binding of the corresponding isozymes to their RACKs. The first C2-derived inhibitory peptides were derived from the C2 domain of PKCβ and were termed βC2-1, βC2-2 and βC2-4. These peptides inhibit PKCβ function when introduced into cells (Ron et al. 1995).

The δRACK- and εRACK-binding sites on the C2 domain of the corresponding isozymes were subsequently identified based on the least homologous sequence between highly homologous isozymes (Chen et al. 2001) and the most conserved sequence in the domain across evolutionarily remote species, such as the sea snail, Aplysia, and rat (Johnson et al. 1996). The peptides derived from these sequences are highly selective inhibitors of translocation and function of the corresponding isozymes in culture and *in vivo* and were found to be useful in a variety of animal models of human diseases. Among these diseases are ischemic cardiac disease (Inagaki et al. 2003a; Inagaki and Mochly-Rosen 2005), tumor angiogenesis (Kim et al. 2008), and heart failure (Inagaki et al. 2008). The peptide inhibitors were found to be safe in humans and are currently used in clinical trials in patients with acute myocardial infarction (Bates et al. 2008) and in patients with postoperative pain. Therefore, identifying the location of intermolecular interaction sites on the PKC C2 domain for the cognate RACK generated useful pharmacological tools and perhaps even drugs to treat patients with diverse acute and chronic diseases.

5.2.3.2 The C2 Domain Also Mediates Many Intramolecular Interactions Between the Domain and Other Domains Within PKC

These intramolecular interactions were first identified using the same rationale for the pseudosubstrate (ψ-substrate) site described above. We reasoned that the RACK-binding sequence in PKC must be unavailable for interaction with the RACK when PKC is inactive and therefore predicted that a sequence within PKC that mimics the PKC-binding site on the RACK is found also in PKC. We predicted that, similar to the homology between the phosphorylation site on substrates and the ψ-substrate site in PKC, the RACK-like (pseudo-RACK) sequence within the enzyme must have an important difference from the PKC-binding site on RACK. In 1994, we identified such a homologous sequence between PKCβII and its cognate binding protein, RACK1 (Ron et al. 1994). The six amino acid stretch in PKCβII, SVEIWD (amino acid 241–246 in PKCβII), had a charge difference from the RACK1 sequence SIKIWD (amino acids 255–260 in RACK1). We subsequently demonstrated that a peptide derived from this region interferes with the intramolecular protein–protein interaction, thus exposing the RACK-binding site on PKC, leading to translocation and activation of PKCβII (Ron and Mochly-Rosen 1995). Such ψ-RACK sites were identified in all the other PKC isozymes and peptides corresponding to these sequences are selective activators of translocation and function of the corresponding isozymes. For example, the ψεRACK peptide selectively causes the translocation of PKCε and not any other PKC isozyme in a variety of species (Chen et al. 2001; Inagaki et al. 2003b, 2005). These peptides appear safe even when given for many weeks in a variety of animal models of human diseases (Inagaki et al. 2005; Koyanagi et al. 2007). Therefore, peptide activators can be identified based on the homology between PKC and its cognate interacting proteins that can be used as pharmacological tools for animal studies and possibly as therapeutics in human diseases.

In contrast to the above rational approach to identify the location of the intra- and intermolecular protein–protein interactions within the C2 domain, a number of other peptide activators derived from the C2 domain were identified using a more systematic search (Brandman et al. 2007). In that study, four activator peptides for PKCε translocation and function were identified. Because the peptides are PKC activators and they are derived from different surfaces of the βC2 sandwich (Brandman et al. 2007), it is likely that they correspond to interaction sites with different domains within that PKC. Further, some of the peptides are highly selective for PKCε; they lead to protection of the heart from ischemic damage, a function of PKCε. Other activator peptides derived from the PKCε C2 domain activate multiple PKC isozymes and lead to PKC-mediated substrate phosphorylation in PKCε

KO cells (Brandman et al. 2007). These peptides are likely to represent regions in the C2 domain that interact with conserved sequences in different PKC isozymes. This more systematic approach of mapping the surface of the C2 domain also lead to the identification of a number of potential peptide inhibitors of PKCε that are derived from sites on the C2 domain unlikely to interact with the RACK. We predict that these peptides represent sites in that domain that interact with select substrates of each PKC isozyme. These may represent STICKs proposed by Jaken and Parker (2000). Further characterization of these peptides and corresponding peptides from other C2 domains is under way.

5.2.4 The V3 Region

The V3 region is the hinge region between the catalytic and regulatory domain. Upon activation, a proteolytic site for a variety of proteases is exposed in that region, leading to the separation of the regulatory domain from the catalytic domain and the generation of constitutively active kinase fragments (Mochly-Rosen and Koshland 1987; Steinberg 2004). The generation of a catalytic fragment occurs also in cells and has been suggested to be critical in processes such as long-term memory (Hernandez et al. 2003) and apoptosis (Emoto et al. 1995). The V3 region is also critical in protein–protein interaction. For example, it interacts with β1 integrin, an interaction that regulates cell migration and chemotaxis (Ng et al. 1999; Parsons et al. 2002).

5.2.5 The C3/C4 Domain

These domains contain the ATP- and substrate-binding sites as well as the catalytic domain of the enzyme. The location of the substrate-binding site was identified by House and Kemp before the crystal structure of that domain was confirmed (House and Kemp 1987). As with any enzyme, the interaction with the substrate is terminated upon its phosphorylation and is thus likely to reflect the binding of the domain to the PKC consensus phosphorylation motif in the substrate protein. Although this intermolecular protein–protein interaction has not been studied in detail and selective regulators of these interactions have not been identified, it may mediate, at least in part, the interaction between PKC isozymes and their STICKs (Jaken and Parker 2000).

5.2.6 The V5 Region

The V5 region contains important posttranslational phosphorylation sites in PKC. Studies by Newton and collaborators showed that a number of serines and threonines in the V5 domain need to be phosphorylated to mature the enzyme; in the un-phosphorylated state, the V5 region occupies the catalytic site of the enzyme and prevents its activation by DAG (Newton 2003). As discussed above, the mature

V5 domain may interact with the C2 domain. Although structural information on this region is not available because it is highly flexible, it is possible that the V5 region acquires a β-sheet-like structure, thus extending the surface of the β sheets of the C2 domain (Kheifets and Mochly-Rosen 2007).

We also confirmed that the V5 region plays a critical role in protein–protein interactions. We found that activated PKCβI and PKCβII are localized to different compartments in cardiac myocytes (Disatnik et al. 1994); whereas activated PKCβII is localized to the perinucleus and plasma membrane, activated PKCβI is found inside the nucleus. The only difference between PKCβI and PKCβII is the last 50 amino acids out of 750 amino acids (Parker et al. 1986). Therefore, we reasoned that the V5 region and, specifically, the least homologous sequences within this region contain the binding sites for the anchoring proteins, RACKs, in each of the cell compartments. We then showed that peptides that correspond to these sequences (6–10 amino acid-long, termed V5-1, V5-3, and V5-5) are selective inhibitors of the translocation and function of the corresponding isozymes (Stebbins and Mochly-Rosen 2001; Kim et al. 2008). Corresponding peptides from the V5 region of PKCα and PKCγ were also found to be selective inhibitors of translocation and function of their corresponding isozymes (Sweitzer et al. 2004).

The role of the V5 region in protein–protein interactions in other PKC isozymes has been confirmed using direct protein–protein interaction studies *in vitro*. For example, we showed that both the V5 region and the C2 region of PKCδ directly bind annexin V (Kheifets et al. 2006).

5.3 The Role of Protein–Protein Interaction of RACK and PKC in Tumorigenesis

PKC is activated by tumor-promoting phorbol esters and its involvement in carcinogenesis was proposed many years ago (Castagna et al. 1982). Its role has since been substantiated in many human cancers, including prostate, breast, ovarian, and colon cancer (Teicher et al. 2002a, b; Koren et al. 2004; Graff et al. 2005). In particular, RACK1 and RACK2 contain seven repeats of WD-40 motifs, which enable them to serve as adaptor platforms to position different molecules in the proximity for efficient signaling (Churchill et al. 2008). Currently, there are several PKC inhibitors and activators used in ongoing clinical trials for cancer treatment (Martiny-Baron and Fabbro 2007). Among these, aurothiomalate (ATM), used in non small cell lung cancer (Fields et al. 2007), was developed based on the protein–protein interaction. Among all the PKC isozymes, the PB1 domain is present only in the atypical PKC isozymes (Parker and Murray-Rust 2004) and ATM is a specific inhibitor of PB1–PB1 domain interaction between PKCι and Par6, an adaptor molecule of this isozyme (Regala et al. 2008). We will not discuss it further here because this is being reviewed extensively in Sect. 5.4 of this series.

Here, we focus on the intermolecular protein–protein interactions involving PKCs and RACKs and will review the functional consequences of these interactions on various stages of tumorigenesis (Fig. 5.2 and summarized in Table 5.1).

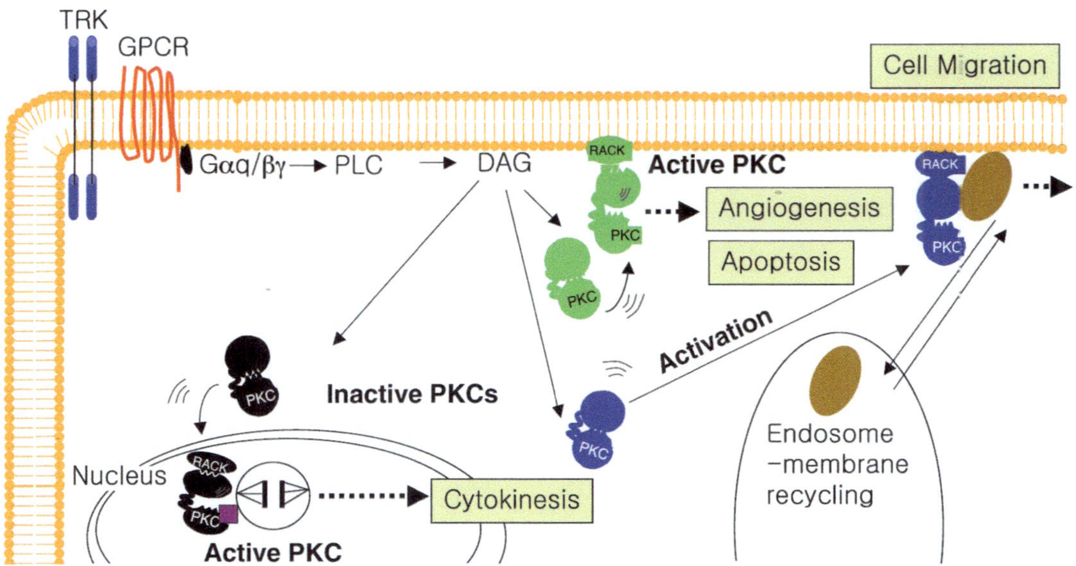

Fig. 5.2 PKC activation involves interruption and induction of several protein–protein interactions. On activation with DAG (and calcium for cPKCs), PKC undergoes conformational change that involves interruption of intramolecular protein–protein interactions, leading to a transitional state. Each PKC isozyme can then translocate to different cellular locations where it establishes new protein–protein interactions through its binding to isozyme-specific RACKs. Intra- and intermolecular protein–protein interactions involving PKC have physiological effects on the tumor growth by regulating cytokinesis, angiogenesis, and migration (*GPCR* G protein-coupled-receptor, *TRK* tyrosine receptor kinase, a *brown circle* and a *purple rectangle* represent proteins that interact with PKC. PKC regulators of these protein–protein interactions, including isozyme-specific inhibitor and activator peptides developed by our lab, can modulate critical events that lead to tumor growth by blocking select protein–protein interactions and the resulting PKC activity

Table 5.1 A summary of a variety of cellular events in tumorigenesis that can be modulated by selective regulators of specific PKC isozymes

PKC/RACK	Tumor type	Interacting molecule	Physiological Effect	References
PKCα	Breast	β1 integrin	Increased migration and chemotaxis	Ng et al. 1999; Parsons et al. 2002
PKCβll	Prostate	Pericentrin	Increased angiogenesis	Kim et al. 2008
PKCδ	Mouse colon	KITENIN	Increased invasion	Kho et al. 2008
PKCε	Squamous cell	Stat3	Increased tumorigenesis	Aziz et al. 2007
PKCε	Breast	Vimentin	Increased integrin recycling	Ivaska et al. 2005
PKCε	HeLa	14–3–3	Normal cell division	Saurin et al. 2008
RACK1	Renal cell	HIF–1α	Decreased angiogenesis	Liu et al. 2007
RACK1	Colon	C–Src	Decreased cell proliferation	Mamidipudi et al. 2008

5.3.1 Proliferation

Accumulating evidence suggests that PKC family members are critical in mediating cytokinesis and cell proliferation (Kiley and Parker 1995; Takahashi et al. 2000; Chen et al. 2004). For example, PKCα has shown to be antiproliferative in intestinal, pancreatic, and mammary cells by inducing G1 arrest, and PKCδ has been found to be antiproliferative, inducing G1 and G2 arrest (Frey et al. 1997; Gavrielides et al. 2004), and also proproliferative depending on the cell types (Grossoni et al. 2007).

5.3.1.1 PKCβII

PKC is necessary for cell division as shown by the deletion of PKC inducing cell cycle arrest in yeast (Levin et al. 1990). Specifically, PKCβII was shown to regulate G2/M transition through phosphorylation of lamin B in eukaryotic cells during cell division (Gokmen-Polar and Fields 1998). Recent functional studies have shown a key role of the interaction of PKCβII-pericentrin, a centrosomal protein (Doxsey et al. 1994; Purohit et al. 1999; Chen et al. 2004), in microtubule organization, spindle assembly, and chromosome segregation. Newton and collaborators showed that C1A domain of PKCβII interacts with pericentrin (amino acids 494–593), and when the interaction between these two proteins is disrupted by increased levels of a fragment of pericentrin that binds PKCβII or PKCβII fragment that binds pericentrin, these lead to mislocalization of PKCβII away from the centrosome and a loss of microtubule anchoring at the centrosome, resulting in cytokinesis failure and aneuploidy in several cell types (Chen et al. 2004).

We recently reported that interaction of PKCβII and pericentrin is critical in the regulation of tumor angiogenesis and tumor growth in human prostate cancer. PKCβII is activated during the growth of prostate cancer in PC-3 xenografts, and inhibition of its activity decreases epithelial cell proliferation, angiogenesis, and tumor growth. Furthermore, tumor endothelial and epithelial cells exhibit abnormal localization, morphology, and increased levels of pericentrin in the xenografts, which were corrected by PKCβII inhibitor peptide, βIIV5-3, developed by our lab. PKCβII coimmunoprecipitated with pericentrin in xenografts *in vivo* and this interaction was significantly higher with βIIV5-3 administration. Also, treatment with conditioned medium from human prostate cancer cells (PC-3) induced similar abnormalities as seen in xenografts and disorganized microtubule organization and actin structures in the pericentrin of tumor endothelial cells *in vitro*, which were normalized by treatment with βIIV5-3. These data suggest that human prostate cancer cells secrete factors that induce pericentrin dysregulation possibly regulated by PKCβII kinase activity. In addition, cell proliferation decreased with siRNA of human PKCβII and pericentrin in PC-3 and tumor endothelial cells. More importantly, we report that in tumor endothelial cells, but not in normal endothelium, increased levels of pericentrin is found in human prostate tumors with Gleason grade 3+, confirming the relevance of pericentrin in tumor-induced angiogenesis in human prostate cancer (Kim et al. 2008). Together, these data suggest that RACK1 may be involved in regulation of cytokinesis or that PKCβII phosphorylation of pericentrin may be critical for regulated cytokinesis. These findings provide strong evidence that protein–protein interaction involving PKCβII regulates cytokinesis in cells.

5.3.1.2 RACK1

RACK1 regulates cell proliferation through its interaction with c-Src (Schechtman and Mochly-Rosen 2001). c-Src is a tyrosine kinase protooncogene that regulates cell growth (Moasser et al. 1999). Using two-hybrid screen, Cartwright laboratory showed that the SH2 domain of Src directly binds RACK1, a protein with multiple WD-40 units (Chang et al. 1998, 2001; Schechtman and Mochly-Rosen 2001). The interactions of c-Src, PKCβII, and RACK1 regulate Src activity. PKCβII association with RACK1 is necessary for c-Src phosphorylation of RACK1 and its binding to RACK1, which in turn results in decreased Src activity (Miller et al. 2004). RACK1-binding reduces c-Src activity by ~50% and results in decreased G1/S entry (Chang et al. 2002; Mamidipudi et al. 2004). This negative regulatory effect of RACK1 through Src on cell proliferation was confirmed by using the PKCβ-specific inhibitor, βC2-4, and the PKCβII-specific inhibitor, βIIV5-3. By binding to RACK1, these peptides interfere with c-Src and RACK1 interaction. This, in turn, increases cell division by inducing cyclin-dependent G1/S transition and inactivating retinoblastoma protein (Mamidipudi et al. 2007). On the other hand, a PKCβII-specific activator peptide, also developed by our lab (Ron and Mochly-Rosen 1995), reversed these effects by inducing interaction of c-Src with RACK1 and decreased cell proliferation (Mamidipudi et al. 2007) in colon cells. These findings indicate that RACK1 is a critical factor in c-Src-mediated cell proliferation. Because RACK1 can interact with multiple proteins with SH2 domains, Src-independent effects on cell proliferation in cancer cannot be ruled out.

5.3.1.3 PKCε

PKCε is generally found to be proproliferative and prosurvival in both normal and cancer cells (Budas et al. 2007a). PKCε levels and activities generally correlate well with tumor progression (Griner and Kazanietz 2007). Previously, PKCε has been linked to squamous cell carcinoma (SCC) development. Transgenic mice overexpressing PKCε are more vulnerable to SCC development by initiation with dimethylbenzanthracene and promotion with 12-*O*-tetradecanoylphorbol 13-acetate (TPA) treatment (Reddig et al. 2000). The importance of PKCε in cell survival is underscored by the recent finding that protein–protein interaction with signal transducers and activators of transcription-3 (Stat3) is required for the development of SCC (Aziz et al. 2007b). The constitutively active form of Stat3 was found in UV irradiation-induced human SCC (Chan et al. 2004a, b) and in a recent study by Aziz et al., PKCε coimmunoprecipitated with Stat3 and phosphorylated Ser 727 of Stat3 was essential for Stat3 DNA binding and transcriptional activity of genes regulating cell cycle progression and cell survival (Aziz et al. 2007a): reduced level of PKCε inhibited PKCε association with Stat3, Stat3 Ser 727 phosphorylation, and Stat3 transcriptional activity. Considering the fact that other constitutively activated Stats are also found in prostate, ovary, breast, head, and neck cancers and in SCC, and because Stats have shared consensus motif in the C-terminal transactivation domain between 720 and 730 (Aziz et al. 2007a), interaction of PKC and other forms of Stats may regulate tumorigenesis in different cell types.

5.3.1.4 PKCε Interaction with 14-3-3-β Regulates Cytokinesis

14-3-3 has been identified 20 years ago as a PKC inhibitory protein by Aitken and collaborators (Aitken et al. 1990). A sequence in 14-3-3 was found to be homologous to another PKC-binding protein called annexin I (ibid) and we showed that a peptide corresponding to this homologous sequence in annexin I inhibits PKC translocation and function when injected to *Xenopus* oocytes (Smith and Mochly-Rosen 1992). Subsequent crystal structure studies demonstrated that the annexin I-like sequence in 14-3-3 corresponds to the α helix in the inner plane of the 14-3-3 dimer, which binds not only PKC but also other protein kinases (Xiao et al. 1995). The role of 14-3-3 in regulating PKC has been further studied in a variety of models. Relevant to this review, a recent study demonstrated that the interaction of PKCε with 14-3-3-β protein is required for completion of cytokinesis (Saurin et al. 2008) in COS-7, HEK293, and HeLa cells. V3 domain of PKCε (amino acids 343–348) was found to interact with 14-3-3 by a yeast two-hybrid screen. This interaction was dependent on phosphorylation in the V3 domain of PKCε at Ser

346 and Ser 368: PKC activator TPA treatment and phosphatase inhibitor increased the interaction and mutation of either site abolished interaction of PKCε with 14-3-3-β. PKCε and 14-3-3-β complex increased in cells undergoing cytokinesis and disruption of this interaction by PKCε-specific siRNAs or by dominant negative 14-3-3 expression induced failure or significant delay in cytokinesis producing double-nucleated cells (Saurin et al. 2008). Because 14-3-3 proteins can bind other protein molecules as well (Bridges and Moorhead 2005), the regulation of cytokinesis by PKCε-independent interactions cannot be excluded.

5.3.2 Angiogenesis

5.3.2.1 RACK1

RACK1 contains multiple WD-40 units that form a seven-bladed propeller structure homologous to G protein β subunit (Csukai et al. 1997; Dell et al. 2002). With its multiple WD-40 repeats, it not only functions as an adaptor for PKCβII but also plays an important role in mediating signaling in endothelial cell growth (Schechtman and Mochly-Rosen 2001). In particular, in cord-forming clones of bovine aortic endothelial cells (BAECs), the levels of RACK1 mRNA and protein were found to be elevated compared to noncord-forming monolayer BAECs.

Also, RACK1 mRNA was found to be highly up-regulated in endothelium and epithelium of human carcinomas (non small cell lung cancer, colon cancer and in breast cancer) compared to noncancerous and nonangiogenic tissues (Berns et al. 2000). This suggests that this 37 kDa protein RACK 1 may play a role in angiogenesis in both tumorigenic and in nontumorigenic tissues. Recently, it was also shown that interaction of RACK1, PKC, and ADAM12 (a disintegrin-like multidomain protein) can induce translocation of ADAM12 to the plasma membrane in liver fibroblasts, where its shedding takes place, showing its potential for increasing liver fibrogenesis and cancer (Bourd-Boittin et al. 2008).

5.3.2.2 PKCβII

One mechanism by which RACK1 controls angiogenesis is through activation of its cognate protein, PKCβII. A role for PKC isozymes in angiogenesis has been demonstrated both *in vitro* and *in vivo* (Montesano and Orci 1985; Takahashi et al. 1999; Das Evcimen and King 2007; Griner and Kazanietz 2007). Specifically, the proangiogenic role of PKCβII has been shown in both mouse and model systems (Chen and LaCasce 2008; Kim et al. 2008). For example, we showed that selective inhibition of PKCβII decreases endothelial cell proliferation and tumor growth in human prostate cancer xenograft models (Kim et al. 2008). PKCβII has also been shown to be important in tumor-induced angiogenesis in hepatocellular carcinoma, an effect mediated by VEGF production and ERK 1/2 activation (Yoshiji et al. 1999). The PKC βII inhibitor enzataurin has been shown to suppress VEGF-induced angiogenesis and cell growth in human colon and renal carcinoma xenografts (Graff et al. 2005), to reduce tumor growth in non small cell lung cancer (Teicher et al. 2001a), breast and ovarian carcinoma xenografts (Teicher et al. 2002b), and hepatocellular and gastric carcinoma xenografts *in vivo* (Teicher et al. 2001b). Furthermore, PKCβ knockout mice showed reduced angiogenic activity in the cornea under hypoxia *vs.* wild type whereas mice overexpressing the protein showed increased angiogenic activity in the cornea (Suzuma et al. 2002).

5.3.2.3 HIF-1α

Another mechanism of regulating angiogenesis can be through interaction of RACK1 with HIF-1α. Recently, in HEK293 and RCC4 cells, RACK1 was found to be competing with HSP90 to bind HIF-1α in an oxygen-independent manner, providing another pathway leading to the degradation of HIF-1α by the proteasome (Liu et al. 2007). In agreement with our data that PKCβII inhibitor peptide reduced angiogenesis in human prostate cancer, this finding suggests that when the interaction of PKCβII with RACK1 is interrupted, this may increase RACK1 availability for interaction with HIF-1α (and possibly other angiogenic proteins) and therefore can induce further degradation of HIF-1α and reduce angiogenesis. RACK1 can interact with multiple proteins with SH2 domains. Therefore, it is possible that both proangiogenic and antiangiogenic proteins interact with RACK1 to balance the angiogenic activity of the tissue. If and how the balance is maintained and dysregulated through protein–protein interaction needs further investigation.

5.3.2.4 PKCδ

Reactive oxygen species (ROS) mediate angiogenic signaling and NADPH oxidase, one of the major sources of ROS in endothelial cells (Ushio-Fukai and Nakamura 2008; Kumar et al. 2008), is therefore emerging as an important signaling mediator of angiogenesis in cancer. Increased NADPH oxidase activities correlate with tumorigenic activity in various cancers (Lim et al. 2005; Lambeth 2007). Further, Nox1, a catalytic subunit of NADPH oxidase, was shown to play a critical role in tumor-induced angiogenesis. Inhibition of Nox1 by siRNA or diphenylene iodonium inhibited synthesis of VEGF mRNA and protein in K-Ras transformed normal rat kidney cells. Mechanistically, Nox1 inhibition decreased ERK-dependent phosphorylation of Sp1 transcriptional factor and its binding to VEGF promoter (Komatsu et al. 2008).

Although PKCδ was found to be antiproliferative and proapoptotic (Murriel et al. 2004; Griner and Kazanietz 2007), PKCδ is the major kinase that promotes angiogenesis and cell survival in several cell types (Gliki et al. 2002; Grossoni et al. 2007; Lee et al. 2007). In relation to angiogenesis, PKCδ was shown to coimmunoprecipitate with NADPH oxidase subunits and to induce NADPH oxidase activation and promote angiogenesis in prostate cancer cell line (Kim et al. unpublished results). This can occur *via* direct phosphorylation of NADPH oxidase subunits by PKCδ; for example, in a human myelomonoblastic leukemia cell line, PKCδ-induced phosphorylation of $p67^{phox}$ subunit of NADPH oxidase increased its activity (Siow et al. 2006). These data suggest that disruption of PKCδ-NADPH oxidase intermolecular protein interactions can negatively modulate tumor-induced angiogenesis. Because PKCδ can be both proangiogenic and proapoptotic depending on the cell type and binding partners available (Griner and Kazanietz 2007), development of therapeutic regulators of PKCδ in cancer needs to be approached with caution.

5.3.3 Migration

PKC activation correlates with cell migration and invasion in many types of tumors and can promote metastasis by interacting with molecules governing the metastatic potential of tumor cells.

5.3.3.1 Internalization of β1 Integrin Is Regulated by Regulatory Domain of PKCα

Integrins are a family of surface receptors that are critical in mediating cell adhesion and directionality of migration by binding to different ligands in the extracellular matrix (ECM) (Parsons et al. 2002). As part of their activation and regulation of cell motility, β1 integrins are internalized into the endosome and retransported to the cell surface (Ivaska et al. 2003, 2005). Activation of PKCα increases the internalization and recycling of β1 integrin. PKCα and β1 integrin physically interact in human breast cancer cells, as shown by colocalization and coimmunoprecipitation studies (Ng et al. 1999). β1 integrin internalization requires dynamin-1 (Ng et al. 1999), a GTPase protein involved in clathrin-coated vesicle pinching and endocytosis (De Camilli et al. 1995). Previously, we showed that dynamin-1 binds PKCβII and RACK1 *in vitro* as shown by coimmunoprecipitation as well as direct *in vitro* binding studies, and this tri-molecular interaction increases the GTPase activity of dynamin-1 and vesicle trafficking (Rodriguez et al. 1999; Schechtman and Mochly-Rosen 2001). Thus, both PKCα and PKCβII may play a role in this process.

5.3.3.2 Association of V3 Region of PKCα with the Cytoplasmic Tail of β1 is Required for Directional Chemotaxis in Human Breast Cancer Cells

The interaction between PKCα-V3 hinge region and β1 integrin has been shown to directly regulate chemotaxis of breast carcinoma cells (Parsons et al. 2002). Introducing a PKCα-V3 binding sequence derived from β1 integrin disrupted interaction between PKCα and β1 integrin and resulted in significantly reduced chemotaxis by epidermal growth factor gradient. These findings suggest that modulators of protein–protein interaction between PKCα and integrins can be used to inhibit cell migration and possibly even invasion and metastasis.

5.3.3.3 PKCα and Src Mediate ErbB2 Signaling

PKCα is also involved in the regulation of cell invasion by interaction with Src (Tan et al. 2006). Overactivity of ErbB2 signaling in human breast cancer is known to increase invasion and reduce patient survival (Emanuel et al. 2008). In breast cancer cells, PKCα activation was critical in mediating ErbB2 signaling; in ErbB2-overexpressing MDA-MB-435 cells or in constitutively active ErbB2-expressing cells, the levels and activities of PKCα (as shown by probing with antiphospho PKCα antibody and by histone phosphorylation) increased whereas ErbB2 kinase defective cells reversed these effects. In ErbB2-overexpressing MDA-MB-435 cells, siRNA of PKCα or cells with dominant negative PKCα resulted in decreased invasion. Also, PKCα interacted with Src as shown by immunoprecipitation and reverse immunoprecipitation, where the interaction was stronger in ErbB2-overexpressing or constitutively active cells. Moreover, activation of PKCα was controlled by Src; with Src inhibitor treatment or with Src dominant-negative mutant, PKCα activity decreased dramatically. Finally, administration of Src inhibitor and/or PKC inhibitor (Gö6976) synergistically decreased urokinase-type plasminogen activator expression and invasion. These results suggest that Src:PKCα interaction mediates ErbB2 signaling in breast cancer cell invasion. Because Src can interact with various other proteins containing SH2 domains as previously mentioned (for example with RACK1), other interacting proteins may also be important in the regulation of cancer cell invasion. Therefore, identifying the sequence in PKCα that interacts with Src may lead to the development of a more fine-tuned modulator of PKCα:Src interaction and may provide a new target for drug development to treat breast cancer invasion. Also, inhibition of PKCα activity by an isozyme-specific inhibitor of PKCα (developed by our laboratory) may regulate this important signaling pathway.

5.3.3.4 Association of PKCε and Vimentin Increases β1 Integrin Recycling and Cell Motility

Vimentin is an intermediate filament protein upregulated during the transition of epithelial to mesenchymal-like cells (Kang and Massague 2004). Fibroblasts defective in vimentin expression showed decreased cell motility as well as wound healing (Eckes et al. 1998, 2000). Phosphorylation of vimentin by PKCε and their interaction were reported to be critical in the recycling of β1 integrin (Ivaska et al. 2005). Mutagenesis of PKCε interacting sites in vimentin and ectopic expression of the variants inhibited efficient recycling of β1 integrin. These findings demonstrate that PKCε regulates cell migration through protein–protein interactions with vimentin and suggest that inhibition of this interaction will provide a new means to interfere with this important step in tumor metastasis. Recently, PKCδ and PKCζ were shown to regulate cell invasion by interacting with KITENIN in rat C6 glioma cells and ZIP/p62 in mouse colon cancer cells, respectively (Huang et al. 2008; Kho et al. 2008), further indicating the role of PKC in cancer cell invasion.

5.3.4 Apoptosis

5.3.4.1 PKCδ and Annexin V

Many types of apoptotic stimuli can induce PKCδ translocation to endoplasmic reticulum and to the mitochondria (Murriel et al. 2004; Qi and Mochly-Rosen 2008). These events lead to cytochrome c release, caspase-3 cleavage, and programmed cell death, or apoptosis (Murriel et al. 2004; Qi et al. 2008). Activation of PKCδ can also trigger the autocrine secretion of death factors thus inducing apoptosis by extrinsic pathways (Humphries et al. 2006; Griner and Kazanietz 2007). We have previously reported that a peptide inhibitor of PKCδ can be derived from a sequence in annexin V, a PKCδ-binding protein (Kheifets et al. 2006). The peptide corresponds to a 6-amino acid sequence in annexin V that is similar to a sequence in PKCδ but is different in one charge from its homologous sequence. This pseudo-annexin V peptide inhibits intramolecular interaction within PKCδ, increases PKCδ binding to annexin V and sequestration of the activated enzyme away from its target on the mitochondria, where the active PKCδ leads to apoptosis and necrosis (Murriel et al. 2004). We further showed when treating a heart with this pseudo-annexin V peptide, it reduces injury induced by myocardial infarction by ~70% (Kheifets et al. 2006). Conversely, treatment with pseudo δRACK, a δPKC-selective activator, increases the damage to heart cells subjected to models of myocardial infarction (Chen et al. 2001). [The development of such PKCδ modulating peptides is reviewed in detail in our recent review (Budas et al. 2007b).] Because resistance to proapoptotic signals is a hallmark of a variety of cancer cells, increasing apoptosis using regulators of protein–protein interactions involving PKCδ may be a useful therapeutic approach.

5.4 Conclusions

The observations that each PKC isozyme translocates to different intracellular locations and the identification of PKC-selective RACKs and other select PKC-binding proteins led to the recognition that an important means to selectively regulate specific PKC isozymes is by regulating select protein–protein interactions. Here, we summarized the rationale for the development of PKC isozyme-specific inhibitors derived from the sequences in PKC that interfere with PKC–RACK intermolecular interaction and some new methods for identifying PKC inhibitors based on C2 intermolecular interactions. We also reviewed our rationale for the development of PKC isozyme-specific activators based on pseudo RACK sequences in PKC. The three dozen PKC regulators developed by our lab are all inhibitors of protein–protein interactions, modulating either intramolecular interactions in PKC or intermolecular interactions between PKC and its cognate proteins. Over 150 studies by different laboratories, using a variety of models of human diseases, showed that these peptides can be used as effective pharmacological tools to identify critical PKC isozymes for different diseases. A variety of these peptides were found to be safe even when given for prolonged periods. These preclinical models led to the use of three of the peptides in clinical trials in human, and one efficacy study in patients with acute myocardial infarction showed sufficient therapeutic effects that led to a large study in thousands of patients. Our laboratory has used several animal models of human cancer to identify the critical PKC isozymes whose inhibition or activation provides benefit (Kim et al. and unpublished results). These and future pre-clinical studies may provide the insight to enable novel cancer drug development.

Acknowledgments We thank Dr. Grant R. Budas for critical reading of the manuscript and apologize to those colleagues whose work was not cited due to space limitations.

References

Aitken, A., Ellis, C. A., Harris, A., Sellers, L. A., & Toker, A. (1990). Kinase and neurotransmitters. *Nature, 344*, 594.

Aziz, M. H., Manoharan, H. T., & Verma, A. K. (2007a). Protein kinase C epsilon, which sensitizes skin to sun's UV radiation-induced cutaneous damage and development of squamous cell carcinomas, associates with Stat3. *Cancer Research, 67*, 1385–1394.

Aziz, M. H., Manoharan, H. T., Sand, J. M., & Verma, A. K. (2007b). Protein kinase Cepsilon interacts with Stat3 and regulates its activation that is essential for the development of skin cancer. *Molecular Carcinogenesis, 46*, 646–653.

Bates, E., Bode, C., Costa, M., Gibson, C. M., Granger, C., Green, C., et al. (2008). Intracoronary KAI-9803 as an adjunct to primary percutaneous coronary intervention for acute ST-segment elevation myocardial infarction. *Circulation, 117*, 886–896.

Berns, H., Humar, R., Hengerer, B., Kiefer, F. N., & Battegay, E. J. (2000). RACK1 is up-regulated in angiogenesis and human carcinomas. *The FASEB Journal, 14*, 2549–2558.

Bourd-Boittin, K., et al. (2008). RACK1, a new ADAM12 interacting protein. Contribution to liver fibrogenesis. *The Journal of Biological Chemistry, 283*, 26000–26009.

Brandman, R., Disatnik, M. H., Churchill, E., & Mochly-Rosen, D. (2007). Peptides derived from the C2 domain of protein kinase C epsilon (epsilon PKC) modulate epsilon PKC activity and identify potential protein-protein interaction surfaces. *The Journal of Biological Chemistry, 282*, 4113–4123.

Bridges, D., & Moorhead, G. B. (2005). 14-3-3 proteins: A number of functions for a numbered protein. *Science's STKE, 2005*, re10.

Budas, G. R., Churchill, E. N., & Mochly-Rosen, D. (2007a). Cardioprotective mechanisms of PKC isozyme-selective activators and inhibitors in the treatment of ischemia-reperfusion injury. *Pharmacological Research, 55*, 523–536.

Budas, G. R., Koyanagi, T., Churchill, E. N., & Mochly-Rosen, D. (2007b). Competitive inhibitors and allosteric activators of protein kinase C isoenzymes: A personal account and progress report on transferring academic discoveries to the clinic. *Biochemical Society Transactions, 35*, 1021–1026.

Castagna, M., Takai, Y., Kaibuchi, K., Sano, K., Kikkawa, U., & Nishizuka, Y. (1982). Direct activation of calcium-activated, phospholipid-dependent protein kinase by tumor-promoting phorbol esters. *The Journal of Biological Chemistry, 257*, 7847–7851.

Chan, K. S., Carbajal, S., Kiguchi, K., Clifford, J., Sano, S., & DiGiovanni, J. (2004a). Epidermal growth factor receptor-mediated activation of Stat3 during multistage skin carcinogenesis. *Cancer Research, 64*, 2382–2389.

Chan, K. S., Carbajal, S., Kiguchi, K., Clifford, J., Sano, S., & DiGiovanni, J. (2004b). Disruption of Stat3 reveals a critical role in both the initiation and the promotion stages of epithelial carcinogenesis. *The Journal of Clinical Investigation, 114*, 720–728.

Chang, B. Y., Chiang, M., & Cartwright, C. A. (2001). The interaction of Src and RACK1 is enhanced by activation of protein kinase C and tyrosine phosphorylation of RACK1. *The Journal of Biological Chemistry, 276*, 20346–20356.

Chang, B. Y., Harte, R. A., & Cartwright, C. A. (2002). RACK1: A novel substrate for the Src protein-tyrosine kinase. *Oncogene, 21*, 7619–7629.

Chang, B. Y., Conroy, K. B., Machleder, E. M., & Cartwright, C. A. (1998). RACK1, a receptor for activated C kinase and a homolog of the beta subunit of G proteins, inhibits activity of src tyrosine kinases and growth of NIH 3T3 cells. *Molecular and Cellular Biology, 18*, 3245–3256.

Chen, Y. B., & LaCasce, A. S. (2008). Enzastaurin. *Expert Opinion on Investigational Drugs, 17*, 939–944.

Chen, L., Hahn, H., Wu, G., Chen, C. H., Liron, T., & Schechtman, D., et al. (2001). Opposing cardioprotective actions and parallel hypertrophic effects of delta PKC and epsilon PKC. *Proceedings of the National Academy of Sciences of the United States of America, 98*, 11114–11119.

Chen, D., Purohit, A., Halilovic, E., Doxsey, S. J., & Newton, A. C. (2004). Centrosomal anchoring of protein kinase C betaII by pericentrin controls microtubule organization, spindle function, and cytokinesis. *The Journal of Biological Chemistry, 279*, 4829–4839.

Churchill, E. N., Qvit, N., & Mochly-Rosen, D. (2009). Rationally designed peptide regulators of protein kinase C. *Trends in Endocrinology and Metabolism, 20*, 25–33.

Churchill, E. N., Murriel, C. L., Chen, C. H., Mochly-Rosen, D., & Szweda, L. I. (2005). Reperfusion-induced translocation of deltaPKC to cardiac mitochondria prevents pyruvate dehydrogenase reactivation. *Circulation Research, 97*, 78–85.

Csukai, M., Chen, C. H., De Matteis, M. A., Mochly-Rosen, D. (1997). The coatomer protein beta'-COP, a selective binding protein (RACK) for protein kinase Cepsilon. *The Journal of Biological Chemistry, 272*, 29200–29206.

Das Evcimen, N., & King, G. L. (2007). The role of protein kinase C activation and the vascular complications of diabetes. *Pharmacological Research, 55*, 498–510.

De Camilli, P., Takei, K., & McPherson, P. S. (1995). The function of dynamin in endocytosis. *Current Opinion in Neurobiology, 5*, 559–565.

Dell, E. J., Connor, J., Chen, S., Stebbins, E. G., Skiba, N. P., Mochly-Rosen, D., et al. (2002). The betagamma subunit of heterotrimeric G proteins interacts with RACK1 and two other WD repeat proteins. *The Journal of Biological Chemistry, 277*, 49888–49895.

Disatnik, M. H., Buraggi, G., & Mochly-Rosen, D. (1994). Localization of protein kinase C isozymes in cardiac myocytes. *Experimental Cell Research, 210*, 287–297.

Doxsey, S. J., Stein, P., Evans, L., Calarco, P. D., & Kirschner, M. (1994). Pericentrin, a highly conserved centrosome protein involved in microtubule organization. *Cell, 76*, 639–650.

Eckes, B., Dogic, D., Colucci-Guyon, E., Wang, N., Maniotis, A. I., Ngber, D., et al. (1998). Impaired mechanical stability, migration and contractile capacity in vimentin-deficient fibroblasts. *Journal of Cell Science, 111*(Pt 13), 1897–1907.

Eckes, B., Colucci-Guyon, E., Smola, H., Nodder, S., Babinet, C., Krieg, T., & Martin, P. (2000). Impaired wound healing in embryonic and adult mice lacking vimentin. *Journal of Cell Science, 113*(Pt 13), 2455–2462.

Edwards, A. S., & Newton, A. C. (1997). Phosphorylation at conserved carboxyl-terminal hydrophobic motif regulates the catalytic and regulatory domains of protein kinase C. *The Journal of Biological Chemistry, 272*, 18382–18390.

Emanuel, S. L., Hughes, T. V., Adams, M., Rugg, C. A., Fuentes-Pesquera, A., Connolly, P. J., et al. (2008). Cellular and in vivo activity of JNJ-28871063, a nonquinazoline pan-ErbB kinase inhibitor that crosses the blood-brain barrier and displays efficacy against intracranial tumors. *Molecular Pharmacology, 73*, 338–348.

Emoto, Y., Manome, Y., Meinhardt, G., Kisaki, H., Kharbanda, S., Robertson, M., et al. (1995). Proteolytic activation of protein kinase C delta by an ICE-like protease in apoptotic cells. *The EMBO Journal, 14*, 6148–6156.

Fields, A. P., Frederick, L. A., & Regala, R. P. (2007). Targeting the oncogenic protein kinase Ciota signalling pathway for the treatment of cancer. *Biochemical Society Transactions, 35*, 996–1000.

Frey, M. R., Saxon, M. L., Zhao, X., Rollins, A., Evans, S. S., & Black, J. D. (1997). Protein kinase C isozyme-mediated cell cycle arrest involves induction of p21(waf1/cip1) and p27(kip1) and hypophosphorylation of the retinoblastoma protein in intestinal epithelial cells. *The Journal of Biological Chemistry, 272*, 9424–9435.

Gavrielides, M. V., Frijhoff, A. F., Conti, C. J., & Kazanietz, M. G. (2004). Protein kinase C and prostate carcinogenesis: Targeting the cell cycle and apoptotic mechanisms. *Current Drug Targets, 5*, 431–443.

Ghosh, S., Xie, W. Q., Quest, A. F., Mabrouk, G. M., Strum, J. C., & Bell, R. M. (1994). The cysteine-rich region of raf-1 kinase contains zinc, translocates to liposomes, and is adjacent to a segment that binds GTP-ras. *The Journal of Biological Chemistry, 269*, 10000–10007.

Gliki, G., Wheeler-Jones, C., & Zachary, I. (2002). Vascular endothelial growth factor induces protein kinase C (PKC)-dependent Akt/PKB activation and phosphatidylinositol 3'-kinase-mediates PKC delta phosphorylation: Role of PKC in angiogenesis. *Cell Biology International, 26*, 751–759.

Gokmen-Polar, Y., & Fields, A. P. (1998). Mapping of a molecular determinant for protein kinase C betaII isozyme function. *The Journal of Biological Chemistry, 273*, 20261–20266.

Graff, J. R., et al. (2005). The protein kinase Cbeta-selective inhibitor, Enzastaurin (LY317615. HCl), suppresses signaling through the AKT pathway, induces apoptosis, and suppresses growth of human colon cancer and glioblastoma xenografts. *Cancer Research, 65*, 7462–7469.

Griner, E. M., & Kazanietz, M. G. (2007). Protein kinase C and other diacylglycerol effectors in cancer. *Nature Reviews. Cancer, 7*, 281–294.

Grossoni, V. C., Falbo, K. B., Kazanietz, M. G., de Kier Joffe, E. D., & Urtreger, A. J. (2007). Protein kinase C delta enhances proliferation and survival of murine mammary cells. *Molecular Carcinogenesis, 46*, 381–390.

Hernandez, A. I., Blace, N., Crary, J. F., Serrano, P. A., Leitges, M., Libien, J. M., et al. (2003). Protein kinase M zeta synthesis from a brain mRNA encoding an independent protein kinase C zeta catalytic domain. Implications for the molecular mechanism of memory. *The Journal of Biological Chemistry, 278*, 40305–40316.

House, C., & Kemp, B. E. (1987). Protein kinase C contains a pseudosubstrate prototope in its regulatory domain. *Science, 238*, 1726–1728.

House, C., Robinson, P. J., & Kemp, B. E. (1989). A synthetic peptide analog of the putative substrate-binding motif activates protein kinase C. *FEBS Letters, 249*, 243–247.

Huang, H. C., Huang, C. Y., Lin-Shiau, S. Y., & Lin, J. K. (2009). Ursolic acid inhibits IL-1beta or TNF-alpha-induced C6 glioma invasion through suppressing the association ZIP/p62 with PKC-zeta and downregulating the MMP-9 expression. *Molecular Carcinogenesis, 48*, 517–531.

Humphries, M. J., Limesand, K. H., Schneider, J. C., Nakayama, K. I., Anderson, S. M., & Reyland, M. E. (2006). Suppression of apoptosis in the protein kinase Cdelta null mouse in vivo. *The Journal of Biological Chemistry, 281*, 9728–9737.

Hurley, J. H., & Meyer, T. (2001). Subcellular targeting by membrane lipids. *Current Opinion in Cell Biology, 13*, 146–152.

Inagaki, K., & Mochly-Rosen, D. (2005). DeltaPKC-mediated activation of epsilonPKC in ethanol-induced cardiac protection from ischemia. *Journal of Molecular and Cellular Cardiology, 39*, 203–211.

Inagaki, K., Chen, L., Ikeno, F., Lee, F. H., Imahashi, K., Bouley, D. M., et al. (2003a). Inhibition of delta-protein kinase C protects against reperfusion injury of the ischemic heart in vivo. *Circulation, 108*, 2304–2307.

Inagaki, K., Hahn, H. S., Dorn, G. W., 2nd & Mochly-Rosen, D. (2003b). Additive protection of the ischemic heart ex vivo by combined treatment with delta-protein kinase C inhibitor and epsilon-protein kinase C activator. *Circulation, 108*, 869–875.

Inagaki, K., Begley, R., Ikeno, F., & Mochly-Rosen, D. (2005). Cardioprotection by epsilon-protein kinase C activation from ischemia: Continuous delivery and antiarrhythmic effect of an epsilon-protein kinase C-activating peptide. *Circulation, 111*, 44–50.

Inagaki, K., Koyanagi, T., Berry, N. C., Sun, L., & Mochly-Rosen, D. (2008). Pharmacological inhibition of epsilon-protein kinase C attenuates cardiac fibrosis and dysfunction in hypertension-induced heart failure. *Hypertension, 51*, 1565–1569.

Inoue, M., Kishimoto, A., Takai, Y., & Nishizuka, Y. (1977). Studies on a cyclic nucleotide-independent protein kinase and its proenzyme in mammalian tissues. II. Proenzyme and its activation by calcium-dependent protease from rat brain. *The Journal of Biological Chemistry, 252*, 7610–7616.

Ivaska, J., Kermorgant, S., Whelan, R., Parsons, M., Ng, T., & Parker, P. J. (2003). Integrin-protein kinase C relationships. *Biochemical Society Transactions, 31*, 90–93.

Ivaska, J., Vuoriluoto, K., Huovinen, T., Izawa, I., Inagaki, M., & Parker, P. J. (2005). PKCepsilon-mediated phosphorylation of vimentin controls integrin recycling and motility. *The EMBO Journal, 24*, 3834–3845.

Jaken, S., & Parker, P. J. (2000). Protein kinase C binding partners. *Bioessays, 22*, 245–254.

Johnson, J. A., Gray, M. O., Chen, C. H., & Mochly-Rosen, D. (1996). A protein kinase C translocation inhibitor as an isozyme-selective antagonist of cardiac function. *The Journal of Biological Chemistry, 271*, 24962–24966.

Kang, Y., & Massague, J. (2004). Epithelial-mesenchymal transitions: Twist in development and metastasis. *Cell, 118*, 277–279.

Keranen, L. M., & Newton, A. C. (1997). Ca2+ differentially regulates conventional protein kinase Cs' membrane interaction and activation. *The Journal of Biological Chemistry, 272*, 25959–25967.

Kheifets, V., & Mochly-Rosen, D. (2007). Insight into intra- and inter-molecular interactions of PKC: Design of specific modulators of kinase function. *Pharmacological Research, 55*, 467–476.

Kheifets, V., Bright, R., Inagaki, K., Schechtman, D., & Mochly-Rosen, D. (2006). Protein kinase C delta (deltaPKC)-annexin V interaction: A required step in deltaPKC translocation and function. *The Journal of Biological Chemistry, 281*, 23218–23226.

Kho, D. H., Bae, J. A., Lee, J. H., Cho, H. J., Cho, S. H., Seo, Y. W., et al. (2009). KITENIN recruits Dishevelled/PKC {delta}to form a functional complex and controls the migration and invasiveness of colorectal cancer cells. *Gut, 58*, 509–519.

Kiley, S. C., & Parker, P. J. (1995). Differential localization of protein kinase C isozymes in U937 cells: Evidence for distinct isozyme functions during monocyte differentiation. *Journal of Cell Science, 108*(Pt 3), 1003–1016.

Kim, J., Choi, Y. L., Vallentin, A., Hunrichs, B. S., Hellerstein, M. K., Peehl, D. M., & Mochly-Rosen, D. (2008). Centrosomal PKCbetaII and pericentrin are critical for human prostate cancer growth and angiogenesis. *Cancer Research, 68*, 6831–6839.

Kohout, S. C., Corbalan-Garcia, S., Torrecillas, A., Gomez-Fernandez, J. C., & Falke, J. J. (2002). C2 domains of protein kinase C isoforms alpha, beta, and gamma: Activation parameters and calcium stoichiometries of the membrane-bound state. *Biochemistry, 41*, 11411–11424.

Komatsu, D., Kato, M., Nakayama, J., Miyagawa, S., & Kamata, T. (2008). NADPH oxidase 1 plays a critical mediating role in oncogenic Ras-induced vascular endothelial growth factor expression. *Oncogene, 27*, 4724–4732.

Koren, R., Ben Meir, D., Langzam, L., Dekel, Y., Konichezky, M., Baniel, J., et al. (2004). Expression of protein kinase C isoenzymes in benign hyperplasia and carcinoma of prostate. *Oncol Rep, 11*, 321–326.

Koyanagi, T., Noguchi, K., Ootani, A., Inagaki, K., Robbins, R. C., & Mochly-Rosen, D. (2007). Pharmacological inhibition of epsilon PKC suppresses chronic inflammation in murine cardiac transplantation model. *Journal of Molecular and Cellular Cardiology, 43*, 517–522.

Kraft, A. S., & Anderson, W. B. (1983). Phorbol esters increase the amount of Ca2+, phospholipid-dependent protein kinase associated with plasma membrane. *Nature, 301*, 621–623.

Kumar, B., Koul, S., Khandrika, L., Meacham, R. B., & Koul, H. K. (2008). Oxidative stress is inherent in prostate cancer cells and is required for aggressive phenotype. *Cancer Research, 68*, 1777–1785.

Lambeth, J. D. (2007). Nox enzymes, ROS, and chronic disease: An example of antagonistic pleiotropy. *Free Radical Biology and Medicine, 43*, 332–347.

Lee, J. W., Park, J. A., Kim, S. H., Seo, J. H., Lim, K. J., Jeong, J. W., et al. (2007). Protein kinase C-delta regulates the stability of hypoxia-inducible factor-1alpha under hypoxia. *Cancer Science, 98*, 1476–1481.

Lehel, C., Olah, Z., Jakab, G., & Anderson, W. B. (1995). Protein kinase C epsilon is localized to the Golgi via its zinc-finger domain and modulates Golgi function. *Proceedings of the National Academy of Sciences of the United States of America, 92*, 1406–1410.

Levin, D. E., Fields, F. O., Kunisawa, R., Bishop, J. M., & Thorner, J. (1990). A candidate protein kinase C gene, PKC1, is required for the S. cerevisiae cell cycle. *Cell, 62*, 213–224.

Lim, S. D., Sun, C., Lambeth, J. D., Marshall, F., Amin, M., Chung, L., et al. (2005). Increased Nox1 and hydrogen peroxide in prostate cancer. *The Prostate, 62*, 200–207.

Liu, Y. V., Baek, J. H., Zhang, H., Diez, R., Cole, R. N., & Semenza, G. L. (2007). RACK1 competes with HSP90 for binding to HIF-1alpha and is required for O(2)-independent and HSP90 inhibitor-induced degradation of HIF-1alpha. *Molecular Cell, 25*, 207–217.

Mamidipudi, V., Zhang, J., Lee, K. C., & Cartwright, C. A. (2004). RACK1 regulates G1/S progression by suppressing Src kinase activity. *Molecular and Cellular Biology, 24*, 6788–6798.

Mamidipudi, V., Dhillon, N. K., Parman, T., Miller, L. D., Lee, K. C., & Cartwright, C. A. (2007). RACK1 inhibits colonic cell growth by regulating Src activity at cell cycle checkpoints. *Oncogene, 26*, 2914–2924.

Martiny-Baron, G., & Fabbro, D. (2007). Classical PKC isoforms in cancer. *Pharmacological Research, 55*, 477–486.

Medkova, M., & Cho, W. (1998). Mutagenesis of the C2 domain of protein kinase C-alpha. Differential roles of Ca2+ ligands and membrane binding residues. *The Journal of Biological Chemistry, 273*, 17544–17552.

Miller, L. D., Lee, K. C., Mochly-Rosen, D., & Cartwright, C. A. (2004). RACK1 regulates Src-mediated Sam68 and p190RhoGAP signaling. *Oncogene, 23*, 5682–5686.

Moasser, M. M., Srethapakdi, M., Sachar, K. S., Kraker, A. J., & Rosen, N. (1999). Inhibition of Src kinases by a selective tyrosine kinase inhibitor causes mitotic arrest. *Cancer Research, 59*, 6145–6152.

Mochly-Rosen, D. (1995). Localization of protein kinases by anchoring proteins: A theme in signal transduction. *Science, 268*, 247–251.

Mochly-Rosen, D., Khaner, H., & Lopez, J. (1991). Identification of intracellular receptor proteins for activated protein kinase C. *Proceedings of the National Academy of Sciences of the United States of America, 88*, 3997–4000.

Mochly-Rosen, D., & Koshland, D. E., Jr. (1987). Domain structure and phosphorylation of protein kinase C. *The Journal of Biological Chemistry, 262*, 2291–2297.

Mochly-Rosen, D., Miller, K. G., Scheller, R. H., Khaner, H., Lopez, J., & Smith, B. L. (1992). p65 fragments, homologous to the C2 region of protein kinase C, bind to the intracellular receptors for protein kinase C. *Biochemistry, 31*, 8120–8124.

Montesano, R., & Orci, L. (1985). Tumor-promoting phorbol esters induce angiogenesis in vitro. *Cell, 42*, 469–477.

Mosior, M., & McLaughlin, S. (1991). Peptides that mimic the pseudosubstrate region of protein kinase C bind to acidic lipids in membranes. *Biophysical Journal, 60*, 149–159.

Murriel, C. L., Churchill, E., Inagaki, K., Szweda, L. I., & Mochly-Rosen, D. (2004). Protein kinase Cdelta activation induces apoptosis in response to cardiac ischemia and reperfusion damage: A mechanism involving BAD and the mitochondria. *The Journal of Biological Chemistry, 279*, 47985–47991.

Nalefski, E. A., & Newton, A. C. (2001). Membrane binding kinetics of protein kinase C betaII mediated by the C2 domain. *Biochemistry, 40*, 13216–13229.
Newton, A. C. (2003). Regulation of the ABC kinases by phosphorylation: Protein kinase C as a paradigm. *Biochemical Journal, 370*, 361–371.
Ng, T., Shima, D., Squire, A., Bastiaens, P. I., Gschmeissner, S., Humphries, M. J., & Parker, P. J. (1999). PKCalpha regulates beta1 integrin-dependent cell motility through association and control of integrin traffic. *The EMBO Journal, 18*, 3909–3923.
Nishizuka, Y. (1986). Studies and perspectives of protein kinase C. *Science, 233*, 305–312.
Ono, Y., Fujii, T., Igarashi, K., Kuno, T., Tanaka, C., Kikkawa, U., & Nishizuka, Y. (1999). Phorbol ester binding to protein kinase C requires a cysteine-rich zinc-fingerlike sequence. *Proceedings of the National Academy of Sciences of the United States of America, 86*, 4868–4871.
Parekh, D. B., Ziegler, W., & Parker, P. J. (2000). Multiple pathways control protein kinase C phosphorylation. *The EMBO Journal, 19*, 496–503.
Parker, P. J., & Murray-Rust, J. (2004). PKC at a glance. *Journal of Cell Science, 117*, 131–132.
Parker, P. J., Coussens, L., Totty, N., Rhee, L., Young, S., Chen, E., et al. (1986). The complete primary structure of protein kinase C – the major phorbol ester receptor. *Science, 233*, 853–859.
Parsons, M., Keppler, M. D., Kline, A., Messent, A., Humphries, M. J., Gilchrist, R., et al. (2002). Site-directed perturbation of protein kinase C – integrin interaction blocks carcinoma cell chemotaxis. *Molecular and Cellular Biology, 22*, 5897–5911.
Pears, C. J., Kour, G., House, C., Kemp, B. E., & Parker, P. J. (1990). Mutagenesis of the pseudosubstrate site of protein kinase C leads to activation. *European Journal of Biochemistry, 194*, 89–94.
Prekeris, R., Mayhew, M. W., Cooper, J. B., & Terrian, D. M. (1996). Identification and localization of an actin-binding motif that is unique to the epsilon isoform of protein kinase C and participates in the regulation of synaptic function. *The Journal of Cell Biology, 132*, 77–90.
Purohit, A., Tynan, S. H., Vallee, R., & Doxsey, S. J. (1999). Direct interaction of pericentrin with cytoplasmic dynein light intermediate chain contributes to mitotic spindle organization. *The Journal of Cell Biology, 147*, 481–492.
Qi, X., & Mochly-Rosen, D. (2008). The PKCdelta–Abl complex communicates ER stress to the mitochondria – an essential step in subsequent apoptosis. *Journal of Cell Science, 121*, 804–813.
Qi, X., Inagaki, K., Sobel, R. A., & Mochly-Rosen, D. (2008). Sustained pharmacological inhibition of deltaPKC protects against hypertensive encephalopathy through prevention of blood-brain barrier breakdown in rats. *The Journal of Clinical Investigation, 118*, 173–182.
Regala, R. P., Thompson, E. A., & Fields, A. P. (2008). Atypical protein kinase C iota expression and aurothiomalate sensitivity in human lung cancer cells. *Cancer Research, 68*, 5888–5895.
Rizo, J., & Sudhof, T. C. (1998). C2-domains, structure and function of a universal Ca2+-binding domain. *The Journal of Biological Chemistry, 273*, 15879–15882.
Rodriguez, M. M., Ron, D., Touhara, K., Chen, C. H., & Mochly-Rosen, D. (1999). RACK1, a protein kinase C anchoring protein, coordinates the binding of activated protein kinase C and select pleckstrin homology domains in vitro. *Biochemistry, 38*, 13787–13794.
Ron, D., Luo, J., & Mochly-Rosen, D. (1995). C2 region-derived peptides inhibit translocation and function of beta protein kinase C in vivo. *The Journal of Biological Chemistry, 270*, 24180–24187.
Ron, D., & Mochly-Rosen, D. (1994). Agonists and antagonists of protein kinase C function, derived from its binding proteins. *The Journal of Biological Chemistry, 269*, 21395–21398.
Ron, D., & Mochly-Rosen, D. (1995). An autoregulatory region in protein kinase C: The pseudo-anchoring site. *Proceedings of the National Academy of Sciences of the United States of America, 92*, 492–496.
Ron, D., Chen, C. H., Caldwell, J., Jamieson, L., Orr, E., & Mochly-Rosen, D. (1994). Cloning of an intracellular receptor for protein kinase C: A homolog of the beta subunit of G proteins. *Proceedings of the National Academy of Sciences of the United States of America, 91*, 839–843.

Saurin, A. T., Durgan, J., Cameron, A. J., Faisal, A., Marber, M. S., & Parker, P. J. (2008). The regulated assembly of a PKCepsilon complex controls the completion of cytokinesis. *Nature Cell Biology, 10*, 891–901.

Schechtman, D., & Mochly-Rosen, D. (2001). Adaptor proteins in protein kinase C-mediated signal transduction. *Oncogene, 20*, 6339–6347.

Schultz, A., Jonsson, J. I., & Larsson, C. (2003). The regulatory domain of protein kinase Ctheta localises to the Golgi complex and induces apoptosis in neuroblastoma and Jurkat cells. *Cell Death and Differentiation, 10*, 662–675.

Siow, Y. L., Au-Yeung, K. K., Woo, & C. W. O, K. (2006). Homocysteine stimulates phosphorylation of NADPH oxidase p47phox and p67phox subunits in monocytes via protein kinase Cbeta activation. *Biochemical Journal, 398*, 73–82.

Smith, B. L., & Mochly-Rosen, D. (1992). Inhibition of protein kinase C function by injection of intracellular receptors for the enzyme. *Biochemical and Biophysical Research Communications, 188*, 1235–1240.

Stebbins, E. G., & Mochly-Rosen, D. (2001). Binding specificity for RACK1 resides in the V5 region of beta II protein kinase C. *The Journal of Biological Chemistry, 276*, 29644–29650.

Steinberg, S. F. (2004). Distinctive activation mechanisms and functions for protein kinase Cdelta. *Biochemical Journal, 384*, 449–459.

Suzuma, K., Takahara, N., Suzuma, I., Isshiki, K., Ueki, K., Leitges, M., et al. (2002). Characterization of protein kinase C beta isoform's action on retinoblastoma protein phosphorylation, vascular endothelial growth factor-induced endothelial cell proliferation, and retinal neovascularization. *Proceedings of the National Academy of Sciences of the United States of America, 99*, 721–726.

Sweitzer, S. M., Wong, S. M., Peters, M. C., Mochly-Rosen, D., Yeomans, D. C., & Kendig, J. J. (2004). Protein kinase C epsilon and gamma: Involvement in formalin-induced nociception in neonatal rats. *The Journal of Pharmacology and Experimental Therapeutics, 309*, 616–625.

Takahashi, T., Ueno, H., & Shibuya, M. (1999). VEGF activates protein kinase C-dependent, but Ras-independent Raf-MEK-MAP kinase pathway for DNA synthesis in primary endothelial cells. *Oncogene, 18*, 2221–2230.

Takahashi, M., Mukai, H., Oishi, K., Isagawa, T., & Ono, Y. (2000). Association of immature hypophosphorylated protein kinase cepsilon with an anchoring protein CG-NAP. *The Journal of Biological Chemistry, 275*, 34592–34596.

Takai, Y., Kishimoto, A., Iwasa, Y., Kawahara, Y., Mori, T., & Nishizuka, Y. (1979a). Calcium-dependent activation of a multifunctional protein kinase by membrane phospholipids. *The Journal of Biological Chemistry, 254*, 3692–3695.

Takai, Y., Kishimoto, A., Kikkawa, U., Mori, T., & Nishizuka, Y. (1979b). Unsaturated diacylglycerol as a possible messenger for the activation of calcium-activated, phospholipid-dependent protein kinase system. *Biochemical and Biophysical Research Communications, 91*, 1218–1224.

Tan, M., Li, P., Sun, M., Yin, G., & Yu, D. (2006). Upregulation and activation of PKC alpha by ErbB2 through Src promotes breast cancer cell invasion that can be blocked by combined treatment with PKC alpha and Src inhibitors. *Oncogene, 25*, 3286–3295.

Teicher, B. A., Menon, K., Alvarez, E., Galbreath, E., Shih, C., & Faul, M. M. (2001a). Antiangiogenic and antitumor effects of a protein kinase Cbeta inhibitor in murine lewis lung carcinoma and human Calu-6 non-small-cell lung carcinoma xenografts. *Cancer Chemotherapy and Pharmacology, 48*, 473–480.

Teicher, B. A., Menon, K., Alvarez, E., Liu, P., Shih, C., & Faul, M. M. (2001b). Antiangiogenic and antitumor effects of a protein kinase C beta inhibitor in human hepatocellular and gastric cancer xenografts. *In Vivo, 15*, 185–193.

Teicher, B. A., Alvarez, E., Menon, K., Esterman, M. A., Considine, E., Shih, C., & Faul, M. M. (2002a). Antiangiogenic effects of a protein kinase Cbeta-selective small molecule. *Cancer Chemotherapy and Pharmacology, 49*, 69–77.

Teicher, B. A., Menon, K., Alvarez, E., Shih, C., & Faul, M. M. (2002b). Antiangiogenic and antitumor effects of a protein kinase Cbeta inhibitor in human breast cancer and ovarian cancer xenografts. *Investigational New Drugs, 20*, 241–251.

Ushio-Fukai, M., & Nakamura, Y. (2008). Reactive oxygen species and angiogenesis: NADPH oxidase as target for cancer therapy. *Cancer Letters, 266*, 37–52.

Vallentin, A., Lo, T. C., & Joubert, D. (2001). A single point mutation in the V3 region affects protein kinase Calpha targeting and accumulation at cell-cell contacts. *Molecular and Cellular Biology, 21*, 3351–3363.

Vattemi, G., Tonin, P., Mora, M., Filosto, M., Morandi, L., Savio, C., et al. (2004). Expression of protein kinase C isoforms and interleukin-1beta in myofibrillar myopathy. *Neurology, 62*, 1778–8262.

Xiao, B., Smerdon, S. J., Jones, D. H., Dodson, G. G., Soneji, Y., Aitken, A., & Gamblin, S. J. (1995). Structure of a 14-3-3 protein and implications for coordination of multiple signalling pathways. *Nature, 376*, 188–191.

Yoshiji, H., Kuriyama, S., Ways, D. K., Yoshii, J., Miyamoto, Y., Kawata, M., et al. (1999). Protein kinase C lies on the signaling pathway for vascular endothelial growth factor-mediated tumor development and angiogenesis. *Cancer Research, 59*, 4413–4418.

Part II
PKC Isozymes in the Control of Cell Function

Chapter 6
Introduction: PKC Isozymes in the Control of Cell Function

Gry Kalstad Lønne and Christer Larsson

Keywords Protein Kinase C • Proliferation • Differentiation • Cell motility • Apoptosis

6.1 Background

Since its discovery more than 30 years ago protein kinase C (PKC) has been implicated in the regulation of essentially any cell function investigated. Many of these assumptions were based on effects obtained by the application of phorbol esters, which for a long time were considered to be specific activators of PKC, and by the use of PKC inhibitors. Some of the conclusions may need to be modified since phorbol esters also target other proteins that contain typical C1 (Kazanietz 2002) domains and the chemical inhibitors have been shown to inhibit other kinases than PKC as well. Furthermore, atypical PKCs are insensitive to phorbol esters and effects of these isoforms will be undetected when phorbol esters are used as PKC activators.

Later research using approaches to more specifically inhibit individual isoforms have revealed that the plethora of PKC effects is mediated by different isoforms in a more or less isoform-specific manner. The picture that emerges from these studies is that there is a remarkable variability between cell types in terms of which cell functions are influenced by a PKC isoform. There are at the most a few examples of mechanisms of an isoform that is general and common for most cell types.

Deletion of a PKC isoform is compatible with life. Mice that lack a PKC isoform are superficially and largely normal, indicating that individual isoforms are not essential for the development of normal and functional tissues and organs.

G.K. Lønne and C. Larsson (✉)
Center for Molecular Pathology, Lund University, Malmö University Hospital, Entr 78, 3rd Floor, SE-205 02, Malmö, Sweden
e-mail: christer.larsson@med.lu.se

M.G. Kazanietz (ed.), *Protein Kinase C in Cancer Signaling and Therapy*,
Current Cancer Research, DOI 10.1007/978-1-60761-543-9_6,

Nevertheless, there is an abundance of indications from in vitro studies that PKC isoforms play important roles for proliferation, differentiation, and survival of a number of cell types. Furthermore, for all isoforms the corresponding knock-out mouse displays some phenotypic abnormalities, particularly when certain cell types are provoked. Thus, PKC isoforms are likely to be crucial for important but more limited parts of differentiation, proliferation, and cell death programs. It is also conceivable that there is redundancy, either that other pathways not involving PKC regulate the same processes or that the most similar PKC isoforms can mediate the same effect.

6.2 Proliferation

Cell proliferation is one process which can be both positively and negatively influenced by PKC isoforms. PKCα is the most studied classical isoform and generally inhibits proliferation. It mediates cell cycle withdrawal (Frey et al. 2000) and cell cycle arrest in G1 by mechanisms including Rb hypophosphorylation, downregulation of cyclin D1, and induction of $p21^{Waf1/Cip1}$ and $p27^{Kip1}$ (Frey et al. 1997; Detjen et al. 2000; Clark et al. 2004). However, it can also be pro-proliferative by mediating transcription of genes involved in cell cycle progression (Soh and Weinstein 2003).

The novel PKC isoforms show different effects on proliferation. The general picture is that PKCδ negatively and PKCε positively regulate proliferation. PKCδ influences cell cycle progression by preventing cells from entering S-phase (Fukumoto et al. 1997; Ashton et al. 1999) and M-phase (Watanabe et al. 1992). However, PKCδ has also shown pro-proliferative effects, conceivably due to its ability to activate the MAPK pathway (Ueda et al. 1996; Keshamouni et al. 2002).

PKCε can promote cell cycle entry by mediating transcriptional regulation of genes involved in G1-S-phase transition (Soh and Weinstein 2003; Bae et al. 2007). It has also been suggested that PKCε interacts with the Ras signaling pathway to promote cell proliferation (Perletti et al. 1998).

There are reports indicating both pro- and antiproliferative effects of PKCη. It enhances cell cycle progression by upregulating G1 cyclins (Fima et al. 2001) and increases proliferation through Akt/mTOR (Aeder et al. 2004) and ERK/Elk1 (Uht et al. 2007) signaling. However, a negative effect on proliferation has been shown for PKCη as well (Ohba et al. 1998).

The atypical PKCs can both favor and repress proliferation. PKCι is required for cell proliferation of glioma cells (Patel et al. 2008) and associates with Cdk7, leading to activation of this protein, which favors cell cycle progression (Acevedo-Duncan et al. 2002). PKCζ reduces proliferation of fibroblasts by interfering with ERK-signaling (Short et al. 2006). However, overexpression and silencing studies have shown pro-proliferative effects of PKCζ as well (Ghosh et al. 2002; Martin et al. 2002).

6.3 Differentiation

There are many examples of PKC isoforms influencing and regulating differentiation programs. One initial finding opening this path was the discovery that phorbol esters, which at the time were basically considered to be tumor promoters, actually stimulate maturation of promyelocytic HL-60 cells (Huberman and Callaham 1979). It was subsequentially clarified that PKCβ isoforms are critical for this effect (Macfarlane and Manzel 1994; Tonetti et al. 1994). Another early discovery was the finding that phorbol esters also induce neuronal differentiation of cultured neuroblastoma cells (Påhlman et al. 1981). However, in terms of neuronal differentiation it is rather PKCε that is the promoting isoform, at least in cell lines that are used as model systems (Hundle et al. 1995). One important feature of a neuronal differentiation program is the establishment of cell polarity with separate axonal and dendritic compartments. Atypical PKC isoforms play a crucial role in this process (Etienne-Manneville and Hall 2001; Nishimura et al. 2004).

Keratinocyte development is another PKC-regulated process, and it is illustrative as the related novel isoforms seem to have unique roles. PKCη is a differentiation-driving isoform and has been suggested to achieve this effect by activation of Fyn (Cabodi et al. 2000) and by inhibition of a cdk2 complex (Kashiwagi et al. 2000). On the other hand, the closest related isoform, PKCε, rather promotes the development of squamous cell carcinoma (Verma et al. 2006). PKCδ, the third novel isoform present in keratinocytes can facilitate differentiation unless it is tyrosine-phosphorylated, which is often the case in transformed cells (Joseloff et al. 2002). On the other hand, when overexpressed, several novel isoforms have the capacity to promote the keratinocyte differentiation program (Efimova et al. 2002).

Classical isoforms have been suggested to counteract each other in intestinal epithelial cell differentiation with PKCα promoting an exit from the cell cycle (Frey et al. 2000) while PKCβII supports a continued proliferation (Murray et al. 1999).

6.4 Morphology and Motility

It has long been recognized that phorbol esters induce morphological changes in a wide range of cell types. Many of these effects have later been confirmed to be mediated via PKC isoforms. In many cases, PKC promotes the formation of protrusions and cell spreading and suppresses stress fibers (Larsson 2006), suggesting that PKC generally counteracts morphological effects that are induced by RhoA. However, there are examples when PKC can stimulate the formation of stress fibers (Woods and Couchman 1992; Massoumi et al. 2002).

PKC also takes part in morphological effects of importance for cellular function. For instance, PKCε stimulates the outgrowth of cellular processes, which in the case of neuronal cells may mature into axons or dendrites (Zeidman et al. 1999). PKC has also been shown to fine-tune the growth direction of growth cones in response to attractants and repellants (Xiang et al. 2002). Another important role for PKC is the regulation of cell polarity which is crucial for the function of a wide range of cell types. This is mediated by atypical PKCs in close conjunction with PAR proteins (Joberty et al. 2000; Lin et al. 2000) and is one of the few examples when a PKC isoform may have an effect that is general.

Most PKC isoforms have been suggested to promote migration whereas there is less evidence for antimigratory effects of PKC. In particular, PKCα has been implicated in promigratory effects in various cell types including endothelial cells (Harrington et al. 1997), breast cancer cells (Ng et al. 1999), and fibrosarcoma cells (Ng et al. 2001).

There does not seem to be a common mechanism for the PKC effects on cellular morphology or motility. PKC can influence cellular receptors and the signaling pathways connecting them with the cytoskeleton. One example is the transport of integrins to and from the plasma membrane which is a PKC-regulated process. The presence of active integrins at the cell surface is central for proper adhesion and a dynamic transport of integrins is crucial for cell motility and adhesion. Both PKCα and PKCε associate with integrins, which is important for the effects (Ng et al. 1999; Ivaska et al. 2002). PKCα also binds other matrix receptors such as syndecan-4. This leads to direct activation of PKCα, which is an important step in the transduction of syndecan-4 effects on the cytoskeleton (Oh et al. 1998). There are also several examples when PKC not only regulates integrins but also takes part in the intracellular pathways induced by integrin activation (Volkov et al. 2001).

PKC isoforms can also directly phosphorylate and thereby functionally modulate components of the cytoskeleton. A common theme for many substrates is that they are accessory proteins that serve as connectors between different cytoskeletal proteins or between the cytoskeleton and the plasma membrane. The MARCKS/GAP43 proteins are examples that belong to the latter category. Upon PKC-mediated phosphorylation MARCKS loses its membrane-binding capacity and dissociates from the plasma membrane (Thelen et al. 1991). This will consequently facilitate dissociation of the actin cytoskeleton from plasma membrane. Another model suggests that the primary function of MARCKS proteins is to sequester phosphatidylinositol 4,5-bisphosphate (Laux et al. 2000). The PKC-mediated phosphorylation of MARCKS would thereby lead to an increase in the free amounts of this lipid, which in turn regulates a number of proteins that modify the cytoskeleton.

Adducin (Ling et al. 1986), which connects actin filaments with spectrin, and fascin (Anilkumar et al. 2003), which bundles actin filaments, are also examples of PKC substrates. In these cases, phosphorylation leads to disruption of the proteins from F-actin. A consequence of PKC action would, in view of the effects on MARCKS, adducin, and fascin, be dissociated F-actin, which thereby may be more prone to structural reformation.

There are also PKC substrates, such as ERM proteins (Ng et al. 2001), whose capacity to bind F-actin is increased as a result of PKC-mediated phosphorylation.

6.5 Apoptosis

PKC isoforms are important regulators of cell death and survival. The classical and atypical PKC subfamilies are generally associated with survival, whereas for the novel isoforms the effects are more isoform-specific. In general, PKCδ is pro-apoptotic and PKCε is antiapoptotic.

Several PKC isoforms, including PKCδ, ε, θ, and ζ, are substrates for caspases, enzymes that execute the programmed cell death, which cleave PKCs in the hinge region, thereby releasing a constitutively active catalytic domain, and which generally have pro-apoptotic activity (Emoto et al. 1995; Datta et al. 1997; Frutos et al. 1999; Hoppe et al. 2001).

PKCδ is the isoform that is strongest associated with pro-apoptotic effects. Proteolytic activation of PKCδ by caspases is directly linked with apoptosis (Emoto et al. 1995; Ghayur et al. 1996). This cofactor-independent activation of PKCδ further activates caspase-3 and function as a positive feedback loop (Anantharam et al. 2002; Blass et al. 2002). Tyrosine phosphorylation (Blass et al. 2002) and localization (DeVries-Seimon et al. 2007; Gomel et al. 2007) of PKCδ appear to determine its apoptotic effect, where nuclear and mitochondrial localizations of PKCδ are the strongest promoters of apoptosis. PKCδ mediates phosphorylation of several nuclear proteins to promote apoptosis (Bharti et al. 1998; Cross et al. 2000; Yoshida et al. 2003). In mitochondria, PKCδ activation can lead to release of cytochrome *c* and decreased mitochondrial membrane potential (Matassa et al. 2001; Sumitomo et al. 2002).

In contrast to PKCδ, PKCε is considered to be antiapoptotic since inhibition or silencing of PKCε in cancer cell lines makes them more susceptible to apoptotic insults (Sivaprasad et al. 2007), and overexpression or activation of PKCε protects against apoptosis (Okhrimenko et al. 2005; Sivaprasad et al. 2007). There is also evidence for PKCε-mediated increase in expression and activity of the pro-survival protein Akt (Okhrimenko et al. 2005; Lu et al. 2006).

Of the classical PKCs, PKCα is the isoform that is mostly studied concerning cell survival and is considered as a survival factor (Ruvolo et al. 1998; Gill et al. 2001). The atypical PKCs appear to be predominantly antiapoptotic (Frutos et al. 1999; Jamieson et al. 1999).

6.6 Concluding Remarks

Although PKC isoforms play critical roles in most cellular functions, their effects are in most cases dependent on context and cell type. The factors upstream or downstream of PKC that determine its functional effects in the specific context are not completely understood. One challenging task for future research is to identify factors that determine and perhaps make it possible to predict the effects a PKC isoform will have on cellular functions.

References

Acevedo-Duncan, M., Patel, R., Whelan, S., & Bicaku, E.(2002). Human glioma PKC-i and PKC-bII phosphorylate cyclin-dependent kinase activating kinase during the cell cycle. *Cell Proliferation*, *35*, 23–36.

Aeder, S.E., Martin, P.M., Soh, J.W., & Hussaini, I.M. (2004). PKC-h mediates glioblastoma cell proliferation through the Akt and mTOR signaling pathways. *Oncogene*, *23*, 062–9069.

Anantharam, V., Kitazawa, M., Wagner, J., Kaul, S., & Kanthasamy, A.G. (2002). Caspase-3-dependent proteolytic cleavage of protein kinase Cd is essential for oxidative stress-mediated dopaminergic cell death after exposure to methylcyclopentadienyl manganese tricarbonyl. *Journal of Neuroscience*, *22*, 1738–1751.

Anilkumar, N., Parsons, M., Monk, R., Ng, T., & Adams, J.C. (2003). Interaction of fascin and protein kinase Ca: A novel intersection in cell adhesion and motility. *EMBO Journal*, *22*, 5390–5402.

Ashton, A.W., Watanabe, G., Albanese, C., Harrington, E.O., Ware, J.A., & Pestell, R.G. (1999). Protein kinase Cd inhibition of S-phase transition in capillary endothelial cells involves the cyclin-dependent kinase inhibitor p27(Kip1). *Journal of Biological Chemistry*, *274*, 20805–20811.

Bae, K-M., Wang, H., Jiang, G., Chen, M.G., Lu, L., & Xiao, L. (2007). Protein kinase Ce is overexpressed in primary human non-small cell lung cancers and functionally required for proliferation of non-small cell lung cancer cells in a p21/Cip1-dependent manner. *Cancer Research*, *67*, 6053–6063.

Bharti, A., Kraeft, S.K., Gounder, M., Pandey, P., Jin, S., Yuan, Z.M., et al. (1998). Inactivation of DNA-dependent protein kinase by protein kinase Cd: Implications for apoptosis. *Molecular and Cellular Biology*, *18*, 6719–6728.

Blass, M., Kronfeld, I., Kazimirsky, G., Blumberg, P.M., & Brodie, C. (2002). Tyrosine phosphorylation of protein kinase Cd is essential for its apoptotic effect in response to etoposide. *Molecular and Cellular Biology*, *22*, 182–195.

Cabodi, S., Calautti, E., Talora, C., Kuroki, T., Stein, P.L., & Dotto, G.P. (2000). A PKC-h/Fyn-dependent pathway leading to keratinocyte growth arrest and differentiation. *Molecular Cell*, *6*, 1121–1129.

Clark, J.A., Black, A.R., Leontieva, O.V., Frey, M.R., Pysz, M.A., Kunneva, L., et al. (2004). Involvement of the ERK signaling cascade in protein kinase C-mediated cell cycle arrest in intestinal epithelial cells. *Journal of Biological Chemistry*, *279*, 9233–9247.

Cross, T., Griffiths, G., Deacon, E., Sallis, R., Gough, M., Watters, D., et al. (2000). PKC-d is an apoptotic lamin kinase. *Oncogene*, *19*, 2331–2337.

Datta, R., Kojima, H., Yoshida, K., & Kufe, D. (1997). Caspase-3-mediated cleavage of protein kinase C q in induction of apoptosis. *Journal of Biological Chemistry*, *272*, 20317–20320.

Detjen, K.M., Brembeck, F.H., Welzel, M., Kaiser, A., Haller, H., Wiedenmann, B., et al. (2000. Activation of protein kinase Ca inhibits growth of pancreatic cancer cells via p21(cip)-mediated G(1) arrest. *Journal of Cell Science*, *113*, 3025–3035.

DeVries-Seimon, T.A., Ohm, A.M., Humphries, M.J., & Reyland, M.E. (2007). Induction of apoptosis is driven by nuclear retention of protein kinase C d. *Journal of Biological Chemistry*, *282*, 22307–22314.

Efimova, T., Deucher, A., Kuroki, T., Ohba, M., & Eckert, R.L. (2002). Novel protein kinase C isoforms regulate human keratinocyte differentiation by activating a p38d mitogen-activated protein kinase cascade that targets CCAAT/enhancer-binding protein a. *Journal of Biological Chemistry*, *277*, 31753–31760.

Emoto, Y., Manome, Y., Meinhardt, G., Kisaki, H., Kharbanda, S., Robertson, M., et al. (1995). Proteolytic activation of protein kinase C d by an ICE-like protease in apoptotic cells. *EMBO Journal*, *14*, 6148–6156.

Etienne-Manneville, S., & Hall, A. (2001). Integrin-mediated activation of Cdc42 controls cell polarity in migrating astrocytes through PKCz. *Cell*, *106*, 489–498.

Fima, E., Shtutman, M., Libros, P., Missel, A., Shahaf, G., Kahana, G., et al. (2001). PKCh enhances cell cycle progression, the expression of G1 cyclins and p21 in MCF-7 cells. *Oncogene*, *20*, 6794–6804.

Frey, M.R., Clark, J.A., Leontieva, O., Uronis, J.M., Black, A.R., & Black, J.D. (2000). Protein kinase C signaling mediates a program of cell cycle withdrawal in the intestinal epithelium. *Journal of Cell Biology*, *151*, 763–778.

Frey, M.R., Saxon, M.L., Zhao, X., Rollins, A., Evans, S.S., & Black, J.D. (1997). Protein kinase C isozyme-mediated cell cycle arrest involves induction of p21(waf1/cip1) and p27(kip1) and hypophosphorylation of the retinoblastoma protein in intestinal epithelial cells. *Journal of Biological Chemistry*, *272*, 9424–9435.

Frutos, S., Moscat, J., & Diaz-Meco, M.T. (1999). Cleavage of zPKC but not l/iPKC by caspase-3 during UV-induced apoptosis. *Journal of Biological Chemistry*, *274*, 10765–10770.

Fukumoto, S., Nishizawa, Y., Hosoi, M., Koyama, H., Yamakawa, K., Ohno, S., et al. (1997). Protein kinase C d inhibits the proliferation of vascular smooth muscle cells by suppressing G1 cyclin expression. *Journal of Biological Chemistry*, *272*, 13816–13822.

Ghayur, T., Hugunin, M., Talanian, R.V., Ratnofsky, S., Quinlan, C., Emoto, Y., et al. (1996). Proteolytic activation of protein kinase C d by an ICE/CED 3-like protease induces characteristics of apoptosis. *Journal of Exprimental Medicine*, *184*, 2399–2404.

Ghosh, P.M., Bedolla, R., Mikhailova, M., & Kreisberg, J.I. (2002). RhoA-dependent murine prostate cancer cell proliferation and apoptosis: role of protein kinase Czeta. *Cancer Research*, *62*, 2630–2636.

Gill, P.K., Gescher, A., & Gant, T.W. (2001). Regulation of MDR1 promoter activity in human breast carcinoma cells by protein kinase C isozymes a and q. *European Journal of Biochemistry*, *268*, 4151–4157.

Gomel, R., Xiang, C., Finniss, S., Lee, H.K., Lu, W., Okhrimenko, H., et al. (2007). The localization of protein kinase Cd in different subcellular sites affects its proapoptotic and antiapoptotic functions and the activation of distinct downstream signaling pathways. *Molecular Cancer Research*, *5*, 627–639.

Harrington, E.O., Loffler, J., Nelson, P.R., Kent, K.C., Simons, M., & Ware, J.A. (1997). Enhancement of migration by protein kinase Ca and inhibition of proliferation and cell cycle progression by protein kinase Cd in capillary endothelial cells. *Journal of Biological Chemistry*, *272*, 7390–7397.

Hoppe, J., Hoppe, V., & Schafer, R. (2001). Selective degradation of the PKC-e isoform during cell death in AKR-2B fibroblasts. *Experimental Cell Research*, *266*, 64–73.

Huberman, E., & Callaham, M.F. (1979). Induction of terminal differentiation in human promyelocytic leukemia cells by tumor-promoting agents. *Proceedings of the National Academy of Sciences of the United States of America*, *76*, 1293–1297.

Hundle, B., McMahon, T., Dadgar, J., & Messing, R.O. (1995). Overexpression of e-protein kinase C enhances nerve growth factor-induced phosphorylation of mitogen-activated protein kinases and neurite outgrowth. *Journal of Biological Chemistry*, *270*, 30134–30140.

Ivaska, J., Whelan, R.D.H., Watson, R., & Parker, P.J. (2002). PKCe controls the traffic of b1 integrins in motile cells. *EMBO Journal*, *21*, 3608–3619.

Jamieson, L., Carpenter, L., Biden, T.J., & Fields, A.P. (1999). Protein kinase Ci activity is necessary for Bcr-Abl-mediated resistance to drug-induced apoptosis. *Journal of Biological Chemistry*, *274*, 3927–3930.

Joberty, G., Petersen, C., Gao, L., & Macara, I.G. (2000). The cell-polarity protein Par6 links Par3 and atypical protein kinase C to Cdc42. *Nature Cell Biology*, *2*, 531–539.

Joseloff, E., Cataisson, C., Aamodt, H., Ocheni, H., Blumberg, P., Kraker, A.J., et al. (2002). Src family kinases phosphorylate protein kinase C delta on tyrosine residues and modify the neoplastic phenotype of skin keratinocytes. *Journal of Biological Chemistry*, *277*, 12318–12323.

Kashiwagi, M., Ohba, M., Watanabe, H., Ishino, K., Kasahara, K., Sanai, Y., et al. (2000). PKCh associates with cyclin E/cdk2/p21 complex, phosphorylates p21 and inhibits cdk2 kinase in keratinocytes. *Oncogene*, *19*, 6334–6341.

Kazanietz, M.G. (2002). Novel "nonkinase" phorbol ester receptors: the C1 domain connection. *Molecular Pharmacology*, *61*, 759–767.

Keshamouni, V.G., Mattingly, R.R., & Reddy, K.B. (2002). Mechanism of 17-b-estradiol-induced Erk1/2 activation in breast cancer cells. A role for HER2 AND PKC-d. *Journal of Biological Chemistry*, *277*, 22558–22565.

Larsson, C. (2006). Protein kinase C and the regulation of the actin cytoskeleton. *Cellular Signalling*, *18*, 276–284.

Laux, T., Fukami, K., Thelen, M., Golub, T., Frey, D., & Caroni, P. (2000). GAP43, MARCKS, and CAP23 modulate PI(4, 5)P(2) at plasmalemmal rafts, and regulate cell cortex actin dynamics through a common mechanism. *Journal of Cell Biology*, *149*, 1455–1472.

Lin, D., Edwards, A.S., Fawcett, J.P., Mbamalu, G., Scott, J.D., & Pawson, T. (2000). A mammalian PAR-3-PAR-6 complex implicated in Cdc42/Rac1 and aPKC signalling and cell polarity. *Nature Cell Biology*, *2*, 540–547.

Ling, E., Gardner, K., & Bennett, V. (1986). Protein kinase C phosphorylates a recently identified membrane skeleton-associated calmodulin-binding protein in human erythrocytes. *Journal of Biological Chemistry*, *261*, 13875–13878.

Lu, D., Huang, J., & Basu, A. (2006). Protein Kinase Ce Activates Protein Kinase B/Akt via DNA-PK to Protect against Tumor Necrosis Factor-a-induced Cell Death. *Journal of Biological Chemistry*, *281*, 22799–22807.

Macfarlane, D.E., & Manzel, L. (1994). Activation of beta-isozyme of protein kinase C (PKC b) is necessary and sufficient for phorbol ester-induced differentiation of HL-60 promyelocytes. Studies with PKC b-defective PET mutant. *Journal of Biological Chemistry*, *269*, 4327–4331.

Martin, P., Duran, A., Minguet, S., Gaspar, M.L., Diaz-Meco, M.T., Rennert, P., et al. (2002). Role of z PKC in B-cell signaling and function. *EMBO Journal*, *21*, 4049–4057.

Massoumi, R., Larsson, C., & Sjölander, A. (2002). Leukotriene D4 induces stress-fibre formation in intestinal epithelial cells via activation of RhoA and PKCd. *Journal of Cell Science*, *115*, 3509–3515.

Matassa, A.A., Carpenter, L., Biden, T.J., Humphries, M.J., & Reyland, M.E. (2001). PKCd is required for mitochondrial-dependent apoptosis in salivary epithelial cells. *Journal of Biological Chemistry*, *276*, 29719–29728.

Murray, N.R., Davidson, L.A., Chapkin, R.S., Clay Gustafson, W., Schattenberg, D.G., & Fields, A.P. (1999). Overexpression of protein kinase C bII induces colonic hyperproliferation and increased sensitivity to colon carcinogenesis. *Journal of Cell Biology*, *145*, 699–711.

Ng, T., Parsons, M., Hughes, W.E., Monypenny, J., Zicha, D., Gautreau, A., et al. (2001). Ezrin is a downstream effector of trafficking PKC-integrin complexes involved in the control of cell motility. *EMBO Journal*, *20*, 2723–2741.

Ng, T., Shima, D., Squire, A., Bastiaens, P.I.H., Gschmeissner, S., Humphries, M.J., et al (1999). PKCa regulates b1 integrin-dependent cell motility through association and control of integrin traffic. *EMBO Journal*, *18*, 3909–3923.

Nishimura, T., Kato, K., Yamaguchi, T., Fukata, Y., Ohno, S., & Kaibuchi, K. (2004). Role of the PAR-3-KIF3 complex in the establishment of neuronal polarity. *Nature Cell Biology*, *6*, 328–334.

Oh, E.S., Woods, A., Lim, S.T., Theibert, A.W., & Couchman, J.R. (1998). Syndecan-4 proteoglycan cytoplasmic domain and phosphatidylinositol 4, 5-bisphosphate coordinately regulate protein kinase C activity. *Journal of Biological Chemistry*, *273*, 10624–10629.

Ohba, M., Ishino, K., Kashiwagi, M., Kawabe, S., Chida, K., Huh, N.H., et al. (1998) Induction of differentiation in normal human keratinocytes by adenovirus-mediated introduction of the eta and delta isoforms of protein kinase C. *Molecular and Cell Biology*, *18*, 5199–5207.

Okhrimenko, H., Lu, W., Xiang, C., Hamburger, N., Kazimirsky, G., & Brodie, C., (2005). Protein Kinase C-e Regulates the Apoptosis and Survival of Glioma Cells. *Cancer Research*, *65*, 7301–7309.

Patel, R., Win, H., Desai, S., Patel, K., Matthews, J.A., & Acevedo-Duncan, M. (2008). Involvement of PKC-iota in glioma proliferation. *Cell Proliferation*, *41*, 122–135.

Perletti, G.P., Concari, P., Brusaferri, S., Marras, E., Piccinini, F., & Tashjian, A.H., Jr. (1998) Protein kinase Ce is oncogenic in colon epithelial cells by interaction with the ras signal transduction pathway. *Oncogene*, *16*, 3345–3348.

Påhlman, S., Odelstad, L., Larsson, E., Grotte, G., & Nilsson, K. (1981). Phenotypic changes of human neuroblastoma cells in culture induced by 12-O-tetradecanoyl-phorbol-13-acetate. *International Journal of Cancer*, *28*, 583–589.

Ruvolo, P.P., Deng, X., Carr, B.K., & May, W.S. (1998). A functional role for mitochondrial protein kinase Ca in Bcl2 phosphorylation and suppression of apoptosis. *Journal of Biological Chemistry*, *273*, 25436–25442.

Short, M.D., Fox, S.M., Lam, C.F., Stenmark, K.R., & Das, M. (2006). Protein kinase Cz attenuates hypoxia-induced proliferation of fibroblasts by regulating MAP kinase phosphatase-1 expression. *Molecular Biology of the Cell*, *17*, 1995–2008.

Sivaprasad, U., Shankar, E., & Basu, A. (2007). Downregulation of Bid is associated with PKCepsilon-mediated TRAIL resistance. *Cell Death and Differentiation*, *14*, 851–860.

Soh, J.W., & Weinstein, I.B. (2003). Roles of specific isoforms of protein kinase C in the transcriptional control of cyclin D1 and related genes. *Journal of Biological Chemistry*, *278*, 34709–34716.

Sumitomo, M., Ohba, M., Asakuma, J., Asano, T., Kuroki, T., Asano, T., et al. (2002). Protein kinase Cd amplifies ceramide formation via mitochondrial signaling in prostate cancer cells. *Journal of Clinical Investigation*, *109*, 827–836.

Thelen, M., Rosen, A., Nairn, A.C. & Aderem, A. (1991). Regulation by phosphorylation of reversible association of a myristoylated protein kinase C substrate with the plasma membrane. *Nature*, *351*, 320–322.

Tonetti, D.A., Henning-Chubb, C., Yamanishi, D.T., & Huberman, E. (1994). Protein kinase C-b is required for macrophage differentiation of human HL-60 leukemia cells. *Journal of Biological Chemistry*, *269*, 23230–23235.

Ueda, Y., Hirai, S., Osada, S., Suzuki, A., Mizuno, K., & Ohno, S. (1996). Protein kinase C activates the MEK-ERK pathway in a manner independent of Ras and dependent on Raf. *Journal of Biological Chemistry*, *271*, 23512–23519.

Uht, R.M., Amos, S., Martin, P.M., Riggan, A.E., & Hussaini, I.M. (2007). The protein kinase C-η isoform induces proliferation in glioblastoma cell lines through an ERK//Elk-1 pathway. *Oncogene*, *26*, 2885–2893

Watanabe, T., Ono, Y., Taniyama, Y., Hazama, K., Igarashi, K., Ogita, K., et al. (1992). Cell division arrest induced by phorbol ester in CHO cells overexpressing protein kinase C-d subspecies. *Proceedings of the National Academy of Sciences of the United States of America*, *89*, 10159–10163.

Verma, A.K., Wheeler, D.L., Aziz, M.H., & Manoharan, H. (2006). Protein kinase Ce and development of squamous cell carcinoma, the nonmelanoma human skin cancer. *Molecular Carcinogenesis*, *45*, 381–388.

Volkov, Y., Long, A., McGrath, S., Ni Eidhin, D., & Kelleher, D. (2001). Crucial importance of PKC-b(I) in LFA-1-mediated locomotion of activated T cells. *Nature Immunology*, *2*, 508–514.

Woods, A., & Couchman, J.R. (1992). Protein kinase C involvement in focal adhesion formation. *Journal of Cell Science*, *101*, 277–290.

Xiang, Y., Li, Y., Zhang, Z., Cui, K., Wang, S., Yuan, X.B., et al. (2002). Nerve growth cone guidance mediated by G protein-coupled receptors. *Nature Neuroscience*, *5*, 843–848.

Yoshida, K., Wang, H-G., Miki, Y., & Kufe, D. (2003). Protein kinase Cd is responsible for constitutive and DNA damage-induced phosphorylation of Rad9. *EMBO Journal*, *22*, 1431–1441.

Zeidman, R., Löfgren, B., Påhlman, S., & Larsson, C. (1999). PKCe, via its regulatory domain and independently of its catalytic domain, induces neurite-like processes in neuroblastoma cells. *Journal of Cell Biology*, *145*, 713–726.

Chapter 7
Regulation and Function of Protein Kinase D Signaling

Enrique Rozengurt

Abstract Protein kinase D (PKD) is an evolutionarily conserved protein kinase with structural, enzymological, and regulatory properties different from the PKC family members. The most distinct characteristics of PKD are the presence of a catalytic domain distantly related to Ca^{2+}-regulated kinases and a pleckstrin homology (PH) domain that regulates enzyme activity. The N-terminal region of PKD also contains a tandem repeat of cysteine-rich, zinc finger-like motifs which confer high affinity binding of phorbol esters and repress catalytic kinase activity. The subsequent identification of PKD2 and PKD3, similar in overall structure and amino acid sequence to PKD, confirmed the notion that PKD is the founding member of a new family of protein kinases, now classified in the mammalian kinome within the Ca^{2+}/calmodulin-dependent protein kinase (CaMK) group. PKD can be activated within intact cells by multiple stimuli acting through receptor-mediated pathways. Rapid PKD activation has been demonstrated in response to G protein-coupled receptor agonists, growth factors, cross-linking of B-cell receptor and T-cell receptor and cellular stress. The phosphorylation of Ser^{744} and Ser^{748} in the activation loop of PKD is critical for its activation. Rapid PKC-dependent PKD activation can be followed by a late, PKC-independent, phase of catalytic activation and phosphorylation induced by agonists of Gq-coupled receptors. Accumulating evidence suggest that PKD plays a role in multiple cellular processes and activities, including signal transduction, chromatin organization, Golgi function, gene expression, immune regulation, cardiac hypertrophy and cell survival, adhesion, motility, polarity, DNA synthesis and proliferation. The studies on regulation and function of PKD reviewed here illustrate a remarkable diversity in both its signal generation and distribution and its potential for complex regulatory interactions with multiple downstream pathways. In conclusion, PKD emerges as a key node in cellular signal transduction.

E. Rozengurt (✉)
Department of Medicine, UCLA School of Medicine, Division of Digestive Diseases and CURE, Digestive Diseases Research Center, David Geffen School of Medicine, University of California, 900 Veteran Avenue, Warren Hall, Room 11-124, Los Angeles, CA 90095-1786, USA
e-mail: erozengurt@mednet.ucla.edu

M.G. Kazanietz (ed.), *Protein Kinase C in Cancer Signaling and Therapy*,
Current Cancer Research, DOI 10.1007/978-1-60761-543-9_7,

Keywords Protein kinase C • G protein-coupled receptors • Phorbol esters • Signal transduction • Intracellular translocation • Cell proliferation

7.1 Introduction

A wide range of external signals, including hormones, neurotransmitters, growth factors, cytokines, chemokines, bioactive lipids and tastants promote the stimulation of the isoforms of the PLC[1] family, including β, γ, δ and ε. PLCs catalyze the hydrolysis of phosphatidylinositol 4,5-biphosphate to produce two second messengers: Ins $(1,4,5)P_3$, which triggers the release of Ca^{2+} from internal stores; and DAG, which elicits cellular responses through a variety of effectors (Brose et al. 2004). The most prominent intracellular targets of DAG are the isoforms of the PKC family, which are differentially expressed in cells and tissues (Newton 1997; Mellor and Parker 1998). PKC isoforms can be classified in three subclasses according to their regulatory properties that are conferred by specific domains located in the NH_2-terminal portion of these proteins. All members of the PKC family, i.e., conventional or classical PKCs (α, βI, βII, γ), novel PKCs (δ, ε, η, θ) and atypical PKCs (ζ, τ), contain a highly conserved catalytic domain and an autoinhibitory pseudosubstrate that maintains these enzymes in an inactive state in the absence of activating second messengers (Mellor and Parker 1998; Kikkawa et al. 1989). Most of the variation between the PKC isoforms occurs in the regulatory domain. The C_1 region of this domain in both conventional and novel PKCs contains a tandem repeat of zinc-finger-like cysteine-rich motifs that confers DAG binding on these PKC isoforms (Hurley et al. 1997). In contrast, atypical PKCs contain a single cysteine-rich motif, and they are not regulated by DAG (Ways et al. 1992; Selbie et al. 1993; Nakanishi et al. 1993). Conventional PKCs have a Ca^{2+}-binding domain (the C_2 region) that allows Ca^{2+} and DAG to act synergistically on these enzymes. The novel and atypical PKCs do not require Ca^{2+} for activation (Nishizuka 1992), but can be modulated by fatty acids (Nishizuka 1995), PtdIns$(3,4,5)P_3$ (Toker and Cantley 1997) and via interaction with the Cdc42-GTP-Par6 complex (Henrique and Schweisguth 2003). The early findings that the potent tumor promoters of the phorbol ester family can substitute for DAG in PKC activation and that PKC is a high-affinity phorbol ester receptor indicated that a major cellular target of the phorbol esters is PKC, and thus established an important link between signal transduction and tumor promotion in multistage carcinogenesis (Kikkawa et al. 1989; Rozengurt et al. 1985; Nishizuka 1988; Weinstein 1988). However, the mechanisms by which PKC-mediated signals are propagated to critical downstream targets remain incompletely understood.

Protein kinase D (PKD), the founding member of a new family of serine/threonine protein kinases and the subject of this chapter, occupies a unique position in the signal transduction pathways initiated by DAG and PKC. As described below, not only is PKD a direct DAG/phorbol ester target, but it also lies downstream of PKCs in a novel signal transduction pathway implicated in the regulation of multiple fundamental biological processes.

7.1.1 The PKD Subfamily Belongs to the CaMK Group

Complementary DNA clones encoding human PKD (initially called atypical PKCμ) and PKD from mouse were identified by two different laboratories in 1994 (Johannes et al. 1994; Valverde et al. 1994). Subsequently, two additional mammalian protein kinases have been identified that share extensive overall homology with PKD (Fig. 7.1), termed PKD2 (Sturany et al. 2001) and PKC_v/PKD3 (Hayashi et al. 1999; Rey et al. 2003b).

The N-terminal regulatory portion of PKD (Fig. 7.1) contains a tandem repeat of zinc finger-like cysteine-rich motifs (termed the cysteine-rich domain, or CRD) highly homologous to domains found in DAG/phorbol ester-sensitive PKCs and in other signaling proteins regulated by DAG, including chimaerins, Ras-GRP, Munc13 and DAG kinases (see Griner and Kazanietz 2007) for review. Accordingly, PKD binds phorbol esters with high affinity via its CRD (Valverde et al. 1994; Van Lint et al. 1995; Wang et al. 2003). The individual cysteine-rich motifs of the CRD (referred to as cys1 and cys2, Fig. 7.1) are functionally dissimilar, with the cys2 motif responsible for the majority of high-affinity [^{3}H] phorbol 12,13-dibutyrate binding, both in vivo and in vitro (Iglesias et al. 1998a; Rey and Rozengurt 2001). As described below, the CRD plays a critical role in mediating PKD translocation to the plasma membrane and nucleus in cells challenged with a variety of stimuli and also represses the catalytic activity of the enzyme (Iglesias and Rozengurt 1999).

Interposed between the CRD and the catalytic domain, PKD also contains a PH domain (Iglesias and Rozengurt 1998). Found in many signal transduction proteins, PH domains bind to membrane lipids as well as to other proteins (reviewed in Cozier et al. 2004). PKD mutants with deletions or with single amino acid substitutions within the PH domain are fully active (Iglesias and Rozengurt 1998; Waldron et al. 1999a), indicating that the PH domain, like the CRD, plays a role in maintaining PKD in an inactive catalytic state.

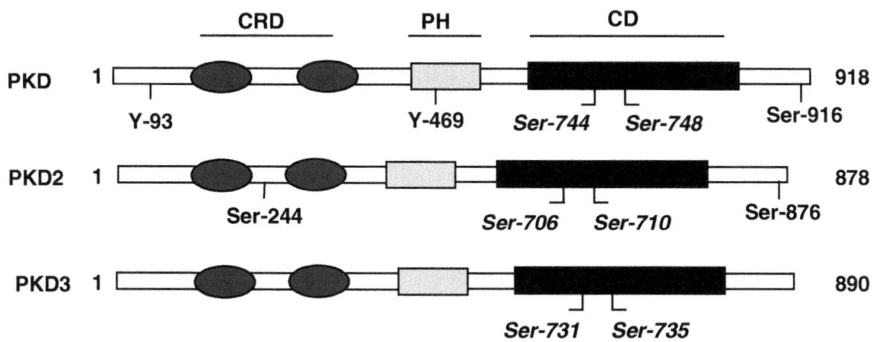

Fig. 7.1 Schematic representation of the members of the PKD family. Numbers correspond to amino acid position. Serine residues within the activation loop of PKDs that became phosphorylated are indicated in *italic*

The initial description of PKD as an atypical isoform of PKC (Johannes et al. 1994) and the inclusion of PKD/PKCμ in reviews concerning the PKC family, which belongs to the AGC group (named for PKA, PKG and PKC) (Newton 1997; Mellor and Parker 1998), contributed to a perception that PKD belongs to the PKC family. However, it was noted from the outset that the catalytic domain of PKD has highest sequence homology with myosin light chain kinase and CaMKs (Valverde et al. 1994). Indeed, the three isoforms of PKD are now classified as a new protein kinase subfamily within the CaMK group, separate from the AGC group (Hanks 2003). This scheme reflects the notion that the evolutionary relationship between protein kinases is most appropriately linked to their respective catalytic domain structures.

Full-length PKD isolated from multiple cell types or tissues exhibits very low catalytic activity (Van Lint et al. 1995), which can be stimulated by phosphatidylserine micelles and either DAG or phorbol esters (Van Lint et al. 1995; Johannes et al. 1995; Matthews et al. 1997). These early studies demonstrated that PKD is a phospholipid/DAG-stimulated serine/threonine protein kinase and implied that PKD represents a novel component of the signal transduction initiated by DAG production in their target cells (Rozengurt et al. 1997).

7.2 Regulation of PKD Activation

7.2.1 Rapid PKD Activation in Intact Cells: A PKC/PKD Phosphorylation Cascade

Subsequent studies, aimed to define the regulatory properties of PKD within intact cells, produced multiple lines of evidence that elucidated a mechanism of PKD activation distinct from the direct stimulation of enzyme activity by DAG/phorbol ester plus phospholipids obtained in vitro. Treatment of intact cells with phorbol esters, cell-permeable DAGs or bryostatin induced a dramatic conversion of PKD from an inactive to an active form, as shown by in vitro kinase assays performed in the absence of lipid co-activators (Matthews et al. 1997; Zugaza et al. 1996). In all these cases, PKD activation was selectively and potently blocked by cell treatment with PKC inhibitors (e.g., GFI, Ro31-8220 and Go6983) that did not directly inhibit PKD catalytic activity (Matthews et al. 1997; Zugaza et al. 1996), suggesting that PKD activation in intact cells is mediated, directly or indirectly, through PKCs. In line with this conclusion, co-transfection of PKD with active mutant forms of "novel" PKCs (PKCs δ, ε, η, θ) resulted in robust PKD activation in the absence of cell stimulation (Waldron et al. 1999a; Zugaza et al. 1996; Yuan et al. 2002; Storz et al. 2004a).

A variety of regulatory peptides, including bombesin, bradykinin, endothelin and vasopressin, or growth factors (e.g., PDGF) also induced PKD activation via a PKC-dependent pathway in intact fibroblasts (Zugaza et al. 1997). These results provided the first evidence indicating the operation of a PKC/PKD signaling cascade in response to receptor-activated pathways.

Subsequently, the functioning of PKC-dependent PKD activation has been extended and further explored in many normal cell types, including fibroblasts (Matthews et al. 1997; Chiu and Rozengurt 2001a; Zhukova et al. 2001a), intestinal and kidney epithelial cells (Rey et al. 2001a; Chiu and Rozengurt 2001b; Chiu et al. 2002; Rey et al. 2004), smooth muscle cells (Abedi et al. 1998), cardiomyocytes (Haworth et al. 2000; Haworth et al. 2004), neuronal cells (Iglesias et al. 2000a; Wang et al. 2004; Song et al. 2007; Poole et al. 2008), human mesenchymal stem cells (Celil and Campbell 2005), osteoblasts (Lemonnier et al. 2004), pancreatic exocrine (Yuan et al. 2008; Berna et al. 2007) and endocrine β cells (Liuwantara et al. 2006), B and T lymphocytes (Sidorenko et al. 1996; Matthews et al. 2000a; Matthews et al. 1999a), mast cells (Matthews et al. 1999b), bone marrow-derived mast cells (Murphy et al. 2007), macrophages (Park et al. 2008) and platelets (Stafford et al. 2003), as well as in a variety of cancer cells, including cells derived from small cell lung carcinoma, ductal pancreatic adenocarcinoma and breast and prostate cancer (Paolucci and Rozengurt 1999; Guha et al. 2002; Mihailovic et al. 2004; Qiang et al. 2004; Jaggi et al. 2005). These studies revealed PKD activation in response to regulatory peptides, including angiotensin, bombesin/gastrin-releasing peptide, cholecystokinin, neurotensin, vasopressin (Zugaza et al. 1997; Chiu and Rozengurt 2001a; Zhukova et al. 2001b; Chiu and Rozengurt 2001b; Chiu et al. 2002; Guha et al. 2002; Zhukova et al. 2001a; Sinnett-Smith et al. 2004), the potent bioactive lipid mediator LPA (Chiu and Rozengurt 2001b; Paolucci et al. 2000; Yuan et al. 2003) and thrombin (Stafford et al. 2003) that act through G_q, G_{12}, G_i and Rho (Chiu and Rozengurt 2001b; Paolucci et al. 2000; Yuan et al. 2003; Yuan et al. 2000; Yuan et al. 2001), as well as polypeptide growth factors, such as PDGF (Zugaza et al. 1997; Abedi et al. 1998; Van Lint et al. 1998), VEGF (Wong and Jin 2005), BMP-2 (Celil and Campbell 2005) and IGF (Celil and Campbell 2005; Qiang et al. 2004), cross-linking of B-cell receptor and T-cell receptor in B and T lymphocytes, respectively (Sidorenko et al. 1996; Matthews et al. 2000a, b; Matthews et al. 1999a), agonist stimulation of TLR2 (Murphy et al. 2007) and TLR9 (Park et al. 2008), aldosterone (McEneaney et al. 2007) and oxidative stress (Waldron and Rozengurt 2000; Waldron et al. 2004; Storz and Toker 2003; Zhang et al. 2005a).

Collectively, these studies demonstrated *rapid* PKC-dependent PKD activation in a broad range of biological systems but did not exclude the possibility of PKD activation through PKC-independent mechanism(s).

7.2.2 PKCs Directly Phosphorylate PKD at the Activation Loop

For many protein kinases, catalytic activity is dependent on the phosphorylation of activating residues located in a region spanning the highly conserved sequences DFG (in kinase subdomain VII) and APE (in kinase subdomain VIII) of the kinase catalytic domain termed the "activation loop" or "activation segment." At least two mechanisms, involving autophosphorylation or transphosphorylation, mediate the phosphorylation of one or more residues within the activation loop leading to stabilization of an active conformation of the catalytic residues (Johnson et al. 1996; Oliver et al. 2007a). Many protein kinases that participate in signal transduction pathways, including those in MAP-kinase cascades (Chang and Karin 2001; Ebisuya et al. 2005; Johnson and Lapadat 2002), are regulated by transphosphorylation of the activation loop mediated by a different upstream kinase. For example, Raf, the earliest identified effector of Ras, transphosphorylates MEK on two key residues in its activation loop, Ser^{217} and Ser^{221}, and thereby stimulates MEK activation (Marais and Marshall 1996). It is also recognized that a substantial number of regulatory protein kinases from different families mediate their own activation loop phosphorylation promoting their catalytic activation. Examples of serine/threonine protein kinases that mediate autoactivation include Aurora A (Eyers et al. 2003), Aurora B (Kelly et al. 2007), CaMK II (Hudmon and Schulman 2002), Chk2 (Oliver et al. 2006), DYRK (Lochhead et al. 2005), GSK-3 (Lochhead et al. 2006), JNK2 (Cui et al. 2005), Mps1 (Mattison et al. 2007; Kang et al. 2007), MTK1/MEKK4 (Miyake et al. 2007) and PAK (Buchwald et al. 2001). In many cases, autophosphorylation occurs by intermolecular autophosphorylation (Oliver et al. 2007b).

Using 2-D tryptic phosphopeptide mapping of metabolically ^{32}P-labeled wild type and mutant PKD forms, two key serine residues in the PKD activation loop, Ser^{744} and Ser^{748} in mouse PKD (Fig. 7.1), were identified (Iglesias et al. 1998b; Waldron et al. 1999b). A PKD mutant with both sites altered to alanine was resistant to activation in response to cell stimulation whereas mutation of Ser^{744} and Ser^{748} to glutamic acid residues, to mimic phosphorylation, generated a constitutively active mutant PKD. Single point mutants in which glutamic acid replaced Ser^{744} and Ser^{748} produced partly activated kinases. The properties of these mutant forms of PKD were consistent with a role of Ser^{744} and Ser^{748} in phosphorylation-dependent activation.

Using an antibody that recognizes PKD phosphorylated at Ser^{744}/Ser^{748}, and a second antibody that detects predominantly PKD phosphorylated at Ser^{748}, PKD activation loop phosphorylation was demonstrated in response to regulatory peptides, expression of heterotrimeric G proteins and oxidative stress in many cell types (Zhukova et al. 2001a; Yuan et al. 2003; Waldron et al. 2004; Brandlin et al. 2002; Storz et al. 2004b; Waldron et al. 2001). In line with the existence of a kinase cascade, Ser^{744} and Ser^{748} also became rapidly phosphorylated in kinase-deficient forms of PKD, indicating that PKD activation depends on transphosphorylation by an upstream kinase (e.g., PKC) (Waldron et al. 2001). Although phosphorylation of other serine (Matthews et al. 1999a; Iglesias et al. 1998b) and tyrosine (Waldron et al. 2004; Storz and Toker 2003) residues are likely to play a role in PKD regulation, it is clear that PKD phosphorylation at Ser^{744} and Ser^{748} is documented in multiple cell types, is triggered by a vast array of stimuli and plays a critical role in PKD activation. However, these initial experiments did not rule out that the activation loop residues, Ser^{744} and Ser^{748}, are phosphorylated independently of each other or via an ordered mechanism involving different isoforms of PKC or additional upstream kinases (see below).

Recent studies in vitro and in vivo examined further the role of PKCε as an upstream kinase in the activation loop phosphorylation of PKD. When incubated in the presence of phosphatidylserine, phorbol ester and ATP, intact PKD autophosphorylated at Ser^{748} rather than on Ser^{744}. In striking contrast, addition of purified PKCε to the incubation mixture induced rapid Ser^{744} and Ser^{748} phosphorylation, concomitant with persistent increase in PKD catalytic activity (Waldron and Rozengurt 2003). A plausible interpretation of these experiments is that PKCε-mediated phosphorylation of Ser^{744} synergizes with lipid co-activators to stimulate catalytic activation that then promotes the autophosphorylation of Ser^{748}. We will return to this point in the next section.

Additional experiments using selective suppression of PKCε: expression in intact cells markedly attenuated activation loop phosphorylation induced by GPCR stimulation (Rey et al. 2004) interfered with PKD activation. Similarly, PKCε has been reported to play a critical role as a PKD kinase in cultured adult rat ventricular myocytes (Haworth et al. 2007). Other investigators implicated PKCδ (Storz et al. 2004a) and PKCθ (Yuan et al. 2002) as an upstream kinase of PKD. Collectively, these studies substantiated the notion that novel PKCs directly activate PKD by activation loop phosphorylation at Ser^{744} and Ser^{748}. Although many studies indicated that novel PKCs preferentially phosphorylate the activation loop of PKD, there is evidence that classic isoforms of PKC can also induce the phosphorylation of these residues (Zugaza et al. 1996; Wong and Jin 2005).

7.2.3 Gq-coupled Receptor Agonists Induce Sequential PKC-dependent and PKC-independent Phases of PKD Activation

In addition to the well-characterized rapid PKC-dependent PKD activation, recent kinetic studies demonstrated that PKD activation in response to Gq-coupled receptor agonists can be dissected into two different phases, consisting of an early PKC-dependent and a late PKC-independent phase of regulation (Jacamo et al. 2008). In particular, PKD autophosphorylation on Ser^{748} was shown to be a major mechanism contributing to late PKD activation loop phosphorylation occurring in cells stimulated by GPCR agonists. This conclusion was supported by several lines of evidence: (1) Catalytic inactivation of PKD by mutation of either Lys^{618} or Asp^{733} produced PKD forms in which the late phase phosphorylation of Ser^{748} was eliminated by the treatment with PKC inhibitors, demonstrating that PKC-independent Ser^{748} phosphorylation requires PKD catalytic activity; (2) Constitutively active PKD generated by the deletion of the PH domain displayed a high level of Ser^{748} (but not of Ser^{744}) phosphorylation in unstimulated cells; (3) The PKC-independent phase of PKD activation induced by bombesin was completely blocked by the substitution of Ser^{748} for Ala. PKC-independent PKD phosphorylation on Ser^{748} is likely to require the recruitment of PKD to DAG-rich microenvironments, as judged by the fact that this phosphorylation is prevented in a PKD mutant (i.e., PKDP287G) with impaired ability to bind DAG (Jacamo et al. 2008). These new results identify a novel mechanism induced by GPCR activation that leads to PKC-independent PKD activation loop autophosphorylation.

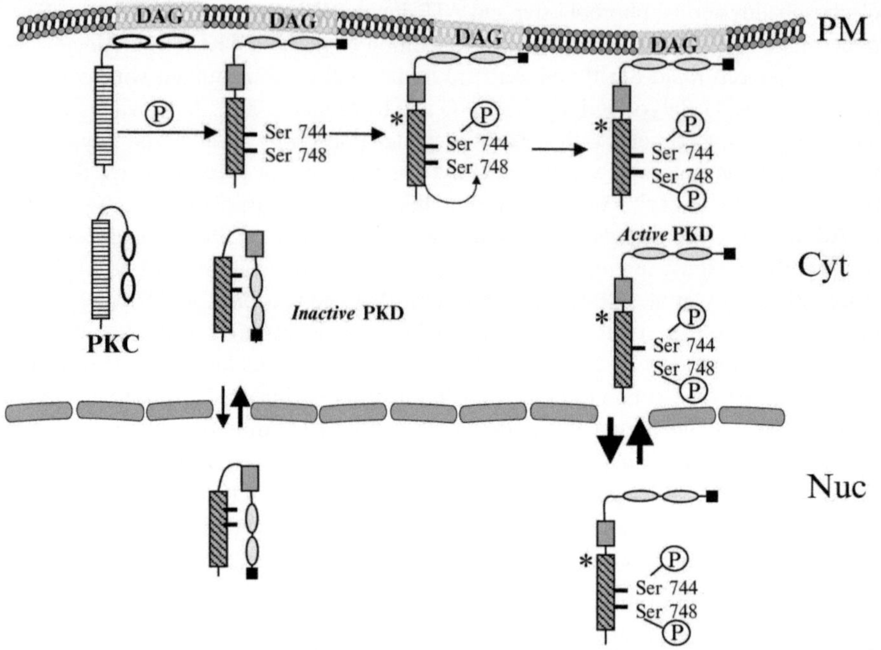

Fig. 7.2 Model of PKDs activation and intracellular distribution regulation. In unstimulated cells, inactive PKD and PKD2 are present in the cytoplasm, whereas PKD3 is continuously shuttling between the cytoplasm and the nucleus at different rates, *i.e.*, faster nuclear import than export. After cell stimulation, PLC-mediated hydrolysis of phosphatidylinositol 4,5-biphosphate (PIP_2) produces DAG at the plasma membrane, which in turn mediates the translocation of inactive PKDs from the cytosol to that cellular compartment. DAG also recruits, and simultaneously activates, novel PKCs to the plasma membrane which mediate the transphosphorylation of PKDs on Ser^{744} (in mouse PKD). DAG and PKC-mediated transphosphorylation of PKD act synergistically to promote PKD catalytic activation and autophosphorylation on Ser^{748}. Active PKDs then dissociate from the plasma membrane and migrate to the cytosol and subsequently into the nuclei while PKD3 increases its rate of nuclear import compared to non-stimulated cells. Upon cessation of agonist-induced cell stimulation, all PKDs return to their steady-state, prior to the cell stimulation. Arrows representing potential pathways leading to the inactivation of PKDs were not included for clarity purposes. Subscripts (PM, Cyt, Nuc.) denote cytosolic, plasma membrane and nuclear localization, respectively; * denotes catalytic active kinases. Arrow direction and thickness represent PKDs directionality and differential rates of transport. The scheme represents primarily the rapid activation of PKDs. In response to Gq-coupled receptors, PKD exhibits a second phase of activation that is not dependent on PKC (see details in the text). The precise mechanism responsible for the switch between PKC-dependent to PKC-independent modes of activation remains incompletely understood but in view of the agonists that trigger this phase, it likely to involve Gq

In the light of these new results, PKD emerges as a unique example of a protein kinase in which the phosphorylation of the key serines, Ser^{744} and Ser^{748}, in its activation loop is regulated by transphosphorylation and autophosphorylation mechanisms. As shown in the scheme illustrated in Fig. 7.2, transphosphorylation by PKC is a major mechanism targeting Ser^{744} and autophosphorylation is a predominant mechanism for Ser^{748}. It is important to emphasize that the pathways leading to the phosphorylation of these residues depend on the time of GPCR stimulation. For example, while PKD phosphorylation on Ser^{744} is mediated entirely by PKC transphosphorylation at early times of bombesin stimulation, a low level of PKC-independent phosphorylation of this residue could be detected consistently at longer times of bombesin stimulation. These findings suggest that, in addition to PKC, another, as yet unidentified, upstream protein kinase (insensitive to GF109203X and Gö6983) contributes to Ser^{744} phosphorylation at later times of Gq-coupled receptor stimulation.

Interestingly, phosphorylation of Ser^{748} was markedly decreased in kinase-dead mutants as well as in the PKD(S744A) mutant even at early times of bombesin stimulation. Therefore, it is conceivable that the elimination of rapid Ser^{748} phosphorylation by PKC inhibitors noted previously by Waldron et al. (Waldron et al. 2001) could be indirect, at least in part, e.g., early PKC-dependent phosphorylation of Ser^{744} could be necessary for PKD catalytic activation and subsequent autophosphorylation on Ser^{748}. In this manner, both autophosphorylation and PKC-mediated transphosphorylation could contribute to early phase Ser^{748} phosphorylation. The results of Jacamo et al. (Jacamo et al. 2008) also indicated that Ser^{748} can be phosphorylated via a PKC-dependent pathway when PKD autophosphorylation is rendered non-functional, e.g., in kinase-dead mutants or in the PKD(S744A) mutant. Collectively, these findings reveal unsuspected complexities and plasticity in the regulation of PKD phosphorylation at the activation loop and emphasize the importance of monitoring the phosphorylation of each residue of the loop at different times of agonist stimulation.

It should be mentioned that in addition to activation loop phosphorylation, the phosphorylation of other serine (Matthews et al. 1999a; Iglesias et al. 1998b) and tyrosine (Waldron et al. 2004; Storz and Toker 2003) residues are likely to play a role in PKD regulation. Nevertheless, it is clear that PKD phosphorylation at Ser^{744} and Ser^{748} is demonstrated in multiple cell types, is triggered by a vast array of stimuli and plays a critical role in PKD activation.

7.2.4 Intracellular Redistributions of PKD, PKD2 and PKD3

The translocation of signaling protein kinases to different cellular compartments is a fundamental process in the regulation of their activity. PKD is present in the cytosol of unstimulated cells (Rey et al. 2001a; Matthews et al. 2000a; Matthews et al. 1999c; Rey et al. 2001b), and to a lesser extent in several intracellular compartments, including Golgi and mitochondria (Liljedahl et al. 2001; Hausser et al. 2002) but rapidly translocates from the cytosol to different subcellular compartments in response to receptor activation, as revealed by real-time imaging of GFP-tagged PKD and immunocytochemistry of expressed or endogenous PKD (Rey et al. 2001a; Matthews et al. 2000a; Matthews et al. 1999c; Rey et al. 2001b; Rey et al. 2003c).

Each translocation step is associated to a particular PKD domain and to rapid and reversible interactions. The first step of PKD translocation is mediated by the cys2 motif of the CRD, which binds to DAG produced at the inner leaflet of the plasma membrane as a result of PLC stimulation (Rey et al. 2001a). Interestingly, it has also been reported that cys2 and flanking sequences can directly bind to activated $G\alpha_q$ (Oancea et al. 2003). In contrast, the cys1 recruits PKD to the Golgi apparatus (Maeda et al. 2001). The second step, i.e., reversible translocation from the plasma membrane to the cytosol, requires the phosphorylation of Ser^{744} and Ser^{748} within the activation loop of PKD (Rey et al. 2001a) leading to its catalytic activation (Rey et al. 2006). Active PKD is then imported, via its cys2 motif, into the nucleus, where it transiently accumulates before being exported to the cytosol through a CRM1-dependent nuclear export pathway that requires the PH domain of PKD (Rey et al. 2001b).

Antigen-receptor engagement of B cells and mast cells induces rapid translocation of PKD from the cytosol to the plasma membrane (Matthews et al. 2000a; Matthews et al. 1999b). The plasma membrane translocation promoted by antigen-receptor engagement is reversible and does not appear to involve the nuclear compartment (Matthews et al. 2000a). These different observations emphasize the notion that in addition to the structural determinants present in PKD, other factors, including cell context, stimulus and scaffolding proteins also influence its intracellular distribution. In this context, the A-kinase anchoring protein (AKAP-Lbc), which possesses Rho-specific guanine nucleotide exchange activity and is linked to $G\alpha_{12/13}$ signaling, forms a multiprotein complex that includes PKD, PKCη and PKA that facilitates PKD translocation and activation (Carnegie et al. 2004). Previous results also demonstrated that $G\alpha_{13}$ and activated Rho promote PKD activation (Yuan et al. 2003; Yuan et al. 2001). These findings support the notion that GPCRs utilize both G_q and $G_{12/13}$ pathways to induce PKD translocation and activation in their target cells. However, the elucidation of the precise contribution of different G proteins to the early and late phases of PKD activation induced by GPCR agonists requires further experimental work.

As in fibroblasts, PKD is cytosolic in unstimulated T cells, but it rapidly polarizes to the immunological synapse in response to antigen/antigen presenting cells (Spitaler et al. 2006). PKD translocation is determined by the accumulation of DAG at the immunological synapse and changes in DAG accessibility of the PKD-CRD. Unstimulated T cells have a uniform distribution of DAG at the plasma membrane, whereas after T cell activation, a gradient of DAG is created with a persistent focus of DAG at the center of the synapse. PKD is only transiently associated with the immune synapse, indicating a fine tuning of PKD responsiveness to DAG by additional regulatory mechanisms (Spitaler et al. 2006). These results reveal the immune synapse as a critical point for DAG and PKD interaction during T cell activation.

PKD2 also undergoes reversible translocation from the cytosol to the plasma membrane in response to GPCR stimulation (Rey et al. 2003a). The reversible translocation of PKD2 requires PKC activity and, as in the case of PKD, it can be prevented by inhibiting the translocation of PKCε (Rey et al. 2004). In gastric cancer AGS cells transfected with the CCK2 receptor, PKD2 has been shown to move into the membrane and subsequently to the nucleus in response to the CCK2 receptor agonist gastrin (von Blume et al. 2007). In addition to activation loop phosphorylation of Ser^{706} and Ser^{710}, PKD2 nuclear accumulation requires phosphorylation on Ser^{244} within the CRD. Casein kinase1 (CK1) δ and CK1ε have been identified as the upstream kinases that phosphorylate PKD2 on Ser^{244}, and thereby are critical for PKD2 nuclear translocation (von Blume et al. 2007).

In contrast to PKD and PKD2, PKD3 is present in both the cytoplasm and the nucleus of unstimulated cells (Rey et al. 2003b). GPCR agonists (e.g., neurotensin) and B-cell antigen receptor engagement induce rapid and reversible plasma membrane translocation of PKD3 (Rey et al. 2003b; Matthews et al. 2003). Subsequently, the rate of PKD3 entry into the nucleus is also enhanced by GPCR activation (Rey et al. 2003b). Real-time imaging of a photoactivatable green fluorescent protein fused to PKD3 revealed that point mutations that render PKD3 catalytically inactive completely prevented its nuclear accumulation (Rey et al. 2006). These results identify a novel function for the kinase activity of PKD3 in promoting its nuclear entry and suggest that the catalytic activity of PKD3 may regulate its nuclear import through autophosphorylation and/or interaction with another protein(s). Further results suggest that the short C-terminal tail of the PKDs plays a role in determining their cytoplasmic/nuclear localization (Papazyan et al. 2006). Collectively, the results imply that the nuclear localization of PKD3, and probably the transient nuclear localization of PKD and PKD2, is governed by its multiple domains. The differences in the intracellular distribution of the different PKD isoenzymes may be related to their ability to execute different functions at different subcellular locations in different cell types.

Although a substantial amount of information is available describing the intracellular distribution of the isoforms of the PKD family during interphase in a variety of cell types, much less is known about their localization during mitosis. Recent results showed that PKD isoforms are phosphorylated within their activation loop in fibroblasts and epithelial cells during mitosis (Papazyan et al. 2008). Activation loop-phosphorylated PKD, PKD2 and PKD3 were found associated with centrosomes, spindles and midbody suggesting that these activated kinases establish dynamic interactions with the mitotic apparatus (Papazyan et al. 2008). These results suggest a link between PKD isozymes and cell division, but the elucidation of the precise role of the PKD family during G_2/M requires further experimental work.

7.2.5 *A Multistep Model of PKD Localization, Phosphorylation and Catalytic Activation*

The studies discussed here suggest a sequential model of PKD activation in response to rapid generation of DAG in the plasma membrane that integrates the spatial and temporal changes in PKD localization with PKD catalytic activity and multisite phosphorylation (Rey et al. 2003b; Rey et al. 2004; Waldron et al. 2004; Waldron et al. 2001; Waldron and Rozengurt 2003; Rey et al. 2001b). The salient features of this model, illustrated in Fig. 7.2, are: (1) In non-stimulated cells, PKD is in a state of very low kinase catalytic activity maintained by the CRD and PH domains, which repress the catalytic activity of the enzyme. The steady-state distribution of inactive PKD results from nucleo-cytoplasmic shuttling in which the rate of nuclear export exceeds its rate of nuclear import (Fig. 7.2). (2) Production of DAG induces CRD-mediated PKD translocation from the cytosol to the plasma membrane, where novel PKCs are also recruited in response to DAG generation. (3) Novel PKCs, allosterically activated by DAG, rapidly transphosphorylate PKD at Ser^{744} which synergizes with DAG in the membrane to stimulate catalytic activation of PKD that then autophosphorylates on Ser^{748}. The phosphorylation of both residues, Ser^{744} and Ser^{748}, stabilizes the activation loop of the PKDs in their active conformation. (4) The phosphorylated and activated PKD dissociates from the plasma membrane and moves to the cytosol and subsequently into the nucleus. Thus, PKD phosphorylation on Ser^{744} and Ser^{748} followed by autophosphorylation on other sites, including Ser^{916}, promotes rapid dissociation of PKD from the plasma membrane into the interior of the cell, including the nucleus (Rey et al. 2006), where it propagates DAG-PKC signals initiated at the cell surface. While this model explains the *rapid* activation of PKD triggered by many stimuli, it should be pointed out that cell stimulation with Gq-coupled agonists initiates a *late* phase of PKC-independent PKD activation that appears driven by PKD autophosphorylation on Ser^{748} and by PKC-independent transphosphorylation on Ser^{744} (Jacamo et al. 2008). More experimental work is needed to define the precise mechanism by which Gq-coupled receptor agonists induce the second phase of PKD activation.

Recent results suggest that a similar model could explain the regulation of the catalytic activity and intracellular distribution of PKD2 and PKD3 in response to agonist-induced DAG generation (Fig. 7.2). In the framework of this model, the steady-state distribution of inactive PKD, PKD2 and PKD3 in the cytosol and nucleus results from the rates of nuclear import and nuclear export (Rey et al. 2001b; Rey et al. 2003a, b). As discussed above for PKD, we envisage that the production of DAG in the plasma membrane triggers changes in localization, phosphorylation and catalytic activation of PKD2 and PKD3, as presented in Fig. 7.2. In this way, a similar mechanism of PKD family activation can potentially generate diverse physiological responses based on the differential distribution of each isoform.

7.3 PKD Function

The multistep model of activation shown in Fig. 7.2 suggests that the PKDs are well positioned to regulate membrane, cytoplasmic and nuclear events. Indeed, it is emerging that the PKDs are implicated in the regulation of a remarkable array of fundamental biological processes, including cell proliferation, survival, polarity, migration and differentiation, membrane trafficking, inflammation and cancer.

7.3.1 PKD in the Regulation of Cell Proliferation

PKD can be activated by multiple growth-promoting GPCR agonists acting through Gq, Gi and G_{12} in a variety of cell types, suggesting that PKD functions in mediating mitogenic signaling (Rozengurt et al. 2005). Indeed, overexpression of either PKD or PKD2 strikingly potentiates the stimulation of DNA synthesis and cell proliferation induced by Gq-coupled receptor agonists in Swiss 3T3 cells (Zhukova et al. 2001a; Sinnett-Smith et al. 2004; Sinnett-Smith et al. 2007). In contrast, overexpression of PKD mutants lacking catalytic activity, failed to promote any enhancement of GPCR-induced mitogenesis. These results indicate that PKD activation plays a critical role in GPCR mitogenic signaling.

A key pathway involved in mitogenic signaling induced by GPCRs is the extracellular-regulated protein kinase (ERK) cascade (Johnson and Lapadat 2002; Rozengurt 1998; Meloche and Pouyssegur 2007). The duration and intensity of ERK pathway activation are of critical importance for determining specific biological outcomes, including proliferation, differentiation and transformation (Marshall 1995; Pouyssegur and Lenormand 2003). ERK signal duration is sensed by the cells through the protein products of immediate early genes, including c-Fos (Murphy et al. 2002; Murphy et al. 2004). When ERK activation is transient, its activity declines before the c-Fos protein accumulates and c-Fos is degraded rapidly. However, when ERK signaling is sustained, c-Fos is phosphorylated by ERK and RSK and its stability is dramatically increased, thereby leading to its accumulation (Sinnett-Smith et al. 2004; Murphy et al. 2002; Murphy et al. 2004).

PKD alters the relative activities of the JNK and ERK pathways, attenuating JNK activation and c-Jun phosphorylation in response to EGF receptor activation (Bagowski et al. 1999; Hurd and Rozengurt 2001) while stimulating the ERK pathway (Sinnett-Smith et al. 2004; Brandlin et al. 2002; Wang et al. 2002; Hurd et al. 2002). For example, the stimulatory effect of PKD on GPCR-induced cell proliferation (Zhukova et al. 2001a) has been linked to its ability to increase the duration of the MEK/ERK/RSK pathway leading to accumulation of immediate gene products, including c-Fos, that stimulate cell cycle progression (Sinnett-Smith et al. 2004).

Although the immediate downstream targets of the PKDs necessary for the transmission of its mitogenic signal have not been fully identified, putative substrates are beginning to emerge. Recently, a number of scaffolding proteins and endogenous inhibitors have been implicated in the regulation of the intensity and duration of the ERK pathway (Kolch 2005). Modeling of the ERK pathway indicates that scaffolds regulate the intensity of pathway activation, whereas inhibitors modulate its duration in response to stimuli (Ebisuya et al. 2005). The activity and subcellular localization of these proteins are also regulated by phosphorylation,

Table 7.1. Identified substrates of the PKD family

PKD Substrate	Residue	Target Sequence −5 −3 0
CREB (human)	**S133**	E I **L** S **R** R P **S** Y R K
DLC-1 (human)	**S327**	S P **V** T **R** T R **S** L S A
HDAC5 (human)	**S498**	R P **L** S **R** T Q **S** S P L
HDAC7 (human)	**S155**	F P **L** R **K** T V **S** E P N
HDAC7 (human)	**S358**	W P **L** S **R** T R **S** E P L
HDAC7 (human)	**S486**	R P **L** S **R** A Q **S** S P A
HPK1 (human)	**S171**	A T **L** A **R** R L **S** F I G
HSP27 (human)	**S82**	R A **L** S **R** Q L **S** S G V
Kidins220 (rat)	**S918**	R T **I** T **R** Q M **S** F D L
Par-1 (human)	**S400**	H K **V** Q **R** S V **S** A N P
Rin1 (human)	**S351**	R P **L** L **R** S M **S** A A F
TNNI (rat)	**S23**	A P **V** R **R** R S **S** A N Y
TLR5 (human)	**S805**	Y Q **L** M **K** H Q **S** I R G
TRPV1 (rat)	**S116**	P R **L** Y D R R **S** I F D

Note that in the majority of cases, PKD phosphorylates a serine surrounded by a sequence characterized by L/V/I at position -5, R/K at position -3. Less strict requirements are seen at other positions.

thereby offering potential new mechanisms for controlling the Raf/MEK/ERK pathway. PKD has been shown to phosphorylate RIN1 (see Table 7.1 for the sequence surrounding the PKD phosphorylation site), a multidomain protein that interferes with the interaction between Ras and Raf, and thereby inhibits ERK activation in its unphosphorylated form (Wang et al. 2002). The phosphorylation of RIN1 at Ser^{351} by PKD induces binding of 14-3-3 proteins that confine RIN1 to the cytosol, thereby preventing it from inhibiting the stimulatory interaction between Ras and Raf-1 (Wang et al. 2002). Although the expression of RIN1 is tissue-specific and therefore unlikely to provide a complete explanation for the effects of PKD on ERK duration in fibroblasts, the mechanism is reminiscent of PKC-mediated phosphorylation of *R*af *K*inase *I*nhibitor *P*rotein (RKIP), a protein that in its unphosphorylated state inhibits Raf-mediated phosphorylation of MEK (Santos et al. 2007). It is plausible that the PKC/PKD pathway phosphorylates and sequentially inactivates different inhibitors of Raf-1 leading to fine regulation of the duration of the active state of the ERK pathway.

7.3.2 Role of PKD in VEGF-induced Endothelial Cell Migration and Proliferation

Recent studies implicated PKD signaling in ERK activation and DNA synthesis in endothelial cells stimulated by vascular endothelial growth factor (VEGF), which is essential for many angiogenic processes both in normal and abnormal conditions

(Wong and Jin 2005). In addition to stimulate activation loop Ser^{744} and Ser^{748} phosphorylation, VEGF, acting via the KDR receptor, also induces PKD phosphorylation on Tyr^{463} (Qin et al. 2006).

Regulation of chromatin accessibility by acetylation/deacetylation of nucleosomal histones is a key mechanism used to modulate gene expression. Class II histone deacetylases (HDACs), including HADCS 5 and 7, regulate chromatin structure by interacting with various transcription factors to repress their transcriptional activity. PKD-mediated phosphorylation of specific residues in class II HDACS (Table 7.1) leads to association with 14-3-3 chaperone proteins, thereby regulating their intracellular distribution in a variety of cell types. Sequestration of HDACs in the cytoplasm presumably relieves target genes from HDAC repressive actions, thereby facilitating gene expression. HDAC7 has been implicated in the regulation of endothelial cells morphology, migration and capacity to form capillary tube-like structures in vitro (Mottet et al. 2007). Treatment of endothelial cells with PMA or VEGF resulted in the exit of HDAC7 from the nucleus through a PKC/PKD pathway (Mottet et al. 2007; Ha et al. 2008a). Further studies indicate that VEGF also stimulates PKD-dependent phosphorylation of HDAC5 at $Ser^{259/498}$ residues, which leads to HDAC5 nuclear exclusion and transcriptional activation (Ha et al. 2008b). It is conceivable that the complex program of gene expression and migration triggered by VEGF in endothelial cells leading to angiogenesis is orchestrated by PKD-mediated phosphorylation of both HADC5 and HDAC7, leading to their nuclear extrusion in these cells. Indeed, it has been recently proposed that PKD is one of the most attractive targets for anti-angiogenic therapies (Altschmied and Haendeler 2008).

7.3.3 PKD and Regulation of Cell Trafficking and Secretion

PKD regulates the budding of secretory vesicles from the trans-Golgi network (Liljedahl et al. 2001; Yeaman et al. 2004). Specifically, inactivation of PKD (e.g., by expression of kinase-deficient mutants of PKD) blocks fission of trans-Golgi network (TGN) transport carriers, inducing the appearance of long tubules filled with cargo. At the TGN, active PKD and PKD2 phosphorylate phosphatidylinositol 4-kinase IIIβ (PI4KIIIβ), a key player required for fission of TGN-to-plasma membrane carriers (Hausser et al. 2005). PI4KIIIb is recruited to the TGN membrane by the small GTPase ARF, and activated by PKD-mediated phosphorylation to generate PI(4)P, which then recruits the machinery that is required for carrier fission (Ghanekar and Lowe 2005).

This process has been implicated in fibroblast locomotion and localized Rac1-dependent leading edge activity (Prigozhina and Waterman-Storer 2004). In agreement with an important role in cell trafficking and motility, PKD also promotes integrin recruitment to newly formed focal adhesions (Woods et al. 2004) and invasiveness of cancer cells (Qiang et al. 2004; Bowden et al. 1999).

Several studies indicate an important role of PKD in secretion in a number of endocrine cell types. PKD has been shown to stimulate the secretion of the gastrointestinal peptide neurotensin (NT) in the human endocrine cell line BON (Li et al. 2004). Further studies determined that the PKD protein substrate Kidins220, [kinase D-interacting substrate of 220 kDa (Iglesias et al. 2000b; Cabrera-Poch et al. 2004) and Table 7.1] mediates NT secretion (Li et al. 2008). Interestingly, the PKD/Kidins220 pathway appears to function downstream of PKD-induced fission of TGN carriers, suggesting that PKD regulates different steps of cell secretion. In addition to PKD, PKD2 has been shown to regulate chromogranin release in BON cells (von Wichert et al. 2008). Other studies indicate that PKDs play a critical role in regulating angiotensin II-mediated cortisol and aldosterone secretion from H295R cells, a human adrenocortical cell line (Romero et al. 2006; Chang et al. 2007). Recent studies using mice deficient in p38δ reveal a novel p38δ-PKD pathway that regulates insulin secretion and survival of pancreatic β cells, suggesting a critical role for PKD in the development of diabetes mellitus (Sumara et al. 2009).

7.3.4 PKD and Neuronal and Epithelial Cell Polarity

Establishing and maintaining cellular polarity is of fundamental importance for the functions of a variety of cell types, including neuronal and epithelial cells. Early neurons develop initial polarity by mechanisms analogous to those used by migrating cells. In line with this notion, PKDs has been shown to play a role in neuronal protein trafficking. In these cells, PKD and PKD2 regulate TGN-derived sorting of dendritic proteins and axon formation, and hence have a role in establishing neuronal polarity (Bisbal et al. 2008; Yin et al. 2008).

In polarized epithelial cells, PKD and PKD2, but not PKD3, specifically regulate the production of TGN carriers destined to the basolateral membrane rather than to the apical membrane and consequently, PKD and PKD2 may play an important role in the generation of epithelial polarity (Yeaman et al. 2004). Another major mechanism involved in establishing cell polarity is mediated by the evolutionary conserved PAR (partitioning-defective) genes (Suzuki and Ohno 2006). The Par-3/Par6/aPKC complex is located at tight junctions, whereas Par-1, a protein kinase, is found in lateral membranes. There is an antagonistic interaction between the Par-3/Par6/aPKC complex and Par-1 mediated by phosphorylation of specific residues that form binding sites for 14-3-3 proteins. Par-1 kinase, activated by mammalian Par-4/LKB1 by phosphorylation of its activation loop, phosphorylates Par-3, thereby destabilizing the complex and removing it from lateral membranes, whereas Par-3/Par6/aPKC phosphorylates Par-1 (on Thr^{595}) to dissociate it from apical plasma membranes (Suzuki and Ohno 2006). Treatment of cells with phorbol-12-myristate-13-acetate (PMA) induced PKD-mediated phosphorylation of Par-1 on a residue (Ser^{400}; see Table 7.1) that promotes Par-1 binding to 14-3-3, thereby promoting its dissociation from the plasma membrane and inhibiting its activity (Watkins et al. 2008). Although these results suggest that PKD may play a role in regulating cell polarity via phosphorylation of Par-1, additional experiments using physiological stimuli rather than PMA are necessary to substantiate this important hypothesis.

7.3.5 *PKD and Regulation of Lymphocyte Function*

A prominent PKC/PKD axis has been demonstrated in B and T lymphocytes (Sidorenko et al. 1996; Matthews et al. 2000a, b; Matthews et al. 1999a). As in other cell types, a recently proposed function for PKD in lymphocytes is the phosphorylation and regulation of class II HDACs (Vega et al. 2004; Dequiedt et al. 2005; Parra et al. 2005). PKD has also been implicated in regulating the functional activity of β1 integrins in T cells via Rap1 (Medeiros et al. 2005). Upon T-cell receptor engagement, PKD stimulated hematopoietic progenitor kinase 1 (HPK1) activity in Jurkat T cells and enhanced HPK1-driven SAPK/JNK and NF-κB activation (Arnold et al. 2005). However, other investigators found that PKD2 is the predominant PKD isoform in T cells (Irie et al. 2006). Expression of PKD2 enhanced interleukin (IL)-2 promoter activity upon stimulation with anti-CD3 mAb in Jurkat T cells, suggesting that PKD2 is involved in IL-2 promoter regulation in response to TCR stimulation (Irie et al. 2006). Using avian DT40 B-cell line lacking PKD and PKD3, Liu et al. concluded that these PKDs are dispensable for proliferation, survival responses and NF-κB transcriptional activity downstream of the B cell antigen receptor (Liu et al. 2007). Apparently, the role of PKD2 was not determined in this system.

The activation of PKD by antigen receptors is a sustained response associated with changes in PKD intracellular location (see above). The function of PKD at these different locations has been probed in an in vivo model using active PKD mutants targeted to either the plasma membrane or the cytosol of pre-T cells of transgenic mice (Marklund et al. 2003). Studies of these mice have shown that PKD can substitute for the pre-T cell receptor and induce both proliferation and differentiation of T cell progenitors in the thymus. Moreover, cellular localization of PKD within a thymocyte is critical; membrane-targeted and cytosolic PKD thus control different facets of pre-T cell differentiation (Marklund et al. 2003). Subsequent studies probed the Rho requirements for the actions of constitutively active PKD mutants localized at the plasma membrane or the cytosol in pre-T cells of transgenic mice. Membrane-localized PKD regulation of pre-T cell differentiation was shown to be Rho-dependent, but the actions of cytosol-localized PKD were not (Mullin et al. 2006). These studies demonstrated that links between PKD and Rho appear to be determined by the cellular location of PKD in T lymphocytes.

7.3.6 *Role of PKD Upstream and Downstream of Toll-like Receptors*

Toll-like receptors (TLRs) have been identified as the primary innate immune receptors. TLRs distinguish between different patterns of pathogens and activate a rapid innate immune response. Recent results implicated PKD in TLR 2, 5 and 9 function in different cell types (Park et al. 2008; Ivison et al. 2007). TLR5, a receptor for bacterial flagellin, is expressed highly in the intestinal mucosa, a major site of exchange with the external environment (Rozengurt and Sternini 2007), and home to an enormous population of bacteria. Ivison et al. suggested that phosphorylation of TLR5 by PKD on Ser^{805} (Table 7.1) may be a necessary proximal event in the response of TLR5 to flagellin, and that this phosphorylation contributes to p38MAPK activation and production of inflammatory cytokines in epithelial cells (Ivison et al. 2007).

Subsequent studies provided evidence that PKD is a downstream target in TLR9 signaling in macrophages (Park et al. 2008) and TLR2 in mouse bone marrow-derived mast cells (Murphy et al. 2007). PKD has been proposed to mediate the increase in expression and release of the chemokine CCL2 (MCP-1) from mast cells (Murphy et al. 2007). Although the precise role of PKD in TLR function remains incompletely understood, these studies provide evidence suggesting that PKD plays a role in the regulation of the innate immune response mediated by this class of pattern recognition receptors.

7.3.7 PKD and Osteoblast Differentiation

Bone morphogenetic proteins (BMPs) are multifunctional growth factors that belong to the transforming growth factor beta (TGFβ) superfamily. BMPs bind to receptor complexes that stimulate multiple intracellular pathways, including the SMADS, leading to a wide range of biological effects in different tissues. In particular, they contribute to the formation of bone and connective tissues by inducing the differentiation of mesenchymal cells into bone-forming cells. Recent studies demonstrated that BMP-2 induces PKD activation through a PKC-independent pathway during osteoblast lineage progression (Lemonnier et al. 2004) and that PKD is required for the effects of BMP-2 on osteoblast differentiation (Celil and Campbell 2005).

More recent studies explored the mechanism of action of the BMP-2/PKD pathway. Runx is a master transcriptional regulator of skeletal biology that plays a critical role in bone cell growth and differentiation, as well as in the structural and functional integrity of skeletal tissue (Stein et al. 2004). Interestingly, HDAC7 associates and represses the activity of Runx2 (Jensen et al. 2008). Further studies demonstrated that BMP-2 induces export of HDAC7 from the nucleus in mesenchymal cells that require Crm1-mediated nuclear export and are associated with increased HDAC7 serine phosphorylation and 14-3-3 binding (Jensen et al. 2009). PKD was shown to form a molecular complex with HDAC7 in a BMP2-enhanced manner, and a constitutively active form of PKD stimulated HDAC7 nuclear export. An important finding was that active PKD inhibited repression of Runx2-mediated transcription by HDAC7 (Jensen et al. 2009). Although other pathways may be involved, these results establish a mechanism by which BMP-2 signaling regulates Runx2 activity via PKD-dependent inhibition of HDAC7 transcriptional repression. The elucidation of the precise mechanism by which BMP-2 induces PKD activation requires further experimental work.

7.3.8 PKD and Heat Shock Proteins

The small heat shock proteins (Hsps), including human Hsp27 and mouse Hsp25, play an important role in the regulation of many cellular functions in response to stress, cytokines, growth factors and GPCR agonists. The level of Hsp27 is markedly increased in many cancer cells and its expression contributes to the malignant properties of these cells, including chemoresistance (Rocchi et al. 2004; Chen et al. 2004; Shin et al. 2005; Xu and Bergan 2006; McCollum et al. 2006; Bruey et al. 2000; Paul et al. 2002; Benn et al. 2002). Many of the functions attributed to Hsp27 require its phosphorylation, especially at Ser-82 (Guay et al. 1997; Geum et al. 2002; Kubisch et al. 2004; Berkowitz et al. 2006; Zheng et al. 2006), a consensus site for PKD-mediated phosphorylation (Table 7.1). Although it is widely recognized that Hsp27 is a substrate of the p38 MAPK/MK2 cascade (Chang and Karin 2001; Widmann et al. 1999), other studies demonstrated that phorbol esters also stimulate the phosphorylation of Hsp27 via a PKC-dependent but p38/MK2-independent pathway (Maizels et al. 1998). However, it has remained unclear whether PKCs directly phosphorylate Hsp27. PKD has been implicated in the phosphorylation of Hsp27 on Ser^{82} in HeLa cells exposed to oxidative stress (Doppler et al. 2005), a condition previously shown to activate PKD (Storz et al. 2004a; Waldron and Rozengurt 2000; Waldron et al. 2004) as well as the p38 MAPK/MK2 cascade. The relative contribution to Hsp27 phosphorylation of these parallel pathways was not evaluated. Human pancreatic cancer PANC-1 cells, which endogenously express PKD and PKD2 (Rey et al. 2003a, b), also express high levels of Hsp27. Knockdown of both PKD and PKD2 virtually abolished neurotensin-induced $Hsp27Ser^{82}$ phosphorylation in PANC-1 cells treated with SB 202190 to eliminate the p38MAPK/MK-2 pathway (Yuan and Rozengurt 2008). These results demonstrate that neurotensin induces Hsp27 phosphorylation on Ser^{82} via simultaneous operation of at least two separate pathways in PANC-1 cells and members of the PKD family play a critical role in mediating one of the pathways. PKD and PKD3 are also required to regulate Hsp27 phosphorylation in DT40 B-cells (Liu et al. 2007). Thus, PKDs function as upstream kinases for Hsp27 in a variety of cell types, in some cases functioning in conjunction with the p38 MAP kinase pathway.

7.3.9 PKD and Pain Transmission via TRPV1

Activation of TRPV1 in response to capsaicin and endogenous ligands, including endocannabinoids or DAG (Woo et al. 2008), leads to Ca^{2+} influx, resulting in membrane depolarization leading to the release of proinflammatory neuropeptides from primary afferent nerve terminals. TRPV1 can be sensitized by several endogenous mediators present in inflammatory conditions, including bradykinin (Tang et al. 2004), ATP (Moriyama et al. 2003), proteases and chemokines

(Zhang et al. 2005b). These agonists are known to bind to Gq-coupled receptors that promote PLC-mediated activation of the PKC/PKD axis. Oxidative stress which activates TRPV1 (Ruan et al. 2005) also leads to PKD activation (Waldron and Rozengurt 2000; Waldron et al. 2004). TRPV1 phosphorylation of several aminoacid residues, including Ser^{117}, Thr^{371}, Ser^{502} and Ser^{801}, (in the human molecule) are known to sensitize the channel to capsaicin, protons and heat (Mohapatra and Nau 2005). PKC has been implicated as one of the upstream kinases (Premkumar and Ahern 2000) and as shown in Table 7.1, Ser^{117} has been identified as a target for PKD (Wang et al. 2004). The other phosphorylation sites in TRPV1 are also consensus target sites for PKDs but the role of these kinases in their phosphorylation has not been investigated. It is plausible that PKC and PKC via PKD sensitize TRPV1 to protons, heat, capsaicin and endogenous ligands by multisite phosphorylation, thus involving the PKC/PKD axis in sensitization to neurogenic inflammation.

7.3.10 PKDs, Inflammation and Oxidative Stress

NK-kB is a key transcription factor that is activated by multiple receptors and regulates the expression of a wide variety of proteins that control innate and adaptive immunity. A number of studies indicate that PKD is a mediator of NF-kB induction in a variety of cells exposed to GPCR agonists or oxidative stress (Storz et al. 2004a; Mihailovic et al. 2004; Storz and Toker 2003; Storz et al. 2004b; Chiu et al. 2007; Song et al. 2009). In view of the increasing recognition of the interplay between inflammation and cancer development, a possible role of PKD in linking these processes is of importance. However, the precise molecular mechanisms remain incompletely understood.

Stimulation of human colonic epithelial NCM460 cells with the GPCR agonist and bioactive lipid lysophosphatidic acid (LPA) led to a rapid and striking activation of PKD2, the major isoform of the PKD family expressed by these cells (Chiu et al. 2007). LPA induced a striking increase in the production of interleukin 8 (IL-8), a potent pro-inflammatory chemokine, and stimulated NF-kB activation. PKD2 gene silencing utilizing small interfering RNAs dramatically reduced LPA-stimulated NF-kB promoter activity and IL-8 production. These results imply that PKD2 mediates LPA-stimulated IL-8 secretion in NCM460 cells through a NF-kB-dependent pathway. PKD2 has also been implicated in mediating NF-κB activation by Bcr-Abl in myeloid leukemia cells (Mihailovic et al. 2004).

NF-κB also plays a critical role in inflammatory and cell death responses during acute pancreatitis. Previous studies demonstrated that the PKC isoforms PKCδ and ε are key regulators of NF-κB activation induced by cholecystokinin-8 (CCK-8), an agonist that induces pancreatitis when administered to rodents at supramaximal doses. PKD has been shown to function as a key downstream target of PKCδ and PKCε in pancreatic acinar cells stimulated by CCK-8 or the cholinergic agonist carbachol (CCh). Furthermore, PKD was necessary for NF-κB activation induced by these GPCR agonists (Yuan et al. 2008). The kinetics of PKD and NF-κB activation during rat pancreatitis showed that both PKD and NF-κB activation were early events during acute pancreatitis and that their time courses of response in vivo were similar (Yuan et al. 2008). These results identify PKD as a novel early point of convergence in the signaling pathways mediating NF-κB activation in pancreatitis, a condition that predisposes to pancreatic cancer.

Since the original finding that oxidative stress induces PKD activation, partly via PKC-mediated activation loop phosphorylation and partly through Src-mediated PKD tyrosine phosphorylation (Waldron and Rozengurt 2000), a number of reports confirmed that PKD is a sensor of oxidative stress (Storz et al. 2004a; Waldron et al. 2004; Storz and Toker 2003; Storz et al. 2004b; Sumara et al. 2009; Song et al. 2009; Storz et al. 2005; Doppler and Storz 2007). Recently, Tyr^{95} in PKD has been identified as a phosphorylation site that is regulated by oxidative stress and generates a binding motif for PKCδ. Oxidative stress-mediated PKCδ/PKD interaction results in PKD activation loop phosphorylation on Ser^{744} and Ser^{748} leading to catalytic activation (Doppler and Storz 2007). A number of studies have shown that PKD opposes the apoptotic effects of oxidative stress in a variety of cells (Sumara et al. 2009; Song et al. 2009; Storz et al. 2005; Singh and Czaja 2007; Storz 2007; Song et al. 2006).

A recent study using pancreatic β cells, demonstrated that stress signals markedly induced TNFAIP3/A20, a zinc finger-containing, immediate-early-response gene with potent antiapoptotic and anti-inflammatory functions (Lee et al. 2000). In fact, A20 is an early NF-κB-responsive gene that encodes a ubiquitin-editing protein that is involved in the negative feedback regulation of NF-κB signaling (Coornaert et al. 2009). Interestingly, other studies demonstrated that PKD induces A20 promoter activity (Liuwantara et al. 2006). It is plausible that PKD initiates not only an inflammatory response via NF-κB, but also stimulates expression of the antiapoptotic and anti-inflammatory A20, as a feedback mechanism that protects cells subject to stress signals, including oxidative stress.

7.3.11 PKD and Cardiac Hypertrophy

Several years after its identification, PKD was shown to be expressed and regulated in ventricular myocytes (Haworth et al. 2000). Treatment of these cells with either PMA or an alpha1-adrenergic receptor (AR) agonist induced rapid PKD activation through PKC-mediated pathways (Haworth et al. 2000). Subsequent studies demonstrated that PKD is implicated as a mediator of cardiac hypertrophy, a condition associated with elevated risk for the development of heart failure, and clarified the mechanism by which PKD exerts such profound influence in the heart (Avkiran et al. 2008). As mentioned above, class II HDACs are direct substrates for PKDs. Vega et al. demonstrated that PKD directly phosphorylates class II HDAC5 (Table 7.1), an enzyme that induces chromatin modifications and suppresses cardiac hypertrophy (Vega et al. 2004).

PKD-mediated phosphorylation of HDCA5 neutralizes its ability to suppress cardiac hypertrophy by triggering CRM1-dependent nuclear export (Vega et al. 2004; Sucharov et al. 2006). The A-Kinase Anchoring Protein (AKAP)-Lbc, which is upregulated in hypertrophic cardiomyocytes, has been proposed to couple PKD activation with the phosphorylation-dependent nuclear export of HDAC5 (Carnegie et al. 2008). In turn, other studies demonstrated that increased myocardial PKD activity induces cardiac troponin I (TNNI) phosphorylation at $Ser^{22/23}$ (Table 7.1) and reduces myofilament Ca^{2+} sensitivity, suggesting that altered PKD activity in disease may impact on contractile function (Cuello et al. 2007). Recent studies demonstrated that mice with cardiac-specific deletion of PKD were viable and showed diminished hypertrophy, fibrosis and fetal gene activation as well as improved cardiac function in response to pressure overload or chronic adrenergic and angiotensin II signaling (Fielitz et al. 2008a).

The cAMP-response element-binding protein (CREB) is activated by phosphorylation on Ser^{133}, and thereby plays a key role in the proliferative and survival responses of a variety of cell types in response to many growth regulatory stimuli. CREB is phosphorylated on Ser^{133} by a number of upstream kinases, but a recent study identified PKD as a cardiac CREB-Ser^{133} kinase (Table 7.1) that can contribute to cardiac remodeling (Ozgen et al. 2008). Additional studies implicate PKD in the altered energy metabolism observed in the diabetic heart (Kim et al. 2008). These studies provide strong support to the notion that PKD functions as a key transducer of stress stimuli involved in pathological cardiac remodeling in vivo.

7.3.12 PKD, Cancer Cell Proliferation and Invasion

Given the widespread role of PKDs in signal transduction, migration, gene expression and proliferation, it is not surprising that PKD signaling has been implicated in a variety of cancer cells, including those originated from the lung, the digestive system, breast and prostate.

GPCR and their ligands have been implicated as autocrine growth factors for small cell lung cancer (SCLC). Earlier results demonstrated the existence of a PKC/PKD pathway in SCLC cell lines and raised the possibility that PKD may be an important mediator of some of the biological responses elicited by PKC activation in SCLC cells (Paolucci and Rozengurt 1999).

The GPCR agonist neurotensin induces PKC-dependent PKD activation (Guha et al. 2002) and translocation (Rey et al. 2003b) and acts as potent growth factor for pancreatic cancer cell lines, including PANC-1 (Ehlers et al. 2000; Ryder et al. 2001; Kisfalvi et al. 2005). As mentioned above, downstream targets of PKD include Hsp27 which contributes to gemcitabine resistance in pancreatic cancer cells (Mori-Iwamoto et al. 2007). Interestingly, PKD is upregulated in pancreatic adenocarcinoma cell lines highly resistant to chemotherapeutic drugs (Trauzold et al. 2003). Preliminary results from our laboratory show that PKD overexpression in PANC-1 cells increases DNA synthesis, cell proliferation and anchorage-independent proliferation (K. Kisvalvi and E. Rozengurt, unpublished results).

A number reports using transgenic or mutant mice have raised the possibility that the PKC/PKD axis plays an important role in the regulation of intestinal epithelial cell proliferation in vivo. For example, mice with transgenic overexpression PKCβII in the intestinal epithelium exhibit hyperproliferation, increased Wnt signaling and an increased susceptibility to azoxymethane-induced preneoplastic lesions in the colon (Murray et al. 1999; Yu et al. 2003). Since PKCβ overexpression stimulates PKD catalytic activation (Zugaza et al. 1996), it is conceivable that PKD mediates, at least in part, the growth-promoting effects of this PKC isoform. Furthermore, in the APC^{min} mouse model of colon cancer, PKD (but not any of the PKCs) exhibited dysplasia-specific nuclear localization, suggesting that this enzyme is activated during adenomatous transformation (Klein et al. 2000). Indeed, our studies demonstrated that active PKD accumulates in the nucleus in response to GPCR stimulation (see above). To clarify the role of PKD in intestinal epithelial cell proliferation in vivo, we generated transgenic mice that express elevated PKD protein in the distal small intestinal and proximal colonic epithelium. We found a significant increase in DNA synthesizing cells in the crypts of the PKD transgenic mice as compared with non-transgenic littermates (our unpublished results). In view of recent results showing that PKD can phosphorylate Par1 (Watkins et al. 2008), it is plausible that PKD plays a role in modulating the polarity and proliferation of epithelial cells in the gut.

Invasive breast cancer cells have the ability to extend membrane protrusions, invadopodia, into the extracellular matrix. These structures are associated with sites of active matrix degradation. The amount of matrix degradation associated with the activity of these membrane protrusions has been shown to directly correlate with invasive potential. PKD has been implicated in the regulation of these structures (Bowden et al. 1999). Indeed, PKD, cortactin and paxillin were co-immunoprecipitated as a complex from invadopodia-enriched membranes. This complex of proteins was not detected in lysates from non-invasive cells that do not form invadopodia (Bowden et al. 1999). These data suggested that the formation of this PKD-containing complex correlates with cellular invasiveness. Increased cellular adhesion to extracellular matrix proteins, such as collagen type IV, often increases metastatic potential. Stimulation of adhesion of human metastatic breast carcinoma cells to collagen type IV in response to arachidonic acid is associated with the activation and translocation of PKD from the cytoplasm to the membrane (Kennett et al. 2004). Additional results indicate that PKD is necessary for the increased adhesion promoted by arachidonic acid. These studies suggest that PKD is an important element in breast tumor cell adhesion and metastasis. However, a recent study using breast cancer tissues as well as cell lines reached a different conclusion (Eiseler et al. 2009). Specifically, loss of PKD expression appears to increase the malignant potential of breast cancer cells. This may be due to the function of PKD as a negative regulator of matrix-metalloproteinases expression (Eiseler et al. 2009). These results suggest that decreased PKD expression may be a marker for invasive breast cancer.

Clearly, more experimental work is needed to define the role of PKD in breast cancer, in particular to elucidate whether PKD could have different roles at different stages of the disease.

An increasing amount of evidence indicates that *DLC1* (*d*eleted in *l*iver *c*ancer), a negative regulator of Rho, is a tumor suppressor gene deleted almost as frequently as p53 in common cancers such as breast, colon and lung (Lahoz and Hall 2008). Recent results show that phorbol ester-induced activation of the PKC/PKD axis stimulates the association of DLC1 with 14-3-3 proteins (Scholz et al. 2009) via phosphorylation involving Ser^{327} and Ser^{431} (Table 7.1). Association with 14-3-3 proteins inhibits DLC1 GAP activity, and thus facilitates signaling by active Rho. The binding to 14-3-3 proteins induced by PKD-mediated phosphorylation is thus a newly discovered mechanism by which DLC1 activity and localization is regulated and compartmentalized. DLC1 is one of 67 Rho GAPs, many of which can act on Rho and are ubiquitously expressed, suggesting that each GAP may make a unique contribution to regulating Rho activity (Lahoz and Hall 2008). This could reflect their different spatial locations, emphasizing a potential important role of PKDs in regulating DLC1. The neutralization of DLC1 function by PKD phosphorylation could represent another mechanism by which PKDs could contribute to the phenotypic transformation of cancer cells.

It is interesting that activated Rho has been implicated in promoting PKD catalytic activation (Yuan et al. 2003; Yuan et al. 2001). If additional results confirm that PKD regulates DLC1 activity, it is possible to envisage an amplification loop involving Rho/PKD/DLC1/Rho. In any case, it will be important to determine whether PKDs regulate DLC1 association with 14-3-3 proteins and its localization in response to receptor-mediated stimuli rather than phorbol esters in a variety of cell types.

7.4 Concluding Remarks

A great deal of progress has been made in understanding the regulatory mechanisms of activation and subcellular localization of PKD and the role of novel PKCs in mediating rapid PKD phosphorylation at the activation loop. As in other phosphorylation cascades, inducible activation loop phosphorylation provides a mechanism of signal integration and amplification. Interestingly, new results uncovered that the regulation of the activation loop phosphorylation of PKD is more complex than previously thought, with the participation of different mechanism at different times, especially in cells stimulated by Gq-coupled receptor agonists.

Recent advances demonstrate an important role of the PKDs in an array of fundamental biological processes, including cell proliferation, motility, polarity, balance of MAP kinase pathways, cardiac hypertrophy, pain transmission, inflammation and cancer. The involvement of PKDs in mediating such a diverse array of normal and abnormal biological activities in different subcellular compartments is likely to depend on the dynamic changes in their spatial and temporal localization, com-

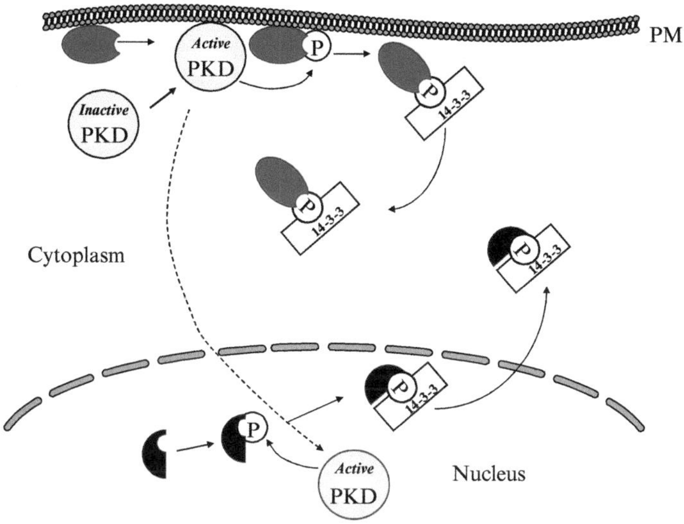

Fig. 7.3 Schematic representation of the mechanism by which PKD modulates intracellular localization of its substrates. In many cases, the phosphorylation of PKD substrates induces binding of 14-3-3 proteins that sequester them to the cytosol, thereby preventing them from acting at the plasma membrane (e.g. RIN1, Par-1, DLC1) or at the nucleus (e.g. HDACS 5 and HDACS 7). An emerging theme is that PKD modulates cell function by altering the subcellular localization of its substrates

bined with its distinct substrate specificity (see Table 7.1). As originally predicted (Rozengurt et al. 1995), it seems that a variety of biological responses attributed originally to PKCs are in fact executed by PKDs. Animal models using PKD transgenics or tissue specific knockout are emerging and will serve to further clarify the function(s) of PKD isoforms in vivo. In this context, it is important to point out that knockout of PKD in mice is embryonic lethal with incomplete penetrance (Fielitz et al. 2008b).

In view of the multifunctional roles of PKD, the search for physiological substrates is gathering pace and already a number of interesting molecules have been identified as PKD targets (summarized in Table 7.1). Interestingly, in many cases PKD-mediated phosphorylation regulates the subcellular localization of the phosphorylated substrate. For example, the phosphorylation of RIN1 on Ser^{351} by PKD induces binding of 14-3-3 proteins that sequester RIN1 to the cytosol, thereby preventing it from inhibiting the stimulatory interaction between Ras and Raf-1 at the membrane (Wang et al. 2002). A similar consequence of PKD phosphorylation leading to changes in substrate localization is critical for HDAC5 and HDAC7. There is evidence suggesting that a similar mode of regulation also functions for Par-1 and DLC1. An emerging theme is that PKD modulates multiple aspects of cell function by altering the subcellular localization of its substrates, either interfering with their membrane or nuclear localization, as shown schematically in Fig. 7.3.

In conclusion, studies on PKD thus far indicate a remarkable diversity of both its signal generation and distribution and its potential for complex regulatory interactions with multiple downstream pathways. It is increasingly apparent that the members of the PKD subfamily are key players in the regulation of cell signaling, organization, migration, inflammation and normal and abnormal cell proliferation. PKD emerges as a valuable target for the development of novel therapeutic approaches in common diseases, including cardiac hypertrophy and cancer.

Acknowledgments The help of Mr. James Sinnett-Smith in the preparation of this chapter is greatly appreciated. Studies from our laboratory presented here were supported in part by National Institutes of Health Grants R01-DK 55003, R0-1DK56930 and P30-DK41301.

References

Abedi, H., Rozengurt, E., & Zachary, I. (1998). Rapid activation of the novel serine/threonine protein kinase, protein kinase D by phorbol esters, angiotensin II and PDGF-BB in vascular smooth muscle cells. *FEBS Letters, 247*, 209–212.

Altschmied, J., & Haendeler, J. (2008). A new kid on the block: PKD1: A promising target for antiangiogenic therapy? *Arteriosclerosis, Thrombosis, and Vascular Biology, 28*, 1689–1690.

Arnold, R., Patzak, I. M., Neuhaus, B., Vancauwenbergh, S., Veillette, A., Van Lint, J., et al. (2005). Activation of hematopoietic progenitor kinase 1 involves relocation, autophosphorylation, and transphosphorylation by protein kinase D1. *Molecular and Cellular Biology, 25*, 2364–2383.

Avkiran, M., Rowland, A. J., Cuello, F., & Haworth, R. S. (2008). Protein kinase d in the cardiovascular system: Emerging roles in health and disease. *Circulation Research, 102*, 157–163.

Bagowski, C. P., Stein-Gerlach, M., Choidas, A., & Ullrich, A. (1999). Cell-type specific phosphorylation of threonines T654 and T669 by PKD defines the signal capacity of the EGF receptor. *EMBO Journal, 18*, 5567–5576.

Benn, S. C., Perrelet, D., Kato, A. C., Scholz, J., Decosterd, I., Mannion, R. J., et al. (2002). Hsp27 upregulation and phosphorylation is required for injured sensory and motor neuron survival. *Neuron, 36*, 45–56.

Berkowitz, P., Hu, P., Warren, S., Liu, Z., Diaz, L. A., & Rubenstein, D. S. (2006). p38MAPK inhibition prevents disease in pemphigus vulgaris mice 10.1073/pnas.0602973103. *Proceedings of the National Academy of Sciences of the United States of America, 103*, 12855–12860.

Berna, M. J., Hoffmann, K. M., Tapia, J. A., Thill, M., Pace, A., Mantey, S. A., et al. (2007). CCK causes PKD1 activation in pancreatic acini by signaling through PKC-delta and PKC-independent pathways. *Biochimica et Biophysica Acta, 1773*, 483–501.

Bisbal, M., Conde, C., Donoso, M., Bollati, F., Sesma, J., Quiroga, S., et al. (2008). Protein kinase d regulates trafficking of dendritic membrane proteins in developing neurons. *Journal of Neuroscience, 28*, 9297–9308.

Bowden, E. T., Barth, M., Thomas, D., Glazer, R. I., & Mueller, S. C. (1999). An invasion-related complex of cortactin, paxillin and PKCmu associates with invadopodia at sites of extracellular matrix degradation. *Oncogene, 18*, 4440–4449.

Brandlin, I., Hubner, S., Eiseler, T., Martinez-Moya, M., Horschinek, A., Hausser, A., et al. (2002). Protein kinase C (PKC)eta-mediated PKC mu activation modulates ERK and JNK signal pathways. *Journal of Biological Chemistry, 277*, 6490–6496.

Brose, N., Betz, A., & Wegmeyer, H. (2004). Divergent and convergent signaling by the diacylglycerol second messenger pathway in mammals. *Current Opinion Neurobiology, 14*, 328–340.

Bruey, J. M., Ducasse, C., Bonniaud, P., Ravagnan, L., Susin, S. A., Diaz-Latoud, C., et al. (2000). Hsp27 negatively regulates cell death by interacting with cytochrome c. *Nature Cell Biology, 2*, 645–652.

Buchwald, G., Hostinova, E., Rudolph, M. G., Kraemer, A., Sickmann, A., Meyer, H. E., et al. (2001). Conformational switch and role of phosphorylation in PAK activation. *Molecular and Cellular Biology, 21*, 5179–5189.

Cabrera-Poch, N., Sanchez-Ruiloba, L., Rodriguez-Martinez, M., & Iglesias, T. (2004). Lipid raft disruption triggers protein kinase C and Src-dependent protein kinase D activation and Kidins220 phosphorylation in neuronal cells. *Journal of Biological Chemistry, 279*, 28592–28602.

Carnegie, G. K., Smith, F. D., McConnachie, G., Langeberg, L. K., & Scott, J. D. (2004). AKAP-Lbc nucleates a protein kinase D activation scaffold. *Molecular Cell, 15*, 889–899.

Carnegie, G. K., Soughayer, J., Smith, F. D., Pedroja, B. S., Zhang, F., Diviani, D., et al. (2008). AKAP-Lbc mobilizes a cardiac hypertrophy signaling pathway. *Molecular Cell, 32*, 169–179.

Celil, A. B., & Campbell, P. G. (2005). BMP-2 and insulin-like growth factor-I mediate osterix (Osx) expression in human mesenchymal stem cells via the MAPK and protein kinase D signaling pathways. *Journal of Biological Chemistry, 280*, 31353–31359.

Chang, H.-W., Chu, T.-S., Huang, H.-Y., Chueh, S.-C., Wu, V.-C., Chen, Y.-M., et al. (2007). Down-regulation of D2 dopamine receptor and increased protein kinase C{micro} phosphorylation in aldosterone-producing adenoma play roles in aldosterone overproduction. *Journal of Clinical Endocrinology and Metabolism, 92*, 1863–1870.

Chang, L., & Karin, M. (2001). Mammalian MAP kinase signalling cascades. *Nature, 410*, 37–40.

Chen, J., Kahne, T., Rocken, C., Gotze, T., Yu, J., Sung, J. J. Y., et al. (2004). Proteome analysis of gastric cancer metastasis by two-dimensional gel electrophoresis and matrix assisted laser desorption/ionization-mass spectrometry for identification of metastasis-related proteins. *Journal of Proteome Research, 3*, 1009–1016.

Chiu, T. T., Leung, W. Y., Moyer, M. P., Strieter, R. M., & Rozengurt, E. (2007). Protein kinase D2 mediates lysophosphatidic acid-induced interleukin 8 production in nontransformed human colonic epithelial cells through NF-{kappa}B. *American Journal of Physiology Cell Physiology, 292*, C767–C777.

Chiu, T., & Rozengurt, E. (2001a). CCK2 (CCKB/gastrin) receptor mediates rapid protein kinase D (PKD) activation through a protein kinase C-dependent pathway. *FEBS Letters, 489*, 101–106.

Chiu, T., & Rozengurt, E. (2001b). PKD in intestinal epithelial cells: Rapid activation by phorbol esters, LPA, and angiotensin through PKC. *American Journal of Physiological Cell Physiology, 280*, C929–C942.

Chiu, T., Wu, S. S., Santiskulvong, C., Tangkijvanich, P., Yee, H. F., Jr., & Rozengurt, E. (2002). Vasopressin-mediated mitogenic signaling in intestinal epithelial cells. *American Journal of Physiology. Cell Physiology, 282*, C434–C450.

Coornaert, B., Carpentier, I., & Beyaert, R. (2009). A20: Central gatekeeper in inflammation and immunity. *Journal of Biological Chemistry, 284*, 8217–8221.

Cozier, G. E., Carlton, J., Bouyoucef, D., & Cullen, P. J. (2004). Membrane targeting by pleckstrin homology domains. *Current Topics in Microbiology and Immunology, 282*, 49–88.

Cuello, F., Bardswell, S. C., Haworth, R. S., Yin, X., Lutz, S., Wieland, T., et al. (2007). Protein kinase D selectively targets cardiac troponin I and regulates myofilament Ca2+ sensitivity in ventricular myocytes. *Circulation Research, 100*, 864–873.

Cui, J., Holgado-Madruga, M., Su, W., Tsuiki, H., Wedegaertner, P., & Wong, A. J. (2005). Identification of a specific domain responsible for JNK2alpha2 autophosphorylation. *Journal of Biological Chemistry, 280*, 9913–9920.

Dequiedt, F., Van Lint, J., Lecomte, E., Van Duppen, V., Seufferlein, T., Vandenheede, J. R., et al. (2005). Phosphorylation of histone deacetylase 7 by protein kinase D mediates T cell receptor-induced Nur77 expression and apoptosis. *Journal of Experimental Medicine, 201*, 793–804.

Doppler, H., & Storz, P. (2007). A novel tyrosine phosphorylation site in protein kinase D contributes to oxidative stress-mediated activation. *Journal of Biological Chemistry, 282*, 31873–31881.

Doppler, H., Storz, P., Li, J., Comb, M. J., & Toker, A. (2005). A phosphorylation state-specific antibody recognizes Hsp27, a novel substrate of protein kinase D. *Journal of Biological Chemistry, 280*, 15013–15019.

Ebisuya, M., Kondoh, K., & Nishida, E. (2005). The duration, magnitude and compartmentalization of ERK MAP kinase activity: Mechanisms for providing signaling specificity. *Journal of Cell Science, 118*, 2997–3002.

Ehlers, R. A., Zhang, Y., Hellmich, M. R., & Evers, B. M. (2000). Neurotensin-mediated activation of MAPK pathways and AP-1 binding in the human pancreatic cancer cell line, MIA PaCa-2. *Biochemical and Biophysical Research Communications, 269*, 704–708.

Eiseler, T., Doppler, H., Yan, I., Goodison, S., & Storz, P. (2009). Protein kinase D1 regulates matrix metalloproteinase expression and inhibits breast cancer cell invasion. *Breast Cancer Research, 11*, R13.

Eyers, P. A., Erikson, E., Chen, L. G., & Maller, J. L. (2003). A novel mechanism for activation of the protein kinase aurora A. *Current Biology, 13*, 691–697.

Fielitz, J., Kim, M.-S., Shelton, J. M., Qi, X., Hill, J. A., Richardson, J. A., et al. (2008a). Requirement of protein kinase D1 for pathological cardiac remodeling. *Proceedings of the National Academy of Sciences, 105*, 3059–3063.

Fielitz, J., Kim, M. S., Shelton, J. M., Qi, X., Hill, J. A., Richardson, J. A., et al. (2008b). Requirement of protein kinase D1 for pathological cardiac remodeling. *Proceedings of the National Academy of Sciences of the United States of America, 105*, 3059–3063.

Geum, D., Son, G. H., & Kim, K. (2002). Phosphorylation-dependent cellular localization and thermoprotective role of heat shock protein 25 in hippocampal progenitor cells 10.1074/jbc.M104396200. *Journal of Biological Chemistry, 277*, 19913–19921.

Ghanekar, Y., & Lowe, M. (2005). Signalling for secretion. *Nature Cell Biology, 7*, 851.

Griner, E. M., & Kazanietz, M. G. (2007). Protein kinase C and other diacylglycerol effectors in cancer. *Nature Reviews Cancer, 7*, 281–294.

Guay, J., Lambert, H., Gingras-Breton, G., Lavoie, J. N., Huot, J., & Landry, J. (1997). Regulation of actin filament dynamics by p38 map kinase-mediated phosphorylation of heat shock protein 27. *Journal of Cell Science, 110*, 357–368.

Guha, S., Rey, O., & Rozengurt, E. (2002). Neurotensin induces protein kinase C-dependent protein kinase D activation and DNA synthesis in human pancreatic carcinoma cell line PANC-1. *Cancer Research, 62*, 1632–1640.

Ha, C. H., Jhun, B. S., Kao, H.-Y., & Jin, Z.-G. (2008a). VEGF stimulates HDAC7 phosphorylation and cytoplasmic accumulation modulating matrix metalloproteinase expression and angiogenesis. *Arteriosclerosis, Thrombosis, and Vascular Biology, 28*, 1782–1788.

Ha, C. H., Wang, W., Jhun, B. S., Wong, C., Hausser, A., Pfizenmaier, K., et al. (2008b). Protein kinase D-dependent phosphorylation and nuclear export of histone deacetylase 5 mediates vascular endothelial growth factor-induced gene expression and angiogenesis. *Journal of Biological Chemistry, 283*, 14590–14599.

Hanks, S. K. (2003). Genomic analysis of the eukaryotic protein kinase superfamily: A perspective. *Genome Biology, 4*, 111.

Hausser, A., Link, G., Bamberg, L., Burzlaff, A., Lutz, S., Pfizenmaier, K., et al. (2002). Structural requirements for localization and activation of protein kinase C mu (PKC mu) at the Golgi compartment. *Journal of Cell Biology, 156*, 65–74.

Hausser, A., Storz, P., Martens, S., Link, G., Toker, A., & Pfizenmaier, K. (2005). Protein kinase D regulates vesicular transport by phosphorylating and activating phosphatidylinositol-4 kinase III[beta] at the Golgi complex. *Nature Cell Biology, 7*, 880.

Haworth, R. S., Cuello, F., Herron, T. J., Franzen, G., Kentish, J. C., Gautel, M., et al. (2004). Protein kinase D is a novel mediator of cardiac troponin I phosphorylation and regulates myofilament function. *Circulation Research, 95*, 1091–1099.

Haworth, R. S., Goss, M. W., Rozengurt, E., & Avkiran, M. (2000). Expression and activity of protein kinase D/protein kinase C mu in myocardium: Evidence for alpha(1)-adrenergic receptor- and protein kinase C-mediated regulation. *Journal of Molecular and Cellular Cardiology, 32*, 1013–1023.

Haworth, R. S., Roberts, N. A., Cuello, F., & Avkiran, M. (2007). Regulation of protein kinase D activity in adult myocardium: Novel counter-regulatory roles for protein kinase C[epsilon] and protein kinase A. *Journal of Molecular and Cellular Cardiology, 43*, 686.

Hayashi, A., Seki, N., Hattori, A., Kozuma, S., & Saito, T. (1999). PKCnu, a new member of the protein kinase C family, composes a fourth subfamily with PKCmu. *Biochimica et Biophysica Acta, 1450*, 99–106.

Henrique, D., & Schweisguth, F. (2003). Cell polarity: The ups and downs of the Par6/aPKC complex. *Current Opinion in Genetics and Development, 13*, 341–350.

Hudmon, A., & Schulman, H. (2002). Neuronal CA2+/calmodulin-dependent protein kinase II: The role of structure and autoregulation in cellular function. *Annual Review of Biochemistry, 71*, 473–510.

Hurd, C., & Rozengurt, E. (2001). Protein kinase D is sufficient to suppress EGF-induced c-Jun Ser 63 phosphorylation. *Biochemical and Biophysical Research Communications, 282*, 404–408.

Hurd, C., Waldron, R. T., & Rozengurt, E. (2002). Protein kinase D complexes with C-Jun N-terminal kinase via activation loop phosphorylation and phosphorylates the C-Jun N-terminus. *Oncogene, 21*, 2154–2160.

Hurley, J. H., Newton, A. C., Parker, P. J., Blumberg, P. M., & Nishizuka, Y. (1997). Taxonomy and function of C1 protein kinase C homology domains. *Protein Science, 6*, 477–480.

Iglesias, T., Cabrera-Poch, N., Mitchell, M. P., Naven, T. J. P., Rozengurt, E., & Schiavo, G. (2000a). Identification and cloning of Kidins220, a novel neuronal substrate of protein kinase D. *Journal of Biological Chemistry, 275*, 40048–40056.

Iglesias, T., Cabrera-Poch, N., Mitchell, M. P., Naven, T. J., Rozengurt, E., & Schiavo, G. (2000b). Identification and cloning of Kidins220, a novel neuronal substrate of protein kinase D. *Journal of Biological Chemistry, 275*, 40048–40056.

Iglesias, T., Matthews, S., & Rozengurt, E. (1998a). Dissimilar phorbol ester binding properties of the individual cysteine-rich motifs of protein kinase D. *FEBS Letters, 437*, 19–23.

Iglesias, T., & Rozengurt, E. (1998). Protein kinase D activation by mutations within its pleckstrin homology domain. *Journal of Biological Chemistry, 273*, 410–416.

Iglesias, T., & Rozengurt, E. (1999). Protein kinase D activation by deletion of its cysteine-rich motifs. *FEBS Letters, 454*, 53–56.

Iglesias, T., Waldron, R. T., & Rozengurt, E. (1998b). Identification of in vivo phosphorylation sites required for protein kinase D activation. *Journal of Biological Chemistry, 273*, 27662–27667.

Irie, A., Harada, K., Tsukamoto, H., Kim, J. R., Araki, N., & Nishimura, Y. (2006). Protein kinase D2 contributes to either IL-2 promoter regulation or induction of cell death upon TCR stimulation depending on its activity in Jurkat cells. *International Immunology, 18*, 1737–1747.

Ivison, S. M., Graham, N. R., Bernales, C. Q., Kifayet, A., Ng, N., Shobab, L. A., et al. (2007). Protein kinase D interaction with TLR5 is required for inflammatory signaling in response to bacterial flagellin. *Journal of Immunology, 178*, 5735–5743.

Jacamo, R., Sinnett-Smith, J., Rey, O., Waldron, R. T., & Rozengurt, E. (2008). Sequential protein kinase C (PKC)-dependent and PKC-independent protein kinase D catalytic activation via Gq-coupled receptors: Differential regulation of activation loop Ser(744) and Ser(748) phosphorylation. *Journal of Biological Chemistry, 283*, 12877–12887.

Jaggi, M., Rao, P. S., Smith, D. J., Wheelock, M. J., Johnson, K. R., Hemstreet, G. P., et al. (2005). E-cadherin phosphorylation by protein kinase D1/protein kinase C{micro} is associated with altered cellular aggregation and motility in prostate cancer. *Cancer Research, 65*, 483–492.

Jensen, E. D., Gopalakrishnan, R., & Westendorf, J. J. (2009). Bone morphogenic protein 2 activates protein kinase D to regulate histone deacetylase 7 localization and repression of Runx2. *Journal of Biological Chemistry, 284*, 2225–2234.

Jensen, E. D., Schroeder, T. M., Bailey, J., Gopalakrishnan, R., & Westendorf, J. J. (2008). Histone deacetylase 7 associates with Runx2 and represses its activity during osteoblast maturation in a deacetylation-independent manner. *Journal of Bone and Mineral Research, 23*, 361–372.

Johannes, F. J., Prestle, J., Dieterich, S., Oberhagemann, P., Link, G., & Pfizenmaier, K. (1995). Characterization of activators and inhibitors of protein kinase C mu. *European Journal of Biochemistry, 227*, 303–307.

Johannes, F. J., Prestle, J., Eis, S., Oberhagemann, P., & Pfizenmaier, K. (1994). PKCu is a novel, atypical member of the protein kinase C family. *Journal of Biological Chemistry, 269*, 6140–6148.

Johnson, G. L., & Lapadat, R. (2002). Mitogen-activated protein kinase pathways mediated by ERK, JNK, and p38 protein kinases. *Science, 298*, 1911–1912.

Johnson, L. N., Noble, M. E., & Owen, D. J. (1996). Active and inactive protein kinases: Structural basis for regulation. *Cell, 85*, 149–158.

Kang, J., Chen, Y., Zhao, Y., & Yu, H. (2007). Autophosphorylation-dependent activation of human Mps1 is required for the spindle checkpoint. *Proceedings of the National Academy of Sciences of the United States of America, 104*, 20232–20237.

Kelly, A. E., Sampath, S. C., Maniar, T. A., Woo, E. M., Chait, B. T., & Funabiki, H. (2007). Chromosomal enrichment and activation of the aurora B pathway are coupled to spatially regulate spindle assembly. *Developmental Cell, 12*, 31–43.

Kennett, S. B., Roberts, J. D., & Olden, K. (2004). Requirement of protein kinase C{micro} activation and calpain-mediated proteolysis for arachidonic acid-stimulated adhesion of MDA-MB-435 human mammary carcinoma cells to collagen type IV. *Journal of Biological Chemistry, 279*, 3300–3307.

Kikkawa, U., Kishimoto, A., & Nishizuka, Y. (1989). The protein kinase C family: Heterogeneity and its implications. *Annual Review Biochemistry, 58*, 31–44.

Kim, M. S., Wang, F., Puthanveetil, P., Kewalramani, G., Hosseini-Beheshti, E., Ng, N., et al. (2008). Protein kinase D is a key regulator of cardiomyocyte lipoprotein lipase secretion after diabetes. *Circulation Research, 103*, 252–260.

Kisfalvi, K., Guha, S., & Rozengurt, E. (2005). Neurotensin and EGF induce synergistic stimulation of DNA synthesis by increasing the duration of ERK signaling in ductal pancreatic cancer cells. *Journal of Cellular Physiology, 202*, 880–890.

Klein, I. K., Ritland, S. R., Burgart, L. J., Ziesmer, S. C., Roche, P. C., Gendler, S. J., et al. (2000). Adenoma-specific alterations of protein kinase C isozyme expression in ApcMIN mice. *Cancer Research, 60*, 2077–2080.

Kolch, W. (2005). Coordinating ERK/MAPK signalling through scaffolds and inhibitors. *Nature Reviews Molecular Cell Biology, 6*, 827–837.

Kubisch, C., Dimagno, M. J., Tietz, A. B., Welsh, M. J., Ernst, S. A., Brandt-Nedelev, B., et al. (2004). Overexpression of heat shock protein Hsp27 protects against cerulein-induced pancreatitis. *Gastroenterology, 127*, 275–286.

Lahoz, A., & Hall, A. (2008). DLC1: A significant GAP in the cancer genome. *Genes Development, 22*, 1724–1730.

Lee, E. G., Boone, D. L., Chai, S., Libby, S. L., Chien, M., Lodolce, J. P., et al. (2000). Failure to regulate TNF-induced NF-kappa B and cell death responses in A20-deficient mice. *Science, 289*, 2350–2354.

Lemonnier, J., Ghayor, C., Guicheux, J., & Caverzasio, J. (2004). Protein kinase C-independent activation of protein kinase D is involved in BMP-2-induced activation of stress mitogen-activated protein kinases JNK and p38 and osteoblastic cell differentiation. *Journal of Biological Chemistry, 279*, 259–264.

Li, J., Chen, L. A., Townsend, C. M., Jr., & Evers, B. M. (2008). PKD1, PKD2, and their substrate Kidins220 regulate neurotensin secretion in the BON human endocrine cell line. *Journal of Biological Chemistry, 283*, 2614–2621.

Li, J., O'Connor, K. L., Hellmich, M. R., Greeley, G. H., Jr., Townsend, C. M., Jr., & Evers, B. M. (2004). The role of protein kinase D in neurotensin secretion mediated by protein kinase C-{alpha}/-{delta} and Rho/Rho kinase. *Journal of Biological Chemistry, 279*, 28466–28474.

Liljedahl, M., Maeda, Y., Colanzi, A., Ayala, I., Van Lint, J., & Malhotra, V. (2001). Protein kinase D regulates the fission of cell surface destined transport carriers from the trans-Golgi network. *Cell, 104*, 409–420.

Liu, P., Scharenberg, A. M., Cantrell, D. A., & Matthews, S. A. (2007). Protein kinase D enzymes are dispensable for proliferation, survival and antigen receptor-regulated NFkappaB activity in vertebrate B-cells. *FEBS Letters, 581*, 1377–1382.

Liuwantara, D., Elliot, M., Smith, M. W., Yam, A. O., Walters, S. N., Marino, E., et al. (2006). Nuclear factor-{kappa}B regulates {beta}-cell death: A critical role for A20 in {beta}-cell protection. *Diabetes, 55*, 2491–2501.

Lochhead, P. A., Kinstrie, R., Sibbet, G., Rawjee, T., Morrice, N., & Cleghon, V. (2006). A chaperone-dependent GSK3beta transitional intermediate mediates activation-loop autophosphorylation. *Molecular Cell, 24*, 627–633.

Lochhead, P. A., Sibbet, G., Morrice, N., & Cleghon, V. (2005). Activation-loop autophosphorylation is mediated by a novel transitional intermediate form of DYRKs. *Cell, 121*, 925–936.

Maeda, Y., Beznoussenko, G. V., Van Lint, J., Mironov, A. A., & Malhotra, V. (2001). Recruitment of protein kinase D to the trans-Golgi network via the first cysteine-rich domain. *EMBO Journal, 20*, 5982–5990.

Maizels, E. T., Peters, C. A., Kline, M., Cutler, R. E., Jr., Shanmugam, M., & Hunzicker-Dunn, M. (1998). Heat-shock protein-25/27 phosphorylation by the delta isoform of protein kinase C. *Biochemical Journal, 332*(Pt 3), 703–712.

Marais, R., & Marshall, C. J. (1996). Control of the ERK MAP kinase cascade by Ras and Raf. *Cancer Surveys, 27*, 101–125.

Marklund, U., Lightfoot, K., & Cantrell, D. (2003). Intracellular location and cell context-dependent function of protein kinase D. *Immunity, 19*, 491–501.

Marshall, C. J. (1995). Specificity of receptor tyrosine kinase signaling: Transient versus sustained extracellular signal-regulated kinase activation. *Cell, 80*, 179–185.

Matthews, S. A., Dayalu, R., Thompson, L. J., & Scharenberg, A. M. (2003). Regulation of protein kinase Cnu by the B-cell antigen receptor. *Journal of Biological Chemistry, 278*, 9086–9091.

Matthews, S., Iglesias, T., Cantrell, D., & Rozengurt, E. (1999a). Dynamic re-distribution of protein kinase D (PKD) as revealed by a GFP-PKD fusion protein: Dissociation from PKD activation. *FEBS Letters, 457*, 515–521.

Matthews, S., Iglesias, T., Cantrell, D., & Rozengurt, E. (1999b). Dynamic re-distribution of protein kinase D (PKD) as revealed by a GFP-PKD fusion protein: Dissociation from PKD activation. *FEBS Letters, 457*, 515–521.

Matthews, S. A., Iglesias, T., Rozengurt, E., & Cantrell, D. (2000a). Spatial and temporal regulation of protein kinase D (PKD). *EMBO Journal, 19*, 2935–2945.

Matthews, S. A., Pettit, G. R., & Rozengurt, E. (1997). Bryostatin 1 induces biphasic activation of protein kinase D in intact cells. *Journal of Biological Chemistry, 272*, 20245–20250.

Matthews, S. A., Rozengurt, E., & Cantrell, D. (1999c). Characterization of serine 916 as an in vivo autophosphorylation site for protein kinase D/protein kinase C mu. *Journal of Biological Chemistry, 274*, 26543–26549.

Matthews, S. A., Rozengurt, E., & Cantrell, D. (2000b). Protein kinase D. A selective target for antigen receptors and a downstream target for protein kinase C in lymphocytes. *Journal of Experimental Medicine, 191*, 2075–2082.

Mattison, C. P., Old, W. M., Steiner, E., Huneycutt, B. J., Resing, K. A., Ahn, N. G., et al. (2007). Mps1 activation loop autophosphorylation enhances kinase activity. *Journal of Biological Chemistry, 282*, 30553–30561.

McCollum, A. K., TenEyck, C. J., Sauer, B. M., Toft, D. O., & Erlichman, C. (2006). Up-regulation of heat shock protein 27 induces resistance to 17-allylamino-demethoxygeldanamycin through a glutathione-mediated mechanism 10.1158/0008-5472.CAN-06-1629. *Cancer Research, 66*, 10967–10975.

McEneaney, V., Harvey, B. J., & Thomas, W. (2007). Aldosterone rapidly activates protein kinase D via a mineralocorticoid receptor/EGFR trans-activation pathway in the M1 kidney CCD cell line. *Journal of Steroid Biochemistry and Molecular Biology, 107*, 180.

Medeiros, R. B., Dickey, D. M., Chung, H., Quale, A. C., Nagarajan, L. R., Billadeau, D. D., et al. (2005). Protein kinase D1 and the beta 1 integrin cytoplasmic domain control beta 1 integrin function via regulation of Rap1 activation. *Immunity, 23*, 213–226.

Mellor, H., & Parker, P. J. (1998). The extended protein kinase C superfamily. *Biochemical Journal, 332*, 281–292.
Meloche, S., & Pouyssegur, J. (2007). The ERK1/2 mitogen-activated protein kinase pathway as a master regulator of the G1- to S-phase transition. *Oncogene, 26*, 3227–3239.
Mihailovic, T., Marx, M., Auer, A., Van Lint, J., Schmid, M., Weber, C., et al. (2004). Protein kinase D2 mediates activation of nuclear factor {kappa}B by Bcr-Abl in Bcr-Abl+human myeloid leukemia cells. *Cancer Research, 64*, 8939–8944.
Miyake, Z., Takekawa, M., Ge, Q., & Saito, H. (2007). Activation of MTK1/MEKK4 by GADD45 through induced N-C dissociation and dimerization-mediated trans autophosphorylation of the MTK1 kinase domain. *Molecular Cell Biology, 27*, 2765–2776.
Mohapatra, D. P., & Nau, C. (2005). Regulation of Ca2+-dependent desensitization in the vanilloid receptor TRPV1 by calcineurin and cAMP-dependent protein kinase. *Journal of Biological Chemistry, 280*, 13424–13432.
Mori-Iwamoto, S., Kuramitsu, Y., Ryozawa, S., Mikuria, K., Fujimoto, M., Maehara, S., et al. (2007). Proteomics finding heat shock protein 27 as a biomarker for resistance of pancreatic cancer cells to gemcitabine. *International Journal of Oncology, 31*, 1345–1350.
Moriyama, T., Iida, T., Kobayashi, K., Higashi, T., Fukuoka, T., Tsumura, H., et al. (2003). Possible involvement of P2Y2 metabotropic receptors in ATP-induced transient receptor potential vanilloid receptor 1-mediated thermal hypersensitivity. *Journal of Neuroscience, 23*, 6058–6062.
Mottet, D., Bellahcene, A., Pirotte, S., Waltregny, D., Deroanne, C., Lamour, V., et al. (2007). Histone deacetylase 7 silencing alters endothelial cell migration, a key step in angiogenesis. *Circulation Research, 101*, 1237–1246.
Mullin, M. J., Lightfoot, K., Marklund, U., & Cantrell, D. A. (2006). Differential requirement for RhoA GTPase depending on the cellular localization of protein kinase D. *Journal of Biological Chemistry, 281*, 25089–25096.
Murphy, T. R., Legere, H. J., III, & Katz, H. R. (2007). Activation of protein kinase D1 in mast cells in response to innate, adaptive, and growth factor signals. *Journal of Immunology, 179*, 7876–7882.
Murphy, L. O., MacKeigan, J. P., & Blenis, J. (2004). A network of immediate early gene products propagates subtle differences in mitogen-activated protein kinase signal amplitude and duration. *Molecular and Cellular Biology, 24*, 144–153.
Murphy, L. O., Smith, S., Chen, R. H., Fingar, D. C., & Blenis, J. (2002). Molecular interpretation of ERK signal duration by immediate early gene products. *Nature Cell Biology, 4*, 556–564.
Murray, N. R., Davidson, L. A., Chapkin, R. S., Clay Gustafson, W., Schattenberg, D. G., & Fields, A. P. (1999). Overexpression of protein kinase C betaII induces colonic hyperproliferation and increased sensitivity to colon carcinogenesis. *Journal of Cell Biology, 145*, 699–711.
Nakanishi, H., Brewer, K. A., & Exton, J. H. (1993). Activation of the zeta isozyme of protein kinase C by phosphatidylinositol 3, 4, 5-trisphosphate. *Journal of Biological Chemistry, 268*, 13–16.
Newton, A. C. (1997). Regulation of protein kinase C. *Current Opinion Cell Biology*, *9*, 161–167.
Nishizuka, Y. (1988). The molecular heterogeneity of protein kinase C and its implications for cellular regulation. *Nature, 334*, 661–665.
Nishizuka, Y. (1992). Intracellular signaling by hydrolysis of phospholipids and activation of protein kinase C. *Science, 258*, 607–614.
Nishizuka, Y. (1995). Protein kinase C and lipid signaling for sustained cellular responses. *FASEB Journal, 9*, 484–496.
Oancea, E., Bezzerides, V. J., Greka, A., & Clapham, D. E. (2003). Mechanism of persistent protein kinase D1 translocation and activation. *Developmental Cell, 4*, 561–574.
Oliver, A. W., Knapp, S., & Pearl, L. H. (2007). Activation segment exchange: A common mechanism of kinase autophosphorylation? Trends in Biochemical Sciences
Oliver, A. W., Knapp, S., & Pearl, L. H. (2007b). Activation segment exchange: A common mechanism of kinase autophosphorylation? *Trends Biochemical Science, 32*, 351–356.
Oliver, A. W., Paul, A., Boxall, K. J., Barrie, S. E., Aherne, G. W., Garrett, M. D., et al. (2006). Trans-activation of the DNA-damage signalling protein kinase Chk2 by T-loop exchange. *EMBO Journal, 25*, 3179–3190.

Ozgen, N., Obreztchikova, M., Guo, J., Elouardighi, H., Dorn, G. W., II, Wilson, B. A., et al. (2008). Protein kinase D links Gq-coupled receptors to cAMP response element-binding protein (CREB)-Ser133 phosphorylation in the heart. *Journal of Biological Chemistry, 283*, 17009–17019.

Paolucci, L., & Rozengurt, E. (1999). Protein kinase D in small cell lung cancer cells: Rapid activation through protein kinase C. *Cancer Research, 59*, 572–577.

Paolucci, L., Sinnett-Smith, J., & Rozengurt, E. (2000). Lysophosphatidic acid rapidly induces protein kinase D activation through a pertussis toxin-sensitive pathway. *American Journal of Physiology. Cell Physiology, 278*, C33–C39.

Papazyan, R., Doche, M., Waldron, R. T., Rozengurt, E., Moyer, M. P., & Rey, O. (2008). Protein kinase D isozymes activation and localization during mitosis. *Experimental Cell Research, 314*, 3057.

Papazyan, R., Rozengurt, E., & Rey, O. (2006). The C-terminal tail of protein kinase D2 and protein kinase D3 regulates their intracellular distribution. *Biochemical and Biophysical Research Communications, 342*, 685–689.

Park, J.-E., Kim, Y.-I., & Yi, A.-K. (2008). Protein kinase D1: A new component in TLR9 signaling. *Journal of Immunology, 181*, 2044–2055.

Parra, M., Kasler, H., McKinsey, T. A., Olson, E. N., & Verdin, E. (2005). Protein kinase D1 phosphorylates HDAC7 and induces its nuclear export after T-cell receptor activation. *Journal of Biological Chemistry, 280*, 13762–13770.

Paul, C., Manero, F., Gonin, S., Kretz-Remy, C., Virot, S., & Arrigo, A.-P. (2002). Hsp27 as a negative regulator of cytochrome c release. *Molecular and Cellular Biology, 22*, 816–834.

Poole, D. P., Amadesi, S., Rozengurt, E., Thacker, M., Bunnett, N. W., & Furness, J. B. (2008). Stimulation of the neurokinin 3 receptor activates protein kinase C{varepsilon} and protein kinase D in enteric neurons. *American Journal of Physiology. Gastrointestinal and Liver Physiology, 294*, G1245–G1256.

Pouyssegur, J., & Lenormand, P. (2003). Fidelity and spatio-temporal control in MAP kinase (ERKs) signalling. *European Journal of Biochemistry, 270*, 3291–3299.

Premkumar, L. S., & Ahern, G. P. (2000). Induction of vanilloid receptor channel activity by protein kinase C. *Nature, 408*, 985–990.

Prigozhina, N. L., & Waterman-Storer, C. M. (2004). Protein kinase D-mediated anterograde membrane trafficking is required for fibroblast motility. *Current Biology, 14*, 88–98.

Qiang, Y. W., Yao, L., Tosato, G., & Rudikoff, S. (2004). Insulin-like growth factor I induces migration and invasion of human multiple myeloma cells. *Blood, 103*, 301–308.

Qin, L., Zeng, H., & Zhao, D. (2006). Requirement of protein kinase D tyrosine phosphorylation for VEGF-A165-induced angiogenesis through its interaction and regulation of phospholipase C{gamma} phosphorylation. *Journal of Biological Chemistry, 281*, 32550–32558.

Rey, O., Papazyan, R., Waldron, R. T., Young, S. H., Lippincott-Schwartz, J., Jacamo, R., et al. (2006). The nuclear import of protein kinase D3 requires its catalytic activity. *Journal of Biological Chemistry, 281*, 5149–5157.

Rey, O., Reeve, J. R., Jr., Zhukova, E., Sinnett-Smith, J., & Rozengurt, E. (2004). G protein-coupled receptor-mediated phosphorylation of the activation loop of protein kinase D: Dependence on plasma membrane translocation and protein kinase Cepsilon. *Journal of Biological Chemistry, 279*, 34361–34372.

Rey, O., & Rozengurt, E. (2001). Protein kinase D interacts with Golgi via its cysteine-rich domain. *Biochemical and Biophysical Research Communications, 287*, 21–26.

Rey, O., Sinnett-Smith, J., Zhukova, E., & Rozengurt, E. (2001a). Regulated nucleocytoplasmic transport of protein kinase D in response to G protein-coupled receptor activation. *Journal of Biological Chemistry, 276*, 49228–49235.

Rey, O., Young, S. H., Cantrell, D., & Rozengurt, E. (2001b). Rapid protein kinase D translocation in response to G protein-coupled receptor activation. Dependence on protein kinase C. *Journal of Biological Chemistry, 276*, 32616–32626.

Rey, O., Yuan, J., & Rozengurt, E. (2003a). Intracellular redistribution of protein kinase D2 in response to G- protein-coupled receptor agonists. *Biochemical and Biophysical Research Communications, 302*, 817–824.

Rey, O., Yuan, J., Young, S. H., & Rozengurt, E. (2003b). Protein kinase C nu/protein kinase D3 nuclear localization, catalytic activation, and intracellular redistribution in response to G protein-coupled receptor agonists. *Journal of Biological Chemistry, 278*, 23773–23785.

Rey, O., Zhukova, E., Sinnett-Smith, J., & Rozengurt, E. (2003c). Vasopressin-induced intracellular redistribution of protein kinase D in intestinal epithelial cells. *Journal of Cellular Physiology, 196*, 483–492.

Rocchi, P., So, A., Kojima, S., Signaevsky, M., Beraldi, E., Fazli, L., et al. (2004). Heat shock protein 27 increases after androgen ablation and plays a cytoprotective role in hormone-refractory prostate cancer 10.1158/0008-5472.CAN-03-3998. *Cancer Research, 64*, 6595–6602.

Romero, D. G., Welsh, B. L., Gomez-Sanchez, E. P., Yanes, L. L., Rilli, S., & Gomez-Sanchez, C. E. (2006). Angiotensin II-mediated protein kinase D activation stimulates aldosterone and cortisol secretion in H295R human adrenocortical cells. *Endocrinology, 147*, 6046–6055.

Rozengurt, E. (1998). Signal transduction pathways in the mitogenic response to G protein-coupled neuropeptide receptor agonists. *Journal of Cellular Physiology, 177*, 507–517.

Rozengurt, E., Rey, O., & Waldron, R. T. (2005). Protein kinase D signaling. *Journal of Biological Chemistry, 280*, 13205–13208.

Rozengurt, E., Rodriguez-Peña, A., & Sinnett-Smith, J. (1985). Signalling mitogenesis in 3T3 cells: Role of Ca2+-sensitive, phospholipid-dependent protein kinase. *Ciba Foundation Symposium, 116*, 66–86.

Rozengurt, E., Sinnett-Smith, J., Van Lint, J., & Valverde, A. M. (1995). Protein kinase D (PKD): A novel target for diacylglycerol and phorbol esters. *Mutation Research, 333*, 153–160.

Rozengurt, E., Sinnett-Smith, J., & Zugaza, J. L. (1997). Protein kinase D: A novel target for diacylglycerol and phorbol esters. *Biochemical Society Transactions, 25*, 565–571.

Rozengurt, E., & Sternini, C. (2007). Taste receptor signaling in the mammalian gut. *Current Opinion in Pharmacology, 7*, 557–562.

Ruan, T., Lin, Y. S., Lin, K.-S., & Kou, Y. R. (2005). Sensory transduction of pulmonary reactive oxygen species by capsaicin-sensitive vagal lung afferent fibres in rats. *Journal of Physiology (London), 565*, 563–578

Ryder, N. M., Guha, S., Hines, O. J., Reber, H. A., & Rozengurt, E. (2001). G protein-coupled receptor signaling in human ductal pancreatic cancer cells: Neurotensin responsiveness and mitogenic stimulation. *Journal of Cellular Physiology, 186*, 53–64.

Santos, S. D. M., Verveer, P. J., & Bastiaens, P. I. H. (2007). Growth factor-induced MAPK network topology shapes Erk response determining PC-12 cell fate. *Nature Cell Biology, 9*, 324–330.

Scholz, R.-P., Regner, J., Theil, A., Erlmann, P., Holeiter, G., Jahne, R., et al. (2009). DLC1 interacts with 14-3-3 proteins to inhibit RhoGAP activity and block nucleocytoplasmic shuttling. *Journal of Cell Science, 122*, 92–102.

Selbie, L. A., Schmitz-Peiffer, C., Sheng, Y., & Biden, T. J. (1993). Molecular cloning and characterization of PKC iota, an atypical isoform of protein kinase C derived from insulin-secreting cells. *Journal of Biological Chemistry, 268*, 24296–24302.

Shin, K. D., Lee, M.-Y., Shin, D.-S., Lee, S., Son, K.-H., Koh, S., et al. (2005). Blocking tumor cell migration and invasion with biphenyl isoxazole derivative KRIBB3, a synthetic molecule that inhibits Hsp27 phosphorylation 10.1074/jbc.M507209200. *Journal of Biological Chemistry, 280*, 41439–41448.

Sidorenko, S. P., Law, C. L., Klaus, S. J., Chandran, K. A., Takata, M., Kurosaki, T., et al. (1996). Protein kinase C mu (PKC mu) associates with the B cell antigen receptor complex and regulates lymphocyte signaling. *Immunity, 5*, 353–363.

Singh, R., & Czaja, M. J. (2007). Regulation of hepatocyte apoptosis by oxidative stress. *Journal of Gastroenterology and Hepatology, 22*(Suppl 1), S45–S48.

Sinnett-Smith, J., Zhukova, E., Hsieh, N., Jiang, X., & Rozengurt, E. (2004). Protein kinase D potentiates DNA synthesis induced by Gq-coupled receptors by increasing the duration of ERK signaling in swiss 3T3 cells. *Journal of Biological Chemistry, 279*, 16883–16893.

Sinnett-Smith, J., Zhukova, E., Rey, O., & Rozengurt, E. (2007). Protein kinase D2 potentiates MEK/ERK/RSK signaling, c-Fos accumulation and DNA synthesis induced by bombesin in Swiss 3T3 cells. *Journal of Cellular Physiology, 211*, 781–790.

Song, J., Li, J., Lulla, A., Evers, B. M., & Chung, D. H. (2006). Protein kinase D protects against oxidative stress-induced intestinal epithelial cell injury via Rho/ROK/PKC-{delta} pathway activation. *American Journal of Physiology. Cell Physiology, 290*, C1469–C1476.

Song, J., Li, J., Qiao, J., Jain, S., Mark Evers, B., & Chung, D. H. (2009). PKD prevents H_2O_2-induced apoptosis via NF-[kappa]B and p38 MAPK in RIE-1 cells. *Biochemical and Biophysical Research Communications, 378*, 610.

Song, M.-J., Wang, Y.-Q., & Wu, G.-C. (2007). Lipopolysaccharide-induced protein kinase D activation mediated by interleukin-1[beta] and protein kinase C. *Brain Research, 1145*, 19.

Spitaler, M., Emslie, E., Wood, C. D., & Cantrell, D. (2006). Diacylglycerol and protein kinase D localization during T lymphocyte activation. *Immunity, 24*, 535–546.

Stafford, M. J., Watson, S. P., & Pears, C. J. (2003). PKD: A new protein kinase C-dependent pathway in platelets. *Blood, 101*, 1392–1399.

Stein, G. S., Lian, J. B., van Wijnen, A. J., Stein, J. L., Montecino, M., Javed, A., et al. (2004). Runx2 control of organization, assembly and activity of the regulatory machinery for skeletal gene expression. *Oncogene, 23*, 4315–4329.

Storz, P. (2007). Mitochondrial ROS-radical detoxification, mediated by protein kinase D. *Trends in Cell Biology, 17*, 13.

Storz, P., Doppler, H., & Toker, A. (2004a). Protein kinase Cdelta selectively regulates protein kinase D-dependent activation of NF-kappaB in oxidative stress signaling. *Molecular and Cell Biology, 24*, 2614–2626.

Storz, P., Doppler, H., & Toker, A. (2004b). Activation loop phosphorylation controls protein kinase D-dependent activation of nuclear factor kappaB. *Molecular Pharmacology, 66*, 870–879.

Storz, P., Doppler, H., & Toker, A. (2005). Protein kinase D mediates mitochondrion-to-nucleus signaling and detoxification from mitochondrial reactive oxygen species. *Molecular and Cellular Biology, 25*, 8520–8530.

Storz, P., & Toker, A. (2003). Protein kinase D mediates a stress-induced NF-kappaB activation and survival pathway. *EMBO Journal, 22*, 109–120.

Sturany, S., Van Lint, J., Mueller, F., Wilda, M., Hameister, H., Hoecker, M., et al. (2001). Molecular cloning and characterization of the human protein kinase D2: A novel member of the protein kinase D family of serine threonine kinases. *Journal of Biological Chemistry, 276*, 3310–3318.

Sucharov, C. C., Langer, S., Bristow, M., & Leinwand, L. (2006). Shuttling of HDAC5 in H9C2 cells regulates YY1 function through CaMKIV/PKD and PP2A. *American Journal of Physiology. Cell Physiology, 291*, C1029–C1037.

Sumara, G., Formentini, I., Collins, S., Sumara, I., Windak, R., Bodenmiller, B., et al. (2009). Regulation of PKD by the MAPK p38delta in insulin secretion and glucose homeostasis. *Cell, 136*, 235–248.

Suzuki, A., & Ohno, S. (2006). The PAR-aPKC system: Lessons in polarity. *Journal of Cell Science, 119*, 979–987.

Tang, H.-B., Inoue, A., Oshita, K., & Nakata, Y. (2004). Sensitization of vanilloid receptor 1 induced by bradykinin via the activation of second messenger signaling cascades in rat primary afferent neurons. *European Journal of Pharmacology, 498*, 37–43.

Toker, A., & Cantley, L. C. (1997). Signalling through the lipid products of phosphoinositide-3-OH kinase. *Nature (London), 387*, 673–676.

Trauzold, A., Schmiedel, S., Sipos, B., Wermann, H., Westphal, S., Roder, C., et al. (2003). PKCmu prevents CD95-mediated apoptosis and enhances proliferation in pancreatic tumour cells. *Oncogene, 22*, 8939–8947.

Valverde, A. M., Sinnett-Smith, J., Van Lint, J., & Rozengurt, E. (1994). Molecular cloning and characterization of protein kinase D: A target for diacylglycerol and phorbol esters with a distinctive catalytic domain. *Proceedings of the National Academy of Sciences of the United States of America, 91*, 8572–8576.

Van Lint, J., Ni, Y., Valius, M., Merlevede, W., & Vandenheede, J. R. (1998). Platelet-derived growth factor stimulates protein kinase D through the activation of phospholipase Cgamma and protein kinase C. *Journal of Biological Chemistry, 273*, 7038–7043.

Van Lint, J. V., Sinnett-Smith, J., & Rozengurt, E. (1995). Expression and characterization of PKD, a phorbol ester and diacylglycerol-stimulated serine protein kinase. *Journal of Biological Chemistry, 270*, 1455–1461.

Vega, R. B., Harrison, B. C., Meadows, E., Roberts, C. R., Papst, P. J., Olson, E. N., et al. (2004). Protein kinases C and D mediate agonist-dependent cardiac hypertrophy through nuclear export of histone deacetylase 5. *Molecular and Cellular Biology, 24*, 8374–8385.

von Blume, J., Knippschild, U., Dequiedt, F., Giamas, G., Beck, A., Auer, A., et al. (2007). Phosphorylation at Ser244 by CK1 determines nuclear localization and substrate targeting of PKD2. *EMBO Journal, 26*, 4619–4633.

von Wichert, G., Edenfeld, T., von Blume, J., Krisp, H., Krndija, D., Schmid, H., et al. (2008). Protein kinase D2 regulates chromogranin A secretion in human BON neuroendocrine tumour cells. *Cellular Signalling, 20*, 925.

Waldron, R. T., Iglesias, T., & Rozengurt, E. (1999a). The pleckstrin homology domain of protein kinase D interacts preferentially with the eta isoform of protein kinase C. *Journal of Biological Chemistry, 274*, 9224–9230.

Waldron, R. T., Iglesias, T., & Rozengurt, E. (1999b). Phosphorylation-dependent protein kinase D activation. *Electrophoresis, 20*, 382–390.

Waldron, R. T., Rey, O., Iglesias, T., Tugal, T., Cantrell, D., & Rozengurt, E. (2001). Activation loop Ser744 and Ser748 in protein kinase D are transphosphorylated in vivo. *Journal of Biological Chemistry, 276*, 32606–32615.

Waldron, R. T., Rey, O., Zhukova, E., & Rozengurt, E. (2004). Oxidative stress induces protein kinase C-mediated activation loop phosphorylation and nuclear redistribution of protein kinase D. *Journal of Biological Chemistry, 279*, 27482–27493.

Waldron, R. T., & Rozengurt, E. (2000). Oxidative stress induces protein kinase D activation in intact cells – involvement of Src and dependence on protein kinase C. *Journal of Biological Chemistry, 275*, 17114–17121.

Waldron, R. T., & Rozengurt, E. (2003). Protein kinase C phosphorylates protein kinase D activation loop Ser744 and Ser748 and releases autoinhibition by the pleckstrin homology domain. *Journal of Biological Chemistry, 278*, 154–163.

Wang, Q. J., Fang, T. W., Yang, D., Lewin, N. E., Van Lint, J., Marquez, V. E., et al. (2003). Ligand structure-activity requirements and phospholipid dependence for the binding of phorbol esters to protein kinase D. *Molecular Pharmacology, 64*, 1342–1348.

Wang, Y., Kedei, N., Wang, M., Wang, Q. J., Huppler, A. R., Toth, A., et al. (2004). Interaction between protein kinase C mu and the vanilloid receptor type 1. *Journal of Biological Chemistry, 279*, 53674–53682.

Wang, Y., Waldron, R. T., Dhaka, A., Patel, A., Riley, M. M., Rozengurt, E., et al. (2002). The RAS effector RIN1 directly competes with RAF and is regulated by 14-3-3 proteins. *Molecular and Cellular Biology, 22*, 916–926.

Watkins, J. L., Lewandowski, K. T., Meek, S. E., Storz, P., Toker, A., & Piwnica-Worms, H. (2008). Phosphorylation of the Par-1 polarity kinase by protein kinase D regulates 14-3-3 binding and membrane association. *Proceedings of the National Academy of Sciences of the United States of America, 105*, 18378–18383.

Ways, D. K., Cook, P. P., Webster, C., & Parker, P. J. (1992). Effect of phorbol esters on protein kinase C-zeta. *Journal of Biological Chemistry, 267*, 4799–4805.

Weinstein, I. B. (1988). The origins of human cancer: Molecular mechanisms of carcinogenesis and their implications for cancer prevention and treatment-twenty-seventh G.H.A. Clowes memorial award lecture. *Cancer Research, 48*, 4135–4143.

Widmann, C., Gibson, S., Jarpe, M. B., & Johnson, G. L. (1999). Mitogen-activated protein kinase: Conservation of a three-kinase module from yeast to human. *Physiological Reviews, 79*, 143–180.

Wong, C., & Jin, Z.-G. (2005). Protein kinase C-dependent protein kinase D activation modulates ERK signal pathway and endothelial cell proliferation by vascular endothelial growth factor. *Journal of Biological Chemistry, 280*, 33262–33269.

Woo, D. H., Jung, S. J., Zhu, M. H., Park, C. K., Kim, Y. H., Oh, S. B., et al. (2008). Direct activation of transient receptor potential vanilloid 1(TRPV1) by diacylglycerol (DAG). *Molecular Pain, 4*, 42.

Woods, A. J., White, D. P., Caswell, P. T., & Norman, J. C. (2004). PKD1/PKCmicro promotes alphavbeta3 integrin recycling and delivery to nascent focal adhesions. *EMBO Journal, 23*, 2531–2543.

Xu, L., & Bergan, R. C. (2006). Genistein inhibits matrix metalloproteinase type 2 activation and prostate cancer cell invasion by blocking the transforming growth factor beta-mediated activation of mitogen-activated protein kinase-activated protein kinase 2-27-kDa heat shock protein pathway 10.1124/mol.106.023861. *Molecular Pharmacology, 70*, 869–877.

Yeaman, C., Ayala, M. I., Wright, J. R., Bard, F., Bossard, C., Ang, A., et al. (2004). Protein kinase D regulates basolateral membrane protein exit from trans-Golgi network. *Nature Cell Biology, 6*, 106–112.

Yin, D. M., Huang, Y. H., Zhu, Y. B., & Wang, Y. (2008). Both the establishment and maintenance of neuronal polarity require the activity of protein kinase D in the Golgi apparatus. *Journal of Neuroscience, 28*, 8832–8843.

Yu, W., Murray, N. R., Weems, C., Chen, L., Guo, H., Ethridge, R., et al. (2003). Role of cyclooxygenase 2 in protein kinase C beta II-mediated colon carcinogenesis. *Journal of Biological Chemistry, 278*, 11167–11174.

Yuan, J. Z., Bae, D., Cantrell, D., Nel, A. E., & Rozengurt, E. (2002). Protein kinase D is a downstream target of protein kinase C theta. *Biochemical and Biophysical Research Communication, 291*, 444–452.

Yuan, J., Lugea, A., Zheng, L., Gukovsky, I., Edderkaoui, M., Rozengurt, E., et al. (2008). Protein kinase D1 mediates NF-kappaB activation induced by cholecystokinin and cholinergic signaling in pancreatic acinar cells. *American Journal of Physiology. Gastrointestinal and Liver Physiology, 295*, G1190–G1201.

Yuan, J., & Rozengurt, E. (2008). PKD, PKD2, and p38 MAPK mediate Hsp27 serine-82 phosphorylation induced by neurotensin in pancreatic cancer PANC-1 cells. *Journal of Cellular Biochemistry, 103*, 648–662.

Yuan, J., Slice, L. W., Gu, J., & Rozengurt, E. (2003). Cooperation of Gq, Gi, and G12/13 in protein kinase D activation and phosphorylation induced by lysophosphatidic acid. *Journal of Biological Chemistry, 278*, 4882–4891.

Yuan, J., Slice, L. W., & Rozengurt, E. (2001). Activation of protein kinase D by signaling through Rho and the alpha subunit of the heterotrimeric G protein G13. *Journal of Biological Chemistry, 276*, 38619–38627.

Yuan, J. Z., Slice, L., Walsh, J. H., & Rozengurt, E. (2000). Activation of protein kinase D by signaling through the alpha subunit of the heterotrimeric G protein G(q). *Journal of Biological Chemistry, 275*, 2157–2164.

Zhang, N., Inan, S., Cowan, A., Sun, R., Wang, J. M., Rogers, T. J., et al. (2005a). A proinflammatory chemokine, CCL3, sensitizes the heat- and capsaicin-gated ion channel TRPV1. *Proceedings of the National Academy of Sciences of the United States of America, 102*, 4536–4541.

Zhang, W., Zheng, S., Storz, P., & Min, W. (2005b). Protein kinase D specifically mediates apoptosis signal-regulating kinase 1-JNK signaling induced by H_2O_2 but not tumor necrosis factor. *Journal of Biological Chemistry, 280*, 19036–19044.

Zheng, C., Lin, Z., Zhao, Z. J., Yang, Y., Niu, H., & Shen, X. (2006). MK2-mediated formation and phosphorylation-regulated dissociation of the signal complex consisting of p38, MK2, Akt and Hsp27 10.1074/jbc.M603622200. *Journal of Biological Chemistry*, M603622200.

Zhukova, E., Sinnett-Smith, J., & Rozengurt, E. (2001a). Protein kinase D potentiates DNA synthesis and cell proliferation induced by bombesin, vasopressin, or phorbol esters in Swiss 3T3 cells. *Journal of Biological Chemistry, 276*, 40298–40305.

Zhukova, E., Sinnett-Smith, J., Wong, H., Chiu, T., & Rozengurt, E. (2001b). CCKB/gastrin receptor mediates synergistic stimulation of DNA synthesis and cyclin D1, D3, and E expression in Swiss 3T3 cells. *Journal of Cellular Physiology, 189*, 291–305.

Zugaza, J. L., Sinnett-Smith, J., Van Lint, J., & Rozengurt, E. (1996). Protein kinase D (PKD) activation in intact cells through a protein kinase C-dependent signal transduction pathway. *EMBO Journal, 15*, 6220–6230.

Zugaza, J. L., Waldron, R. T., Sinnett-Smith, J., & Rozengurt, E. (1997). Bombesin, vasopressin, endothelin, bradykinin, and platelet-derived growth factor rapidly activate protein kinase D through a protein kinase C-dependent signal transduction pathway. *Journal of Biological Chemistry, 272*, 23952–23960.

Chapter 8
PKC and Control of the Cell Cycle

Jennifer D. Black

Abstract Members of the PKC family have been widely implicated in control of cell proliferation. Consistent with this role, PKC signaling can negatively or positively modulate the cell cycle at multiple stages, including cell cycle entry and exit, progression through G_1 and S phases, and transit through the G_1 and G_2 checkpoints. The cell cycle-specific effects of PKCs are dependent on the timing and duration of PKC activation, the specific PKC isozyme(s) involved, and the cellular context. Various cell cycle regulatory molecules, including cyclins, cyclin-dependent kinases (Cdks), and Cdk inhibitors, have been implicated in PKC-induced cell cycle effects, with $p21^{Waf1/Cip1}$ and cyclin D1 emerging as key targets of PKC control. $p21^{Waf1/Cip1}$ can be targeted by distinct PKC isozymes at different stages of the cell cycle to control both the $G_1 \rightarrow S$ and $G_2 \rightarrow M$ transitions, while cyclin D1 expression is modulated by transcriptional or translational mechanisms to regulate progression through G1 phase. PKC signaling can also phosphorylate lamin B to promote nuclear lamina disassembly and $G_2 \rightarrow M$ progression. Although understanding of the specific functions of individual PKC isozymes in regulation of the cell cycle remains a major challenge for the future, accumulated evidence indicates that PKCα and δ can either inhibit or promote $G_1 \rightarrow S$ and $G_2 \rightarrow M$ progression in a highly context-dependent manner, while PKCβII and ε are predominantly cell cycle stimulatory, and PKCη is generally inhibitory. Elucidation of the complex mechanisms underlying PKC isozyme-mediated control of the cell cycle is critical for the development of novel anticancer therapies targeting individual PKCs or their downstream effectors.

J.D. Black (✉)
Department of Pharmacology and Therapeutics, Roswell Park Cancer Institute,
Elm and Carlton Streets, Buffalo, NY 14263, USA
e-mail: jennifer.black@roswellpark.org

M.G. Kazanietz (ed.), *Protein Kinase C in Cancer Signaling and Therapy*,
Current Cancer Research, DOI 10.1007/978-1-60761-543-9_8,

Keywords PKC • Cell cycle • G1 progression • G1 and G2/M arrest • Cell cycle exit • G2/M progression • Senescence • S phase • Cyclin D1 • p21Waf1/Cip1 • PKCalpha • PKCδ • PKCβII • PKC… • PKCθ • PKCζ and ι

8.1 Introduction

The involvement of members of the protein kinase C (PKC) family in regulation of cell growth and cell cycle progression is well established. Early studies supported a role for PKC signaling in growth stimulation and mitogenesis (Dicker and Rozengurt 1978; Rozengurt 1986; Takuwa et al. 1988), findings that were consistent with the identification of the enzyme as a major cellular receptor for the phorbol ester class of tumor promoters (Castagna et al. 1982; Kikkawa et al. 1983; Leach et al. 1983). However, it was soon recognized that PKC-mediated pathways can also potently promote cell growth arrest and differentiation in a wide variety of cellular systems (Huang and Ives 1987; Kariya et al. 1987; Rovera et al. 1979; Yamamoto et al. 1988). The contrasting effects of PKC signaling on cell growth are reflected in both positive and negative control of cell cycle transitions, which appear to be dependent on the cellular context, the timing and duration of PKC activation during the cell cycle, and the specific PKC isozymes involved. Initial studies on PKC-mediated cell cycle-specific effects were extensively described in several comprehensive reviews published in the late 1990s and in 2000 (Black 2000; Fishman et al. 1998; Livneh and Fishman 1997). Here, we discuss major advances and new concepts that have emerged during the past eight years.

By the late 1990s, accumulated evidence pointed to PKCs as key regulators of the cell cycle at two stages, in G_1 phase and at the G_2/M transition (Fishman et al. 1998; Livneh and Fishman 1997). Limited evidence also supported a role for these enzymes in control of cell cycle entry and exit (Vrana et al. 1998; Wang et al. 2000; Frey et al. 2000). Most of the mechanistic information available was related to negative regulation of these transitions (Black 2000). As discussed below, it is now clear that PKCs can negatively and positively regulate the cell cycle at multiple points, including S-phase, in an isozyme-dependent manner. Signal strength and duration can play a determining role in outcome. Various cell cycle regulatory molecules [cyclins, cyclin-dependent kinases (Cdks), and Cdk inhibitors (CKIs)] have been implicated in PKC's effects, with $p21^{Waf1/Cip1}$ and cyclin D1 emerging as key targets of PKC control. Notably, the cell cycle effects of PKCs are highly context-dependent. A single PKC isozyme can exert opposite effects on a specific cell cycle target in different cellular systems, and distinct PKC isozymes can modulate the same target in different phases of the cell cycle to produce divergent cell cycle-specific responses. Activation of PKCs in one phase of the cell cycle can lead to effects in a different phase, and a single PKC can inhibit one cell cycle transition while stimulating another. Regulation of cell cycle progression by the PKC enzyme system thus exhibits a high degree of complexity. In order to take advantage of PKC-mediated cell cycle effects for the development of novel anticancer therapies, an in-depth understanding of underlying mechanisms is clearly essential.

8.1.1 Regulation of Cell Cycle Progression

8.1.1.1 The Mammalian Cell Cycle Machinery

According to the classical view, progression through the cell cycle is a highly coordinated process that is driven by heterodimeric enzymes consisting of a catalytic subunit, or Cdk, and a regulatory cyclin subunit (Sherr and Roberts 2004). Activation of Cdks is dependent on association with partner cyclins, which are generally transiently expressed during the cell cycle. The key cell cycle checkpoints, the $G_1 \rightarrow S$ and $G_2 \rightarrow M$ transitions, are controlled by specific cyclin/Cdk complexes. Progression through early G_1 depends on Cdk4 and Cdk6, which are activated by association with one of the three D-type cyclins (D1, D2, D3), while transit through late G_1 and into S phase involves Cdk2, regulated by cyclin E. Synthesis of the D-type cyclins and early G_1 progression are dependent on growth factors. Major targets of activated Cdk4,6/cyclin D and Cdk2/cyclin E complexes are members of the pocket protein family, the retinoblastoma protein (pRb), p107, and p130 (Cobrinik 2005). In their hypophosphorylated state, pocket proteins act as repressors of E2F transcription factors (Trimarchi and Lees 2002), preventing the expression of genes necessary for DNA replication (Du 2006). Phosphorylation by Cdk4,6/cyclin D at a subset of available sites relieves repression of E2F, leading to upregulation of cyclin E. Functional Cdk2/cyclin E complexes complete pocket protein phosphorylation, promoting release of E2F and enabling a wave of transcriptional activity essential for S phase progression. These events drive cells through the restriction point, after which the cell cycle progresses in the absence of growth factors. When mitogenic signals are not strong enough to enhance Cdk activity and inactivate pRb, cells exit the cell cycle and acquire a reversible nonreplicative state, quiescence or G_0, which is associated with changes in the phosphorylation and expression levels of pocket protein family members (Classon and Dyson 2001; Grana et al. 1998). Levels of p130 phosphoforms 1 and 2 markedly increase during cell cycle withdrawal, coinciding with the accumulation of p130/E2F complexes. p107, on the other hand, is not expressed in quiescent cells, while levels of pRb are generally similar in quiescent and cycling cells.

Cdk2/cyclin E activity has also been implicated in initiating DNA replication by facilitating loading of licensing factors onto origins of replication (Malumbres and Barbacid 2005). Once cells enter S phase, Cdk2/cyclin E activity is inhibited by rapid proteasomal degradation of cyclin E, thus avoiding DNA rereplication (Hwang and Clurman 2005). Continued inactivation of pRb also allows the transcription of genes necessary for subsequent phases of the cell cycle, including cyclin A and cyclin B. A-type cyclins accumulate during late G_1 and S phase and associate with Cdk2, resulting in phosphorylation of numerous proteins believed to be required for completion and exit from S phase (e.g., pRb, E2F1, cdc6, $p21^{Waf1/Cip1}$). The G2/M transition is regulated by Cdk1 (Cdc2) in association with cyclins

A and B (Malumbres and Barbacid 2005; Sanchez and Dynlacht 2005). At the end of S-phase, cyclin A associates with Cdk1 to promote phosphorylation of several proteins involved in DNA replication and cell cycle progression. During G_2, cyclin A is degraded, while B-type cyclins are actively synthesized; Cdk1 thus associates with cyclin B to trigger mitosis.

Multiple mechanisms control the activity of cyclin/Cdk complexes, including degradation of cyclins, positive and negative phosphorylation events, and association with CKIs. Positive phosphorylation of Cdks is mediated by Cdk activating kinase (CAK or Cdk7/cyclin H) (Fisher and Morgan 1994), while negative phosphorylation involves the kinases Wee1 and Myt1. Inhibitory phosphorylation is removed by Cdc25 phosphatases, a necessary step for full kinase activity. Cdk inhibition is also achieved by CKIs, which bind to cyclin/Cdk complexes and render them inactive (Sherr and Roberts 1995). The Cip/Kip CKIs, which include $p21^{Waf1/Cip1}$, $p27^{Kip1}$, and p57, inhibit cyclin E-, cyclin A- and cyclin B-dependent activity (i.e., Cdk2 and Cdk1). Members of the INK4 family (p15, p16, p18, and p19), on the other hand, are specific inhibitors of Cdk4 and Cdk6. Notably, $p21^{Waf1/Cip1}$ and $p27^{Kip1}$ act as positive regulators of cyclin D-dependent kinases (Sherr and Roberts 1999).

Cell-cycle progression can be blocked at the $G_1 \rightarrow S$ and $G_2 \rightarrow M$ checkpoints as well as in S phase and mitosis. Inhibitory pathways usually activate $p21^{Waf1/Cip1}$ or other CKIs, which block Cdk2 and/or Cdk1 activity. The balance between mitogenic, antimitogenic, apoptotic, and stress response signals ultimately determines cell fate and the ability to proliferate.

8.1.1.2 A Minimal Model of Cell Cycle Control

It should be noted that recent work with gene-targeted mice has revealed considerable redundancy within the classical model of cell cycle regulation. It is now clear that Cdks and cyclins have overlapping functions and that only a limited subset of these molecules is absolutely required for control of cell cycle progression (Malumbres and Barbacid 2005; Hochegger et al. 2008; Berthet and Kaldis 2007; Malumbres 2005). Genetic evidence has demonstrated that (a) Cdk4, Cdk6, and Cdk2 are not required for the mitotic cell cycle (Santamaria et al. 2007); (b) Cdk4/Cdk6 are not essential for mitogen-induced entry into the cell cycle; and (c) the inhibitory and tumor suppressor activities of the Cip/Kip CKIs can occur in the absence of Cdk2. However, a strict requirement for these molecules has been noted in certain specialized cells: Cdk2 is essential for the meiotic cell cycle in germ cells, Cdk4 is required in the pancreas, and Cdk6 is critical for erythropoiesis.

On the basis of these findings, the following minimal model for cell cycle control has emerged (Hochegger et al. 2008): any combination of Cdk1 or Cdk2, partnered with nuclear cyclin E or cyclin A, is sufficient to trigger S phase. Completion of S phase and entry into mitosis probably requires cyclin A, while cyclin B is essential for raising the activity of Cdk1 above the threshold level required for mitosis. It has been proposed that the difference between interphase and mitotic Cdks is not related to substrate specificity but rather to differential localization and a higher activity threshold for mitosis than interphase (Hochegger et al. 2008; Stern and Nurse 1996). Although this is a rapidly evolving field, and many questions remain unanswered, it is important to bear this model in mind when interpreting the cell cycle-specific effects of PKC isozymes in different biological systems, and evaluating their potential use as therapeutic targets in cancer.

8.2 Cell Cycle-Specific Effects of PKC Signaling: Regulation by Timing and Duration of PKC Activation

Multiple points in the cell cycle are targets for regulation by PKC signaling. Studies in a wide variety of cell types (e.g., epithelial, endothelial, hematopoietic, smooth muscle, and neuronal) have identified PKCs as important positive and negative modulators of cell cycle entry and exit (Vrana et al. 1998; Frey et al. 2000; Wang et al. 1998; Santiago-Walker et al. 2005), the G_1 and G_2 checkpoints (Black 2000; Fishman et al. 1998; Livneh and Fishman 1997), as well as transit through S phase (Harrington et al. 1997; Kinzel et al. 1980; Oliva et al. 2008). This section highlights accumulated evidence on the importance of timing and duration of PKC signaling in determining PKC-mediated cell cycle-specific responses.

Many of the studies addressing PKC-mediated regulation of the cell cycle have used pharmacological agonists as tools to mimic competence factors and modulate PKC activity directly. Examples include phorbol esters such as phorbol 12-myristate 13-acetate [PMA; also known as 12-O-tetradecanoylphorbol-13-acetate (TPA)] and phorbol 12,13-dibutyrate (PDBu), the macrocyclic lactone bryostatin, or membrane-permeant diacylglycerol (DAG) analogs. Despite the inherent limitations associated with the use of these agents (Griner and Kazanietz 2007), including lack of specificity for individual PKC family members, the existence of "non-PKC" targets for these molecules, and their ability to promote PKC downregulation, a large body of information on the cell cycle effects of PKCs has been generated using this approach. Addition of pharmacological PKC agonists to asynchronously growing cell populations has been shown to promote a biphasic cell cycle blockade in G_0/G_1 and G_2/M in a variety of nontransformed and transformed cell types [e.g., HeLa cells (Kinzel et al. 1980), melanoma cells (Coppock et al. 1992), intestinal epithelial cells (Frey et al. 1997; Clark et al. 2004), HL60 myelocytic leukemia cells (Millard et al. 1997), pancreatic cancer cells (Salabat et al. 2006), MCF7 breast cancer cells (Barboule et al. 1999), lung cancer cells (Oliva et al. 2008)]. Cells that are in G_1 remain in G_1 or exit the cell cycle into G_0, while cells that have progressed through the $G_1 \rightarrow S$ transition complete S phase and arrest in G_2. Recent studies using synchronized cell populations have identified additional effects of PKCs in S phase. For example, delayed S phase progression was noted in PMA-treated HeLa cells and NSCLC cells (Kinzel et al. 1980; Oliva et al. 2008), as well as in PKCδ overexpressing microvascular endothelial cells (Harrington et al. 1997). In addition, increased activity of PKCδ, but not PKCα or ε, in quiescent thyroid epithelial cells stimulated G_1/S phase progression prior to promoting cell cycle arrest and caspase-dependent apoptosis in S phase (Santiago-Walker et al. 2005). It appears that PKC activation does not generally modulate progression through M phase (Coppock et al. 1992; Arita et al. 1998; Kosaka et al. 1996; Barth and Kinzel 1994), although PKCδ-overexpressing Chinese Hamster Ovary cells were shown to arrest in telophase in response to phorbol ester treatment (Watanabe et al. 1992) and antimitotic effects of this isozyme were noted in 3Y1 murine fibroblasts (Kitamura et al. 2003).

Studies in synchronized cell populations have highlighted the importance of timing and duration of PKC signaling in determining cell cycle-specific responses (e.g., Kinzel et al. 1980; Kosaka et al. 1996; Zhou et al. 1993). Early work in vascular endothelial cells demonstrated an interesting bidirectional regulation of $G_1 \rightarrow S$ progression, depending on timing of PKC stimulation (Zhou et al. 1993, 1994). Short-term activation of PKC in early G_1 by PDBu potentiated $G_1 \rightarrow S$ progression, while activation in mid-to-late G_1 prevented entry of cells into S phase. The ability of PKC signaling to promote cell cycle progression in early G_1 phase is consistent with evidence that phorbol esters and DAG analogs can mimic the action of growth factors and induce the expression of immediate early genes such as *c-fos* and *c-myc* (Hug and Sarre 1993; Olashaw and Pledger 1988). In a series of elegant studies (Balciunaite et al. 2000; Balciunaite and Kazlauskas 2001, 2002), Balciunaite and Kazlauskas further demonstrated that the natural PKC agonist/growth factor PDGF stimulates PKC/PKCε activity in HepG2 hepatoma cells at two distinct times, within 10 min of addition and between 5 and 9 h of treatment. The first phase of PKC activity was dispensable for cell cycle progression, while late activity was required for PDGF-dependent S phase transit and DNA synthesis. DAG analogs were shown to recapitulate the cell cycle effects of PDGF signaling. Interestingly, the same effects were observed in serum-stimulated fibroblasts, although PKCδ was identified as the requisite isozyme in these cells (Kitamura et al. 2003). Additional studies by Kazlauskas and colleagues demonstrated that the strength/duration of late phase PKC activation determined the ability of PKC signaling to promote or inhibit progression into S phase (Balciunaite and Kazlauskas 2002). Thus, in contrast to the effects of DAG, which is rapidly metabolized by the cell, addition of the relatively stable and longer lasting PKC agonist PMA in mid-to-late G_1 phase inhibited PDGF-induced DNA synthesis in HepG2 cells. The role of signal duration in the cell cycle-specific effects of late G_1 phase PKC activation is further highlighted by the fact that DAG can inhibit cell cycle progression in various cell types, as long as it is added repeatedly to compensate for its rapid metabolism (Bi and Mamrack 1994; Kosaka et al. 1993; Sasaguri et al. 1993; Zezula et al. 1997).

The importance of timing and duration of PKC activation is not restricted to G_1 phase. Using human umbilical vein endothelial cells, Kosaka et al. (1996) demonstrated that activation of PKC signaling in G_2, but not in M phase, can promote G_2/M arrest. An interesting recent study by Kazanietz et al. in NSCLC cells further showed that G_2/M blockade can also result from stimulation of PKC in S phase, but not in G1 (Oliva et al. 2008). This study introduced the paradigm that activation of

PKC in one phase of the cell cycle can lead to effects in a different phase, and provided the first evidence that sustained activation of PKC in S phase can lead to an irreversible G_2/M arrest and induction of a senescence program.

8.3 Mechanisms of PKC-Mediated Cell Cycle Control

The cell cycle-specific effects of PKC signaling can involve transcriptional, translational, and/or posttranslational modulation. Numerous studies have explored the molecular mechanisms underlying PKC-mediated regulation of G_1→S and G_2→M progression, and the importance of p21$^{Waf1/Cip1}$ and cyclin D1 as targets of PKC control is now clear (see Figs. 8.1 and 8.2). Although information is more limited, several key reports are beginning to provide an understanding of the pathways involved in PKC-mediated cell cycle withdrawal into G_0 and induction of differentiation. On the other hand, effects of PKC in S and M phase remain poorly defined at a mechanistic level.

8.3.1 PKC Regulation of G1→S Phase Progression

8.3.1.1 Effects on Pocket Proteins and E2F Transcription Factors

Studies in a wide variety of systems have highlighted the ability of PKC signaling to modulate the activity of members of the pocket protein family, i.e., pRb, p130, and p107, key regulators of G_0/G_1→S progression (De Falco 2006). pRb was first identified as a target of PKC control by Zhou et al. (1993) in vascular endothelial cells. In some systems, PKC activation promotes pRb phosphorylation, concomitant with potentiation of cell cycle progression (Zhou et al. 1993). However, in most cases, the effect is inhibitory, resulting in G_1 arrest (Frey et al. 1997, 2000; Zhou et al. 1993; Fukumoto et al. 1997; Livneh et al. 1996; Sasaguri et al. 1996; Whyte and Eisenman 1992; Nakagawa et al. 2005; Afrasiabi et al. 2008). PKC signaling has also been shown to regulate the phosphorylation state and expression levels of the pRb-related proteins p107 and p130 in some cell types (Frey et al. 2000; Tibudan et al. 2002) (see below). Consistent with a role in pocket protein regulation, limited evidence supports the ability of PKCs to affect the expression and activity of members of the E2F family of transcription factors (Oliva et al. 2008; Zhou et al. 1994; Nakaigawa et al. 1996; Zhang and Chellappan 1996; Saunders et al. 1998). PKC signaling is a potent regulator of E2F1 mRNA and protein levels in several systems. For example, bimodal regulation of E2F1 message levels was reported in human umbilical vein endothelial cells, dependent on the timing of PKC activation by PMA in G_1 phase (Zhou et al. 1994). PMA-induced growth arrest was associated with rapid destabilization of E2F1 mRNA in normal human keratinocytes (Saunders et al. 1998), and reduced levels of E2F1 mRNA, accompanied by accumulation of E2F5/p130 complexes, was observed during phorbol ester-induced differentiation of U937 cells (Zhang and Chellappan 1996).

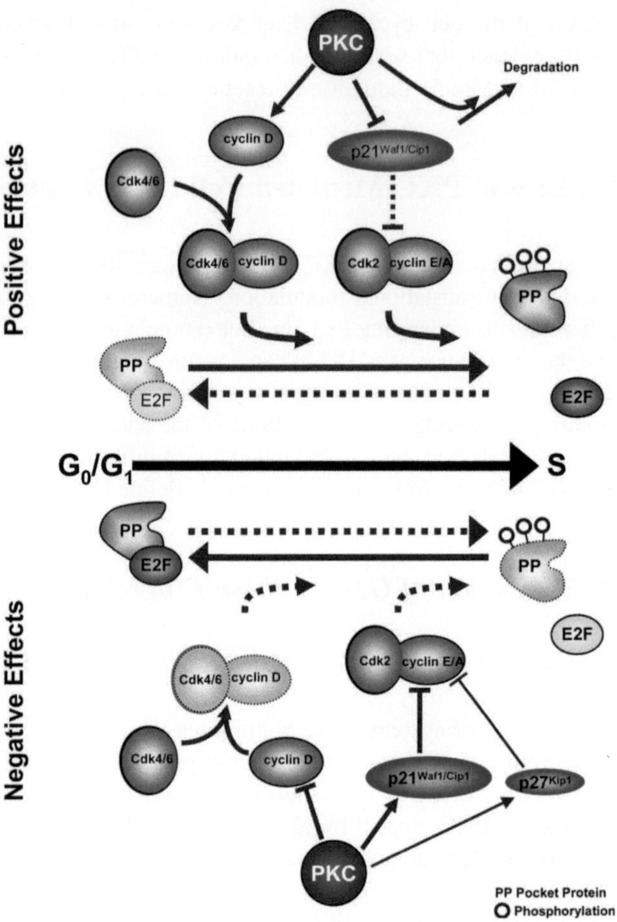

Fig. 8.1 Model of the molecular mechanisms underlying PKC-mediated regulation of $G_1 \rightarrow S$ progression. The upper part of the figure illustrates PKC-mediated cell cycle stimulatory pathways (positive effects), while the lower portion depicts PKC-regulated cell cycle inhibitory events (negative effects). PKC activation, usually in early G_1, can lead to increased levels of cyclin D1 and/or destabilization of $p21^{Waf1/Cip1}$ protein, resulting in hyperphosphorylation of pocket proteins and stimulation of $G_1 \rightarrow S$ progression. Activation of PKC in mid-to-late G1 induces rapid down-regulation of cyclin D1 and/or robust induction of Cip/Kip CKIs, usually $p21^{Waf1/Cip1}$. These effects result in inhibition of cdk4/6 and cdk2 activity, respectively, activation of pocket proteins and G_0/G_1 arrest

8.3.1.2 Control of Cdk Activity

Cdk2: The Role of $p21^{Waf1/Cip1}$

PKC-mediated regulation of pocket protein phosphorylation involves alterations in Cdk activity. A number of reports support a role for PKCs in regulation of the activity of Cdk2 (Frey et al. 2000; Zhou et al. 1993; Zezula et al. 1997; Livneh et al. 1996;

Sasaguri et al. 1996; Ashton et al. 1999; Hamada et al. 1996; Coppock et al. 1995), partnered with either cyclin E (Frey et al. 2000; Zezula et al. 1997; Livneh et al. 1996; Ashton et al. 1999) or cyclin A (Frey et al. 2000; Ashton et al. 1999). PKC-induced effects are generally inhibitory, although bimodal regulation of Cdk2 was observed during the $G_0 \rightarrow S$ transition in vascular endothelial cells (Zhou et al. 1993) and enhanced Cdk2 activity was observed in PKCδ overexpressing thyroid epithelial cells (Santiago-Walker et al. 2005). Analysis of a variety of systems indicates that modulation of Cdk2 activity is not usually the result of alterations in partner cyclin or Cdk expression. Although PKC-induced alterations in cyclin E/A levels have been noted in some cell types, it is likely that these changes are a consequence of PKC-induced cell cycle synchronization, rather than a cause of Cdk2 inhibition (Frey et al. 2000; Arita et al. 1998; Zhou et al. 1994; Kosaka et al. 1993; Zezula et al. 1997; Fukumoto et al. 1997; Sasaguri et al. 1996; Nakagawa et al. 2005; Coppock et al. 1995). For example, inhibitory effects of PKC signaling on cyclin A expression are likely secondary to PKC-induced pocket protein activation and inhibition of E2F activity, rather than a direct effect on cyclin A promoter activity (Nakagawa et al. 2005). This notion is further supported by the demonstration that PKC-induced inhibition of Cdk2 can occur under conditions in which cyclin E and cyclin A are not limiting (Frey et al. 2000; Livneh et al. 1996; Ashton et al. 1999; Coppock et al. 1995; Asiedu et al. 1997). With the exception of a study in thyroid cells overexpressing PKCδ (Santiago-Walker et al. 2005), there is no evidence that PKC signaling can affect Cdk2 expression (Black 2000), although PKC-mediated modulation of the activating phosphorylation at Thr160 has been reported in some systems (Hamada et al. 1996; Coppock et al. 1995; Asiedu et al. 1997, 1995). Alterations in CAK (Cdk7/cyclin H) activity (Hamada et al. 1996; Asiedu et al. 1995; Acevedo-Duncan et al. 2002; Coppock and Nathanson 1993) and PKC-induced Thr160 dephosphorylation (Asiedu et al. 1995, 1997; Kashiwagi et al. 2000) have been implicated in mediating the effect. As discussed below, these alterations are likely a consequence of PKC-induced increased levels of $p21^{Waf1/Cip1}$.

Accumulated evidence increasingly points to Cip/Kip CKIs, particularly $p21^{Waf1/Cip1}$, as key mediators of PKC-induced inhibition of Cdk2 activity. It is well established that PKC signaling induces the expression of $p21^{Waf1/Cip1}$ in a wide variety of cell types (Black 2000; Fishman et al. 1998; Livneh and Fishman 1997; Frey et al. 2000; Griner and Kazanietz 2007; Salabat et al. 2006; Nakagawa et al. 2005; Gavrielides et al. 2004; Cerda et al. 2006; Slosberg et al. 1999; Sugibayashi et al. 2001; Lin et al. 2002; Cabodi et al. 2000; Detjen et al. 2000). Induction is rapid and robust, generally transient, and can occur by several distinct mechanisms including increased transcription (Zezula et al. 1997; Zeng 1996; Akashi et al. 1999; Deeds et al. 2003), RNA stabilization (Akashi et al. 1999; Deeds et al. 2003; Park et al. 2001), altered translation (Zezula et al. 1997), and enhanced stability of the protein (Zezula et al. 1997). Transcriptional control appears to be mediated by Sp1/Sp3 transcription factors (Biggs et al. 1996; Prowse et al. 1997; Schavinsky-Khrapunsky et al. 2003; Sakaguchi et al. 2004; Traore et al. 2005). PKC-induced upregulation of $p21^{Waf1/Cip1}$ is almost always p53-independent (Zeng 1996; Akashi et al. 1999; Zhang et al. 1995; Todd and Reynolds 1998) and generally involves activation of the ERK/MAPK signaling pathway (Clark et al. 2004; Salabat et al. 2006; Zezula et al. 1997; Lin et al. 2002; Akashi et al. 1999; Liu et al. 1996; Esposito et al. 1997). PKC agonists promote increased association of $p21^{Waf1/Cip1}$ with Cdk2/cyclin E (Frey et al. 2000; Livneh et al. 1996; Ashton et al. 1999; Coppock et al. 1995; Asiedu et al. 1997) and Cdk2/ cyclin A (Frey et al. 2000) complexes, leading to attenuation of Cdk2 activity and cell cycle blockade. The importance of $p21^{Waf1/Cip1}$ in mediating PKC-induced G_1 arrest has been demonstrated directly using several approaches. For example, in contrast to its effects in wild-type mouse embryo fibroblasts (MEFs), PMA was unable to block PDGF-induced DNA synthesis in MEFs isolated from $p21^{Waf1/Cip1}$-null embryos (Balciunaite and Kazlauskas 2002). In addition, recent studies using $p21^{Waf1/Cip1}$ siRNA have confirmed a requirement for $p21^{Waf1/Cip1}$ in PKCδ-induced inhibition of G1→S progression in lung adenocarcinoma cells (Nakagawa et al. 2005). Similar effects have been observed in coronary smooth muscle cells (Bowles et al. 2007), where testosterone promoted cell cycle arrest via PKCδ-mediated induction of $p21^{Waf1/Cip1}$. Although upregulation of $p21^{Waf1/Cip1}$ generally results in cell cycle blockade, it should be noted that accumulation of this CKI can be required to promote rather than inhibit cell cycle progression in some systems. Antisense strategies revealed that PKCα-mediated induction of $p21^{Waf1/Cip1}$ is necessary for cyclin/Cdk complex formation and increased proliferation in glioma cells (Besson and Yong 2000).

PKC signaling can also lead to a reduction in $p21^{Waf1/Cip1}$ levels and accelerated mitogenesis (Walker et al. 2006). Recent studies in MEFs identified a role for PKCδ in mediating posttranscriptional destabilization of $p21^{Waf1/Cip1}$ via a proteasome-dependent mechanism, an effect that was associated with G_1→S progression. Loss of PKCδ, on the other hand, increased $p21^{Waf1/Cip1}$ levels and reduced entry into S phase, effects not observed in $p21^{Waf1/Cip1}$-null cells. Destabilization of $p21^{Waf1/Cip1}$ can also be induced by PKCζ, as demonstrated in HeLa cells, where the effect is dependent on PDK1 (Scott et al. 2002). In addition, inhibition of PKCε signaling in NSCLC cells by expression of kinase inactive, dominant negative enzyme led to p53-independent induction of $p21^{Waf1/Cip1}$ and cell growth arrest; similar effects were noted in fibroblasts following combined loss of PKCα and θ (Deeds et al. 2003). Finally, there is evidence that PKCs can mediate phosphorylation of $p21^{Waf1/Cip1}$ to regulate its stability, activity, and/or localization (Kashiwagi et al. 2000; Scott et al. 2002; Agell et al. 2006; Rodriguez-Vilarrupla et al. 2005).

PKC activation can also lead to increased expression of the CKI $p27^{Kip1}$ [11, 107], although induction is generally delayed compared with that of $p21^{Waf1/Cip1}$ (Black 2000; Frey et al. 1997, 2000; Tibudan et al. 2002; Asiedu et al. 1997). Limited evidence indicates that $p27^{Kip1}$ can be the only Cip/Kip CKI induced in response to PKC activation in some systems (e.g., Fukumoto et al. 1997; Asiedu et al. 1997). Although the mechanism(s) underlying PKC-induced accumulation of $p27^{Kip1}$ have not been extensively studied, posttranscriptional or posttranslational mechanisms appear to be involved (Asiedu et al. 1997). $p27^{Kip1}$ has been shown to accumulate in both cyclin E- and cyclin A-Cdk2 complexes, indicating that the molecule could potentially mediate PKC-induced suppression of Cdk2 activity in some cases (Frey et al. 2000; Asiedu et al. 1997). However, the role of $p27^{Kip1}$ in the cell cycle-specific effects of PKC signaling remains unclear based on its inability to mediate PKC-induced cell cycle arrest in the absence of $p21^{Waf1/Cip1}$ in some cell types (e.g., Vrana et al. 1998; Wang et al. 1998). The timing of $p27^{Kip1}$ induction suggests that it is a downstream event of $p21^{Waf1/Cip1}$ expression involved in initiation and/or maintenance of differentiation rather than in mediating cell cycle arrest (Frey et al. 1997, 2000; Wang et al. 1998; Tibudan et al. 2002; Yamamoto et al. 1999).

Cdk4/6: The Role of Cyclin D1 Modulation

Another important PKC target in G_1 phase appears to be the activity of Cdk4/6. Unlike Cdk2, which is largely modulated by PKC-induced alterations in the levels of the CKI $p21^{Waf1/Cip1}$ (see above), Cdk4/6 activity appears to be controlled by PKC-induced changes in partner cyclin expression. With the exception of a study in 293 T and HCT116 cells demonstrating PKC-dependent suppression of p18 promoter activity by PMA (Matsuzaki et al. 2004) and recent evidence that PKC can mediate induction of p15 and p16 in HepG2 cells (Wen-Sheng and Jun-Ming 2005; Wu and Hsu 2001), there is little support for the ability of PKC signaling to regulate members of the INK4 family of CKIs. In contrast, there is extensive evidence for PKC-mediated control of cyclin D1 expression. PKC regulation of cyclin D1 levels can be negative or positive, depending on the cell type and PKC isozyme profile. A reduction in cyclin D1 levels by PKC signaling has been noted in nontransformed intestinal epithelial cells (Frey et al. 2000, 2004; Clark et al. 2004; Hizli et al. 2006; Guan et al. 2007), colon cancer cells with restored expression of PKCα (Pysz et al. 2009) or PKCδ (Cerda et al. 2006), as well as in PKCδ overexpressing vascular smooth muscle cells (Fukumoto et al. 1997), primary bovine airway smooth muscle cells (Page et al. 2002), and NIH3T3 cells (Soh and Weinstein 2003). Conversely, loss of PKCδ activity has been shown to result in increased levels of cyclin D1 in colon cancer cells (Cerda et al. 2006) and bovine airway smooth muscle cells (Page et al. 2002). Interestingly, the bone remodeling peptide hormone, PTH-related protein, was recently shown to inhibit cyclin D1 expression and to markedly reduce Cdk4/6-cyclin D1 activity in differentiated osteoblasts via a PKC-dependent mechanism (Datta et al. 2005). Similarly, testosterone-induced G1 arrest in coronary smooth muscle cells was found to be associated with PKCδ-mediated downregulation of cyclin D1 (Bowles et al. 2007).

PKC-mediated inhibition of cyclin D1 expression can occur at the level of transcription or translation. Transcriptional blockade has been noted in several systems (e.g., Pysz et al. 2009; Page et al. 2002; Soh and Weinstein 2003). In bovine airway smooth muscle cells, PKCδ appears to attenuate cyclin D1 promoter activity via complex regulation of three distinct *cis*-acting promoter elements in a region -22 basepairs from the transcription start site (CRE/ATF2 and Ets enhancer sites, and an NF-κB suppressor site) (Page et al. 2002). In contrast, transcriptional inhibition of cyclin D1 by PKCα in colon cancer cells appears to involve promoter elements between −1745 and −163, highlighting the context-dependence of the effect (Pysz et al. 2009). Our recent studies in intestinal epithelial cells have further demonstrated that PKC can inhibit cyclin D1 expression via activation of the translational repressor 4E-BP1 and blockade of cap-dependent translation initiation (Hizli et al. 2006). Activation of 4E-BP1 involves PKCα and occurs via a phosphoinositide 3-kinase/Akt-independent, protein phosphatase 2A-dependent mechanism (Guan et al. 2007). PKCα promotes association of hypophosphorylated/active 4E-BP1 with the mRNA cap-binding protein eIF4E, accompanied by sequestration of cyclin D1 mRNA in 4E-BP1-associated complexes, thus inhibiting cyclin D1 protein synthesis (Hizli et al. 2006).

PKC signaling is also able to markedly enhance cyclin D1 expression in several cell types (Frey et al. 2000; Zhou et al. 1994; Nakagawa et al. 2005; Soh and Weinstein 2003; Li and Weinstein 2006; Grossoni et al. 2007; Yan and Wenner 2001; Huang et al. 1995; Mann et al. 1997). Cyclin D1 upregulation has been observed in response to treatment with pharmacological PKC activators, including phorbol esters (e.g., Frey et al. 2000; Nakagawa et al. 2005; Yan and Wenner 2001; Huang et al. 1995; Mann et al. 1997), bryostatin (Pysz et al. Unpublished data), or diacylglycerol analogs (added repeatedly during the course of the experiment) (Black et al. Unpublished data), as well as following exposure to physiological PKC agonists such as the neurohormone arginine vasopressin or the transmembrane protein polycystin-1 (Manzati et al. 2005; He et al. 2008). Exposure of adult rat cardiac fibroblasts to arginine vasopressin increased cyclin D1 levels and stimulated $G_0/G_1 \rightarrow S$ progression via a PKC-dependent pathway (He et al. 2008). Similarly, polycystin-1 enhanced HEK293 $G_0/G_1 \rightarrow S$ transit via activation of PKCα and upregulation of cyclin D1 and D3 (Manzati et al. 2005). Several members of the PKC family have been implicated in the effect, with overexpression of PKCα (Soh and Weinstein 2003), βI/βII (Li and Weinstein 2006), δ (Grossoni et al. 2007), ε (Soh and Weinstein 2003) , or η (Fima et al. 2001) resulting in hyperinduction of cyclin D1 in different cell types. The ERK/MAPK pathway appears to be a critical downstream player in PKC-induced cyclin D1 upregulation (Grossoni et al. 2007; He et al. 2008; Matsumoto et al. 2006), which generally results in hyperphosphorylation of pRb and enhanced cell cycle progression (Soh and Weinstein 2003; Li and Weinstein 2006; Grossoni et al. 2007; Yan and Wenner 2001; Huang et al. 1995; Fima et al. 2001). It should be noted, however, that cyclin D1 hyperinduction can also occur in cells undergoing PKC-induced cell cycle arrest or differentiation (Frey et al. 2000; Zhou et al. 1994; Nakagawa et al. 2005; Matsumoto et al. 2006), likely reflecting contrasting effects of individual isozymes activated at the same time, and pointing to the dominant effects of $p21^{Waf1/Cip1}$ induction and Cdk2 inhibition on cell cycle regulation in some cells (e.g., Nakagawa et al. 2005; Matsumoto et al. 2006). Interestingly, the opposite effect has also been observed, highlighting the complexity of PKC-mediated cell cycle regulation. For example, in MCF7 cells overexpressing PKCη under the control of a tetracycline responsive inducible promoter, the inhibitory effects of $p21^{Waf1/Cip1}$ were overcome by increased levels of cyclin D1, resulting in enhanced cell growth (Fima et al. 2001). PKC signaling has also been reported to regulate the compartmentalization of cyclin D1; activation of PKC induced rapid translocation of cyclin D1 to the nucleus in NIH3T3 cells (Lin et al. 2000).

PKC-induced increased expression of cyclin D1 appears to occur via transcriptional mechanism(s). Using a series of 5'-deleted cyclin D1 promoter constructs, Weinstein and colleagues demonstrated a role for an AP-1 enhancer element at position −194 upstream of the transcription start site (Soh and Weinstein 2003; Li and Weinstein 2006). AP-1 activity appears to mediate stimulation of the cyclin D1 promoter by PKCα/ε overexpression in NIH3T3 cells (Soh and Weinstein 2003) and by PKC βI/βII overexpression in MCF-7 cells (Li and Weinstein 2006). In contrast, our studies in nontransformed intestinal epithelial cells point to the involvement of a proximal 163-base region, indicating that PKC-mediated transcriptional control of the cyclin D1 promoter is highly cell type-dependent (Hao et al. Unpublished data).

8.3.2 PKC Regulation of G_2/M Progression

PKC signaling has been implicated in both negative and positive regulation of G_2/M phase progression (Fig. 8.2). As discussed above, negative regulation of $G_2 \rightarrow M$ transit can be achieved when PKC is activated in S or G_2 phase, but not in G_1 phase, and occurs even when PKC agonists are added near the end of G_2, pointing to rapid engagement of inhibitory mechanisms (Oliva et al. 2008; Arita et al. 1998; Kosaka et al. 1996; Barth and Kinzel 1994). PKC agonist-induced cell cycle arrest in G_2 is generally transient (Frey et al. 1997; Arita et al. 1998; Kosaka et al. 1996; Barth and Kinzel 1994), likely as a result of depletion of requisite PKC isozyme(s). Consistent with this notion, studies by Kazanietz and colleagues (Oliva et al. 2008) have recently demonstrated that sustained activation of PKCα, initially triggered in S phase, can lead to irreversible cell cycle arrest in G_2/M and induction of a senescence program (also see Cozzi et al. 2006). A key target of PKC-mediated negative regulation of the $G_2 \rightarrow M$ transition appears to be the Cdk1/cyclin B complex (Barboule et al. 1999; Arita et al. 1998; Kosaka et al. 1996; Barth and Kinzel 1994). Modulation of Cdk1/cyclin B activity is not generally associated with alterations in levels of cyclin B or Cdk1, although loss of cyclin B1 was observed in NSCLC cells undergoing senescence (Oliva et al. 2008). Instead, PKC-mediated suppression of this complex appears to be the result of (a) downregulation of the phosphatase Cdc25, which prevents dephosphorylation of Tyr15 on Cdk1, thereby preventing activation of the kinase (Arita et al. 1998; Kosaka et al. 1996; Barth and Kinzel 1994; Barth et al. 1996), or (b) induction of $p21^{Waf1/Cip1}$, likely via an ERK/MAPK-dependent mechanism (e.g., Oliva et al. 2008; Barboule et al. 1999; Arita et al. 1998; Tchou et al. 1996; Dangi et al. 2006). These mechanisms may, in fact, be related, since $p21^{Waf1/Cip1}$-mediated inhibition of Cdk2/cyclin A activity has been linked to downregulation of Cdc25 (Guadagno and Newport 1996; Niculescu et al. 1998). It should be noted that PKCα-induced irreversible G_2/M arrest and senescence in NSCLC cells depends on sustained $p21^{Waf1/Cip1}$ upregulation, which may explain why the G_2/M effects of PKC signaling are transient in many of the systems examined (e.g., Coppock et al. 1992; Frey et al. 1997).

PKC signaling can also play a role in positive regulation of the G_2/M transition. In a series of studies, Fields and colleagues demonstrated that PKCβII signaling is required for entry into mitosis in human erythroleukemia cells (Thompson and

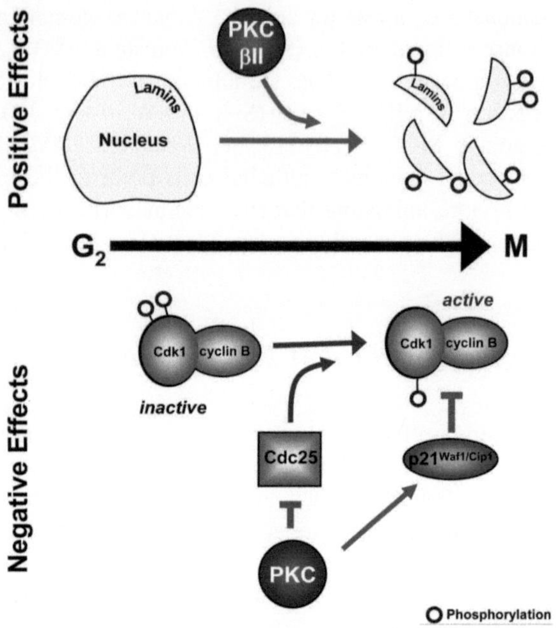

Fig. 8.2 Model of PKC-mediated regulation of the $G_2 \rightarrow M$ transition. PKC can stimulate $G_2 \rightarrow M$ progression by promoting nuclear lamina disassembly (upper portion of the figure). Alternatively, PKC activation can inhibit this transition by blocking Cdc25-mediated dephosphorylation of Cdk1 on Thr14 and Tyr15, a requisite step in Cdk1 activation. PKC signaling can also promote the accumulation of $p21^{Waf/Cip}$, which inhibits Cdk1/cyclin B complex activity

Fields 1996; Goss et al. 1994; Walker et al. 1995; Murray and Fields 1998). During G_2 phase, PKCβII is activated at the nuclear periphery by phosphatidylglycerol, leading to phosphorylation of lamin B and nuclear lamina disassembly. Inhibition of PKCβII by chelerythine chloride leads to profound G_2 arrest in this system. Interestingly, chelerythrine-induced cell cycle arrest does not involve inhibition of Cdk1/cyclin B activity, indicating that PKCβII and Cdk1 act in distinct pathways to regulate G2→M progression in these leukemic cells (Thompson and Fields 1996).

8.3.3 PKC Regulation of Cell Cycle Entry and Exit

The ability of PKC signaling to promote $G_0 \rightarrow G_1$ progression has been noted in several cell types (Santiago-Walker et al. 2005; Chiu et al. 2002, 2003), although the underlying mechanisms and key players have yet to be defined. PKC signaling has also been shown to promote cell cycle exit and differentiation in a number of systems, including intestinal epithelial cells, keratinocytes, PKC-overexpressing fibroblasts, and leukemic cell lines (Black 2000). It is well established that cell differentiation requires an irreversible cell cycle exit that is dominant under optimal growth conditions (Yee et al. 1998; Miller et al. 2007). Thus, prior to induction of tissue-specific gene expression, differentiation-inducing signals activate mechanisms of cell cycle arrest that lead to cell cycle withdrawal into G_0. It has been proposed that quiescence and differentiation-inducing signals are recognized by the restriction point machinery, leading to inhibition of Cdk activity, likely by rapid induction of CKIs in combination with other mechanisms (Miller et al. 2007; Mayol and Grana 1998). Cell cycle exit appears to involve a coordinated program of cell cycle regulatory events including induction of $p21^{Waf1/Cip1}$ and/or $p27^{Kip1}$, hypophosphorylation of the pocket proteins pRb and p107, inactivation of E2F-dependent transcription, downregulation of p107 protein, accumulation of p130 phosphoforms 1 and 2, and predominance of E2F4/p130 and E2F5/p130 complexes (Grana et al. 1998; De Falco 2006; Garriga et al. 1998; Smith et al. 1996). Other events associated with cell cycle withdrawal include rapid downregulation of D-type cyclins (Zwijsen et al. 1996; Diehl et al. 1997) and disappearance of DNA replication licensing factors such as Cdc6 (Fujita 1999). Studies in leukemia cells (Vrana et al. 1998; Wang et al. 1998; Zhang and Chellappan 1996), nontransformed intestinal epithelial cells (Frey et al. 2000), and keratinocytes (Tibudan et al. 2002) indicate that PKC family members are capable of activating a complete program of regulatory events associated with cell cycle exit. For example, work in our laboratory has shown that activation of PKCα in intestinal epithelial cells results in rapid downregulation of cyclin D1 and differential induction of $p21^{Waf1/Cip1}$ and $p27^{Kip1}$, thus targeting all of the major G_1/S Cdk complexes (Frey et al. 2000). These events are associated with coordinated alterations in the expression and phosphorylation of the pocket proteins p107, pRb, and p130, including downregulation of p107, hypophosphorylation of pRb, and accumulation of phosphoforms 1 and 2 of p130. Cell cycle arrest was also accompanied by loss of cdc6. A similar program was triggered by PKCα in keratinocytes induced to differentiate in suspension culture or by treatment with phorbol ester (Tibudan et al. 2002) and in PMA-treated pancreatic cancer cells (Detjen et al. 2000). Studies in several systems have demonstrated the ability of PKC signaling to promote loss of E2F1 mRNA and protein (Zhang and Chellappan 1996; Saunders et al. 1998), and phorbol ester-induced differentiation of U937 cells is associated with the appearance of E2F5/p130 complexes (Zhang and Chellappan 1996). The differentiation-inducing properties of PKCs are well established in many systems, particularly in keratinocytes (Denning 2004) and hematopoietic cells (Hocevar et al. 1992; Murray et al. 1993; Harris and Ralph 1985). Since cell cycle withdrawal appears to be a prerequisite for cell differentiation, it is likely that PKCs can trigger a program of cell cycle withdrawal in many systems.

8.4 Cell Cycle-Specific Effects of Individual PKC Family Members

While significant advances have been made in defining the mechanisms underlying PKC-mediated regulation of the cell cycle, a major challenge for the future remains understanding the specific function(s) of individual PKC isozymes. Progress has been hindered by the fact that many studies have relied on nonselective pharmacological PKC agonists, PKC isozyme overexpression strategies, or PKC inhibitors of questionable specificity (Griner and Kazanietz 2007). While RNA interference technology may provide a more informative approach, potential limitations include the need for a high level of silencing (e.g., >80% Pysz et al. Unpublished data; Cameron et al. 2008) and evidence that knockdown of one PKC isozyme can affect accumulation of other members of the family (Pysz et al. Unpublished data). The following sections discuss current understanding of PKC isozyme-specific cell cycle regulation.

Early experiments involving overexpression or selective activation of individual PKCs highlighted the opposing effects of PKC isozymes on cell proliferation. In a series of elegant studies, Weinstein and colleagues demonstrated that increased expression of PKCβI (Housey et al. 1988) or PKCε (Cacace et al. 1993) in rat embryo fibroblasts resulted in enhanced growth or transformation, while PKCα (Borner et al. 1991) led to marked suppression of cell proliferation. Similarly, overexpression of PKCε in mouse NIH3T3 cells resulted in decreased doubling time and increased saturation density, while overexpression of PKCδ produced the opposite phenotype (Mischak et al. 1993). The ability of PKCε to enhance cell growth and induce neoplastic transformation was also observed in epithelial cells of the colon (Perletti et al. 1998), while PKCδ showed growth suppressive effects in the same cells (Perletti et al. 1999). In erythroleukemia cells, PKCβII was shown to be essential for cell growth, while PKCα was implicated in control of cytostasis and megakaryocytic differentiation (Murray et al. 1993). Taken together, these initial findings generally supported a role for PKCα, δ, and η in negative regulation of cell cycle progression and/or differentiation, and for PKCβII and ε in growth stimulation (Black 2000). However, it was soon recognized that the actions of individual isozymes could be highly dependent on cellular context, with opposite effects seen in different biological systems (Murray et al. 1993; Housey et al. 1988; Choi et al. 1990; Gamard et al. 1994). As discussed below, more recent studies support this complexity, with each member of the PKC family shown to be capable of promoting or suppressing cell cycle progression in different cell types. Notably, in certain cases, a single enzyme can have opposite effects within the same cell, with the outcome depending on timing of stimulation and cell cycle phase (Kitamura et al. 2003).

8.4.1 Context-Dependent Cell Cycle-Specific Effects of PKCα and PKCδ

8.4.1.1 PKCα

PKCα has antiproliferative and differentiation-inducing effects in several cell types (Black 2000), including intestinal epithelial cells (Frey et al. 1997, 2000), keratinocytes (Tibudan et al. 2002), mammary epithelial cells (Slosberg et al. 1999), melanoma cells (Niles 2003), pancreatic cancer cells (Detjen et al. 2000), and leukemia cells (Hocevar et al. 1992; Murray et al. 1993), among others. The enzyme has been reported to inhibit both $G_1 \rightarrow S$ and $G_2 \rightarrow M$ progression, and to promote G_0 exit. Early studies by Sasaguri and colleagues demonstrated that downregulation of PKCα and ε in porcine aortic smooth muscle cells by preincubation with PMA not only prevented PDBu- and DiC_8-induced G_1 arrest but also accelerated entry into S phase (Sasaguri et al. 1993). Use of selective pharmacological inhibitors, antisense technology, or siRNA has since confirmed the role of PKCα in cell cycle arrest in other systems (Frey et al. 1997, 2000; Oliva et al. 2008; Clark et al. 2004; Detjen et al. 2000; Wen-Sheng and Jun-Ming 2005; Scaglione-Sewell et al. 1998). PKCα-mediated $G_1 \rightarrow S$ blockade appears to involve downregulation of cyclin D1 (Detjen et al. 2000; Hizli et al. 2006; Guan et al. 2007) and/or induction of the Cip/Kip CKIs $p21^{Waf1/Cip1}$ (Black 2000; Frey et al. 1997, 2000; Clark et al. 2004; Tibudan et al. 2002; Slosberg et al. 1999; Detjen et al. 2000; Abraham et al. 1998) and $p27^{Kip1}$ (Frey et al. 1997, 2000; Tibudan et al. 2002; Detjen et al. 2000) in diverse cell types. Activation of the enzyme in S phase results in delayed S phase progression and irreversible, $p21^{Waf1/Cip1}$-dependent blockade in G_2/M, associated with induction of a senescence program (Oliva et al. 2008). Interestingly, expression of bovine PKCα in *Saccharomyces cerevisiae* resulted in accumulation of cells in G_2/M phase and inhibition of chromosome segregation, cytokinesis, and septum formation (Sprowl et al. 2007). The effect may involve activation of a subunit of the PP2A phosphatase complex (cdc55), a component of the mitotic spindle checkpoint.

Growth stimulatory effects of PKCα have also been reported in several cell types, including glioma cells (Besson and Yong 2000; Mandil et al. 2001), osteoblasts (Lampasso et al. 2002), chick embryo hepatocytes (Alisi et al. 2004), hepatocellular carcinoma cells (Wu et al. 2008), and myoblasts (Buitrago et al. 2003). PKCα is necessary and sufficient to promote cell cycle progression in glioma cells via a $p21^{Waf1/Cip1}$-dependent mechanism (Besson and Yong 2000). Thyroid hormone-induced $G_1 \rightarrow S$ progression in hepatocytes appears to be mediated by PKCα, and involves increased levels of cyclin D1 and Cdk4 as well as enhanced cyclinE/A complex activity (Alisi et al. 2004). PKCα has also been reported to stimulate the ERK/MAPK pathway in various cell types (Schonwasser et al. 1998; Shatos et al. 2008) and to enhance cyclin D1 levels in hepatocellular carcinoma cells (Wu et al. 2008).

Consistent with the contrasting growth regulatory effects of PKCα activation described above, physiological agonists can either upregulate or decrease PKCα levels to produce growth arrest in different systems. In addition, the enzyme can mediate opposing cell cycle-specific effects of these agents dependent on context. For example, while all-*trans* retinoic acid (ATRA) inhibited $G_1 \rightarrow S$ progression in SKRB-3 breast cancer cells by decreasing PKCα expression and ERK/MAPK activity (Nakagawa et al. 2003), PKCα was found to be required for ATRA-induced growth arrest in T-47D breast cancer cells (Cho et al. 1997). PKCα appears to mediate both the growth-inhibitory (Chen et al. 1999; Bikle et al. 2001) and growth-stimulatory (Buitrago et al. 2003) effects of vitamin D in different systems. Similarly, the enzyme has been implicated in transforming growth factor-β (TGF-β)-induced growth arrest (Sakaguchi et al. 2004) as well as proliferation (Chow et al. 2008). It should be noted, however, that morphological and/or biochemical analysis of murine and human tissues has demonstrated that PKCα is strongly associated with cell membranes of postmitotic intestinal epithelial cells (Frey et al. 2000; Saxon et al. 1994) and epidermal keratinocytes (Tibudan et al. 2002), pointing to translocation and activation of the enzyme in association with growth arrest and differentiation in vivo. Furthermore, consistent with a growth suppressor role, PKCα is lost early during intestinal carcinogenesis (Black 2001) and PKCα knockout mice show increased proliferative activity within intestinal crypts and spontaneous intestinal adenoma formation. Importantly, PKCα-deficient $Apc^{Min/+}$ mice develop more aggressive tumors and exhibit reduced survival relative to PKCα-expressing littermates (Oster and Leitges 2006).

8.4.1.2 PKCδ

As is the case for PKCα, the cell cycle-specific effects of PKCδ exhibit a high degree of complexity and cell type variability. The enzyme has been reported to inhibit cell cycle progression in G_1 (Fukumoto et al. 1997; Nakagawa et al. 2005; Ashton et al. 1999; Cerda et al. 2006) and G_2/M phase (Watanabe et al. 1992) in response to pharmacological activators or physiological PKCδ agonists such as inositol hexaphosphate (IP6) (Vucenik et al. 2005), ATRA (Kambhampati et al. 2003), interferons (Uddin et al. 2002), and testosterone (Bowles et al. 2007). PKCδ signaling directly or indirectly targets G_1 cyclins (cyclins D1, E, and A) and/or Cip/Kip CKIs ($p21^{Waf1/Cip1}$ and/or $p27^{Kip1}$) to block $G_1 \rightarrow S$ phase progression in several cell types (Vrana et al. 1998; Fukumoto et al. 1997; Nakagawa et al. 2005; Ashton et al. 1999; Cerda et al. 2006). Recent studies using specific siRNAs have confirmed a requirement for PKCδ and $p21^{Waf1/Cip1}$ upregulation in phorbol ester-induced G_1 arrest in lung adenocarcinoma cells (Nakagawa et al. 2005). On the other hand, IP6 blocks G_1 progression in MCF-7 cells via PKCδ-dependent $p27^{Kip1}$ induction and pRb hypophosphorylation, consistent with several studies linking PKCδ signaling to $p27^{Kip1}$ induction (Fukumoto et al. 1997; Ashton et al. 1999). Inhibitory effects of PKCδ in G_2/M have been reported in CHO cells (Watanabe et al. 1992), where stable overexpression of the enzyme results in accumulation of cells in telophase in response to PMA. PKCδ also has a negative effect on mitosis in 3Y1 murine fibroblasts (Kitamura et al. 2003).

While PKCδ signaling has been linked to inhibition of cell proliferation in many systems, increasing evidence points to an additional role of the enzyme in positive regulation of the cell cycle (Kitamura et al. 2003; Jackson and Foster 2004; Cho et al. 2004). For example, PKCδ is required for insulin-like growth factor 1-induced proliferation (Czifra et al. 2006) and can stimulate $G_0/G_1 \rightarrow S$ progression in several cell types (Santiago-Walker et al. 2005; Kitamura et al. 2003; Grossoni et al. 2007). The growth stimulatory effects of the enzyme can be associated with increased expression of G_1 cyclins, including cyclin D1 (Grossoni et al. 2007; Black et al. Unpublished data), cyclin E, and cyclin A (Santiago-Walker et al. 2005), posttranscriptional destabilization of $p21^{Waf1/Cip1}$ (Walker et al. 2006), enhanced Cdk2 expression and activity (Santiago-Walker et al. 2005), and increased E2F promoter activity in late G_1 phase (Nakaigawa et al. 1996). In many cases, PKCδ-induced mitogenesis involves activation of the ERK/MAPK pathway (Grossoni et al. 2007; Jackson and Foster 2004). It has been proposed that the opposing effects of PKCδ on cell cycle progression are regulated by differential tyrosine phosphorylation of the enzyme (Steinberg 2004; Acs et al. 2000). In this regard, mutation of Tyr155 on PKCδ to phenylalanine dramatically altered the effects of the enzyme on growth of NIH3T3 cells: while the wild-type enzyme inhibited NIH3T3 cell growth, the mutant promoted growth and tumorigenicity of the cells (Acs et al. 2000). Studies with PKCδ/ε chimeras have identified the carboxy-terminal 51 amino acids of PKCδ as critical for the cell cycle promoting effects of the enzyme (Kitamura et al. 2003). Interestingly, the entire PKCδ molecule appears to be required for suppression of mitosis (Acs et al. 1997), suggesting that specific regions mediate differential association with substrates or binding proteins to modulate different stages of the cell cycle.

8.4.2 Cell Cycle Stimulatory Roles of PKCβII and PKCε

8.4.2.1 PKCβII

The cell cycle-specific effects of PKCβII appear to be largely stimulatory. As discussed above, studies by Fields and colleagues have established a role for this isozyme in promoting the $G_2 \rightarrow M$ transition in leukemia cells (Thompson and Fields 1996; Goss et al. 1994; Walker et al. 1995; Murray and Fields 1998). Consistent with this role, Newton et al. (Chen et al. 2004) have implicated PKCβII in regulation of cytokinesis, via interaction with the centrosomal scaffold protein pericentrin. Cell cycle promoting effects of PKCβII have also been noted in G_1. PKCβII increases cyclin D1 levels in breast cancer cells via a transcriptional mechanism involving AP-1 (Li and Weinstein 2006) and promotes pRb phosphorylation in retinal endothelial cells (Suzuma et al. 2002). The enzyme also appears to phosphorylate CAK and stimulate its activity in glioma cells (Acevedo-Duncan et al. 2002). It should be noted, however, that PKCβII has also been reported to inhibit cell proliferation and induce differentiation in some cell types [e.g., dendritic cells (Cejas et al. 2005), leukemia cells (Yoshida et al. 2003)]. Thus, as is the case for PKCα and PKCδ, it appears that the cell cycle regulatory effects of PKCβII may be complex and context-dependent.

8.4.2.2 PKCε

Like PKCβII, PKCε generally mediates proproliferative responses, although its effects appear to be predominantly in G_1/S rather than G_2/M (Balciunaite and Kazlauskas 2001; Graham et al. 2000). The enzyme is rapidly activated by PDGF in fibroblasts and has been implicated in mediating PDGF-induced $G_0/G_1 \rightarrow S$ progression in these cells (Balciunaite and Kazlauskas 2001). Loss of PKCε activity in NSCLC cells is associated with induction of $p21^{Waf1/Cip1}$, prolonged $G_1 \rightarrow S$ transition in response to serum stimulation, and reduced activation of Cdk2 complexes (Bae et al. 2007), pointing to a role of the enzyme in suppressing $p21^{Waf1/Cip1}$ accumulation to facilitate cell cycle progression. PKCε can also induce cyclin D1 transcriptional activity and upregulate cyclin D1 and cyclin E protein, leading to enhanced transit through G_1 (Soh and Weinstein 2003). Although PKCε is generally downregulated during differentiation (e.g., Yang et al. 2003), the enzyme promotes adipogenic commitment and is essential for terminal differentiation of 3T3-F442A preadipocytes (Webb et al. 2003).

8.4.3 Complex Regulation of the Cell Cycle by PKCη

PKCη also appears to have context-dependent effects on cell cycle progression. Expression of the enzyme in NIH3T3 cells led to induction of $p21^{Waf1/Cip1}$ and $p27^{Kip1}$, decreased cyclin E-associated kinase activity, hypophosphorylation of pRb, and growth arrest (Livneh et al. 1996). In keratinocytes, PKCη has been implicated in negative control of $G_1 \rightarrow S$ progression via association with cyclin E/Cdk2/$p21^{Waf1/Cip1}$ complexes and inhibition of Cdk2 activity (Ishino et al. 1998). PKCη overexpressing NIH3T3 undergo adipocyte differentiation in response to adipogenic hormones (Livneh et al. 1996), an effect that is consistent with the differentiation-inducing properties of the enzyme in other cell types, including keratinocytes and B-cells (Cabodi et al. 2000). Studies in keratinocytes have further demonstrated a requirement for PKCη in Vitamin D- and calcium-induced squamous cell differentiation (Ohba et al. 1998). These findings are consistent with evidence that PKCη is predominantly expressed in differentiated suprabasal layers of squamous epithelia (Kashiwagi et al. 2002; Breitkreutz et al. 2007), in the uppermost granular layer of the epidermis (Breitkreutz et al. 2007), and in postmitotic intestinal epithelial cells (Osada et al. 1993). In contrast, however, overexpression of PKCη in MCF-7 breast cancer cells upregulated cyclin D and cyclin E levels and promoted a redistribution of $p21^{Waf1/Cip1}$ and $p27^{Kip1}$ from Cdk2 to Cdk4 complexes, resulting in stimulation of cell growth (Fima et al. 2001).

8.4.4 Role of Atypical PKC Isozymes in Control of Cell Cycle Progression

Understanding of the cell cycle-specific effects of atypical PKCs, PKCζ and PKCι, lags behind that of other members of the PKC family. However, emerging evidence supports a role for these PKC family members in stimulation of cell cycle progression. Expression of PKCζ or activation of endogenous atypical PKCs by insulin or

PDK1 in HCT116 colon cancer cells led to proteasome-dependent degradation of p21$^{Waf1/Cip1}$ protein (Scott et al. 2002), an effect that was associated with phosphorylation of p21$^{Waf1/Cip1}$ at Ser146. Consistent with a cell cycle stimulatory role of PKCζ, keratin-induced blockade of HaCaT keratinocyte cell cycle progression involved inhibition of PKCζ activity, a reduction in cyclin D1 and cyclin E levels, and pRb hypophosphorylation (Paramio et al. 2001). In addition, transcriptional activation of cyclin D1 by oncogenic Ras required PKCζ and ERK/MAPK activity in mouse mammary epithelial cells (Kampfer et al. 2001). Exciting studies by Fields and colleagues have recently identified PKCι as an oncogene which is required for the transformed growth of various human cancer cell types (Fields and Regala 2007). However, with the exception of a potential role in regulation of CAK phosphorylation and activity in glioma cells (Acevedo-Duncan et al. 2002), its cell cycle targets remain undefined.

8.5 Conclusions and New Paradigms

Significant advances have been made in recent years in our understanding of PKC-mediated regulation of the cell cycle. It is clear that PKC signaling can positively or negatively regulate cell cycle progression at multiple points (Fig. 8.3). In addition to controlling cell cycle entry and exit, PKC-mediated pathways can regulate passage through G_1 and S phases as well as transit through the G_1 and G_2 checkpoints. It is also clear that the timing and duration of PKC activation, as well as the specific PKC isozyme(s) targeted, play a determining role in cell cycle-specific responses. Effects are also highly dependent on cellular context. Short-term PKC activation stimulates cell cycle progression in some systems, while prolonged activation induces cell cycle arrest in a variety of cell types (Zhou et al. 1993; Balciunaite et al. 2000; Balciunaite and Kazlauskas 2002). Sustained activation of PKC can promote irreversible cell cycle blockade and differentiation (Tibudan et al. 2002) or senescence (Oliva et al. 2008). It appears that the cell cycle stage during which PKC is activated determines how the cell integrates signaling events. In this regard, Kazanietz and colleagues have introduced the novel paradigm that activation of PKCs in one phase of the cell cycle can lead to effects in a different phase, as shown in NSCLC cells (Oliva et al. 2008) and thyroid cells (Santiago-Walker et al. 2005).

Perhaps not surprisingly, Cdk activity is a major target of PKC modulation: Cdk4/6 and Cdk2 in G_1 phase and Cdk1 in G_2. The CKI p21$^{Waf1/Cip1}$ has been established as a key player in the cell cycle-specific effects of PKCs, mediating PKC-induced control of both the $G_0/G_1 \rightarrow S$ and $G_2 \rightarrow M$ transitions. The notion that control of these transitions involves a common target, previously proposed based on the strong parallels between the cell cycle-specific effects of PKC signaling and p21$^{Waf1/Cip1}$ in a wide variety of cell types (Black 2000; Niculescu et al. 1998), was recently tested and confirmed directly using RNA interference technology (Oliva et al. 2008; Nakagawa et al. 2005) and p21$^{Waf1/Cip1}$-null cells (Balciunaite and Kazlauskas 2002).

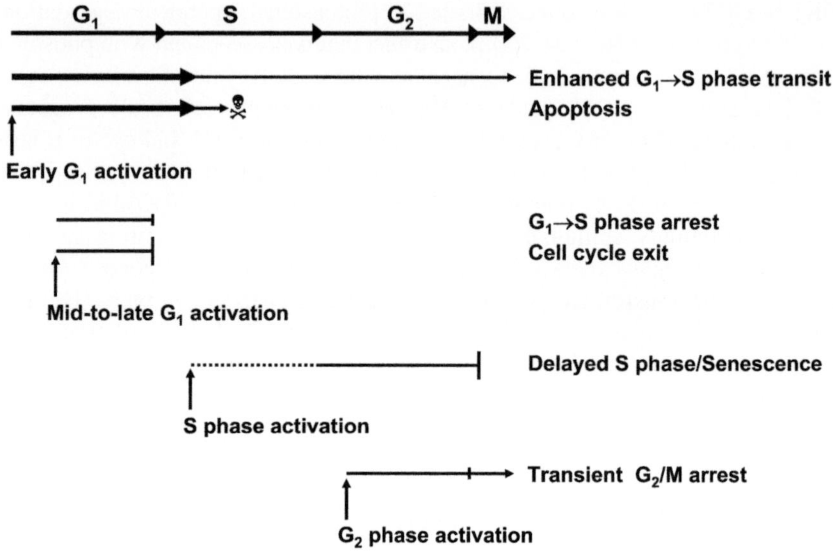

Fig. 8.3 The timing of PKC activation plays a determining role in PKC-mediated cell cycle-specific responses. Activation of PKCs in early G1 usually has a stimulatory effect on $G_1 \rightarrow S$ progression, while mid-to-late G_1 phase activation results in G_0/G_1 arrest/differentiation. In some cases, late G1 phase PKC activity is required for cell cycle progression. Increased activity of PKCδ can promote $G_1 \rightarrow S$ progression followed by arrest in S phase and induction of apoptosis. PKCα activation in S phase delays S phase progression and promotes irreversible cell cycle blockade in G_2/M, in association with senescence. G_2 stimulation of PKC generally results in transient G_2/M arrest, an effect seen even when stimulation occurs at the end of G_2

Studies in NSCLC cells also led to the recognition that distinct PKCs can trigger different responses via a common mediator depending on the phase of the cell cycle in which they are activated (Oliva et al. 2008; Nakagawa et al. 2005). Thus, activation of PKCδ in mid-to-late G1 phase resulted in $p21^{Waf1/Cip1}$-mediated $G_1 \rightarrow S$ arrest, while activation of PKCα in S phase promoted $p21^{Waf1/Cip1}$-dependent G_2/M blockade. A common target may also mediate opposing cell cycle effects in different systems, as exemplified by the ability of PKCδ to promote $G_1 \rightarrow S$ progression in fibroblasts via destabilization of $p21^{Waf1/Cip1}$ protein (Walker et al. 2006).

PKC-mediated modulation of cdk4/6 appears to involve regulation of cyclin D1 expression. It is now well established that PKC isozymes can use transcriptional or translational mechanisms to modulate the levels of this key mitogenic molecule and thereby regulate G_1 progression. Our studies on PKC-mediated regulation of cyclin D1 also provided the first evidence that PKC signaling can regulate the activity of the translational repressor 4E-BP1 to inhibit cap-dependent translation initiation (Hizli et al. 2006; Guan et al. 2007).

Elucidation of the cell cycle-specific effects of individual members of the PKC family remains a major challenge for the future. A large number of studies have focused on PKCα and δ, which exhibit negative or positive regulation of cell cycle progression in a highly context-dependent manner. The complexity of the cell cycle-specific effects of these molecules is well exemplified by studies in 3Y1 fibroblasts, where PKCδ stimulates G_1/S progression while potently inhibiting mitosis (Kitamura et al. 2003). Fewer studies have addressed the roles of other members of the PKC family. Thus, although PKCβII and ε generally appear to play a cell cycle stimulatory role, a few studies point to additional complexity. In view of the inherent drawbacks in current approaches used to understand PKC isozyme-specific functions (i.e., overexpression studies, use of nonselective agonists and inhibitors, analysis of transformed cell lines), it may be helpful to focus future studies on gaining a better understanding of the expression and activation of these molecules in unperturbed tissue systems, and using this information to guide subsequent mechanistic analyses. In seeking deeper insight into the biological functions of individual PKCs, it will also be critical to understand the regulatory inputs and downstream events induced by PKC activation, and to identify target proteins that are modulated by PKCs in vivo.

Acknowledgments I would like to thank past and present members of my laboratory for their contributions to the study of PKC and control of cell cycle progression. I also thank Drs. Adrian Black and Debora Kramer for critical reading of the manuscript and Dr. Adrian Black and Margaret Frey for the artwork. I apologize to many colleagues whose work was not cited due to space limitations. Work in my laboratory is supported by NIH grants DK54909, DK60632, and CA16056.

References

Abraham, C., Scaglione-Sewell, B., Skarosi, S. F., Qin, W., Bissonnette, M., & Brasitus, T. A. (1998). Protein kinase C alpha modulates growth and differentiation in Caco-2 cells. *Gastroenterology, 114*, 503–509.

Acevedo-Duncan, M., Patel, R., Whelan, S., & Bicaku, E. (2002). Human glioma PKC-iota and PKC-betaII phosphorylate cyclin-dependent kinase activating kinase during the cell cycle. *Cell Prolif, 35*, 23–36.

Acs, P., Beheshti, M., Szallasi, Z., Li, L., Yuspa, S. H., & Blumberg, P. M. (2000). Effect of a tyrosine 155 to phenylalanine mutation of protein kinase cdelta on the proliferative and tumorigenic properties of NIH 3T3 fibroblasts. *Carcinogenesis, 21*, 887–891.

Acs, P., Wang, Q. J., Bogi, K., Marquez, A. M., Lorenzo, P. S., Biro, T., et al. (1997). Both the catalytic and regulatory domains of protein kinase C chimeras modulate the proliferative properties of NIH 3T3 cells. *J Biol Chem, 272*, 28793–28799.

Afrasiabi, E., Ahlgren, J., Bergelin, N., & Tornquist, K. (2008). Phorbol 12-myristate 13-acetate inhibits FRO anaplastic human thyroid cancer cell proliferation by inducing cell cycle arrest in G1/S phase: evidence for an effect mediated by PKCdelta. *Mol Cell Endocrinol, 292*, 26–35.

Agell, N., Jaumot, M., Rodriguez-Vilarrupla, A., Brun, S., Abella, N., Canela, N., et al. (2006). The diverging roles of calmodulin and PKC in the regulation of p21 intracellular localization. *Cell Cycle, 5*, 3–6.

Akashi, M., Osawa, Y., Koeffler, H. P., & Hachiya, M. (1999). p21WAF1 expression by an activator of protein kinase C is regulated mainly at the post-transcriptional level in cells lacking p53: important role of RNA stabilization. *Biochem J, 337*(Pt 3), 607–616.

Alisi, A., Spagnuolo, S., Napoletano, S., Spaziani, A., & Leoni, S. (2004). Thyroid hormones regulate DNA-synthesis and cell-cycle proteins by activation of PKCalpha and p42/44 MAPK in chick embryo hepatocytes. *J Cell Physiol, 201*, 259–265.

Arita, Y., Buffolino, P., & Coppock, D. L. (1998). Regulation of the cell cycle at the G2/M boundary in metastatic melanoma cells by 12-O-tetradecanoyl phorbol-13-acetate (TPA) by blocking p34cdc2 kinase activity. *Exp Cell Res, 242*, 381–390.

Ashton, A. W., Watanabe, G., Albanese, C., Harrington, E. O., Ware, J. A., & Pestell, R. G. (1999). Protein kinase Cdelta inhibition of S-phase transition in capillary endothelial cells involves the cyclin-dependent kinase inhibitor p27(Kip1). *J Biol Chem, 274*, 20805–20811.

Asiedu, C., Biggs, J., & Kraft, A. S. (1997). Complex regulation of CDK2 during phorbol ester-induced hematopoietic differentiation. *Blood, 90*, 3430–3437.

Asiedu, C., Biggs, J., Lilly, M., & Kraft, A. S. (1995). Inhibition of leukemic cell growth by the protein kinase C activator bryostatin 1 correlates with the dephosphorylation of cyclin-dependent kinase 2. *Cancer Res, 55*, 3716–3720.

Bae, K. M., Wang, H., Jiang, G., Chen, M. G., Lu, L., & Xiao, L. (2007). Protein kinase C epsilon is overexpressed in primary human non-small cell lung cancers and functionally required for proliferation of non-small cell lung cancer cells in a p21/Cip1-dependent manner. *Cancer Res, 67*, 6053–6063.

Balciunaite, E., Jones, S., Toker, A., & Kazlauskas, A. (2000). PDGF initiates two distinct phases of protein kinase C activity that make unequal contributions to the G0 to S transition. *Curr Biol, 10*, 261–267.

Balciunaite, E., & Kazlauskas, A. (2001). Early phosphoinositide 3-kinase activity is required for late activation of protein kinase Cepsilon in platelet-derived-growth-factor-stimulated cells: evidence for signalling across a large temporal gap. *Biochem J, 358*, 281–285.

Balciunaite, E., & Kazlauskas, A. (2002). The timing and extent of activation of diacylglycerol-responsive protein kinase-cs determines their ability to inhibit or promote platelet-derived growth factor-dependent DNA synthesis. *Exp Cell Res, 281*, 167–174.

Barboule, N., Lafon, C., Chadebech, P., Vidal, S., & Valette, A. (1999). Involvement of p21 in the PKC-induced regulation of the G2/M cell cycle transition. *FEBS Lett, 444*, 32–37.

Barth, H., Hoffmann, I., Klein, S., Kaszkin, M., Richards, J., & Kinzel, V. (1996). Role of cdc25-C phosphatase in the immediate G2 delay induced by the exogenous factors epidermal growth factor and phorbolester. *J Cell Physiol, 168*, 589–599.

Barth, H., & Kinzel, V. (1994). Phorbol ester TPA rapidly prevents activation of p34cdc2 histone H1 kinase and concomitantly the transition from G2 phase to mitosis in synchronized HeLa cells. *Exp Cell Res, 212*, 383–388.

Berthet, C., & Kaldis, P. (2007). Cell-specific responses to loss of cyclin-dependent kinases. *Oncogene, 26*, 4469–4477.

Besson, A., & Yong, V. W. (2000). Involvement of p21(Waf1/Cip1) in protein kinase C alpha-induced cell cycle progression. *Mol Cell Biol, 20*, 4580–4590.

Bi, N., & Mamrack, M. D. (1994). PMA inhibits the growth of human fibroblasts after the induction of immediate-early genes. *Exp Cell Res, 212*, 105–112.

Biggs, J. R., Kudlow, J. E., & Kraft, A. S. (1996). The role of the transcription factor Sp1 in regulating the expression of the WAF1/CIP1 gene in U937 leukemic cells. *J Biol Chem, 271*, 901–906.

Bikle, D. D., Ng, D., Tu, C. L., Oda, Y., & Xie, Z. (2001). Calcium- and vitamin D-regulated keratinocyte differentiation. *Mol Cell Endocrinol, 177*, 161–171.

Black, J. D. (2000). Protein kinase C-mediated regulation of the cell cycle. *Front Biosci, 5*, D406–D423.

Black, J. D. (2001). Protein kinase C isozymes in colon carcinogenesis: guilt by omission. *Gastroenterology, 120*, 1868–1872.

Black, A. R., Curry, K. J., & Black, J. D. Unpublished data.

Borner, C., Filipuzzi, I., Weinstein, I. B., & Imber, R. (1991). Failure of wild-type or a mutant form of protein kinase C-alpha to transform fibroblasts. *Nature, 353*, 78–80.

Bowles, D. K., Maddali, K. K., Dhulipala, V. C., & Korzick, D. H. (2007). PKCdelta mediates anti-proliferative, pro-apoptic effects of testosterone on coronary smooth muscle. *Am J Physiol Cell Physiol, 293*, C805–C813.

Breitkreutz, D., Braiman-Wiksman, L., Daum, N., Denning, M. F., & Tennenbaum, T. (2007). Protein kinase C family: on the crossroads of cell signaling in skin and tumor epithelium. *J Cancer Res Clin Oncol, 133*, 793–808.

Buitrago, C. G., Pardo, V. G., de Boland, A. R., & Boland, R. (2003). Activation of RAF-1 through Ras and protein kinase Calpha mediates 1alpha, 25(OH)2-vitamin D3 regulation of the mitogen-activated protein kinase pathway in muscle cells. *J Biol Chem, 278*, 2199–2205.

Cabodi, S., Calautti, E., Talora, C., Kuroki, T., Stein, P. L., & Dotto, G. P. (2000). A PKC-eta/Fyn-dependent pathway leading to keratinocyte growth arrest and differentiation. *Mol Cell, 6*, 1121–1129.

Cacace, A. M., Guadagno, S. N., Krauss, R. S., Fabbro, D., & Weinstein, I. B. (1993). The epsilon isoform of protein kinase C is an oncogene when overexpressed in rat fibroblasts. *Oncogene, 8*, 2095–2104.

Cameron, A. J., Procyk, K. J., Leitges, M., & Parker, P. J. (2008). PKC alpha protein but not kinase activity is critical for glioma cell proliferation and survival. *Int J Cancer, 123*, 769–779.

Castagna, M., Takai, Y., Kaibuchi, K., Sano, K., Kikkawa, U., & Nishizuka, Y. (1982). Direct activation of calcium-activated, phospholipid-dependent protein kinase by tumor-promoting phorbol esters. *J Biol Chem, 257*, 7847–7851.

Cejas, P. J., Carlson, L. M., Zhang, J., Padmanabhan, S., Kolonias, D., Lindner, I., et al. (2005). Protein kinase C betaII plays an essential role in dendritic cell differentiation and autoregulates its own expression. *J Biol Chem, 280*, 28412–28423.

Cerda, S. R., Mustafi, R., Little, H., Cohen, G., Khare, S., Moore, C., et al. (2006). Protein kinase C delta inhibits Caco-2 cell proliferation by selective changes in cell cycle and cell death regulators. *Oncogene, 25*, 3123–3138.

Chen, A., Davis, B. H., Bissonnette, M., Scaglione-Sewell, B., & Brasitus, T. A. (1999). 1, 25-Dihydroxyvitamin D(3) stimulates activator protein-1-dependent Caco-2 cell differentiation. *J Biol Chem, 274*, 35505–35513.

Chen, D., Purohit, A., Halilovic, E., Doxsey, S. J., & Newton, A. C. (2004). Centrosomal anchoring of protein kinase C betaII by pericentrin controls microtubule organization, spindle function, and cytokinesis. *J Biol Chem, 279*, 4829–4839.

Chiu, T., Santiskulvong, C., & Rozengurt, E. (2003). ANG II stimulates PKC-dependent ERK activation, DNA synthesis, and cell division in intestinal epithelial cells. *Am J Physiol Gastrointest Liver Physiol, 285*, G1–G11.

Chiu, T., Wu, S. S., Santiskulvong, C., Tangkijvanich, P., Yee, H. F., Jr., & Rozengurt, E. (2002). Vasopressin-mediated mitogenic signaling in intestinal epithelial cells. *Am J Physiol Cell Physiol, 282*, C434–C450.

Cho, C. H., Lee, C. S., Chang, M., Jang, I. H., Kim, S. J., Hwang, I., et al. (2004). Localization of VEGFR-2 and PLD2 in endothelial caveolae is involved in VEGF-induced phosphorylation of MEK and ERK. *Am J Physiol Heart Circ Physiol, 286*, H1881–H1888.

Cho, Y., Tighe, A. P., & Talmage, D. A. (1997). Retinoic acid induced growth arrest of human breast carcinoma cells requires protein kinase C alpha expression and activity. *J Cell Physiol, 172*, 306–313.

Choi, P. M., Tchou-Wong, K. M., & Weinstein, I. B. (1990). Overexpression of protein kinase C in HT29 colon cancer cells causes growth inhibition and tumor suppression. *Mol Cell Biol, 10*, 4650–4657.

Chow, J. Y., Dong, H., Quach, K. T., Van Nguyen, P. N., Chen, K., & Carethers, J. M. (2008). TGF-beta mediates PTEN suppression and cell motility through calcium-dependent PKC-alpha activation in pancreatic cancer cells. *Am J Physiol Gastrointest Liver Physiol, 294*, G899–G905.

Clark, J. A., Black, A. R., Leontieva, O. V., Frey, M. R., Pysz, M. A., Kunneva, L., et al. (2004). Involvement of the ERK signaling cascade in protein kinase C-mediated cell cycle arrest in intestinal epithelial cells. *J Biol Chem, 279*, 9233–9247.

Classon, M., & Dyson, N. (2001). p107 and p130: versatile proteins with interesting pockets. *Exp Cell Res, 264*, 135–147.

Cobrinik, D. (2005). Pocket proteins and cell cycle control. *Oncogene, 24*, 2796–2809.

Coppock, D. L., Buffolino, P., Kopman, C., & Nathanson, L. (1995). Inhibition of the melanoma cell cycle and regulation at the G1/S transition by 12-O-tetradecanoylphorbol-13-acetate (TPA) by modulation of CDK2 activity. *Exp Cell Res, 221*, 92–102.

Coppock, D. L., & Nathanson, L. (1993). Positive and negative growth regulation in melanoma: growth factors, intracellular signalling, and the cell cycle. *Cancer Treat Res, 65*, 279–296.

Coppock, D. L., Tansey, J. B., & Nathanson, L. (1992). 12-O-tetradecanoylphorbol-13-acetate induces transient cell cycle arrest in G1 and G2 in metastatic melanoma cells: inhibition of phosphorylation of p34cdc2. *Cell Growth Differ, 3*, 485–494.

Cozzi, S. J., Parsons, P. G., Ogbourne, S. M., Pedley, J., & Boyle, G. M. (2006). Induction of senescence in diterpene ester-treated melanoma cells via protein kinase C-dependent hyperactivation of the mitogen-activated protein kinase pathway. *Cancer Res, 66*, 10083–10091.

Czifra, G., Toth, I. B., Marincsak, R., Juhasz, I., Kovacs, I., Acs, P., et al. (2006). Insulin-like growth factor-I-coupled mitogenic signaling in primary cultured human skeletal muscle cells and in C2C12 myoblasts. A central role of protein kinase Cdelta. *Cell Signal, 18*, 1461–1472.

Dangi, S., Chen, F. M., & Shapiro, P. (2006). Activation of extracellular signal-regulated kinase (ERK) in G2 phase delays mitotic entry through p21CIP1. *Cell Prolif, 39*, 261–279.

Datta, N. S., Chen, C., Berry, J. E., & McCauley, L. K. (2005). PTHrP signaling targets cyclin D1 and induces osteoblastic cell growth arrest. *J Bone Miner Res, 20*, 1051–1064.

De Falco, G. (2006). Comes F. *Simone C. pRb: master of differentiation. Coupling irreversible cell cycle withdrawal with induction of muscle-specific transcription. Oncogene, 25*, 5244–5249.

Deeds, L., Teodorescu, S., Chu, M., Yu, Q., & Chen, C. Y. (2003). A p53-independent G1 cell cycle checkpoint induced by the suppression of protein kinase C alpha and theta isoforms. *J Biol Chem, 278*, 39782–39793.

Denning, M. F. (2004). Epidermal keratinocytes: regulation of multiple cell phenotypes by multiple protein kinase C isoforms. *Int J Biochem Cell Biol, 36*, 1141–1146.

Detjen, K. M., Brembeck, F. H., Welzel, M., Kaiser, A., Haller, H., Wiedenmann, B., et al. (2000). Activation of protein kinase Calpha inhibits growth of pancreatic cancer cells via p21(cip)-mediated G(1) arrest. *J Cell Sci, 113*(Pt 17), 3025–3035.

Dicker, P., & Rozengurt, E. (1978). Stimulation of DNA synthesis by tumour promoter and pure mitogenic factors. *Nature, 276*, 723–726.

Diehl, J. A., Zindy, F., & Sherr, C. J. (1997). Inhibition of cyclin D1 phosphorylation on threonine-286 prevents its rapid degradation via the ubiquitin-proteasome pathway. *Genes Dev, 11*, 957–972.

Du, W. (2006). *Pogoriler J. Retinoblastoma family genes. Oncogene, 25*, 5190–5200.

Esposito, F., Cuccovillo, F., Vanoni, M., Cimino, F., Anderson, C. W., Appella, E., et al. (1997). Redox-mediated regulation of p21(waf1/cip1) expression involves a post-transcriptional mechanism and activation of the mitogen-activated protein kinase pathway. *Eur J Biochem, 245*, 730–737.

Fields, A. P., & Regala, R. P. (2007). Protein kinase C iota: human oncogene, prognostic marker and therapeutic target. *Pharmacol Res, 55*, 487–497.

Fima, E., Shtutman, M., Libros, P., Missel, A., Shahaf, G., Kahana, G., et al. (2001). PKCeta enhances cell cycle progression, the expression of G1 cyclins and p21 in MCF-7 cells. *Oncogene, 20*, 6794–6804.

Fisher, R. P., & Morgan, D. O. (1994). A novel cyclin associates with MO15/CDK7 to form the CDK-activating kinase. *Cell, 78*, 713–724.

Fishman, D. D., Segal, S., & Livneh, E. (1998). The role of protein kinase C in G1 and G2/M phases of the cell cycle (review). *Int J Oncol, 12*, 181–186.

Frey, M. R., Clark, J. A., Bateman, N. W., Kazanietz, M. G., Black, A. R., & Black, J. D. (2004). Cell cycle- and protein kinase C-specific effects of resiniferatoxin and resiniferonol 9, 13, 14-ortho-phenylacetate in intestinal epithelial cells. *Biochem Pharmacol, 67*, 1873–1886.

Frey, M. R., Clark, J. A., Leontieva, O., Uronis, J. M., Black, A. R., & Black, J. D. (2000). Protein kinase C signaling mediates a program of cell cycle withdrawal in the intestinal epithelium. *J Cell Biol, 151*, 763–778.

Frey, M. R., Saxon, M. L., Zhao, X., Rollins, A., Evans, S. S., & Black, J. D. (1997). Protein kinase C isozyme-mediated cell cycle arrest involves induction of p21(waf1/cip1) and p27(kip1) and hypophosphorylation of the retinoblastoma protein in intestinal epithelial cells. *J Biol Chem, 272*, 9424–9435.

Fujita, M. (1999). Cell cycle regulation of DNA replication initiation proteins in mammalian cells. *Front Biosci, 4*, D816–D823.

Fukumoto, S., Nishizawa, Y., Hosoi, M., Koyama, H., Yamakawa, K., Ohno, S., et al. (1997). Protein kinase C delta inhibits the proliferation of vascular smooth muscle cells by suppressing G1 cyclin expression. *J Biol Chem, 272*, 13816–13822.

Gamard, C. J., Blobe, G. C., Hannun, Y. A., & Obeid, L. M. (1994). Specific role for protein kinase C beta in cell differentiation. *Cell Growth Differ, 5*, 405–409.

Garriga, J., Limon, A., Mayol, X., Rane, S. G., Albrecht, J. H., Reddy, E. P., et al. (1998). Differential regulation of the retinoblastoma family of proteins during cell proliferation and differentiation. *Biochem J, 333*(Pt 3), 645–654.

Gavrielides, M. V., Frijhoff, A. F., Conti, C. J., & Kazanietz, M. G. (2004). Protein kinase C and prostate carcinogenesis: targeting the cell cycle and apoptotic mechanisms. *Curr Drug Targets, 5*, 431–443.

Goss, V. L., Hocevar, B. A., Thompson, L. J., Stratton, C. A., Burns, D. J., & Fields, A. P. (1994). Identification of nuclear beta II protein kinase C as a mitotic lamin kinase. *J Biol Chem, 269*, 19074–19080.

Graham, M. A., Rawe, I., Dartt, D. A., & Joyce, N. C. (2000). Protein kinase C regulation of corneal endothelial cell proliferation and cell cycle. *Invest Ophthalmol Vis Sci, 41*, 4124–4132.

Grana, X., Garriga, J., & Mayol, X. (1998). Role of the retinoblastoma protein family, pRB, p107 and p130 in the negative control of cell growth. *Oncogene, 17*, 3365–3383.

Griner, E. M., & Kazanietz, M. G. (2007). Protein kinase C and other diacylglycerol effectors in cancer. *Nat Rev Cancer, 7*, 281–294.

Grossoni, V. C., Falbo, K. B., & Kazanietz, M. G. (2007). de Kier Joffe ED, Urtreger AJ. Protein kinase C delta enhances proliferation and survival of murine mammary cells. *Mol Carcinog, 46*, 381–390.

Guadagno, T. M., & Newport, J. W. (1996). Cdk2 kinase is required for entry into mitosis as a positive regulator of Cdc2-cyclin B kinase activity. *Cell, 84*, 73–82.

Guan, L., Song, K., Pysz, M. A., Curry, K. J., Hizli, A. A., Danielpour, D., et al. (2007). Protein kinase C-mediated down-regulation of cyclin D1 involves activation of the translational repressor 4E-BP1 via a phosphoinositide 3-kinase/Akt-independent, protein phosphatase 2A-dependent mechanism in intestinal epithelial cells. *J Biol Chem, 282*, 14213–14225.

Hamada, K., Takuwa, N., Zhou, W., Kumada, M., & Takuwa, Y. (1996). Protein kinase C inhibits the CAK-CDK2 cyclin-dependent kinase cascade and G1/S cell cycle progression in human diploid fibroblasts. *Biochim Biophys Acta, 1310*, 149–156.

Hao, F., Black, A. R., & Black, J. D. Unpublished data.

Harrington, E. O., Loffler, J., Nelson, P. R., Kent, K. C., Simons, M., & Ware, J. A. (1997). Enhancement of migration by protein kinase Calpha and inhibition of proliferation and cell cycle progression by protein kinase Cdelta in capillary endothelial cells. *J Biol Chem, 272*, 7390–7397.

Harris, P., & Ralph, P. (1985). Human leukemic models of myelomonocytic development: a review of the HL-60 and U937 cell lines. *J Leukoc Biol, 37*, 407–422.

He, Y. P., Zhao, L. Y., Zheng, Q. S., Liu, S. W., Zhao, X. Y., Lu, X. L., et al. (2008). Arginine vasopressin stimulates proliferation of adult rat cardiac fibroblasts via protein kinase C-extracellular signal-regulated kinase 1/2 pathway. *Sheng Li Xue Bao, 60*, 333–340.

Hizli, A. A., Black, A. R., Pysz, M. A., & Black, J. D. (2006). Protein kinase C alpha signaling inhibits cyclin D1 translation in intestinal epithelial cells. *J Biol Chem, 281*, 14596–14603.

Hocevar, B. A., Morrow, D. M., Tykocinski, M. L., & Fields, A. P. (1992). Protein kinase C isotypes in human erythroleukemia cell proliferation and differentiation. *J Cell Sci, 101*(Pt 3), 671–679.

Hochegger, H., Takeda, S., & Hunt, T. (2008). Cyclin-dependent kinases and cell-cycle transitions: does one fit all? *Nat Rev Mol Cell Biol, 9*, 910–916.

Housey, G. M., Johnson, M. D., Hsiao, W. L., O'Brian, C. A., Murphy, J. P., Kirschmeier, P., et al. (1988). Overproduction of protein kinase C causes disordered growth control in rat fibroblasts. *Cell, 52*, 343–354.

Huang, T. S., Duyster, J., & Wang, J. Y. (1995). Biological response to phorbol ester determined by alternative G1 pathways. *Proc Natl Acad Sci USA, 92*, 4793–4797.

Huang, C. L., & Ives, H. E. (1987). Growth inhibition by protein kinase C late in mitogenesis. *Nature, 329*, 849–850.

Hug, H., & Sarre, T. F. (1993). Protein kinase C isoenzymes: divergence in signal transduction? *Biochem J, 291*(Pt 2), 329–343.

Hwang, H. C., & Clurman, B. E. (2005). Cyclin E in normal and neoplastic cell cycles. *Oncogene, 24*, 2776–2786.

Ishino, K., Ohba, M., Kashiwagi, M., Kawabe, S., Chida, K., & Kuroki, T. (1998). Phorbol ester-induced G1 arrest in BALB/MK-2 mouse keratinocytes is mediated by delta and eta isoforms of protein kinase C. *Jpn J Cancer Res, 89*, 1126–1133.

Jackson, D. N., & Foster, D. A. (2004). The enigmatic protein kinase Cdelta: complex roles in cell proliferation and survival. *FASEB J, 18*, 627–636.

Kambhampati, S., Li, Y., Verma, A., Sassano, A., Majchrzak, B., Deb, D. K., et al. (2003). Activation of protein kinase C delta by all-trans-retinoic acid. *J Biol Chem, 278*, 32544–32551.

Kampfer, S., Windegger, M., Hochholdinger, F., Schwaiger, W., Pestell, R. G., Baier, G., et al. (2001). Protein kinase C isoforms involved in the transcriptional activation of cyclin D1 by transforming Ha-Ras. *J Biol Chem, 276*, 42834–42842.

Kariya, K., Fukumoto, Y., Tsuda, T., Kawahara, Y., Fukuzaki, H., Yamamoto, T., et al. (1987). Inhibition of DNA synthesis by phorbol esters through protein kinase C in cultured rabbit aortic smooth muscle cells. *FEBS Lett, 217*, 69–73.

Kashiwagi, M., Ohba, M., Chida, K., & Kuroki, T. (2002). Protein kinase C eta (PKC eta): its involvement in keratinocyte differentiation. *J Biochem, 132*, 853–857.

Kashiwagi, M., Ohba, M., Watanabe, H., Ishino, K., Kasahara, K., Sanai, Y., et al. (2000). PKCeta associates with cyclin E/cdk2/p21 complex, phosphorylates p21 and inhibits cdk2 kinase in keratinocytes. *Oncogene, 19*, 6334–6341.

Kikkawa, U., Takai, Y., Tanaka, Y., Miyake, R., & Nishizuka, Y. (1983). Protein kinase C as a possible receptor protein of tumor-promoting phorbol esters. *J Biol Chem, 258*, 11442–11445.

Kinzel, V., Richards, J., & Stohr, M. (1980). Tumor promoter TPA mimics irradiation effects on the cell cycle of HeLa cells. *Science, 210*, 429–431.

Kitamura, K., Mizuno, K., Etoh, A., Akita, Y., Miyamoto, A., Nakayama, K., et al. (2003). The second phase activation of protein kinase C delta at late G1 is required for DNA synthesis in serum-induced cell cycle progression. *Genes Cells, 8*, 311–324.

Kosaka, C., Sasaguri, T., Ishida, A., & Ogata, J. (1996). Cell cycle arrest in the G2 phase induced by phorbol ester and diacylglycerol in vascular endothelial cells. *Am J Physiol, 270*, C170–C178.

Kosaka, C., Sasaguri, T., Masuda, J., Zen, K., Shimokado, K., Yokota, T., et al. (1993). Protein kinase C-mediated inhibition of cyclin A expression in human vascular endothelial cells. *Biochem Biophys Res Commun, 193*, 991–998.

Lampasso, J. D., Marzec, N., Margarone, J., 3rd, & Dziak, R. (2002). Role of protein kinase C alpha in primary human osteoblast proliferation. *J Bone Miner Res, 17*, 1968–1976.

Leach, K. L., James, M. L., & Blumberg, P. M. (1983). Characterization of a specific phorbol ester aporeceptor in mouse brain cytosol. *Proc Natl Acad Sci U S A, 80*, 4208–4212.

Li, H., & Weinstein, I. B. (2006). Protein kinase C beta enhances growth and expression of cyclin D1 in human breast cancer cells. *Cancer Res, 66*, 11399–11408.

Lin, S. Y., Liang, Y. C., Ho, Y. S., Tsai, S. H., Pan, S., & Lee, W. S. (2002). Involvement of both extracellular signal-regulated kinase and c-jun N-terminal kinase pathways in the 12-O-tetradecanoylphorbol-13-acetate-induced upregulation of p21(Cip1) in colon cancer cells. *Mol Carcinog, 35*, 21–28.

Lin, X., Nelson, P., & Gelman, I. H. (2000). SSeCKS, a major protein kinase C substrate with tumor suppressor activity, regulates G(1)–>S progression by controlling the expression and cellular compartmentalization of cyclin D. *Mol Cell Biol, 20*, 7259–7272.

Liu, Y., Martindale, J. L., Gorospe, M., & Holbrook, N. J. (1996). Regulation of p21WAF1/CIP1 expression through mitogen-activated protein kinase signaling pathway. *Cancer Res, 56*, 31–35.

Livneh, E., & Fishman, D. D. (1997). Linking protein kinase C to cell-cycle control. *Eur J Biochem, 248*, 1–9.

Livneh, E., Shimon, T., Bechor, E., Doki, Y., Schieren, I., & Weinstein, I. B. (1996). Linking protein kinase C to the cell cycle: ectopic expression of PKC eta in NIH3T3 cells alters the expression of cyclins and Cdk inhibitors and induces adipogenesis. *Oncogene, 12*, 1545–1555.

Malumbres, M. (2005). Revisiting the "Cdk-centric" view of the mammalian cell cycle. *Cell Cycle, 4*, 206–210.

Malumbres, M., & Barbacid, M. (2005). Mammalian cyclin-dependent kinases. *Trends Biochem Sci, 30*, 630–641.

Mandil, R., Ashkenazi, E., Blass, M., Kronfeld, I., Kazimirsky, G., Rosenthal, G., et al. (2001). Protein kinase Calpha and protein kinase Cdelta play opposite roles in the proliferation and apoptosis of glioma cells. *Cancer Res, 61*, 4612–4619.

Mann, D. J., Higgins, T., Jones, N. C., & Rozengurt, E. (1997). Differential control of cyclins D1 and D3 and the cdk inhibitor p27Kip1 by diverse signalling pathways in Swiss 3T3 cells. *Oncogene, 14*, 1759–1766.

Manzati, E., Aguiari, G., Banzi, M., Manzati, M., Selvatici, R., Falzarano, S., et al. (2005). The cytoplasmic C-terminus of polycystin-1 increases cell proliferation in kidney epithelial cells through serum-activated and Ca(2+)-dependent pathway(s). *Exp Cell Res, 304*, 391–406.

Matsumoto, E., Hatanaka, M., Bohgaki, M., & Maeda, S. (2006). PKC pathway and ERK/MAPK pathway are required for induction of cyclin D1 and p21Waf1 during 12-o-tetradecanoylphorbol 13-acetate-induced differentiation of myeloleukemia cells. *Kobe J Med Sci, 52*, 181–194.

Matsuzaki, Y., Takaoka, Y., Hitomi, T., Nishino, H., & Sakai, T. (2004). Activation of protein kinase C promotes human cancer cell growth through downregulation of p18(INK4c). *Oncogene, 23*, 5409–5414.

Mayol, X., & Grana, X. (1998). The p130 pocket protein: keeping order at cell cycle exit/re-entrance transitions. *Front Biosci, 3*, d11–d24.

Millard, S. S., Yan, J. S., Nguyen, H., Pagano, M., Kiyokawa, H., & Koff, A. (1997). Enhanced ribosomal association of p27(Kip1) mRNA is a mechanism contributing to accumulation during growth arrest. *J Biol Chem, 272*, 7093–7098.

Miller, J. P., Yeh, N., Vidal, A., & Koff, A. (2007). Interweaving the cell cycle machinery with cell differentiation. *Cell Cycle, 6*, 2932–2938.

Mischak, H., Goodnight, J. A., Kolch, W., Martiny-Baron, G., Schaechtle, C., Kazanietz, M. G., et al. (1993). Overexpression of protein kinase C-delta and -epsilon in NIH 3T3 cells induces opposite effects on growth, morphology, anchorage dependence, and tumorigenicity. *J Biol Chem, 268*, 6090–6096.

Murray, N. R., Baumgardner, G. P., Burns, D. J., & Fields, A. P. (1993). Protein kinase C isotypes in human erythroleukemia (K562) cell proliferation and differentiation. Evidence that beta II protein kinase C is required for proliferation. *J Biol Chem, 268*, 15847–15853.

Murray, N. R., & Fields, A. P. (1998). Phosphatidylglycerol is a physiologic activator of nuclear protein kinase C. *J Biol Chem, 273*, 11514–11520.

Nakagawa, S., Fujii, T., Yokoyama, G., Kazanietz, M. G., Yamana, H., & Shirouzu, K. (2003). Cell growth inhibition by all-trans retinoic acid in SKBR-3 breast cancer cells: involvement of protein kinase Calpha and extracellular signal-regulated kinase mitogen-activated protein kinase. *Mol Carcinog, 38*, 106–116.

Nakagawa, M., Oliva, J. L., Kothapalli, D., Fournier, A., Assoian, R. K., & Kazanietz, M. G. (2005). Phorbol ester-induced G1 phase arrest selectively mediated by protein kinase Cdelta-dependent induction of p21. *J Biol Chem, 280*, 33926–33934.

Nakaigawa, N., Hirai, S., Mizuno, K., Shuin, T., Hosaka, M., & Ohno, S. (1996). Differential effects of overexpression of PKC alpha and PKC delta/epsilon on cellular E2F activity in late G1 phase. *Biochem Biophys Res Commun, 222*, 95–100.

Niculescu, A. B., 3rd, Chen, X., Smeets, M., Hengst, L., Prives, C., & Reed, S. I. (1998). Effects of p21(Cip1/Waf1) at both the G1/S and the G2/M cell cycle transitions: pRb is a critical determinant in blocking DNA replication and in preventing endoreduplication. *Mol Cell Biol, 18*, 629–643.

Niles, R. M. (2003). Vitamin A (retinoids) regulation of mouse melanoma growth and differentiation. *J Nutr, 133*, 282S–286S.

Ohba, M., Ishino, K., Kashiwagi, M., Kawabe, S., Chida, K., Huh, N. H., et al. (1998). Induction of differentiation in normal human keratinocytes by adenovirus-mediated introduction of the eta and delta isoforms of protein kinase C. *Mol Cell Biol, 18*, 5199–5207.

Olashaw, N. E., & Pledger, W. J. (1988). Cellular mechanisms regulating proliferation. *Adv Second Messenger Phosphoprotein Res, 22*, 139–173.

Oliva, J. L., Caino, M. C., Senderowicz, A. M., & Kazanietz, M. G. (2008). S-Phase-specific activation of PKC alpha induces senescence in non-small cell lung cancer cells. *J Biol Chem, 283*, 5466–5476.

Osada, S., Hashimoto, Y., Nomura, S., Kohno, Y., Chida, K., Tajima, O., et al. (1993). Predominant expression of nPKC eta, a Ca(2+)-independent isoform of protein kinase C in epithelial tissues, in association with epithelial differentiation. *Cell Growth Differ, 4*, 167–175.

Oster, H., & Leitges, M. (2006). Protein kinase C alpha but not PKCzeta suppresses intestinal tumor formation in ApcMin/+ mice. *Cancer Res, 66*, 6955–6963.

Page, K., Li, J., Corbit, K. C., Rumilla, K. M., Soh, J. W., Weinstein, I. B., et al. (2002). Regulation of airway smooth muscle cyclin D1 transcription by protein kinase C-delta. *Am J Respir Cell Mol Biol, 27*, 204–213.

Paramio, J. M., Segrelles, C., Ruiz, S., & Jorcano, J. L. (2001). Inhibition of protein kinase B (PKB) and PKCzeta mediates keratin K10-induced cell cycle arrest. *Mol Cell Biol, 21*, 7449–7459.

Park, J. W., Jang, M. A., Lee, Y. H., Passaniti, A., & Kwon, T. K. (2001). p53-independent elevation of p21 expression by PMA results from PKC-mediated mRNA stabilization. *Biochem Biophys Res Commun, 280*, 244–248.

Perletti, G. P., Concari, P., Brusaferri, S., Marras, E., Piccinini, F., & Tashjian, A. H., Jr. (1998). Protein kinase Cepsilon is oncogenic in colon epithelial cells by interaction with the ras signal transduction pathway. *Oncogene, 16*, 3345–3348.

Perletti, G. P., Marras, E., Concari, P., Piccinini, F., & Tashjian, A. H., Jr. (1999). PKCdelta acts as a growth and tumor suppressor in rat colonic epithelial cells. *Oncogene, 18*, 1251–1256.

Prowse, D. M., Bolgan, L., Molnar, A., & Dotto, G. P. (1997). Involvement of the Sp3 transcription factor in induction of p21Cip1/WAF1 in keratinocyte differentiation. *J Biol Chem, 272*, 1308–1314.

Pysz, M. A., Hizli, A. A., Curry, K. J., Black, A. R., & Black, J. D. Unpublished data.

Pysz, M. A., Leontieva, O. V., Bateman, N. W., Uronis, J. M., Curry, K. J., Threadgill, D. W., et al. (2009). PKCa tumor suppression in the intestine is associated with transcriptional and translational inhbition of cyclin D1. *Exp Cell Res, 315*, 1415–1428.

Rodriguez-Vilarrupla, A., Jaumot, M., Abella, N., Canela, N., Brun, S., Diaz, C., et al. (2005). Binding of calmodulin to the carboxy-terminal region of p21 induces nuclear accumulation via inhibition of protein kinase C-mediated phosphorylation of Ser153. *Mol Cell Biol, 25*, 7364–7374.

Rovera, G., Santoli, D., & Damsky, C. (1979). Human promyelocytic leukemia cells in culture differentiate into macrophage-like cells when treated with a phorbol diester. *Proc Natl Acad Sci U S A, 76*, 2779–2783.
Rozengurt, E. (1986). Early signals in the mitogenic response. *Science, 234*, 161–166.
Sakaguchi, M., Miyazaki, M., Sonegawa, H., Kashiwagi, M., Ohba, M., Kuroki, T., et al. (2004). PKCalpha mediates TGFbeta-induced growth inhibition of human keratinocytes via phosphorylation of S100C/A11. *J Cell Biol, 164*, 979–984.
Salabat, M. R., Ding, X. Z., Flesche, J. B., Ujiki, M. B., Robin, T. P., Talamonti, M. S., et al. (2006). On the mechanisms of 12-O-tetradecanoylphorbol-13-acetate-induced growth arrest in pancreatic cancer cells. *Pancreas, 33*, 148–155.
Sanchez, I., & Dynlacht, B. D. (2005). New insights into cyclins, CDKs, and cell cycle control. *Semin Cell Dev Biol, 16*, 311–321.
Santamaria, D., Barriere, C., Cerqueira, A., Hunt, S., Tardy, C., Newton, K., et al. (2007). Cdk1 is sufficient to drive the mammalian cell cycle. *Nature, 448*, 811–815.
Santiago-Walker, A. E., Fikaris, A. J., Kao, G. D., Brown, E. J., Kazanietz, M. G., & Meinkoth, J. L. (2005). Protein kinase C delta stimulates apoptosis by initiating G1 phase cell cycle progression and S phase arrest. *J Biol Chem, 280*, 32107–32114.
Sasaguri, T., Ishida, A., Kosaka, C., Nojima, H., & Ogata, J. (1996). Phorbol ester inhibits the phosphorylation of the retinoblastoma protein without suppressing cyclin D-associated kinase in vascular smooth muscle cells. *J Biol Chem, 271*, 8345–8351.
Sasaguri, T., Kosaka, C., Hirata, M., Masuda, J., Shimokado, K., Fujishima, M., et al. (1993). Protein kinase C-mediated inhibition of vascular smooth muscle cell proliferation: the isoforms that may mediate G1/S inhibition. *Exp Cell Res, 208*, 311–320.
Saunders, N. A., Dicker, A. J., Jones, S. J., & Dahler, A. L. (1998). E2F1 messenger RNA is destabilized in response to a growth inhibitor in normal human keratinocytes but not in a squamous carcinoma cell line. *Cancer Res, 58*, 1646–1649.
Saxon, M. L., Zhao, X., & Black, J. D. (1994). Activation of protein kinase C isozymes is associated with post-mitotic events in intestinal epithelial cells in situ. *J Cell Biol, 126*, 747–763.
Scaglione-Sewell, B., Abraham, C., Bissonnette, M., Skarosi, S. F., Hart, J., Davidson, N. O., et al. (1998). Decreased PKC-alpha expression increases cellular proliferation, decreases differentiation, and enhances the transformed phenotype of CaCo-2 cells. *Cancer Res, 58*, 1074–1081.
Schavinsky-Khrapunsky, Y., Huleihel, M., Aboud, M., & Torgeman, A. (2003). TPA activates p21WAF-1 promoter in human T-cells through its second most upstream Sp1 site. *Biochem Biophys Res Commun, 304*, 696–700.
Schonwasser, D. C., Marais, R. M., Marshall, C. J., & Parker, P. J. (1998). Activation of the mitogen-activated protein kinase/extracellular signal-regulated kinase pathway by conventional, novel, and atypical protein kinase C isotypes. *Mol Cell Biol, 18*, 790–798.
Scott, M. T., Ingram, A., & Ball, K. L. (2002). PDK1-dependent activation of atypical PKC leads to degradation of the p21 tumour modifier protein. *EMBO J, 21*, 6771–6780.
Shatos, M. A., Hodges, R. R., Bair, J. A., Lashkari, K., & Dartt, D. A. (2008). PKC{alpha} Stimulates Conjunctival Goblet Cell Proliferation Using Extracellular Regulated Kinase 1/2. *Invest Ophthalmol Vis Sci, 49*(8), 629–636.
Sherr, C. J., & Roberts, J. M. (1995). Inhibitors of mammalian G1 cyclin-dependent kinases. *Genes Dev, 9*, 1149–1163.
Sherr, C. J., & Roberts, J. M. (1999). CDK inhibitors: positive and negative regulators of G1-phase progression. *Genes Dev, 13*, 1501–1512.
Sherr, C. J., & Roberts, J. M. (2004). Living with or without cyclins and cyclin-dependent kinases. *Genes Dev, 18*, 2699–2711.
Slosberg, E. D., Klein, M. G., Yao, Y., Han, E. K., Schieren, I., & Weinstein, I. B. (1999). The alpha isoform of protein kinase C mediates phorbol ester-induced growth inhibition and p21cip1 induction in HC11 mammary epithelial cells. *Oncogene, 18*, 6658–6666.
Smith, E. J., Leone, G., DeGregori, J., Jakoi, L., & Nevins, J. R. (1996). The accumulation of an E2F-p130 transcriptional repressor distinguishes a G0 cell state from a G1 cell state. *Mol Cell Biol, 16*, 6965–6976.

Soh, J. W., & Weinstein, I. B. (2003). Roles of specific isoforms of protein kinase C in the transcriptional control of cyclin D1 and related genes. *J Biol Chem, 278*, 34709–34716.

Sprowl, J. A., Villeneuve, D. J., Guo, B., Young, A. J., Hembruff, S. L., & Parissenti, A. M. (2007). Changes in expression of cell wall turnover genes accompany inhibition of chromosome segregation by bovine protein kinase C alpha expression in Saccharomyces cerevisiae. *Cell Biol Int, 31*, 1160–1172.

Steinberg, S. F. (2004). Distinctive activation mechanisms and functions for protein kinase Cdelta. *Biochem J, 384*, 449–459.

Stern, B., & Nurse, P. (1996). A quantitative model for the cdc2 control of S phase and mitosis in fission yeast. *Trends Genet, 12*, 345–350.

Sugibayashi, R., Shimizu, T., Suzuki, T., Yamamoto, N., Hamada, H., & Takeda, K. (2001). Upregulation of p21(WAF1/CIP1) leads to morphologic changes and esterase activity in TPA-mediated differentiation of human prostate cancer cell line TSU-Pr1. *Oncogene, 20*, 1220–1228.

Suzuma, K., Takahara, N., Suzuma, I., Isshiki, K., Ueki, K., Leitges, M., et al. (2002). Characterization of protein kinase C beta isoform's action on retinoblastoma protein phosphorylation, vascular endothelial growth factor-induced endothelial cell proliferation, and retinal neovascularization. *Proc Natl Acad Sci U S A, 99*, 721–726.

Takuwa, N., Takuwa, Y., & Rasmussen, H. (1988). Stimulation of mitogenesis and glucose transport by 1-monooleoylglycerol in Swiss 3T3 fibroblasts. *J Biol Chem, 263*, 9738–9745.

Tchou, W. W., Rom, W. N., & Tchou-Wong, K. M. (1996). Novel form of p21(WAF1/CIP1/SDI1) protein in phorbol ester-induced G2/M arrest. *J Biol Chem, 271*, 29556–29560.

Thompson, L. J., & Fields, A. P. (1996). betaII protein kinase C is required for the G2/M phase transition of cell cycle. *J Biol Chem, 271*, 15045–15053.

Tibudan, S. S., Wang, Y., & Denning, M. F. (2002). Activation of protein kinase C triggers irreversible cell cycle withdrawal in human keratinocytes. *J Invest Dermatol, 119*, 1282–1289.

Todd, C., & Reynolds, N. J. (1998). Up-regulation of p21WAF1 by phorbol ester and calcium in human keratinocytes through a protein kinase C-dependent pathway. *Am J Pathol, 153*, 39–45.

Traore, K., Trush, M. A., George, M., Jr., Spannhake, E. W., Anderson, W., & Asseffa, A. (2005). Signal transduction of phorbol 12-myristate 13-acetate (PMA)-induced growth inhibition of human monocytic leukemia THP-1 cells is reactive oxygen dependent. *Leuk Res, 29*, 863–879.

Trimarchi, J. M., & Lees, J. A. (2002). Sibling rivalry in the E2F family. *Nat Rev Mol Cell Biol, 3*, 11–20.

Uddin, S., Sassano, A., Deb, D. K., Verma, A., Majchrzak, B., Rahman, A., et al. (2002). Protein kinase C-delta (PKC-delta) is activated by type I interferons and mediates phosphorylation of Stat1 on serine 727. *J Biol Chem, 277*, 14408–14416.

Vrana, J. A., Saunders, A. M., Chellappan, S. P., & Grant, S. (1998). Divergent effects of bryostatin 1 and phorbol myristate acetate on cell cycle arrest and maturation in human myelomonocytic leukemia cells (U937). *Differentiation, 63*, 33–42.

Vucenik, I., Ramakrishna, G., Tantivejkul, K., Anderson, L. M., & Ramljak, D. (2005). Inositol hexaphosphate (IP6) blocks proliferation of human breast cancer cells through a PKCdelta-dependent increase in p27Kip1 and decrease in retinoblastoma protein (pRb) phosphorylation. *Breast Cancer Res Treat, 91*, 35–45.

Walker, J. L., Castagnino, P., Chung, B. M., Kazanietz, M. G., & Assoian, R. K. (2006). Post-transcriptional destabilization of p21cip1 by protein kinase C in fibroblasts. *J Biol Chem, 281*, 38127–38132.

Walker, S. D., Murray, N. R., Burns, D. J., & Fields, A. P. (1995). Protein kinase C chimeras: catalytic domains of alpha and beta II protein kinase C contain determinants for isotype-specific function. *Proc Natl Acad Sci U S A, 92*, 9156–9160.

Wang, Z., Su, Z. Z., Fisher, P. B., Wang, S., VanTuyle, G., & Grant, S. (1998). Evidence of a functional role for the cyclin-dependent kinase inhibitor p21(WAF1/CIP1/MDA6) in the recip-

rocal regulation of PKC activator-induced apoptosis and differentation in human myelomonocytic leukemia cells. *Exp Cell Res, 244*, 105–116.
Wang, Z., Wang, S., Fisher, P. B., Dent, P., & Grant, S. (2000). Evidence of a functional role for the cyclin-dependent kinase inhibitor p21CIP1 in leukemic cell (U937) differentiation induced by low concentrations of 1-beta-D-arabinofuranosylcytosine. *Differentiation, 66*, 1–13.
Watanabe, T., Ono, Y., Taniyama, Y., Hazama, K., Igarashi, K., Ogita, K., et al. (1992). Cell division arrest induced by phorbol ester in CHO cells overexpressing protein kinase C-delta subspecies. *Proc Natl Acad Sci U S A, 89*, 10159–10163.
Webb, P. R., Doyle, C., & Anderson, N. G. (2003). Protein kinase C-epsilon promotes adipogenic commitment and is essential for terminal differentiation of 3T3-F442A preadipocytes. *Cell Mol Life Sci, 60*, 1504–1512.
Wen-Sheng, W., & Jun-Ming, H. (2005). Activation of protein kinase C alpha is required for TPA-triggered ERK (MAPK) signaling and growth inhibition of human hepatoma cell HepG2. *J Biomed Sci, 12*, 289–296.
Whyte, P., & Eisenman, R. N. (1992). Dephosphorylation of the retinoblastoma protein during differentiation of HL60 cells. *Biochem Cell Biol, 70*, 1380–1384.
Wu, T. T., Hsieh, Y. H., Hsieh, Y. S., & Liu, J. Y. (2008). Reduction of PKC alpha decreases cell proliferation, migration, and invasion of human malignant hepatocellular carcinoma. *J Cell Biochem, 103*, 9–20.
Wu, W. S., & Hsu, H. Y. (2001). Involvement of p-15(INK4b) and p-16(INK4a) gene expression in saikosaponin a and TPA-induced growth inhibition of HepG2 cells. *Biochem Biophys Res Commun, 285*, 183–187.
Yamamoto, H., Soh, J. W., Shirin, H., Xing, W. Q., Lim, J. T., Yao, Y., et al. (1999). Comparative effects of overexpression of p27Kip1 and p21Cip1/Waf1 on growth and differentiation in human colon carcinoma cells. *Oncogene, 18*, 103–115.
Yamamoto, T., Tsuda, T., Hamamori, Y., Nishimura, N., & Takai, Y. (1988). Inhibition of DNA synthesis by protein kinase C-activating phorbol esters in NIH/3T3 cells. *J Biochem, 104*, 53–56.
Yan, S., & Wenner, C. E. (2001). Modulation of cyclin D1 and its signaling components by the phorbol ester TPA and the tyrosine phosphatase inhibitor vanadate. *J Cell Physiol, 186*, 338–349.
Yang, L. C., Ng, D. C., & Bikle, D. D. (2003). Role of protein kinase C alpha in calcium induced keratinocyte differentiation: defective regulation in squamous cell carcinoma. *J Cell Physiol, 195*, 249–259.
Yee, A. S., Shih, H. H., & Tevosian, S. G. (1998). New perspectives on retinoblastoma family functions in differentiation. *Front Biosci, 3*, D532–D547.
Yoshida, M., Heike, Y., Ohno, S., Ikekawa, T., & Wakasugi, H. (2003). Involvement of PKC betaII in anti-proliferating action of a new antitumor compound gnidimacrin. *Int J Cancer, 105*, 601–606.
Zeng, Y. X. (1996). el-Deiry WS. Regulation of p21WAF1/CIP1 expression by p53-independent pathways. *Oncogene, 12*, 1557–1564.
Zezula, J., Sexl, V., Hutter, C., Karel, A., Schutz, W., & Freissmuth, M. (1997). The cyclin-dependent kinase inhibitor p21cip1 mediates the growth inhibitory effect of phorbol esters in human venous endothelial cells. *J Biol Chem, 272*, 29967–29974.
Zhang, Y., & Chellappan, S. P. (1996). Transcriptional activation and expression of DP transcription factors during cell cycle and TPA-induced U937 differentiation. *Mol Cell Differentiation, 4*, 297–316.
Zhang, W., Grasso, L., McClain, C. D., Gambel, A. M., Cha, Y., Travali, S., et al. (1995). p53-independent induction of WAF1/CIP1 in human leukemia cells is correlated with growth arrest accompanying monocyte/macrophage differentiation. *Cancer Res, 55*, 668–674.
Zhou, W., Takuwa, N., Kumada, M., & Takuwa, Y. (1993). Protein kinase C-mediated bidirectional regulation of DNA synthesis, RB protein phosphorylation, and cyclin-dependent kinases in human vascular endothelial cells. *J Biol Chem, 268*, 23041–23048.

Zhou, W., Takuwa, N., Kumada, M., & Takuwa, Y. (1994). E2F1, B-myb and selective members of cyclin/cdk subunits are targets for protein kinase C-mediated bimodal growth regulation in vascular endothelial cells. *Biochem Biophys Res Commun, 199*, 191–198.

Zwijsen, R. M., Klompmaker, R., Wientjens, E. B., Kristel, P. M., van der Burg, B., & Michalides, R. J. (1996). Cyclin D1 triggers autonomous growth of breast cancer cells by governing cell cycle exit. *Mol Cell Biol, 16*, 2554–2560.

Chapter 9
PKC and the Control of Apoptosis

Mary E. Reyland and Andrew P. Bradford

Abstract Cells respond to a wide range of cellular toxins by inducing a suicide pathway known as apoptosis. Apoptosis plays a critical role in development, tissue remodeling, and in the removal of damaged and genetically altered cells. Activation of cell death pathways must be tightly regulated as too much or too little apoptosis can have drastic consequences for tissue homeostasis and for disease initiation and progression. While the mechanism of how apoptosis is executed has been extensively studied, little is known about how other signaling pathways influence the decision to die. In this regard, members of the protein kinase C (PKC) family of serine/threonine protein kinases are emerging as important modulators of the apoptotic response. Here we will discuss the role of specific PKC isoforms in regulating cell death, and address how alterations in the expression or activity of some of these kinases contribute to human diseases.

Keywords Disease • Cell death • Protein kinase C • Signal transduction • Cancer

9.1 Introduction

Over 35 years ago, Kerr, Wyllie, and Currie described a form of cell death characterized by blebbing of the cell membrane, condensation of nuclear chromatin, and the appearance of small extracellular bodies containing fragments of the nucleus and subcellular organelles (Crawford et al. 1972; Kerr et al. 1972; Wyllie et al.

M.E. Reyland (✉)
Department of Craniofacial Biology, School of Dental Medicine, Anschutz Medical Campus, University of Colorado Denver, 13001 E 17th Place, Aurora, CO 80045, USA
e-mail: Mary.Reyland@UCdenver.edu

A.P. Bradford
The Department of Obstetrics and Gynecology, School of Medicine, Anschutz Medical Campus, University of Colorado Denver, 13001 E 17th Place, Aurora, CO 80045, USA

M.G. Kazanietz (ed.), *Protein Kinase C in Cancer Signaling and Therapy*, Current Cancer Research, DOI 10.1007/978-1-60761-543-9_9,

1972, 1973; Kerr 2002). This process was recognized and originally referred to as "shrinkage necrosis" as the subcellular organelles such as mitochondria and ribosomes appeared to be intact (Kerr 1971), and was later coined "apoptosis" after the Greek "dropping off." A critical observation of these early studies was that this mode of cell death appeared to follow a predetermined program; hence it also came to be known as "programmed cell death." Kerr and colleagues concluded that this process regulates "the size of cell populations under both normal and pathological conditions" (Kerr 2002). Nonapoptotic cell death programs such as autophagy have also been the focus of intense study recently. Interestingly, autophagy and apoptosis both appear to contribute to processes such as development and chemotherapeutic cell death, suggesting cross-talk between these seemingly distinct modes of cell death (Gozuacik and Kimchi 2007; Levine et al. 2008).

The protein kinase C (PKC) family of serine/threonine protein kinases consists of 11 isoforms that regulate a wide variety of biological functions including cell proliferation and differentiation, cell survival and cell death (Reyland 2007). Many isoforms including PKCα, −β, −δ, −ε, and −ζ show widespread tissue expression, while others such as PKCγ and −η have a more tissue specific expression pattern (Wetsel et al. 1992). Despite overlapping expression patterns and common substrates, many of the functions of PKC appear to be isoform-specific. This may be achieved in part by changes in subcellular localization which facilitate interaction with specific signaling modules. A clear role in apoptosis and/or cell survival has been demonstrated for a subset of PKC isoforms, in particular PKCα, −δ, −ε, and −ζ. The function of these isoforms may vary with cell type, suggesting that the distinct composition of PKC isoforms in a cell determines the ultimate response. PKC isoforms are also well integrated into both proliferation and apoptotic signaling networks; hence the specific "wiring" of other regulatory pathways in the cell may also contribute to signal transduction by this family of kinases. Given their central role in proliferation and apoptosis, it is not surprising that expression or activation of PKC isoforms is altered in some human diseases, particularly cancer. This review will focus on the contribution of specific PKC isoforms to the regulation of cell survival and apoptosis.

9.2 Apoptosis and Human Disease

Early studies suggested a role for apoptosis during normal development, in endocrine tissues upon hormone withdrawal, and in breast carcinomas (Kerr and Searle 1972; Kerr et al. 1972). More recent studies from mice lacking specific components of the apoptotic cascade show that disruption of caspase-3, 7, 8 or 9, or Bcl-2 results in either embryonic or perinatal death, indicating a critical role in development (Varfolomeev et al. 1998; Zheng and Flavell 2000; Ranger et al. 2001). It is now appreciated that in addition to physiological stimuli, cells respond to a wide range of cellular toxins by inducing this suicide pathway. Moreover, too much or too little apoptosis can have drastic consequences for tissue homeostasis and for disease initiation and progression. In the nervous system, increased apoptosis may result in the inappropriate loss of cells and contribute to neurodegenerative disorders such as Alzheimer's, Parkinson's, and Huntington's disease (Ekshyyan and Aw 2004). In acute disorders such as stroke and myocardial infarction, apoptosis contributes to tissue loss, and blocking apoptosis in ischemic tissue is a promising approach for diminishing damage (Churchill et al. 2008). Likewise, apoptosis is thought to be the main mechanism by which pancreatic beta cells are destroyed in patients with type I diabetes mellitus and may contribute to reduced β-cell volume in type-II diabetes (Hayashi and Faustman 2003; Liadis et al. 2005; Lee and Pervaiz 2007). Alterations in apoptosis also may underlie autoimmune diseases such as systemic lupus erythematous and Hashimoto's thyroiditis (Eguchi 2001; Nagata 2006; Wang and Baker 2007).

Too little apoptosis may result in the failure to remove defective or unwanted cells and facilitate development of cancer (Evan and Vousden 2001). Genetic disruption of key apoptotic mediators is common in human tumor cells, and correlations between the expression of specific apoptotic markers and clinical outcome underscore the relevance of this pathway to cancer biology (Hanahan and Weinberg 2000; Johnstone et al. 2002; Konstantinidou et al. 2002; Rogers et al. 2002; Yip and Reed 2008). The antiapoptotic protein, Bcl-2, was first identified as a result of a chromosomal translocation in nonHodgkin's lymphoma, which results in a dramatic increase in Bcl-2 transcription (Tsujimoto et al. 1985). Overexpression of Bcl-2, and Bcl-2 family members, has since been shown to contribute to many human tumors (Yip and Reed 2008). Curiously, human tumors utilize a variety of mechanisms to achieve this goal, including Bcl-2 gene amplification, gene hypomethylation, and elimination of micro-RNAs that normally suppress Bcl-2 expression (Cimmino et al. 2005; Yip and Reed 2008). Inactivating mutations in the pro-apoptotic protein Bax, are also observed in human cancers including a subset of human colon cancers, and experimentally loss of Bax is associated with increased cancer cell growth in vivo and in vitro (Rampino et al. 1997; Shibata et al. 1999; Ionov et al. 2000). As most chemotherapeutic drugs depend on Bcl-2/Bax for cell death, acquired defects in apoptosis may also hamper the treatment of many tumors (Debatin et al. 2002). For instance, loss of Bax in glioblastoma multiforme tumors results in resistance to apoptotic stimuli in vitro (Cartron et al. 2003). Understanding the molecular mechanisms that underlie the apoptotic response is hence critical for the design of more effective therapeutic approaches.

9.3 Molecular Mechanisms of Apoptosis

Tissue homeostasis in multicellular organisms requires a balance between cell proliferation, cell differentiation, and cell death. While the signals that induce apoptotic cell death are likely to depend on the specific context (i.e., development versus DNA damage), execution of the apoptotic pathway appears to rely on a common set of biochemical mediators which are highly conserved from nematodes to mammals (Ellis and Horvitz 1986; Kroemer 1997). Critical genes in the apoptotic pathway were identified first by Horvitz and colleagues in their seminal studies in *C. elegans*, for which they were awarded the Nobel prize in 2002 (Metzstein et al. 1998; Horvitz 2003). Sulston and Horvitz mapped the fate of all 1,090 cells in *C. elegans* and discovered that 131 of these cells were fated to die by apoptosis (Sulston and Horvitz 1977). Subsequent studies in *C. elegans* identified genes that are required for apoptosis as their mutation blocked cell death (Metzstein et al. 1998). Mammalian homologues of these genes have been found and include members of the Bcl-2 family of pro- and antiapoptotic proteins, cysteine-dependent aspartate-directed (caspase) proteases and APAF-1, a regulator of caspase activation (Kuwana and Newmeyer 2003; Riedl and Shi 2004; Li and Yuan 2008).

Two pathways for activation of apoptosis have been defined based on the initiating signals and upstream apoptotic effectors (see Fig. 9.1; Adams 2003). Activation of "effector" caspases including caspase 3, 6, and 7, and cleavage of cellular proteins, is common to both pathways and is responsible for hallmarks of apoptosis such as chromatin condensation and cell blebbing (Wolf and Green 1999; Riedl and Shi 2004). In the receptor-mediated ("extrinsic") pathway, ligand binding to death receptors such as tumor necrosis factor-alpha (TNF-α), Fas, and TRAIL receptors leads to the formation of signaling complexes which activate caspases and lead to cell death (reviewed in Thorburn 2004). Key to this pathway is formation of the

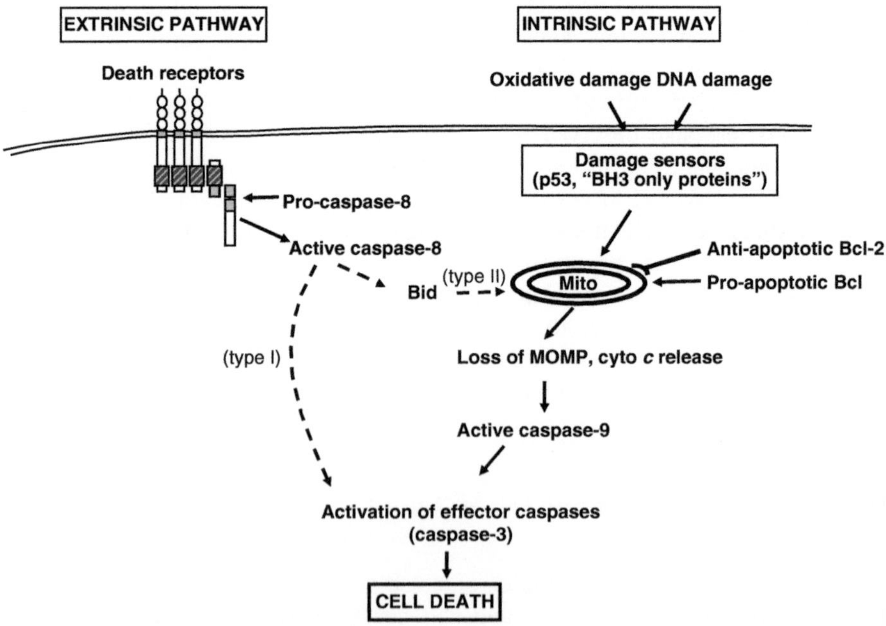

Fig. 9.1 Intrinsic and extrinsic apoptotic pathways. Apoptosis can be activated through the extrinsic/death receptor-dependent pathway, or the intrinsic/mitochondrial-dependent pathway. Both pathways converge to activate a common set of effector caspases. See text for details

death-inducing signaling complex (DISC), which recruits and activates the initiating caspase, pro-caspase-8, through an auto-catalytic mechanism (Wang and El-Deiry 2003). In some cells (type I cells), activation of caspase-8 leads to activation of effector caspases independent of the mitochondria. However, in other cell types (type-II cells), the mitochondrion is necessary for the amplification of apoptotic signals initiated by death receptor engagement. This is thought to occur chiefly through cleavage of Bid, a member of the Bcl-2 family, which can then promote the release of apoptogenic proteins from the mitochondria (Sheridan and Martin 2008).

Chemotherapeutic agents and other types of cell stress activate the intrinsic apoptotic pathway. In the case of genotoxin or oncogene-induced cell stress, the apoptotic response often, but not always, requires stabilization and activation of p53 (Chipuk and Green 2006). p53 is arguably the most frequently mutated gene in human cancer, and its activation has a multitude of effects on apoptotic signaling, including transcriptional activation of pro-apoptotic proteins such as Bax, death receptors, caspases, and Apaf-1 (Pietsch et al. 2008; Riley et al. 2008). A common outcome of cell damage induced apoptosis and death receptor induced apoptosis in type-II cells, is the loss of mitochondrial membrane potential (MOMP) resulting in release of cytochrome *c*. Cytochrome *c* released from the mitochondria, together with Apaf-1, ATP, and pro-caspase-9, forms the "apoptosome" and leads to activation of caspase-9 and cell execution. This is generally thought to be the "commitment" step in apoptosis, and it is tightly regulated by the Bcl-2 family of pro- and anti-apoptotic proteins (Sheridan and Martin 2008; Youle and Strasser 2008). In the absence of an apoptotic signal, antiapoptotic Bcl-2 proteins bind to and neutralize pro-apoptotic Bax and Bak. Apoptotic stimuli alleviate suppression of Bax/Bak, allowing these proteins to oligomerize at the mitochondria membrane resulting in loss of MOMP, cytochrome *c* release, and caspase activation. The key intermediates in this process are a subclass of pro-apoptotic Bcl-2 proteins known as "BH3" only proteins. These include Bim, Bid, Bik, PUMA, Noxa, and Bad, which act as apical damage sensors and are thought to function by antagonizing the interaction of pro-survival Bcl-2 proteins with Bax/Bak (Huang and Strasser 2000; Gelinas and White 2005). As the ratio of pro- to antiapoptotic Bcl-2 proteins at the mitochondria is an important determinant of cell fate, these protein–protein interactions are tightly regulated by a number of mechanisms including phosphorylation and protein sequestration (Cory et al. 2003; Gelinas and White 2005).

9.4 PKC and the Control of Apoptosis

In addition to the canonical apoptotic proteins described above, protein kinase pathways can regulate apoptotic signaling directly, through phosphorylation of apoptotic proteins, or indirectly, via regulation of transcription of pro- or antiapoptotic genes (Utz and Anderson 2000). Critical signaling pathways in apoptosis include members of the mitogen-activated kinase pathways (MAPK), including c-Jun-N-terminal-kinase (JNK) and p38, the phosphoinositide 3-kinase/AKT (PI3K/AKT) pathway, the janus-kinase-signal transducer and activator of transcription (JAK-STAT) pathway, and isoforms of PKC (Stephanou et al. 2002; Franke 2007; Reyland 2007; Dhanasekaran and Reddy 2008). In some instances, a hierarchy of these pathways has been described such as the PKCδ-dependent activation of JNK in response to DNA damage (Yoshida et al. 2002; Humphries et al. 2006).

The central role of the PKC family in cell survival and apoptosis suggests that specific isoforms may function as molecular sensors, promoting cell survival or cell death depending on environmental cues. These potential dual functions have added significant complexity to our understanding of the role of specific PKC isoforms in these pathways. Early approaches to defining the role of PKC in apoptosis utilized phorbol-12-myristate-13-acetate (PMA), an activator of the conventional (PKCα, −β and −γ) and novel (PKCδ, −ε, −η), inhibition of PKC by pharmacological agents, and expression of dominant negative forms of these kinases. Studies using PMA showed that activation of PKC typically blocks death-receptor induced apoptosis, although in some studies PMA appeared to sensitize cells to this apoptotic pathway (Reyland et al. 2000; Gomez-Angelats and Cidlowski 2001; Ito et al. 2001; Sarker et al. 2001; Herrant et al. 2002, 2003; Yin et al. 2005). Treatment of cells with PMA likewise blocks apoptosis in response to irradiation and oxidative stress in Jurkat and HL-60 cells (Haimovitz-Friedman et al. 1994; Zhuang et al. 2001), although it induces apoptosis in salivary epithelial cells and prostate cancer cells (Reyland et al. 2000; Fujii et al. 2000; Yin et al. 2005). As PMA both activates and then down-regulates PKC, these discrepant findings may reflect differences in the kinetics of these processes in different cells, or in expression of specific PKC isoforms in these cells.

The complexity and functional redundancy of PKC isoforms have prompted the development of isoform specific tools such as “knockout” and transgenic mice and siRNA to define the function of these individual kinases in the apoptotic pathway. Below we will discuss what is currently known about the contribution of specific isoforms of PKC to apoptosis, and how signal transduction by specific PKC isoforms integrates with other molecular regulators to promote or inhibit apoptosis and modulate the development and progression of cancers.

9.4.1 Contribution of Conventional Isoforms of PKC

The classical PKCs (α, β and γ) have typically been linked to the development, maintenance, and progression of malignancies. PKCα and β are considered critical regulators of cell survival, proliferation, migration, and invasion in a variety of tumors and tissues (Gutcher et al. 2003; Mackay and Twelves 2003; Lahn et al. 2004; Koivunen et al. 2006; Griner and Kazanietz 2007; Martiny-Baron and Fabbro 2007). However, it is clear that isoform specific functions of PKCs are dependent on cell type and context, such that PKCα and β can exert opposite effects to either promote or repress tumorigenesis.

9.4.1.1 PKCα

Numerous studies have implicated PKCα as a regulator of cell survival and sensitivity to apoptotic stimuli (Dempsey et al. 2000; Gutcher et al. 2003; Teicher 2006). Depletion or inhibition of PKCα has been shown to induce apoptosis in salivary epithelial cells, glioblastoma cells and bladder, endometrial and melanoma cancer cells (Dooley et al. 1998; Whelan and Parker 1998; Mandil et al. 2001; Matassa et al. 2003; Jorgensen et al. 2005; Haughian et al. 2006). PKCα also inhibits heregulin induced apoptosis in breast cancer cells (Le et al. 2001) and protects endothelial cells against radiation induced apoptosis (Haimovitz-Friedman et al. 1994). In contrast, overexpression or activation of PKCα results in apoptosis in androgen-dependent prostate cancer cells (Powell et al. 1996), and PKCα mediates caspase-3 activation and cytochrome *c* release in cisplatin-induced programmed cell death in renal cells (Nowak 2002). PKCα-dependent effects on apoptosis may be mediated by loss of proliferative or survival signals, such as Raf, MAPK (ERK), and Akt (Kolch et al. 1993; Ueda et al. 1996; Li et al. 1999, 2006; Partovian and Simons 2004), or by direct targeting of the apoptotic machinery (Gutcher et al. 2003). PKCα has been shown to colocalize with the mitochondrial protein Bcl-2 and increase phosphorylation of serine 70, thereby stabilizing and enhancing the antiapoptotic functions of Bcl-2 (Deng et al. 1998; Ruvolo et al. 1998; Jiffar et al. 2004). Accordingly, COS cells depleted of PKCα undergo apoptosis, concomitant with down regulation of Bcl-2 expression (Whelan and Parker 1998), while in hematopoetic cells, PKCα-dependent activation of Raf/Akt signaling results in phosphorylation and inactivation of the pro-apoptotic Bcl-2 family member Bad (Majewski et al. 1999).

While early evidence indicated that PKCα was a mitogenic kinase, promoting cell proliferation, subsequent reports have demonstrated antiproliferative actions of PKCα in a number of cell types (Black 2000; Gavrielides et al. 2004; Griner and Kazanietz 2007). Over expression of PKCα increases proliferation of fibroblasts, breast cancer, and glioma cells (Ways et al. 1995; Mandil et al. 2001; Soh and Weinstein 2003; Cameron et al. 2008) and inhibition or depletion of PKCα suppressed growth of lung cancer cells (Yin et al. 2003) and hepatocellular carcinoma cells (Wu et al. 2008). However, PKCα has been shown to inhibit proliferation of intestinal epithelial (Frey et al. 2000), pancreatic (Detjen et al. 2000), melanoma (Krasagakis et al. 2004), and mammary gland cells (Slosberg et al. 1999). In intestinal cells, PKCα is thought to function as a tumor suppressor in that knockout of PKCα increased the number of spontaneous intestinal neoplasms and enhanced tumorigenesis in the APC mutant mouse colon cancer model (Oster and Leitges 2006). PKCα effects on proliferation appear to be mediated primarily by modulation of the Ras/raf/MAPK and/or Akt signaling pathways (Kolch et al. 1993; Li et al. 1999; Partovian and Simons 2004; Li and Weinstein 2006) and regulation of expression of cyclin D1 and the cyclin-dependent kinase inhibitors p21 and p27 (Detjen et al. 2000; Clark et al. 2004; Frey et al. 2004; Guan et al. 2007). In the intestinal epithelium, PKCα induced cell cycle arrest is linked to increased expression of p21 and p27, decreased Rb phosphorylation, and sustained activation of MAPK (Frey et al. 1997; Clark et al. 2004). Similarly, PKCα-dependent G2/M arrest and senescence in nonsmall cell lung cancer (NSCLC) cells is a result of increased p21 levels (Oliva et al. 2008; Xiao et al. 2008), and upregulation of p21 and p27, combined with decreased cyclin D1 expression, is thought to underlie PKCα-mediated antiproliferative effects in hepatoma and pancreatic cancer cells (Detjen et al. 2000; Frey et al. 2000; Guan et al. 2007; Wu et al. 2008). PKCα also suppresses cyclin D1 translation in intestinal epithelium via protein phosphatase 2A catalyzed dephosphorylation and activation of the translational repressor 4E-BP1 (Hizli et al. 2006; Guan et al. 2007). The dual role of p21 as both a cell cycle inhibitor and an oncogenic, pro-proliferative/antiapoptotic protein (Blagosklonny 2002; Coqueret 2003; Child and Mann 2006; Abukhdeir and Park 2008) may in part explain the differential roles of PKCα in the cell type specific activation/inhibition of proliferation and apoptosis (Besson and Yong 2000).

PKCα has also been implicated in increased cell motility and invasion in bladder, renal, colon, and breast cancer cells (Ways et al. 1995; Engers et al. 2000; Masur et al. 2001; Parsons et al. 2002; Podar et al. 2002; Koivunen et al. 2004; Tan et al. 2006). However, such studies frequently were based on nonspecific PKC inhibitors and the functional role of specific PKC isozymes in invasion and metastasis remains to be established (Koivunen et al. 2006; Griner and Kazanietz 2007). PKCα effects on cell invasion and migration may be mediated by interaction with β1 integrins (Ng et al. 1999; Parsons et al. 2002), disruption of adherens junctions (Masur et al. 2001; Koivunen et al. 2004), or regulation of the expression and secretion of extracellular matrix remodeling proteins such as matrix metalloprotease MMP9 and components of the uroplasminogen activator receptor (uPAR) pathway (Liu et al. 2002; Sliva et al. 2002).

Overexpression of PKCα has been observed in a variety of human tumors, including hepatocellular, bladder, prostate and endometrial cancers (Koren et al. 2000, 2004; Tsai et al. 2000; Fournier et al. 2001; Langzam et al. 2001; Varga et al. 2004). In contrast, PKCα is down regulated in basal cell, colon, and ovarian tumors (Kahl-Rainer et al. 1994; Neill et al. 2003; Weichert et al. 2003). Elevated PKCα in breast cancers correlated with resistance to tamoxifen therapy and a role for PKCα in the development of an estrogen receptor (ER) negative, and selective ER modifier (SERM)-resistant phenotype in breast cancer cells has been proposed (Tonetti et al. 2000; Lahn et al. 2004; Assender et al. 2007). However, other studies show decreased PKCα expression in breast cancers correlating with increasing tumor grade (Ainsworth et al. 2004; Kerfoot et al. 2004). Mutations in PKCα are rare but have been detected in a fraction of unusually invasive aggressive pituitary tumors and follicular thyroid adenomas and carcinomas. In both cases, a point mutation (D294G) in a GDE motif within the hinge region of PKCα impairs translocation to the plasma membrane and consequent substrate phosphorylation, but the functional role of PKCα mutations in the development and progression of these endocrine malignancies is not known (Alvaro et al. 1993; Prevostel et al. 1997; Zhu et al. 2005).

9.4.1.2 PKCβ

Like PKCα, PKCβI and βII have been implicated in both the promotion and suppression of apoptosis and cell survival. Over expression of PKCβII protects small cell lung cancer (SCLC) cells against c-myc induced apoptosis (Barr et al. 1997) and antisense down regulation of PKCβII enhances ara-c mediated cell death in HL-60 leukemia cells, concomitant with a reduction in Bcl-2 expression (Whitman et al. 1997). Interestingly, selective activation of PKCβI induced apoptosis in these cells (Macfarlane and Manzel 1994), implying that the two splice variants may have opposing roles in cell survival. PKCβ knockout mice also indicate a critical role for the kinase in activation of NFκB-dependent survival pathways in B lymphocytes (Su et al. 2002). PKCβII is overexpressed in chronic myelogenous leukemia (CML; Abrams et al. 2007), and membrane-localized expression of PKCβII is a predictor of poor responsiveness to chemotherapy and decreased survival in patients with large B-cell lymphoma (Espinosa et al. 2006).

In contrast to the tumor suppressive role of PKCα in the intestine (Griner and Kazanietz 2007), increased PKCβII levels are observed in colon cancer (Gokmen-Polar et al. 2001) and PKCβ has been shown to mediate increased proliferation and invasion of intestinal cancer cells (Schwartz et al. 1993; Sauma et al. 1996; Murray et al. 1999; Jiang et al. 2004). Overexpression of PKCβII in mouse colon induced hyperproliferation and enhanced sensitivity to azoxymethane tumorigenesis, while PKCβ null mice exhibited a corresponding resistance to carcinogen induced colon cancer (Murray et al. 1999; Liu et al. 2004). PKCβII is also required for invasion of colon cancer cells acting via a Ras and MEK-dependent pathway, upstream of the atypical PKCι (Zhang et al. 2004).

PKCβ is overexpressed in prostate and pancreatic tumors (Koren et al. 2004; El-Rayes et al. 2008) but down regulated in bladder cancer (Koren et al. 2000; Langzam et al. 2001; Varga et al. 2004). Loss of PKCβ is also observed in melanoma cells but this is considered a result of melanocyte differentiation and not thought to play a role in tumorigenesis (Gilhooly et al. 2001). Metastatic hepatocellular carcinoma cell lines selectively upregulate PKCβ, relative to the other classical PKCs and inhibition or RNAi targeting of PKCβ suppressed cell migration and invasion (Guo et al. 2009). Expression of constitutively active forms of PKCβI and βII increased proliferation of breast cancer cells and resulted in an AP-1-dependent increase in cyclin D1 promoter activity and protein levels, while cells expressing dominant negative PKCβ mutants showed growth inhibition and decreased cyclin D1 levels (Li and Weinstein 2006). PKCβ1 also increased cell proliferation in neuroblastoma cells and PKCβ inhibitors enhanced sensitivity to chemotherapeutic agents (Svensson et al. 2000).

In addition to its role in cancer cell survival, proliferation, and invasion, PKCβ is also implicated in angiogenesis (Griner and Kazanietz 2007; Ma and Rosen 2007; Martiny-Baron and Fabbro 2007). PKCβ is a mediator of VEGF signaling, and its inhibition results in decreased endothelial cell proliferation and a reduction in tumor neovascularization (Xia et al. 1996; Yoshiji et al. 1999). Inhibition of PKCβ impaired VEGF-dependent tumor growth and angiogenesis in mouse xenograft models of hepatocellular and colon carcinomas (Yoshiji et al. 1999; Teicher et al. 2001b). Antiangiogenic activity of the PKCβ specific inhibitor enzastaurin has been demonstrated in a variety of cancers (Teicher et al. 2001a, b, 2002) and is currently in clinic trials as a primary or adjunct chemotherapeutic agent in both solid tumors and hematologic malignancies (Ma and Rosen 2007).

9.4.1.3 PKCγ

Few studies have addressed the role of PKCγ in cancer and apoptosis. PKCγ null mice exhibit modest neurological deficits in learning, memory, and motor coordination, consistent with the predominant expression of this isozyme in neuronal tissues, but do not indicate a role for PKCγ in cell survival or tumorigenesis (Abeliovich et al. 1993a, b). PKCγ may indirectly modulate survival and apoptsis in neuronal cells and lens epithelia, due to its role in oxidative stress and control of the formation and function of gap junctions (Lin et al. 2007; Lin and Takemoto 2005). ER negative breast cancer cell exhibit higher levels of PKCα and PKCγ (Morse-Gaudio et al. 1998) and overexpression of PKCγ in immortalized murine mammary epithelial cells conferred growth in soft agar and tumor formation in nude mice (Mazzoni et al. 2003). However, a specific role for PKCγ in breast cancer remains to be established (Martiny-Baron and Fabbro 2007). In contrast, increased expression of PKCγ and down regulation of PKCα was associated with TNF-α induced growth arrest of pancreatic cancer cells (Franz et al. 1996) suggesting that like PKCα and PKCβ, PKCγ may have tissue specific effects on cell survival, proliferation, and tumorigenesis. Accordingly, PKCγ was shown to be a positive prognostic factor in a small subset of Burkitt's lymphoma cases but has been implicated in Rac-dependent migration and invasion of colon carcinoma cells (Kamimura et al. 2004; Parsons and Adams 2008).

9.4.2 Contribution of Novel Isoforms of PKC

The novel PKC isoforms (PKCδ, ε, θ and η) regulate diverse cellular responses including cell migration, proliferation, cell death, and secretion. Two isoforms in this family, PKCδ and PKCε, have emerged as important regulators of apoptosis and cell survival while PKCθ and PKCη do not appear to play an important role in these processes. PKCδ activation is a hallmark of many apoptotic inducers and is essential for promoting the apoptotic response of these agents (Reyland 2007). In contrast, PKCε enhances cell proliferation, inhibits apoptosis, and altered PKCε expression/activation is associated with cell transformation in vitro and tumor promotion in vitro (Jansen et al. 2001; Verma et al. 2006). In some instances, such as in response to cardiac ischemia/reperfusion, cell death and survival are determined by the balance of these PKC signaling pathways (Churchill et al. 2008). Curiously, both isoforms are caspase substrates; however, caspase cleavage of PKCδ results in a pro-apoptotic signal, while caspase cleave of PKCε generates an active, antiapoptotic form of PKCε (Ghayur et al. 1996; Koriyama et al. 1999; Basu et al. 2002).

9.4.2.1 PKCε

PKCε was identified as an antiapoptotic signal based on studies which showed that it contributes to the resistance of tumor cells to cell death induced by TRAIL and chemotherapeutic drugs (Mayne and Murray 1998; Ding et al. 2002; Yonekawa and Akita 2008). In vascular endothelial cells, glioma cells, and MCF-7 breast cancer cells, PKCε suppresses TRAIL/TNF-α induced apoptosis (Okhrimenko et al. 2005a; Sivaprasad et al. 2006; Steinberg et al. 2007). PKCε mediates TRAIL resistance of MCF-7 cells by decreasing pro-apoptotic Bid expression and increasing levels of the pro-survival Bcl-2 (Sivaprasad et al. 2006; Steinberg et al. 2007; Shankar et al. 2008). Most of these studies place PKCε activation of the pro-survival kinase, Akt, upstream of regulation of pro and antiapoptotic Bcl-2 proteins (Basu and Sivaprasad 2007; Steinberg et al. 2007; Shankar et al. 2008). However, in glioma cells, TRAIL-induced suppression of the pro-survival kinase, Akt, was abolished by overexpression of PKCε, in the absence of changes in the expression of Bcl-2 or Bax (Okhrimenko et al. 2005a). In addition to Akt activation, PKCε has also been shown to enhance survival and promote cell transformation through activation of the NF-κB pathway and Ras/Raf pathways (Cacace et al. 1996; Perletti et al. 1998; Piiper et al. 2003; Catley et al. 2004).

PKCε appears to function as a bonafide oncogene based on studies that show it is able to transform rodent cells (Cacace et al. 1996, 1998). Suppression of apoptosis has been linked to the ability of PKCε to promote tumorigenesis in animal models, and changes in PKCε expression are associated with tumor progression in humans (Knauf et al. 1999; Sharif and Sharif 1999; Tachado et al. 2002; Wu et al. 2002; McJilton et al. 2003; Wheeler et al. 2004; Verma et al. 2006). Human prostate carcinoma cells frequently show increased expression of PKCε, and this correlates with conversion from an androgen-dependent to an androgen-independent state (Cornford et al. 1999). PKCε was shown to be required for resistance to apoptosis in prostate tumors cells, and resistance correlated with binding of PKCε to the pro-apoptotic protein, Bax (McJilton et al. 2003). Likewise, PKCε has been shown to play a role in the development of skin cancer in mice and humans (Jansen et al. 2001; Verma et al. 2006). Overexpression of PKCε in the skin sensitizes mice to the development of squamous cell carcinoma by increasing TNFα production and suppressing apoptosis of UV-irradiated skin cells (Wheeler et al. 2003, 2004). Activation of STAT-3 may also contribute to PKCε induced skin tumors in mice (Aziz et al. 2007). While these studies demonstrate a role for PKCε in the pathogenesis of skin cancer in the transgenic mouse model, further studies are needed to address the contribution of this isoform to human cancer.

9.4.2.2 PKCδ

Studies in vitro and in PKCδ null mice have identified a role for this isoform in immune regulation and in the control of cell proliferation and apoptosis (Leitges et al. 2001a; Miyamoto et al. 2002; Mecklenbrauker et al. 2004). PKCδ is a negative regulator of cell cycle progression through inhibition of cyclin D1 and p21 (Toyoda et al. 1998; Shanmugam et al. 2001; Wakino et al. 2001; Santiago-Walker et al. 2005). Furthermore, a myriad of studies show that PKCδ activity is required for apoptosis induced by irradiation, genotoxins, and oxidative stress (Reyland et al. 1999; Majumder et al. 2000, 2001; Matassa et al. 2001). In spite of these known functions of PKCδ in vitro, mice in which the PKCδ gene has been "knocked-out" develop normally, suggesting that PKCδ is not required for proliferation or apoptosis during development or for tissue homeostasis. However, PKCδ null mice do have an increased B-cell population and develop autoimmunity by 6–12 months of age (Leitges et al. 2002; Miyamoto et al. 2002). Studies in this mouse model support a role for PKCδ in stress induced apoptosis, as cell death in response to irradiation is suppressed in vitro, and smooth muscle and epithelial cells cultured from these mice are resistant to multiple apoptotic stimuli (Leitges et al. 2002; Humphries et al. 2006). Consistent with a role in cell death, loss of PKCδ has been associated with cell transformation (Watanabe et al. 1992; Mischak et al. 1993; Lu et al. 1997; Toyoda et al. 1998; Acs et al. 2000). In human cancer, reduced PKCδ expression correlates with increasing tumor grade in human squamous cell and endometrial carcinomas (D'Costa et al. 2006; Reno et al. 2008).

In contrast to its proposed pro-apoptotic function, in some transformed and tumor cells PKCδ appears to promote cell survival and suppress apoptosis, suggesting that PKCδ signaling maybe "re-wired" in this context to promote proliferation (Li et al. 1998; Kiley et al. 1999a, b; Wert and Palfrey 2000; Peluso et al. 2001; Kilpatrick et al. 2002; Okhrimenko et al. 2005b; Grossoni et al. 2007). Studies from Jaken and coworkers show that PKCδ expression increases as rat embryo fibroblasts and mammary tumor cells become more transformed, and that PKCδ is required for the anchorage-independent growth of mammary tumor cells (Liao et al. 1994; Kiley et al. 1999a, b). Furthermore, in human breast cancer, increased PKCδ mRNA correlated with reduced overall patient survival, suggesting a role for PKCδ in breast cancer progression (McKiernan et al. 2008). Consistent with a pro-proliferation function, PKCδ activates ERK downstream of the epidermal growth factor (EGF) receptor and collaborates with the hedgehog pathway to activate ERK signaling and the pro-proliferative transcription factor, GLI (Riobo et al. 2006). PKCδ also contributes to signal transduction downstream of the insulin growth factor-1 (IGF-1) receptor in some tumor cells and to insulin-induced keratinocyte proliferation (Ueda et al. 1996; Li et al. 1998; Datta et al. 2000; Keshamouni et al. 2002; Fan et al. 2005; Mingo-Sion et al. 2005; Gartsbein et al. 2006). Thus, while in normal cells PKCδ negatively regulates cell proliferation and cell death, some tumor cells appear to redirect PKCδ to activate pro-proliferative signals and promote transformation (Jackson and Foster 2004).

Activation of PKCδ by Apoptotic Signals

Studies from a variety of labs show that inhibition of PKCδ suppresses "downstream" apoptotic signals such as caspase activation and DNA fragmentation, as well as "upstream" apoptotic events such as loss of MOMP, suggesting that PKCδ contributes to both apoptotic initiation and amplification (Reyland et al. 1999; Leitges et al. 2001a; Matassa et al. 2001; Humphries et al. 2006; Reyland 2007). As PKCδ is a ubiquitously expressed kinase, its ability to activate apoptotic pathways must be tightly regulated. Our studies suggest that activation of PKCδ and caspase-3 occurs rapidly in response to apoptotic stimuli and that both proteins accumulate in the nucleus under these conditions (DeVries-Seimon et al. 2007). If activation of PKCδ and apoptosis are temporally linked, is there a cell damage signal that activates both pathways? One potential candidate is p53, a common mediator of many stress/DNA damage responses (Chipuk and Green 2006; Helton and Chen 2007). Many apoptotic agents induce phosphorylation and stabilization of the p53 protein, resulting in an increase in p53-dependent transcription of pro-apoptotic proteins. In addition to its role in transcription, p53 can also act at the mitochondria to directly affect MOMP by interacting with Bcl-2 family members (Erster et al. 2004; Fuster et al. 2007; Wolff et al. 2008). Some studies indicate a role for PKCδ in regulating p53 transcriptional activation in response to genotoxins and oxidative stress (Johnson et al. 2002; Ryer et al. 2005; Liu et al. 2007; Yamaguchia et al. 2007). Yoshida's group has reported PKCδ mediated phosphorylation of p53 on serine 43, an event required for p53 transcriptional activation (Yoshida et al. 2006). In vascular smooth muscle cells, overexpression of PKCδ resulted in activation of p53 both at the transcriptional and posttranscriptional levels and depletion of p53 prevented PKCδ-induced apoptosis in these cells (Ryer et al. 2005). PKCδ has also been shown to regulate p53 phosphorylation at serine 20 in response to oxidative stress, via activation of the IKKα pathway (Yamaguchia et al. 2007). A recent study showed that PKCδ can function as a coactivator by binding to and regulating the interaction of the transcription factor Btf with the p53 promoter (Liu et al. 2007). In contrast, studies from the Reyland lab show no difference in p53 stabilization or target gene expression in irradiated or genotoxin treated salivary epithelial cells from PKCδ WT and PKCδ null mice (Humphries et al. 2006). Furthermore, based on microarray analysis, with the exception of p21, p53 targets do appear to be differentially expressed in irradiated salivary epithelial cells from PKCδ WT and PKCδ null mice (Ohm and Reyland, unpublished data). Thus, while PKCδ may contribute to the apoptotic response via activation of p53, evidence that it regulates the transcription of p53-dependent pro-apoptotic genes is lacking.

In addition to p53, other substrates of PKCδ in apoptotic cells include the DNA repair and checkpoint proteins Rad9, topoisomerase IIα and DNA-dependent protein kinase (DNA-PK). PKCδ phosphorylates the checkpoint protein, Rad9, both constitutively and in response to DNA damage (Yoshida et al. 2003). Rad9 phosphorylation promotes formation of the Rad9-Hus1-Rad1 complex, which is a critical component of G2/M DNA checkpoint control (Yoshida et al. 2003; Yoshida 2007). Similar studies show that PKCδ is required for increased expression and activation of topoisomerase IIα in response to DNA damage (Yoshida et al. 2006). PKCδ has also been suggested to suppress activation of the repair enzyme, DNA-PK, in response to DNA damage (Bharti et al. 1998). PKCδ phosphorylation of DNA-PK inhibits the latters binding to DNA, resulting in suppression of DNA double strand break repair (Bharti et al. 1998). Some reports suggest that PKCδ may directly target and inactivate the apoptosis machinery. PKCδ phosphorylation of Rad9 enhances the interaction of the BH3 domain of Rad9 with Bcl-2, potentiating apoptosis (Yoshida et al. 2003). PKCδ also promotes apoptosis by suppressing phosphorylation of the pro-apoptotic protein, Bad, and by phosphorylating and targeting the antiapoptotic protein Mcl-1 for degradation (Murriel et al. 2004; Sitailo et al. 2006). Finally, PKCδ has been shown to directly phosphorylate and activate capsase-3 in response to apoptotic signals (Voss et al. 2005).

PKCδ is a ubiquitously expressed kinase; thus its ability to activate apoptosis must be tightly regulated in order to prevent inappropriate cell death. Studies from the Reyland lab show that, under basal conditions, PKCδ is largely cytoplasmic and that nuclear retention of PKCδ commits a cell to apoptosis. Events that regulate nuclear import and accumulation of PKCδ appear to be temporally coordinated in response to apoptotic agents (see Fig. 9.2). We propose that transduction of the "death" signal to PKCδ occurs through activation of a tyrosine kinase and phosphorylation of PKCδ on specific tyrosine residues (Okhrimenko et al. 2005b; Humphries et al. 2008; Lomonaco et al. 2008). Tyrosine phosphorylated PKCδ then

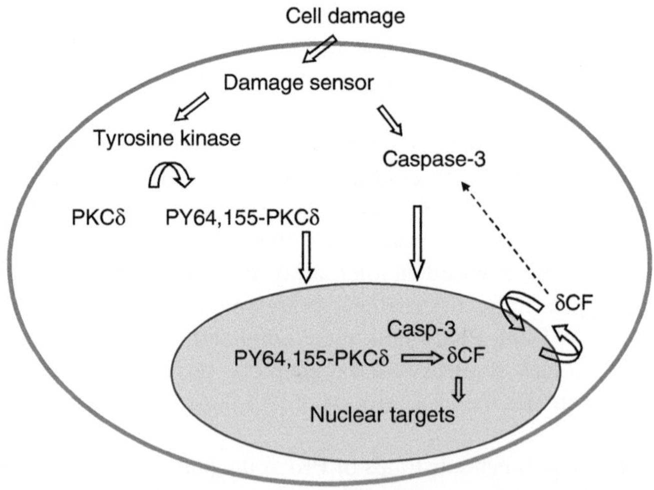

Fig. 9.2 Activation of PKCδ in response to apoptotic signals. Under basal conditions PKCδ is retained in the cytoplasm; however, in response to cell damage signals it accumulates in the nucleus. Activation of tyrosine kinase(s) downstream of damage sensors results in tyrosine phosphorylation of PKCδ in the regulatory domain and facilitates nuclear accumulation of PKCδ. Active capsase-3 also translocates to the nucleus, resulting in cleavage of PKCδ and generation of δCF. Early in apoptosis, δCF accumulates in the nucleus where studies suggest that it regulates targets involved in the cell damage response. As apoptosis progresses, δCF can also be found in the cytoplasm where it may function to regulate mitochondrial apoptotic events

translocates to the nucleus, where it is cleaved by caspase to generate the PKCδ catalytic fragment (δCF; DeVries-Seimon et al. 2007). As caspase cleavage removes the regulatory domain of the kinase, cleaved PKCδ is constitutively active and localized to the nucleus (Matassa et al. 2001). Although the mechanism of how nuclear PKCδ promotes apoptotic signaling is not well understood, most nuclear substrates of PKCδ are involved in DNA damage sensing and/or repair; hence it is likely that nuclear PKCδ facilitates transduction of these signals to the apoptotic machinery.

Tyrosine Phosphorylation of PKCδ

Multiple studies indicate that phosphorylation of PKCδ on tyrosine regulates stimulus specific functions of PKCδ (Konishi et al. 1997; Blass et al. 2002; Humphries et al. 2008; Lomonaco et al. 2008). c-Abl, Src, and Lyn tyrosine kinases have been shown to mediate phosphorylation of PKCδ in cells treated with genotoxins or H_2O_2 (Kharbanda et al. 1997; Zang et al. 1997; Sun et al. 2000). PKCδ is phosphorylated on Y64 and Y187 in glioma cells treated with etoposide (Blass et al. 2002) and on Y311, Y332, and Y512 in response to H_2O_2 (Konishi et al. 1997). In glioma cells, tyrosine phosphorylation of PKCδ is not required for its nuclear import; however, in parotid C5 cells, tyrosine phosphorylation of PKCδ at Y64 and Y155 regulate the nuclear translocation and pro-apoptotic function of PKCδ (Humphries et al. 2008). As glioblastoma is a highly aggressive tumor, these differences may reflect the "re-wiring" of PKCδ regulation in transformed cells.

Nuclear Localization of PKCδ

PKCδ translocates to the nucleus in response to many apoptotic agents (Yuan et al. 1998; DeVries et al. 2002). A bipartite nuclear localization sequence in the catalytic domain of PKCδ is required for nuclear import; however, PKCδ is largely cytoplasmic in the absence of an apoptotic signal. A second, apoptosis specific, signal appears to be required in order to activate the pro-apoptotic functions of PKCδ (DeVries-Seimon et al. 2007). Much evidence indicates that this second signal is relayed by damage induce tyrosine kinases that phosphorylate PKCδ in the regulatory domain. Once phosphorylated, PKCδ rapidly accumulates in the nucleus, suggesting that tyrosine phosphorylation may be necessary to facilitate importin binding (Humphries et al. 2008). Retention of tyrosine phosphorylated PKCδ in the nucleus is transient and may depend upon the coordinated import of active caspase-3 (DeVries-Seimon et al. 2007). Caspase cleavage of PKCδ promotes nuclear accumulation; thus, the regulatory domain may function to retain PKCδ in the cytoplasm in the absence of an apoptotic signal. In support of this, tyrosine residues important for nuclear import appear to be exclusively in the regulatory domain.

Caspase Cleavage of PKCδ

Early studies by Emoto et al. showed that irradiation activated a 40 kD myelin basic protein kinase that was subsequently identified as a stable, proteolytically cleaved, yet catalytically competent fragment of PKCδ (Emoto et al. 1996). Caspase cleavage of PKCδ occurs in the hinge region of the kinase and effectively separates the regulatory domain from the catalytic domain, resulting in the release of a constitutively active catalytic fragment (δCF; Matassa et al. 2001). Expression of δCF is sufficient to induce cell death, and Sitailo et al. have shown that expression of δCF is associated with activation of the pro-apoptotic protein, Bax, and cytochrome *c* release (Sitailo et al. 2004). Furthermore, a caspase resistant mutant of PKCδ protects keratinocytes from UV-induced apoptosis (D'Costa and Denning 2005). Studies from the Reyland lab indicate that PKCδ is cleaved in the nucleus, and that nuclear import of PKCδ and activated caspase-3 are temporally linked (DeVries-Seimon et al. 2007). These studies also suggest that it is the nuclear accumulation of PKCδ, and not caspase cleavage *per se*, which is critical for the apoptotic response as targeting a caspase-resistant mutant of PKCδ to the nucleus is also able to induce apoptosis (DeVries-Seimon et al. 2007). Although initially largely nuclear, in the later stages of apoptosis δCF also can be found in the cytoplasm, consistent with studies that a role for PKCδ at the mitochondria in apoptotic cells (Majumder et al. 2000; Sitailo et al. 2004).

9.4.3 Contribution of Atypical Isoforms of PKC

The atypical isoforms, PKCι (its murine counterpart PKCλ) and PKCζ, have been shown to be critical components of cell survival signal transduction pathways, downstream of PI3K (Akimoto et al. 1996; Cataldi et al. 2003). Atypical PKCs also suppress apoptosis by activation of pro-survival NFκB and MAPK signaling (Berra et al. 1993; Diaz-Meco et al. 1993). PKCι phosphorylation and subsequent degradation of IKKαβ is thought to play a role in NFκB activation and survival of androgen-independent (DU-145) prostate cancer cells (Win and Acevedo-Duncan 2008). In addition, the atypical PKCs may directly target mediators and regulators of apoptotic signaling pathways. In NSCLC cells, evidence suggests that PKCζ functions as a Bax kinase; phosphorylation resulting in cytoplasmic sequestration and inhibition of the pro-apoptotic functions of this Bcl-2 family member (Xin et al. 2007). Expression of a dominant negative, kinase inactive PKCζ construct resulted in increased Bax expression and downregulation of the antiapoptotic Bcl-2 in leukemic cells (Filomenko et al. 2002). PKCζ may also phosphorylate FADD, inhibiting formation of DISC and thereby conferring resistance to Fas mediated apoptosis in leukemic cells (de Thonel et al. 2001). Finally, PKCζ can protect against UV-induced apoptosis by inhibition of acid sphingomyolinase-dependent production of ceramide (Charruyer et al. 2007).

Consistent with their role in cell survival, atypical PKCs are frequently activated or upregulated in response to apoptotic stimuli and are thought to mediate protective signals activated in tumor cells exposed to cytotoxic agents. Thus, expression of PKCζ and PKCι is enhanced by UV and ceramide (Berra et al. 1997; Wang et al. 2005) and the atypical PKCs are activated by caspase cleavage releasing catalytic fragments, which are subsequently degraded by the proteosome (Smith et al. 2000, 2003; Garin et al. 2007). Conversely, p38 MAP kinase mediates nitric oxide induced apoptosis by interaction with the regulatory domain of PKCζ to inhibit autophosphorylation of PKCζ required for its activation (Kim et al. 2002).

PKCζ suppresses Fas-induced apoptosis in Jurkat cells and inhibition of PKCζ in leukemia cells potentiated the apoptotic effects of etoposide and TNF-α (Filomenko et al. 2002; Leroy et al. 2005). PKCζ overexpression also inhibited topoisomerase II activity and drug induced cytotoxicity in the U937 monocytic leukemia cell line (Plo et al. 2002). PKCι plays a similar role in hematologic malignancies, conferring resistance to cytotoxic agents in human leukemia cells and is required for Bcr-Abl mediated resistance to chemotherapy (Murray and Fields 1997; Jamieson et al. 1999). Indeed PKCι has been identified as a human oncogene, genomically amplified in a number of cancers, and may be the more important of the atypical PKCs in tumorigensis and suppression of apoptosis (Regala et al. 2005, 2008; Fields et al. 2007; Fields and Regala 2007). PKCι is overexpressed in benign and malignant meningiomas and gliomas and may also be required for cell proliferation (Patel et al. 2008).

The prostate apoptotic response (PAR4) 4 protein interacts with and inhibits the atypical PKCs, suppressing activation of NFκB, thereby promoting apoptosis (Diaz-Meco et al. 1993, 1996, 1999; Leitges et al. 2001b). PAR4 null mice exhibit enhanced Ras-induced lung tumorigenesis, consistent with the role of PAR4 as a tumor suppressor (Joshi et al. 2008). Activated PKCζ, due to the loss of PAR4 inhibition of the atypical PKC, was shown to directly phosphorylate and activate Akt in this model (Diaz-Meco and Moscat 2008; Joshi et al. 2008). In contrast, PKCζ and PKCλ have been shown to inactivate Akt by binding to the pleckstrin homology domain in breast cancer and COS cells (Doornbos et al. 1999; Mao et al. 2000). These data suggest that the functional role of PKCζ in cell survival and apoptosis may be modulated by interaction with specific partners that regulate its activity and subcellular location.

Atypical PKCs may also be pro-apoptotic in some cell types. Overexpression of PKCζ in Caco-2 colon cancer cells inhibited cell proliferation and growth in soft agar but enhanced apoptosis (Mustafi et al. 2006). Consistent with a growth-inhibitory, pro-apoptotic role for PKCζ this atypical isoform is down regulated in human colonic adenocarcinoma (Mustafi et al. 2006). Similarly, in contrast to the oncogenic role of PKCι in lung cancer (Fields and Regala 2007), PKCζ deficient mice exhibit enhanced Ras-induced lung tumorigenesis, implying a role for PKCζ as a tumor suppressor (Galvez et al. 2009). However, inhibition of PKCζ suppressed chemotaxis in human NSCLC cells, suggesting a role for PKCζ in promotion of lung metastasis (Liu et al. 2009). PKCζ was also required for EGF-induced chemotaxis of breast cancer cells and motility of invasive pancreatic adenocarcinoma cells (Laudanna et al. 2003; Sun et al. 2005). Paradoxically, PKCζ appears to function as both a tumor suppressor (Galvez et al. 2009) and tumor promoter

(Diaz-Meco and Moscat 2008), in the same mouse model of Ras-induced lung cancer. Tumorigenic actions of PKCζ appear to be mediated by regulation of Akt survival signaling, in the context of a complex with PAR4 (Diaz-Meco and Moscat 2008). In contrast, lung carcinogenesis in PKCζ deficient mice is attributed to elevated interleukin 6 (IL-6) an essential factor in the ability of Ras-transformed cells to grow under nutrient deprived conditions (Galvez et al. 2009). PKCζ in complex with the scaffold protein p62 represses IL-6 promoter acetylation and activation (Galvez et al. 2009). Thus, interaction of atypical PKCs with specific protein adaptors may mediate distinct functional responses (Moscat and Diaz-Meco 2000).

9.5 Conclusions

As discussed above, current evidence implicates the PKC family as critical regulators of cell proliferation, survival, and apoptosis in a variety of cell types and tumors. PKC kinases can act "upstream" as mediators of growth factor and cytokine signaling, as well as "downstream" via regulation of the activity and stability of transcription factors and components of apoptotic/proliferative pathways by phosphorylation. Thus, PKCs exhibit both direct and indirect effects on the extrinsic and intrinsic apoptotic pathway machinery. Consistent with their critical role as regulators of cell growth and apoptosis, alterations in the expression and/or activity of specific members of the PKC family are correspondingly associated with the pathogenesis of human diseases including cancer, making them attractive therapeutic targets. However, it is important to note that classical, novel, and atypical forms of PKC can exhibit both positive and negative effects on cell survival and either enhance or suppress the effects of chemotherapeutic agents in a cell and context specific manner. Indeed, it is likely that the balance of expression and activity of pro-apoptotic and pro-survival PKC isoforms is a key determinant in the response of tumor cells to cytotoxic and/or proliferative agents. Moreover, the levels and activation profile of distinct PKC isoforms may be altered in response to such stimuli or cellular transformation. Thus, understanding the molecular mechanisms by which specific PKC isoforms regulate and respond to cellular apoptotic or survival signaling pathways, and the relative expression and functional contributions of distinct PKCs in a given tissue or cell type, are critical elements in the targeting of PKCs and identification of new therapeutic targets in the treatment of cancer. Development of isoform specific activators and inhibitors of PKCs will also be an essential component in PKC-dependent apoptosis-based treatment strategies.

References

Abeliovich, A., Chen, C., Goda, Y., Silva, A. J., Stevens, C. F., & Tonegawa, S. (1993a). Modified hippocampal long-term potentiation in PKC-gamma-mutant mice. *Cell, 7*, 1253–1262.

Abeliovich, A., Paylor, R., Chen, C., Kim, J. J., Wehner, J. M., & Tonegawa, S. (1993b). PKC-gamma mutant mice exhibit mild deficits in spatial and contextual learning. *Cell, 75*, 1263–1271.

Abrams, S. T., Lakum, T., Lin, K., Jones, G. M., Treweeke, A. T., Farahani, M., et al. (2007). B-cell receptor signaling in chronic lymphocytic leukemia cells is regulated by overexpressed active protein kinase C beta. *Blood, 109*, 1193–1201.

Abukhdeir, A. M., & Park, B. H. (2008). P21 and p27: Roles in carcinogenesis and drug resistance. *Expert Reviews in Molecular Medicine, 10*, e19.

Acs, P., Beheshti, M., Szallasi, Z., Li, L., Yuspa, S. H., & Blumberg, P. M. (2000). Effect of a tyrosine 155 to phenylalanine mutation of protein kinase C delta on the proliferative and tumorigenic properties of NIH 3T3 fibroblasts. *Carcinogenesis, 21*, 887–891.

Adams, J. M. (2003). Ways of dying: Multiple pathways to apoptosis. *Genes and Development, 17*, 2481–2495.

Ainsworth, P. D., Winstanley, J. H., Pearson, J. M., Bishop, H. M., & Garrod, D. R. (2004). Protein kinase C alpha expression in normal breast, ductal carcinoma in situ and invasive ductal carcinoma. *European Journal of Cancer, 40*, 2269–2273.

Akimoto, K., Mizuno, K., Osada, S., Hirai, S., Tanuma, S., Suzuki, K., et al. (1994). A new member of the third class in the protein kinase C family, PKC lambda, expressed dominantly in an undifferentiated mouse embryonal carcinoma cell line and also in many tissues and cells. *The Journal of Biological Chemistry, 269*, 12677–12683.

Alvaro, V., Levy, L., Dubray, C., Roche, A., Peillon, F., Querat, B., et al. (1993). Invasive human pituitary tumors express a point-mutated alpha-protein kinase-C. *Journal of Clinical Endocrinology and Metabolism, 77*, 1125–1129.

Assender, J. W., Gee, J. M., Lewis, I., Ellis, I. O., Robertson, J. F., & Nicholson, R. I. (2007). Protein kinase C isoform expression as a predictor of disease outcome on endocrine therapy in breast cancer. *Journal of Clinical Pathology, 60*, 1216–1221.

Aziz, M. H., Manoharan, H. T., Church, D. R., Dreckschmidt, N. E., Zhong, W., Oberley, T. D., et al. (2007). Protein kinase C epsilon interacts with signal transducers and activators of transcription 3 (stat3), phosphorylates stat3ser727, and regulates its constitutive activation in prostate cancer. *Cancer Research, 67*, 8828–8838.

Barr, L. F., Campbell, S. E., & Baylin, S. B. (1997). Protein kinase C-beta 2 inhibits cycling and decreases c-myc-induced apoptosis in small cell lung cancer cells. *Cell Growth and Differentiation, 8*, 381–392.

Basu, A., Lu, D., Sun, B., Moor, A. N., Akkaraju, G. R., & Huang, J. (2002). Proteolytic activation of protein kinase C-epsilon by caspase-mediated processing and transduction of antiapoptotic signals. *The Journal of Biological Chemistry, 277*, 41850–41856.

Basu, A., & Sivaprasad, U. (2007). Protein kinase C epsilon makes the life and death decision. *Cellular Signalling, 19*, 1633–1642.

Berra, E., Diaz-Meco, M. T., Dominguez, I., Municio, M. M., Sanz, L., Lozano, J., et al. (1993). Protein kinase C zeta isoform is critical for mitogenic signal transduction. *Cell, 74*, 555–563.

Berra, E., Municio, M. M., Sanz, L., Frutos, S., Diaz-Meco, M. T., & Moscat, J. (1997). Positioning atypical protein kinase C isoforms in the uv-induced apoptotic signaling cascade. *Molecular and Cellular Biology, 17*, 4346–4354.

Besson, A., & Yong, V. W. (2000). Involvement of p21(waf1/cip1) in protein kinase C alpha-induced cell cycle progression. *Molecular and Cellular Biology, 20*, 4580–4590.

Bharti, A., Kraeft, S.-K., Gounder, M., Pandey, P., Jin, S., Yuan, Z.-M., et al. (1998). Inactivation of DNA-dependent protein kinase by protein kinase C delta: Implications for apoptosis. *Molecular and Cellular Biology, 18*, 6719–6728.

Black, J. D. (2000). Protein kinase C-mediated regulation of the cell cycle. *Frontiers in Bioscience, 5*, D406–D423.

Blagosklonny, M. V. (2002). Are p27 and p21 cytoplasmic oncoproteins? *Cell Cycle, 1*, 391–393.

Blass, M., Kronfeld, I., Kazimirsky, G., Blumberg, P., & Brodie, C. (2002). Tyrosine phosphorylation of protein kinase C delta is essential for its apoptotic effect in response to etoposide. *Molecular and Cellular Biology, 22*, 182–195.

Cacace, A. M., Ueffing, M., Han, E. K.-H., Marmè, D., & Weinstein, I. (1998). Overexpression of PKC epsilon in R6 fibroblasts causes increased production of active TGFbeta. *Journal of Cellular Physiology, 175*, 314–322.

Cacace, A. M., Ueffing, M., Philipp, A., Han, E. K.-H., Kolch, W., & Weinstein, I. B. (1996). PKC epsilon functions as an oncogene by enhancing activation of the raf kinase. *Oncogene, 13*, 2517–2626.

Cameron, A. J., Procyk, K. J., Leitges, M., & Parker, P. J. (2008). PKC alpha protein but not kinase activity is critical for glioma cell proliferation and survival. *International Journal of Cancer, 123*, 769–779.

Cartron, P.-F., Juin, P., Oliver, L., Martin, S., Meflah, K., & Vallette, F. M. (2003). Nonredundant role of bax and bak in bid-mediated apoptosis. *Molecular and Cellular Biology, 23*, 4701–4712.

Cataldi, A., Di Pietro, R., Centurione, L., Rapino, M., Santavenere, E., Garaci, F., et al. (2003). Engagement of PI-3-kinase mediated protein kinase C zeta activation in protecting friend cells from ionizing radiation-induced apoptosis. *International Journal of Oncology, 22*, 129–135.

Catley, M. C., Cambridge, L. M., Nasuhara, Y., Ito, K., Chivers, J. E., Beaton, A., et al. (2004). Inhibitors of protein kinase C (PKC) prevent activated transcription: Role of events downstream of NF-κB DNA binding. *The Journal of Biological Chemistry, 279*, 18457–18466.

Charruyer, A., Jean, C., Colomba, A., Jaffrazou, J.-P., Quillet-Mary, A., Laurent, G., et al. (2007). PKC iota protects against uv-c-induced apoptosis by inhibiting acid sphingomyelinase-dependent ceramide production. *Biochemical Journal, 405*, 77–83.

Child, E. S., & Mann, D. J. (2006). The intricacies of p21 phosphorylation: Protein/protein interactions, subcellular localization and stability. *Cell Cycle, 5*, 1313–1319.

Chipuk, J., & Green, D. R. (2006). Dissecting p53-dependent apoptosis. *Cell Death and Differentiation, 13*, 994–1002.

Churchill, E., Budas, G., Vallentin, A., Koyanagi, T., & Mochly-Rosen, D. (2008). PKC isozymes in chronic cardiac disease: Possible therapeutic targets? *Annual Review of Pharmacology and Toxicology, 48*, 569–599.

Cimmino, A., Calin, G. A., Fabbri, M., Iorio, M. V., Ferracin, M., Shimizu, M., et al. (2005). Mir-15 and mir-16 induce apoptosis by targeting bcl2. *Proceedings of the National Academy of Science of the United States of America, 102*, 13944–13949.

Clark, J. A., Black, A. R., Leontieva, O. V., Frey, M. R., Pysz, M. A., Kunneva, L., et al. (2004). Involvement of the ERK signaling cascade in protein kinase C-mediated cell cycle arrest in intestinal epithelial cells. *The Journal of Biological Chemistry, 279*, 9233–9247.

Coqueret, O. (2003). New roles for p21 and p27 cell-cycle inhibitors: A function for each cell compartment? *Trends in Cell Biology, 13*, 65–70.

Cornford, P., Evans, J., Dodson, A., Parsons, K., Woolfenden, A., Neoptolemos, J., et al. (1999). Protein kinase C isoenzyme patterns characteristically modulated in early prostate cancer. *American Journal of Pathology, 154*, 137–144.

Cory, S., Huang, D. C., & Adams, J. M. (2003). The bcl-2 family: Roles in cell survival and oncogenesis. *Oncogene, 22*, 8590–8607.

Crawford, A. M., Kerr, J. F., & Currie, A. R. (1972). The relationship of acute mesodermal cell death to the teratogenic effects of 7-OHM-12-MBA in the foetal rat. *British Journal of Cancer, 26*, 498–503.

D'Costa, A. M., & Denning, M. F. (2005). A caspase-resistant mutant of PKC-delta protects keratinocytes from uv-induced apoptosis. *Cell Death and Differentiation, 12*, 224–232.

D'Costa, A., Robinson, J., Maududi, T., Chaturvedi, V., Nickoloff, B. J., & Denning, M. F. (2006). The proapoptotic tumor suppressor protein kinase C-delta is lost in human squamous cell carcinomas. *Oncogene, 25*, 375–386.

Datta, K., Nambudripad, R., Pal, S., Zhou, M., Cohen, H. T., & Mukhopadhyay, D. (2000). Inhibition of insulin-like growth factor-1-mediated cell signaling by the von hippel-lindau gene product in renal cancer. *The Journal of Biological Chemistry, 275*, 20700–20706.

de Thonel, A., Bettaieb, A., Jean, C., Laurent, G., & Quillet-Mary, A. (2001). Role of protein kinase C zeta isoform in Fas resistance of immature myeloid KG1a leukemic cells. *Blood, 98*, 3770–3777.

Debatin, K. M., Poncet, D., & Kroemer, G. (2002). Chemotherapy: Targeting the mitochondrial cell death pathway. *Oncogene, 21*, 8786–8803.

Dempsey, E. C., Newton, A. C., Mochly-Rosen, D., Fields, A. P., Reyland, M. E., Insel, P. A., et al. (2000). Protein kinase C isozymes and the regulation of diverse cell responses. *American Journal of Physiology, 279*, L429–L438.
Deng, X., Ito, T., Carr, B., Mumby, M., & May, W. S., Jr. (1998). Reversible phosphorylation of bcl2 following interleukin 3 or bryostatin 1 is mediated by direct interaction with protein phosphatase 2a. *The Journal of Biological Chemistry, 273*, 34157–34163.
Detjen, K., Brembeck, F., Welzel, M., Kaiser, A., Haller, H., Wiedenmann, B., et al. (2000). Activation of protein kinase C alpha inhibits growth of pancreatic cancer cells via p21(cip)-mediated G(1) arrest. *Journal of Cell Science, 113*, 3025–3035.
DeVries, T. A., Neville, M. C., & Reyland, M. E. (2002). Nuclear import of PKC delta is required for apoptosis: Identification of a novel nuclear import sequence. *EMBO Journal, 21*, 6050–6060.
DeVries-Seimon, T. A., Ohm, A. M., Humphries, M. J., & Reyland, M. E. (2007). Induction of apoptosis is driven by nuclear retention of protein kinase C delta. *The Journal of Biological Chemistry, 282*, 22307–22314.
Dhanasekaran, D., & Reddy, E. (2008). JNK signaling in apoptosis. *Oncogene, 27*, 6245–6251.
Diaz-Meco, M., Berra, E., Municio, M., Sanz, L., Lozano, J., Dominguez, I., et al. (1993). A dominant negative protein kinase Cζ subspecies blocks NF-κB activation. *Molecular and Cellular Biology, 13*, 4770–4775.
Diaz-Meco, M. T., Lallena, M. J., Monjas, A., Frutos, S., & Moscat, J. (1999). Inactivation of the inhibitory κB protein kinase/nuclear factor κB pathway by par-4 expression potentiates tumor necrosis factor alpha-induced apoptosis. *The Journal of Biological Chemistry, 274*, 19606–19612.
Diaz-Meco, M. T., & Moscat, J. (2008). AKT regulation and lung cancer: A novel role and mechanism of action for the tumor suppressor par-4. *Cell Cycle, 7*, 2817–2820.
Diaz-Meco, M., Municio, M., Frutos, S., Sanchez, P., Lazano, J., Sanz, L., et al. (1996). The product of *par*-4, a gene induced during apoptosis, interacts selectively with the atypical isoforms of protein kinase C. *Cell, 86*, 777–786.
Ding, L., Wang, H., Lang, W., & Xiao, L. (2002). Protein kinase C-epsilon promotes survival of lung cancer cells by suppressing apoptosis through dysregulation of the mitochondrial caspase pathway. *The Journal of Biological Chemistry, 277*, 35305–35313.
Dooley, N. P., Baltuch, G. H., Groome, N., Villemure, J. G., & Yong, V. W. (1998). Apoptosis is induced in glioma cells by antisense oligonucleotides to protein kinase C alpha and is enhanced by cycloheximide. *Neuroreport, 9*, 1727–1733.
Doornbos, R. P., Theelen, M., van der Hoeven, P. C., van Blitterswijk, W. J., Verkleij, A. J., & van Bergen en Henegouwen, P. M. (1999). Protein kinase C zeta is a negative regulator of protein kinase B activity. *The Journal of Biological Chemistry, 274*, 8589–8596.
Eguchi, K. (2001). Apoptosis in autoimmune diseases. *Internal Medicine, 40*, 275–284.
Ekshyyan, O., & Aw, T. Y. (2004). Apoptosis: A key in neurodegenerative disorders. *Current Neurovascular Research, 1*, 355–371.
Ellis, H., & Horvitz, H. R. (1986). Genetic control of programmed cell death in the nematode c. Elegans. *Cell, 44*, 817–829.
El-Rayes, B. F., Ali, S., Philip, P. A., & Sarkar, F. H. (2008). Protein kinase C: A target for therapy in pancreatic cancer. *Pancreas, 36*, 346–352.
Emoto, Y., Kisaki, H., Manome, Y., Kharbanda, S., & Kufe, D. (1996). Activation of protein kinase C delta in human myeloid leukemia cells treated with 1-β-d-arabinofuranosylcytosine. *Blood, 87*, 1990–1996.
Engers, R., Mrzyk, S., Springer, E., Fabbro, D., Weissgerber, G., Gernharz, C. D., et al. (2000). Protein kinase C in human renal cell carcinomas: Role in invasion and differential isoenzyme expression. *British Journal of Cancer, 82*, 1063–1069.
Erster, S., Mihara, M., Kim, R. H., Petrenko, O., & Moll, U. M. (2004). In vivo mitochondrial p53 translocation triggers a rapid first wave of cell death in response to DNA damage that can precede p53 target gene activation. *Molecular and Cellular Biology, 24*, 6728–6741.
Espinosa, I., Briones, J., Bordes, R., Brunet, S., Martino, R., Sureda, A., et al. (2006). Membrane PKC-beta 2 protein expression predicts for poor response to chemotherapy and survival in patients with diffuse large B-cell lymphoma. *Annals of Hematology, 85*, 597–603.

Evan, G., & Vousden, K. (2001). Proliferation, cell cycle and apoptosis in cancer. *Nature, 411*, 342–348.

Fan, C., Katsuyama, M., & Yabe-Nishimura, C. (2005). Protein kinase C delta mediates up-regulation of nox1, a catalytic subunit of nadph oxidase, via transactivation of the EGF receptor: Possible involvement of PKC delta in vascular hypertrophy. *Biochemical Journal, 390*, 761–767.

Fields, A. P., Frederick, L. A., & Regala, R. P. (2007). Targeting the oncogenic protein kinase C iota signaling pathway for the treatment of cancer. *Biochemical Society Transactions, 35*, 996–1000.

Fields, A. P., & Regala, R. P. (2007). Protein kinase C iota: Human oncogene, prognostic marker and therapeutic target. *Pharmacological Research, 55*, 487–497.

Filomenko, R., Poirson-Bichat, F., Billerey, C., Belon, J. P., Garrido, C., Solary, E., et al. (2002). Atypical protein kinase C zeta as a target for chemosensitization of tumor cells. *Cancer Research, 62*, 1815–1821.

Fournier, D. B., Chisamore, M., Lurain, J. R., Rademaker, A. W., Jordan, V. C., & Tonetti, D. A. (2001). Protein kinase C alpha expression is inversely related to ER status in endometrial carcinoma: Possible role in AP-1-mediated proliferation of ER-negative endometrial cancer. *Gynecologic Oncology, 81*, 366–372.

Franke, T. (2007). PI3K/AKT:Getting it right matters. *Oncogene, 27*, 6473–6488.

Franz, M. G., Norman, J. G., Fabri, P. J., & Gower, W. R., Jr. (1996). Differentiation of pancreatic ductal carcinoma cells associated with selective expression of protein kinase C isoforms. *Annals of Surgical Oncology, 3*, 564–569.

Frey, M. R., Clark, J. A., Leontieva, O., Uronis, J. M., Black, A. R., & Black, J. D. (2000). Protein kinase C signaling mediates a program of cell cycle withdrawal in the intestinal epithelium. *Journal of Cell Biology, 151*, 763–778.

Frey, M. R., Saxon, M. L., Zhao, X., Rollins, A., Evans, S. S., & Black, J. D. (1997). Protein kinase C isozyme-mediated cell cycle arrest involves induction of p21(waf1/cip1) and p27(kip1) and hypophosphorylation of the retinoblastoma protein in intestinal epithelial cells. *The Journal of Biological Chemistry, 272*, 9424–9435.

Frey, M. R., Clark, J. A., Bateman, N. W., Kazanietz, M. G., Black, A. R., & Black, J. D. (2004). Cell cycle- and protein kinase C-specific effects of resiniferatoxin and resiniferonol 9,13,14-ortho-phenylacetate in intestinal epithelial cells. *Biochemical Pharmacology, 67*, 1873–1886.

Fujii, T., Garcia-Bermejo, M. L., Bernabo, J. L., Caamano, J., Ohba, M., Kuroki, T., et al. (2000). Involvement of protein kinase C delta (PKCdelta) in phorbol ester-induced apoptosis in lncap prostate cancer cells. Lack of proteolytic cleavage of PKC delta. *The Journal of Biological Chemistry, 275*, 7574–7582.

Fuster, J. J., Sanz-Gonzalez, S. M., Moll, U. M., & Andres, V. (2007). Classic and novel roles of p53: Prospects for anticancer therapy. *Trends in Molecular Medicine, 13*, 192–199.

Galvez, A. S., Duran, A., Linares, J. F., Pathrose, P., Castilla, E. A., Abu-Baker, S., et al. (2009). Protein kinase C zeta represses the interleukin-6 promoter and impairs tumorigenesis *in vivo*. *Molecular and Cellular Biology, 29*, 104–115.

Garin, G., Abe, J., Mohan, A., Lu, W., Yan, C., Newby, A. C., et al. (2007). Flow antagonizes TNF-alpha signaling in endothelial cells by inhibiting caspase-dependent PKC zeta processing. *Circulation Research, 101*, 97–105.

Gartsbein, M., Alt, A., Hashimoto, K., Nakajima, K., Kuroki, T., & Tennenbaum, T. (2006). The role of protein kinase C delta activation and STAT3 ser727 phosphorylation in insulin-induced keratinocyte proliferation. *Journal of Cell Science, 119*, 470–481.

Gavrielides, M. V., Frijhoff, A. F., Conti, C. J., & Kazanietz, M. G. (2004). Protein kinase C and prostate carcinogenesis: Targeting the cell cycle and apoptotic mechanisms. *Current Drug Targets, 5*, 431–443.

Gelinas, C., & White, E. (2005). BH3-only proteins in control: Specificity regulates mcl-1 and bak-mediated apoptosis. *Genes and Development, 19*, 1263–1268.

Ghayur, T., Hugunin, M., Talanian, R. V., Ratnofsky, S., Quinlan, C., Emoto, Y., et al. (1996). Proteolytic activation of protein kinase C delta by an ice/ced 3-like protease induces characteristics of apoptosis. *Journal of Experimental Medicine, 184*, 2399–2404.

Gilhooly, E. M., Morse-Gaudio, M., Bianchi, L., Reinhart, L., Rose, D. P., Connolly, J. M., et al. (2001). Loss of expression of protein kinase C beta is a common phenomenon in human malignant melanoma: A result of transformation or differentiation? *Melanoma Research, 11*, 355–369.

Gokmen-Polar, Y., Murray, N. R., Velasco, M. A., Gatalica, Z., & Fields, A. P. (2001). Elevated protein kinase C beta II is an early promotive event in colon carcinogenesis. *Cancer Research, 61*, 1375–1381.

Gomez-Angelats, M., & Cidlowski, J. A. (2001). Protein kinase C regulates Fadd recruitment and death-inducing signaling complex formation in Fas/cd95-induced apoptosis. *The Journal of Biological Chemistry, 276*, 44944–44952.

Gozuacik, D., & Kimchi, A. (2007). Autophagy and cell death. *Current Topics in Developmental Biology, 78*, 217–245.

Griner, E. M., & Kazanietz, M. G. (2007). Protein kinase C and other diacylglycerol effectors in cancer. *Nature Reviews. Cancer, 7*, 281–294.

Grossoni, V., Falbo, K., Kazanietz, M., de Kier Joffé, E., & Urtreger, A. (2007). Protein kinase C delta enhances proliferation and survival of murine mammary cells. *Molecular Carcinogenesis, 46*, 381–390.

Guan, L., Song, K., Pysz, M. A., Curry, K. J., Hizli, A. A., Danielpour, D., et al. (2007). Protein kinase C-mediated down-regulation of cyclin D1 involves activation of the translational repressor 4e-BP1 via a phosphoinositide 3-kinase/AKT-independent, protein phosphatase 2a-dependent mechanism in intestinal epithelial cells. *The Journal of Biological Chemistry, 282*, 14213–14225.

Guo, K., Li, Y., Kang, X., Sun, L., Cui, J., Gao, D., et al. (2009). Role of PKC beta in hepatocellular carcinoma cells migration and invasion in vitro: A potential therapeutic target. *Clinical & Experimental Metastasis, 26*, 189–195.

Gutcher, I., Webb, P. R., & Anderson, N. G. (2003). The isoform-specific regulation of apoptosis by protein kinase C. *Cellular and Molecular Life Sciences, 60*, 1061–1070.

Haimovitz-Friedman, A., Balaban, N., McLoughlin, M., Ehleiter, D., Michaeli, J., Vlodavsky, I., et al. (1994). Protein kinase C mediates basic fibroblast growth factor protection of endothelial cells against radiation-induced apoptosis. *Cancer Research, 54*, 2591–2597.

Hanahan, D., & Weinberg, R. (2000). The hallmarks of cancer. *Cell, 100*, 57–70.

Harper, N., Hughes, M. A., Farrow, S. N., Cohen, G. M., & MacFarlane, M. (2003). Protein kinase C modulates tumor necrosis factor-related apoptosis-inducing aigand-induced apoptosis by targeting the apical events of death receptor signaling. *The Journal of Biological Chemistry, 278*, 44338–44347.

Haughian, J. M., Jackson, T. A., Koterwas, D. M., & Bradford, A. P. (2006). Endometrial cancer cell survival and apoptosis is regulated by protein kinase C alpha and delta. *Endocrine Related Cancer, 13*, 1251–1267.

Hayashi, T., & Faustman, D. L. (2003). Role of defective apoptosis in type 1 diabetes and other autoimmune diseases. *Recent Progress in Hormone Research, 58*, 131–153.

Helton, S., & Chen, X. (2007). P53 modulation of the DNA damage response. *Journal of Cellular Biochemistry, 100*, 883–896.

Herrant, M., Liuciano, F., Loubat, A., & Auberger, P. (2002). The protective effect of phorbol esters on Fas-mediated apoptosis in T cells. Transcriptional and postranscriptional regulation. *Oncogene, 21*, 4957–4968.

Hizli, A. A., Black, A. R., Pysz, M. A., & Black, J. D. (2006). Protein kinase C alpha signaling inhibits cyclin D1 translation in intestinal epithelial cells. *The Journal of Biological Chemistry, 281*, 14596–14603.

Horvitz, H. R. (2003). Nobel lecture. Worms, life and death. *Bioscience Reports, 23*, 239–303.

Huang, D. C., & Strasser, A. (2000). BH3-only proteins-essential initiators of apoptotic cell death. *Cell, 103*, 839–842.

Humphries, M. J., Limesand, K. H., Schneider, J. C., Nakayama, K. I., Anderson, S. M., & Reyland, M. E. (2006). Suppression of apoptosis in the protein kinase C delta null mouse in vivo. *The Journal of Biological Chemistry, 281*, 9728–9737.

Humphries, M. J., Ohm, A. M., Schaack, J., Adwan, T. S., & Reyland, M. E. (2008). Tyrosine phosphorylation regulates nuclear translocation of PKC delta. *Oncogene, 27*, 3045–3053.

Ionov, Y., Yamamoto, H., Krajewski, S., Reed, J. C. & Perucho, M. (2000). Mutational inactivation of the proapoptotic gene bax confers selective advantage during tumor clonal evolution. *Proceedings of the National Academy of Science of the United States of America, 97*, 10872–10877.

Ito, Y., Mishra, N., Yoshida, K., Kharbanda, S., Saxena, S., & Kufe, D. (2001). Mitochondrial targeting of JNK/SAPK in the phorbol ester response of myeloid leukemia cells. *Oncogene, 8*, 794–800.

Jackson, D. N., & Foster, D. A. (2004). The enigmatic protein kinase C delta: Complex roles in cell proliferation and survival. *FASEB Journal, 18*, 627–636.

Jamieson, L., Carpenter, L., Biden, T. J., & Fields, A. P. (1999). Protein kinase C iota activity is necessary for bcr-abl-mediated resistance to drug-induced apoptosis. *The Journal of Biological Chemistry, 274*, 3927–3930.

Jansen, A. P., Verwiebe, E. G., Dreckschmidt, N. E., Wheeler, D. L., Oberley, T. D., & Verma, A. K. (2001). Protein kinase C-epsilon transgenic mice: A unique model for metastatic squamous cell carcinoma. *Cancer Research, 61*, 808–812.

Jiang, X. H., Tu, S. P., Cui, J. T., Lin, M. C., Xia, H. H., Wong, W. M., et al. (2004). Antisense targeting protein kinase C alpha and beta1 inhibits gastric carcinogenesis. *Cancer Research, 64*, 5787–5794.

Jiffar, T., Kurinna, S., Suck, G., Carlson-Bremer, D., Ricciardi, M. R., Konopleva, M., et al. (2004). PKC alpha mediates chemoresistance in acute lymphoblastic leukemia through effects on bcl2 phosphorylation. *Leukemia, 18*, 505–512.

Johnson, C. L., Lu, D., Huang, J., & Basu, A. (2002). Regulation of p53 stabilization by DNA damage and protein kinase C. *Molecular Cancer Therapeutics, 1*, 861–867.

Johnstone, R., Ruefli, A., & Lowe, S. (2002). Apoptosis: A link between cancer genetics and chemotherapy. *Cell, 108*, 135–144.

Jorgensen, K., Skrede, M., Cruciani, V., Mikalsen, S. O., Slipicevic, A., & Florenes, V. A. (2005). Phorbol ester phorbol-12-myristate-13-acetate promotes anchorage-independent growth and survival of melanomas through MEK-independent activation of ERK1/2. *Biochemical and Biophysical Research Communications, 329*, 266–274.

Joshi, J., Fernandez-Marcos, P. J., Galvez, A., Amanchy, R., Linares, J. F., Duran, A., et al. (2008). Par-4 inhibits AKT and suppresses ras-induced lung tumorigenesis. *EMBO Journal, 27*, 2181–2193.

Kahl-Rainer, P., Karner-Hanusch, J., Weiss, W., & Marian, B. (1994). Five of six protein kinase C isoenzymes present in normal mucosa show reduced protein levels during tumor development in the human colon. *Carcinogenesis, 15*, 779–782.

Kamimura, K., Hojo, H., & Abe, M. (2004). Characterization of expression of protein kinase C isozymes in human b-cell lymphoma: Relationship between its expression and prognosis. *Pathology International, 54*, 224–230.

Kerfoot, C., Huang, W., & Rotenberg, S. A. (2004). Immunohistochemical analysis of advanced human breast carcinomas reveals downregulation of protein kinase C alpha. *Journal of Histochemistry and Cytochemistry, 52*, 419–422.

Kerr, J. F. R. (1971). Shrinkage necrosis: A distinct mode of cellular death. *Journal Pathology, 105*, 13–20.

Kerr, J. F. R. (2002). History of the events leading to the formulation of the apoptosis concept. *Toxicology, 181*, 471–474.

Kerr, J. F. R., & Searle, J. (1972). Digestion of cellular fragments within the phagolysosoes in carcinoma cells. *Journal of Pathology, 108*, 55–58.

Kerr, J. F. R., Wyllie, A. H., & Currie, A. R. (1972) Apoptosis: A basic biological phenomenon with wide-ranging implications in tissue kinetics. *British Journal of Cancer, 26*, 239–257.

Keshamouni, V. G., Mattingly, R. R., & Reddy, K. B. (2002). Mechanism of 17-beta-estradiol-induced ERK1/2 activation in breast cancer cells. *The Journal of Biological Chemistry, 277*, 22558–22565.

Kharbanda, S., Pandey, P., Jin, S., Inoue, S., Bharti, A., Yuan, Z.-M., et al. (1997). Functional interaction between DNA-PK and c-abl in response to DNA damage. *Nature, 386*, 732–735.

Kiley, S. C., Clark, K. J., Duddy, S., Welch, D., & Jaken, S. (1999). Increased protein kinase C delta in mammary tumor cells: Relationship to transformation and metastatic progression. *Oncogene, 18*, 6748–6757.

Kiley, S. C., Clark, K. J., Goodnough, M., Welch, D. R., & Jaken, S. (1999). Protein kinase C delta involvement in mammary tumor cell metastasis. *Cancer Research, 59*, 3230–3238.

Kilpatrick, L. E., Lee, J. Y., Haines, K. M., Campbell, D. E., Sullivan, K. E., & Korchak, H. M. (2002). A role for PKC-delta and PI 3-kinase in TNF-alpha-mediated antiapoptotic signaling in the human neutrophil. *American Journal of Physiology. Cell Physiology, 283*, C48–C57.

Kim, G.-Y., Mercer, S. E., Ewton, D. Z., Yan, Z., Jin, K., & Friedman, E. (2002). The stress-activated protein kinases p38alpha and JNK1 stabilize p21cip1 by phosphorylation. *The Journal of Biological Chemistry, 277*, 29792–29802.

Knauf, J. A., Elisei, R., Mochly-Rosen, D., Liron, T., Chen, X.-N., Gonsky, R., et al. (1999). Involvement of protein kinase C epsilon in thyroid cell death. A truncated chimeric PKC epsilon cloned from a thyroid cancer cell line protects thyroid cells from apoptosis. *The Journal of Biological Chemistry, 274*, 23414–23425.

Koivunen, J., Aaltonen, V., Koskela, S., Lehenkari, P., Laato, M., & Peltonen, J. (2004). Protein kinase C alpha/beta inhibitor GO6976 promotes formation of cell junctions and inhibits invasion of urinary bladder carcinoma cells. *Cancer Research, 64*, 5693–5701.

Koivunen, J., Aaltonen, V., & Peltonen, J. (2006). Protein kinase C (PKC) family in cancer progression. *Cancer Letters, 235*, 1–10.

Kolch, W., Heidecker, G., Kochs, G., Hummel, R., Vahidi, H., Mischak, H., et al. (1993). Protein kinase C alpha activates raf-1 by direct phosphorylation. *Nature, 364*, 249–252.

Konishi, H., Tanaka, M., Takemura, Y., Hiderori, H., Ono, Y., Kikkawa, U., et al. (1997). Activation of protein kinase C by tyrosine phosphorylation in response to H_2O_2. *Proceedings of the National Academy of Science of the United States of America, 94*, 11233–11237.

Konstantinidou, A., Korkolopoulou, P., & Patsouris, E. (2002). Apoptotic markers for tumor recurrence:A minireview. *Apoptosis, 7*, 461–470.

Koren, R., Ben Meir, D., Langzam, L., Dekel, Y., Konichezky, M., Baniel, J., et al. (2004). Expression of protein kinase C isoenzymes in benign hyperplasia and carcinoma of prostate. *Oncology Reports, 11*, 321–326.

Koren, R., Langzam, L., Paz, A., Livne, P. M., Gal, R., & Sampson, S. R. (2000). Protein kinase C (PKC) isoenzymes immunohistochemistry in lymph node revealing solution-fixed, paraffin-embedded bladder tumors. *Applied Immunohistochemistry & Molecular Morphology, 8*, 166–171.

Koriyama, H., Kouchi, Z., Umeda, T., Saido, T. C., Momoi, T., Ishiura, S., et al. (1999). Proteolytic activation of protein kinase C delta and epsilon by caspase-3 in U937 cells during chemotherapeutic agent-induced apoptosis. *Cellular Signalling, 11*, 831–838.

Krasagakis, K., Lindschau, C., Fimmel, S., Eberle, J., Quass, P., Haller, H., et al. (2004). Proliferation of human melanoma cells is under tight control of protein kinase C alpha. *Journal of Cellular Physiology, 199*, 381–387.

Kroemer, G. (1997). The proto-oncogene bcl-2 and its role in regulating apoptosis. *Nature Medicine, 3*, 614–620.

Kuwana, T., & Newmeyer, D. D. (2003). Bcl-2-family proteins and the role of mitochondria in apoptosis. *Current Opinion in Cell Biology, 15*, 691–699.

Lahn, M., Kohler, G., Sundell, K., Su, C., Li, S., Paterson, B. M., et al, (2004). Protein kinase C alpha expression in breast and ovarian cancer. *Oncology, 67*, 1–10.

Langzam, L., Koren, R., Gal, R., Kugel, V., Paz, A., Farkas, A., et al. (2001). Patterns of protein kinase C isoenzyme expression in transitional cell carcinoma of bladder. Relation to degree of malignancy. *American Journal of Clinical Pathology, 116*, 377–385.

Laudanna, C., Sorio, C., Tecchio, C., Butcher, E. C., Bonora, A., Bassi, C., et al. (2003). Motility analysis of pancreatic adenocarcinoma cells reveals a role for the atypical zeta isoform of protein kinase C in cancer cell movement. *Laboratory Investigation, 83*, 1155–1163.

Le, X. F., Marcelli, M., McWatters, A., Nan, B., Mills, G. B., O'Brian, C. A., et al. (2001). Heregulin-induced apoptosis is mediated by down-regulation of bcl-2 and activation of caspase-7 and is potentiated by impairment of protein kinase C alpha activity. *Oncogene, 20*, 8258–8269.

Lee, S., & Pervaiz, S. (2007). Apoptosis in the pathophysiology of diabetes mellitus. *International Journal of Biochemistry and Cell Biology, 39*, 497–504.

Leitges, M., Gimborn, K., Elis, W., Kalesnikoff, J., Hughes, M. R., Krystal, G., et al. (2002). Protein kinase C-delta is a negative regulator of antigen-induced mast cell degranulation. *Molecular and Cellular Biology, 22*, 3970–3980.

Leitges, M., Mayr, M., Braun, U., Mayr, U., Li, C., Pfister, G., et al. (2001). Exacerbated vein graft arteriosclerosis in protein kinase C delta-null mice. *Journal of Clinical Investigation, 108*, 1505–1512.

Leitges, M., Sanz, L., Martin, P., Duran, A., Braun, U., Garcia, J. F., et al. (2001). Targeted disruption of the zeta PKC gene results in the impairment of the NF-kB pathway. *Molecular Cell, 8*, 771–780.

Leroy, I., de Thonel, A., Laurent, G., & Quillet-Mary, A. (2005). Protein kinase C zeta associates with death inducing signaling complex and regulates fas ligand-induced apoptosis. *Cellular Signalling, 17*, 1149–1157.

Levine, B., Sinha, S., & Kroemer, G. (2008). Bcl-2 family members: Dual regulators of apoptosis and autophagy. *Autophagy, 4*, 600–606.

Li, W., Jiang, Y.-X., Zhang, J., Soon, L., Flechner, L., Kapoor, V., et al. (1998). Protein kinase C-delta is an important signaling molecule in insulin-like growth factor 1 receptor-mediated cell transformation. *Molecular and Cellular Biology, 18*, 5888–5898.

Li, L., Sampat, K., Hu, N., Zakari, J., & Yuspa, S. H. (2006). Protein kinase C negatively regulates AKT activity and modifies uvc-induced apoptosis in mouse keratinocytes. *The Journal of Biological Chemistry, 281*, 3237–3243.

Li, H., & Weinstein, I. B. (2006). Protein kinase C beta enhances growth and expression of cyclin D1 in human breast cancer cells. *Cancer Research, 66*, 11399–11408.

Li, J., & Yuan, J. (2008). Caspases in apoptosis and beyond. *Oncogene, 27*, 6194–6206.

Li, W., Jiang, Y.-X., Zhang, J., Soon, L., Flechner, L., Kapoor, V., et al. (1998). Protein kinase C-delta is an important signaling molecule in insulin-like growth factor receptor-mediated cell transformation. *Molecular and Cellular Biology, 18*, 5888–5898.

Liadis, N., Murakami, K., Eweida, M., Elford, A. R., Sheu, L., Gaisano, H. Y., et al. (2005). Caspase-3-dependent beta-cell apoptosis in the initiation of autoimmune diabetes mellitus. *Molecular and Cellular Biology, 25*, 3620–3629.

Liao, L., Ramsay, K., & Jaken, S. (1994). Protein kinase C isozymes in progressively transformed rat embryo fibroblasts. *Cell Growth and Differentiation, 5*, 1185–1194.

Lin, D., & Takemoto, D., J. (2005). Oxidative activation of protein kianse C gamma through the c1 domain. Effects on gap junctions. *The Journal of Biological Chemistry, 280*, 13682–13693.

Lin, D., Shanks, D., Prakash, O., & Takemoto, D. J. (2007). Protein kinase C gamma mutations in the c1b domain cause caspase-3-linked apoptosis in lens epithelial cells through gap junctions. *Experimental Eye Research, 85*, 113–122.

Liu, H., Zhang, H. Y., Zhu, X., Shao, Z., & Yao, Z. (2002). Preconditioning blocks cardiocyte apoptosis: Role of K(ATP) channels and PKC-epsilon. *American Journal of Physiology-Heart and Circulatory Physiology 282*, H1380–H1386.

Liu, H., Lu, Z.-G., Miki, Y. & Yoshida, K. (2007). Protein kinase C delta induces transcription of the tp53 tumor suppressor gene by controlling death-promoting factor btf in the apoptotic response to DNA damage. *Molecular and Cellular Biology*, *27*, 8480–8491.

Liu, Y., Su, W., Thompson, E. A., Leitges, M., Murray, N. R., & Fields, A. P. (2004). Protein kinase C betaII regulates its own expression in rat intestinal epithelial cells and the colonic epithelium in vivo. *The Journal of Biological Chemistry*, *279*, 45556–45563.

Liu, Y., Wang, B., Wang, J., Wan, W., Sun, R., & Zhao, Y., et al. (2009). Down-regulation of pkczeta expression inhibits chemotaxis signal transduction in human lung cancer cells. *Lung Cancer, 63*, 210–218.

Lomonaco, S. L., Kahana, S., Blass, M., Brody, Y., Okhrimenko, H., Xiang, C., et al. (2008). Phosphorylation of protein kinase c{delta} on distinct tyrosine residues induces sustained activation of erk1/2 via down-regulation of mkp-1: Role in the apoptotic effect of etoposide. *The Journal of Biological Chemistry, 283*, 17731–17739.

Lu, Z., Hornia, A., Jiang, Y.-W., Zang, Q., Ohno, S. & Foster, D. (1997). Tumor promotion by depleting cells of protein kinase C delta. *Molecular and Cellular Biology, 17*, 3418–3428.

Ma, S., & Rosen, S. T. (2007). Enzastaurin. *Current Opinion in Oncology, 19*, 590–595.

Macfarlane, D., & Manzel, L. (1994). Activation of beta-isozyme of protein kinase C (pkc beta) is necessary and sufficient for phorbol ester-induced differentiation of hl-60 promyelocytes. Studies with pkc beta-defective pet mutant. *The Journal of Biological Chemistry, 269*, 4327–23484.

Mackay, H. J., & Twelves, C. J. (2003). Protein kinase C: A target for anticancer drugs? *Endocrine-Related Cancer, 10*, 389–396.

Majewski, M., Nieborowska-Skorska, M., Salomoni, P., Slupianek, A., Reiss, K., Trotta, R., et al. (1999). Activation of mitochondrial raf-1 is involved in the antiapoptotic effects of AKT. *Cancer Research, 59*, 2815–2819.

Majumder, P., Mishra, N., Sun, X., Bharti, A., Kharbanda, S., Saxena, S., et al. (2001). Targeting of protein kinase C delta to mitochondria in the oxidative stress response. *Cell Death and Differentiation, 12*, 465–470.

Majumder, P. K., Pandey, P., Sun, X., Cheng, K., Datta, R., Saxena, S., et al. (2000). Mitochondrial translocation of protein kinase C delta in phorbol ester-induced cytochrome c release and apoptosis. *The Journal of Biological Chemistry, 275*, 21793–21796.

Mandil, R., Ashkenazi, E., Blass, M., Kronfeld, I., Kazimirsky, G., Rosenthal, G., et al. (2001). Protein kinase C alpha and protein kinase C delta play opposite roles in the proliferation and apoptosis of glioma cells. *Cancer Research, 61*, 4612–4619.

Mao, M., Fang, X., Lu, Y., Lapushin, R., Bast, R. C., Jr., & Mills, G. B. (2000). Inhibition of growth-factor-induced phosphorylation and activation of protein kinase b/akt by atypical protein kinase C in breast cancer cells. *The Biochemistry Journal, 352*(Pt 2), 475–482.

Martiny-Baron, G., & Fabbro, D. (2007). Classical PKC isoforms in cancer. *Pharmacological Research, 55*, 477–486.

Masur, K., Lang, K., Niggemann, B., Zanker, K. S., & Entschladen, F. (2001). High pkc alpha and low e-cadherin expression contribute to high migratory activity of colon carcinoma cells. *Molecular Biology of the Cell, 12*, 1973–1982.

Matassa, A., Carpenter, L., Biden, T., Humphries, M. & Reyland, M. (2001). PKC delta is required for mitochondrial dependent apoptosis in salivary epithelial cells. *The Journal of Biological Chemistry, 276*, 29719–29728.

Matassa, A. A., Kalkofen, R. L., Carpenter, L., Biden, T. J., & Reyland, M. E. (2003). Inhibition of pkcalpha induces a pkcdelta-dependent apoptotic program in salivary epithelial cells. *Cell Death and Differentiation, 10*, 269–277.

Mayne, G. C. & Murray, A. W. (1998). Evidence that protein kinase cepsilon mediates phorbol ester inhibition of calphostin c- and tumor necrosis factor-α-induced apoptosis in u937 histiocytic lymphoma cells. *The Journal of Biological Chemistry, 273*, 24115–24121.

Mazzoni, E., Adam, A., Bal de Kier Joffe, E., & Aguirre-Ghiso, J. A. (2003). Immortalized mammary epithelial cells overexpressing protein kinase C gamma acquire a malignant phenotype and become tumorigenic in vivo. *Molecular Cancer Research, 1*, 776–787.

McJilton M. A., Wescott, G.G., Wu, D., Foreman, T. L., Gregory, C. W., Weidner, D. A., et al. (2003). Protein kinase C epsilon interacts with bax and promotes survival of human prostate cancer cells. *Oncogene, 22*, 7958–7968.

McKiernan, E., O'Brien, K., Grebenchtchikov, N., Geurts-Moespot, A., Sieuwerts, A., Martens, J., et al. (2008). Protein kinase C expression in breast cancer as measured by real-time PCR, western blotting and ELISA. *British Journal of Cancer, 99*, 1644–1650.

Mecklenbrauker, I., Kalled, S. L., Leitges, M., Mackay, F., & Tarakhovsky, A. (2004). Regulation of B-cell survival by baff-dependent PKC delta-mediated nuclear signaling. *Nature, 431*, 456–461.

Metzstein, M., Stanfield, G., & Horvitz, H. (1998). Genetics of programmed cell death in c. Elegans: Past, present and future. *Trends in Genetics, 14*, 410–416.

Mingo-Sion, A. M., Ferguson, H. A., Koller, E., Reyland, M. E., & Van Den Berg, C. L. (2005). PKC delta and mTOR interact to regulate stress and IGF-1 induced IRS-1 ser312 phosphorylation in breast cancer cells. *Breast Cancer Research and Treatment, 91*, 259–269.

Mischak, H., Goodnight, J. A., Kolch, W., Martiny-Baron, G., Schaechtle, C., Kazanietz, M. G., et al. (1993). Overexpression of protein kinase C-delta and -epsilon in NIH 3T3 cells induces opposite effects on growth, morphology, anchorage dependence, and tumorigenicity. *The Journal of Biological Chemistry, 268*, 6090–6096.

Miyamoto, A., Nakayama, K., Imaki, H., Hirose, S., Jiang, Y., Abe, M., et al. (2002). Increased proliferation of B cells and auto-immunity in mice lacking protein kinase C delta. *Nature, 416*, 865–869.

Morse-Gaudio, M., Connolly, J. M., & Rose, D. P. (1998). Protein kinase C and its isoforms in human breast cancer cells: Relationship to the invasive phenotype. *International Journal of Oncology, 12*, 1349–1354.

Moscat, J., & Diaz-Meco, M. T. (2000). The atypical protein kinase C's. Functional specificity mediated by specific protein adapters. *EMBO Reports, 1*, 399–403.

Murray, N. R., Davidson, L. A., Chapkin, R. S., Clay Gustafson, W., Schattenberg, D. G., & Fields, A. P. (1999). Overexpression of protein kinase C betaII induces colonic hyperproliferation and increased sensitivity to colon carcinogenesis. *The Journal of Cell Biology, 145*, 699–711.

Murray, N. R., & Fields, A. P. (1997). Atypical protein kinase C iota protects human leukemia cells against drug-induced apoptosis. *The Journal of Biological Chemistry, 272*, 27521–27524.

Murriel, C. L., Churchill, E., Inagaki, K., Szweda, L. I., & Mochly-Rosen, D. (2004). Protein kinase C delta activation induces apoptosis in response to cardiac ischemia and reperfusion damage: A mechanism involving bad and the mitochondria. *The Journal of Biological Chemistry, 279*, 47985–47991.

Mustafi, R., Cerda, S., Chumsangsri, A., Fichera, A., & Bissonnette, M. (2006). Protein kinase-zeta inhibits collagen i-dependent and anchorage-independent growth and enhances apoptosis of human Caco-2 cells. *Molecular Cancer Research, 4*, 683–694.

Nagata, S. (2006). Apoptosis and autoimmune diseases. *IUBMB Life, 58*, 358–362.

Neill, G. W., Ghali, L. R., Green, J. L., Ikram, M. S., Philpott, M. P., & Quinn, A. G. (2003). Loss of protein kinase C alpha expression may enhance the tumorigenic potential of gli1 in basal cell carcinoma. *Cancer Research, 63*, 4692–4697.

Ng, T., Shima, D., Squire, A., Bastiaens, P. I., Gschmeissner, S., Humphries, M. J., et al. (1999). Pkcalpha regulates beta1 integrin-dependent cell motility through association and control of integrin traffic. *EMBO Journal, 18*, 3909–3923.

Nowak, G. (2002). Protein kinase c-alpha and erk1/2 mediate mitochondrial dysfunction, decreases in active na+ transport, and cisplatin-induced apoptosis in renal cells. *The Journal of Biological Chemistry, 277*, 43377–43388.

Okhrimenko, H., Lu, W., Xiang, C., Hamburger, N., Kazimirsky, G., & Brodie, C. (2005a). Protein kinase c-epsilon regulates the apoptosis and survival of glioma cells. *Cancer Research, 65*, 7301–7309.

Okhrimenko, H., Lu, W., Xiang, C., Ju, D., Blumberg, P. M., Gomel, R., et al. (2005b). Roles of tyrosine phosphorylation and cleavage of protein kinase c{delta} in its protective effect against tumor necrosis factor-related apoptosis inducing ligand-induced apoptosis. *The Journal of Biological Chemistry, 280*, 23643–23652.

Oliva, J. L., Caino, M. C., Senderowicz, A. M., & Kazanietz, M. G. (2008). S-phase-specific activation of pkcalpha induces senescence in non-small cell lung cancer cells. *The Journal of Biological Chemistry, 283*, 5466–5476.

Oster, H., & Leitges, M. (2006). Protein kinase C alpha but not PKC zeta suppresses intestinal tumor formation in ApcMin/+ mice. *Cancer Research, 66*, 6955–6963.

Parsons, M., & Adams, J. C. (2008). Rac regulates the interaction of fascin with protein kinase C in cell migration. *Journal of Cell Science, 121*, 2805–2813.

Parsons, M., Keppler, M. D., Kline, A., Messent, A., Humphries, M. J., Gilchrist, R., et al. (2002). Site-directed perturbation of protein kinase c-integrin interaction blocks carcinoma cell chemotaxis. *Molecular and Cellular Biology, 22*, 5897–5911.

Partovian, C., & Simons, M. (2004). Regulation of protein kinase B/AKT activity and ser473 phosphorylation by protein kinase C alpha in endothelial cells. *Cellular Signalling, 16*, 951–957.

Patel, R., Win, H., Desai, S., Patel, K., Matthews, J. A., & Acevedo-Duncan, M. (2008). Involvement of PKC-iota in glioma proliferation. *Cell Proliferation, 41*, 122–135.

Peluso, J. J., Pappalardo, A., & Fernandez, G. (2001). Basic fibroblast growth factor maintains calcium homeostasis and granulosa cell viability by stimulating calcium efflux via a PKCdelta-dependent pathway. *Endocrinology, 142*, 4203–4211.

Perletti, G. P., Concari, P., Brusaferri, S., Marras, E., Piccinini, F., & Tashjian, A. H. (1998). Protein kinase C epsilon is oncogenic in colon epithelial cells by interaction with the ras signal transduction pathway. *Oncogene, 16*, 3345–3348.

Pietsch, E., Sykes, S., McMahon, S., & Murphy, M. (2008). The p53 family and programmed cell death. *Oncogene, 27*, 6507–6521.

Piiper, A., Elez, R., You, S.-J., Kronenberger, B., Loitsch, S., & Roche, S., et al. (2003). Cholecystokinin stimulates extracellular signal-regulated kinase through activation of the epidermal growth factor receptor, yes, and protein kinase c. Signal amplification at the level of raf by activation of protein kinase C epsilon. *The Journal of Biological Chemistry, 278*, 7065–7072.

Plo, I., Hernandez, H., Kohlhagen, G., Lautier, D., Pommier, Y., & Laurent, G. (2002). Overexpression of the atypical protein kinase C zeta reduces topoisomerase II catalytic activity, cleavable complexes formation, and drug-induced cytotoxicity in monocytic u937 leukemia cells. *The Journal of Biological Chemistry, 277*, 31407–31415.

Podar, K., Tai, Y. T., Lin, B. K., Narsimhan, R. P., Sattler, M., Kijima, T., et al. (2002). Vascular endothelial growth factor-induced migration of multiple myeloma cells is associated with beta 1 integrin- and phosphatidylinositol 3-kinase-dependent PKC alpha activation. *The Journal of Biological Chemistry, 277*, 7875–7881.

Powell, C. T., Brittis, N. J., Stec, D., Hug, H., Heston, W. D., & Fair, W. R. (1996). Persistent membrane translocation of protein kinase C alpha during 12-O-tetradecanoylphorbol-13-acetate-induced apoptosis of lncap human prostate cancer cells. *Cell Growth and Differentiation, 7*, 419–428.

Prevostel, C., Martin, A., Alvaro, V., Jaffiol, C., & Joubert, D. (1997). Protein kinase C alpha and tumorigenesis of the endocrine gland. *Hormone Research, 47*, 140–144.

Rampino, N., Yamamoto, H., Ionov, Y., Li, Y., Sawai, H., Reed, J. C., et al. (1997). Somatic frameshift mutations in the bax gene in colon cancers of the microsatellite mutator phenotype. *Science, 275*, 967–969.

Ranger, A. M., Malynn, B. A., & Korsmeyer, S. J. (2001). Mouse models of cell death. *Nature Genetics, 28*, 113–118.

Regala, R. P., Thompson, E. A., & Fields, A. P. (2008). Atypical protein kinase C iota expression and aurothiomalate sensitivity in human lung cancer cells. *Cancer Research, 68*, 5888–5895.

Regala, R. P., Weems, C., Jamieson, L., Khoor, A., Edell, E. S., Lohse, C. M., et al. (2005). Atypical protein kinase C iota is an oncogene in human non-small cell lung cancer. *Cancer Research, 65*, 8905–8911.

Reno, E. M., Haughian, J. M., Dimitrova, I. K., Jackson, T. A., Shroyer, K. R., & Bradford, A. P. (2008). Analysis of protein kinase C delta (PKCdelta) expression in endometrial tumors. *Human Pathology, 39*, 21–29.

Reyland, M. E. (2007). In R. Srivastava (Ed.), PKC and apoptosis. *Apoptosis, cell signaling, and human disease: Molecular mechanisms.* (Vol, 2: 31–55). Totowa, NJ: Humana.

Reyland, M. E., Anderson, S. M., Matassa, A. A., Barzen, K. A., & Quissell, D. O. (1999b). Protein kinase C delta is essential for etoposide-induced apoptosis in salivary gland acinar cells. *The Journal of Biological Chemistry, 274*, 19115–19123.

Reyland, M., Barzen, K., Anderson, S., Quissell, D., & Matassa, A. (2000). Activation of pkc is sufficient to induce an apoptotic program in salivary gland acinar cells. *Cell Death and Differentiation, 7*, 1200–1209.

Riedl, S. J., & Shi, Y. (2004). Molecular mechanisms of caspase regulation during apoptosis. *Nature Reviews. Molecular Cell Biology, 9*, 897–907.

Riley, T., Sontag, E., Chen, P., & Levine, A. (2008). Transcriptional control of human p53-regulated genes. *Nature Reviews. Molecular Cell Biology, 9*, 402–412.

Riobo, N. A., Haines, G. M., & Emerson, C. P., Jr. (2006). Protein kinase C-{delta} and mitogen-activated protein/extracellular signal-regulated kinase-1 control gli activation in hedgehog signaling. *Cancer Research, 66*, 839–845.

Rogers, C., Loveday, R., Drew, P., & Greenman, J. (2002). Molecular prognostic indicators in breast cancer. *European Journal of Surgical Oncology, 28*, 467–478.

Ruvolo, P. P., Deng, X., Carr, B. K., & May, W. S. (1998). A functional role for mitochondrial protein kinase calpha in bcl2 phosphorylation and suppression of apoptosis. *The Journal of Biological Chemistry*, *273*, 25436–25442.

Ryer, E. J., Sakakibara, K., Wang, C., Sarkar, D., Fisher, P. B., Faries, P. L., et al. (2005). Protein kinase C delta induces apoptosis of vascular smooth muscle cells through induction of the tumor suppressor p53 by both p38-dependent and p38-independent mechanisms. *The Journal of Biological Chemistry*, *280*, 35310–35317.

Santiago-Walker, A. E., Fikaris, A. J., Kao, G. D., Brown, E. J., Kazanietz, M. G. & Meinkoth, J. L. (2005). Protein kinase C delta stimulates apoptosis by initiating G1 phase cell cycle progression and S phase arrest. *The Journal of Biological Chemistry*, *280*, 32107–32114.

Sarker, M., Ruiz-Ruiz, C., & López-Rivas, A. (2001). Activation of protein kinase C inhibits trail-induced caspases activation, mitochondrial events and apoptosis in a human leukemic t cell line. *Cell Death and Differentiation, 2*, 172–181.

Sauma, S., Yan, Z., Ohno, S., & Friedman, E. (1996). Protein kinase C beta 1 and protein kinase C beta 2 activate p57 mitogen-activated protein kinase and block differentiation in colon carcinoma cells. *Cell Growth and Differentiation*, *7*, 587–594.

Schwartz, G. K., Jiang, J., Kelsen, D., & Albino, A. P. (1993). Protein kinase C: A novel target for inhibiting gastric cancer cell invasion. *Journal of the National Cancer Institute*, *85*, 402–407.

Shankar, E., Sivaprasad, U., & Basu, A. (2008). Protein kinase C epsilon confers resistance of mcf-7 cells to trail by akt-dependent activation of hdm2 and downregulation of p53. *Oncogene, 27*, 3957–3966.

Shanmugam, M., Krett, N. L., Maizels, E. T., Murad, F. M., Rosen, S. T., & Hunzicker-Dunn, M. (2001). A role for protein kinase C delta in the differential sensitivity of MCF-7 and MDA-MB 231 human breast cancer cells to phorbol ester-induced growth arrest and p21waf1/cip1 induction. *Cancer Letters, 172*, 43–53.

Sharif, T., & Sharif, M. (1999). Overexpression of protein kinase C epsilon in astroglial brain tumor derived cell lines and primary tumor samples. *International Journal of Oncology, 15*, 237–243.

Sheridan, C., & Martin, S. (2008). Commitment in apoptosis: Slightly dead but mostly alive. *Trends in Cell Biology, 18*, 353–357.

Shibata, M.-A., Liu, M.-L., Knudson, M. C., Shibata, E., Yoshidome, K., Bandey, T., et al. (1999). Haploid loss of bax leads to accelerated mammary tumor development in Cc3(1)/SV40-TAg transgenic mice: Reduction in protective apoptotic response at the preneoplastic stage. *EMBO Journal, 18*, 2692–2701.

Sitailo, L. A., Tibudan, S. S., & Denning, M. F. (2004). Bax activation and induction of apoptosis in human keratinocytes by the protein kinase C delta catalytic domain. *The Journal of Investigative Dermatology, 123*, 434–443.

Sitailo, L. A., Tibudan, S. S., & Denning, M. F. (2006). The protein kinase C delta catalytic fragment targets mcl-1 for degradation to trigger apoptosis. *The Journal of Biological Chemistry, 281*, 29703–29710.

Sivaprasad, U., Shankar, E., & Basu, A. (2006). Downregulation of Bid is associated with PKC epsilon-mediated trail resistance. *Cell Death and Differentiation, 14*, 851–860.

Silva, D., English, D., Lyons, D., & Lloyd, F.P. Jr. (2002). Protein kinase C induces motility of breast cancers by upregulating secretion of urokinase-type plasminogen activator through activation of AP-1 and NF-kappaB. *Biochemical and Biophysical Research Communications, 11*, 552–557.
Slosberg, E. D., Klein, M. G., Yao, Y., Han, E. K., Schieren, I., & Weinstein, I. B. (1999). The alpha isoform of protein kinase c mediates phorbol ester-induced growth inhibition and p21cip1 induction in HC11 mammary epithelial cells. *Oncogene 18*, 6658–6666.
Smith, L., Chen, L., Reyland, M. E., DeVries, T. A., Talanian, R. V., Omura, S., et al. (2000). Activation of atypical protein kinase C zeta by caspase processing and degradation by the ubiquitin-proteasome system. *The Journal of Biological Chemistry, 275*, 40620–40627.
Smith, L., Wang, Z., & Smith, J. B. (2003). Caspase processing activates atypical protein kinase C zeta by relieving autoinhibition and destabilizes the protein. Biochemical Journal, *375*, 663–671.
Soh, J. W., & Weinstein, I. B. (2003). Roles of specific isoforms of protein kinase C in the transcriptional control of cyclin d1 and related genes. *The Journal of Biological Chemistry, 278*, 34709–34716.
Steinberg, R., Harari, O. A., Lidington, E. A., Boyle, J. J., Nohadani, M., Samarel, A. M., et al. (2007). A protein kinase C epsilon-anti-apoptotic kinase signaling complex protects human vascular endothelial cells against apoptosis through induction of bcl-2. *The Journal of Biological Chemistry, 282*, 32288–32297.
Stephanou, A., Scarabelli, T. M., Townsend, P. A., Bell, R., Yellon, D., Knight, R. A., et al. (2002). The carboxy-terminal activation domain of the stat-1 transcription enhances ischemia/reperfusion-induced apoptosis in cardiac myocytes. *FASEB Journal, 16*, 1841–1843.
Su, T. T., Guo, B., Kawakami, Y., Sommer, K., Chae, K., Humphries, L. A., et al. (2002). PKC-beta controls I kappa B kinase lipid raft recruitment and activation in response to bcr signaling. *Nature Immunology, 3*, 780–786.
Sulston, J. E., & Horvitz, H. R. (1977). Post-embryonic cell lineages of the nematode, caenorhabditis elegans. *Developmental Biology, 56*, 110–156.
Sun, R., Gao, P., Chen, L., Ma, D., Wang, J., Oppenheim, J. J., et al. (2005). Protein kinase C zeta is required for epidermal growth factor-induced chemotaxis of human breast cancer cells. *Cancer Research, 65*, 1433–1441.
Sun, X., Wu, F., Datta, R., Kharbanda, S., & Kufe, D. (2000). Interaction between protein kinase C delta and the c-abl tyrosine kinase in the cellular response to oxidative stress. *The Journal of Biological Chemistry, 275*, 7470–7473.
Svensson, K., Zeidman, R., Troller, U., Schultz, A., & Larsson, C. (2000). Protein kinase C beta1 is implicated in the regulation of neuroblastoma cell growth and proliferation. *Cell Growth and Differentiation, 11*, 641–648.
Tachado, S., Mayhew, M., Wescott, G., Foreman, T., Goodwin, C., McJilton, M., et al. (2002). Regulation of tumor invasion and metastasis in protein kinase C epsilon-transformed NIH3T3 fibroblasts. *Journal of Cellular Biochemistry, 85*, 785–797.
Tan, M., Li, P., Sun, M., Yin, G., & Yu, D. (2006). Upregulation and activation of pkc by erbb2 through src promotes breast cancer cell invasion that can be blocked by combined treatment with pkc and src inhibitors. *Oncogene, 25*, 3286–3295.
Teicher, B. A. (2006). Protein kinase C as a therapeutic target. *Clinical Cancer Research, 12*, 5336–5345.
Teicher, B. A., Menon, K., Alvarez, E., Galbreath, E., Shih, C., & Faul, M. (2001). Antiangiogenic and antitumor effects of a protein kinase C beta inhibitor in human T98G glioblastoma multiforme xenografts. *Clin Cancer Res, 7*: 634–640.
Teicher, B. A., Menon, K., Alvarez, E., Galbreath, E., Shih, C., Faul, M. M. (2001). Antiangiogenic and antitumor effects of a protein kinase C beta inhibitor in murine lewis lung carcinoma and human calu-6 non-small-cell lung carcinoma xenografts. *Cancer Chemotherapy and Pharmacology, 48*, 473–480.
Teicher, B. A., Menon, K., Alvarez, E., Shih, C., & Faul, M. M. (2002). Antiangiogenic and antitumor effects of a protein kinase C beta inhibitor in human breast cancer and ovarian cancer xenografts. *Investigational New Drugs, 20*, 241–251.

Thorburn, A. (2004) Death receptor-induced cell killing. *Cellular Signalling, 16*, 139–144.

Tonetti, D. A., Chisamore, M. J., Grdina, W., Schurz, H., & Jordan, V. C. (2000). Stable transfection of protein kinase C alpha cDNA in hormone-dependent breast cancer cell lines. *British Journal of Cancer, 83*, 782–791.

Toyoda, M., Gotoh, N., Handa, H., & Shibuya, M. (1998). Involvement of map kinase-independent protein kinase C signaling pathway in the EGF-induced p21(waf1/cip1) expression and growth inhibition of A431 cells. *Biochemical and Biophysical Research Communications, 250*, 430–435.

Tsai, J. H., Hsieh, Y. S., Kuo, S. J., Chen, S. T., Yu, S. Y., Huang, C. Y., et al. (2000). Alteration in the expression of protein kinase C isoforms in human hepatocellular carcinoma. *Cancer Letters, 161*, 171–175.

Tsujimoto, Y., Cossman, J., Jaffe, E., & Croce, C. M. (1985). Involvement of the bcl-2 gene in human follicular lymphoma. *Science, 228*, 1440–1443.

Ueda, Y., Hirai, S.-I., Osada, S.-I., Suzuki, A., Mizuno, K., & Ohno, S. (1996). Protein kinase C delta activates the MEK-ERK pathway in a manner independent of ras and dependent on raf. *The Journal of Biological Chemistry, 271*, 23512–23519.

Utz, P. J., & Anderson, P. (2000). Life and death decisions: Regulation of apoptosis by proteolysis of signaling molecules. *Cell Death and Differentiation, 7*, 589–602.

Varfolomeev, E. E., Schuchmann, M., Luria, V., Chiannikulchai, N., Beckmann, J. S., Mett, I. L., et al. (1998). Targeted disruption of the mouse caspase 8 gene ablates cell death induction by the tnf receptors, fas/apo1, and dr3 and is lethal prenatally. *Immunity, 9*, 267–276.

Varga, A., Czifra, G., Tallai, B., (2004). *European Urology, 46*, 462–465.

Verma, A. K., Wheeler, D. L., Aziz, M. H., & Manoharan, H. (2006). Protein kinase C epsilon and development of squamous cell carcinoma, the nonmelanoma human skin cancer. *Molecular Carcinogenesis, 45*, 381–388.

Voss, O. H., Kim, S., Wewers, M. D., & Doseff, A. I. (2005). Regulation of monocyte apoptosis by the protein kinase C delta-dependent phosphorylation of caspase-3. *The Journal of Biological Chemistry, 280*, 17371–17379.

Wakino, S., Kintscher, U., Liu, Z., Kim, S., Yin, F., Ohba, M., et al. (2001). Peroxisome proliferator-activated receptor gamma ligands inhibit mitogenic induction of p21cip1 by modulating the protein kinase C delta pathway in vascular smooth muscle cells. *The Journal of Biological Chemistry, 276*, 47650–47657.

Wang, S., & Baker, J. (2007) The role of apoptosis in thyroid autoimmunity. *Thyroid, 17*, 975–979.

Wang, S., & El-Deiry, W. (2003). Trail and apoptosis induction by tnf-family death receptors. *Oncogene. 22*, 8628–8633.

Wang, G., Silva, J., Krishnamurthy, K., Tran, E., Condie, B. G., & Bieberich, E. (2005). Direct binding to ceramide activates protein kinase C zeta before the formation of a pro-apoptotic complex with par-4 in differentiating stem cells. *The Journal of Biological Chemistry, 280*, 26415–26424.

Watanabe, T., Ono, Y., Taniyama, Y., Hazama, K., Igarashi, K., Ogita, K., et al. (1992). Cell division arrest induced by phorbol ester in cho cells overexpressing protein kinase C-delta subspecies. *Proceedings of the National Academy of Sciences of the United States of America, 89*, 10159–10163.

Ways, D. K., Kukoly, C. A., deVente, J., Hooker, J. L., Bryant, W. O., Posekany, K. J., et al. (1995). MCF-7 breast cancer cells transfected with protein kinase C-alpha exhibit altered expression of other protein kinase C isoforms and display a more aggressive neoplastic phenotype. *The Journal of Clinical Investigation, 95*, 1906–1915.

Weichert, W., Gekeler, V., Denkert, C., Dietel, M., & Hauptmann, S. (2003). Protein kinase C isoform expression in ovarian carcinoma correlates with indicators of poor prognosis. *International Journal of Oncology, 23*, 633–639.

Wert, M., & Palfrey, C. (2000). Divergence in the anti-apoptotic signalling pathways used by nerve growth factor and basic fibroblast growth factor (bFGF) in PC12 cells: Rescue by bFGF involves protein kinase C delta. *Biochemical Journal, 352*, 175–182.

Wetsel, W. C., Khan, W. A., Merchenthaler, I., Rivera, H., Halpern, A. E., Phung, H. M., et al. (1992). Tissue and cellular distribution of the extended family of protein kinase C isoenzymes. *The Journal of Cell Biology, 117*, 121–133.

Wheeler, D. L., Martin, K. E., Ness, K. J., Li, Y., Dreckschmidt, N. E., Wartman, M., et al. (2004). Protein kinase C epsilon is an endogenous photosensitizer that enhances ultraviolet radiation-induced cutaneous damage and development of squamous cell carcinomas. *Cancer Research, 64*, 7756–7765.

Wheeler, D. L., Ness, K. J., Oberley, T. D., & Verma, A. K. (2003). Protein kinase C epsilon is linked to 12-O-tetradecanoylphorbol-13-acetate-induced tumor necrosis factor-alpha ectodomain shedding and the development of metastatic squamous cell carcinoma in protein kinase Cepsilon transgenic mice. *Cancer Research, 63*, 6547–6555.

Whelan, R. D., & Parker, P. J. (1998). Loss of protein kinase C function induces an apoptotic response. *Oncogene, 16*, 1939–1944.

Whitman, S. P., Civoli, F., & Daniel, L. W. (1997). Protein kinase C beta II activation by 1-beta-d-arabinofuranosylcytosine is antagonistic to stimulation of apoptosis and bcl-2alpha down-regulation. *The Journal of Biological Chemistry, 272*, 23481–23484.

Win, H. Y., & Acevedo-Duncan, M. (2008). Atypical protein kinase C phosphorylates IKKalphabeta in transformed non-malignant and malignant prostate cell survival. *Cancer Letters, 270*(2), 302–311.

Wolf, B., & Green, D. (1999). Suicidal tendencies: Apoptotic cell death by caspase family proteinases. *The Journal of Biological Chemistry, 274*, 20049–20052.

Wolff, S., Erster, S., Palacios, G., & Moll, U. M. (2008). P53's mitochondrial translocation and MOMP action is independent of puma and bax and severely disrupts mitochondrial membrane integrity. *Cell Research, 18*, 733–744.

Wu, D., Foreman, T. L., Gregory, C. W., McJilton, M. A., Wescott, G. G., Ford, O. H., et al. (2002). Protein kinase C epsilon has the potential to advance the recurrence of human prostate cancer. *Cancer Research, 62*, 2423–2429.

Wu, T., Hsieh, Y.-H., Hsieh, Y.-S., & Liu, J. (2008). Reduction of PKC alpha decreases cell proliferation, migration, and invasion of human malignant hepatocellular carcinoma. *Journal of Cellular Biochemistry, 103*, 9–20.

Wyllie, A. H., Kerr, J. F., & Currie, A. R. (1972). Cellular events in the adrenal cortex following ACTH deprivation. *Journal of Pathology, 106*, Pix.

Wyllie, A. H., Kerr, J. F., & Currie, A. R. (1973). Cell death in the normal neonatal rat adrenal cortex. *Journal of Pathology, 111*, 255–261.

Xia, P., Aiello, L. P., Ishii, H., Jiang, Z. Y., Park, D. J., Robinson, G. S., et al. (1996). Characterization of vascular endothelial growth factor's effect on the activation of protein kinase C, its isoforms, and endothelial cell growth. *The Journal of Clinical Investigation, 98*, 2018–2026.

Xiao, L., Caino, M. C., von Burstin, V. A., Oliva, J. L., & Kazanietz, M. G. (2008). Phorbol ester-induced apoptosis and senescence in cancer cell models. *Methods in Enzymology, 446*, 123–139.

Xin, M., Gao, F., May, W. S., Flagg, T., & Deng, X. (2007). Protein kinase czeta abrogates the proapoptotic function of bax through phosphorylation. *The Journal of Biological Chemistry, 282*, 21268–21277.

Yamaguchi, T., Miki, Y., & Yoshida, K. (2007). Protein kinase C delta activates IkappaB-kinase alpha to induce the p53 tumor suppressor in response to oxidative stress. *Cellular Signalling, 19*, 2088–2097.

Yin, L., Bennani-Baiti, N., & Powell, C. T. (2005). Phorbol ester-induced apoptosis of C4–2 cells requires both a unique and a redundant protein kinase C signaling pathway. *Molecular and Cellular Biology, 280*, 5533–5541.

Yin, J. Q., Gao, J., Shao, R., Tian, W. N., Wang, J., & Wan, Y. (2003). SiRNA agents inhibit oncogene expression and attenuate human tumor cell growth. *Journal of Experimental Therapeutics & Oncology, 3*, 194–204.

Yip, K., & Reed, J. (2008). Bcl-2 family proteins and cancer. *Oncogene, 27*, 6398–6406.

Yonekawa, H., & Akita, Y. (2008). Protein kinase C epsilon: The mitochondria-mediated signaling pathway. *FEBS Journal, 275*, 4005–4013.
Yoshida, K. (2007). PKC delta signaling: Mechanisms of DNA damage response and apoptosis. *Cellular Signalling, 19*, 892–901.
Yoshida, K., Liu, H., & Miki, Y. (2006). Protein kinase C delta regulates ser46 phosphorylation of p53 tumor suppressor in the apoptotic response to DNA damage. *The Journal of Biological Chemistry, 281*, 5734–5740.
Yoshida, K., Miki, Y., & Kufe, D. (2002). Activation of SAPK/JNK signaling by protein kinase C delta in response to DNA damage. *The Journal of Biological Chemistry, 277*, 48372–48378.
Yoshida, K., Wang, H., Miki, Y. & Kufe, D. (2003). Protein kinase C delta is responsible for constitutive and DNA damage induced phosphorylation of rad9. *EMBO Journal, 22*, 1431–1441.
Yoshiji, H., Kuriyama, S., Ways, D. K., Yoshii, J., Miyamoto, Y., Kawata, M., et al. (1999). Protein kinase C lies on the signaling pathway for vascular endothelial growth factor-mediated tumor development and angiogenesis. *Cancer Research, 59*, 4413–4418.
Youle, R. J., & Strasser, A. (2008). The bcl-2 protein family: Opposing activities that mediate cell death. *Nature Review. Molecular Cell Biology, 9*, 47–59.
Yuan, Z.-M., Utsugisawa, T., Ishiko, T., Nakada, S., Huang, Y., Kharbanda, S., et al. (1998). Activation of protein kinase C delta by the c-abl tyrosine kinase in response to ionizing radiation. *Oncogene, 16*, 1643–1648.
Zang, Q., Lu, Z., Curto, M., Barile, N., Shalloway, D., & Foster, D. A. (1997). Association between v-src and protein kinase C delta in v-src-transformed fibroblasts. *The Journal of Biological Chemistry, 272*, 13275–13280.
Zhang, J., Anastasiadis, P. Z., Liu, Y., Thompson, E. A., & Fields, A. P. (2004). Protein kinase C (PKC) betaII induces cell invasion through a ras/MEK-, PKC iota/rac 1-dependent signaling pathway. *The Journal of Biological Chemistry, 279*, 22118–22123.
Zheng, T. S., & Flavell, R. A. (2000). Divinations and suprises: Genetic analysis of caspase function in mice. *Experimental Cell Research, 256*, 67–73.
Zhu, Y., Dong, Q., Tan, B. J., Lim, W. G., Zhou, S., & Duan, W. (2005). The PKC alpha-D294G mutant found in pituitary and thyroid tumors fails to transduce extracellular signals. *Cancer Research, 65*, 4520–4524.
Zhuang S., Demirs, J. T., & Kochevar, I. E. (2001). Protein kinase C inhibits singlet oxygen-induced apoptosis by decreasing caspase-8 activation. *Oncogene, 20*, 6764–6776.

Chapter 10
Atypical PKCs, NF-κB, and Inflammation

Maria T. Diaz-Meco and Jorge Moscat

Abstract Complex biological processes, such as inflammation and cancer, rely on the crosstalk and activation of distinct cellular networks that control gene expression. Central to this process is the transcription factor NF-κB. In this chapter, we focus on the role of the atypical protein kinase C (PKC) isoforms (aPKC) in inflammation and cancer through the activation of this transcription factor in vivo. The aPKCs are key members of a network of kinases, adapters and regulators, such as p62 and Par-4, which confer functional plasticity and specificity to this critical pathway. Here, we summarize the molecular mechanisms that govern that network, learnt from the knock-out mouse models, to unravel its role and function in vivo.

Keywords aPKC • PKCζ • PKCι • p62 • Sequestosome • Par-4 • NF-κB • Inflammation • Cancer

10.1 Introduction

The atypical protein kinase C (aPKC) subfamily of kinases is composed of two members, PKCζ and PKCλ/ι. PKCλ is the mouse homolog of the human PKCι. The two aPKC isoforms are highly related, sharing an overall amino acid identity of 72% (Nishizuka 1995). The conservation in their sequences is most striking in the catalytic domain, which is also conserved among other PKC isotypes that belong to the classical and novel subfamilies. In contrast, the regulatory domain of the aPKC subfamily diverts from other members of the PKC family; it has only one zinc finger, whereas the other PKCs have two (Nishizuka 1995). Through the zinc-finger domain, the aPKCs bind Par-4, a negative regulator of their enzymatic activity.

M.T. Diaz-Meco (✉) and J. Moscat
Department of Cancer and Cell Biology, University of Cincinnati College of Medicine, 3125 Eden Ave, Cincinnati, OH 45267, USA
e-mail: maria.diazmeco@uc.edu

M.G. Kazanietz (ed.), *Protein Kinase C in Cancer Signaling and Therapy*, Current Cancer Research, DOI 10.1007/978-1-60761-543-9_10,

Like the novel PKCs, the aPKCs lack the characteristic C2 domain that is present in the classical isoforms. These important structural differences may explain why the aPKCs are insensitive to Ca^{2+}, diacylglycerol, and phorbol esters, which are potent activators of the other isoforms (Nishizuka 1995).

The recent identification of the protein interaction domain PB1, present at the N-terminus of the aPKCs, has opened new avenues to explore the roles of these kinase isoforms by looking for adapters and regulators that could shed light on their functions. It is well known that the PKCs are kinases that display little selectivity in vitro and in vivo. This invokes the need for cellular mechanisms to confer functional specificity while preserving the capacity for crosstalk, which is necessary for the regulation of complex biological processes. Because the aPKCs have been implicated in diverse cellular functions, the existence of different adapter proteins that serve to provide the required selectivity has been hypothesized (Moscat and Diaz-Meco 2000). In this regard, the PB1s are dimerization/oligomerization domains present in adapter and scaffold proteins, as well as in kinases, and serve to organize platforms that ensure specificity and fidelity during cellular signaling. The PB1 domains are named after the prototypical domains found in Phox and Bem1p, which mediate polar, heterodimeric interactions. The PB1 domains comprise about 80 amino acid residues and are grouped into three types: type I (or type A), type II (or type B), and type I/II (or type AB). The type I domain group contains a conserved acidic DX(D/E)GD segment (called the OPCA motif) that interacts with a conserved lysine residue from a type II domain. Type I includes the PB1 domains of p40phox, MEK5, and Nbr1, whereas type II occurs in p67phox, Par-6, MEKK2, and MEKK3. The type I/II PB1 domain, containing both the OPCA motif and the invariant lysine, is present in the aPKCs and in p62 (also known as sequestosome-1) (Moscat et al. 2006a; Sumimoto et al. 2007). Type I and type II PB1 domains interact with each other in a front-to-back manner resulting

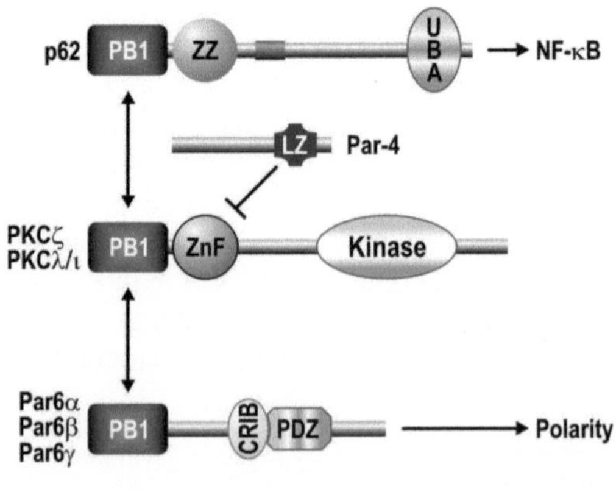

Fig. 10.1 The atypical PKC network. The atypical PKC isoforms, PKCζ and PKC λ/ι, establish a network of protein interactions with adapters proteins (such as p62 and Par-6) binding through the PB1 domain and with regulators (such as Par-4) through their zinc-finger domain. PB1–PB1 interactions confer specificity to the actions of the aPKCs. The interaction with p62 allocates the aPKCs in the NF-κB pathway, whereas through Par-6 the aPKCs regulate cell polarity

in heterodimers in which acidic residues on the OPCA motif form salt bridges with basic residues of the type II PB1 domain. Of note, two-hybrid screenings in yeast identified p62 and Par-6 as selective adapters for the aPKCs (Macara 2004; Moscat and Diaz-Meco 2000; Ohno 2001; Puls et al. 1997; Sanchez et al. 1998). Par-6 has been shown to be central to the control of cell polarity and, through its PB1 domain, allocates the aPKCs specifically in polarity-related functions. On the other hand, the p62/aPKC signaling platform plays a critical role in NF-κB activation. p62 interacts with PKCζ and PKCλ/ι, but not with any of the other closely related PKC family members. It is not a substrate and does not seem to significantly affect the intrinsic kinase activity of PKCζ or of PKCλ/ι. Moreover, it harbors a number of domains that support its role as a scaffold in aPKC signaling. Thus, the formation of aPKC complexes with different adapters, scaffold proteins, and regulators, such as Par-6, p62, and Par-4, serves to confer specificity and plasticity to the actions of these kinases and to establish a signaling network. However, the factors that determine which complex is formed at a given time remain to be identified (Fig. 10.1).

10.2 NF-κB Activation: A Key Event in Inflammation

Complex biological processes, such as inflammation, rely on crosstalk between apparently disparate signaling pathways. In the case of inflammation, the pathological condition is brought about when distinct cellular networks controlled by molecular interactions are set in motion, ultimately resulting in the activation of gene expression programs for cytokines and chemokines. The transcription factor NF-κB (nuclear factor κB) is central to this process. NF-κB consists of protein transcriptional complexes that are critically involved in the control of inflammation-related functions associated with the innate and adaptive responses of the immune system. It also plays a role in repression of apoptosis in cancer and other physiological conditions (Karin 1998; Karin et al. 2002; Li and Verma 2002). NF-κB complexes are dimers of various combinations of Rel proteins, which include RelA (p65), c-Rel, RelB, NF-κB1 (p105), and NF-κB2 (p100) (Ghosh and Karin 2002), which add an important layer of selectivity depending on the different complexes that are present in a given tissue or cell type. The proteolytic processing of NF-κB1 and NF-κB2 generates p50 and p52, respectively. The genetic inactivation of each of these proteins in mice has revealed the existence of specificities and redundancies in different cells and functions of the immune system (Li and Verma 2002; Silverman and Maniatis 2001).

In the so-called canonical pathway, NF-κB is retained in the cytosol of unstimulated cells by the IκB inhibitor proteins, which are degraded upon cell activation by a number of stimuli. These include TNFα, IL-1, and bacterial lipopolysaccharide (LPS) in fibroblasts and macrophages, and activation of the T-cell receptor (TCR) and B-cell receptor (BCR) in lymphocytes (Chen and Greene 2004; Li and Verma 2002). This leads to the release and subsequent nuclear translocation of NF-κB. The degradation of IκB takes place after its ubiquitination, and is carried out by the proteasome system (Ghosh and Karin 2002). The triggering event in this pathway is the phosphorylation of IκB by the IKK complex, which is composed of two catalytic subunits (IKKα and

IKKβ) and a scaffold protein named NEMO, IKKγ, or IKKAP. Genetic evidence demonstrates that IKKβ and IKKγ are ubiquitously required for IκB phosphorylation (Ghosh and Karin 2002), whereas IKKα seems to be necessary only in mammary gland epithelial cells (Cao et al. 2001). However, IKKα plays other roles in the NF-κB pathways related to its ability to control histone H3 phosphorylation (Anest et al. 2003; Yamamoto et al. 2003), and through the activation of an alternative noncanonical NF-κB cascade. The activation of this alternative cascade is initiated by the processing of NF-κB2/p100, which is critical for the BAFF and lymphotoxin-β receptor signaling pathways that control B cell maturation and the development of secondary lymphoid organs (Claudio et al. 2002; Dejardin et al. 2002; Kayagaki et al. 2002; Senftleben et al. 2001; Xiao et al. 2001). In addition, a role for IKKα has been proposed as a negative regulator of NF-κB through its ability to phosphorylate the Ser536 in RelA, which apparently controls the stability of this NF-κB subunit in macrophages, accelerating their removal from inflammatory gene promoters (Lawrence et al. 2005). In this regard, it is important to note that the nuclear translocation of NF-κB is not sufficient to drive transcription. It is also necessary that the RelA subunit be phosphorylated on at least two residues: Ser276 (Zhong et al. 2002; 1997; 1998) and Ser311 (Duran et al. 2003). These phosphorylations recruit the transcriptional coactivators CBP and p300 to regulate full κB-dependent gene expression.

10.2.1 The aPKCs in NF-κB Activation

The similarity between the aPKC isoforms, PKCλ/ι and PKCζ, and the lack of specific genetic and biochemical tools have, until recently, hampered the effort to assign unique functions to the individual isoforms. Many studies have used antibodies that do not discern between the two kinases, and the function of the two aPKCs is abrogated by the same pseudosubstrate inhibitor (Dominguez et al. 1992; 1993). Moreover, the overexpression of dominant-negative and active forms of the two proteins does not necessarily discriminate specific functions for each isoform, as these manipulations may impinge on pathways other than those with physiological relevance to each aPKC isotype. However, the genetic inactivation of these isoforms is starting to shed light on their specific roles. The fact that PKCλ/ι knockout (KO) mice are embryonic lethal at early stages, probably due to defects in cell polarity (Soloff et al. 2004), whereas the PKCζ KO are born in Mendelian ratios (Leitges et al. 2001) was a first indication of the different and specific functions that each of these kinases might play in vivo.

The aPKCs have been implicated as important mediators in the control of cell survival through the activation of NF-κB (Diaz-Meco et al. 1993; Moscat and Diaz-Meco 2000; Moscat et al. 2003). Indeed, the genetic inactivation of PKCζ in mice supports a key role of this isoform in the activation of NF-κB. PKCζ deficiency impairs NF-κB at two levels (Leitges et al. 2001). In lung, where PKCζ is especially abundant, this kinase is required for the activation of IKK in vivo. While, in other systems, such as embryo fibroblasts, endothelial cells, and B cells (Anrather et al. 1999; Martin et al. 2002), PKCζ controls the phosphorylation of the RelA subunit of the NF-κB complex, enabling its interaction with the transcriptional coactivator CBP and subsequent gene expression (Duran et al. 2003). Therefore, depending on the system, PKCζ can be considered an IKK kinase or may act downstream of IKK by controlling the transcriptional activity of the NF-κB complex. Evidence for this latter mechanism of action comes from ^{32}P metabolic labeling experiments in which the phosphorylation of RelA that is normally induced by TNFα and IL-1 is severely ablated in PKCζ-deficient cells (Leitges et al. 2001). Further in vitro studies demonstrated that PKCζ directly interacts with NF-κB once IκB has been degraded upon cell stimulation, and that it directly phosphorylates Ser311 in the proximity of the Rel-homology domain of RelA (Duran et al. 2003). Cell culture studies from our laboratory demonstrated that phosphorylation of Ser311 and Ser276 are required for efficient recruitment of the transcriptional coactivator CBP and of the transcriptional machinery (Duran et al. 2003). Phosphorylation of Ser276 is controlled by PKA (Zhong et al. 1997) and Misk1 (Vermeulen et al. 2003), which further illustrates the complexity of NF-κB activation, even after it has been released from the IκB molecule (Fig. 10.2).

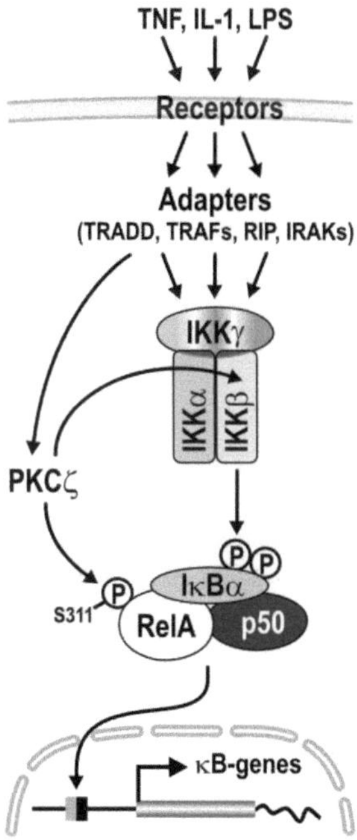

Fig. 10.2 Role of PKCζ in NF-κB activation. The binding of different ligands to their respective receptors in the plasma membrane triggers the recruitment of specific adapters for each receptor that orchestrate the formation of a signalosome complex that includes two catalytic (IKKα and IKKβ) and one regulatory subunit (IKKγ). This complex phosphorylates IκB, which is subsequently ubiquitinated and degraded through the proteasome system, releasing NF-κB (the more classical components of which are RelA-p50 heterodimers), which is free now to translocate to the nucleus and interact with elements in the promoter of inflammatory and survival genes harboring κB-elements in their promoters. RelA has to be phosphorylated to be fully functional. PKCζ phosphorylates RelA at Serine 311, an important residue for recruiting the CBP coactivator complex. Depending on tissue specificity PKCζ could also act as an IKK kinase

Compared with PKCζ, less is known about the in vivo role of PKCλ/ι because of its requirement in development. That is, PKCλ/ι-deficient mice die by embryonic day 9.5, with defects and abnormalities in development detected as early as day 6.5 (Soloff et al. 2004). This phenotype is in agreement with that found for the disrupted expression of the aPKC orthologs in *C. elegans* (Tabuse et al. 1998), *Xenopus* (Dominguez et al. 1992), and *Drosophila* (Betschinger et al. 2003; Cai et al. 2003; Cox et al. 2001; Wodarz et al. 1999). Functional KOs in these organisms result in early embryonic lethality due to defects in polarity and asymmetric cell division. An initial attempt to address the role of PKCλ/ι deficiency in cytokine-mediated cellular activation was studied in chimeras in which PKCλ/ι-deficient embryonic stem cells were combined with C57BL/6 or Rag2-deficient blastocysts. Cell lines derived from these chimeric animals showed no defects in NF-κB activation, as judged by the unimpaired degradation of IκB and induction of an NF-κB–luciferase reporter construct in response to TNFα treatment (Soloff et al. 2004). In addition, no abnormalities were found in T cell development or T cell activation (Soloff et al. 2004). However, another approach involving tissue-specific conditional PKCλ/ι deficient mice would be more helpful in elucidating the in vivo role of this atypical isoform. In fact, preliminary experiments from our laboratory, in which PKCλ/ι was specifically deleted in activated T-cells, demonstrated that this isoform plays a role in the NF-κB pathway upon T-cell activation (Moscat et al., unpublished observations).

10.2.2 The Regulation of the Inflammatory Response by the aPKCs

The canonical NF-κB pathway is essential in the control of fetal liver survival. This was demonstrated in IKKβ and RelA KO mice, which die of liver apoptosis during gestation in a TNFα-dependent manner (Karin 1998). Surprisingly, recent results using a liver-specific conditional IKKβ KO mouse demonstrated, in LPS-challenged mice, that the loss of NF-κB does not sensitize hepatocytes to apoptosis induced by circulating TNFα. However, injection of concanavalin A (ConA) produced massive hepatocyte apoptosis through a cell-bound TNFa-mediated mechanism involving both TNF receptors 1 and 2, as well as the sustained activation of JNK (Maeda et al. 2003). ConA-induced liver injury is also an excellent model of T cell-mediated hepatitis (Tiegs et al. 1992) in which the release of cytokines affects liver cells through the STAT signaling cascades. In this regard, recent evidence from mice in which IL-4 or Stat6 has been genetically inactivated demonstrates that ConA injection induces hepatitis through an IL-4/Stat6 pathway that upregulates IL-5 and eotaxin levels, which in turn triggers the recruitment of leukocytes, thus inducing hepatitis (Jaruga et al. 2003). IL-4 is a Th2 cytokine that activates the tyrosine phosphorylation of Stat6 through a Jak1/Jak3-dependent mechanism promoting its homodimerization and nuclear translocation (Ho and Glimcher 2002; O'Shea et al. 2002; Shuai and Liu 2003). Therefore, it seems that, while the IL-4/Stat6 cascade plays a pro-inflammatory role in ConA-induced hepatitis, NF-κB exerts a protective function. This is also of interest with regard to

PKCζ signaling, as T cells from mice deficient in Par-4, an inhibitor of the aPKCs, overproduce IL-4 when chronically challenged through the TCR (Lafuente et al. 2003). Therefore, in addition to regulating NF-κB, PKCζ could conceivably play an important role in the IL-4 signaling pathway (Fig. 10.2). Interestingly, the genetic inactivation of PKCζ in mice inhibits ConA-induced NF-κB activation, similar to what has been reported for the liver-specific IKKβ–/– conditional KO mice. But, surprisingly, and in contrast to the IKKβ mutant mouse, PKCζ KO mice were resistant to ConA-induced hepatitis, as determined by several parameters (Duran et al. 2004a). This could be interpreted in two ways: either NF-κB activated by PKCζ does not protect the liver from the toxic actions of T cell-mediated hepatitis, or PKCζ may be required for ConA-induced liver apoptosis through a mechanism that is independent of its role in NF-κB activation. Further analysis of this system revealed that PKCζ was, in fact, an important mediator in the activation of Jak1 and Stat6 and the consequent induction of IL-5 and eotaxin, which are critical for the recruitment of eosinophils that are central to the induction of liver damage (Duran et al. 2004a). Of potential importance is the fact that adoptive transfer experiments with liver cells demonstrated that PKCζ is necessary for NKT cells to induce liver damage in ConA-treated mice (Duran et al. 2004a). Also, in Par-4-deficient mice, in which aPKC and NF-κB activities are enhanced (Garcia-Cao et al. 2005; 2003; Lafuente et al. 2003), ConA-induced hepatitis was increased despite enhanced liver NF-κB activation. Collectively, these results suggest that, of the two main pathways activated by PKCζ (i.e., NF-κB and Stat6, Fig. 10.3), the Stat6 cascade seems to be predominant in complex inflammatory situations like T cell-mediated hepatitis.

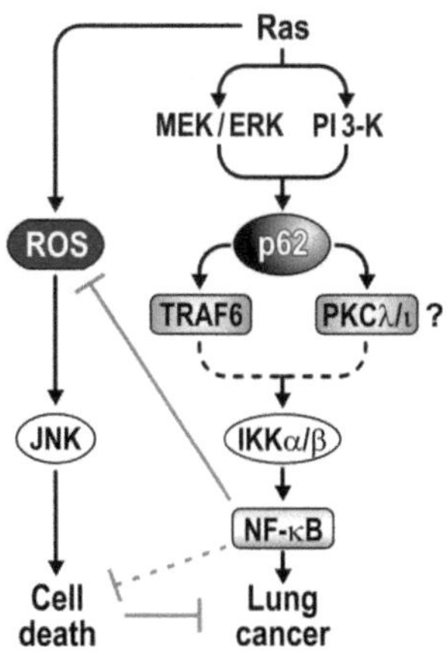

Fig. 10.3 Model of the molecular mechanism of p62 signaling in Ras-induced lung cancer. The Ras oncogene induces p62 levels through the activation of the PI-3K and MEK pathways. p62 oligomerization regulates TRAF6 ubiquitination, which leads to the activation of the IKK complex. This process could be mediated by the aPKC isoform PKCλ/ι. The activation of the IKK complex results in the activation of NF-κB, which provides a survival signal that promotes cancer. On the other hand, Ras activates also a cell death signal that is mediated by the production of ROS and JNK activation. This apoptotic cascade is counteracted by NF-κB

The fact that PKCζ is important in IL-4 signaling suggests its involvement in other pathologies, such as asthma, in which this cytokine plays a relevant role. Asthma is a chronic lung inflammatory disease with increased prevalence in developed countries. The pathology of asthma is associated with aberrant activation of CD4+ lymphocytes differentiated along the T helper (Th) 2 lineage (Luster and Tager 2004). Naïve CD4+ Th cells can differentiate in response to antigen stimulation into two distinct subsets of effector cells, Th1 and Th2, which display distinct cytokine profiles and immune regulatory functions (Mosmann and Coffman 1989). Th1 cells mainly produce interferon-γ (IFN-γ) and interleukin-2 (IL-2), and are essential for cell-mediated immune responses against intracellular pathogens. Th2 cells produce a different set of cytokines, including IL-4, IL-5, IL-10, and IL-13, and are relevant in the control of humoral immunity and allergies (Shuai and Liu 2003). IL-4 is important for the induction and maintenance of differentiated Th2 cells and for B cell immunoglobulin isotype switching to IgE (Paul and Seder 1994). Consistent with this, adult PKCζ KO mice are unable to mount an optimal immune response (Martin et al. 2002), suggesting alterations in lymphocyte function. Although the humoral reaction to a T-independent antigen was reduced in PKCζ KO mice, probably as a consequence of its role in B cell function, the major defects were found in mice challenged with a T-dependent antigen, specifically in the levels of IgG1, IgG2a, and IgG2b (Martin et al. 2002). Basal IgE levels were also dramatically reduced in PKCζ KO mice, as compared to WT controls (Martin et al. 2002). This indicates that some kind of T-cell alteration, possibly in the Th2 lineage, might be produced by the loss of PKCζ. Surprisingly, while the ability of B cells to proliferate in response to BCR challenge was reproducibly impaired in the PKCζ-deficient mice, no major alterations have been observed in the proliferation of naïve T cells (Martin et al. 2002). However, these observations could be explained by the fact that PKCζ is important in IL-4 signaling and suggest that this kinase may be critical for the regulation of Th2 function and asthma. In fact, PKCζ levels were increased during Th2, but not Th1, differentiation of CD4+ T cells. In addition, the loss of PKCζ impaired the secretion of Th2 cytokines in vitro and in vivo, as well as Jak1 activation, and the nuclear translocation and tyrosine phosphorylation of Stat6, essential downstream targets of IL-4 signaling. Moreover, PKCζ KO mice displayed dramatic inhibition of ovalbumin (OVA)-induced allergic airway disease, strongly suggesting that PKCζ might be a good candidate for a novel therapeutic target in asthma. Adoptive transfer experiments confirmed the critical role of PKCζ in the function of Th2 cells in vivo, demonstrating that the loss of PKCζ in resident lung cells does not significantly contribute to OVA-induced airway inflammation (Martin et al. 2005).

With regard to the other aPKC isoform, recent results from our laboratory demonstrated that PKCλ/ι also plays an essential role in Th2 establishment and allergic airway disease. In these studies, PKCλ/ι$^{fl/fl}$ mice (Farese et al. 2007) were crossed with CreOX40 mice in which the expression of Cre was under the control of the Tnfrsf4 locus (Zhu et al. 2004). OX40 is almost exclusively expressed in activated T cells, especially CD4+ cells, upon stimulation (Zhu et al. 2004). In this mutant, mouse line PKCλ/ι was expressed at normal levels in immature thymocytes and naïve T cells and was only deleted upon T cell activation. This strategy is advantageous in that it avoids embryonic lethality and prevents potential confounding effects resulting from the deletion of PKCλ/ι during development or in resting cells. This approach has been used previously to specifically delete the GATA3 gene in activated T cells during Th2 differentiation experiments (Zhu et al. 2004). Ex vivo experiments with PKCλ/ι-deficient T cells demonstrated that this kinase is required for the activation of transcription factors, such as NF-κB, NFATc1, and GATA-3, which are critical for adequate Th2 cell function and differentiation. We also provided the first genetic evidence, by using conditional KO cells, that PKCλ/ι is essential for T cell polarity, an event that has been suggested to be relevant to T cell function (Krummel and Macara 2006). Considered collectively, these results suggest that defects in cell polarity caused by the lack of PKCλ/ι in activated T cells, along with alterations in gene-expression programs, are responsible for the defects in Th2 cytokine production detected in the ex vivo experiments, and for the impaired lung inflammatory response observed in the PKCλ/ι mutant mice when challenged with an allergic stimulus. This is an important observation as it has been previously shown that defects in cell polarity, at least in T cells, were consistently linked to increased T cell proliferation and skewing of T cell differentiation toward the Th1 lineage (Yeh et al. 2008). However, these data are consistent with a model according to which the inactivation of different polarity proteins would have different cellular consequences, most probably owing to their association with different signaling complexes.

10.2.3 Role of the aPKC Inhibitor Par-4 in NF-κB Signaling

Par-4 was originally identified as a gene upregulated in prostate cancer cell lines undergoing apoptosis following androgen withdrawal (Sells et al. 1994). Subsequently, our laboratory identified Par-4 as a negative regulator of the aPKC isoforms (Diaz-Meco et al. 1996). This observation was of particular importance because PKCζ and PKCλ/ι are relevant pro-inflammatory molecules through their ability to regulate NF-κB (Diaz-Meco et al. 1999; Moscat and Diaz-Meco 2000; Moscat et al. 2003). In fact, multiple studies independently demonstrated that overexpression of Par-4 leads to inhibition of NF-κB, thus potentiating TNFα-induced cell death (Barradas et al. 1999; Nalca et al. 1999). The available data support a model according to which the interaction of Par-4 with the zinc-finger region of the aPKC regulatory domain leads to the inhibition of PKCζ and PKCλ/ι activity and the consequent reduction of NF-κB activity in Par-4-overexpressing cells (Moscat et al. 2006b). In this regard, the loss of Par-4 in embryo fibroblasts leads to the hyperactivation of PKCζ and of NF-κB transcriptional activity (Garcia-Cao et al. 2003). Consistent with this, the NF-κB-dependent antiapoptotic protein XIAP is expressed at significantly elevated levels in the Par-4-null cells, which correlates with reduced caspase-3 activation and apoptosis (Garcia-Cao et al. 2003).

These observations are relevant because they suggest that PKCζ is a bona fide physiologically relevant target of Par-4. In addition, Par-4 and PKCζ KO mice display opposite phenotypes in vivo in the immune system (Lafuente et al. 2003; Martin et al. 2002). Thus, whereas PKCζ–/– mice have impaired B cell proliferation and function (Martin et al. 2002), Par-4–/– mice have increased B cell and T cell proliferation (Lafuente et al. 2003). Also, Par-4–/– T cells hyperproduce the Th2 cytokine IL-4 (Duran et al. 2004a), whereas PKCζ–/– T cells show impaired Th2 polarization and IL-4 secretion in vivo (Martin et al. 2005).

10.2.4 Role of the aPKC Adapter p62 in NF-κB

p62 interacts with PKCζ and PKCλ/ι through their PB1 domains, but not with any of the other closely related PKC family members. It is not a substrate and does not seem to significantly affect the intrinsic kinase activity of PKCζ or PKCλ/ι. p62 was shown to be required for NF-κB signaling in several systems (Sanz et al. 2000; 1999; Wooten et al. 2001), including *Drosophila* in which a functionally relevant homologue termed Ref(2)P was identified (Avila et al. 2002). This protein has an overall structure very similar to that of p62, suggesting that, in partnership with PKCζ, it may be a critical mediator of NF-κB in *Drosophila* cells.

The NF-κB pathway is remarkably conserved in *Drosophila*, controlling not only development but also the innate immune response (Hoffmann 2003). Thus, the RelA homologs in *Drosophila*, dorsal-related immunity factor (Dif) and Dorsal, have been shown to be necessary for the synthesis of the antimicrobial peptide drosomycin in response to the activation of the Toll pathway by fungal pathogens. This pathway involves the adapter Tube, the kinase Pelle, and the phosphorylation, and subsequent degradation of the IκB homolog, Cactus. Parallel to this pathway, there is another one in *Drosophila* that involves the kinase dTAK1, which serves to control the degradation of Relish. Relish is the fly homolog of NF-κB1/NF-κB2 and is required for the synthesis of diptericin in response to bacterial infection. Interestingly, knocking down PKCζ with RNAi in *Drosophila* cells inhibits drosomycin expression but not that of diptericin, indicating that PKCζ is located specifically in the Toll antifungal pathway (Avila et al. 2002). Of note, PKCζ downmodulation does not affect Cactus or relish degradation, but does inhibit drosomycin transcriptional activity. Therefore, in *Drosophila*, as in mammalian cells, PKCζ is necessary for NF-κB transcriptional activity. In this regard, PKCζ is capable of phosphorylating Dif, the fly homolog of RelA, which suggests a high degree of conservation of the role of PKCζ in the NF-κB pathway. In fact, the knockdown of Ref(2)P with RNAi in *Drosophila* cells leads to impaired drosomycin expression (Avila et al. 2002; Goto et al. 2003). This indicates a high degree of conservation for the role of PKCζ in the control of NF-κB transcriptional activity downstream of IκB degradation, which reinforces the importance of this kinase in the regulation of NF-κB and the innate immune response. On the other hand, Par-6 has been shown, via genetic manipulations, to be critically implicated in the control of cell polarity in *C. elegans* and *Drosophila* (Macara 2004; Ohno 2001). Although genetic data has yet to be produced to prove the role of the aPKCs and Par-6 in different aspects of mammalian cell polarity, overexpression analyses have implicated the Par-6/aPKC complex in the control of the epithelial–mesenchymal transition (Ozdamar et al. 2005), T cell (Ludford-Menting et al. 2005) and neuronal polarity (Shi et al. 2003), and cell polarity in migrating astrocytes (Etienne-Manneville and Hall 2003), among other functions. Therefore, the formation of aPKC complexes with different adapters and scaffold proteins serves to confer specificity and plasticity to the actions of these kinases. The mechanistic link between p62 and NF-κB received further support when it was shown that TRAF6 interacts with p62 (Sanz et al. 2000). This is particularly relevant because TRAF6 is an important intermediary in the IL-1, NGF, and LPS signaling pathways controlling NF-κB through still not totally clarified mechanisms likely involving K-63 ubiquitination of IKKγ (Sun et al. 2004). Ref(2)P interacts not only with *Drosophila* PKCζ, but also with the fly homologue of TRAF6 (dTRAF2), again reinforcing the conservation of function in this pathway (Avila et al. 2002). In mammalian cells p62 also interacts with RIP, which mediates NF-κB activation in response to TNFα (Sanz et al. 1999). Both interactions seem to be physiologically relevant since downregulation of p62 levels with antisense constructs leads to a significant reduction of NF-κB activation in IL-1 or TNFα-activated cells. In vivo experiments using p62-deficient mice show a clear impairment in osteoclastogenesis in response to injections of the calciotropic hormone PTHrP. Also, osteoclast precursors from p62–/– mice respond poorly to RANK-L in cell cultures, and are unable to produce a sustained NF-κB response (Duran et al. 2004b). This suggests that p62 can be considered essential in the control of osteoclastogenesis and bone remodeling. Whether this is due to its ability to interact with the aPKCs or whether it is related to the preferential binding of p62 to K63-linked ubiquitinated proteins is a matter for future research. Nonetheless, these observations highlight the importance of specificity during cell signaling through specific adapters that serve to restrict a kinase's action, but that simultaneously allow enough crosstalk between pathways to regulate complex biological events such as inflammation.

Of note, p62 has been shown to be required for the sustained phase of NF-κB activation during T-cell differentiation, a process that is critical in asthma and other allergic diseases (Martin et al. 2006). Interestingly, as in osteoclasts, p62 levels are induced upon T cell differentiation (Martin et al. 2006), suggesting that p62 is necessary to control biochemical events required for proper differentiation in different cell systems. The loss of p62 in T cells impairs their ability to produce Th2 cytokines ex vivo, but the activation of the IL-4/Jak1/Stat6 cascade was not affected by the loss of p62 (Martin et al. 2006). In addition, experiments using the OVA-induced allergic airway inflammation model clearly demonstrated that p62 is required for an optimal lung inflammatory response (Martin et al. 2006). Therefore, p62, like PKCζ, emerges as an important component of the signaling cascades regulating Th2 function and asthma, but acting through a different mechanism.

10.3 NF-κB and Cancer

10.3.1 Introduction

There is growing evidence for a crucial connection between inflammation and cancer, and for the role of NF-κB as an essential molecular link joining the two processes. NF-κB activation is an important step in integrating multiple stress stimuli and regulating innate and adaptive immune responses seen in states of inflammation (Karin et al. 2006). Moreover, inflammatory conditions are often associated with or precede cancer. Many chronic infectious diseases are, in fact, associated with the development of cancer, with approximately 15% of cancer burden linked to chronic infections and the accompanying inflammatory reaction, and with 15–20% of cancer deaths arising from preventable infections (Naugler and Karin 2008). Similarly, many noninfectious inflammatory conditions increase the risk of cancer and promote carcinogenesis (Karin 2006). The common denominator in these conditions is the presence of chronic inflammation, invariably associated with the activation of NF-κB and its effector pathways.

Tumor progression depends on disruption of the normally fine-tuned balance between cell growth, apoptosis, and survival. Oncogenes trigger alterations in all three of these properties and, depending on the specifics of those alterations, can shift the equilibrium toward more or less aggressive forms of cancer. NF-κB is central to this process because it controls the expression of a number of genes that play essential roles in cell survival and angiogenesis (Karin 1998; Karin and Lin 2002; Li and Verma 2002). However, NF-κB also controls critical aspects of the innate and adaptive immune response, which complicates the interpretation of some data that have been generated on this question. There is an abundant literature on this topic for recent reviews see (Karin et al. 2002; 2006), and the following examples are provided to illustrate the complexity of this problem. The selective KO of the NF-κB pathway in colon epithelial cells prevented tumor progression in a model of inflammation-modulated colon cancer (Greten et al. 2004). In contrast, the inactivation of NF-κB in the epidermis increased tumorigenesis in skin in the apparent absence of inflammation and likely due to increased JNK activation (Zhang et al. 2004). Likewise, NF-κB inhibition in hepatocytes enhanced liver cancer in a model of diethylnitrosamine-induced hepatocarcinogenesis in mice (Maeda et al. 2005). A potential interpretation of this intriguing observation could be related to liver-specific compensatory proliferation in response to the increased cell death in NF-κB–deficient hepatocytes. This could activate an inflammatory response orchestrated by liver macrophages (Kupffer cells) that secrete proliferative cytokines favoring the growth and development of the surviving hepatocytes, which, in turn, could result in increased tumorigenesis (Maeda et al. 2005). In this example, JNK, which is increased due to the lack of NF-κB, was essential for the enhanced tumorigenicity, despite its well-established role as pro-apoptotic kinase (Chang et al. 2006; Kamata et al. 2005). This could be because the ablation of JNK in this system blocked the compensatory proliferation induced by the loss of NF-κB activity in hepatocytes (Sakurai et al. 2006), thus preventing the inflammatory response and the increased proliferation of the surviving hepatocytes provoked by the Kupffer cells. Clearly, all of these findings suggest that the role played by NF-κB and its regulators in tumor initiation and progression will greatly depend on the tissue and the cell type involved. As general paradigms have not yet emerged on this question, it is essential to understand the function of the different signaling cascades in specific tumor types to advance our comprehension of the etiopathogenesis of cancer.

10.3.2 The aPKC Network in the NF-κB Cancer Paradigm: Its Role in Lung Cancer

Aberrant activation of NF-κB is strongly associated with cancer. That is, NF-κB is abnormally activated in many kinds of tumors, including pancreatic cancer, breast cancer, gastric carcinoma, prostate cancer, and lung cancer. This, and NF-κB's known antiapoptotic activity, suggests that NF-κB is a logical therapeutic target for research on cancer treatment. Additionally, it has been shown that NF-κB is an important player in preventing apoptotic death after DNA-damaging treatments, such as chemotherapy and irradiation, clearly affecting the efficacy of these treatments.

10.3.2.1 The aPKC Adapter, p62, is Critical for Ras-Induced Lung Cancer

Lung cancer is the leading cause of cancer deaths throughout the world, and current treatments do not lead to a cure for most patients with this type of neoplasia. Targeted antitumor therapies are likely to prove more effective, but their development will require a better understanding of the signaling cascades involved. Ras oncogenes are frequently mutated in human cancers where they play an unquestionably important role in the genesis and progression of the disease (Downward 2003). In lung adenocarcinomas, mutations in Ras are present in at least 25% of cases (Bos 1989), suggesting that the components of Ras-related signaling pathways are promising candidates for therapeutic targets in lung cancer treatment. The ability of Ras to promote cell survival is essential for the suppression of apoptosis associated with Ras-induced transformation and, therefore, for cancer initiation and progression (Hanahan and Weinberg 2000). Previous studies showed that Ras activates NF-κB, which is important for cell survival and tumor transformation because it suppresses p53-independent Ras-induced cell death (Mayo et al. 1997). This is a key event because the final outcome of the oncogenic process is determined by a finely tuned balance between cell survival and death (Luo et al. 2005). However, the mechanism by which Ras controls NF-κB activation remains unclear. IKK activation by Ras requires the combined action of extracellular signal-regulated kinase (ERK) and Akt signals, although how these translate into IKK activation has not been determined (Arsura et al. 2000). Recent results show that the aPKC adapter protein p62 is a crucial molecule linking Ras to NF-κB activation and that it is required for Ras-induced tumorigenesis (Duran et al. 2008). This was demonstrated using a lung-specific inducible model of oncogenic Ras (Fisher et al. 2001) crossed to p62 KO mice. Of note, in the p62 KO mice, Ras was no longer able to induce tumor formation. This is most likely a cell-autonomous effect, as p62 was also required for NF-κB and cell survival during Ras-induced transformation in embryo fibroblasts. Interestingly, Ras had the ability to regulate p62 levels, and in accordance with this, p62 was found to be overexpressed in human cancer samples (Duran et al. 2008).

The mechanism through which p62 channels Ras signals to NF-κB is mediated by IKK activation, as Ras-induced IKK activity was blocked in p62 KO mice and this was facilitated by the ubiquitination of the adapter TRAF6. How TRAF6 ubiquitination regulates IKK activation is not completely clear yet. TRAF6 is an E3 ubiquitin-ligase that promotes K63-type polyubiquitination of many proteins, including itself (Chen 2005), and creates a number of docking sites that are necessary for IKK activation. Also, it has been shown that p62 is critically involved in the oligomerization of TRAF6 (Sanz et al. 2000; Wooten et al. 2005), which is an essential step for these polyubiquitinations (Chen 2005). In addition to K63 polyubiquitination, IKK activation correlates with the phosphorylation of both of its catalytic subunits (IKKα and IKKβ) at their activation loops (Ghosh and Karin 2002). Recent results suggest that these are separate required events that, independently, are not sufficient to trigger IKK activity (Grabiner et al. 2007; Misra et al. 2007; Su et al. 2005). In fact, p62 deficiency has an impact on both processes: it significantly reduces the ability of Ras to induce IKKα/β phosphorylation, and completely abolishes the polyubiquitination of TRAF6.

This mediation of Ras signaling relies on the ability of p62 to impinge on the NF-κB cascade and on the subsequent expression of survival genes. Among the different antiapoptotic signals delivered by NF-κB is the expression of genes involved in ROS scavenging, such as FHC (Pham et al. 2004). Interestingly, the loss of p62 impaired FHC expression, and provided a likely explanation for why Ras transformation leads to higher ROS levels in p62 KO cells than in WT. Higher ROS levels translate into enhanced JNK activity in the absence of p62. Elevated JNK activity induces the production of more ROS (Nakano et al. 2006), activating a positive-feedback cycle that is deleterious for the viability of the transformed cell and explains the lack of cell toxicity in JNK-deficient Ras-transformed cells (Ventura et al. 2004). These results indicate that the p62 adapter is a crucial step in the activation of NF-κB by the Ras oncogenic pathway to ensure survival of the transformed cell (Fig. 10.3).

10.3.2.2 The Role of Par-4 in Lung Cancer

A number of previous cell culture studies have suggested that Par-4 could be a negative regulator of tumor progression, at least in vitro (Ranganathan and Rangnekar 2005). This was further reinforced by in vivo studies of Par-4 KO mice showing reduced lifespan, enhanced benign tumor development, and low-frequency carcinogenesis (Garcia-Cao et al. 2005). It has also been shown that Par-4-null mice develop spontaneous benign neoplasias in hormone-dependent tissues, including prostate (Garcia-Cao et al. 2005). In addition, it has been shown that Par-4 is downregulated in approximately 40% of human endometrial carcinomas, human prostate carcinomas, and human lung adenocarcinomas (Joshi et al. 2008; Moreno-Bueno et al. 2007). The loss of Par-4 in these tumors corresponds to Par-4 distribution in normal tissue. That is, Par-4 is highly expressed in prostate, endometrium, and lung. As PKCζ expression is also very abundant in lung tissue, this suggests that the PKCζ/Par-4 complex may play a role in the normal physiology of the lung and in lung pathology. Consistent with the role of Par-4 as a negative regulator of NF-κB, analysis of Par-4-deficient lungs revealed increased activation of NF-κB (Joshi et al. 2008), but more importantly, this was reduced to normal levels in PKCζ/Par-4 DKO mice. This in vivo genetic evidence strongly suggests a role for PKCζ as the target of Par-4 in the NF-κB pathway. Our recent data also demonstrate that the loss of Par-4 dramatically enhances Ras-induced lung carcinoma formation in association with enhanced NF-κB activity, and unveiled an unanticipated role for Par-4 as an indirect inhibitor of Akt, in vitro and in vivo, through downmodulation of PKCζ (Joshi et al. 2008). That is, Par-4-deficient mice had higher levels of activated Akt in alveolar and airway epithelial cells, and this was inhibited in the PKCζ/Par-4 DKO mice (Joshi et al. 2008). Together, these observations identify Par-4 as a tumor suppressor in the NF-κB and Akt pathways in lung cancer.

10.3.2.3 PKCζ, a Novel Tumor Suppressor in Lung Cancer

There is compelling evidence supporting a role for PKCζ in Ras-induced lung cancer: The aPKC adapter p62 is required for lung carcinogenesis through its ability to activate the NF-κB inflammatory and survival pathway (Duran et al. 2008); p62 binds the aPKC isoform PKCζ (Moscat et al. 2006a; Sanz et al. 2000); and PKCζ has been implicated in the regulation of NF-κB, and, moreover, has been suggested to be relevant for Ras-induced oncogenesis in co-transfection and overexpression experiments (Berra et al. 1993; Diaz-Meco et al. 1994). However, until recently there has been no report of an in vivo test, at the organismal level, of the role of PKCζ in cancer. Unexpectedly, recent results demonstrated that the loss of PKCζ enhances Ras-induced lung carcinogenesis in vivo despite the fact that expression of NF-κB-dependent genes was dramatically reduced in Ras-expressing PKCζ–/– lungs as compared to WT, which is in keeping with PKCζ's proposed role in regulating NF-κB transcriptional activity through the phosphorylation of the Rel A subunit (Galvez et al. 2009). The enhanced tumorigenic response found in Ras-expressing PKCζ–/– lungs was due to an increase in IL-6 levels, even in the presence of high levels of IFNγ and IL-12 and reduced IL-4 levels, which are hallmarks of an M1-type immunological antitumor response (Galvez et al. 2009). IL-6 has been shown to be essential for tumor progression in several systems (Amsen et al. 2004; Gao et al. 2001; Grivennikov and Karin 2008; Naugler et al. 2007; Sansone et al. 2007), and a recent study has shown that Ras-induced IL-6 production promotes tumorigenesis via the paracrine induction of angiogenesis (Ancrile et al. 2007). Consistent with this, the loss of PKCζ enhances Ras-induced angiogenesis in lung tumors. However, in addition to that role, IL-6 secretion is essential to providing the Ras transformants with a cell-autonomous competitive advantage with respect to growth under nutrient-scarce conditions, a situation that the cancer cell encounters during tumor development. Therefore, in the context of PKCζ deficiency, the ability of Ras to induce IL-6 production is enhanced, which promotes tumorigenesis through two different mechanisms: increased angiogenesis and an enhanced ability to grow under nutrient-limiting conditions. These results unveiled a previously unrecognized role of PKCζ as a negative regulator of lung cancer through its ability to control IL-6 production by transformed cells. In addition, these results indicate a new layer of selectivity differentiating the two aPKC isoforms and their roles in cancer. Interestingly, and in contrast to PKCζ, PKCλ/ι is required for Ras-induced lung cancer. Thus, in xenograft experiments involving nude mice, injection of Ras-expressing PKCλ/ι-deficient cells, generated by infection of PKCλ/ι$^{fl/fl}$ cells with Cre adenovirus, led to reduced tumor size in vivo {Moscat et al., unpublished results}, indicating that PKCλ/ι acts as an oncogene in contrast with the tumor suppressor activity of its closely related isotype PKCζ. These results are consistent with previous observations implicating PKCλ/ι in lung cancer (Fields and Murray 2008). PKCλ/ι is highly overexpressed at both the mRNA and protein level in the majority (70%) of primary NSCLC tumors and the level of PKCλ/ι expression in tumors is highly predictive of poor patient survival (Regala et al. 2005). However, the precise mechanism by which PKCλ/ι channels Ras signals deserves further research. Previous studies have implicated Rac as a downstream target of PKCλ/ι (Regala et al. 2005), but more definitive in vivo analyses are needed to determine where the functional specificity of the two aPKCs resides, and which molecular mechanisms account for these opposite functions.

References

Amsen, D., Blander, J. M., Lee, G. R., Tanigaki, K., Honjo, T., & Flavell, R. A. (2004). Instruction of distinct CD4 T helper cell fates by different notch ligands on antigen-presenting cells. *Cell, 117*, 515–526.

Ancrile, B., Lim, K. H., & Counter, C. M. (2007). Oncogenic Ras-induced secretion of IL6 is required for tumorigenesis. *Genes and Development, 21*, 1714–1719.

Anest, V., Hanson, J. L., Cogswell, P. C., Steinbrecher, K. A., Strahl, B. D., & Baldwin, A. S. (2003). A nucleosomal function for IkappaB kinase-alpha in NF-kappaB-dependent gene expression. *Nature, 423*, 659–663.

Anrather, J., Csizmadia, V., Soares, M. P., & Winkler, H. (1999). Regulation of NF-kappaB RelA phosphorylation and transcriptional activity by p21(ras) and protein kinase Czeta in primary endothelial cells. *The Journal of Biological Chemistry, 274*, 13594–13603.

Arsura, M., Mercurio, F., Oliver, A. L., Thorgeirsson, S. S., & Sonenshein, G. E. (2000). Role of the IkappaB kinase complex in oncogenic Ras- and Raf-mediated transformation of rat liver epithelial cells. *Molecular and Cellular Biology, 20*, 5381–5391.

Avila, A., Silverman, N., Diaz-Meco, M. T., & Moscat, J. (2002). The Drosophila atypical protein kinase C-ref(2)p complex constitutes a conserved module for signaling in the toll pathway. *Molecular and Cellular Biology, 22*, 8787–8795.

Barradas, M., Monjas, A., Diaz-Meco, M.T., Serrano, M., & Moscat, J. (1999). The downregulation of the pro-apoptotic protein Par-4 is critical for Ras-induced survival and tumor progression. *The EMBO Journal*, 18, 6362–6369.

Berra, E., Diaz-Meco, M. T., Dominguez, I., Municio, M. M., Sanz, L., Lozano, J., et al. (1993). Protein kinase C zeta isoform is critical for mitogenic signal transduction. *Cell, 74*, 555–563.

Betschinger, J., Mechtler, K., & Knoblich, J. A. (2003). The Par complex directs asymmetric cell division by phosphorylating the cytoskeletal protein Lgl. *Nature, 422*, 326–330.

Bos, J. L. (1989). Ras oncogenes in human cancer: A review. *Cancer Research, 49*, 4682–4689.

Cai, Y., Yu, F., Lin, S., Chia, W., & Yang, X. (2003). Apical complex genes control mitotic spindle geometry and relative size of daughter cells in Drosophila neuroblast and pI asymmetric divisions. *Cell, 112*, 51–62.

Cao, Y., Bonizzi, G., Seagroves, T. N., Greten, F. R., Johnson, R., Schmidt, E. V., et al. (2001). IKKalpha provides an essential link between RANK signaling and cyclin D1 expression during mammary gland development. *Cell, 107*, 763–775.

Chang, L., Kamata, H., Solinas, G., Luo, J. L., Maeda, S., Venuprasad, K., et al. (2006). The E3 ubiquitin ligase itch couples JNK activation to TNFalpha-induced cell death by inducing c-FLIP(L) turnover. *Cell, 124*, 601–613.

Chen, Z. J. (2005). Ubiquitin signalling in the NF-kappaB pathway. *Nature Cell Biology, 7*, 758–765.

Chen, L. F., & Greene, W. C. (2004). Shaping the nuclear action of NF-kappaB. *Nature Reviews. Molecular Cell Biology, 5*, 392–401.

Claudio, E., Brown, K., Park, S., Wang, H., & Siebenlist, U. (2002). BAFF-induced NEMO-independent processing of NF-kappaB2 in maturing B cells. *Nature Immunology, 3*, 958–965.

Cox, D. N., Seyfried, S. A., Jan, L. Y., & Jan, Y. N. (2001). Bazooka and atypical protein kinase C are required to regulate oocyte differentiation in the Drosophila ovary. *Proceedings of the National Academy of Sciences of the United States of America, 98*, 14475–14480.

Dejardin, E., Droin, N. M., Delhase, M., Haas, E., Cao, Y., Makris, C., et al. (2002). The lymphotoxin-beta receptor induces different patterns of gene expression via two NF-kappaB pathways. *Immunity, 17*, 525–535.

Diaz-Meco, M. T., Berra, E., Municio, M. M., Sanz, L., Lozano, J., Dominguez, I., et al. (1993). A dominant negative protein kinase C zeta subspecies blocks NF-kappa B activation. *Molecular and Cellular Biology, 13*, 4770–4775.

Diaz-Meco, M. T., Lallena, M. J., Monjas, A., Frutos, S., & Moscat, J. (1999). Inactivation of the Inhibitory kappaB protein kinase/nuclear factor kappaB pathway by Par-4 expression potentiates tumor necrosis factor alpha-induced apoptosis. *The Journal of Biological Chemistry, 274*, 19606–19612.

Diaz-Meco, M. T., Lozano, J., Municio, M. M., Berra, E., Frutos, S., Sanz, L., et al. (1994). Evidence for the in vitro and in vivo interaction of Ras with protein kinase C zeta. *The Journal of Biological Chemistry, 269*, 31706–31710.

Diaz-Meco, M. T., Municio, M. M., Frutos, S., Sanchez, P., Lozano, J., Sanz, L., et al. (1996). The product of par-4, a gene induced during apoptosis, interacts selectively with the atypical isoforms of protein kinase C. *Cell, 86*, 777–786.

Dominguez, I., Diaz-Meco, M. T., Municio, M. M., Berra, E., Garcia de Herreros, A., Cornet, M. E., et al. (1992). Evidence for a role of protein kinase C zeta subspecies in maturation of Xenopus laevis oocytes. *Molecular and Cellular Biology, 12*, 3776–3783.

Dominguez, I., Sanz, L., Arenzana-Seisdedos, F., Diaz-Meco, M. T., Virelizier, J. L., & Moscat, J. (1993). Inhibition of protein kinase C zeta subspecies blocks the activation of an NF-kappa B-like activity in Xenopus laevis oocytes. *Molecular and Cellular Biology, 13*, 1290–1295.

Downward, J. (2003). Targeting RAS signalling pathways in cancer therapy. *Nature Reviews. Cancer, 3*, 11–22.

Duran, A., Diaz-Meco, M. T., & Moscat, J. (2003). Essential role of RelA Ser311 phosphorylation by zetaPKC in NF-kappaB transcriptional activation. *The EMBO Journal, 22*, 3910–3918.

Duran, A., Linares, J. F., Galvez, A. S., Wikenheiser, K., Flores, J. M., Diaz-Meco, M. T., et al. (2008). The signaling adaptor p62 is an important NF-kappaB mediator in tumorigenesis. *Cancer Cell, 13*, 343–354.

Duran, A., Rodriguez, A., Martin, P., Serrano, M., Flores, J. M., Leitges, M., et al. (2004a). Crosstalk between PKCzeta and the IL4/Stat6 pathway during T-cell-mediated hepatitis. *The EMBO Journal, 23*, 4595–4605.

Duran, A., Serrano, M., Leitges, M., Flores, J. M., Picard, S., Brown, J. P., et al. (2004b). The atypical PKC-interacting protein p62 is an important mediator of RANK-activated osteoclastogenesis. *Developmental Cell, 6*, 303–309.

Etienne-Manneville, S., & Hall, A. (2003). Cell polarity: Par6, aPKC and cytoskeletal crosstalk. *Current Opinion in Cell Biology, 15*, 67–72.

Farese, R. V., Sajan, M. P., Yang, H., Li, P., Mastorides, S., Gower, W. R., Jr., et al. (2007). Muscle-specific knockout of PKC-lambda impairs glucose transport and induces metabolic and diabetic syndromes. *The Journal of Clinical Investigation, 117*, 2289–2301.

Fields, A. P., & Murray, N. R. (2008). Protein kinase C isozymes as therapeutic targets for treatment of human cancers. *Advances in Enzyme Regulation, 48*, 166–178.

Fisher, G. H., Wellen, S. L., Klimstra, D., Lenczowski, J. M., Tichelaar, J. W., Lizak, M. J., et al. (2001). Induction and apoptotic regression of lung adenocarcinomas by regulation of a K-Ras transgene in the presence and absence of tumor suppressor genes. *Genes and Development, 15*, 3249–3262.

Galvez, A. S., Duran, A., Linares, J. F., Pathrose, P., Castilla, E. A., Abu-Baker, S., et al. (2009). Protein kinase Czeta represses the interleukin-6 promoter and impairs tumorigenesis in vivo. *Molecular and Cellular Biology, 29*, 104–115.

Gao, T., Toker, A., & Newton, A. C. (2001). The carboxyl terminus of protein kinase c provides a switch to regulate its interaction with the phosphoinositide-dependent kinase, PDK-1. *The Journal of Biological Chemistry, 276*, 19588–19596.

Garcia-Cao, I., Duran, A., Collado, M., Carrascosa, M. J., Martin-Caballero, J., Flores, J. M., et al. (2005). Tumour-suppression activity of the proapoptotic regulator Par4. *EMBO Reports, 6*, 577–583.

Garcia-Cao, I., Lafuente, M., Criado, L., Diaz-Meco, M., Serrano, M., & Moscat, J. (2003). Genetic inactivation of Par4 results in hyperactivation of NF-kB and impairment of JNK and p38. *EMBO Reports, 4*, 307–312.

Ghosh, S., & Karin, M. (2002). Missing pieces in the NF-kappaB puzzle. *Cell, 109*(Suppl), S81–S96.

Goto, A., Blandin, S., Royet, J., Reichhart, J. M., & Levashina, E. A. (2003). Silencing of Toll pathway components by direct injection of double-stranded RNA into Drosophila adult flies. *Nucleic Acids Research, 31*, 6619–6623.

Grabiner, B. C., Blonska, M., Lin, P. C., You, Y., Wang, D., Sun, J., et al. (2007). CARMA3 deficiency abrogates G protein-coupled receptor-induced NF-{kappa}B activation. *Genes and Development, 21*, 984–996.

Greten, F. R., Eckmann, L., Greten, T. F., Park, J. M., Li, Z. W., Egan, L. J., et al. (2004). IKKbeta links inflammation and tumorigenesis in a mouse model of colitis-associated cancer. *Cell, 118*, 285–296.

Grivennikov, S., & Karin, M. (2008). Autocrine IL-6 signaling: A key event in tumorigenesis? *Cancer Cell, 13*, 7–9.

Hanahan, D., & Weinberg, R. A. (2000). The hallmarks of cancer. *Cell, 100*, 57–70.

Ho, I. C., & Glimcher, L. H. (2002). Transcription: Tantalizing times for T cells. *Cell, 109*(Suppl), S109–S120.

Hoffmann, J. A. (2003). The immune response of Drosophila. *Nature, 426*, 33–38.

Jaruga, B., Hong, F., Sun, R., Radaeva, S., & Gao, B. (2003). Crucial role of IL-4/STAT6 in T cell-mediated hepatitis: Up-regulating eotaxins and IL-5 and recruiting leukocytes. *Journal of Immunology, 171*, 3233–3244.

Joshi, J., Fernandez-Marcos, P. J., Galvez, A., Amanchy, R., Linares, J. F., Duran, A., et al. (2008). Par-4 inhibits Akt and suppresses Ras-induced lung tumorigenesis. *The EMBO Journal, 27*, 2181–2193.

Kamata, H., Honda, S., Maeda, S., Chang, L., Hirata, H., & Karin, M. (2005). Reactive oxygen species promote TNFalpha-induced death and sustained JNK activation by inhibiting MAP kinase phosphatases. *Cell, 120*, 649–661.

Karin, M. (1998). The NF-kappa B activation pathway: Its regulation and role in inflammation and cell survival. *The Cancer Journal from Scientific American, 4*(Suppl 1), S92–S99.

Karin, M. (2006). Nuclear factor-[kappa]B in cancer development and progression. *Nature, 441*, 431–436.

Karin, M., Cao, Y., Greten, F. R. and Li, Z. W. (2002). NF-kappaB in cancer: From innocent bystander to major culprit. *Nature Reviews. Cancer, 2*, 301–310.

Karin, M., Lawrence, T., & Nizet, V. (2006). Innate immunity gone awry: Linking microbial infections to chronic inflammation and cancer. *Cell, 124*, 823–835.

Karin, M., & Lin, A. (2002). NF-kappaB at the crossroads of life and death. *Nature Immunology, 3*, 221–227.

Kayagaki, N., Yan, M., Seshasayee, D., Wang, H., Lee, W., French, D. M., et al. (2002). BAFF/BLyS receptor 3 binds the B cell survival factor BAFF ligand through a discrete surface loop and promotes processing of NF-kappaB2. *Immunity, 17*, 515–524.

Krummel, M. F., & Macara, I. (2006). Maintenance and modulation of T cell polarity. *Nature Immunology, 7*, 1143–1149.

Lafuente, M. J., Martin, P., Garcia-Cao, I., Diaz-Meco, M. T., Serrano, M., & Moscat, J. (2003). Regulation of mature T lymphocyte proliferation and differentiation by Par-4. *The EMBO Journal, 22*, 4689–4698.

Lawrence, T., Bebien, M., Liu, G. Y., Nizet, V., & Karin, M. (2005). IKKalpha limits macrophage NF-kappaB activation and contributes to the resolution of inflammation. *Nature, 434*, 1138–1143.

Leitges, M., Sanz, L., Martin, P., Duran, A., Braun, U., Garcia, J. F., et al. (2001). Targeted disruption of the zetaPKC gene results in the impairment of the NF-kappaB pathway. *Molecular Cell, 8*, 771–780.

Li, Q., & Verma, I. M. (2002). NF-kappaB regulation in the immune system. *Nature Reviews. Immunology, 2*, 725–734.

Ludford-Menting, M. J., Oliaro, J., Sacirbegovic, F., Cheah, E. T., Pedersen, N., Thomas, S. J., et al. (2005). A network of PDZ-containing proteins regulates T cell polarity and morphology during migration and immunological synapse formation. *Immunity, 22*, 737–748.

Luo, J. L., Kamata, H., & Karin, M. (2005). IKK/NF-kappaB signaling: Balancing life and death – a new approach to cancer therapy. *The Journal of Clinical Investigation, 115*, 2625–2632.

Luster, A. D., & Tager, A. M. (2004). T-cell trafficking in asthma: Lipid mediators grease the way. *Nature Reviews. Immunology, 4*, 711–724.

Macara, I. G. (2004). Par proteins: Partners in polarization. *Current Biology, 14*, R160–R162.

Maeda, S., Chang, L., Li, Z. W., Luo, J. L., Leffert, H., & Karin, M. (2003). IKKbeta is required for prevention of apoptosis mediated by cell-bound but not by circulating TNFalpha. *Immunity, 19*, 725–737.

Maeda, S., Kamata, H., Luo, J. L., Leffert, H., & Karin, M. (2005). IKKbeta couples hepatocyte death to cytokine-driven compensatory proliferation that promotes chemical hepatocarcinogenesis. *Cell, 121*, 977–990.

Martin, P., Diaz-Meco, M. T., & Moscat, J. (2006). The signaling adapter p62 is an important mediator of T helper 2 cell function and allergic airway inflammation. *The EMBO Journal, 25*, 3524–3533.

Martin, P., Duran, A., Minguet, S., Gaspar, M. L., Diaz-Meco, M. T., Rennert, P., et al. (2002). Role of zeta PKC in B-cell signaling and function. *The EMBO Journal, 21*, 4049–4057.

Martin, P., Villares, R., Rodriguez-Mascarenhas, S., Zaballos, A., Leitges, M., Kovac, J., et al. (2005). Control of T helper 2 cell function and allergic airway inflammation by PKC{zeta}. *Proceedings of the National Academy of Sciences of the United States of America, 102*, 9866–9871.

Mayo, M. W., Wang, C. Y., Cogswell, P. C., Rogers-Graham, K. S., Lowe, S. W., Der, C. J., et al. (1997). Requirement of NF-kappaB activation to suppress p53-independent apoptosis induced by oncogenic Ras. *Science, 278*, 1812–1815.

Misra, R. S., Russell, J. Q., Koenig, A., Hinshaw-Makepeace, J. A., Wen, R., Wang, D., et al. (2007). Caspase-8 and c-FLIPL associate in lipid rafts with NF-kappa B adaptors during T cell activation. *The Journal of Biological Chemistry, 282*, 19365–19374.

Moreno-Bueno, G., Fernandez-Marcos, P. J., Collado, M., Tendero, M. J., Rodriguez-Pinilla, S. M., Garcia-Cao, I., et al. (2007). Inactivation of the candidate tumor suppressor par-4 in endometrial cancer. *Cancer Research, 67*, 1927–1934.

Moscat, J., & Diaz-Meco, M. T. (2000). The atypical protein kinase Cs. Functional specificity mediated by specific protein adapters. *EMBO Reports, 1*, 399–403.

Moscat, J., Diaz-Meco, M. T., Albert, A., & Campuzano, S. (2006a). Cell signaling and function organized by PB1 domain interactions. *Molecular Cell, 23*, 631–640.

Moscat, J., Diaz-Meco, M. T., & Rennert, P. (2003). NF-kappaB activation by protein kinase C isoforms and B-cell function. *EMBO Reports, 4*, 31–36.

Moscat, J., Rennert, P., & Diaz-Meco, M. T. (2006b). PKCzeta at the crossroad of NF-kappaB and Jak1/Stat6 signaling pathways. *Cell Death and Differentiation, 13*, 702–711.

Mosmann, T. R., & Coffman, R. L. (1989). TH1 and TH2 cells: Different patterns of lymphokine secretion lead to different functional properties. *Annual Review of Immunology, 7*, 145–173.

Nakano, H., Nakajima, A., Sakon-Komazawa, S., Piao, J. H., Xue, X., & Okumura, K. (2006). Reactive oxygen species mediate crosstalk between NF-kappaB and JNK. *Cell Death and Differentiation, 13*, 730–737.

Nalca, A., Qiu, S. G., El-Guendy, N., Krishnan, S., & Rangnekar, V. M. (1999). Oncogenic Ras sensitizes cells to apoptosis by Par-4. *The Journal of Biological Chemistry, 274*, 29976–29983.

Naugler, W. E., & Karin, M. (2008). NF-kappaB and cancer-identifying targets and mechanisms. *Current Opinion in Genetics and Development, 18*, 19–26.

Naugler, W. E., Sakurai, T., Kim, S., Maeda, S., Kim, K., Elsharkawy, A. M., et al. (2007). Gender disparity in liver cancer due to sex differences in MyD88-dependent IL-6 production. *Science, 317*, 121–124.

Nishizuka, Y. (1995). Protein kinase C and lipid signaling for sustained cellular responses. *The FASEB Journal, 9*, 484–496.

O'Shea, J. J., Gadina, M., & Schreiber, R. D. (2002). Cytokine signaling in 2002: New surprises in the Jak/Stat pathway. *Cell, 109*(Suppl), S121–S131.

Ohno, S. (2001). Intercellular junctions and cellular polarity: The PAR–aPKC complex, a conserved core cassette playing fundamental roles in cell polarity. *Current Opinion in Cell Biology, 13*, 641–648.

Ozdamar, B., Bose, R., Barrios-Rodiles, M., Wang, H. R., Zhang, Y., & Wrana, J. L. (2005). Regulation of the polarity protein Par6 by TGFbeta receptors controls epithelial cell plasticity. *Science, 307*, 1603–1609.

Paul, W. E., & Seder, R. A. (1994). Lymphocyte responses and cytokines. *Cell, 76*, 241–251.

Pham, C. G., Bubici, C., Zazzeroni, F., Papa, S., Jones, J., Alvarez, K., et al. (2004). Ferritin heavy chain upregulation by NF-kappaB inhibits TNFalpha-induced apoptosis by suppressing reactive oxygen species. *Cell, 119*, 529–542.

Puls, A., Schmidt, S., Grawe, F., & Stabel, S. (1997). Interaction of protein kinase C zeta with ZIP, a novel protein kinase C-binding protein. *Proceedings of the National Academy of Sciences of the United States of America, 94*, 6191–6196.

Ranganathan, P., & Rangnekar, V. M. (2005). Regulation of cancer cell survival by Par-4. *Annals of the New York Academy of Sciences, 1059*, 76–85.

Regala, R. P., Weems, C., Jamieson, L., Khoor, A., Edell, E. S., Lohse, C. M., et al. (2005). Atypical protein kinase C iota is an oncogene in human non-small cell lung cancer. *Cancer Research, 65*, 8905–8911.

Sakurai, T., Maeda, S., Chang, L., & Karin, M. (2006). Loss of hepatic NF-kappa B activity enhances chemical hepatocarcinogenesis through sustained c-Jun N-terminal kinase 1 activation.

Proceedings of the National Academy of Sciences of the United States of America, 103, 10544–10551.

Sanchez, P., De Carcer, G., Sandoval, I. V., Moscat, J., & Diaz-Meco, M. T. (1998). Localization of atypical protein kinase C isoforms into lysosome-targeted endosomes through interaction with p62. *Molecular and Cellular Biology, 18*, 3069–3080.

Sansone, P., Storci, G., Tavolari, S., Guarnieri, T., Giovannini, C., Taffurelli, M., et al. (2007). IL-6 triggers malignant features in mammospheres from human ductal breast carcinoma and normal mammary gland. *The Journal of Clinical Investigation, 117*, 3988–4002.

Sanz, L., Diaz-Meco, M. T., Nakano, H., & Moscat, J. (2000). The atypical PKC-interacting protein p62 channels NF-kappaB activation by the IL-1-TRAF6 pathway. *The EMBO Journal, 19*, 1576–1586.

Sanz, L., Sanchez, P., Lallena, M. J., Diaz-Meco, M. T., & Moscat, J. (1999). The interaction of p62 with RIP links the atypical PKCs to NF-kappaB activation. *The EMBO Journal, 18*, 3044–3053.

Sells, S. F., Wood, D. P., Jr., Joshi-Barve, S. S., Muthukumar, S., Jacob, R. J., Crist, S. A., et al. (1994). Commonality of the gene programs induced by effectors of apoptosis in androgen-dependent and -independent prostate cells. *Cell Growth and Differentiation, 5*, 457–466.

Senftleben, U., Cao, Y., Xiao, G., Greten, F. R., Krahn, G., Bonizzi, G., et al. (2001). Activation by IKKalpha of a second, evolutionary conserved, NF-kappa B signaling pathway. *Science, 293*, 1495–1499.

Shi, S. H., Jan, L. Y., & Jan, Y. N. (2003). Hippocampal neuronal polarity specified by spatially localized mPar3/mPar6 and PI 3-kinase activity. *Cell, 112*, 63–75.

Shuai, K., & Liu, B. (2003). Regulation of JAK-STAT signalling in the immune system. *Nature Reviews. Immunology, 3*, 900–911.

Silverman, N., & Maniatis, T. (2001). NF-kappaB signaling pathways in mammalian and insect innate immunity. *Genes and Development, 15*, 2321–2342.

Soloff, R. S., Katayama, C., Lin, M. Y., Feramisco, J. R., & Hedrick, S. M. (2004). Targeted deletion of protein kinase C lambda reveals a distribution of functions between the two atypical protein kinase C isoforms. *Journal of Immunology, 173*, 3250–3260.

Su, H., Bidere, N., Zheng, L., Cubre, A., Sakai, K., Dale, J., et al. (2005). Requirement for caspase-8 in NF-kappaB activation by antigen receptor. *Science, 307*, 1465–1468.

Sumimoto, H., Kamakura, S., & Ito, T. (2007). Structure and function of the PB1 domain, a protein interaction module conserved in animals, fungi, amoebas, and plants. *Sci. STKE*, 2007 (401) re6, [DOI: 10.1126/stke.4012007re6].

Sun, L., Deng, L., Ea, C. K., Xia, Z. P., & Chen, Z. J. (2004). The TRAF6 ubiquitin ligase and TAK1 kinase mediate IKK activation by BCL10 and MALT1 in T lymphocytes. *Molecular Cell, 14*, 289–301.

Tabuse, Y., Izumi, Y., Piano, F., Kemphues, K. J., Miwa, J., & Ohno, S. (1998). Atypical protein kinase C cooperates with PAR-3 to establish embryonic polarity in Caenorhabditis elegans. *Development, 125*, 3607–3614.

Tiegs, G., Hentschel, J., & Wendel, A. (1992). A T cell-dependent experimental liver injury in mice inducible by concanavalin A. *The Journal of Clinical Investigation, 90*, 196–203.

Ventura, J. J., Cogswell, P., Flavell, R. A., Baldwin, A. S., Jr., & Davis, R. J. (2004). JNK potentiates TNF-stimulated necrosis by increasing the production of cytotoxic reactive oxygen species. *Genes and Development, 18*, 2905–2915.

Vermeulen, L., De Wilde, G., Damme, P. V., Vanden Berghe, W., & Haegeman, G. (2003). Transcriptional activation of the NF-{kappa}B p65 subunit by mitogen- and stress-activated protein kinase-1 (MSK1). *The EMBO Journal, 22*, 1313–1324.

Wodarz, A., Ramrath, A., Kuchinke, U., & Knust, E. (1999). Bazooka provides an apical cue for Inscuteable localization in Drosophila neuroblasts. *Nature, 402*, 544–547.

Wooten, M. W., Geetha, T., Seibenhener, M. L., Babu, J. R., Diaz-Meco, M. T. and Moscat, J. (2005). The p62 scaffold regulates nerve growth factor-induced NF-kappaB activation by influencing TRAF6 polyubiquitination. *The Journal of Biological Chemistry, 280*, 35625–35629.

Wooten, M. W., Seibenhener, M. L., Mamidipudi, V., Diaz-Meco, M. T., Barker, P. A., & Moscat, J. (2001). The atypical protein kinase C-interacting protein p62 is a scaffold for NF-kappaB activation by nerve growth factor. *The Journal of Biological Chemistry, 276*, 7709–7712.

Xiao, G., Harhaj, E. W., & Sun, S. C. (2001). NF-kappaB-inducing kinase regulates the processing of NF-kappaB2 p100. *Molecular Cell, 7*, 401–409.

Yamamoto, Y., Verma, U. N., Prajapati, S., Kwak, Y. T., & Gaynor, R. B. (2003). Histone H3 phosphorylation by IKK-alpha is critical for cytokine-induced gene expression. *Nature, 423*, 655–659.

Yeh, J. H., Sidhu, S. S., & Chan, A. C. (2008). Regulation of a late phase of T cell polarity and effector functions by Crtam. *Cell, 132*, 846–859.

Zhang, J. Y., Green, C. L., Tao, S., & Khavari, P. A. (2004). NF-kappaB RelA opposes epidermal proliferation driven by TNFR1 and JNK. *Genes and Development, 18*, 17–22.

Zhong, H., May, M. J., Jimi, E., & Ghosh, S. (2002). The phosphorylation status of nuclear NF-kappa B determines its association with CBP/p300 or HDAC-1. *Molecular Cell, 9*, 625–636.

Zhong, H., SuYang, H., Erdjument-Bromage, H., Tempst, P., & Ghosh, S. (1997). The transcriptional activity of NF-kappaB is regulated by the IkappaB-associated PKAc subunit through a cyclic AMP-independent mechanism. *Cell, 89*, 413–424.

Zhong, H., Voll, R. E., & Ghosh, S. (1998). Phosphorylation of NF-kappa B p65 by PKA stimulates transcriptional activity by promoting a novel bivalent interaction with the coactivator CBP/p300. *Molecular Cell, 1*, 661–671.

Zhu, J., Min, B., Hu-Li, J., Watson, C. J., Grinberg, A., Wang, Q., et al. (2004). Conditional deletion of Gata3 shows its essential function in T(H)1-T(H)2 responses. *Nature Immunology, 5*, 1157–1165.

Part III
PKC Isozymes in Cancer

Chapter 11
Introduction: PKC and Cancer

Marcelo G. Kazanietz

Keywords Protein kinase C • Cancer development • Mitogenesis • Survival

Despite our extensive understanding of the biochemical regulation of PKC isozymes and the well-established roles for phorbol esters as tumor promoters, the contribution of specific PKC isozymes in cancer progression is unclear. A vast number of studies have thoroughly assigned key roles for PKC isozymes in proliferation, survival, apoptosis, differentiation, and malignant transformation, and their involvement as mediators of growth factor receptor responses is unquestionable. Whether PKC isozymes can be defined as oncogenes or tumor suppressors, on the other hand, is controversial.

PKC isozymes are known to exert overlapping, different, and even opposite biological functions, particularly in the context of mitogenesis, transformation, and metastasis. There is a great degree of cell type specificity that probably relates to the differential relative expression of PKC isozymes and/or their different access to intracellular compartments where substrates are located. For example, PKCα mediates antiproliferative signaling in a number of cell types including intestinal, pancreatic, and mammary cells, and it has a tumor suppressor role in the intestine (Detjen et al. 2000; Frey et al. 2000; Sun and Rotenberg 1999). However, in other models, PKCα drives proliferative and tumorigenic responses, similarly to PKCβ (Jiang et al. 2004; Sharma et al. 2007; Wu et al. 2008). Regarding the novel PKCδ and PKCε isozymes, they have in most cases a "yin–yang" relationship and mediate opposite responses both in normal and cancer cells (Griner and Kazanietz 2007). For example, studies revealed that while PKCδ overexpression leads to growth arrest in fibroblasts, PKCε overexpresssion transforms both fibroblasts and colon epithelial cells. PKCε overexpression is sufficient to confer tumorigenic properties

M.G. Kazanietz (✉)
Department of Pharmacology, University of Pennsylvania School of Medicine,
1256 Biomedical Research Building II/III, 421 Curie Blvd., Philadelphia, PA 19104-6160, USA
e-mail: marcelog@upenn.edu

M.G. Kazanietz (ed.), *Protein Kinase C in Cancer Signaling and Therapy*,
Current Cancer Research, DOI 10.1007/978-1-60761-543-9_11,

to NIH 3T3 cells when inoculated into nude mice (Mischak et al. 1993, Perletti et al. 1996; Perletti et al. 1999). PKCε has also been linked to survival in various cancer cell types through the activation of Akt, Bax, or other pro-survival molecules (Lu et al. 2006; McJilton et al. 2003; Okhrimenko et al. 2005), and it can signal to mitogenesis via Raf/MEK/ERK and cyclin D1 induction (Kampfer et al. 2001; Soh and Weinstein 2003; Schönwasser et al. 1998; Slupsky et al. 2007). On the other hand, PKCδ is mostly a negative regulator of the cell cycle or drives apoptotic responses triggered by a variety of stimuli, including phorbol esters and chemotherapeutic drugs, and has been linked to the DNA damage response (Gonzalez-Guerrico and Kazanietz 2005; Yoshida 2007; Fukumoto et al. 1997; Nakagawa et al. 2005). Positive roles for PKCδ in proliferation have also been shown in some models such as mammary cell lines (Kiley et al. 1999; Grossoni et al. 2007). The last years have witnessed major advances in establishing roles for PKC isozymes in Wnt and Hedgehog signaling (Riobo et al. 2006; Dissanayake and Weeraratna 2008; Cai et al. 2009; Koyanagi et al. 2009). Conceivably, PKC activation may have profound impact on cancer stem cell biology, an area that has not been exhaustively studied. There is a great need to ascertain the roles for individual PKC isozymes in different cancer types as well as to dissect the molecular basis for the heterogeneity in PKC isozyme function in cancer models.

The pattern of expression of PKC isozymes is profoundly altered in various types of cancers, potentially reflecting their involvement in disease progression (Griner and Kazanietz 2007). For example, a significant reduction in PKCβ expression together with a reciprocal increase in expression of PKCε has been found in human prostate cancer specimens (Cornford et al. 1999). High-grade human prostate tumors express very high PKCε levels relative to low-grade tumors or benign prostatic hyperplasia (Aziz et al. 2007). Notably, de-regulation of PKCε expression or activation has been reported in other cancer types, such as in lung, breast, and thyroid cancer (Knauf et al. 1999; Pan et al. 2005; Bae et al. 2007). All in all, PKCε represents an attractive target for cancer therapy. It is unclear at this stage whether PKCε overexpression as that observed in human prostate or lung tumors has any causal relationship with the initiation and/or progression of the disease, largely because of the limited number of animal models developed so far. Some interesting advances have been made using skin cancer animal models. Studies have shown that PKCδ skin transgenic mice are resistant to tumor promotion in a typical DMBA-phorbol ester paradigm, while PKCε skin transgenic mice are highly susceptible to develop metastatic squamous cell carcinoma in the skin (Reddig et al. 1999; Jansen et al. 2001). The development of conditional and inducible PKC mouse models should greatly help to establish the involvement of specific PKC isozymes in the initiation and maintenance of the malignant phenotype as well as potential cooperations with oncogenic inputs such as Ras and PI3K mutations, or Pten deficiency. Likewise, despite the multiple *in vitro* studies suggesting important roles for PKC isozymes in migration, metalloprotease secretion, and angiogenesis (Griner and Kazanietz 2007), models to address these issues in an *in vivo* context need to be developed.

The next series of chapters will focus on the analysis of PKC isozymes in cancer models and pathways that are highly relevant for cancer development. Skin, breast, lung, and prostate cancer will be presented as paradigms, but the involvement of PKC isozymes in many other cancers is also well established.

References

Aziz, M. H., Manoharan, H. T., Church, D. R., Dreckschmidt, N. E., Zhong, W., Oberley, T. D., et al. (2007). Protein kinase C epsilon interacts with signal transducers and activators of transcription 3 (Stat3), phosphorylates Stat3Ser727, and regulates its constitutive activation in prostate cancer. *Cancer Research, 67*, 8828–8838.

Bae, K. M., Wang, H., Jiang, G., Chen, M. G., Lu, L., & Xiao, L. (2007). Protein kinase C epsilon is overexpressed in primary human non-small cell lung cancers and functionally required for proliferation of non-small cell lung cancer cells in a p21/Cip1-dependent manner. *Cancer Research, 67*, 6053–6063.

Cai, Q., Li, J., Gao, T., Xie, J., & Evers, B. M. (2009). Protein kinase C delta negatively regulates Hedgehog signaling by inhibition of Gli1 activity. *J Biol Chem, 284*, 2150–2158.

Cornford, P., Evans, J., Dodson, A., Parsons, K., Woolfenden, A., Neoptolemos, J., et al. (1999). Protein kinase C isoenzyme patterns characteristically modulated in early prostate cancer. *The American Journal of Pathology, 154*, 137–144.

Detjen, K. M., Brembeck, F. H., Welzel, M., Kaiser, A., Haller, H., Wiedenmann, B., et al. (2000). Activation of protein kinase Calpha inhibits growth of pancreatic cancer cells via p21(cip)-mediated G(1) arrest. *Journal of Cell Science, 113*, 3025–3035.

Dissanayake, S. K., & Weeraratna, A. T. (2008). Detecting PKC phosphorylation as part of the Wnt/calcium pathway in cutaneous melanoma (2008). *Methods in Molecular Biology, 468*, 157–172.

Frey, M. R., Clark, J. A., Leontieva, O., Uronis, J. M., Black, A. R., & Black, J. D. (2000). Protein kinase C signaling mediates a program of cell cycle withdrawal in the intestinal epithelium. *The Journal of Cell Biology, 151*, 763–778.

Fukumoto, S., Nishizawa, Y., Hosoi, M., Koyama, H., Yamakawa, K., Ohno, S., et al. (1997). Protein kinase C delta inhibits the proliferation of vascular smooth muscle cells by suppressing G1 cyclin expression. *The Journal of Biological Chemistry, 272*, 13816–13822.

Gonzalez-Guerrico, A. M., & Kazanietz, M. G. (2005). Phorbol ester-induced apoptosis in prostate cancer cells via autocrine activation of the extrinsic apoptotic cascade: a key role for protein kinase C delta. *The Journal of Biological Chemistry, 280*, 38982–38991.

Griner, E. M., & Kazanietz, M. G. (2007). Protein kinase C and other diacylglycerol effectors in cancer. *Nature Reviews Cancer, 7*, 281–294.

Grossoni, V. C., Falbo, K. B., Kazanietz, M. G., de Kier Joffe, E. D., & Urtreger, A. J. (2007). Protein kinase C delta enhances proliferation and survival of murine mammary cells. *Molecular Carcinogenesis, 46*, 381–390.

Jansen, A. P., Verwiebe, E. G., Dreckschmidt, N. E., Wheeler, D. L., Oberley, T. D., & Verma, A. K. (2001). Protein kinase C-epsilon transgenic mice: a unique model for metastatic squamous cell carcinoma. *Cancer Research, 61*, 808–812.

Jiang, X. H., Tu, S. P., Cui, J. T., Lin, M. C. M., Xia, H. H. X., Wong, W. M., et al. (2004). Antisense targeting protein kinase C alpha and beta1 inhibits gastric carcinogenesis. *Cancer Research, 64*, 5787–5794.

Kampfer, S., Windegger, M., Hochholdinger, F., Schwaiger, W., Pestell, R. G., Baier, G., et al. (2001). Protein kinase C isoforms involved in the transcriptional activation of cyclin D1 by transforming Ha-Ras. *The Journal of Biological Chemistry, 276*, 42834–42842.

Kiley, S. C., Clark, K. J., Duddy, S. K., Welch, D. R., & Jaken, S. (1999). Increased protein kinase C delta in mammary tumor cells: relationship to transformation and metastatic progression. *Oncogene, 18*, 6748–6757.

Knauf, J. A., Elisei, R., Mochly-Rosen, D., Liron, T., Chen, X.-N., Gonsky, R., et al. (1999). Involvement of protein kinase Cepsilon (PKCepsilon) in thyroid cell death. A truncated chimeric PKCepsilon cloned from a thyroid cancer cell line protects thyroid cells from apoptosis. *The Journal of Biological Chemistry, 274*, 23414–23425.

Koyanagi, M., Iwasaki, M., Haendeler, J., Leitges, M., Zeiher, A. M., & Dimmeler, S. (2009). Wnt5a increases cardiac gene expressions of cultured human circulating progenitor cells via a PKC delta activation. *PLoS ONE, 4*, e5765.

Lu, D., Huang, J., & Basu, A. (2006). Protein kinase C epsilon activates protein kinase B/Akt via DNA-PK to protect against tumor necrosis factor-alpha-induced cell death. *The Journal of Biological Chemistry, 281*, 22799–22807.

McJilton, M. A., Van Sikes, C., Wescott, G. G., Wu, D., Foreman, T. L., Gregory, C. W., et al. (2003). Protein kinase Cepsilon interacts with Bax and promotes survival of human prostate cancer cells. *Oncogene, 22*, 7958–7968.

Mischak, H., Goodnight, J. A., Kolch, W., Martiny-Baron, G., Schaechtle, C., Kazanietz, M. G., et al. (1993). Overexpression of protein kinase C-delta and -epsilon in NIH 3T3 cells induces opposite effects on growth, morphology, anchorage dependence, and tumorigenicity. *The Journal of Biological Chemistry, 268*, 6090–6096.

Nakagawa, M., Oliva, J. L., Kothapalli, D., Fournier, A., Assoian, R. K., & Kazanietz, M. G. (2005). Phorbol ester-induced G1 phase arrest selectively mediated by protein kinase C delta-dependent induction of p21. *The Journal of Biological Chemistry, 280*, 33926–33934.

Okhrimenko, H., Lu, W., Xiang, C., Hamburger, N., Kazimirsky, G., & Brodie, C. (2005). Protein kinase C-epsilon regulates the apoptosis and survival of glioma cells. *Cancer Research, 65*, 7301–7309.

Pan, Q., Bao, L. W., Kleer, C. G., Sabel, M. S., Griffith, K. A., Teknos, T. N., et al. (2005). Protein kinase C epsilon is a predictive biomarker of aggressive breast cancer and a validated target for RNA interference anticancer therapy. *Cancer Research, 65*, 8366–8371.

Perletti, G. P., Folini, M., Lin, H. C., Mischak, H., Piccinini, F., & Tashjian, A. H., Jr. (1996). Overexpression of protein kinase C epsilon is oncogenic in rat colonic epithelial cells. *Oncogene, 12*, 847–854.

Perletti, G. P., Marras, E., Concari, P., Piccinini, F., & Tashjian, A. H., Jr. (1999). PKC delta acts as a growth and tumor suppressor in rat colonic epithelial cells. *Oncogene, 18*, 1251–1256.

Reddig, P. J., Dreckschmidt, N. E., Ahrens, H., Simsiman, R., Tseng, C.-P., Zou, J., et al. (1999). Transgenic mice overexpressing protein kinase Cdelta in the epidermis are resistant to skin tumor promotion by 12-O-tetradecanoylphorbol-13-acetate. *Cancer Research, 59*, 5710–5718.

Riobo, N. A., Haines, G. M., & Emerson, C. P., Jr. (2006). Protein kinase C-delta and mitogen-activated protein/extracellular signal-regulated kinase-1 control GLI activation in hedgehog signaling. *Cancer Research, 66*, 839–845.

Schönwasser, D. C., Marais, R. M., Marshall, C. J., & Parker, P. J. (1998). Activation of the mitogen-activated protein kinase/extracellular signal-regulated kinase pathway by conventional, novel, and atypical protein kinase C isotypes. *Molecular and Cellular Biology, 18*, 790–798.

Sharma, G. D., Kakazu, A., & Bazan, H. E. (2007). Protein kinase C alpha and epsilon differentially modulate hepatocyte growth factor-induced epithelial proliferation and migration. *Experimental Eye Research, 85*, 289–297.

Slupsky, J. R., Kamiguti, A. S., Harris, R. J., Cawley, J. C., & Zuzel, M. (2007). Central role of protein kinase Cepsilon in constitutive activation of ERK1/2 and Rac1 in the malignant cells of hairy cell leukemia. *The American Journal of Pathology, 170*, 745–754.

Soh, J. W., & Weinstein, I. B. (2003). Roles of specific isoforms of protein kinase C in the transcriptional control of cyclin D1 and related genes. *The Journal of Biological Chemistry, 278*, 34709–34716.

Sun, X. G., & Rotenberg, S. A. (1999). Overexpression of protein kinase Calpha in MCF-10A human breast cells engenders dramatic alterations in morphology, proliferation, and motility. *Cell Growth & Differentiation, 10*, 343–352.

Wu, T. T., Hsieh, Y. H., Hsieh, Y. S., & Liu, J. Y. (2008). Reduction of PKC alpha decreases cell proliferation, migration, and invasion of human malignant hepatocellular carcinoma. *Journal of Cellular Biochemistry, 103*, 9–20.

Yoshida, K. (2007). PKCdelta signaling: mechanisms of DNA damage response and apoptosis. *Cellular Signalling, 19*, 892–901.

Chapter 12
Protein Kinase C, p53, and DNA Damage

Kiyotsugu Yoshida

Abstract The cellular response to genotoxic stress that damages DNA includes cell cycle arrest, activation of DNA repair, and in the event of irreparable damage, induction of apoptosis. However, the signals that determine cell fate, that is, survival or apoptosis, are largely unknown. The protein kinase C (PKC) has been implicated in many important cellular processes, including regulation of apoptotic cell death. In particular, δ isoform of PKC (PKCδ) contributes to the induction of apoptosis in response to DNA damage. When cells encounter genotoxic stress, certain sensors for DNA lesions activate PKCδ. PKCδ is then proteolytically activated by caspase-3, and the cleaved catalytic fragment translocates to the nucleus and induces apoptosis. Importantly, nuclear targeting of PKCδ is essential for induction of apoptosis. In this regard, PKCδ regulates transcription by phosphorylating various transcription factors, including the p53 tumor suppressor that is critical for cell cycle arrest and apoptosis in response to DNA damage. These findings collectively support a pivotal role for PKCδ in the induction of apoptosis with significant impact. This review is focused on the current views regarding the regulation of cell fate by PKCδ signaling and p53 in response to DNA damage.

Keywords Apoptosis • DNA damage • Nuclear targeting • PKC • p53 • Phosphorylation

12.1 Introduction

Genotoxic stress that damages DNA induces cell cycle arrest, activation of DNA repair, and in the event of irreparable damage, induction of apoptosis. The decision of cells either to repair DNA lesions and continue through the cell cycle or to

K. Yoshida (✉)
Medical Research Institute, Tokyo Medical and Dental University, Tokyo 113-8510, Japan
e-mail: yos.mgen@mri.tmd.ac.jp

M.G. Kazanietz (ed.), *Protein Kinase C in Cancer Signaling and Therapy*,
Current Cancer Research, DOI 10.1007/978-1-60761-543-9_12,

undergo apoptosis is relevant to the incidence of mutagenesis and, subsequently, carcinogenesis. In this context, incomplete repair of DNA damage prior to replication or mitosis can result in the accumulation of heritable genetic changes. Therapeutic anticancer treatments that use DNA-damaging agents must strike a balance between induction of repair and apoptosis in order to maximize the therapeutic effect. However, the nature of the cellular signaling response that determines cell survival or cell death is far from being understood. Certain insights have been derived from the finding that diverse isozymes of the protein kinase C (PKC) family are activated in response to DNA damage. PKC signal transduction pathway regulates cell fate following genotoxic stress (Yoshida 2007a, 2008a). More importantly, recent studies have shown that a certain isozyme of PKC controls function of the p53 tumor suppressor to induce apoptosis. In the past 10 years, understanding of the cellular mechanisms of apoptosis mediated by PKC has advanced considerably, and the primary focus of this review is to provide an overview of PKC and p53, its mode of action, and its physiological role in genotoxic stress-induced apoptosis.

12.2 Protein Kinase C

The protein kinase C (PKC) family of serine-threonine kinases was first described as a calcium-activated, phospholipid-dependent serine/threonine protein kinase (Takai et al. 1977). PKC is activated diacylglycerol (DAG) hydrolyzed from phosphatidylinositol (PI) by phospholipase C (PLC) under a different cell-signaling system (Nishizuka 1984, 1988, 1992, 1995). It has attracted attention as an intracellular receptor for tumor-promotor phorbol esters, such as 12-O-tetradecanoyl-13-phorbol acetate (TPA) (Niedel et al. 1983). Although PKC had been recognized as a protein kinase, subsequent studies have revealed that it belongs to a family of serine/threonine-specific protein kinases and is activated by diverse stimuli and participates in a variety of cellular processes, such as growth, differentiation, apoptosis, and cellular senescence (Casabona 1997; Clemens et al. 1992; Goodnight et al. 1994; Hofmann 1997; Hug and Sarre 1993; Nishizuka 1984, 1988; 1992, 1995). PKC consists of at least 11 isozymes (α, βI, βII, γ, δ, ε, ζ, η, θ, ι/λ, and μ) with selective tissue distribution, activators, and substrates. PKC isozymes have been categorized into three groups: (1) the classical/conventional PKCs (cPKCs: α, βI, βII γ), which are calcium-dependent and activated by DAG; (2) the novel PKCs (nPKCs: δ, ε, θ, μ), which are calcium-independent and activated by DAG; and (3) the atypical PKCs (aPKCs: ζ, λ), which are calcium-independent and not activated by DAG (Casabona 1997; Goodnight et al. 1994; Hug and Sarre 1993; Nishizuka 1988, 1992, 1995). The cell-specific expression and subcellular localization of individual PKC isozymes indicate important isozyme-specific functions. To elucidate these functions, it will be necessary to study in vitro and/or in vivo the individual features of each isozyme, such as expression, posttranslational modification, substrate specificity, subcellular localization, and signaling cross-talk with other proteins.

Moreover, the involvement of a PKC isozyme in a signaling pathway resulting in a specific cellular response can be investigated by several distinct methods such as overexpression of the enzyme or inhibition of enzyme expression or activity.

12.3 PKC and Apoptosis in Response to Genotoxic Stress

Novel PKCδ, θ, and μ are substrates for the effector caspase-3, and proteolytic activation of these novel PKCs has been associated with cell death (Datta et al. 1997; Emoto et al. 1995; Endo et al. 2000). However, recent studies have shown that PKC acts upstream of caspases to regulate cell death. For example, PKC activators enhanced caspase activation, whereas an inhibitor of PKC prevented caspase activation in response to DNA damage (Basu et al. 2001). In particular, studies with PKCδ$^{-/-}$ mice suggest that PKCδ plays pivotal roles in the regulation of cell proliferation and apoptosis (Humphries et al. 2006; Leitges et al. 2001). PKCδ is activated by a variety of stimuli including ionizing radiation, anticancer agents, reactive oxygen species (ROS), ultraviolet radiation, growth factors, and cytokines (Carpenter et al. 2002; Chen et al. 1999; Denning et al. 1996; Konishi et al. 2001; Reyland et al. 1999; Yoshida and Kufe 2001; Yoshida et al. 2002). Molecular mechanisms such as tyrosine phosphorylation and proteolytic cleavage by caspase-3 are of importance for understanding the proapoptotic role for PKCδ activation. PKC isozymes have long been implicated in the growth factor signal transduction pathway (Nishizuka 1992). By contrast, activation of PKCδ is associated with inhibition of cell cycle progression and down-regulation of PKCδ is linked to tumor promotion, suggesting that PKCδ may have a negative effect on cell survival (Lu et al. 1997; Watanabe et al. 1992). In many cases, the growth-inhibitory effects of PKCδ have been linked to changes in the expression of factors that influence cell cycle progression. Furthermore, PKCδ plays an essential role in the genotoxic stress response leading to apoptotic cell death in various cell types (Brodie and Blumberg 2003; Reyland 2007; Yoshida 2007a). In addition, cells derived from PKCδ$^{-/-}$ mice were shown to be defective in mitochondria-dependent apoptosis (Humphries et al. 2006; Leitges et al. 2001). These findings collectively support our proposition of a progenotoxic role of PKCδ. Mechanistically, PKCδ is activated in response to numerous cellular stimuli by various mechanisms, including membrane translocation (Joseloff et al. 2002; Wang et al. 1999), protein–protein interaction (Benes et al. 2005), tyrosine phosphorylation (Denning et al. 1996; Kaul et al. 2005), and proteolytic cleavage (Emoto et al. 1995; Ghayur et al. 1996; Yoshida 2007a; Yoshida et al. 2003). The translocation of PKCδ to different subcellular compartments and/or proteolytic cleavage can be induced by ceramide, TNFα, UV irradiation, ionizing radiation, oxidative stress, and etoposide (DeVries et al. 2002; Majumder et al. 2000; Matassa et al. 2001; Reyland et al. 1999; Yamaguchi et al. 2007b; Yoshida 2007a; Yoshida et al. 2002, 2003, 2006a, b). Importantly, recent studies have shown that DNA-damage-induced PKCδ activation is in part dependent on Ataxia telangiectasia mutated (ATM) (Yoshida et al. 2003). While ATM activates c-Abl, and c-Abl activates PKCδ, a potential explanation is that DNA damage induces an ATM→c-Abl→PKCδ pathway (Yoshida 2007b; Yoshida and Miki 2005; Yoshida et al. 2005). Alternatively, ATM may directly activate PKCδ in the DNA damage response. In any case, nuclear targeting of PKCδ is required for ATM-mediated full activation of PKCδ in the DNA damage response.

12.4 Nuclear Translocation of PKCδ During Apoptotic Responses

Translocation of PKCδ into the nucleus has been demonstrated in various cells (Blass et al. 2002; DeVries et al. 2002; DeVries-Seimon et al. 2007; Eitel et al. 2003; Scheel-Toellner et al. 1999; Yoshida et al. 2003; Yuan et al. 1998). Recent study demonstrated that PKCδ translocates into the nucleus following exposure of cells with 1-β-D-arabinofuranosylcytosine (ara-C) (Yoshida et al. 2003). Moreover, pretreatment with rottlerin attenuates nuclear targeting of PKCδ (Yoshida et al. 2003), suggesting that its kinase activity is required for nuclear translocation. A putative nuclear localization signal has been identified at the C-terminus of the catalytic domain of PKCδ (DeVries et al. 2002). Numerous PKCδ targets and substrates, including the p53 tumor suppressor, are nuclear proteins that function in apoptotic cell death.

12.5 Role for p53 in Response to DNA Damage

The tumor suppressor protein p53 plays a central role in mediating stress and DNA damage-induced growth arrest and apoptosis (Vogelstein et al. 2000). The p53 protein regulates normal responses to DNA damage and other forms of genotoxic stress and is a key element in maintaining genomic stability (Vogelstein et al. 2000). In fact, the p53 tumor suppressor gene is the most frequently inactivated gene in human malignancy (Nigro et al. 1989). The level of p53 protein is largely undetectable in normal cells but rapidly increases in response to a variety of stress signals. The mechanism by which the p53 protein is stabilized is not completely understood, but posttranslational modification plays a pivotal role (Shieh et al. 1997). Mutations in the p53 gene are frequently associated with the formation of human cancer; however, the p53 pathway can be also derailed by numerous oncogenic proteins (Oren et al. 2002). Mice engineered to have the p53 gene knocked out develop tumors at an increased rate (Donehower et al. 1992). It is plausible that many agents may inhibit the p53 pathway as part of the road toward tumor promotion. However, the mechanisms of action of many chemical agents that promote tumor development have not been elucidated. With the central role of p53 in mind, it is logical to presume that agents that promote tumor formation might block the p53 pathway. Importantly, p53 is regulated primarily through posttranslational modifications, especially phosphorylation, and the accumulation of p53 is the first step in response to cellular stress (Oren 1999). The *mdm2* gene is a transcriptional target of p53, and once synthesized, the MDM2 protein can bind to p53 at its NH2 terminus leading to its rapid degradation through the ubiquitination and proteasome-mediated pathway (Kubbutat and Vousden 1998; Oren 1999; Ryan et al. 2001). In response to DNA damage, p53 becomes phosphorylated at multiple sites at the NH2 terminus, thereby inhibiting MDM2 binding (Burns and El-Deiry 1999; Canman et al. 1998; Kubbutat and Vousden 1998; Oren 1999; Ryan et al. 2001; Siliciano et al. 1997). As a result, p53 degradation does not occur and p53 accumulates. p53 can also be phosphorylated at its COOH-terminal regulatory domain, which influences its DNA binding (Meek 1998). It has been reported recently that constitutive phosphorylation of p53 by PKC at its COOH-terminal domain can lead to its degradation via the ubiquitination and proteasome-mediated pathway (Chernov et al. 2001). Treatment of mouse or human fibroblasts with PKC inhibitors, such as H7 or bisindolylmaleimide I, inhibited COOH-terminal phosphorylation of p53 and increased accumulation of p53 without affecting the formation of the p53-MDM2 complex (Chernov et al. 2001). However, PKC inhibitors were unable to increase accumulation of p53 in HPV-positive HeLa cells (Chernov et al. 1998, 2001).

12.6 Regulation of p53 by PKCδ

The p53 tumor suppressor is activated in the cellular response to genotoxic stress. Transactivation of p53 target genes dictates cell cycle arrest and DNA repair or induction of apoptosis. Recent studies have demonstrated that PKCδ controls expression of the p53 at the transcriptional and posttranslational levels.

12.6.1 Transcriptional Regulation

Recent reports document that PKCδ transactivates p53 expression at the transcriptional level (Abbas et al. 2004; Liu et al. 2007; Yoshida 2008a). The tumor-promoting phorbol ester 12-O-tetradecanoylphorbol-13-acetate (TPA) prevents DNA damage-induced up-regulation of p53 by down-regulating PKCδ. TPA promotes tumor formation in a variety of mice and tissue culture models, and this has been associated with the down-regulation of PKC (Hansen et al. 1990). TPA is known to activate but then down-regulate the diacylglycerol-dependent PKC isoforms (Fournier and Murray 1987; Hansen et al. 1990). Previous studies demonstrated that the tumor-promoting activities of TPA are mediated at least in part by down-regulating PKCδ (Lu et al. 1997). Moreover, transgenic mice overexpressing PKCδ in their epidermis are resistant to tumor promotion by TPA (Reddig et al. 1999). Previous studies have suggested that TPA can inhibit the DNA damage-mediated induction of p53 (Magnelli et al. 1995). Moreover, other studies with protein kinase inhibitors have suggested that PKCδ regulates the p53 pathway (Ghosh et al. 1999). Regulation of p53 in response to stress most commonly occurs by preventing ubiquitination and degradation of the p53 protein. By contrast, suppression of p53 expression by inhibition of PKCδ is caused by the inhibition of p53 synthesis, not increased degradation of p53 protein. Inhibiting PKCδ blocks both basal transcription of the human p53 gene and initiation of transcription from the human p53 promoter. The DNA damage-induced increase in p53 accumulation is dramatically inhibited by pretreatment of cells with the tumor promoter TPA. In addition, the PKCδ inhibitor, rottlerin, is also able to block the DNA damage-mediated induction of p53. More importantly, pretreatment of cells with TPA or treatment with rottlerin results in the inhibition of basal p53 transcription. In this regard, accumulation of p53 could not be achieved by any means, including proteosome inhibition, after TPA or rottlerin treatment, because p53 transcription is blocked. It is thus conceivable that the tumor-suppressing effects of PKCδ are mediated at least in part through activating p53 transcription. Repression of the p53 promoter has been suggested as a mechanism for tumor promotion (Raman et al. 2000; Stuart et al. 1995). Damaged genes in tumor cells are usually thought to be the mechanistic drivers toward oncogenesis. However, down-regulation of endogenous genes, specifically tumor suppressors, may be also a key regulatory mechanism resulting in tumor promotion. A transcriptional repression mechanism for tumor promotion by TPA predicts that agents that interfere with the activity of PKCδ may inhibit p53 responses.

Recent study also demonstrated that PKCδ induces the promoter activity of p53 through the p53 core promoter element (CPE-p53) and that such induction is enhanced in response to DNA damage. Upon exposure to genotoxic stress, PKCδ activates and interacts with the death-promoting transcription factor Btf (Bcl-2-associated transcription factor) to co-occupy CPE-p53. Inhibition of PKCδ activity decreases the affinity of Btf for CPE-p53, thereby reducing p53 expression at both the mRNA and protein levels. In concert with these results, disruption of Btf-mediated p53 transcription by RNA interference leads to suppression of p53-mediated apoptosis following genotoxic stress. These findings provide evidence that activation of p53 transcription by PKCδ triggers p53-dependent apoptosis in response to DNA damage (Fig. 12.1) (Liu et al. 2007).

12.6.2 Posttranslational Regulation

Recent study demonstrated that both PKCδ and IKKα, but not IKKβ, translocate to the nucleus in response to oxidative stress (Yamaguchi et al. 2007a, b). PKCδ interacts with and activates IKKα. Importantly, the data suggest that, upon exposure to oxidative stress, PKCδ-mediated IKKα activation does not contribute to NF-κB activation; instead, nuclear IKK regulates p53 transcription activity by phosphorylation at Ser20. These findings collectively support a novel mechanism

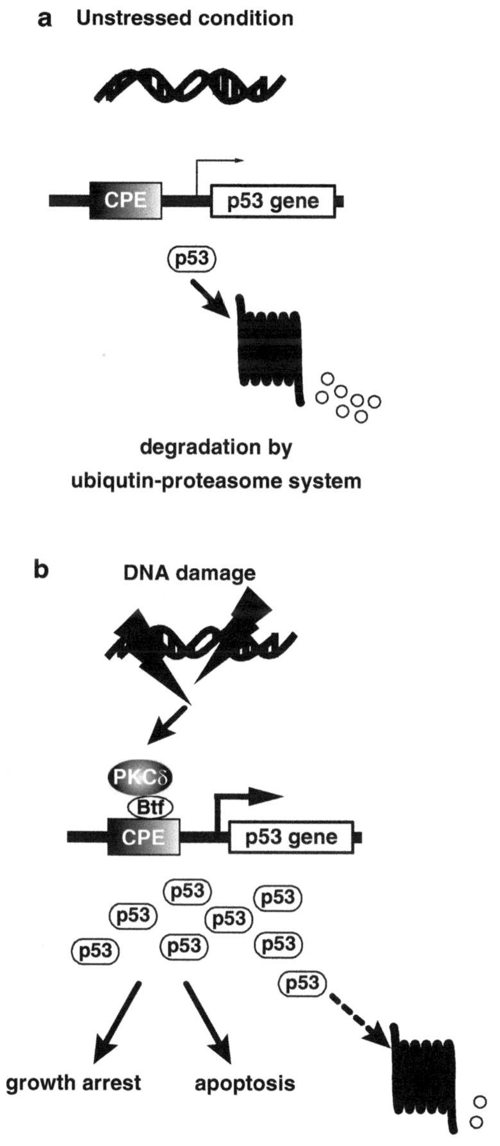

Fig. 12.1 Schematic depiction of the regulation for p53 gene transcription by PKCδ and Btf. (**a**) In control cells, expressed p53 is immediately degraded by MDM2-mediated ubiquitination and proteasome system. (**b**) Upon exposure to genotoxic stress, activated PKCδ induces Btf for co-occupancy to CPE-p53, thereby up-regulating the expression of p53 at mRNA levels. p53 is also stabilized at protein levels by abrogating ubiquitin-dependent degradation

in which the PKCδ→IKKα signaling pathway contributes to ROS-induced p53 activation. Recent studies have also revealed that phosphorylation of p53 on Ser46 is associated with the induction of p53AIP1 expression, resulting in the commitment of the cell fate to apoptotic cell death (Matsuda et al. 2002;

Oda et al. 2000; Taira et al. 2007; Yoshida 2008b). Moreover, upon exposure to genotoxic stress, p53DINP1 is expressed and then recruits a kinase(s) to p53 that specifically phosphorylates Ser46 (Okamura et al. 2001). Our recent data showed that PKCδ is involved in phosphorylation of p53 on Ser46 (Yoshida et al. 2006a). PKCδ-mediated phosphorylation was required for the interaction of PKCδ with p53. The results also demonstrated that p53DINP1 associates with PKCδ upon exposure to genotoxic agents. In concert with these results, PKCδ potentiates p53-dependent apoptosis by Ser46 phosphorylation in response to genotoxic stress. These findings indicate that PKCδ regulates p53 to induce apoptotic cell death in the cellular response to DNA damage (Yoshida et al. 2006a). We also recently found that PKCδ regulates MDM2 expression by controlling

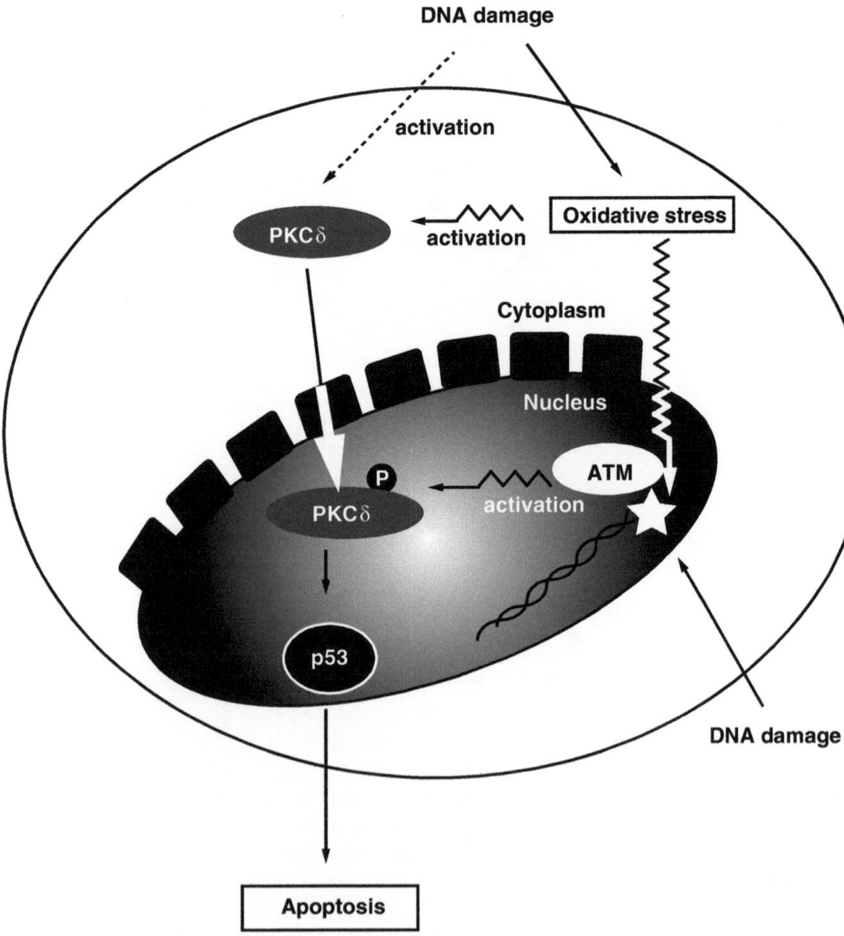

Fig. 12.2 A hypothetical schema for nuclear targeting of PKCδ in response to DNA damage. Following DNA damage, PKCδ translocates from the cytoplasm into the nucleus. In addition, some genotoxic stress also exerts cytoplasmic oxidative stress to activate PKCδ. In the nucleus, PKCδ is activated by ATM, then induces apoptosis in a p53-dependent manner

Akt-mediated phosphorylation. As a result, PKCδ could control p53 expression indirectly by modulating MDM2 function in response to DNA damage (Hew HC and Yoshida K unpublished observation).

12.7 Concluding Remarks

PKCδ plays an essential role in the regulation of apoptosis in response to a large and diverse array of apoptotic stimuli. PKCδ is thus a proapoptotic kinase activated by multiple mechanisms, including subcellular translocation and proteolysis. The proteolytic activation of PKCδ is also important not only in activating the downstream apoptotic cascade including p53, but also in amplifying upstream caspase signaling. Most of the studies mentioned above suggest that the role for PKCδ in the induction of apoptosis involves its caspase-dependent cleavage and the regulation of p53. However, functional control of p53 by PKCδ remains largely obscure. In this context, thorough investigation coupled with PKCδ and p53 should be accelerated from multiple aspects. In the encounter with DNA damage, ATM contributes to various cellular responses, such as growth arrest, transcription, DNA repair, and apoptosis. Importantly, genotoxic stress-induced PKCδ is controlled under ATM, suggesting the notion that establishment of the ATM→PKCδ→p53 signaling cascade could confer new mechanistic light on how PKCδ functions as a proapoptotic kinase in the nucleus (Fig. 12.2) (Yoshida 2007a, Yoshida 2008a). While deregulation of the PKCδ signaling pathway can contribute to anticancer drug resistance (Meinhardt et al. 1999), there is little understanding of how the PKCδ signaling pathway is affected when cancer cells acquire resistance to chemotherapeutic drugs. Considering the importance of PKCδ in DNA damage-induced apoptosis, a thorough understanding of how it regulates apoptosis should benefit cancer therapy. Moreover, novel PKCδ-based therapy may well be used in combination with other agents to create synergism and help prevent the development of drug resistance.

Acknowledgments This work was supported by grants from the Ministry of Education, Science and Culture of Japan, Kato Memorial Bioscience Foundation, and the Cell Science Research Foundation.

References

Abbas, T., White, D., Hui, L., Yoshida, K., Foster, D. A., & Bargonetti, J. (2004). Inhibition of human p53 basal transcription by down-regulation of protein kinase Cδ. *The Journal of Biological Chemistry*, *279*, 9970–9977.

Basu, A., Woolard, M. D., & Johnson, C. L. (2001). Involvement of protein kinase C-δ in DNA damage-induced apoptosis. *Cell Death and Differentiation*, *8*, 899–908.

Benes, C. H., Wu, N., Elia, A. E., Dharia, T., Cantley, L. C., & Soltoff, S. P. (2005). The C2 domain of PKCδ is a phosphotyrosine binding domain. *Cell*, *121*, 271–280.

Blass, M., Kronfeld, I., Kazimirsky, G., Blumberg, P. M., & Brodie, C. (2002). Tyrosine phosphorylation of protein kinase Cδ is essential for its apoptotic effect in response to etoposide. *Molecular and Cellular Biology*, *22*, 182–195.

Brodie, C., & Blumberg, P. M. (2003). Regulation of cell apoptosis by protein kinase Cδ. *Apoptosis*, *8*, 19–27.

Burns, T. F., & El-Deiry, W. S. (1999). The p53 pathway and apoptosis. *Journal of Cellular Physiology*, *181*, 231–239.

Canman, C. E., Lim, D. S., Cimprich, K. A., Taya, Y., Tamai, K., Sakaguchi, K., et al. (1998). Activation of the ATM kinase by ionizing radiation and phosphorylation of p53. *Science*, *281*, 1677–1679.

Carpenter, L., Cordery, D., & Biden, T. J. (2002). Inhibition of protein kinase Cδ protects rat INS-1 cells against interleukin-1β and streptozotocin-induced apoptosis. *Diabetes*, *51*, 317–324.

Casabona, G. (1997). Intracellular signal modulation: A pivotal role for protein kinase C. *Progress in Neuro-psychopharmacology & Biological Psychiatry*, *21*, 407–425.

Chen, N., Ma, W., Huang, C., & Dong, Z. (1999). Translocation of protein kinase Cε and protein kinase Cδ to membrane is required for ultraviolet B-induced activation of mitogen-activated protein kinases and apoptosis. *The Journal of Biological Chemistry*, *274*, 15389–15394.

Chernov, M. V., Bean, L. J., Lerner, N., & Stark, G. R. (2001). Regulation of ubiquitination and degradation of p53 in unstressed cells through C-terminal phosphorylation. *The Journal of Biological Chemistry*, *276*, 31819–31824.

Chernov, M. V., Ramana, C. V., Adler, V. V., & Stark, G. R. (1998). Stabilization and activation of p53 are regulated independently by different phosphorylation events. *Proceedings of the National Academy of Sciences of the United States of America*, *95*, 2284–2289.

Clemens, M. J., Trayner, I., & Menaya, J. (1992). The role of protein kinase C isoenzymes in the regulation of cell proliferation and differentiation. *Journal of Cell Science*, *103*(Pt 4), 881–887.

Datta, R., Kojima, H., Yoshida, K., & Kufe, D. (1997). Caspase-3-mediated cleavage of protein kinase Cθ in induction of apoptosis. *The Journal of Biological Chemistry*, *272*, 20317–20320.

Denning, M. F., Dlugosz, A. A., Threadgill, D. W., Magnuson, T., & Yuspa, S. H. (1996). Activation of the epidermal growth factor receptor signal transduction pathway stimulates tyrosine phosphorylation of protein kinase Cδ. *The Journal of Biological Chemistry*, *271*, 5325–5331.

DeVries, T. A., Neville, M. C., & Reyland, M. E. (2002). Nuclear import of PKCδ is required for apoptosis: Identification of a novel nuclear import sequence. *The EMBO Journal*, *21*, 6050–6060.

DeVries-Seimon, T. A., Ohm, A. M., Humphries, M. J., & Reyland M. E. (2007). Induction of apoptosis is driven by nuclear retention of protein kinase Cδ. *The Journal of Biological Chemistry*, *282*, 22307–22314.

Donehower, L. A., Harvey, M., Slagle, B. L., McArthur, M. J., Montgomery, C. A. Jr, Butel, J. S., & Bradley, A. (1992). Mice deficient for p53 are developmentally normal but susceptible to spontaneous tumours. *Nature*, *356*, 215–221.

Eitel, K., Staiger, H., Rieger, J., Mischak, H., Brandhorst, H., Brendel, M.D., et al. (2003). Protein kinase Cδ activation and translocation to the nucleus are required for fatty acid-induced apoptosis of insulin-secreting cells. *Diabetes*, *52*, 991–997.

Emoto, Y., Manome, Y., Meinhardt, G., Kisaki, H., Kharbanda, S., Robertson, M., Ghayur, T., et al. (1995). Proteolytic activation of protein kinase Cδ by an ICE-like protease in apoptotic cells. *The EMBO Journal*, *14*, 6148–6156.

Endo, K., Oki, E., Biedermann, V., Kojima, H., Yoshida, K., Johannes, F. J., et al. (2000). Proteolytic cleavage and activation of protein kinase Cμ by caspase-3 in the apoptotic response of cells to 1-β-D-arabinofuranosylcytosine and other genotoxic agents. *The Journal of Biological Chemistry*, *275*, 18476–18481.

Fournier, A., & Murray, A. W. (1987). Application of phorbol ester to mouse skin causes a rapid and sustained loss of protein kinase C. *Nature*, *330*, 767–769.

Ghayur, T., Hugunin, M., Talanian, R. V., Ratnofsky, S., Quinlan, C., Emoto, Y., et al. (1996). Proteolytic activation of protein kinase Cδ by an ICE/CED 3-like protease induces characteristics of apoptosis. *The Journal of Experimental Medicine*, *184*, 2399–2404.

Ghosh, J. C., Suzuki, K., Kodama, S., & Watanabe, M. (1999). Effects of protein kinase inhibitors on the accumulation kinetics of p53 protein in normal human embryo cells following X-irradiation. *Journal of Radiation Research (Tokyo)*, *40*, 23–37.

Goodnight, J., Mischak, H., & Mushinski, J. F. (1994). Selective involvement of protein kinase C isozymes in differentiation and neoplastic transformation. *Advances in Cancer Research*, *64*, 159–209.

Hansen, L. A., Monteiro-Riviere, N. A., & Smart, R. C. (1990). Differential down-regulation of epidermal protein kinase C by 12-O-tetradecanoylphorbol-13-acetate and diacylglycerol: Association with epidermal hyperplasia and tumor promotion. *Cancer Research*, *50*, 5740–5745.

Hofmann, J. (1997). The potential for isoenzyme-selective modulation of protein kinase C. *The FASEB Journal*, *11*, 649–669.

Hug, H., & Sarre, T. F. (1993). Protein kinase C isoenzymes: Divergence in signal transduction? *The Biochemical Journal*, *291*(Pt 2), 329–343.

Humphries, M. J., Limesand, K. H., Schneider, J. C., Nakayama, K. I., Anderson, S. M., & Reyland, M. E. (2006). Suppression of apoptosis in the protein kinase Cδ null mouse in vivo. *The Journal of Biological Chemistry*, *281*, 9728–9737.

Joseloff, E., Cataisson, C., Aamodt, H., Ocheni, H., Blumberg, P., Kraker, A. J., & Yuspa, S. H. (2002). Src family kinases phosphorylate protein kinase Cδ on tyrosine residues and modify the neoplastic phenotype of skin keratinocytes. *The Journal of Biological Chemistry*, *277*, 12318–12323.

Kaul, S., Anantharam, V., Yang, Y., Choi, C. J., Kanthasamy, A., & Kanthasamy, A. G. (2005). Tyrosine phosphorylation regulates the proteolytic activation of protein kinase Cδ in dopaminergic neuronal cells. *The Journal of Biological Chemistry*, *280*, 28721–28730.

Konishi, H., Yamauchi, E., Taniguchi, H., Yamamoto, T., Matsuzaki, H., Takemura, Y., et al. (2001). Phosphorylation sites of protein kinase Cδ in H_2O_2-treated cells and its activation by tyrosine kinase in vitro. *Proceedings of the National Academy of Sciences of the United States of America*, *98*, 6587–6592.

Kubbutat, M. H., & Vousden, K. H. (1998). Keeping an old friend under control: Regulation of p53 stability. *Molecular Medicine Today*, *4*, 250–256.

Leitges, M., Mayr, M., Braun, U., Mayr, U., Li, C., Pfister, G., et al. (2001). Exacerbated vein graft arteriosclerosis in protein kinase Cδ-null mice. *The Journal of Clinical Investigation*, *108*, 1505–1512.

Liu, H., Lu, Z. G., Miki, Y., & Yoshida, K. (2007). Protein kinase Cδ induces transcription of the TP53 tumor suppressor gene by controlling death-promoting factor Btf in the apoptotic response to DNA damage. *Molecular and Cellular Biology*, *2*, 8480–8491.

Lu, Z., Hornia, A., Jiang, Y. W., Zang, Q., Ohno, S., & Foster, D. A. (1997). Tumor promotion by depleting cells of protein kinase Cδ. *Molecular and Cellular Biology*, *17*, 3418–3428.

Magnelli, L., Cinelli, M., & Chiarugi, V. (1995). Phorbol esters attenuate the expression of p53 in cells treated with doxorubicin and protect TS-P53/K562 from apoptosis. *Biochemical and Biophysical Research Communications*, *215*, 641–645.

Majumder, P. K., Pandey, P., Sun, X., Cheng, K., Datta, R., Saxena, S., et al. (2000). Mitochondrial translocation of protein kinase Cδ in phorbol ester-induced cytochrome c release and apoptosis. *The Journal of Biological Chemistry*, *275*, 21793–21796.

Matassa, A. A., Carpenter, L., Biden, T. J., Humphries, M. J., & Reyland, M. E. (2001). PKCδ is required for mitochondrial-dependent apoptosis in salivary epithelial cells. *The Journal of Biological Chemistry*, *276*, 29719–29728.

Matsuda, K., Yoshida, K., Taya, Y., Nakamura, K., Nakamura, Y., & Arakawa, H. (2002). p53AIP1 regulates the mitochondrial apoptotic pathway. *Cancer Research*, *62*, 2883–2889.

Meek, D. W. (1998). Multisite phosphorylation and the integration of stress signals at p53. *Cellular Signalling*, *10*, 159–166.

Meinhardt, G., Roth, J., Totok, G., Auner, H., Emmerich, B., & Hass, R. (1999). Signaling defect in the activation of caspase-3 and PKCδ in human TUR leukemia cells is associated with resistance to apoptosis. *Experimental Cell Research*, *247*, 534–542.

Niedel, J. E., Kuhn, L. J., & Vandenbark, G. R. (1983). Phorbol diester receptor copurifies with protein kinase C. *Proceedings of the National Academy of Sciences of the United States of America*, *80*, 36–40.

Nigro, J. M., Baker, S. J., Preisinger, A. C., Jessup, J. M., Hostetter, R., Cleary, K., et al. (1989). Mutations in the p53 gene occur in diverse human tumour types. *Nature*, *342*, 705–708.

Nishizuka, Y. (1984). The role of protein kinase C in cell surface signal transduction and tumor promotion. *Nature*, *308*, 693–698.

Nishizuka, Y. (1988). The molecular heterogeneity of protein kinase C and its implications for cellular regulation. *Nature*, *334*, 661–665.

Nishizuka, Y. (1992). Intracellular signaling by hydrolysis of phospholipids and activation of protein kinase C. *Science*, *258*, 607–614.

Nishizuka, Y. (1995). Protein kinase C and lipid signaling for sustained cellular responses. *The FASEB Journal*, *9*, 484–496.

Oda, K., Arakawa, H., Tanaka, T., Matsuda, K., Tanikawa, C., Mori, T., et al. (2000). p53AIP1, a potential mediator of p53-dependent apoptosis, and its regulation by Ser-46-phosphorylated p53. *Cell*, *102*, 849–862.

Okamura, S., Arakawa, H., Tanaka, T., Nakanishi, H., Ng, C. C., Taya, Y., et al. (2001). p53DINP1, a p53-inducible gene, regulates p53-dependent apoptosis. *Molecular Cell*, *8*, 85–94.

Oren, M. (1999). Regulation of the p53 tumor suppressor protein. *The Journal of Biological Chemistry*, *274*, 36031–36034.

Oren, M., Damalas, A., Gottlieb, T., Michael, D., Taplick, J., Leal, J. F., et al. (2002). Regulation of p53: Intricate loops and delicate balances. *Biochemical Pharmacology*, *64*, 865–871.

Raman, V., Martensen, S. A., Reisman, D., Evron, E., Odenwald, W. F., Jaffee, E., et al. (2000). Compromised HOXA5 function can limit p53 expression in human breast tumours. *Nature*, *405*, 974–978.

Reddig, P. J., Dreckschmidt, N. E., Ahrens, H., Simsiman, R., Tseng, C. P., Zou, J., et al. (1999). Transgenic mice overexpressing protein kinase Cδ in the epidermis are resistant to skin tumor promotion by 12-O-tetradecanoylphorbol-13-acetate. *Cancer Research*, *59*, 5710–5718.

Reyland, M. E. (2007). Protein kinase Cδ and apoptosis. *Biochemical Society Transactions*, *35*, 1001–1004.

Reyland, M. E., Anderson, S. M., Matassa, A. A., Barzen, K. A., & Quissell, D. O. (1999). Protein kinase Cδ is essential for etoposide-induced apoptosis in salivary gland acinar cells. *The Journal of Biological Chemistry*, *274*, 19115–19123.

Ryan, K. M., Phillips, A. C., & Vousden, K. H. (2001). Regulation and function of the p53 tumor suppressor protein. *Current Opinion in Cell Biology*, *13*, 332–337.

Scheel-Toellner, D., Pilling, D., Akbar, A. N., Hardie, D., Lombardi, G., Salmon, M., & Lord, J.M. (1999). Inhibition of T cell apoptosis by IFN-β rapidly reverses nuclear translocation of protein kinase C-δ. *European Journal of Immunology*, *29*, 2603–2612.

Shieh, S. Y., Ikeda, M., Taya, Y., & Prives, C. (1997). DNA damage-induced phosphorylation of p53 alleviates inhibition by MDM2. *Cell*, *91*, 325–334.

Siliciano, J. D., Canman, C. E., Taya, Y., Sakaguchi, K., Appella, E., & Kastan, M. B. (1997). DNA damage induces phosphorylation of the amino terminus of p53. *Genes & Development*, *11*, 3471–3481.

Stuart, E. T., Haffner, R., Oren, M., & Gruss, P. (1995). Loss of p53 function through PAX-mediated transcriptional repression. *The EMBO Journal*, *14*, 5638–5645.

Taira, N., Nihira, K., Yamaguchi, T., Miki, Y., & Yoshida, K. (2007). DYRK2 is targeted to the nucleus and controls p53 via Ser46 phosphorylation in the apoptotic response to DNA damage. *Molecular Cell*, *25*, 725–738.

Takai, Y., Kishimoto, A., Inoue, M., & Nishizuka, Y. (1977). Studies on a cyclic nucleotide-independent protein kinase and its proenzyme in mammalian tissues. I. Purification and characterization of an active enzyme from bovine cerebellum. *The Journal of Biological Chemistry*, *252*, 7603–7609.

Vogelstein, B., Lane, D., & Levine, A. J. (2000). Surfing the p53 network. *Nature*, *408*, 307–310.

Wang, Q. J., Bhattacharyya, D., Garfield, S., Nacro, K., Marquez, V. E., & Blumberg, P. M. (1999). Differential localization of protein kinase Cδ by phorbol esters and related compounds using a fusion protein with green fluorescent protein. *The Journal of Biological Chemistry, 274*, 37233–37239.

Watanabe, T., Ono, Y., Taniyama, Y., Hazama, K., Igarashi, K., Ogita, K., et al. (1992). Cell division arrest induced by phorbol ester in CHO cells overexpressing protein kinase C-δ subspecies. *Proceedings of the National Academy of Sciences of the United States of America, 89*, 10159–10163.

Yamaguchi, T., Kimura, J., Miki, Y., & Yoshida, K. (2007a). The deubiquitinating enzyme USP11 controls an IKKα-p53 signaling pathway in response to TNFα. *The Journal of Biological Chemistry, 282*, 33943–33948.

Yamaguchi, T., Miki, Y., & Yoshida, K. (2007b). Protein kinase Cδ activates IκB-kinase α to induce the p53 tumor suppressor in response to oxidative stress. *Cellular Signalling, 19*, 2088–2097.

Yoshida, K. (2007a). PKCδ signaling: Mechanisms of DNA damage response and apoptosis. *Cellular Signalling, 19*, 892–901.

Yoshida, K. (2007b). Regulation for nuclear targeting of the Abl tyrosine kinase in response to DNA damage. *Advances in Experimental Medicine and Biology, 604*, 155–165.

Yoshida, K. (2008a). Nuclear trafficking of pro-apoptotic kinases in response to DNA damage. *Trends in Molecular Medicine, 14*, 305–313.

Yoshida, K. (2008b). Role for DYRK family kinases on regulation of apoptosis. *Biochemical Pharmacology, 76*, 1389–1394.

Yoshida, K., & Kufe, D. (2001). Negative regulation of the SHPTP1 protein tyrosine phosphatase by protein kinase Cδ in response to DNA damage. *Molecular Pharmacology, 60*, 1431–1438.

Yoshida, K., Liu, H., & Miki, Y. (2006a). Protein Kinase Cδ regulates Ser46 phosphorylation of p53 tumor suppressor in the apoptotic response to DNA damage. *The Journal of Biological Chemistry, 281*, 5734–5740.

Yoshida, K., & Miki, Y. (2005). Enabling death by the Abl tyrosine kinase: Mechanisms for nuclear shuttling of c-Abl in response to DNA damage. *Cell Cycle, 4*, 777–779.

Yoshida, K., Miki, Y., & Kufe, D. (2002). Activation of SAPK/JNK signaling by protein kinase Cδ in response to DNA damage. *The Journal of Biological Chemistry, 27*, 48372–48378.

Yoshida, K., Wang, H. G., Miki, Y., & Kufe, D. (2003). Protein kinase Cδ is responsible for constitutive and DNA damage-induced phosphorylation of Rad9. *The EMBO Journal, 22*, 1431–1441.

Yoshida, K., Yamaguchi, T., Natsume, T., Kufe, D., & Miki, Y. (2005). JNK phosphorylation of 14-3-3 proteins regulates nuclear targeting of c-Abl in the apoptotic response to DNA damage. *Nature Cell Biology, 7*, 278–285.

Yoshida, K., Yamaguchi, T., Shinagawa, H., Taira, N., Nakayama, K. I., & Miki, Y. (2006b). Protein kinase Cδ activates topoisomerase IIα to induce apoptotic cell death in response to DNA damage. *Molecular and Cellular Biology, 26*, 3414–3431.

Yuan, Z. M., Utsugisawa, T., Ishiko, T., Nakada, S., Huang, Y., Kharbanda, S., et al. (1998). Activation of protein kinase Cδ by the c-Abl tyrosine kinase in response to ionizing radiation. *Oncogene, 16*, 1643–1648.

Chapter 13
PKCs as Mediators of the Hedgehog and Wnt Signaling Pathways

Natalia A. Riobo

Abstract The Hedgehog and Wnt signaling pathways play critical roles in patterning, proliferation and survival during embryonic development. Abnormal activation of these pathways in adult organisms is associated with a wide rage of neoplasias. In this chapter we provide a concise description of the two signaling pathways and their role in cancer and discuss the current knowledge of the contribution of different PKC isozymes in the Hedgehog and Wnt pathways.

Keywords PKC • PKCδ • Hedgehog • Wnt • Non-canonical signaling • Cancer

Abbreviations

CE	Convergent extension
CK-1	Casein kinase-1
DAG	Diacylglycerol
Fz	Frizzled
GI	Gastrointestinal
GSK-3	Glycogen synthase kinase-3
JNK	c-Jun N-terminal kinase
Hh	Hedgehog
MEF	Mouse embryonic fibroblast
MEK-1	Mitogen extracellular signal-activated kinase 1
PCP	Planar cell polarity
PKC	Protein kinase C
PLC	Phospholipase C
PMA	Phorbol myristate acetate

N.A. Riobo (✉)
Department of Biochemistry and Molecular Biology, Thomas Jefferson University, Bluemle Life Sciences Building, Suite 922, 233 S. 10th St, Philadelphia, PA 19107, USA
e-mail: natalia.riobo@jefferson.edu

M.G. Kazanietz (ed.), *Protein Kinase C in Cancer Signaling and Therapy*, Current Cancer Research, DOI 10.1007/978-1-60761-543-9_13,

PTCH Patched
SMO Smoothened

13.1 Introduction

The Hedgehog (Hh) and Wnt signaling pathways are essential for embryonic development; impairment of those pathways is lethal and even a subtle dysregulation results in significant pattern abnormalities and in growth defects. Moreover, deregulated hyperactivation of these pathways in adult organisms is often found in association with cancer.

The *hedgehog (hh)* gene was discovered in a *Drosophila melanogaster* screening for segment polarity mutants (Nüsslein-Volhard and Wieschaus 1980). Further screening of mutants with similar phenotypes and genetic complementation analysis led to the identification of additional components of the signal transduction pathway. Homologs were later found in vertebrate model organisms, which were shown to contain a large number of homologs of the secreted Hh proteins, as well as a more complex receptor and intracellular signaling network than *Drosophila* (Riobo et al. 2006a; Riobo and Manning 2007).

The *wnt-1* gene was discovered in 1982 as a proto-oncogene activated by integration of mouse mammary tumor virus (MMTV) in mammary tumors (Nusse and Varmus 1982). With the identification of the *Drosophila* segment polarity gene *wingless* (*wg*) as the ortholog of Wnt-1 (Cabrera et al. 1987; Rijsewijk et al. 1987), it became clear that Wnt genes are important regulators of many developmental processes and cancer (Nusse and Varmus 1992; Cadigan and Nusse 1996). Up to date, about 20 Wnt genes have been identified in humans, all encoding a secreted glycoprotein with an almost invariant motif of 23 cysteines named cysteine-rich domain.

The pattern formation role of Hh and Wnt is essential in mammals (although it is conserved in other vertebrate groups like fish, amphibians, and birds), as well as their function in balancing proliferation and survival of embryonic tissues during development. As already mentioned, in adult mammals alterations of these two pathways are very frequently found in many cancer types (Beachy et al. 2004). Hedgehog signaling is upregulated in basal cell carcinoma, medulloblastoma, rhabdomyosarcoma; the majority of GI-tract derived carcinomas, prostate, adenocarcinomas some lymphomas, and lung cancers. Dysregulation of Wnt signaling has been frequently found as the underlying cause of colon cancer and as a common event in breast cancer. There is a very straightforward association of Hh or Wnt with cancer in those tissues in which the same pathways are active during embryonic development.

The family of calcium-dependent protein kinases (PKCs) is implicated in numerous cell responses, and the Hh and Wnt pathways are not the exception. Understanding the mediators that transmit these signals and their crosstalk with other oncogenic factors is of critical importance in the development of more effective therapeutic approaches for cancer. Excellent reviews have been published on Hh and Wnt signaling; therefore, this chapter introduces the pathways only for background purposes and focuses on the role of PKC in those cascades.

13.2 Overview of the Hedgehog Signaling Pathway

The *hedgehog* gene is represented in vertebrates by three highly homologous isoforms encoded by separate genes: Sonic Hedgehog (Shh), Indian Hedgehog (Ihh), and Desert Hedgehog (Dhh). The use of powerful genetic models, spontaneous mutations and the availability of specific pathway inhibitors such as cyclopamine have been instrumental to define numerous functions of the Hedgehog ligands during embryonic development, adult homeostasis, and disease. During vertebrate development, Hh has functions as a morphogen in the dorso-ventral patterning of the neural tube, in the anterior-posterior patterning of the limbs, as a mitogen for neuronal precursors, as an inducer of tissue remodeling processes like vasculogenesis, angiogenesis, and branching morphogenesis, and as a pro-survival factor in all those tissues (McMahon et al. 2003).

In adult animals, Shh has a critical role in stem cell maintenance due to normal cell replacement and in tissue regeneration after injury (Beachy et al. 2004). When deregulated, Hh signaling promotes tumor formation and growth and inhibits apoptosis very potently. In fact, blockade of the Hh pathway with specific pharmacological agents significantly increases cancer cell death and, in many cases, leads to significant tumor regression.

The three Hh isoforms are synthesized as precursor proteins of ~45 kDa which enter the secretory pathway and undergo a unique set of posttranslational modifications (Mann and Beachy 2004). First, the precursors are cleaved through an autocatalytic process to generate the ~19 kDa N-terminal signaling fragment (N-Hh) with a covalently attached cholesterol moiety in the COOH-terminus. To this date, this modification has not been found in any other mammalian protein. Subsequently, N-Hh is palmitoylated at its N-terminus generating the most potent Hh derivative, sometimes called N-Hhp, but usually referred to simply as Hh. This dual lipidated molecule forms a multimeric complex through the hydrophobic groups which is soluble in aqueous environments and allows signaling at a distance from the source (Chen et al. 2004).

The receptors for Hh proteins are two 12-transmembrane (12-TM) proteins named Patched1 and Patched2 (abbreviated as Ptc/PTCH in mice and humans, respectively). In the absence of Hh proteins, PTCH-1 catalytically inhibits the activity of a 7-TM protein called Smoothened (SMO) (Fig. 13.1). The capacity of PTCH-2 to repress SMO appears to be less significant. Binding of the Hh proteins to Ptc1/PTCH-1 results in pathway activation by de-repression of SMO. While the binding affinity of Shh, Ihh, and Dhh is in the same range, and is similar between PTCH-1 and PTCH-2, their potency to promote SMO activation is very distinct, with Shh > Ihh >> Dhh. The basis for this difference in potency is unknown. Upon activation, SMO promotes stabilization and nuclear translocation of the Gli family of transcription factors (Gli1, 2 and 3). The signal transduction pathways employed by SMO to activate Gli family have been the focus of much research, but many questions still remain. One of the more recent areas of research in Hh signaling is the role of active transport of the pathway components in and out of the primary

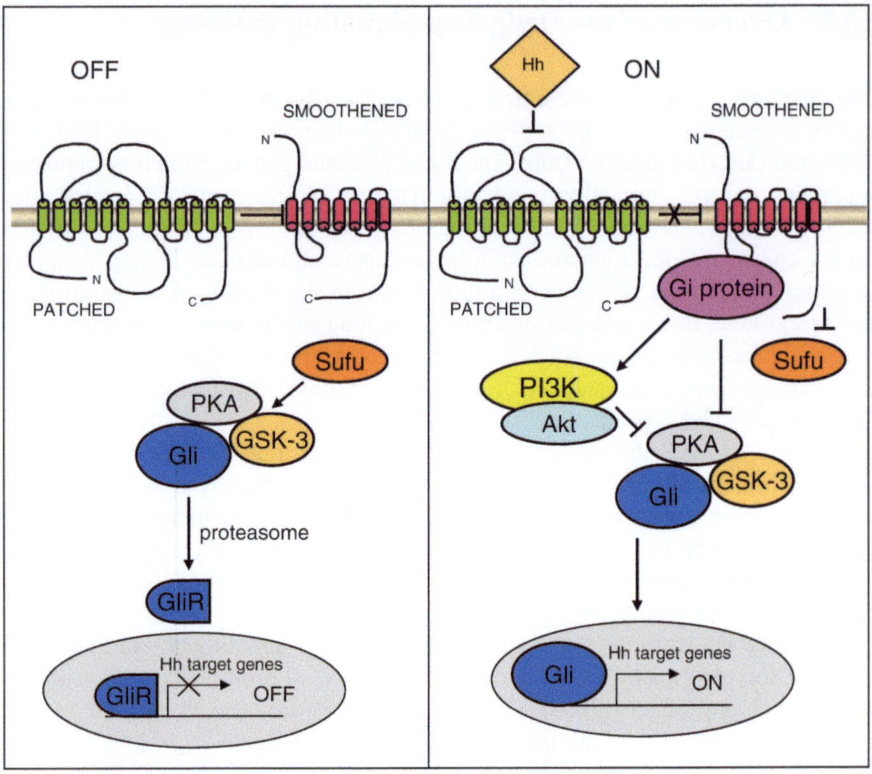

Fig. 13.1 *The Hedgehog signaling pathway*. In the absence of Hh proteins, Patched inhibits the activity of Smoothened and allows Suppressor of Fused (Sufu) to promote phosphorylation of the Gli transcription factors by PKA and GSK-3. Phosphorylation targets Gli to the proteasome for partial destruction, generating a repressor of transcription (GliR). Binding of Hhs to Patched relieves Smoothened repression. Smoothened inhibits Sufu activity and engages Gi proteins to prevent Gli phosphorylation, allowing intact Gli factors to accumulate in the cell nucleus and initiate transcription of target genes. Additional abbreviations: PKA, cAMP-dependent protein kinase; PI3K, phosphoinositide 3-kinase; GSK-3, glycogen synthase kinase 3

cilium, a single organelle in all mammalian cells that serves as a signaling antenna (Rohatgi et al. 2007). Localization of some proteins of the pathway in the primary cilium appears to be a *sine qua non* requirement for the engagement of Gli-transcriptional responses.

SMO belongs to the superfamily of G protein-coupled receptors, and we have demonstrated that it couples to heterotrimeric Gi proteins by virtue of its constitutive activity (Riobo et al. 2006c; Ogden et al. 2008). Activation of Gi proteins by Shh is critical for activation of the Gli-transcriptional responses in some cell types, and it is probably linked to the engagement of the phosphoinositide 3-kinase (PI3K)/Akt by Shh. In mouse fibroblasts and many Hh-dependent human cancer cell types, activation of the PI3K/Akt axis is necessary but not sufficient for activation of Gli transcriptional responses. Akt appears to prevent or reduce PKA-mediated Gli degradation by the proteasome (Riobo et al. 2006b).

SMO activity also represses another central inhibitory protein of the pathway named Suppressor of Fused (Sufu). This protein prevents nuclear localization of full length Gli isoforms by actively promoting nuclear export (Barnfield et al. 2005; Methot and Basler 2000). Loss of Sufu leads to congenital defects akin to those of loss of Patched-1, and *sufu*$^{+/-}$ mice have a high incidence of skin and cerebellar cancer (Svard et al. 2006).

Hh signaling can be effectively blocked by cyclopamine and related compounds, which bind to the hydrophobic core of SMO (Chen et al. 2002a, b). Conversely, purmorphamine and SAG act as SMO agonists, bypassing the requirement of Hedgehog/PTCH for Gli activation (Chen et al. 2002b; Sinha and Chen 2006). Elevation of cAMP levels with forskolin also results in an indirect repression of Gli by activation of PKA (Wang et al. 2000).

The three Gli isoforms contain five zinc-finger DNA-binding domains that recognize a conserved hexameric sequence known as Gli binding site (GBS). Their N-terminal domains are the most divergent regions. Gli1 functions as a strong transcriptional activator; its expression is induced by Gli2 and/or Gli3-mediated transcription, so that Gli1 expression can be used as a convenient readout of the state of Hh pathway activation. Human Gli1 can be phosphorylated *in* vitro and *in* vivo by cAMP-dependent protein kinase (PKA); however, the role of PKA in the regulation of Gli1 activity is not clear.

Gli2 and Gli3 share a high degree of sequence similarity, including activator and repressor domains, and multiple clusters of phosphorylation sites. At least in Gli-luciferase-reporter assays in cultured cells, Gli2 exhibits a stronger transcriptional activity than Gli3, but weaker than Gli1. Gli2 is constitutively expressed, but its stability is tightly regulated through phosphorylation-targeted proteasomal degradation (Riobo et al. 2006b). The phosphorylated Gli2 protein is recognized by the ubiquitin ligase β-TrCP, ubiquitinated and partially degraded by the proteasome to generate Gli2-R (a 78 kDa transcriptional repressor fragment of Gli2) (Pan et al. 2006). Both Gli2 processing and frank degradation are inhibited by Shh signaling *in* vivo, but the mechanisms involved are just starting to be defined.

Gli3 mainly functions as a transcriptional repressor that is suppressed through Hh signaling. The full-length 190 kDa Gli3 protein is efficiently processed in the absence of Hh to generate Gli3-R (the 83 kDa transcriptional repressor fragment) (Wang et al. 2000). Gli3-R localizes to the cell nucleus where it binds Gli-responsive elements to prevent ectopic induction. Gli3 undergoes phosphorylation by PKA, and likely by GSK-3 and CK-1, which are conserved between Gli2 and Gli3 proteins. The PKA-dependent processing of Gli3 in the developing limb is antagonized by Shh signaling (Wang et al. 2000). Mutations of Gli3 in Greig's cephalopolysyndactyly syndrome in humans produce a range of limb patterning malformations due to inefficient Gli3 regulation as an activator/repressor of transcription (Kalff-Suske et al. 1999).

13.3 Hedgehog Signaling in Cancer

Genetic studies in mice and analysis of sporadic and familiar mutations in humans have revealed that loss-of-function of Ptc1/PTCH-1 or gain-of-function of SMO (usually by W539L or S537N point mutations) result almost exclusively in the development of rhabdomyosarcoma, basal cell carcinoma and medulloblastoma (Goodrich et al. 1997; Hahn et al. 1998; Xie et al. 1998; Mao et al. 2006). Specific expression of constitutively active SMO-M2 (also known as SMO-A1) in cerebellar granule neuron precursors results also in high incidence of medulloblastoma, which depends on subsequent activation of Notch signaling for survival (Hatton et al. 2008; Hallahan et al. 2004).

Induction of basal cell carcinoma has been observed with high penetrance by epidermal overexpression of SMO-M2, Gli1, or Gli2 (Mao et al. 2006; Nilsson et al. 2000; Grachtchouk et al. 2000). It is therefore clear that cerebellar granule neuron precursors, epidermal basal cells, and skeletal muscle side population cells are exquisitely sensitive to hyperactivation of the Gli-transcriptional pathway, which is referred to by some as "canonical Hh pathway". It is worth noting that, in the case of SMO activating mutations, activation of the canonical pathway leads to the induction of PTCH-1 in the absence of Hh proteins, so that PTCH-1 is active and SMO is active at the same time due to its gain of function. In contrast, upregulation of Shh and Ihh is highly associated with epithelial adenocarcinomas of endoderm-derived organs, such as the GI tract, lung, breast, and prostate, as shown in Fig. 13.2 (Berman et al. 2003). In the latter context, while the canonical pathway is activated and PTCH-1 expression is similarly induced, an uncharacterized PTCH-1 growth inhibitory activity is repressed by continuous binding of ligand followed by internalization of the receptor/ligand complex. Direct evidence of this dichotomy was further provided with a transgenic mice model of inducible ubiquitous expression of the SMO-M2 oncogene. Induction of widespread SMO-M2 expression in those mice led to the development of rhabdomyosarcoma and basal cell carcinoma in 100% of the animals and a high incidence of medulloblastoma (~40%), while in the GI-tract there were hyperplastic areas but no cancer lesions and prostate adenocarcinoma was not detected, although the transgene was highly expressed in those epithelia (Hatton et al. 2008). This observation suggests that the canonical Hh pathway is not sufficient to sustain GI tract and prostate cancer. Although Gli-target genes are invariably upregulated through the canonical Hedgehog pathway as a result of Shh/Ihh upregulation, loss-of-function of PTCH-1 and gain-of-function of SMO, the association of either type of mutations with different cancer types suggests that new functions have been evolutionary gained in both PTCH and SMO, and that those functions can be regulated by binding of Hh ligands.

Despite the heterogenic etiology, most primary cancer cells derived from medulloblastoma (Berman et al. 2002; Sanchez and Ruiz i Altaba 2005), GI tract-derived neoplasias (Berman et al. 2003), prostate (Karhadkar et al. 2004) and small-cell lung carcinoma (Watkins et al. 2003) are sensitive to cyclopamine or other

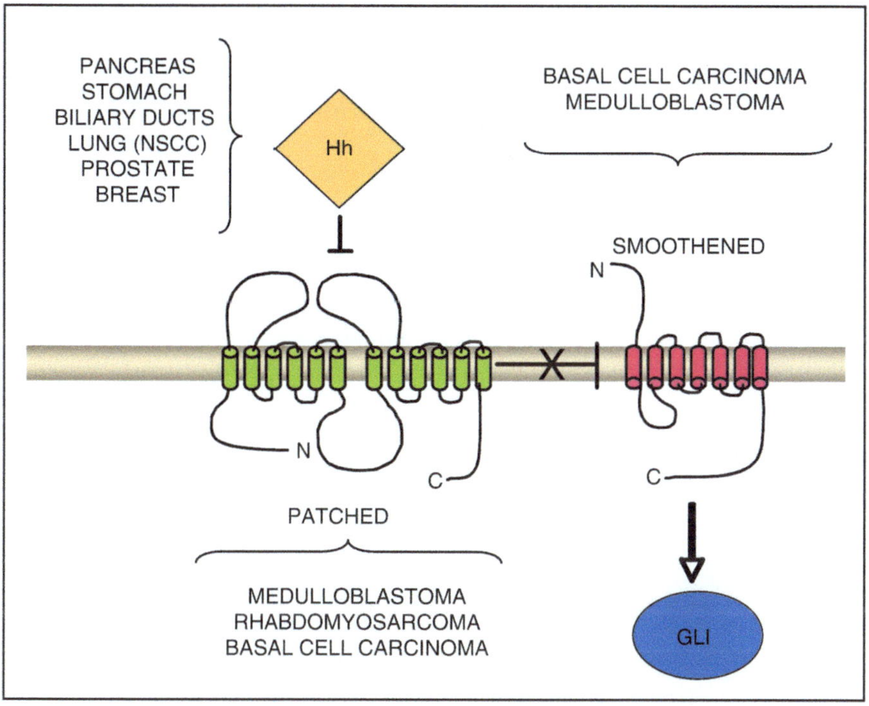

Fig. 13.2 *Common alterations in the Hedgehog pathway in human cancer*. Overexpression of Hh ligands is found in numerous epithelial carcinomas, while loss of function or loss of heterozygosity of Patched-1 or gain of function mutations in Smoothened underlie the development of different cancer types

SMO antagonists *in* vitro and *in* vivo when transplanted as xenografts in nude mice, showing decreased proliferation and/or increased apoptosis. Cyclopamine sensitivity has been nevertheless questioned in prostate cancer cells (Zhang et al. 2007; Varjosalo et al. 2006). Despite some negative observations, the efficacy of small molecule inhibitors of SMO in cancer is currently being tested in clinical trials, with very promising preliminary results.

13.4 PKC as a Mediator of Hedgehog Signaling

The protein kinase C family of Ser/Thr kinases is subdivided into three subfamilies: the classical, novel, and atypical PKCs (cPKC, nPKC, and aPKC, respectively). cPKC is activated by Ca^{2+} and diacylglycerol (DAG), nPKC is activated by DAG but not by Ca^{2+}, and aPKC is not activated by any of these molecules (Kikkawa et al. 1989; Newton 1997). Recently, we and others reported that protein kinase C, in particular the novel PKCδ isoform, is a necessary mediator of Hh signaling for the activation of Gli-dependent transcription. The first evidence came from observations by Neill et al. (2003), who showed that in 293T cells, PKCδ potentiates Gli1 transcriptional activity while PKCα has an inhibitory effect. We subsequently found that, in NIH 3T3 fibroblasts, the phorbol ester PMA–a diacylglycerol-like compound known to bind to the C1 domains of classical and novel PKCs–induced endogenous Gli target genes and a Gli-luciferase reporter within 6 h in a Hh ligand-independent manner (Riobo et al. 2006d). The effect of PMA on Gli-dependent transcription was abolished by rottlerin, a PKCδ isoform-selective inhibitor, and was prevented by PKC downregulation by prolonged PMA exposure. Additional supporting observations for a role of PKCδ in Shh signal transduction include: (1) downregulation of all phorbol ester-sensitive PKCs by prolonged PMA exposure abolished Gli activation, (2) selective inhibition of classical PKCs (PKCα and -βI) with Gö6976 did not prevent the rapid transcriptional effects of PMA, and (3) the only non-classical PKC isoform expressed in NIH 3T3 fibroblasts is PKCδ. Although rottlerin targets other non-PKC enzymes in addition to PKCδ (Soltoff 2007), in the context of those additional independent observations, the effect of rottlerin indicated that PKCδ is a component of the Shh signal transduction pathway. Work by Lauth and colleagues confirmed the requirement of PKCδ and the lack of involvement of classical PKC isozymes, for Hh signaling in NIH 3T3 and in mesenchymal precursors C3H10T1/2 cells stimulated, in this case, by the SMO agonist SAG (Lauth et al. 2007). The cancer cell lines DU145 (prostate) and PANC1 (pancreas) also show repression of the Hh pathway after PKC downregulation, while 22Rv1 prostate cancer cells are insensitive to PKC downregulation. Coincidentally, hyperactivation of Gli in the 22Rv1 cell line is the result of an unknown and atypical defect in the Hh pathway, since Gli activity is only partially inhibited by activation of PKA-mediated Gli degradation with forskolin. These authors also established that PKC inhibition does not disrupt the formation of primary cilia. Interestingly, the spontaneous Gli activation that is found in $Sufu^{-/-}$ MEFs is also reduced by PKC downregulation, placing the requirement of PKCδ downstream of Sufu (Lauth et al. 2007).

The positive effects of PKCδ on Gli transcriptional activity were not limited to phorbol ester-mediated activation. We demonstrated that PKCδ is also a mediator of Shh-initiated Gli activation, since both rottlerin and PKCs downregulation by prolonged PMA treatment impair Gli-luciferase activity while, conversely, the classical PKC inhibitor Gö6976 increases Gli-reporter activity. In addition, activation of Gli induced by Shh or PMA was prevented by U0126 and PD98059, which led us to speculate that MEK-1 was a necessary downstream mediator of PKCδ. Moreover, overexpression of a constitutively active MEK-1 allele in the absence of Shh is sufficient to increase basal Gli activation by ~5-fold, but co-expression with Gli1 or Gli2 leads to a striking synergistic response (1,000-fold above baseline). This synergism of MEK-1 and Gli has been reported as the underlying mechanism for hyperactivation of Gli-target genes in the gastric cancer cell lines AGS, MKN1, MKN45, MKN74, and SH101-P4 (Seto et al. 2009). Apparently, MEK-1 is a necessary downstream mediator of PKCδ but also affects a PKC-independent pathway to stimulate Gli transcriptional activity. This hypothesis is supported by the observation that depletion of PKCs still inhibits endogenous target gene activation by the SMO agonist SAG even in the presence of overexpressed constitutively active MEK-1 (Lauth et al. 2007).

In sharp contrast to these findings, another group recently reported that PKCα stimulates Gli1 activity and PKCδ is inhibitory (Cai et al. 2009). They co-transfected Gli1 and an activated PKCα mutant into NIH3T3 fibroblasts and found that PKCα increased Gli-luciferase activity. However, in an experimental setting in which one would expect a 50–100 fold Gli-luciferase induction simply by Gli1 overexpression, they report a grim 2–2.5 fold activation by the positive control, raising a warning on the validity of those results. Similarly, they show that wild-type but not kinase-dead PKCδ dose dependently reduces this low Gli1-mediated Gli-luciferase activity, but that is accompanied by a clear reduction of Gli1 expression levels, which could account for the apparent inhibitory activity. On the other hand, the kinase-dead PKCδ that does not inhibit Gli1 activity does not reduce Gli1 expression, and indeed the expression level of the kinase-dead PKCδ is much lower than of wild-type PKCδ or the constitutively active mutant. This mutual expression co-dependence, which could be an overexpression artifact or the result of the experimental design, precludes the good interpretation of the results. Therefore, given the existence of four independent reports of positive effects of PKCδ on Gli activity and the pointed problems with the article of Cai et al., I strongly support the notion that PKCδ is indeed necessary for Hedgehog signaling and a positive modulator of Gli activity.

Using the C3H10T1/2 mesenchymal precursor cell line, Dweyr et al. demonstrated that some particular oxysterols act at the level of SMO or immediately upstream to promote activation of the Hh pathway (Dwyer et al. 2007). In that study, the authors also found that rottlerin and downregulation of PKCs by prolonged PMA treatment both inhibit Shh- and oxysterol-mediated Gli1 and Ptch-1 induction in C3H10T1/2 cells, indicating that the role of PKCδ in the Hh pathway is not confined to fibroblasts. In support of the latter argument, a pan-PKC inhibitor (bisindolylmaleimide I) abolishes embryonic stem cells proliferation induced by Shh, although an isoform-specific analysis was not conducted in this study (Heo et al. 2007).

13.4.1 Overview of Wnt Signaling Components

The Wnt pathway is also evolutionary conserved and, like the Hedgehog pathway, it has critical roles during embryonic development in body axis formation and tissue homeostasis. Wnt signals promote proliferation but can also control cell fate determination and terminal differentiation in a tissue- and temporal-specific manner. There are 19 mammalian Wnt isoform proteins and 10 variants of the receptor Frizzled (Fz), giving 190 possible Wnt/Fz combinations that contribute to tissue-specific responses and also to the engagement of differential signaling pathways (Van Amerongen et al. 2008). For example, Wnt3 mediates body axis formation in vertebrates in a β-catenin-dependent pathway; animals deficient in Wnt3 or β-catenin fail to gastrulate. This function of Wnt3 can be reconstituted by exogenous Wnt1, Wnt3a, and Wnt8 but not by other Wnts.

Wnts are cysteine-rich secreted proteins modified by N-glycosylation and palmitoylation in a way similar to the Hh proteins. Palmitoylation occurs in the ER and is catalyzed by the porcupine homolog protein; *porcupine* deficiency is phenotypically identical to the loss of *wingless* in Drosophila (Mikels and Nusse 2006). Like Hhs, Wnt proteins are mostly insoluble, and require the action of several proteins to be transported away from the source. Some of those proteins involved in secretion and transport are: WLS/evi, Dally (a GPI-anchored heparan sulfate proteoglycan), and heparan sulfatases like Sulf1 and Sulf2 (Ai et al. 2006). Wnts bind to Fz receptors, 7-transmembrane proteins of the GPCR superfamily, in a quasi-combinatorial fashion. For engagement of the canonical Wnt pathway (see below), the single pass membrane low-density lipoprotein family member arrow/LRP5/6 act as a necessary co-receptor (Wehrli et al. 2000). In addition to the Fz receptors, high affinity binding of Wnts to the tyrosine kinase receptors Ryk and Ror2 has been documented, but the signaling mechanisms downstream of these receptors remain unknown (Mikels and Nusse 2006).

Accessibility of the Wnt ligands to their receptors is limited by the extracellular antagonists Dickkopf (Dkk), WIF, Cerberus, and soluble Frizzled-related proteins (SFRPs), which compete with the receptors for Wnt (reviewed by Kawano and Kypta 2003).

Following binding of Wnt to the Frizzled receptors, three different signaling cascades can be engaged: the canonical Wnt/β-catenin pathway, the non-canonical planar cell polarity (PCP) pathway, and the non-canonical Wnt/Ca^{2+} pathway (Fig. 13.3).

13.4.2 The Canonical Wnt/β-Catenin Pathway

In the canonical pathway, the "canonical Wnts" (Wnt1, Wnt3a, Wnt8, and Wnt8b) bind to some Fz receptors and to LRP5/6, leading to phosphorylation of the cytoplasmic domain of LRP5/6 and the recruitment of the scaffold proteins Axin and Dishevelled to the co-receptor complex. When the canonical pathway is not engaged, Axin forms part of a β-catenin cytoplasmic destruction complex. The destruction complex is composed of, at least, the scaffold tumor suppressor proteins Axin and Adenomatous Polyposis Coli (APC), the kinases GSK-3 and CK-1 and β-catenin. These kinases phosphorylate the N-terminal domain of β-catenin, which labels the protein for β-TrCP E3 ligase-mediated proteasomal degradation (Kimelman and Xu 2006). Note that GSK-3, CK-1, and β-TrCP are also part of the Gli2 and Gli3 processing machinery.

In the presence of canonical Wnts, the destruction complex is disassembled, which prevents β-catenin degradation and allows it to accumulate in the nucleus. Nuclear β-catenin acts as a transcriptional co-activator via its association with TCF/Lef proteins. In the absence of Wnt signals, TCF factors work as transcriptional repressors in complex with Groucho/Grg/TLE proteins. β-catenin interacts with the N-terminal portion of TCF, physically displacing Groucho, and converting it into a

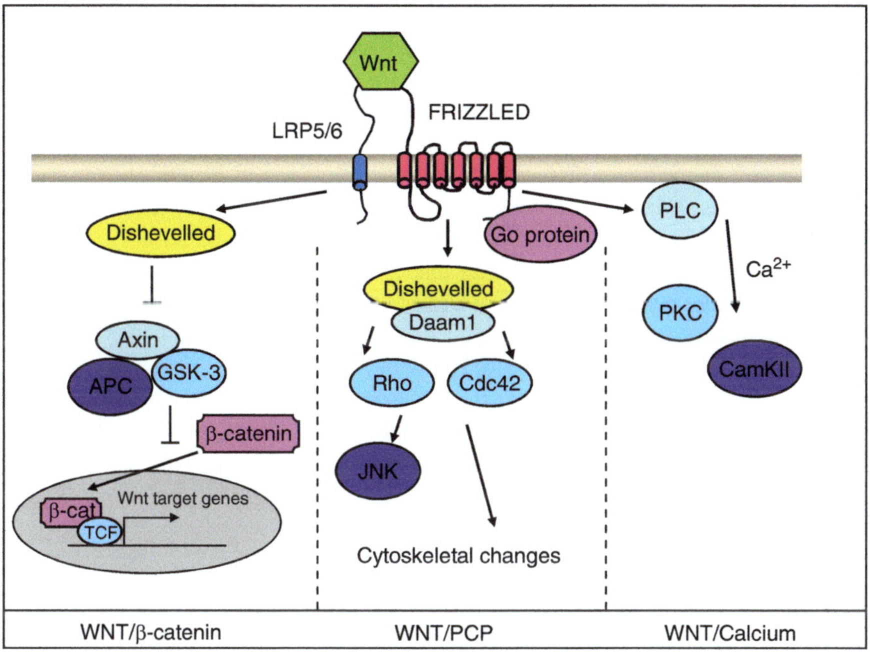

Fig. 13.3 *The Wnt signaling pathway*. Binding of Wnt ligands to Frizzled receptors and LRP5/6 co-receptor activates the Wnt canonical pathway (Wnt/β-catenin); while LRP5/6 is not engaged for activation of the non-canonical Wnt pathways (Wnt/PCP and Wnt/Calcium). Abbreviations: APC, Adenomatous Polyposis Coli; TCF, T cell factor; JNK, c-Jun N-terminal kinase; PLC, phospholipase C; PKC, protein kinase C; GSK-3, glycogen synthase kinase 3; PCP, planar cell polarity

transcriptional activator (Clevers 2006). A negative regulatory step at the level of the TCF/β-catenin complex is the phosphorylation of TCF by the MAP kinase-related protein kinase NLK/Nemo, which reduces the complex affinity for DNA and, therefore, inhibits the expression of Wnt target genes (Ishitani et al. 1999).

The Wnt target genes regulated by the β-catenin/TCF complex are cell-type specific. The most relevant for cancer progression are c-myc and cyclin D1, which are shared with the Hh pathway. The most ubiquitous Wnt target genes, frequently used to evaluate the degree of the canonical pathway activation, are Axin2 and SP5 (Clevers 2006). Wnt signaling is regulated by an autoinhibitory loop, since the expression of negative regulators such as Fzs, LRPs, HSPGs, Axin2, and TCF/Lef are all controlled by the β-catenin/TCF complex.

13.4.3 Canonical Wnt Signaling in Cancer

The role of Wnt signaling in cancer was first evidenced by the finding of a recessive mutation of the APC gene in the hereditary cancer syndrome Familiar Adenomatous Polyposis (FAP) (Kinzler et al. 1991; Nishisho et al. 1991). FAP patients develop numerous colon polyps, many of which progress into malignant colon carcinoma. Indeed, in sporadic colon cancer, bi-allelic loss of APC is a very common event (Kinzler and Vogelstein 1996). Inactivation of APC renders the destruction complex inactive, leading to stabilization of β-catenin, and transcription of Wnt target genes by the β-catenin/Tcf4 complex (the intestinal TCF family member). In a small fraction of colon carcinomas that lack APC mutations, Axin2 is mutated (Liu et al. 2000) or β-catenin itself is mutated at the N-terminal domain, which harbors the phosphorylation motifs required for its degradation (Morin et al. 1997). There is a hotspot S45 mutation site in β-catenin that is a CK-1 phosphorylation site, which controls β-catenin stability during non-canonical Wnt pathway activation (see below for details on this pathway).

Mutations of Axin2 have been also found in hepatocellular carcinoma, and oncogenic β-catenin mutations underlie the development of some hair follicle tumors, such as pilomatricomas and trichofolliculomas (Clevers 2006). In both chronic and acute myeloid leukemia, the Wnt pathway is hyperactivated, but the underlying cause is still unknown, since no mutations have been found in these cells.

13.4.4 The Wnt Planar Cell Polarity Pathway

Planar cell polarity (PCP) is the process of reorganization of protein complexes in cells within the plane of a single layered sheet of cells, which occurs orthogonal to the apical-basal axis (James et al. 2008). This process is very well studied in the Drosophila wing, in which sensory hairs of every individual epithelial cell is oriented in a particular direction. Genetic studies in Drosophila have identified genes that function to establish PCP in the wing; among them several belong to the fly Wingless pathway, such as Frizzled and Dishevelled, but this process is independent of β-catenin. In vertebrates, the most studied β-catenin-independent Wnt-induced process is convergent extension (CE) during embryonic gastrulation, in which cell migration along the anterior-posterior axis is coordinated with precise cell intercalation. Both processes are mediated by a β-catenin-independent Wnt pathway that is commonly referred to as the Wnt PCP pathway or the Wnt/JNK pathway (Kühl 2002; Tada et al. 2002).

The PCP pathway utilizes Wnt5/Wnt11, Fz2/Fz4/Fz6, Dishevelled, and specific proteins like Diego/Diversin, Strabismus/Vangl2, Flamingo/Celsr, and Prickle (James et al. 2008). Diversin inhibits the canonical Wnt pathway and promotes signaling of the Wnt/JNK pathway. The name given to this protein reflects its opposite function in the two branches of the Wnt pathway. Biochemical analyses revealed that Diversin recruits CK-Iε to the destruction complex and promotes degradation of β-catenin by phosphorylation at S45, a cancer hotspot mutation. Through epistasis experiments in frogs and zebrafish, Diversin was situated downstream of Dishevelled and CKIε, and upstream of GSK3β and β-catenin. Diversin influences gastrulation movements in zebrafish embryos, which are controlled by the Wnt/JNK pathway (Yamanaka et al. 2002).

During early embryogenesis in vertebrates, the narrowing and the lengthening of the embryonic axis and neural plate are driven by CE. During gastrulation and neurulation, cells elongate medio-laterally and produce cytoplasmic protrusions which enable cells to move directionally and to intercalate with other neighboring cells, resulting in convergence in one plane and extension of the cell mass in the perpendicular direction (Keller 2002). To generate polarized protrusions, cells must undergo dramatic cytoskeletal changes. Frizzled receptor function is required for proper CE movements and interference with endogenous Fz7 and Fz8 inhibits CE in frog embryos (Deardorff et al. 1998; Djiane et al. 2000). The non-canonical Wnt ligands that have been shown to participate in CE in zebrafish and Xenopus are Wnt11 and Wnt5 (Moon et al. 1993; Heisenberg et al. 2000). Dishevelled controls cellular cytoskeletal rearrangements during gastrulation movements. Dishevelled is a modular protein that has three conserved domains: Dishevelled-Axin (DIX), PSD95-Disc Large-ZO1 (PDZ) and Dishevelled-EGL10-Pleckstrin (DEP). The DEP domain is crucial for the function of the PCP pathway possibly by interaction with Daam1, which links Dishevelled to the RhoA GTPase (Habas et al. 2001). There is strong evidence that RhoA small GTPases mediate cellular polarization following stimulation of the non-canonical Wnt pathway, which subsequently activates Jun kinase (Wnt–JNK pathway) (Yamanaka et al. 2002; Habas et al. 2003).

Although many evidences suggest that PKC is involved in the Wnt signaling pathway, the exact role that PKC plays in this pathway is not well understood. A search for PKC genes affecting the non-canonical Wnt signaling pathways led to the identification and functional analyses of *Xenopus* PKCδ, which belongs to the nPKC subfamily. PKCδ was shown to be essential for convergent extension at least in part through regulation of Dishevelled function in the Wnt/JNK pathway (Kinoshita et al. 2003). When a dominant-negative PKCδ mRNA is introduced into Xenopus blastopores, a striking defect in gastrulation is observed similar to the lack of Xwnt11 and Xfz7, two components of the Wnt non-canonical pathway. Dominant negative PKCα or PKCβ mRNA injection does not inhibit gastrulation. Controls were done to rule out that the gastrulation defect was due to impaired mesodermal specification. In support of a role for PKCδ in the CE process during gastrulation, morpholino-mediated depletion of the two PKCδ paralogs in Xenopus embryos (Xenopus species have a duplicated genome) (Chen et al. 1988) elicited the same phenotype that the dominant negative PKCδ. Overexpression of the Fz7 receptor in Xenopus embryos induced PKCδ translocation to the plasma membrane in association with Dishevelled. The interaction between Dishevelled and PKCδ appears to be stable, independent of the activation of PKC, but Fz7 signaling seems to re-localize the complex from the cytoplasm to the plasma membrane. It is very likely that this complex is of fundamental importance for non-canonical Wnt signaling, since depletion of PKCδ impairs Dishevelled translocation to the membrane compartment and hyperphosphorylation in response to Fz7 overexpression. PKCδ depletion also prevented activation of JNK, a downstream effector of Dishevelled in the Wnt PCP pathway, in response to Fz7. Moreover, phorbol esters promote receptor-independent Dishevelled translocation and JNK activation (Kinoshita et al. 2003).

It is remarkable that the same PKC isoform is involved in the Hh pathway and that sustained activation of PKCδ with phorbol esters is sufficient to activate both the canonical Hh pathway and the PCP Wnt pathway. However, other PKC isoforms seem to mediate some non-canonical Wnt signals. For instance, defects in tissue separation during *Xenopus* gastrulation by loss of Fz7 function can be rescued by overexpression of PKCα (Winklbauer et al. 2001), and activated PKCα is able to phosphorylate Dishevelled *in* vitro (Kühl et al. 2001). PKCs are also implicated in the Xwnt11 signaling pathway for *Xenopus* cardiogenesis (Pandur et al. 2002) and in the Dwnt4 pathway for *Drosophila* ovarian morphogenesis (Cohen et al. 2002).

Some evidences suggest that the requirement of classical or novel PKCs for Wnt signaling is not extended to the Wnt-β-catenin pathway. Injection of Wnt8 into Xenopus embryos leads to the induction of siamois and Xnr3, mesodermal markers. However, when Wnt8 was coinjected with PKCδ morpholino, the induction of siamois and Xnr3 was not inhibited, neither was the secondary axis formation activity of Wnt8 (Kinoshita et al. 2003). Therefore, although PKCδ is required for the Wnt/JNK pathway, it may not be necessary for the canonical Wnt pathway, which is independent of the membrane relocalization of Dishevelled.

13.4.5 The Wnt/Calcium Signaling Pathway

The Wnt/Calcium pathway is another β-catenin-independent Wnt signaling event. Early on it was demonstrated that Wnt5a, Wnt11, and Fz2 activate the phospholipase C (PLC) pathway and double the frequency of Ca^{2+} transients through the release of inositol-3-phosphate in zebrafish embryos (Slusarski et al. 1997a, b). This increase in intracellular calcium stimulates the activities of two calcium-sensitive proteins: calcium/calmodulin-dependent kinase II (CamKII) and PKC. Engagement of the Wnt/Calcium pathway by specific Wnt/Fz combinations is believed to occur in part by the use of a different co-receptors instead of LRP5/6: Knypek (Topczewski et al. 2001) and Ror2 (Hikasa et al. 2002). Based on the selectivity for different intracellular responses, the Frizzled receptors can be grouped as follows: Fzl1, Fzl7, and Fzl8 activate the β-catenin-dependent transcriptional response; Fzl2, Fzl3, Fzl4, and Fzl6 stimulate CamKII and PKC but not β-catenin (Kohn and Moon 2005). The simultaneous generation of DAG and the rise in Ca^{2+} levels induce PKC activation. In fact, overexpression of non-canonical Fz2, Fz7, or Wnt5 causes the translocation of PKCα from the cytoplasm to the plasma membrane in *Xenopus* embryos, while the β-catenin activator Fz1 does not (Sheldahl et al. 1999; Medina et al. 2000). Observations suggest that Fz7 might activate PKCδ through DAG on the plasma membrane, although there is no direct evidence that activation of the Wnt pathway produces DAG. However, heterotrimeric G proteins have been implicated in Wnt signal transduction (Liu et al. 1999a, b, 2001). Taken together, these findings suggest that Fz7 probably activates PKCδ through a heterotrimeric G protein that produces DAG.

A model was proposed in that DAG activates PKCδ on the membrane, and PKCδ phosphorylates Dishevelled directly. However, Dishevelled is known to interact with other kinases, such as CK-1 and -2, Par-1, and PAK1/MuSK (Willert et al. 1997; Sun et al. 2001; Luo et al. 2002). PKCδ may regulate such protein kinases and thus indirectly regulate Dishevelled phosphorylation. It would be interesting to examine whether PKCδ phosphorylates Dishevelled directly, and to elucidate the role of Dishevelled phosphorylation in its localization and in the activation of downstream signaling. Identification of the phosphorylation sites in Dishevelled upon Fz7 stimulation will shed light onto this complex topic.

13.4.6 Heterotrimeric G Proteins in PKC Activation by Wnts

The Frizzled family of Wnt receptors belong structurally to the 7-TM (also known as GPCR) superfamily, and together with SMO they constitute a separate group. Indeed, several evidences suggest that Frizzled signals not only through Dishevelled, but also independently through engagement of heterotrimeric G proteins (Gαβγ), like SMO.

Early on, Slusarski et al. reported that Wnt5a, through activation of Fz2, induces calcium release through the Gβγ subunit of Gi proteins, since pertussis toxin, GDP-β-S and Gαt blocked the calcium increase (Slusarski et al. 1997a). In another elegant study, a chimeric receptor between the β-adrenergic receptor ligand binding domain and Fz2 (a non-canonical Fz that is the Wnt5a receptor) was used to control calcium release with an adrenergic agonist (isoproterenol). Expression and activation of this chimera in F9 teratocarcinoma cells induced the formation of primitive endoderm and calcium transients (Liu et al. 1999a). Primitive endoderm formation was blocked by interference with Gi proteins by pertussis toxin, by depletion of Gαo (a Gi member) or Gβ2 with antisense oligodeoxynucleotides, and by inhibitors of PKC (bisindolylmaleimide I) and MEK1/2 (PD98059). In comparison, the same group performed studies with overexpressed Fz1 (a canonical Fz that responds to Wnt8) in F9 cells. The findings indicate that primitive endoderm formation induced by canonical Wnt8/Fz1 is sensitive not only to Gαo, but also to Gαq, and further downstream, to PKC and MEK (Liu et al. 1999b). Moreover, a chimeric receptor containing the intracellular loops of canonical Fz1 on a β-adrenergic backbone, induced β-catenin stabilization, primitive endoderm formation, and transcriptional activation of a Tcf/LEF reporter construct when stimulated with isoproterenol. These effects were abolished by pertussis toxin, indicating Gi proteins involvement, and by depletion of Gαo (a Gi family member) and Gαq (Liu et al. 2001). These important observations suggest that G proteins are integral part of the Wnt pathway, both canonical and non-canonical, as we have found to be the case for the Hedgehog signaling pathway. Indeed, the role of Gαo is evolutionary conserved since it is immediately downstream of Frizzled in Drosophila Wnt/β-catenin and planar cell polarity pathways (Katanaev et al. 2005).

Some Wnt isoforms, known to activate β-catenin-independent pathways, initiate cytoskeletal rearrangements mediated by activation of the small GTPase Rho and by JNK. In animal cap ectoderm, Cdc42 activity also increases as a response to Wnt11 expression. This increase is inhibited by pertussis toxin, or sequestration of free Gβγ subunits. Activation of Cdc42 is also produced by the expression of bovine Gβ1 and Gγ2. This process is abolished by a PKC inhibitor, while phorbol ester treatment of ectodermal explants activates Cdc42 in a PKC-dependent way, implicating PKC downstream of Gβγ (Penzo-Mendez et al. 2003). Therefore, the Wnt PCP pathway, operating during convergent extension at gastrulation, also requires G protein signaling.

Another example of non-canonical Wnt signaling, in which heterotrimeric G proteins and PKCδ are involved is the stimulation of bone formation by Wnt3a in ST2 cells (Tu et al. 2007). While Gi proteins are not required for Wnt3a-dependent bone formation, inhibition of Gαq activation by competition with a Gαq C-terminal peptide abolished Wnt3a response and translocation of PKCδ to the plasma membrane. A PLC inhibitor also prevented the induction of bone formation by Wnt3a, suggesting that Wnt3a engages Gq for stimulation of PLC and PKCδ during bone formation.

13.5 Concluding Remarks

The PKC family of protein kinases participates in the signal transduction of numerous growth factors, including the Hedgehog and Wnt pathways. In both signaling cascades, PKC seems to mediate both canonical – transcription-dependent – and non-canonical aspects of the pathways. Remarkably, PKCδ plays a very unique role among the other PKC isoforms in Hedgehog and Wnt signaling. The essential role of PKCδ in Hedgehog- and Wnt-dependent cells positions it as a potential target for inhibition of cancer cell growth.

References

Ai, X., Do, A.T., Kusche-Gullberg, M., Lindahl, U., Lu, K., Emerson, C.P. Jr. (2006). Substrate specificity and domain functions of extracellular heparan sulfate 6-O-endosulfatases, QSulf1 and QSulf2. *J Biol Chem, 281*,4969–4976.

Barnfield, P.C., Zhang, X., Thanabalasingham, V., Yoshida, M., Hui, C.C. (2005). Negative regulation of Gli1 and Gli2 activator function by Suppresor of fused through multiple mechanisms. *Differentiation, 73*,397–405.

Beachy, P.A., Karhadkar, S.S., Berman, D.M. (2004). Tissue repair and stem cell renewal in carcinogenesis. *Nature, 432*,324–331.

Berman, D.M., Karhadkar, S.S., Hallahan, A.R., Pritchard, J.I., Eberhart, C.G., Watkins, D.N., et al. (2002). Medulloblastoma growth inhibition by hedgehog pathway blockade. *Science, 297*,1559–1561.

Berman, D.M., Karhadkar, S.S., Maitra, A., Montes De Oca, R., Gerstenblith, M.R., Briggs, K., et al. (2003). Widespread requirement for Hedgehog ligand stimulation in growth of digestive tract tumours. *Nature, 425*,846–851.

Cabrera, C.V., Alonso, M.C., Johnston, P., Phillips, R.G., Lawrence, P.A. (1987). Phenocopies induced with antisense RNA identify the *wingless* gene. *Cell, 50*,659–63 .

Cadigan, K.M., Nusse, R. (1996). *wingless* signaling in the *Drosophila* eye and embryonic epidermis. *Development 122*,2801–2812.

Cai, Q., Li, J., Gao, T., Xie, J., Evers, B.M. (2009). Protein kinase C delta negatively regulates Hedgehog signaling by inhibition of Gli1 activity. *J Biol Chem, 284*,2150–2158.

Chen, K.H., Peng, Z.G., Lavu, S., Kung, H.F. (1988). Molecular cloning and sequence analysis of two distinct types of Xenopus leavis protein kinase C. *Second Messengers Phosphoproteins, 12*,251–260.

Chen, J.K., Taipale, J., Cooper, M.K., Beachy, P.A. (2002). Inhibition of Hedgehog signaling by direct binding of cyclopamine to Smoothened. *Genes Dev*, 16,2743–2748.

Chen, J.K., Taipale, J., Young, K.E., Maiti, T., Beachy, P.A. (2002). Small molecule modulation of Smoothened activity. *Proc Natl Acad Sci USA*, 99,14071–14076.

Chen, M.H., Li, Y.J., Kawakami, T., Xu, S.M., Chuang, P.T. (2004). Palmitoylation is required for the production of a soluble multimeric Hedgehog protein complex and long-range signaling in vertebrates. *Genes Dev, 18*,641–659

Clevers, H. (2006). Wnt/beta-catenin signaling in development and disease. *Cell 127*,469–480.

Cohen, E.D., Mariol, M.C., Wallace, R.M., Weyers, J., Kamberov, Y.G., Pradel, J., et al. (2002). DWnt4 regulates cell movement and focal adhesion kinase during Drosophila ovarian morphogenesis. *Dev Cell 2*,437–448.

Deardorff, M.A., Tan, C., Conrad, L.J., Klein, P.S. (1998). Frizzled-8 is expressed in the Spemann organizer and plays a role in early morphogenesis. *Development, 125*, 2687–2700.

Djiane, A., Riou, J., Umbhauer, M., Boucaut, J., Shi, D. (2000). Role of frizzled 7 in the regulation of convergent extension movements during gastrulation in *Xenopus* laevis. *Development, 127*, 3091–3100.

Dwyer, J.R., Sever, N., Carlson, M., Nelson, S.F., Beachy, P.A., Parhami, F. (2007). Oxysterols are novel activators of the hedgehog signaling pathway in pluripotent mesenchymal cells. *J Biol Chem, 282*,8959–8968.

Goodrich, L.V., Milenkovi , L., Higgins, K.M., Scott, M.P. (1997). Altered neural cell fates and medulloblastoma in mouse patched mutants. *Science, 277*,1109–1113.

Grachtchouk, M., Mo, R., Yu, S., Zhang, X., Sasaki, H., Hui, C.C., Dlugosz, A.A. (2000). Basal cell carcinomas in mice overexpressing Gli2 in skin. *Nat Genet, 24*,216–217

Habas, R., Kato, Y., He, X. (2001). Wnt/Frizzled activation of Rho regulates vertebrate gastrulation and requires a novel Formin homology protein Daam1. *Cell, 107*,843–854.

Habas, R., Dawid, I.B., He, X. (2003). Coactivation of Rac and Rho by Wnt/Frizzled signaling is required for vertebrate gatrulation. *Genes Dev, 17*,295–309.

Hahn,H.,Wojnowski,L.,Zimmer,A.M.,Hall,J.,Miller,G.,Zimmer,A.(1998).Rhabdomyosarcomas and radiation hypersensitivity in a mouse model of Gorlin syndrome. *Nat Med, 4*,619–622.

Hallahan, A.R., Pritchard, J.I., Hansen, S., Benson, M., Stoeck, J., Hatton, B.A., et al. (2004). The SmoA1 mouse model reveals that notch signaling is critical for the growth and survival of sonic hedgehog-induced medulloblastomas. *Cancer Res, 64*,7794–7800.

Hatton, B.A., Villavicencio, E.H., Tsuchiya, K.D., Pritchard, J.I., Ditzler, S., Pullar, B., et al. (2008). The Smo/Smo model: hedgehog-induced medulloblastoma with 90% incidence and leptomeningeal spread. *Cancer Res, 68*,1768–1776.

Heisenberg, C.P., Tada, M., Rauch, G.J., Saude, L., Concha, M.L., Geisler, R., et al.(2000). Silberblick/Wnt11 mediates convergent extension movements during zebrafish gastrulation. *Nature, 405*,76–81.

Heo, J.S., Lee, M.Y., Han, H.J. (2007). Sonic Hedgehog stimulates mouse embryonic stem cell proliferation by cooperation of Ca2+/protein kinase C and epidermal growth factor receptor as well as Gli1 activation. *Stem Cells, 25*,3069–3080.

Hikasa, H., Shibata, M., Hiratani, I., Taira, M. (2002). The Xenopus receptor tyrosine kinase Xror2 modulates morphogenetic movements of the axial mesoderm and neuroextoderm via Wnt signaling. *Development, 129*,5227–5239.

Ishitani, T., Ninomiya-Tsuji, J., Nagai, S., Nishita, M., Meneghini, M., Barker, N., et al. (1999). The TAK1-NLK-MAPK-related pathway antagonizes signaling between beta-catenin and transcription factor TCF. *Nature, 399*,798–802.

James, R.G., Conrad, W.H., Moon, R.T. (2008). Chapter 10: b-catenin-independent Wnt pathways: signals, core proteins, and effectors. *In Wnt signaling: Pathway methods and mammalian models, 468,* 131-144. Ed. Elizabeth Vincan.

Kalff-Suske, M., Wild, A., Topp, J., Wessling, M., Bornholdt, D., Engel, H., et al. (1999). Point mutations throughout the GLI3 gene cause Greig cephalopolysyndactyly syndrome. *Hum Mol Genet, 8*,1769–1777.

Karhadkar, S.S., Bova, G.S., Abdallah, N., Dhara, S., Gardner, D., Maitra, A., et al. (2004) . Hedgehog signalling in prostate regeneration, neoplasia and metastasis. *Nature, 431*,707–712.

Katanaev, V.L., Ponzielli, R., Semeriva, M., Tomlinson, A. (2005). Trimeric G protein-dependent frizzled signaling in Drosophila. *Cell,120*,111–122.

Kawano, Y., Kypta, R. (2003). Secreted antagonists of the Wnt signaling pathway. *J Cell Sci, 116*,2627–2634.

Keller, R. (2002). Shaping the vertebrate body plan by polarized embryonic cell movements. *Science, 298*,1950–1954

Kikkawa, U., Kishimoto, A., Nishizuka, Y. (1989). The protein kinase C family: heterogeneity and its implications. *Annu Rev Biochem, 58*,31–44.

Kimelman, D., Xu, W. (2006). b-Catenin destruction complex: insights and questions from a structural perspective. *Oncogene, 25*,7482–7491.

Kinoshita, N., Hidekazu, I., Miyakoshi, A., Ueno, N. (2003). PKCd is essential for Dishevelled function in a noncanonical Wnt pathway that regulates Xenopus convergent extension movements. *Genes Dev, 17*,1663–1676.

Kinzler, K.W., Nilbert, M.C., Su, L.K., Vogelstein, B., Bryan, T.M., Levy, D.B., et al. (1991). Identification of FAP locus genes from chromosome 5q21. *Science, 253*,661–665

Kinzler, K.W., Vogelstein, B. (1996). Lessons from hereditary colorectal cancer. *Cell, 87*,159–170.

Kohn, A.D., Moon, R.T. (2005). Wnt and calcium signaling: b-catenin-independent pathways. *Cell Calcium, 38*,439–446.

Kühl, M., Geis, K., Sheldahl, L.C., Pukrop, T., Moon, R.T., Wedlich, D. (2001). Antagonistic regulation of convergent extension movements in Xenopus by Wnt/beta-catenin and Wnt/Ca2+ signaling. *Mech Dev, 106*, 61–76.

Kühl, M. (2002). Non-canonical Wnt signaling in Xenopus: regulation of axis formation and gastrulation. Semin *Cell Dev Bio, l13*,243–24.

Lauth, M., Bergstrom, A., Toftgard, R. (2007). Phorbol esters inhibit the Hedgehog signaling pathway downstream of Suppressor of Fused, but upstream of Gli. *Oncogene, 26*,5163–5168.

Liu, X., Liu, T., Slusarski, D.C., Yang-Snyder, J., Moon, R.T., Wang, H. (1999a). Activation of a Frizzled-2/b-adrenergic chimera promotes Wnt signaling and differentiation of mouse F9 teratocarcinoma cells via Gao and Gat. *Proc Natl acad Sci USA, 96*,14383–14388.

Liu, T., Liu, X., Wang, H., Moon, R.T., Malbon, C.C. (1999b). Activation of rat frizzled-1 promotes Wnt signaling and differentiation of mouse F9 teratocarcinoma cells via pathways that require Galpha(q) and Galpha(o) function. *J Biol Chem, 274*,33539–33544.

Liu, W., Dong, X., Mai, M., Seelan, R.S., Taniguchi, K., Krishnadath, K.K., et al. (2000) . Mutations in AXIN2 cause colorectal cancer with defective mismatch repair by activating beta-catenin/TCF signaling. *Nat Genet, 26*,146–147.

Liu, T., DeCostanzo, A.J., Liu, X., Wang, H.Y., Hallagan, S., Moon, R.T., Malbon, C.C. (2001). G protein signaling from activated rat frizzled-1 to the beta-catenin-Lef-Tcf pathway. *Science, 292*,1718–1722.

Luo, Z.G., Wang, Q., Zhou, J.Z., Wang, J., Luo, Z., Liu, M., et al. (2002). Regulation of AchR clustering by Dishevelled interacting with MuSK and PAK1. *Neuron, 35*,489–505.

Mann, R.K., Beachy, P.A. (2004). Novel lipid modifications of secreted protein signals. *Annu Rev Biochem 73*:891–923.

Mao, J., Ligon, K.L., Rakhlin, E.Y., Thayer, S.P., Bronson, R.T., Rowitch, D., McMahon, A.P. (2006). A novel somatic mouse model to survey tumorigenic potential applied to the Hedgehog pathway. *Cancer Res, 66*,10171–10178.

McMahon, A.P., Ingham, P.W., Tabin, C.J. (2003). Developmental roles and clinical significance of hedgehog signaling. *Curr Top Dev Biol 53*,1–14.

Medina, A., Reintsch, W., Steinbeisser, H. (2000). Xenopus frizzled 7 can act in canonical and non-canonical Wnt signaling pathways: implications on early patterning and morphogenesis. *Mech Dev, 92*,227–237.

Methot, N., Basler, K. (2000). Suppresor of fused opposes hedgehog signal transduction by impeding nuclear accumulation of the activator form of Cubitus interruptus. *Development, 127*,4001–4010.

Mikels, A.J., Nusse, R. (2006). Wnts as ligands: processing, secretion and reception. *Oncogene, 25*,7461–7468.

Moon, R.T., Campbell, R.M., Christian, J.L., McGrew, L.L., Shih, J., Fraser, S. (1993). Xwnt-5A: a maternal Wnt that affects morphogenetic movements after overexpression in embryos of Xenopus leavis. *Development, 11*,97–111.

Morin, P.J., Sparks, A.B., Korinek, V., Barker, N., Clevers, H., Vogelstein, B., et al. (1997). Activation of beta-catenin-Tcf signaling in colon cancer by mutations in beta-catenin or APC. *Science, 275*,1787–1790.

Neill, G.W., Ghali, L.R., Green, J.L., Philpott, M.P., Quinn, A.G. (2003). Loss of protein kinase C alpha expression may enhance the tumorigenic potential of Gli1 in basal cell carcinoma. *Cancer Res, 63*,4692–4697

Newton, A.C. (1997). Regulation of protein kinase C. *Curr Opin Cell Bio,l 9*,161–167.

Nilsson, M., Undèn, A.B., Krause, D.,Malmqwist, U., Raza, K., Zaphiropoulos, P.G., Toftgård, R. (2000). Induction of basal cell carcinomas and trichoepitheliomas in mice overexpressing GLI-1. *Proc Natl Acad Sci USA, 97*,3438–3443.

Nishisho, I., Nakamura, Y., Miyoshi, Y., Miki, Y., Ando, H., Horii, A., et al. (1991). Mutations of chromosome 5q21 genes in FAP and colorectal cancer patients. *Science, 253*,665–669.

Nusse, R., Varmus, H.E. (1982). Many tumors induced by the mouse mammary tumor virus contain a provirus integrated in the same region of the host genome. *Cell, 31*,99–109.

Nusse, R., Varmus, H.E. (1992). Wnt genes. Cell, 69,1073–1087.

Nüsslein-Volhard, C., Wieschaus, E. (1980). Mutations affecting segment number and polarity in Drosophila. *Nature, 287*,795–801.

Ogden, S.K., Fei, D.L., Schilling, N.S., Ahmed, Y.F., Hwa, J., Robbins, D.J. (2008) G protein Galphai functions immediately downstream of Smoothened in Hedgehog signaling. *Nature, 456*,967–970.

Pan, Y., Bai, C.B., Joyner, A.L., Wang, B. (2006). Sonic hedgehog signaling regulates Gli2 transcriptional activity by suppressing its processing and degradation. *Mol Cell Biol, 26*,3365–3377.

Pandur, P., Läsche, M., Eisenberg, L.M., Kühl, M. (2002). Wnt-11 activation of a non-canonical Wnt signalling pathway is required for cardiogenesis. *Nature*, *418*,636–641.

Penzo-Mendez, A., Umbhauer, M., Djiane, A., Boucaut, J.C., Riou, J.F. (2003). Activation of Gbetagamma signaling downstream of Wnt-11/Xfz7 regulates Cdc42 activity during Xenopus gastrulation. *Dev Biol*, 257:302–314.

Rijsewijk, F., Schuermann, M., Wagenaar, E., Parren, P., Weigel, D., Nusse, R. (1987). The Drosophila homolog of the mouse mammary oncogene int-1 is identical to the segment polarity gene wingless. *Cell 50*,649–57.

Riobo, N.A., Lu, K., Emerson, C.P. Jr. (2006a). Hedgehog signal transduction: signal integration and crosstalk in development and cancer. *Cell Cycle*, *5*,1612–1615.

Riobo, N.A., Lu, K., Ai, X., Haines, G.M., Emerson, C.P. Jr. (2006b). Phosphoinositide 3-kinase and Akt are essential for Sonic Hedgehog signaling. *Proc Natl Acad Sci USA*, *103*, 4505–4510

Riobo, N.A., Saucy, B., DiLizzio, C., Manning, D.R. (2006c). Activation of heterotrimeric G proteins by Smoothened. *Proc Natl Acad Sci USA*, *103*,12607–12612.

Riobo, N.A., Haines, G.M., Emerson, C.P. Jr. (2006d). Protein kinase C-delta and mitogen-activated protein/extracellular signal-regulated kinase-1 control GLI activation in Hedgehog signaling. *Cancer Res*, *66*,839–845.

Riobo, N.A., Manning, D.R. (2007). Pathways of signal transduction employed by vertebrate Hedgehogs. *Biochem J*, *403*,369–379.

Rohatgi, R., Milenkovic, L., Scott, M.P. (2007). Patched1 regulates hedgehog signaling at the primary cilium. *Science, 317*,372–376.

Sanchez, P., Ruiz i Altaba, A. (2005). In vivo inhibition of endogenous brain tumors through systemic interference of Hedgehog signaling in mice. *Mech Dev, 122*,223–230.

Seto, M., Ohta, M., Asaoka, Y., Ikenoue, T., Tada, M., Miyabayashi, K., et al. (2009). Regulation of hedgehog signaling by the mitogen-activated protein kinase cascade in gastric cancer. *Mol Carcinog*, *48*,703–12.

Sheldahl, L.C., Park, M., Malbon, C.C., Moon, R.T. (1999). Protein kinase C is differentially stimulated by Wnt and Frizzled homologs in a G-protein-dependent manner. *Curr Biol, 9*,695–8.

Sinha, S., Chen, J.K. (2006). Purmorphamine activates the Hedgehog pathway by targeting Smoothened. *Nat Chem Biol, 2*,29–30.

Slusarski, D.C., Yang-Snyder, J., Busa, W.B., Moon, R.T. (1997). Modulation of embryonic intracellular Ca2+ signaling by Wnt5a. *Dev Biol, 182*,114–120.

Slusarski, D.C., Corces, V.G., Moon, R.T. (1997). Interaction of Wnt and a Frizzled homologue triggers a G protein-linked phosphatidylinositol signaling. *Nature, 390*,410–413.

Soltoff, S.P. (2007). Rottlerin: an inappropriate and ineffective inhibitor of PKCdelta. *Trends Pharmacol Sci, 28*,453–458.

Svard, J., Heby-Henricson, K., Persson-Lek, M., Rozell, B., Bergström, A., Ericson, J., et al. (2006). Genetic elimination of Suppressor of fused reveals an essential repressor function in the mammalian Hedgehog signaling pathway. *Dev Cell, 10*,187–197.

Sun, T.Q., Lu, B., Feng, J.J., Reinhard, C., Jan, Y.N., Fantl, W.J., et al. (2001). PAR-1 is a Dishevelled-associated kinase and a positive regulator of Wnt signaling. *Nat Cell Biol, 3*,628–636.

Tada, M., Concha, M.L., Heisenberg, C.P. (2002). Non-canonical Wnt signalling and regulation of gastrulation movements. Semin *Cell Dev Biol, 13*,251–260.

Topczewski, J., Sepich, D.S., Myers, D.C., Walker, C., Amores, A., Lele, Z., et al. (2001). The zebrafish glypican knypek controls cell polarity during gastrulation movements of convergent extension. *Dev Cel ,11*,251–264.

Tu, X., Joeng, K.S., Nakayama, K.I., Nakayama, K., Rajagopal, J., Carroll, T.J., et al. (2007) Noncanonical Wnt signaling through G protein-linked PKCdelta activation promotes bone formation. *Dev Cel, 112*,113–127.

Varjosalo, M., Li, S.P., Taipale, J. (2006). Divergence of Hedgehog signal transduction mechanism between *Drosophila* and mammals. *Dev Cell, 10*,177–186.

Van Amerongen, R., Mikels, A., Nusse, R. (2008). Alternative Wnt signaling is initiated by distinct receptors. *Sci Signal 1*,re9.

Wang, B., Fallon, J.F., Beachy, P.A. (2000). Hedgehog-regulated processing of Gli3 produces an anterior/posterior repressor gradient in the developing vertebrate limb. *Cell 100*,423–434.

Watkins, D.N., Berman, D.M., Burkholder, S.G., Wang, B., Beachy, P.A., Baylin, S.B. (2003). Hedgehog signalling within airway epithelial progenitors and in small-cell lung cancer. *Nature, 422*,313–317.

Wehrli, M., Dougan, S.T., Caldwell, K., O'Keefe, L., Schwartz, S., Vaizel-Ohayon, D., et al. (2000). Arrow encodes an LDL-receptor-related protein essential for Wingless signaling. *Nature, 407*,527–530.

Willert, K., Brink, M., Wodarz, A., Varmus, H., Nusse, R. (1997). Casein kinase 2 associates with and phosphorylates dishevelled. *EMBO J, 16*,3089–3096.

Winklbauer, R., Medina, A., Swain, R.K., Steinbeisser, H. (2001). Frizzled-7 signalling controls tissue separation during Xenopus gastrulation. *Nature*, *413*,856–860.

Xie, J., Murone, M., Luoh, S.M., Ryan, A., Gu, Q., Zhang, C., et al. (1998). Activating Smoothened mutations in sporadic basal-cell carcinoma. *Nature, 391*,90–92.

Yamanaka, H., Moriguchi, T., Masuyama, N., Kusakabe, M., Hanafusa, H., Takada, R., et al. (2002). JNK functions in the non-canonical Wnt pathway to regulate convergent extension movements in vertebrates. *EMBO Rep, 3*,69–75.

Zhang, J., Lipinski, R., Shaw, A., Gipp, J., Bushman, W. (2007). Lack of demonstrable autocrine hedgehog signaling in human prostate cancer cell lines. *J Urol, 177*,1179–1185.

Chapter 14
PKC–PKD Interplay in Cancer

Q. Jane Wang

Abstract The family of protein kinase D (PKD) serine/threonine kinases is a novel diacylglycerol (DAG) receptor and an immediate target of protein kinase C (PKC). The PKC/PKD pathway regulates many important biological processes in response to growth factor receptor and G-protein-coupled receptor activation. Recent studies have linked PKD to hyperproliferative disorders and cancer in several organs. Aberrant expression and activity of PKD have been demonstrated in malignant tumors and are associated with tumor progression. PKD has been implicated in neoplastic transformation and tumor metastasis by modulating tumor cell proliferation, survival, migration, invasion, and, potentially, angiogenesis. Important downstream targets of PKC/PKD in these processes have been identified. Furthermore, selective targeting of the PKC/PKD signaling in cancer is now possible with the discovery of potent and selective small molecule inhibitors of PKD. Thus, the PKC/PKD pathway may contribute to cancer development and represent an emerging target for cancer therapy.

Keywords Protein kinase D • Protein kinase C • Cancer • Diacylglycerol • Signal transduction • Kinase inhibitor • Cancer therapy

14.1 Protein Kinase D (PKD) Is a Novel Receptor of Diacylglycerol (DAG) and Phorbol Esters

DAG is a key second messenger in cells. It is generated through lipid hydrolysis by phospholipase C (PLC) that is activated in response to G-protein-coupled receptors (GPCRs) or growth factor receptors. Phorbol esters, the natural products from plants and potent tumor promoters in mouse skin, are pharmacological analogs of DAG

Q.J. Wang (✉)
Department of Pharmacology and Chemical Biology, University of Pittsburgh
School of Medicine, E1354 BST, Pittsburgh, PA 15261, USA
e-mail: qjw1@pitt.edu

M.G. Kazanietz (ed.), *Protein Kinase C in Cancer Signaling and Therapy*,
Current Cancer Research, DOI 10.1007/978-1-60761-543-9_14,

(Blumberg 1988). DAG and phorbol esters target a variety of structurally and functionally divergent proteins, named "DAG receptors," to regulate a variety of fundamental cellular responses (Colon-Gonzalez and Kazanietz 2006; Griner and Kazanietz 2007). There are at least six families of DAG receptors, including protein kinase C (PKC), PKD, chimaerin – the Rac GTPase-activating protein, the Ras guanyl nucleotide-releasing protein (RasGRP), Munc13, and DAG kinase (DGKβ and γ) (Brose and Rosenmund 2002; Fang et al. 2006; Ferrannini et al. 1999; Yang and Kazanietz 2003). All DAG receptors share a highly conserved structure motif-C1 domain that binds DAG and phorbol esters with high affinity (Colon-Gonzalez and Kazanietz 2006).

PKC is the first and the largest family of DAG receptors identified so far. It belongs to a large family of serine/threonine kinases that are divided into three subfamilies on the basis of their activator/co-activator requirements, the DAG- and Ca^{2+}-dependent classical PKCs (α, βI/βII, γ), the DAG-dependent but Ca^{2+}-independent novel PKCs (δ, ε, η, θ), and the Ca^{2+}- and DAG-independent atypical PKCs (ζ, λ/ι) (Bell and Burns 1991; Davidson-Moncada et al. 2002; Nishizuka 1992). PKC as the primary DAG receptor regulates a plethora of biological responses. Isoforms of PKC act in a cell type-, isoform-, and stimulus-specific manner. However, the basis underlying the differential effects of PKC isoforms is not fully understood. Downstream targets of PKC, such as PKD, may be important in determining the signaling specificity of PKC.

PKD is a novel family of serine/threonine kinases and DAG receptors (Rozengurt et al. 2005; Wang 2006). Three isoforms have been identified. The first isoform, PKD1, was identified in 1994 (Johannes et al. 1994; Valverde et al. 1994), followed by the discovery of two other isoforms, PKD2 (Sturany et al. 2001) and PKD3 (Hayashi et al. 1999). Members of PKD are highly homologous and show broad tissue distribution. The structure of PKD can be divided into the N-terminal regulatory region and the C-terminal catalytic region. The catalytic domain of PKD is highly homologous to Ca^{2+}/calmodulin-dependent kinases (CaMKs). Therefore, PKD has been classified as a subfamily of the CaMK superfamily (Manning et al. 2002). The regulatory region of PKD contains a C1 domain that binds DAG/phorbol esters, an acidic region, followed by a pleckstrin homology (PH) domain. The PH domain exerts an inhibitory effect on overall kinase activity, possibly acting as an autoinhibitory domain for PKD (Iglesias and Rozengurt 1998; Waldron and Rozengurt 2003) (Fig. 14.1).

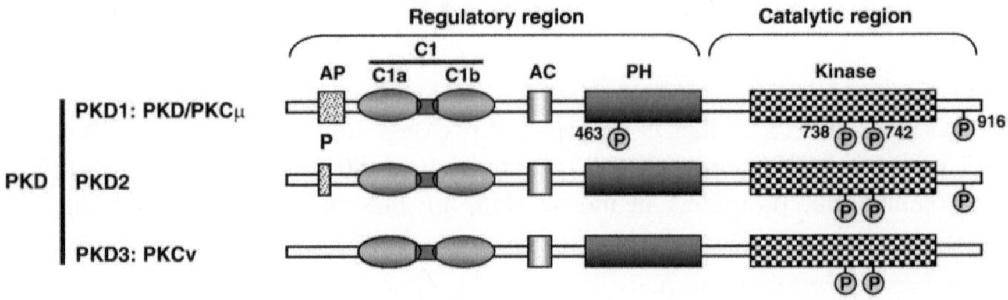

Fig. 14.1 A schematic diagram of the structure of the PKD family. C1, DAG-binding domain that contains C1a and C1b domains; *AP* apolar region; *P* proline-rich region; *AC* acidic domain; *PH* pleckstrin-homology domain; *Kinase* catalytic domain. Key phosphorylation sites in PKD are marked: Tyr 463, an Abl phosphorylation site; Ser 738 and 742, PKC phosphorylation sites; Ser 916, an autophosphorylation site

14.2 The Basis of the Canonical PKC/PKD Pathway

Although the C1 domain in the structure of PKD mediates the direct binding of DAG/phorbol esters, unlike most DAG receptors where the binding of DAG drives enzyme activation, the binding of DAG/phorbol esters to PKD *per se* has little impact on its kinase activity; rather the activity of PKD is controlled by phosphorylation through PKC (Rozengurt et al. 2005; Zugaza et al. 1996). It has been demonstrated over a decade ago by the Rozengurt group that PKD is activated in intact cells through a PKC-dependent mechanism (Zugaza et al. 1996). Subsequent studies from the same group show that the DAG-responsive PKC isoforms, predominantly the novel PKCs, can phosphorylate the two conserved serine residues in the activation loop of PKD, which relieves PKD from repression by the PH domain, leading to PKD activation (Waldron and Rozengurt 2003). This forms the molecular basis of a canonical PKC/PKD pathway that has been demonstrated in many cellular systems. The activation of the PKC/PKD pathway couples to a specific set of biological responses including protein transport, epigenetic regulation of gene expression, cell growth, proliferation, survival, and immune responses, etc. (Rozengurt et al. 2005). Deregulation of this pathway has been implicated in many pathological conditions and diseases including pathological cardiac remodeling and cancer.

The PKC/PKD pathway can be activated by a range of stimuli including GPCR agonists such as mitogenic neuropeptides (bombesin, neurotensin, etc.) (Paolucci and Rozengurt 1999; Zugaza et al. 1997), angiotensin II (Tan et al. 2004), lysophosphatidic acid (Paolucci et al. 2000), thrombin (Tan et al. 2003), endothelin-I (Zugaza et al. 1997), certain growth factors such as platelet-derived growth factor (Van Lint et al. 1998) and vascular endothelial growth factor (VEGF) (Wong and Jin 2005), oxidative stress (Storz and Toker 2003b; Waldron and Rozengurt 2000), and phorbol esters (Van Lint et al. 1995). The activation of the PKC/PKD pathway by GPCR agonists and growth factors is mediated through PLCs, while the activation by oxidative stress additionally involves tyrosine kinase Src and Abl.

PKD, upon activation by PKC, is unique in that activated PKD is mobile and can translocate to different subcellular locations to propagate and execute signals from DAG/PKC (Rozengurt et al. 2005). Among sites of redistribution, translocation to the nucleus is most evident and of particular interest. The activation of PKD caused by mitogenic neuropeptides, phorbol esters, and growth factors couples with transient nuclear translocation of PKD (Auer et al. 2005; Rey et al. 2003a, b).

14.3 The Major PKC/PKD-Regulated Pathways

Several major PKC/PKD-mediated signaling pathways have been identified and intensely studied (Fig. 14.2). Direct targets of PKD in these pathways have been demonstrated. One of the best characterized pathways is in the trans-Golgi network

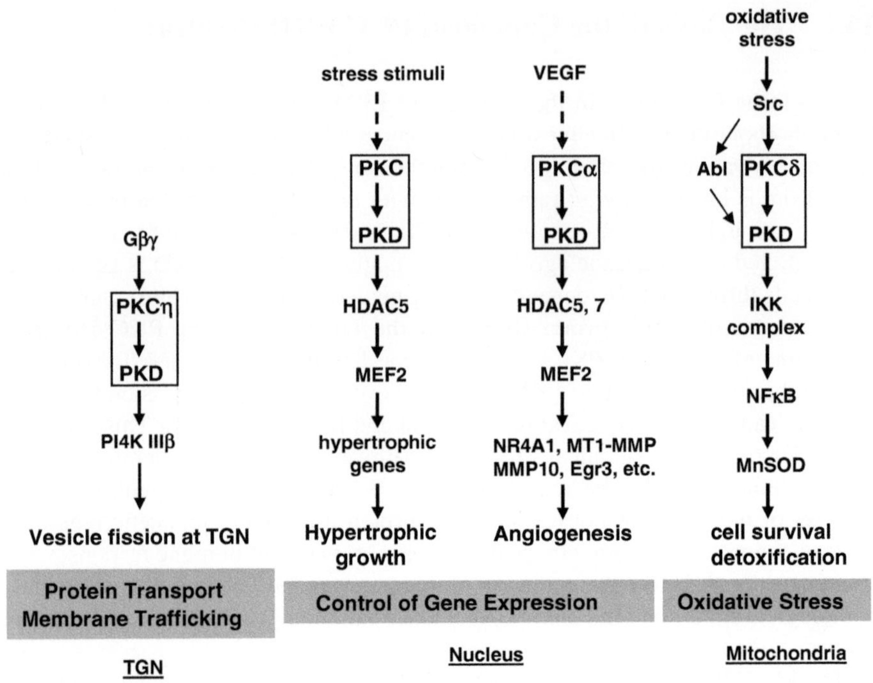

Fig. 14.2 The schemes of major PKD-mediated signaling pathways and biological responses

(TGN), where PKD plays an important role in maintaining proper Golgi structure and regulating protein transport from the Golgi to the plasma membrane (Jamora et al. 1999; Yeaman et al. 2004). The action of PKD in TGN is dependent on its catalytic activity since a kinase-dead PKD blocks the formation of transport vesicles from the Golgi to the cell surface (Liljedahl et al. 2001). PKD is activated in this pathway by the Gβγ subunit, which through PLCγ/PKCη/PKD regulates protein transport. At least one downstream target of PKD has been identified, namely phosphatidylinositol-4 kinase IIIβ (PI4KIIIβ) (Diaz Anel and Malhotra 2005; Hausser et al. 2005).

In the nucleus, a unique PKC/PKD pathway has been identified through modulating the class IIa histone deacetylases (HDAC 4, 5, 7, 9), which are important in the epigenetic control of gene expression (Vega et al. 2004). Histone acetylation and deacetylation are fundamental mechanisms that regulate chromatin structure and gene expression. Histone acetyltransferases catalyze the acetylation of histones, while the reverse reaction is catalyzed by HDACs. Class IIa HDACs are capable of interaction with certain transcription factors that are responsive to extracelluar signals and regulate their transcriptional activities. PKD, once activated by extracellular stimuli, is capable of phosphorylating several conserved serine residues in the N-terminus of HDACs, which creates docking sites for 14-3-3 chaperone proteins and driving the nuclear exclusion of these HDACs and resulting in derepression of gene expression (Bossuyt et al. 2008; Harrison et al. 2006; Vega et al. 2004). The myocyte enhancer factor-2 (MEF2) is one of the key downstream transcription factors targeted for repression by class IIa HDACs. PKD-mediated derepression of MEF2 allows the expression of many genes, including hypertrophic genes, VEGF-regulated genes and the nuclear receptor Nur77 (Avkiran et al. 2008).

The regulation of HDACs by PKD has been implicated in cardiac remodeling. PKD can be activated by a variety of hypertrophic stress stimuli in PKC-dependent and -independent mechanisms in cardiac myocytes (Harrison et al. 2006). Activated PKD relieves MEF2 repression by phosphorylating HDAC5 and promotes pathological cardiac remodeling (Harrison et al. 2006; Vega et al. 2004). The role of PKD1 in cardiac remodeling has been demonstrated in multiple rodent models of pathological cardiac remodeling. Ectopic expression of constitutively active PKD1 in mouse heart leads to dilated cardiomyopathy (Harrison et al. 2006). Conversely, cardiac-specific knockout of PKD1 in mice blunts cardiac hypertrophy, fibrosis caused by pressure overload, chronic adrenergic stimulation, and angiotensin II treatment and improved cardiac function, indicating an essential role of PKD in stress-induced pathological cardiac remodeling *in vivo* (Fielitz et al. 2008). Thus, PKD1 may represent a promising drug target of multiple heart diseases.

The regulation of class IIa HDACs by PKD has also been implicated in VEGF signaling in the vascular endothelial cells (Altschmied and Haendeler 2008). VEGF is the most prominent stimulator of angiogenesis in endothelial cells, a process that is linked to numerous vascular disorders and cancer. VEGF promotes angiogenesis by stimulating the proliferation, migration, and survival of endothelial cells. PKD1 has been shown to be activated by VEGF through PLCγ/PKCα in endothelial cells (Wong and Jin 2005). Activated PKD1, similar to that in the heart, phosphorylates class IIa HDACs, triggering their transport from the nucleus to the cytoplasm, and derepressing MEF2 transcriptional activity. At least two major class IIa HDACs are involved in this process, HDAC5 (Ha et al. 2008b) and HDAC7 (Ha et al. 2008a). HDAC7 in particular is expressed exclusively in endothelial cells. It maintains vascular integrity by repressing matrix metalloprotease 10 (MMP10) expression in endothelial cells. Recent studies indicate that VEGF stimulates HDAC7 phosphorylation and nuclear exclusion by activating PKD1, leading to the expression of MMPs, MTI-MMP, and MMP10 (Ha et al. 2008a). Other MEF2-regualted genes involved in angiogenesis have also been identified such as an orphan nuclear receptor – NR4A1 (Ha et al. 2008b) and early growth response 3 (Egr3) (Liu et al. 2008). Functionally, it has been shown that HDAC7 phosphorylation by PKD1 is essential for VEGF-induced endothelial cell proliferation/migration and tube formation/microvessel sprouting in a mouse aorta ring assay (Ha et al. 2008a). Hence, PKD is a critical component of VEGF-induced angiogenesis and a potential drug target for angiogenesis-related diseases, such as macular degeneration, diabetic retinopathy, and cancer.

Oxidative stress also activates a distinct PKD signaling pathway that involves Src-Abl and PKCδ, leading to the activation of NFκB transcription factor. Activation of PKD by oxidative stress promotes cell survival (Storz and Toker

2003b). At the mitochondria, PKD1 is a sensor of mitochondria reactive oxygen species (ROS) and relays signals between the mitochondria and the nucleus, resulting in mitochondria detoxification (Storz et al. 2005). Thus, PKD is also a potential drug target in ROS-associated pathophysiological processes.

In summary, there are several important PKC/PKD signaling pathways that regulate unique biological processes including protein trafficking, epigenetic gene expression, and oxidative stress signaling. Deregulation of these pathways could contribute to pathological conditions and cancer. Most of these pathways have been linked to cancer development; for example, HDACs are promising targets of anticancer therapy and HDAC inhibitors are in clinical trials for various cancers (Abbas and Gupta 2008; Al-Janadi et al. 2008; Hausser et al. 2005). Similarly, VEGF signaling plays essential roles in tumor angiogenesis, and agents that target components of the VEGF pathway have demonstrated proven clinical benefits (Chung and Stadler 2008).

14.4 The PKC/PKD Pathway in Cancer

A growing body of evidence supports a role of PKD in hyperproliferative disorders and cancer. Studies have demonstrated deregulated PKC/PKD pathways in several cancers and have linked them to tumor cell proliferation, survival, migration, and invasion. The downstream targets of PKC/PKD in these biological processes have also emerged (Fig. 14.3). Additionally, increasing evidence demonstrates isotype- and tumor type-specific functions of PKD isoforms in cancer. Here, we will discuss each potential role, the underlying signaling mechanisms, and targets of the PKC/PKD pathway in cancer.

14.4.1 Proliferation

PKD plays a key role in transducing mitogenic signals and promoting cell proliferation induced by growth factors, mitogenic GPCR agonists, and phorbol ester tumor promoters in normal and cancer cells (Rozengurt et al. 2005; Wang 2006). PKD1 promotes DNA synthesis and cell proliferation stimulated by GPCR agonists including bombesin, vasopressin, and neurotensin in Swiss 3T3 cells (Sinnett-Smith et al. 2004; Zhukova et al. 2001) and in human pancreatic carcinoma cells (Guha et al. 2002). Similar effects have been demonstrated for PKD2 (Sinnett-Smith et al. 2007) and PKD3 (Chen et al. 2008). Although PKD-mediated mitogenic signaling may be important in normal cell physiology, aberrant activation of the PKC/PKD pathway and upregulation of its signaling components could lead to abnormal growth and cancer. In this regard, aberrant activity and expression of PKD have been demonstrated in tumors originated from the skin, pancreas, and prostate. Changes of PKD expression and activity alter proliferative properties of

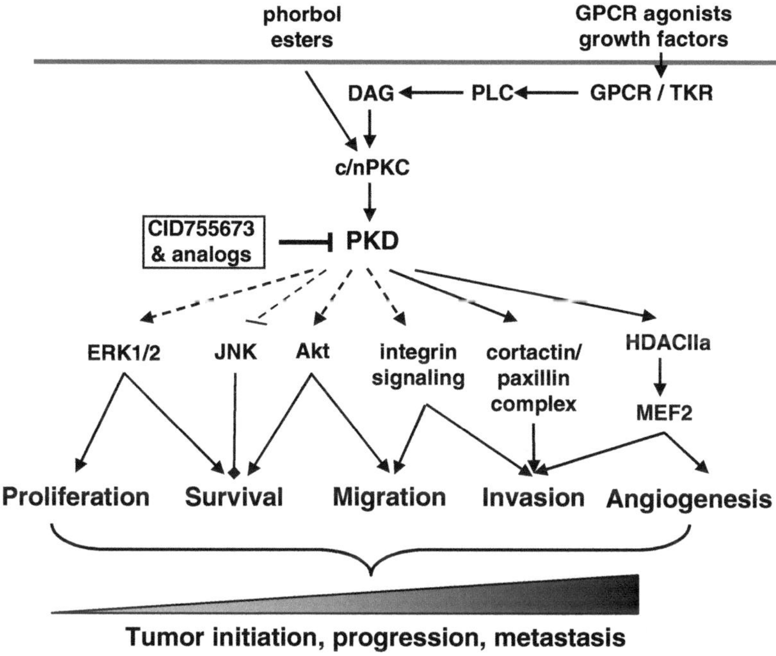

Fig. 14.3 A schematic representation of the PKC/PKD pathway and its potential targets in tumor initiation, progression, and metastasis

the tumor cells (Bollag et al. 2004; Chen et al. 2008; Seufferlein 2002). In the normal skin, the expression of PKD is restricted primarily to the stratum basalis, the proliferative epidermal compartment. PKD1 expression and activity decrease with differentiation and differentiation-induced growth arrest, and increase with proliferation and induction of proliferative phenotypes in primary mouse keratinocytes (Bollag et al. 2004; Rennecke et al. 1999). Thus, the downregulation of PKD may be required for normal keratinocyte maturation and its upregulation could potentially lead to hyperproliferative cell phenotype and possibly cancer (Bollag et al. 2004). This is further supported by the findings that PKD1 is elevated and aberrantly distributed in neoplastic mouse keratinocytes and hyperproliferative human skin disorders including basal cell carcinoma and psoriasis (Rennecke et al. 1999; Ristich et al. 2006). In the pancreas, PKD1 expression is markedly elevated in pancreatic tumor tissues. Overexpression of PKD1 strongly enhances cell proliferation and associates with resistance to apoptosis (Trauzold et al. 2003). In the prostate, our recent study has demonstrated elevated PKD3 in malignant human prostate tumor tissues and aberrant nuclear accumulation of PKD3 in high grade tumors, correlating to hyperactive state of this PKD isoform (Chen et al. 2008). Overexpression of PKD3 in prostate cancer cells promotes cell survival and proliferation, while depleting PKD3 by siRNA inhibits cell proliferation and cell cycle progression (Chen et al. 2008). Aberrant PKD activity and expression in various tumors and their proproliferative/prosurvival properties may fit to the "oncogene addition" theory (Weinstein 2002), implying that tumor cells could addict to the upregulated/hyperactive PKD signaling for growth and survival, providing the basis for targeting PKD for tumor suppression.

The proproliferative effects of PKD are in part mediated through modulating extracellular signal-regulated kinases (ERK1/2) activity. PKD was found to selectively activate the Raf/MEK1/ERK1/2 signaling (Hausser et al. 2001). It modulates both the magnitude and duration of ERK activation, though the later effect appears more prominent. Overexpression of wild-type PKD, but not the kinase-dead, leads to sustained MEK1/ERK1/2 activation and DNA synthesis in response to bombesin and vasopressin stimulation in Swiss 3T3 cells, which can be abrogated by a MEK1 inhibitor (Sinnett-Smith et al. 2004). In addition to GPCR agonists, the VEGF-stimulated endothelial cell proliferation is also mediated through the PKC/PKD pathway (PKCα/PKD1) and the activation of ERK1/2 in endothelial cells (Wong and Jin 2005). Although it remains to be determined the mechanisms through which PKD activates ERK1/2, a potential mediator has been identified as RIN1, a guanyl nucleotide exchange factor for Ras. PKD1 has been shown to phosphorylate and inhibit RIN1, leading to enhanced Ras and downstream ERK1/2 activity (Wang et al. 2002).

14.4.2 Survival and Apoptosis

The proproliferative effect of PKD in general coincides with its prosurvival and antiapoptotic functions in tumor cells. It has been shown that overexpression of PKD1 leads to increased cell survival and the induction of antiapoptotic proteins, c-FLIPL and survivin, in pancreatic adenocarcinoma cells (Trauzold et al. 2003). Although most studies are focused on PKD1, the prosurvival effect has been described for other PKD isoforms, as examples, overexpression of PKD3 confers resistance to phorbol 12-myristate 13-acetate (PMA)-induced apoptosis in prostate cancer cells (Chen et al. 2008). Of particular relevance to cancer development, PKD is a key mediator of cell survival response induced by oxidative stress (Storz and Toker 2003a, b). Increased ROS, a condition otherwise known as oxidative stress, due to internal (mitochondria, metabolic process, inflammation) or external (environment) factors plays an essential role in several cellular processes associated with the etiology and development of many cancers (Lambeth 2007). Oxidative stress activates PKD1, which in turn promotes cell survival by activating the NF-κB signaling (Song et al. 2009; Storz and Toker 2003b; Waldron and Rozengurt 2000). This is by far one of the best characterized PKC/PKD pathways. A unique feature of this pathway is the phosphorylation of PKD1 by Src and Abl tyrosine kinases, which primes PKD1 for full activation by PKCδ (Storz et al. 2003, 2004; Waldron and Rozengurt 2000). The activation of NF-κB could be a general mechanism for PKC/PKD to promote cell survival since the protection of cells from TNF-induced apoptosis by PKD1 is also mediated through induction of NF-κB-regulated prosurvival genes (Johannes et al. 1998). Additionally, lost of cell–cell contact also triggers PKD1-mediated NF-κB activation, which may be relevant to epithelial tumor cell survival (Cowell et al. 2009). Besides NF-κB, PKC/PKD may target additional pathways including the downregulation of JNK and p38 and/or the upregulation of EKR1/2 and Akt to promote cell survival (Buder-Hoffmann et al. 2009; Chen et al. 2008; Song et al. 2009; Wang et al. 2004). Among them, the suppression of the JNK signaling is especially compelling. JNK plays a crucial role in mediating apoptotic signaling induced by stress factors, inflammatory cytokines, and genotoxic agents. PKD was first found to attenuate epidermal growth factor (EGF) signaling to JNK (Bagowski et al. 1999; Hurd and Rozengurt 2001). Later, it was found that PKD formed complex with JNK and phosphorylated alternative sites in N-terminus of c-Jun (now identified as S58) to suppress the phosphorylation of c-Jun by JNK and modulate AP-1 transcriptional activity (Hurd et al. 2002; Waldron et al. 2007).

An interesting observation is that PKD1 can be cleaved by caspase-3 during apoptosis induced by 1-β-D-arabinofuranosylcytosine (ara-C) and other genotoxic agents (cisplatin, etoposide, and doxorubicin) in tumor cells (U-937 or A431) (Endo et al. 2000; Vantus et al. 2004). The cleavage generates active PKD1 catalytic fragments (59 and 62 kDa in case of doxorubicin treatment) (Vantus et al. 2004). However, although the cleaved catalytic fragment of PKD1 is not sufficient to induce apoptosis when overexpressed, it sensitizes tumor cells to DNA damage-induced apoptosis (Endo et al. 2000). This finding suggests that increased PKD1 expression may benefit chemotherapy by enhancing chemosensitivity of tumor cells, implying that PKD may be pro-apoptotic under certain conditions. Assessing the functional relevance of the cleavage at an endogenous level, such as selective blockade of the cleavage, may bring more insights.

14.4.3 Adhesion, Migration, and Invasion

Cell adhesion, migration, and invasion play important roles in tumor initiation, progression, and metastasis. For a cancer cell to metastasize, it is required to escape from the primary tumor, enter the circulation, arrest in the microcirculation, invade and grow in a secondary tissue compartment. This complex multistep process is dependent on many properties of a tumor cell including adhesion, migration, invasion, activities of proteases, survival and growth, and many regulatory components need to act in a highly concerted manner to drive a cell to move and invade. PKD plays an essential role in the control of cell motility and invasion. Accumulating evidence indicate that PKD promotes cell motility at multiple levels, and isoforms of PKD are differentially implicated. In fibroblasts, PKD regulates cell migration through at least two important processes, trafficking and integrin signaling. It has been demonstrated that PKD1 modulates fibroblast locomotion and localized Rac1-dependent leading edge activity by affecting anterograde membrane traffic from the TGN to the plasma membrane (Prigozhina and Waterman-Storer 2004). PKD1 also has been shown to regulate cell motility by promoting αvβ3 integrin recycling and delivery to nascent focal adhesions (Woods et al. 2004). These functions of PKD coincide with its major role as a protein/membrane trafficking regulator. Besides regulating integrin recycling, PKD1 has been shown to directly promote integrin activation by binding to the β1 integrin subunit cytoplasmic domain to regulate the activation and membrane translocation of GTPase Rap1, an integrin regulator, in T cells (Medeiros et al. 2005).

In tumor cells, PKD1 was found to colocalize with cortactin and paxillin in invasive breast cancer cells at invadopodia, the structure that associates with sites of active extracellular matrix degradation, which directly correlates with the invasive potential of the tumor cell (Bowden et al. 1999). However, in contrast to its positive effect in fibroblast motility, recent studies indicate that PKD1 is a negative regulator of tumor cell migration and invasion, which is coupled to its reduced expression in tumors. In gastric cancer, PKD1 was found to be silenced in 73% of gastric cancer cell lines due to hypermethylation. 59% of primary gastric tumor samples exhibit a two-fold decrease in PKD1 expression compared with the normal tissue counterparts, correlating to the high frequency of PKD1 hypermethylation in these tumor samples. Functionally, depletion of PKD1 by siRNA promotes gastric tumor cell invasion (Kim et al. 2008). Consistent with these findings, in breast cancer, PKD1 is detected at high levels in ductal epithelial cells of normal human breast tissue but reduced in more than 95% of analyzed human invasive breast tumors. Similar to that in the gastric tumor cells, silence of PKD1 is caused by DNA hypermethylation, and loss of PKD1 expression associates with enhanced invasiveness of breast cancer cells. The silence of PKD1 is coupled to increased MMP expression, implying the negative regulation of MMPs by PKD1 (Eiseler et al. 2009). In prostate cancer, PKD1 has also been shown to be downregulated in androgen-independent prostate cancer (Jaggi et al. 2003) and overexpression of PKD1 associates with enhanced cellular aggregation and reduced motility by binding and phosphorylating E-cadherin in prostate cancer cells (Jaggi et al. 2005). A common underlying message from these studies is that PKD1 silencing by hypermethylation may be one of the early events that predispose an individual to certain cancers. More studies are required to explain the intricate effects of PKD1 on cell migration and invasion in different cellular systems, though it has been postulated that loss of PKD1 in certain cancer could potentially relieve the negative regulatory influence of αvβ3 on α5β1 trafficking or alter the Rho–ROCK signaling, a pathway that has been implicated tumor cell invasion (Croft et al. 2004; Sahai and Marshall 2003; White et al. 2007). Compared to PKD1, little is known of the roles of PKD2 and PKD3 in migratory and invasive potentials of tumor cells. However, a positive role on cell proliferation and invasion has been demonstrated for PKD2 in human carcinoid BON cell line (Jackson et al. 2006), suggesting differential effects of PKD isoforms in tumor cell invasion.

14.4.4 Other Potential Roles in Tumor Development

In addition to direct regulatory effects on tumor cell growth, survival, migration, and invasion, PKD could potentially contribute to tumor progression and metastasis by promoting angiogenesis. As described previously, PKD is a key transducer of

VEGF-induced angiogenesis (Altschmied and Haendeler 2008), which is essential for the development of tumor vasculature that supports the growth and metastasis of tumors. This potential role of PKD in tumor development should be evaluated in future studies.

14.5 Advances in Targeting the PKC/PKD Pathway by Small Molecules

Since the discovery of the first PKD isoform in 1994 (Johannes et al. 1994; Valverde et al. 1994), no PKD-specific inhibitors have been reported until recently (Sharlow et al. 2008). The lack of a potent and selective inhibitor for PKD has greatly impeded analysis of PKD in biological processes and targeting PKD in pathological conditions. Several kinase inhibitors have been reported to inhibit PKD, including staurosporine (IC_{50}=40 nM), staurosporine-derived compound K252a (IC_{50}=7 nM) and Gö6976 (IC_{50}=20 nM) (Gschwendt et al. 1996), and isoquinoline sulfonamide H89 (IC_{50}=0.5 μM) (Johannes et al. 1995). However, these inhibitors generally have many targets and are not suitable for dissecting PKD-specific pathways or for therapeutic application. The PKD-sensitive indolocarbazole Gö6976 in combination with Gö6983, a pan-PKC inhibitor that inhibits PKD poorly, has been used widely in the past as an alternative to dissect PKD-mediated cellular processes in intact cells despite the fact that it is foremost known as a PKC inhibitor that preferentially inhibits classical PKC isoforms at single digit nanomolar concentrations (Martiny-Baron et al. 1993). This inhibitor clearly lacks the specificity to selectively block PKD signaling let alone for therapeutic applications. Although other compounds such as *trans*-3,4′,5-trihydroxystilbene (resveratrol), an antioxidant and chemopreventive agent, has also been demonstrated to inhibit PKD at IC_{50} 200 μM *in vitro* and 800 μM *in vivo*, these agents are not suitable PKD-selective inhibitors.

The apparent shortage of an effective PKD ablative agent urged us to launch a major HTS campaign to search for selective small molecule inhibitors of PKD, which ultimately leads to the recent discovery of the first potent and selective PKD inhibitor CID755673 (Sharlow et al. 2008). This compound is identified by HTS of the Pittsburgh Molecular Libraries Screening Center's (PMLSC) 196,173 member library using an immobilized metal affinity phosphochemical (IMAP)-based fluorescence polarization (FP) assay developed specifically for PKD1. CID755673 is a pan-PKD inhibitor with potency in the submicromolar concentrations and is considerably superior in specificity to all reported PKD inhibitors. In intact cells, it blocks the activation of endogenous PKD1 and inhibits several known biological actions of PKD1. In particular, this novel PKD inhibitor suppresses proliferation, migration, and invasion of prostate cancer cells. CID755673 is not competitive with ATP or substrate (unpublished data) for enzyme inhibition, implying unique mechanisms of action (Sharlow et al. 2008). In summary, this novel agent shows great promises for further development into an effective therapeutic agent that selectively targets the PKC/PKD signaling.

14.6 Concluding Remarks

It has now been well established that the classical and novel PKC isoforms signal through PKD to regulate a range of fundamental cellular processes. Recent studies have demonstrated potential roles of the PKC/PKD pathway at all stages of tumor development. Aberrant PKD expression and activity have been demonstrated and associated with tumor progression. Thus, the PKC/PKD pathway has emerged as a novel target for cancer therapy. In the future, more efforts should be devoted to dissect the specific roles of PKD isoforms at different stages of tumor development, to advance cellular studies to *in vivo* animal models of cancer and to exploit the possibility of targeting the PKC/PKD pathway therapeutically using potent and selective novel small molecule inhibitors of PKD.

Acknowledgments Work in our laboratory is supported by grants from the National Institutes of Health.

References

Abbas, A., & Gupta, S. (2008). The role of histone deacetylases in prostate cancer. *Epigenetics, 3*, 300–309.

Al-Janadi, A., Chandana, S. R., & Conley, B. A. (2008). Histone deacetylation: An attractive target for cancer therapy? *Drugs in R&D, 9*, 369–383.

Altschmied, J., & Haendeler, J. (2008). A new kid on the block: PKD1: A promising target for antiangiogenic therapy? *Arteriosclerosis, Thrombosis, and Vascular Biology, 28*, 1689–1690.

Auer, A., von Blume, J., Sturany, S., von Wichert, G., Van Lint, J., Vandenheede, J., et al. (2005). Role of the regulatory domain of protein kinase D2 in phorbol ester binding, catalytic activity, and nucleocytoplasmic shuttling. *Molecular Biology of the Cell, 16*, 4375–4385.

Avkiran, M., Rowland, A. J., Cuello, F., & Haworth, R. S. Ullrich (2008). Protein kinase d in the cardiovascular system: Emerging roles in health and disease. *Circulation Research, 102*, 157–163.

Bagowski, C. P., Stein-Gerlach, M., Choidas, A., & Ullrich, A. (1999). Cell-type specific phosphorylation of threonines T654 and T669 by PKD defines the signal capacity of the EGF receptor. *The EMBO Journal, 18*, 5567–5576.

Bell, R. M., & Burns, D. J. (1991). Lipid activation of protein kinase C. *The Journal of Biological Chemistry, 266*, 4661–4664.

Blumberg, P. M. (1988). Protein kinase C as the receptor for the phorbol ester tumor promoters: Sixth Rhoads memorial award lecture. *Cancer Research, 48*, 1–8.

Bollag, W. B., Dodd, M. E., & Shapiro, B. A. (2004). Protein kinase D and keratinocyte proliferation. *Drug News & Perspectives, 17*, 117–126.

Bossuyt, J., Helmstadter, K., Wu, X., Clements-Jewery, H., Haworth, R. S., Avkiran, M. et al. (2008). Ca2+/calmodulin-dependent protein kinase II{delta} and protein kinase D overexpression reinforce the histone deacetylase 5 redistribution in heart failure. *Circulation Research, 102*, 695–702.

Bowden, E. T., Barth, M., Thomas, D., Glazer, R. I., & Mueller, S. C. (1999). An invasion-related complex of cortactin, paxillin and PKCmu associates with invadopodia at sites of extracellular matrix degradation. *Oncogene, 18*, 4440–4449.

Brose, N., & Rosenmund, C. (2002). Move over protein kinase C, you've got company: Alternative cellular effectors of diacylglycerol and phorbol esters. *Journal of Cell Science, 115*, 4399–4411.

Buder-Hoffmann, S. A., Shukla, A., Barrett, T. F., MacPherson, M. B., Lounsbury, K. M., & Mossman, B. T. (2009). A protein kinase Cdelta-dependent protein kinase D pathway modulates ERK1/2 and JNK1/2 phosphorylation and Bim-associated apoptosis by asbestos. *American Journal of Pathology, 174*, 449–459.

Chen, J., Deng, F., Singh, S. V., & Wang, Q. J. (2008). Protein kinase D3 (PKD3) contributes to prostate cancer cell growth and survival through a PKCepsilon/PKD3 pathway downstream of Akt and ERK 1/2. *Cancer Research, 68*, 3844–3853.

Chung, E. K., & Stadler, W. M. (2008). Vascular endothelial growth factor pathway – targeted therapy as initial systemic treatment of patients with renal cancer. *Clinical Genitourinary Cancer, 6*, s22–s28.

Colon-Gonzalez, F., & Kazanietz, M. G. (2006). C1 domains exposed: From diacylglycerol binding to protein–protein interactions. *Biochimica et Biophysica Acta, 1761*, 827–837.

Cowell, C. F., Yan, I. K., Eiseler, T., Leightner, A. C., Doppler, H., & Storz, P. (2009). Loss of cell-cell contacts induces NF-kappaB via RhoA-mediated activation of protein kinase D1. *Journal of Cellular Biochemistry, 106*(4), 714–728.

Croft, D. R., Sahai, E., Mavria, G., Li, S., Tsai, J., Lee, W. M., et al. (2004). Conditional ROCK activation in vivo induces tumor cell dissemination and angiogenesis. *Cancer Research, 64*, 8994–9001.

Davidson-Moncada, J. K., Lopez-Lluch, G., Segal, A. W., & Dekker, L. V. (2002). Involvement of protein kinase D in Fc gamma-receptor activation of the NADPH oxidase in neutrophils. *Biochemical Journal, 363*, 95–103.

Diaz Anel, A. M., & Malhotra, V. (2005). PKCeta is required for beta1gamma2/beta3gamma2- and PKD-mediated transport to the cell surface and the organization of the Golgi apparatus. *The Journal of Cell Biology, 169*, 83–91.

Eiseler, T., Doeppler, H., Yan, I. K., Goodison, S., & Storz, P. (2009). Protein Kinase D1 regulates MMP expression and inhibits breast cancer cell invasion. *Breast Cancer Research, 11*, R13.

Endo, K., Oki, E., Biedermann, V., Kojima, H., Yoshida, K., Johannes, F. J., et al. (2000). Proteolytic cleavage and activation of protein kinase C [micro] by caspase-3 in the apoptotic response of cells to 1-beta -D-arabinofuranosylcytosine and other genotoxic agents. *The Journal of Biological Chemistry, 275*, 18476–18481.

Fang, C. X., Yang, X., Sreejayan, N., & Ren, J. (2006). Acetaldehyde promotes rapamycin-dependent activation of p70(S6K) and glucose uptake despite inhibition of Akt and mTOR in dopaminergic SH-SY5Y human neuroblastoma cells. *Exp Neurol, 203*, 196–204.

Ferrannini, E., Galvan, A. Q., Gastaldelli, A., Camastra, S., Sironi, A. M., Toschi, E., et al. (1999). Insulin: New roles for an ancient hormone. *European Journal of Clinical Investigation, 29*, 842–852.

Fielitz, J., Kim, M. S., Shelton, J. M., Qi, X., Hill, J. A., Richardson, J. A., et al. (2008). Requirement of protein kinase D1 for pathological cardiac remodeling. *Proceedings of the National Academy of Sciences of the United States of America, 105*, 3059–3063.

Griner, E. M., & Kazanietz, M. G. (2007). Protein kinase C and other diacylglycerol effectors in cancer. *Nature Reviews Cancer, 7*, 281–294.

Gschwendt, M., Dieterich, S., Rennecke, J., Kittstein, W., Mueller, H. J., & Johannes, F. J. (1996). Inhibition of protein kinase C mu by various inhibitors. Differentiation from protein kinase c isoenzymes. *FEBS Letters, 392*, 77–80.

Guha, S., Rey, O., & Rozengurt, E. (2002). Neurotensin induces protein kinase C-dependent protein kinase D activation and DNA synthesis in human pancreatic carcinoma cell line PANC-1. *Cancer Research, 62*, 1632–1640.

Ha, C. H., Jhun, B. S., Kao, H. Y., & Jin, Z. G. (2008). VEGF stimulates HDAC7 phosphorylation and cytoplasmic accumulation modulating matrix metalloproteinase expression and angiogenesis. *Arteriosclerosis, Thrombosis, and Vascular Biology, 28*, 1782–1788.

Ha, C. H., Wang, W., Jhun, B. S., Wong, C., Hausser, A., Pfizenmaier, K., et al. (2008). Protein kinase D-dependent phosphorylation and nuclear export of histone deacetylase 5 mediates vascular endothelial growth factor-induced gene expression and angiogenesis. *The Journal of Biological Chemistry, 283*, 14590–14599.

Harrison, B. C., Kim, M. S., van Rooij, E., Plato, C. F., Papst, P. J., Vega, R. B., et al. (2006). Regulation of cardiac stress signaling by protein kinase d1. *Molecular and Cellular Biology, 26*, 3875–3888.

Hausser, A., Storz, P., Hubner, S., Braendlin, I., Martinez-Moya, M., Link, G., et al. (2001). Protein kinase C mu selectively activates the mitogen-activated protein kinase (MAPK) p42 pathway. *FEBS Letters, 492*, 39–44.

Hausser, A., Storz, P., Martens, S., Link, G., Toker, A., & Pfizenmaier, K. (2005). Protein kinase D regulates vesicular transport by phosphorylating and activating phosphatidylinositol-4 kinase IIIbeta at the Golgi complex. *Nature Cell Biology, 7*, 880–886.

Hayashi, A., Seki, N., Hattori, A., Kozuma, S., & Saito, T. (1999). PKCnu, a new member of the protein kinase C family, composes a fourth subfamily with PKCmu. *Biochimica et Biophysica Acta, 1450*, 99–106.

Hurd, C., & Rozengurt, E. (2001). Protein kinase D is sufficient to suppress EGF-induced c-Jun Ser 63 phosphorylation. *Biochemical and Biophysical Research Communications, 282*, 404–408.

Hurd, C., Waldron, R. T., & Rozengurt, E. (2002). Protein kinase D complexes with C-Jun N-terminal kinase via activation loop phosphorylation and phosphorylates the C-Jun N-terminus. *Oncogene, 21*, 2154–2160.

Iglesias, T., & Rozengurt, E. (1998). Protein kinase D activation by mutations within its pleckstrin homology domain. *The Journal of Biological Chemistry, 273*, 410–416.

Jackson, L. N., Li, J., Chen, L. A., Townsend, C. M., & Evers, B. M. (2006). Overexpression of wild-type PKD2 leads to increased proliferation and invasion of BON endocrine cells. *Biochemical and Biophysical Research Communications, 348*, 945–949.

Jaggi, M., Rao, P. S., Smith, D. J., Hemstreet, G. P., & Balaji, K. C. (2003). Protein kinase C mu is down-regulated in androgen-independent prostate cancer. *Biochemical and Biophysical Research Communications, 307*, 254–260.

Jaggi, M., Rao, P. S., Smith, D. J., Wheelock, M. J., Johnson, K. R., Hemstreet, G. P., et al. (2005). E-cadherin phosphorylation by protein kinase D1/protein kinase C{mu} is associated with altered cellular aggregation and motility in prostate cancer. *Cancer Research, 65*, 483–492.

Jamora, C., Yamanouye, N., Van Lint, J., Laudenslager, J., Vandenheede, J. R., Faulkner, D. J., et al. (1999). Gbetagamma-mediated regulation of Golgi organization is through the direct activation of protein kinase D. *Cell, 98*, 59–68.

Johannes, F. J., Horn, J., Link, G., Haas, E., Siemienski, K., Wajant, H., et al. (1998). Protein kinase Cmu downregulation of tumor-necrosis-factor-induced apoptosis correlates with enhanced expression of nuclear-factor-kappaB-dependent protective genes. *European Journal of Biochemistry, 257*, 47–54.

Johannes, F. J., Prestle, J., Dieterich, S., Oberhagemann, P., Link, G., & Pfizenmaier, K. (1995). Characterization of activators and inhibitors of protein kinase C mu. *European Journal of Biochemistry, 227*, 303–307.

Johannes, F. J., Prestle, J., Eis, S., Oberhagemann, P., & Pfizenmaier, K. (1994). PKCu is a novel, atypical member of the protein kinase C family. *The Journal of Biological Chemistry, 269*, 6140–6148.

Kim, M., Jang, H. R., Kim, J. H., Noh, S. M., Song, K. S., Cho, J. S., et al. (2008). Epigenetic inactivation of protein kinase D1 in gastric cancer and its role in gastric cancer cell migration and invasion. *Carcinogenesis, 29*, 629–637.

Lambeth, J. D. (2007). Nox enzymes, ROS, and chronic disease: An example of antagonistic pleiotropy. *Free Radical Biology and Medicine, 43*, 332–347.

Liljedahl, M., Maeda, Y., Colanzi, A., Ayala, I., Van Lint, J., & Malhotra, V. (2001). Protein kinase D regulates the fission of cell surface destined transport carriers from the trans-Golgi network. *Cell, 104*, 409–420.

Liu, D., Evans, I., Britton, G., & Zachary, I. (2008). The zinc-finger transcription factor, early growth response 3, mediates VEGF-induced angiogenesis. *Oncogene, 27*, 2989–2998.

Manning, G., Whyte, D. B., Martinez, R., Hunter, T., & Sudarsanam, S. (2002). The protein kinase complement of the human genome. *Science, 298*, 1912–1934.

Martiny-Baron, G., Kazanietz, M. G., Mischak, H., Blumberg, P. M., Kochs, G., Hug, H., et al. (1993). Selective inhibition of protein kinase C isozymes by the indolocarbazole Go 6976. *The Journal of Biological Chemistry, 268*, 9194–9197.

Medeiros, R. B., Dickey, D. M., Chung, H., Quale, A. C., Nagarajan, L. R., Billadeau, D. D., et al. (2005). Protein kinase D1 and the beta 1 integrin cytoplasmic domain control beta 1 integrin function via regulation of Rap1 activation. *Immunity, 23*, 213–226.

Nishizuka, Y. (1992). Intracellular signaling by hydrolysis of phospholipids and activation of protein kinase C. *Science, 258*, 607–614.

Paolucci, L., & Rozengurt, E. (1999). Protein kinase D in small cell lung cancer cells: Rapid activation through protein kinase C. *Cancer Research, 59*, 572–577.

Paolucci, L., Sinnett-Smith, J., & Rozengurt, E. (2000). Lysophosphatidic acid rapidly induces protein kinase D activation through a pertussis toxin-sensitive pathway. *American Journal of Physiology. Cell physiology, 278*, C33–C39.

Prigozhina, N. L., & Waterman-Storer, C. M. (2004). Protein kinase D-mediated anterograde membrane trafficking is required for fibroblast motility. *Current Biology, 14*, 88–98.

Rennecke, J., Rehberger, P. A., Furstenberger, G., Johannes, F. J., Stohr, M., Marks, F., et al. (1999). Protein-kinase-Cmu expression correlates with enhanced keratinocyte proliferation in normal and neoplastic mouse epidermis and in cell culture. *International Journal of Cancer, 80*, 98–103.

Rey, O., Yuan, J., Young, S. H., & Rozengurt, E. (2003). Protein kinase C nu/protein kinase D3 nuclear localization, catalytic activation, and intracellular redistribution in response to G protein-coupled receptor agonists. *The Journal of Biological Chemistry, 278*, 23773–23785.

Rey, O., Zhukova, E., Sinnett-Smith, J., & Rozengurt, E. (2003). Vasopressin-induced intracellular redistribution of protein kinase D in intestinal epithelial cells. *Journal of Cellular Physiology, 196*, 483–492.

Ristich, V. L., Bowman, P. H., Dodd, M. E., & Bollag, W. B. (2006). Protein kinase D distribution in normal human epidermis, basal cell carcinoma and psoriasis. *British Journal of Dermatology, 154*, 586–593.

Rozengurt, E., Rey, O., & Waldron, R. T. (2005). Protein kinase D signaling. *The Journal of Biological Chemistry, 280*, 13205–13208.

Sahai, E., & Marshall, C. J. (2003). Differing modes of tumour cell invasion have distinct requirements for Rho/ROCK signalling and extracellular proteolysis. *Nature Cell Biology, 5*, 711–719.

Seufferlein, T. (2002). Novel protein kinases in pancreatic cell growth and cancer. *International Journal of Gastrointestinal Cancer, 31*, 15–21.

Sharlow, E. R., Giridhar, K. V., Lavalle, C. R., Chen, J., Leimgruber, S., Barrett, R., et al. (2008). Potent and selective disruption of protein kinase d functionality by a benzoxoloazepinolone. *The Journal of Biological Chemistry, 283*, 33516–33526.

Sinnett-Smith, J., Zhukova, E., Hsieh, N., Jiang, X., & Rozengurt, E. (2004). Protein kinase D potentiates DNA synthesis induced by Gq-coupled receptors by increasing the duration of ERK signaling in swiss 3T3 cells. *The Journal of Biological Chemistry, 279*, 16883–16893.

Sinnett-Smith, J., Zhukova, E., Rey, O., & Rozengurt, E. (2007). Protein kinase D2 potentiates MEK/ERK/RSK signaling, c-Fos accumulation and DNA synthesis induced by bombesin in Swiss 3T3 cells. *Journal of Cellular Physiology, 211*, 781–790.

Song, J., Li, J., Qiao, J., Jain, S., Mark Evers, B., & Chung, D. H. (2009). PKD prevents H2O2-induced apoptosis via NF-kappaB and p38 MAPK in RIE-1 cells. *Biochemical and Biophysical Research Communications, 378*, 610–614.

Storz, P., Doppler, H., Johannes, F. J., & Toker, A. (2003). Tyrosine phosphorylation of protein kinase D in the pleckstrin homology domain leads to activation. *The Journal of Biological Chemistry, 278*, 17969–17976.

Storz, P., Doppler, H., & Toker, A. (2004). Protein kinase Cdelta selectively regulates protein kinase D-dependent activation of NF-kappaB in oxidative stress signaling. *Molecular and Cellular Biology, 24*, 2614–2626.

Storz, P., Doppler, H., & Toker, A. (2005). Protein kinase D mediates mitochondrion-to-nucleus signaling and detoxification from mitochondrial reactive oxygen species. *Molecular and Cellular Biology, 25*, 8520–8530.

Storz, P., & Toker, A. (2003a). NF-kappaB signaling – an alternate pathway for oxidative stress responses. *Cell Cycle, 2*, 9–10.

Storz, P., & Toker, A. (2003b). Protein kinase D mediates a stress-induced NF-kappaB activation and survival pathway. *The EMBO Journal, 22*, 109–120.

Sturany, S., Van Lint, J., Muller, F., Wilda, M., Hameister, H., Hocker, M., et al. (2001). Molecular cloning and characterization of the human protein kinase D2. A novel member of the protein kinase D family of serine threonine kinases. *The Journal of Biological Chemistry, 276*, 3310–3318.

Tan, M., Xu, X., Ohba, M., Ogawa, W., & Cui, M. Z. (2003). Thrombin rapidly induces protein kinase D phosphorylation, and protein kinase C delta mediates the activation. *The Journal of Biological Chemistry, 278*, 2824–2828.

Tan, M., Xu, X., Ohba, M., Ogawa, W., & Cui, M. Z. (2004). Angiotensin II-induced protein kinase D activation is regulated by protein kinase Cdelta and mediated via the angiotensin II type 1 receptor in vascular smooth muscle cells. *Arteriosclerosis, Thrombosis, and Vascular Biology, 24*, 2271–2276.

Trauzold, A., Schmiedel, S., Sipos, B., Wermann, H., Westphal, S., Roder, C., et al. (2003). PKCmu prevents CD95-mediated apoptosis and enhances proliferation in pancreatic tumour cells. *Oncogene, 22*, 8939–8947.

Valverde, A. M., Sinnett-Smith, J., Van Lint, J., & Rozengurt, E. (1994). Molecular cloning and characterization of protein kinase D: A target for diacylglycerol and phorbol esters with a distinctive catalytic domain. *Proceedings of the National Academy of Sciences of the United States of America, 91*, 8572–8576.

Van Lint, J., Ni, Y., Valius, M., Merlevede, W., & Vandenheede, J. R. (1998). Platelet-derived growth factor stimulates protein kinase D through the activation of phospholipase Cgamma and protein kinase C. *The Journal of Biological Chemistry, 273*, 7038–7043.

Van Lint, J. V., Sinnett-Smith, J., & Rozengurt, E. (1995). Expression and characterization of PKD, a phorbol ester and diacylglycerol-stimulated serine protein kinase. *The Journal of Biological Chemistry, 270*, 1455–1461.

Vantus, T., Vertommen, D., Saelens, X., Rykx, A., De Kimpe, L., Vancauwenbergh, S., et al. (2004). Doxorubicin-induced activation of protein kinase D1 through caspase-mediated proteolytic cleavage: Identification of two cleavage sites by microsequencing. *Cellular Signalling, 16*, 703–709.

Vega, R. B., Harrison, B. C., Meadows, E., Roberts, C. R., Papst, P. J., Olson, E. N., et al. (2004). Protein kinases C and D mediate agonist-dependent cardiac hypertrophy through nuclear export of histone deacetylase 5. *Molecular and Cellular Biology, 24*, 8374–8385.

Waldron, R. T., & Rozengurt, E. (2000). Oxidative stress induces protein kinase D activation in intact cells. Involvement of Src and dependence on protein kinase C. *The Journal of Biological Chemistry, 275*, 17114–17121.

Waldron, R. T., & Rozengurt, E. (2003). Protein kinase C phosphorylates protein kinase D activation loop Ser744 and Ser748 and releases autoinhibition by the pleckstrin homology domain. *The Journal of Biological Chemistry, 278*, 154–163.

Waldron, R. T., Whitelegge, J. P., Faull, K. F., & Rozengurt, E. (2007). Identification of a novel phosphorylation site in c-jun directly targeted in vitro by protein kinase D. *Biochemical and Biophysical Research Communications, 356*, 361–367.

Wang, Q. J. (2006). PKD at the crossroads of DAG and PKC signaling. *Trends in Pharmacological Sciences, 27*, 317–323.

Wang, Y., Schattenberg, J. M., Rigoli, R. M., Storz, P., & Czaja, M. J. (2004). Hepatocyte resistance to oxidative stress is dependent on protein kinase C-mediated down-regulation of c-Jun/AP-1. *The Journal of Biological Chemistry, 279*, 31089–31097.

Wang, Y., Waldron, R. T., Dhaka, A., Patel, A., Riley, M. M., Rozengurt, E., et al. (2002). The RAS effector RIN1 directly competes with RAF and is regulated by 14-3-3 proteins. *Molecular and Cellular Biology, 22*, 916–926.

Weinstein, I. B. (2002). Cancer. Addiction to oncogenes – the Achilles heal of cancer. *Science, 297*, 63–64.
White, D. P., Caswell, P. T., & Norman, J. C. (2007). alpha v beta3 and alpha5beta1 integrin recycling pathways dictate downstream Rho kinase signaling to regulate persistent cell migration. *The Journal of Cell Biology, 177*, 515–525.
Wong, C., & Jin, Z. G. (2005). Protein kinase C-dependent protein kinase D activation modulates ERK signal pathway and endothelial cell proliferation by vascular endothelial growth factor. *The Journal of Biological Chemistry, 280*, 33262–33269.
Woods, A. J., White, D. P., Caswell, P. T., & Norman, J. C. (2004). PKD1/PKCmu promotes alphavbeta3 integrin recycling and delivery to nascent focal adhesions. *The EMBO Journal, 23*, 2531–2543.
Yang, C., & Kazanietz, M. G. (2003). Divergence and complexities in DAG signaling: Looking beyond PKC. *Trends in Pharmacological Sciences, 24*, 602–608.
Yeaman, C., Ayala, M. I., Wright, J. R., Bard, F., Bossard, C., Ang, A., et al. (2004). Protein kinase D regulates basolateral membrane protein exit from trans-Golgi network. *Nature Cell Biology, 6*, 106–112.
Zhukova, E., Sinnett-Smith, J., & Rozengurt, E. (2001). Protein kinase D potentiates DNA synthesis and cell proliferation induced by bombesin, vasopressin, or phorbol esters in Swiss 3T3 cells. *The Journal of Biological Chemistry, 276*, 40298–40305.
Zugaza, J. L., Sinnett-Smith, J., Van Lint, J., & Rozengurt, E. (1996). Protein kinase D (PKD) activation in intact cells through a protein kinase C-dependent signal transduction pathway. *The EMBO Journal, 15*, 6220–6230.
Zugaza, J. L., Waldron, R. T., Sinnett-Smith, J., & Rozengurt, E. (1997). Bombesin, vasopressin, endothelin, bradykinin, and platelet-derived growth factor rapidly activate protein kinase D through a protein kinase C-dependent signal transduction pathway. *The Journal of Biological Chemistry, 272*, 23952–23960.

Chapter 15
Transgenic Mouse Models to Investigate Functional Specificity of Protein Kinase C Isoforms in the Development of Squamous Cell Carcinoma, a Nonmelanoma Human Skin Cancer

Ajit K. Verma

Abstract The multistage model of mouse skin carcinogenesis is a useful system in which biochemical events unique to initiation, promotion, or progression steps of carcinogenesis can be studied and related to cancer formation. 12-O-tetradecanoylphorbol-13-acetate (TPA), a component of croton oil, is a potent mouse skin tumor promoter (CRC Critical Reviews in Toxicology 2:419–443, 1974; The Journal of Investigative Dermatology Symposium Proceedings 1:147–150, 1996; Pharmacology and Therapeutics 54:63–128, 1992). A major breakthrough in understanding the mechanism of TPA tumor promotion has been the identification of protein kinase C (PKC), as its receptor. PKC, which is ubiquitous in eukaryotes, is a major intracellular receptor for TPA (Nature Reviews Cancer 7:281–294, 2007). PKC forms part of the signal transduction system involving the turnover of inositol phospholipids and is activated by DAG, which is produced as a consequence of this turnover (Nature Reviews Cancer 7:281–294, 2007). PKC represents a family of phospholipid-dependent serine/threonine kinases (Nature Reviews Cancer 7:281–294, 2007; Biochemical Journal 332:281–292, 1998; Chemical Reviews 101:2353–2364, 2001; Advances in Pharmacology 44:91–145, 1998). At least six PKC isoforms (α, δ, ε, η, μ, ζ) are expressed in both human and mouse skin (International Journal of Biochemistry and Cell Biology 36:1141–1146, 2004). Evidence is presented in this chapter, using genetic approach in intact mouse skin in vivo, indicating that: (1) PKC isoforms exhibit functional specificity in skin cancer induction, (2) PKCε mediates the development of skin cancer and (3) PKCε signal transduction pathway to development of skin cancer involves Stat3 interaction and expression of proinflammatory cytokine TNFα.

Keywords Protein kinase C • Skin • Mitogenesis • Transgenic mice

---

A.K. Verma (✉)
Department of Human Oncology, School of Medicine and Public Health,
University of Wisconsin, Madison, WI 53792, USA
e-mail: akverma@wisc.edu

M.G. Kazanietz (ed.), *Protein Kinase C in Cancer Signaling and Therapy*,
Current Cancer Research, DOI 10.1007/978-1-60761-543-9_15,

15.1 Introduction

Knowledge about the molecular mechanisms involved in the genesis of cancer is essential for the identification of cancer targets in early diagnosis as well as for the rational design of agents to prevent and/or treat cancer. The multistep model of mouse skin carcinogenesis has been on the forefront of the identification of irreversible genetic events of initiation and progression and epigenetic events of tumor promotion (Boutwell 1974; Yuspa et al. 1996; DiGiovanni 1992) (Fig. 15.1). The initiation can be accomplished by a single exposure to a sufficiently small dose of a carcinogen, and this step is rapid and irreversible. Given that humans are constantly exposed to environmental carcinogens, and that even the human diet contains nitrites and nitrates, which are converted in the gut to the potent carcinogens nitrosamines, one may believe the initiation step of carcinogenesis is inevitable. Promotion of tumor formation requires a repeated and prolonged exposure to a promoter, and that tumor promotion is reversible, at least in the early stages. An understanding of the promotion step of carcinogenesis is essential for human cancer prevention. There are two common protocols to induce skin cancer in mice: (1) initiation with 7,12-dimethylbenz[*a*]anthracene (DMBA) – promotion with TPA and (2) by the complete carcinogenesis using repeated exposure to ultraviolet

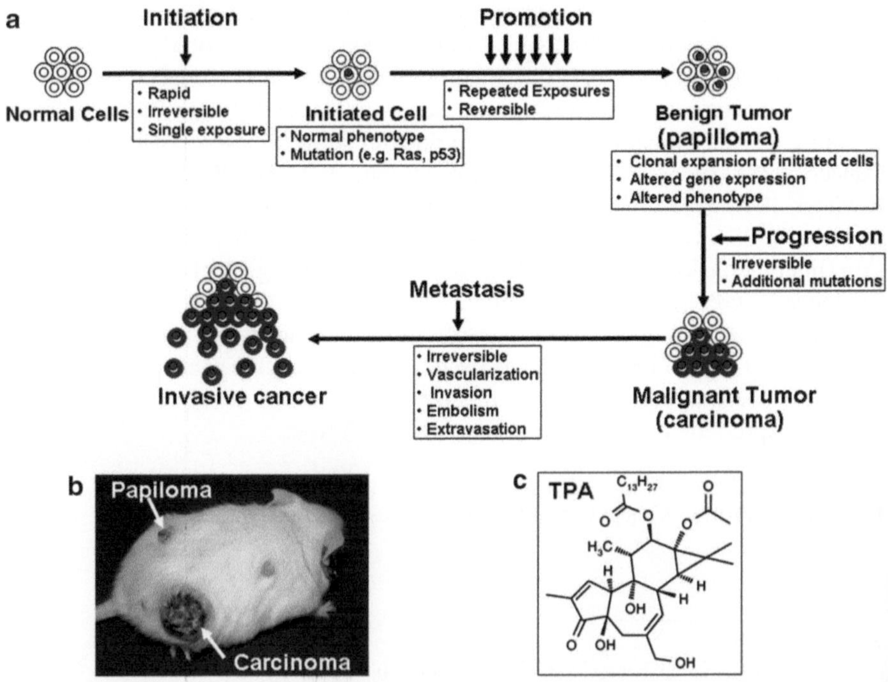

Fig. 15.1 (**a**) Multistep mouse skin carcinogenesis; (**b**) Mouse exhibiting skin papillomas and carcinomas elicited by the initiation/promotion protocol; (**c**) TPA structure

radiation (UVR) (Aziz et al. 2006). Both TPA and the tumor promotion component of UVR carcinogenesis involve clonal expansion of initiated cells as the result of aberrant expression of genes altered during tumor initiation (Wheeler et al. 2004, 2005). TPA and UVR have been reported to alter the expression of genes regulating inflammation, cell growth, and differentiation. Specific examples include upregulation of the expression of p21 (WAF1/C1P1), p53, AP-1 activation, ornithine decarboxylase, cyclooxygenase-2 (COX2), cytokines, and growth factors (Aziz et al. 2006; Wheeler et al. 2004, 2005; Reddig et al. 1999, 2000; Jansen et al. 2001a, b). We found that the development of skin cancer, either by TPA promotion or UVR, involves PKC activation, as a common converging point (Aziz et al. 2006; Wheeler et al. 2004, 2005; Reddig et al. 2000). Evidence is reviewed in this chapter, using PKC transgenic mouse models, indicating that PKC isozymes exhibit functional specificity in skin cancer induction.

15.2 Results and Discussion

15.2.1 Generation of PKC-Overexpressing Transgenic Mice

PKC isoforms in skin are differentially expressed in proliferative (basal layer) and nonproliferative compartments (spinous, granular, cornified layers), which exhibit divergence in their roles in the regulation of epidermal cell proliferation, differentiation, and apoptosis. Immunocytochemical localization of PKC isoforms indicate that PKCα is found in the membranes of suprabasal cells in the spinous and granular layers. PKCε is mostly localized in the proliferative basal layers. PKCη is localized exclusively in the granular layer. PKCδ is detected throughout the epidermis (Denning 2004). To evaluate the distinct role that each individual PKC isoform plays in vivo in mouse skin carcinogenesis, we generated transgenic mice overexpressing an epitope-tagged PKC (T7-PKC) under the control of the human keratin 14 promoter (Reddig et al. 1999, 2000; Jansen et al. 2001a, b). K5/K14 are expressed in the basal layer of the epidermis, which contains epidermal stem cells and transient amplifying cells (Vassar et al. 1989). The human K14 promoter was used to direct expression of mouse PKC cDNA (α, δ, ε) to basal epidermal keratinocytes of the mouse skin (Fig. 15.2). The human K14 promoter has been previously shown to direct high-level expression of various transgenes to the mouse basal epidermis (Vassar et al. 1989). To facilitate the five differentiations between the endogenous PKC and the exogenous transgene, a T7 bacteriophage epitope was subcloned to the amino terminus of PKC (T7-PKC). The transgenic vector (pG3Z-K14-T7-PKC) was initially confirmed to be functional by Western blot and immunocomplex kinase assays of transiently transfected cells. The linear 5.6 kb K14-T7-PKC expression cassette was removed from pG3Z-K14-T7-PKC by endonuclease restriction digestion with EheI and purified before injection into the pronuclei of fertilized FVB/n eggs (Reddig et al. 1999, 2000; Jansen et al. 2001a, b).

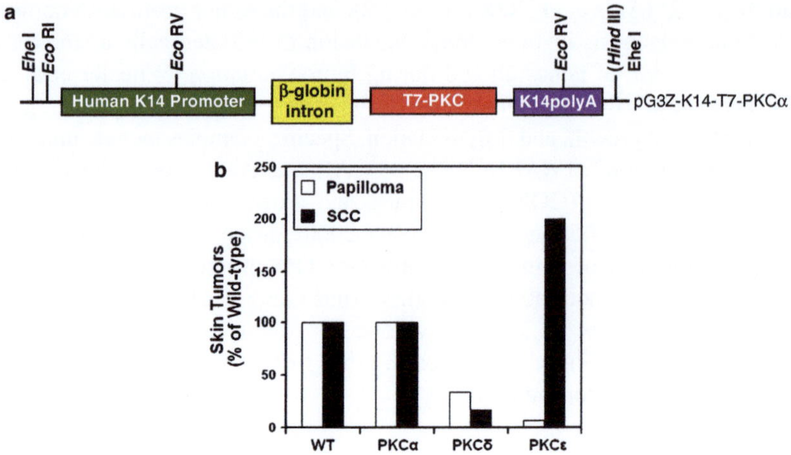

Fig. 15.2 (**a**) PKC expression vectors; (**b**) Susceptibility of PKC transgenic mice to skin carcinogenesis

PKCα overexpressing transgenic mice were also generated by Wang and Smart (1999). In their findings, PKCα expression was targeted using K5 promoter (Wang and Smart 1999).

15.2.2 Susceptibility of PKC Transgenic Mice to Skin Carcinogenesis

15.2.2.1 PKCα Transgenic Mice

K5-PKCα mice exhibited normal keratinocyte growth and differentiation in the epidermis. However, a single topical treatment with TPA resulted in an inflammatory response characterized by edema and extensive epidermal infiltration of neutrophils in the epidermis. Compared to TPA-treated wild-type mice, the epidermis of TPA-treated K5-PKCα mice displayed increased expression of COX-2, the neutrophil chemotactic factor macrophage inflammatory protein-2 (MIP-2) mRNA and the proinflammatory cytokine TNFα mRNA but not IL-6 or IL-1α mRNA (Wang and Smart 1999). Cataisson et al. (2005) also reported that transgenic mice overexpressing K5-PKCα in the skin exhibit severe intra-epidermal neutrophilic inflammation and keratinocyte apoptosis when treated topically with TPA. Activation of PKCα increases the production of TNFα and the transcription of chemotactic factors (MIP-2, KC, S100A8/A9), vascular endothelial growth factor, and GM-CSF in K5-PKCα keratinocytes (Cataisson et al. 2005).

To determine whether K5-PKCα mice display an altered response to TPA-promotion, DMBA-initiated K5-PKCα mice and wild-type mice were promoted with TPA. No differences in papilloma incidence or multiplicity were observed between K5-PKCα mice and wild-type littermates. A lack of effect of PKCα over-expression on skin tumor promotion by TPA was confirmed by Jansen et al. using K14-PKCα transgenic mice (Jansen et al. 2001a). These results (Jansen et al. 2001a; Wang and Smart 1999; Cataisson et al. 2005) indicate that the overexpression of PKCα in the epidermis increases the expression of specific proinflammatory mediators and induces cutaneous inflammation but has little to no effect on TPA tumor promotion.

15.2.2.2 PKCδ Transgenic Mice

PKCδ transgenic mice were generated by Reddig et al. (1999). Two mouse lines with high (line 16) and low (line 37) T7-PKCδ expression levels were generated. Immunoblots of the total extracts probed with the anti-PKCδ antibody displayed an eightfold increase in PKC protein levels in line 16 mice and a twofold increase in line 37 mice when compared with the endogenous level of PKCδ protein in wild-type littermates. Neither transgenic line 16 nor line 37 exhibited any significant phenotypic abnormalities. The results of mouse skin tumor promotion with the T7-PKCδ mice were dramatic. The T7-PKCδ line 16 mice averaged a 73% reduction in papilloma burden for both male and female mice. The appearance of tumors on the T7-PKCδ line 16 mice was also delayed by 4 weeks on average. The carcinoma incidence was also reduced in the line 16 mice. The T7-PKCδ line 37 mice exhibited lower levels of T7-PKCδ protein and activity compared with line 16 and did not display any alterations in sensitivity to mouse skin tumor promotion. The inhibition of tumor promotion in the line 16 mice and the lack of alterations in the line 37 mice imply that the inhibition of papilloma formation by treatment with DMBA/TPA requires a threshold level of PKCδ activity (Reddig et al. 1999).

The ability of PKCδ to suppress papilloma formation implies that its activation may block epidermal keratinocyte proliferation and/or transformation. These results are consistent with the role of PKCδ in in vitro cell culture studies that have shown PKCδ to be an inhibitor of cell growth and transformation. Overexpression of PKCδ inhibited the growth of fibroblasts, vascular smooth muscle cells, and endothelial cells by delaying passage through different phases of the cell cycle, depending on the cell type. Inhibition of vascular smooth muscle cell proliferation by elevated PKCδ levels correlated with decreased expression of cyclins D1 and E. In the mouse epidermis, TPA-induced proliferation correlated with upregulation of both cyclin D1 and cyclin E. Additionally, the homozygous deletion of cyclin D1 reduced the papilloma burden in response to DMBA-TPA treatment of the mouse skin. These proteins may be important mediators of the TPA response mediated by PKCδ in the mouse epidermis (reviewed in Reddig et al. 1999).

15.2.2.3 PKCε Transgenic Mice

PKCε transgenic mice were also generated by Reddig et al. (2000). Three independent mouse lines that overexpressed the T7-PKCε in their epidermis were produced. The three independent lines 206, 224, and 215 exhibited a 3-, 6-, and 18-fold elevation, respectively, in the level of PKCε immunoreactive protein. Line 215 exhibited a 19-fold greater phosphatidylserine and TPA stimulated kinase activity than line 224. Line 206 exhibited a low basal T7-PKCε activity, which failed to be stimulated by phosphatidylserine and TPA. All of the line 215 transgenic mice (F0 to the F2 generation) displayed phenotypic changes in the skin. The phenotypic changes progressed gradually, starting around 4–5 months of age, with mild dryness of the tail accompanied by hair loss and inflammation at the base of the tail. Hyperproliferation and ulceration of the affected regions were observed around 7–8 months of age. The hyperproliferative epidermis from the affected regions exhibited an expansion of the suprabasal epidermal cells. Inflammation and/or ulceration were also observed in the dorsal skin, the ears, and around the eyes. The line 215 mice, which expressed the highest level of PKCε were evaluated for sensitivity to mouse skin tumor promotion by TPA. Tumors were elicited by the initiation (DMBA, 100 nmol)-promotion (TPA, 5 nmol/twice weekly) protocol. The papilloma burden was reduced by 95–96% for male and female T7-PKCε mice compared to wild-type controls. However, carcinomas developed rapidly in the T7-PKCε mice treated with DMBA and TPA. These carcinomas appeared to form independently of prior papilloma development (Reddig et al. 2000). Similarly, epidermal PKCε level was observed to dictate the susceptibility of transgenic mice to the development of papilloma-independent SCC by repeated exposure to UVR (Wheeler et al. 2005; Reddig et al. 2000).

15.2.2.4 SCC Developed in PKCε Transgenic Mice is Metastatic and Originates from Hair Follicle

The papilloma-independent carcinomas, which develop in PKCε transgenic mice, arise from the hair follicle and have increased metastatic potential (Jansen et al. 2001b). The difference in metastatic potential and the different origin of malignancy provided support for the hypothesis that papilloma-independent carcinomas in PKCε transgenic mice were pathologically distinct from wild-type mouse carcinomas. Although the papilloma-independent carcinomas appeared to originate from the hair follicle, it was possible that the origin of the tumor was not within the hair follicle. The hair follicle might have been the easiest pathway for invasion. However, this did not appear to be the case because we observed neoplastic cells arising only from the hair follicle and not the epidermis. By harvesting PKCε transgenic and wild-type mice after 8 weeks of DMBA + TPA or DMBA + acetone treatments, we identified possible premalignant areas in PKCε transgenic mice as early as 8 weeks after DMBA + TPA treatment. The premalignant lesions originated within the hair follicle (Jansen et al. 2001b).

The metastatic potential of a transformed keratinocyte appeared to inversely correlate with the differentiation potential of that keratinocyte in the limited number of tumors studied to date. This conclusion was based on the location of invasion and pathological categorization of PKCε transgenic mouse carcinoma compared with wild-type mouse carcinoma. Bulge keratinocytes are located near the sebaceous gland within the hair follicle. Evidence suggests that these cells appear to be the stem or progenitor cells for both the hair follicle and epidermis and, therefore, would be in a less-differentiated state than epidermal cells (Jansen et al. 2001b). These properties may increase the metastatic potential of these cells. The carcinomas of PKCε transgenic mice that led to metastases were also less differentiated than carcinomas from wild-type mice. Evidence suggested that malignant cells that invaded from the hair follicle were less differentiated and had a higher metastatic potential than cells that invaded from the epidermis.

15.2.2.5 Possible Mechanisms by Which PKCε Sensitizes Skin to the Development of SCC

PKCε, when get activated either via direct binding to TPA or indirectly by UVR treatment, mediates two potential signals leading to inhibition of apoptosis (Basu and Sivaprasad 2007; Verma et al. 2006) and induction of cell proliferation (Wheeler et al. 2004, 2005).

15.2.2.6 PKCε Overexpression in Transgenic Mice Inhibits UVR-Induced Formation of Sunburn Cells

PKCε overexpression in transgenic mice, as compared with their wild-type littermates, reduced the appearance of sunburn cells. Sunburn cells are DNA-damaged keratinocytes undergoing apoptosis (Hill et al. 1999; Lu et al. 2007; Ziegler et al. 1994). UVR is a complete carcinogen, which both initiates and promotes carcinogenesis. UVR initiates carcinogenesis by directly damaging DNA (Berton et al. 1997; de Gruijl 1999; Kunisada et al. 2007), which results in the induction of p53 protein (Ziegler et al. 1994; Berton et al. 1997). The p53 protein transactivates $p21^{WAF1/CIP1}$ inducing cell cycle arrest to allow DNA repair. If the damage is not repaired, p53-dependent apoptosis is triggered to erase the DNA damage. The p53-dependent apoptosis of UV-damaged normal cells (sunburn cells) is prevented due to p53 mutation. Thus, these mutated cells can clonally expand to form SCC after subsequent UVR exposures. In this context, it is notable that mice deficient in p53 have reduced sunburn cell formation and increased susceptibility to UVR-induced skin carcinogenesis (Li et al. 1998). These findings indicate that apoptosis inhibition may be an important component of the mechanism of UVR-induced skin carcinogenesis.

The Fas pathway is important in eliminating DNA-damaged cells both by augmenting p53-mediated apoptosis (Muller et al. 1998) and by inducing apoptosis when p53 has been mutated (Rossi and Gaidano 2003). In Fas-mediated apoptosis, the homotrimeric Fas ligand binds to the Fas receptor, inducing it to trimerize within the membrane (de Gruijl 1999). UVR is also able to activate the Fas receptor independently of its ligand by inducing aggregation of the receptor, possibly through disruption of the plasma membrane (Kulms et al. 2002). After the Fas receptor trimerizes, the intracellular death domain of the receptor binds to Fas-associated with death domain (FADD), forming the death-inducing signaling complex (DISC) (Rossi and Gaidano 2003; Chinnaiyan et al. 1995). FADD then induces the autocatalytic cleavage of initiator caspases 8 or 10, followed by the cleavage of the effector caspases. The executioner caspases cause the cleavage of structural proteins, such as poly(ADP-ribose)polymerases (PARP), leading to membrane blebbing, degradation of nuclear proteins leading to nuclear collapse, fragmentation of nuclear DNA, and finally cell death. FADD is a common adaptor protein in both Fas and TNFR-mediated apoptosis (Gaur and Aggarwal 2003; Sheikh and Huang 2003a, b; Thorburn 2004). We determined the effects of PKCε overexpression in transgenic mice on the UVR-induced Fas- and TNFR-mediated apoptotic pathways (Verma et al. 2006). We found that the inhibition of UVR-induced sunburn cell formation in PKCε transgenic mice may be the result of the inhibition of the expression of the components of Fas/Fas-L (Fas/Fas-L and FADD) and TNFα/TNFR1 (TNFα/TNFR1, FADD)-mediated apoptotic pathways. These results indicate that UVR-induced activated PKCε mediates potential signal that facilitates accumulation of UVR-induced DNA-damaged keratinocytes (preneoplastic cells) to form SCC via inhibition of FADD, component of the death-inducing signaling complex (Verma et al. 2006).

As discussed in the foregoing section that PKCε-mediated proliferation signals include Stat3 activation (Aziz et al. 2007) and TNFα expression (Wheeler et al. 2004, 2005).

15.2.2.7 PKCε Mediates UVR-Induced Activation of Stat3

STATs comprise a family of seven [Stat1 (α and β splice isoforms), Stat2 and Stat3 (α and β isoforms), Stat4, Stat5a, Stat5b, and Stat6] latent transcription factors which reside in the cytoplasm and are encoded by seven distinct genes (Quesnelle et al. 2007; Kortylewski and Yu 2007). STATs are activated through tyrosine phosphorylation by a wide variety of growth factors (e.g., EGF, PDGF), and cytokines (e.g., IL-6), which act through intrinsic receptor tyrosine kinases (Klampfer 2006; Hodge et al. 2005; Kortylewski et al. 2005; Nikitakis et al. 2004; Vinkemeier 2004; Stephanou and Latchman 2005). Tyrosine phosphorylation enables STAT homo- or heterodimerization via reciprocal interactivation between the conserved Src homology 2 (SH2) domain of one monomer and the phosphorylated tyrosine of the other. The dimerized STATs then localize to the nucleus where they bind specific DNA targets and induce the transcription of specific genes (e.g., *c-myc*, *cyclin* D1, cyclin *E*, *cdc25A*, *Bcl-2* and *Bcl-xL*) (Hodge et al. 2005; Kortylewski et al. 2005; Nikitakis et al. 2004). All STATs have similar DNA binding elements (Berton et al. 1997).

STATs exhibit functional divergence in their roles in oncogenesis. Stat3 and Stat5 promote cell survival while Stat1 has been associated with growth inhibitory effects (Stephanou and Latchman 2005; Akira 2000). Constitutively activated STATs, in particular Stat3, have been found in a number of human cancers (e.g., SCCs, head and neck, breast, ovary, prostate, lung) (Rivat et al. 2005; Clevenger 2004; Huang et al. 2005; Alvarez et al. 2005; Burke et al. 2001). Since naturally occurring Stat3 mutations have not been observed, constitutive activation of Stat3 appears to be mediated by aberrant growth factor signaling (Klampfer 2006; Akira 2000; Rivat et al. 2005; Clevenger 2004; Huang et al. 2005; Alvarez et al. 2005; Burke et al. 2001). The physiological role of each individual STAT protein was evaluated using knockout mice. In contrast to other STAT-deficient mice, which were viable, Stat3-deficient mice die during early embryogenesis (Akira 2000).

The pioneering work of DiGiovanni and his associates about the role of EFGR-mediated Stat3 activation in skin carcinogenesis is noteworthy (Chan et al. 2004a, b; Kataoka et al. 2008; Sano et al. 2005). In their findings, activation of STATs (Stat1, 3, and 5) is an essential component of mechanism of mouse skin tumor promotion by diverse tumor promoters. Tumor promoter-induced activation of STATs is mediated by EGFR. Furthermore, Stat3 is constitutively activated in both skin papillomas and carcinomas (Chan et al. 2004a, b). Disruption of Stat3 prevents development of skin tumors elicited by DMBA initiation and TPA promotion (Kataoka et al. 2008). Reports by Sano et al. link activated Stat3 to keratinocyte survival, and to keratinocyte proliferation following UVR (Sano et al. 2005). Constitutive activation of Stat3 is observed in UVR-induced human or mouse squamous cell carcinomas (Chan et al. 2004a, b).

PKCε may impart sensitivity to UVR carcinogenesis via its association with Stat3, the transcriptional factor, which is constitutively activated in both mouse and human SCC (Aziz et al. 2007).

PKCε overexpression, but not PKCδ overexpression, in mouse epidermis stimulated UVR-induced phosphorylation of Stat3 at both tyrosine 705 and serine 727 residues (Aziz et al. 2007). The transcriptional activity of Stat3 involves its dimerization, nuclear-translocation, DNA binding, and recruitment of transcriptional coactivators (Klampfer 2006; Vinkemeier 2004). Stat1, Stat3, and Stat4 share a consensus motif between 720 and 730 in C-terminal transactivation domain in which the serine (serine 727 in Stat3) residue is the target for phosphorylation (Turkson et al. 1998; Li and Shaw 2004; Decker and Kovarik 2000). Evidence indicates that cooperation of both tyrosine and serine phosphorylation is necessary for full activation of Stat3 (Li and Shaw 2004). Ser727 phosphorylation of Stat3 is required for transactivation by association with CREB binding protein p300 (Wang et al. 2005). The constitutive phosphorylation of Stat3 of both tyrosine 705 and Serine 727 residues may be essential components of the mechanism by which PKCε mediate sensitivity to UVR carcinogenesis (Wheeler et al. 2004, 2005).

The mechanism by which PKCε associates and mediate the phosphorylation of Stat3Ser727 is unclear. A few motifs in the signal transducing proteins are known to activate Stat3. For example, Erk has been reported to be involved in Stat3Ser727 phosphorylation through YSTV motif. The YVNV motif in the HGF receptor and the YXXC motif in G-CSF receptor has been reported to serve as the docking sites for Stat3 (Abe et al. 2001). IFNg receptor was shown to cause phosphorylation on Ser727 of Stat1 through its YDKP docking motif. The YXXQ motif is known to activate Stat3 in a variety of signal-transducing receptors including LiF-R, G-CSF-R, leptin-R and IL-10-R. Besides offering as a docking site for Stat3, the YXXQ motif in gp130 is also important for serine phosphorylation of Stat3 (Abe et al. 2001). It is notable that mouse PKCε has three repeats of the YXXQ motif (regions from 176 to 179, 199 to 202, and 468 to 471). Two of the motifs occur in the TPA-binding region and the third may bind and facilitate the serine phosphorylation of Stat3.

The results indicate that PKCε-mediated Stat3Ser727 phosphorylation may be an important component of the mechanism by which PKCε imparts sensitivity to UVR-induced development of SCC. The role of Stat3Ser727 phosphorylation in UVR-induced activation of Stat3 transcriptional activity can be explored using Stat3Ser727Ala knockin mice. There are two reports explaining the generation of genetically engineered Stat (Stat1 and Stat3) serine mutant knockin mice (Varinou et al. 2003; Shen et al. 2004). Both strains of mice with knockin mutations are viable, normal, and fertile (Varinou et al. 2003; Shen et al. 2004). Varinou et al. showed, using a Stat1Ser727 to alanine knockin mouse, that phosphorylation of the Stat1 transactivation domain is required for Stat1 regulated transcriptional activity (Varinou et al. 2003). Similarly, Shen et al. have shown using knockin mouse models that Stat3Ser727 plays an essential role in postnatal survival and growth (Shen et al. 2004).

15.2.3 TNFα is Linked to PKCε-Induced Development of SCC

TNFα is a potent proinflammatory cytokine that is produced by a multitude of cell types including macrophages, lymphocytes, monocytes, fibroblasts, and keratinocytes (Hunt et al. 1992; Old 1985). This molecule was originally discovered as a cytotoxic cytokine for tumor cells and for its ability to cause necrosis of transplanted tumors (Black et al. 1997). Mature murine TNFα consists of 156 amino acids (157 in humans) and is translated with a 79 amino acid (76 in humans) long precursor sequence. For TNFα to exert its pleiotropic inflammatory responses at distant sites from its synthesis, it must be cleaved from the membrane in a process called ectodomain shedding. A specific enzyme called Tumor Necrosis Factor Alpha Convertase (TACE) cleaves proTNFα in response to extracellular stimuli. The cloning of TACE (human and porcine) revealed it to be a member of "A disintegrin and metalloprotease" or ADAM family of proteins (Moss et al. 1997). The TACE protein is a multidomain, type I transmembrane protein that includes a zinc-dependent catalytic domain. The TNFα protein has six domains: prodomain, catalytic domain, disintegrin domain, cysteine-rich domain, transmembrane domain, and the cytoplasmic domain. The prodomain contains a cysteine residue that interacts with a zinc molecule in the catalytic domain. This interaction must be displaced for TACE activation and is believed to be mediated by reactive oxygen species (ROS). Upon its release, TNFα exerts its biological effects by trimerizing and binding to two distinct receptors, TNFR1 and TNFR2. Binding of TNFα induces trimerization of each of these receptors, which then recruit several signaling proteins to the cytoplasmic membrane. With the ability to activate two distinct receptors and to recruit different receptor signaling complexes, TNFα has the ability to regulate a vast array of cellular responses including cellular inflammation, immunity, cell proliferation, differentiation, and apoptosis (Komori et al. 1993; Wheeler et al. 2003).

Evidence indicates that TNFα is linked to skin tumor promotion by TPA and UVR (Komori et al. 1993; Wheeler et al. 2003; Starcher 2000; Suganuma et al. 1999; Moore et al. 1999; Arnott et al. 2004). Experiments using tumor promoters of the okadaic acid class have provided evidence that TNFα is the central mediator of tumor promotion in the mouse skin. These experiments indicated that TNFα shed from the initiated cell or various tissues surrounding the initiated lesion can induce clonal expansion and transformation of initiated cells. This work led to the development of in vivo mouse models, which have further implicated TNFα as the key cytokine for tumor promotion in the mouse skin. Using either the 2-stage model of carcinogenesis or UVR, mice deficient for TNFα or either of its receptors render the mice resistant to skin tumor formation (Komori et al. 1993; Wheeler et al. 2003; Starcher 2000; Suganuma et al. 1999; Moore et al. 1999; Arnott et al. 2004).

PKCε transgenic mice elicit elevated both serum and epidermal TNFα levels during skin tumor promotion either by TPA or UVR and this increase is linked to the development of SCC (Wheeler et al. 2004, 2005). A single topical application of TPA (5 nmol) to the skin, as early as 2.5 h after treatment, result in a significant ($p<0.01$) increase (twofold) in epidermal TNFα and more than a sixfold increase in ectodomain shedding of TNFα into the serum of PKCε transgenic mice relative to their wild-type littermates. Furthermore, this TPA-stimulated TNFα shedding is proportional to the level of expression of PKCε in the epidermis. Using the TNF alpha converting enzyme (TACE) inhibitor, TAPI-1, TPA-stimulated TNFα shedding can be completely prevented in PKCε transgenic mice and isolated keratinocytes. These results indicate that PKCε signal transduction pathways to TPA-stimulated TNFα ectodomain shedding are mediated by TACE, a transmembrane metalloprotease. Using the superoxide dismutase mimetic CuDIPs and the glutathione reductase mimetic ebselen, TPA-stimulated TNFα shedding from PKCε transgenic mice can be completely attenuated, implying the role of reactive oxygen species (Wheeler et al. 2003).

To determine whether TNFα is a critical intermediate component in PKCε -mediated squamous cell carcinoma formation, bigenic mice were created by cross breeding K14-PKCε 215 transgenic mice with TNFα knockout mice. The bigenic mice were then used for the two-step DMBA-TPA skin tumor promotion protocol. TNFα deficiency significantly inhibited (~50%) the development of SCC in PKCε transgenic mice.

TNFα-deficient PKCε transgenic mice were also evaluated for UVR-induced cutaneous damage. Deletion of TNFα gene in PKCε transgenic mice inhibited

UVR-induced release of TNFα and inflammation (Wheeler et al. 2004). The dorsal skin of UVR-exposed PKCε transgenic mice exhibited severely hyperplastic interfollicular epidermis with alternating regions of ulceration associated with severe scarring. The scar tissue also contained remarkable amounts of inflammatory infiltrate. In addition, the skin histopathology exhibited a disorganization of the hair follicle and hyperplasia of the bulb region. The PKCε transgenic-TNFα knockout skin was intact and showed significant reduction in both the interfollicular and follicular hyperplasia, relative to the PKCε transgenic mice (Wheeler et al. 2004).

15.2.3.1 PKCε-Induced Release of TNFα May Stimulate Proliferation of Keratinocyte Stem Cells, SCC Precursor Cells

In adult skin, each hair follicle contains a reservoir of stem cells, which can be mobilized to regenerate the new follicle with each hair cycle and to reepithelialize epidermis during wound repair (Tumbar et al. 2004). Several lines of evidence suggest that epithelial stem cells reside in the bulge (Tumbar et al. 2004; Kangsamaksin et al. 2007; Chebotaev et al. 2007; Trempus et al. 2003; Faurschou et al. 2007). Epidermal stem cells in the mouse hair follicle are known to be the precursor cells for SCC in the mouse skin (Tumbar et al. 2004; Kangsamaksin et al. 2007; Chebotaev et al. 2007; Trempus et al. 2003; Faurschou et al. 2007). Stem cells, unlike transit amplifying cells, are slowly cycling, and thus seem probable target cells. Moreover, stem cells may retain those mutations and pass them on to their progeny (Tumbar et al. 2004).

Morris et al. (2000) demonstrated that label retaining cells (LRCs) retain carcinogen-DNA adducts, another property characteristic of potential initiated cells. Morris et al. (2000) also determined the contribution of follicular and interfollicular stem cells to the induction of skin papillomas and carcinomas. Both follicular and interfollicular stem cells contributed to the development of papillomas. However, only follicular stem cells were linked to the development of carcinomas.

It has recently become possible to isolate living hair follicle stem and progenitor cells from mouse skin because of the discovery of cell surface properties that facilitate enrichment (Gerdes and Yuspa 2005; Lavker and Sun 2000; Liu et al. 2003). The cell surface markers CD34, K15, and α6-integrin mark mouse hair follicle bulge cells, which have attributes of stem cells, including quiescence and multipotency. About 30% of bulge keratinocytes are putative stem cells; however, there is no immunostaining for stem cells specifically. CD34 will mark all bulge keratinocytes. α6-integrin will mark all basal keratinocytes in the hair follicle and interfollicular epidermis. LRCs in the bulge region are putative SCs. SCC in PKCε transgenic mice appears to develop and invade from the hair follicle (Jansen et al. 2001a). However, the mechanism by which hair follicle stem cells are activated and induced to proliferate is unknown. This remains to be determined whether PKCε-mediated UVR-induced release of TNFα promote the proliferation of hair follicle putative stem cells.

15.3 Summary and Conclusion

PKC, a family of phospholipid-dependent serine/threonine kinase, constitutes a component of signaling network involved in numerous cell functions (Griner and Kazanietz 2007; Mellor and Parker 1998; Newton 2001; Mochly-Rosen and Kauvar 1998; Denning 2004; Aziz et al. 2006). PKC isoforms exhibit functional diversity (Griner and Kazanietz 2007; Denning 2004; Aziz et al. 2006; Wheeler et al. 2004, 2005; Reddig et al. 1999, 2000; Jansen et al. 2001a, b). Disregulation of PKC activity has been associated with malignant transformation (Griner and Kazanietz 2007; Basu and Sivaprasad 2007). Molecular genetic experiments involving transgenic mouse model indentified PKCε as a key mediator of induction of SCC, a nonmelanoma human skin cancer (Aziz et al. 2006; Wheeler et al. 2004, 2005; Verma et al. 2006). PKCε associates with Stat3 and regulates UVR-induced activation of Stat3 (Aziz et al. 2007). Stat3 activation regulates the expression of TNFα, which is linked to the development of SCC (Kataoka et al. 2008). Epidermal stem cells, which reside in the bulge region of hair follicle are known to be the precursor cells for SCC in the mouse skin (Morris et al. 2000). The possibility that PKCε activation is an initial signal in the activation and proliferation of these slow cycling stem cells remains to be defined (Fig. 15.3).

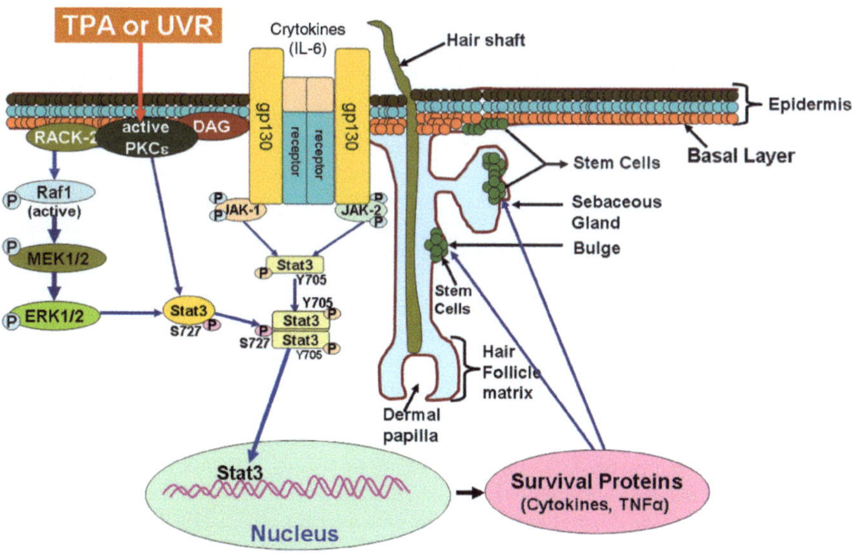

Fig. 15.3 Proposed model illustrating how PKCε mediates the development of carcinomas by TPA or UVR. This model assumes that PKCε activation mediates Stat3 activation and nuclear translocation, which leads to induction of survival proteins such as TNFε. Subsequently, TNFε may stimulate proliferation of hair follicle Stem cells, the proposed carcinoma precursor cells

References

Abe, K., Hirai, M., Mizuno, K., Higashi, N., Sekimoto, T., Miki, T., et al. (2001). The YXXQ motif in gp 130 is crucial for STAT3 phosphorylation at Ser727 through an H7-sensitive kinase pathway. *Oncogene, 20*, 3464–3474.

Akira, S. (2000). Roles of STAT3 defined by tissue-specific gene targeting. *Oncogene, 19*, 2607–2611.

Alvarez, J. V., Febbo, P. G., Ramaswamy, S., Loda, M., Richardson, A., & Frank, D. A. (2005). Identification of a genetic signature of activated signal transducer and activator of transcription 3 in human tumors. *Cancer Research, 65*, 5054–5062.

Arnott, C. H., Scott, K. A., Moore, R. J., Robinson, S. C., Thompson, R. G., & Balkwill, F. R. (2004). Expression of both TNF-alpha receptor subtypes is essential for optimal skin tumour development. *Oncogene, 23*, 1902–1910.

Aziz, M. H., Manoharan, H. T., & Verma, A. K. (2007). Protein kinase C epsilon, which sensitizes skin to sun's UV radiation-induced cutaneous damage and development of squamous cell carcinomas, associates with Stat3. *Cancer Research, 67*, 1385–1394.

Aziz, M. H., Wheeler, D. L., Bhamb, B., & Verma, A. K. (2006). Protein kinase C delta overexpressing transgenic mice are resistant to chemically but not to UV radiation-induced development of squamous cell carcinomas: A possible link to specific cytokines and cyclooxygenase-2. *Cancer Research, 66*, 713–722.

Basu, A., & Sivaprasad, U. (2007). Protein kinase Cepsilon makes the life and death decision. *Cellular Signalling, 19*, 1633–1642.

Berton, T. R., Mitchell, D. L., Fischer, S. M., & Locniskar, M. F. (1997). Epidermal proliferation but not quantity of DNA photodamage is correlated with UV-induced mouse skin carcinogenesis. *The Journal of Investigative Dermatology, 109*, 340–347.

Black, R. A., Rauch, C. T., Kozlosky, C. J., Peschon, J. J., Slack, J. L., Wolfson, M. F., et al. (1997). A metalloproteinase disintegrin that releases tumour-necrosis factor-alpha from cells. *Nature, 385*, 729–733.

Boutwell, R. K. (1974). The function and mechanism of promoters of carcinogenesis. *CRC Critical Reviews in Toxicology, 2*, 419–443.

Burke, W. M., Jin, X., Lin, H. J., Huang, M., Liu, R., Reynolds, R. K., et al. (2001). Inhibition of constitutively active Stat3 suppresses growth of human ovarian and breast cancer cells. *Oncogene, 20*, 7925–7934.

Cataisson, C., Pearson, A. J., Torgerson, S., Nedospasov, S. A., & Yuspa, S. H. (2005). Protein kinase C alpha-mediated chemotaxis of neutrophils requires NF-kappa B activity but is independent of TNF alpha signaling in mouse skin in vivo. *Journal of Immunology, 174*, 1686–1692.

Chan, K. S., Carbajal, S., Kiguchi, K., Clifford, J., Sano, S., & DiGiovanni, J. (2004). Epidermal growth factor receptor-mediated activation of Stat3 during multistage skin carcinogenesis. *Cancer Research, 64*, 2382–2389.

Chan, K. S., Sano, S., Kiguchi, K., Anders, J., Komazawa, N., Takeda, J., et al. (2004). Disruption of Stat3 reveals a critical role in both the initiation and the promotion stages of epithelial carcinogenesis. *The Journal of Clinical Investigation, 114*, 720–728.

Chebotaev, D. V., Yemelyanov, A. Y., Lavker, R. M., & Budunova, I. V. (2007). Epithelial cells in the hair follicle bulge do not contribute to epidermal regeneration after glucocorticoid-induced cutaneous atrophy. *The Journal of Investigative Dermatology, 127*, 2749–2758.

Chinnaiyan, A. M., O'Rourke, K., Tewari, M., & Dixit, V. M. (1995). FADD, a novel death domain-containing protein, interacts with the death domain of Fas and initiates apoptosis. *Cell, 81*, 505–512.

Clevenger, C. V. (2004). Roles and regulation of stat family transcription factors in human breast cancer. *American Journal of Pathology, 165*, 1449–1460.

de Gruijl, F. R. (1999). Skin cancer and solar UV radiation. *European Journal of Cancer, 35*, 2003–2009.

Decker, T., & Kovarik, P. (2000). Serine phosphorylation of STATs. *Oncogene, 19*, 2628–2637.
Denning, M. F. (2004). Epidermal keratinocytes: Regulation of multiple cell phenotypes by multiple protein kinase C isoforms. *International Journal of Biochemistry and Cell Biology, 36*, 1141–1146.
DiGiovanni, J. (1992). Multistage carcinogenesis in mouse skin. *Pharmacology and Therapeutics, 54*, 63–128.
Faurschou, A., Haedersdal, M., Poulsen, T., & Wulf, H. C. (2007). Squamous cell carcinoma induced by ultraviolet radiation originates from cells of the hair follicle in mice. *Experimental Dermatology, 16*, 485–489.
Gaur, U., & Aggarwal, B. B. (2003). Regulation of proliferation, survival and apoptosis by members of the TNF superfamily. *Biochemical Pharmacology, 66*, 1403–1408.
Gerdes, M. J., & Yuspa, S. H. (2005). The contribution of epidermal stem cells to skin cancer. *Stem Cell Reviews, 1*, 225–231.
Griner, E. M., & Kazanietz, M. G. (2007). Protein kinase C and other diacylglycerol effectors in cancer. *Nature Reviews Cancer, 7*, 281–294.
Hill, L. L., Ouhtit, A., Loughlin, S. M., Kripke, M. L., Ananthaswamy, H. N., & Owen-Schaub, L. B. (1999). Fas ligand: A sensor for DNA damage critical in skin cancer etiology. *Science, 285*, 898–900.
Hodge, D. R., Hurt, E. M., & Farrar, W. L. (2005). The role of IL-6 and STAT3 in inflammation and cancer. *European Journal of Cancer, 41*, 2502–2512.
Huang, H. F., Murphy, T. F., Shu, P., Barton, A. B., & Barton, B. E. (2005). Stable expression of constitutively-activated STAT3 in benign prostatic epithelial cells changes their phenotype to that resembling malignant cells. *Molecular Cancer, 4*, 2.
Hunt, J. S., Chen, H. L., Hu, X. L., Chen, T. Y., & Morrison, D. C. (1992). Tumor necrosis factor-alpha gene expression in the tissues of normal mice. *Cytokine, 4*, 340–346.
Jansen, A. P., Dreckschmidt, N. E., Verwiebe, E. G., Wheeler, D. L., Oberley, T. D., & Verma, A. K. (2001). Relation of the induction of epidermal ornithine decarboxylase and hyperplasia to the different skin tumor-promotion susceptibilities of protein kinase C alpha, -delta and -epsilon transgenic mice. *International Journal of Cancer, 93*, 635–643.
Jansen, A. P., Verwiebe, E. G., Dreckschmidt, N. E., Wheeler, D. L., Oberley, T. D., & Verma, A. K. (2001). Protein kinase C-epsilon transgenic mice: A unique model for metastatic squamous cell carcinoma. *Cancer Research, 61*, 808–812.
Kangsamaksin, T., Park, H. J., Trempus, C. S., & Morris, R. J. (2007). A perspective on murine keratinocyte stem cells as targets of chemically induced skin cancer. *Molecular Carcinogenesis, 46*, 579–584.
Kataoka, K., Kim, D. J., Carbajal, S., Clifford, J. L., & DiGiovanni, J. (2008). Stage-specific disruption of Stat3 demonstrates a direct requirement during both the initiation and promotion stages of mouse skin tumorigenesis. *Carcinogenesis, 29*, 1108–1114.
Klampfer, L. (2006). Signal transducers and activators of transcription (STATs): Novel targets of chemopreventive and chemotherapeutic drugs. *Current Cancer Drug Targets, 6*, 107–121.
Komori, A., Yatsunami, J., Suganuma, M., Okabe, S., Abe, S., Sakai, A., et al. (1993). Tumor necrosis factor acts as a tumor promoter in BALB/3T3 cell transformation. *Cancer Research, 53*, 1982–1985.
Kortylewski, M., Jove, R., & Yu, H. (2005). Targeting STAT3 affects melanoma on multiple fronts. *Cancer and Metastasis Reviews, 24*, 315–327.
Kortylewski, M., & Yu, H. (2007). Stat3 as a potential target for cancer immunotherapy. *Journal of Immunotherapy, 30*, 131–139.
Kulms, D., Dussmann, H., Poppelmann, B., Stander, S., Schwarz, A., & Schwarz, T. (2002). Apoptosis induced by disruption of the actin cytoskeleton is mediated via activation of CD95 (Fas/APO-1). *Cell Death and Differentiation, 9*, 598–608.
Kunisada, M., Kumimoto, H., Ishizaki, K., Sakumi, K., Nakabeppu, Y., & Nishigori, C. (2007). Narrow-band UVB induces more carcinogenic skin tumors than broad-band UVB through the formation of cyclobutane pyrimidine dimer. *The Journal of Investigative Dermatology, 127*, 2865–2871.

Lavker, R. M., & Sun, T. T. (2000). Epidermal stem cells: Properties, markers, and location. *Proceedings of the National Academy of Sciences of the United States of America, 97*, 13473–13475.

Li, L., & Shaw, P. E. (2004). A STAT3 dimer formed by inter-chain disulphide bridging during oxidative stress. *Biochemical and Biophysical Research Communications, 322*, 1005–1011.

Li, G., Tron, V., & Ho, V. (1998). Induction of squamous cell carcinoma in p53-deficient mice after ultraviolet irradiation. *The Journal of Investigative Dermatology, 110*, 72–75.

Liu, Y., Lyle, S., Yang, Z., & Cotsarelis, G. (2003). Keratin 15 promoter targets putative epithelial stem cells in the hair follicle bulge. *The Journal of Investigative Dermatology, 121*, 963–968.

Lu, Y. P., Nolan, B., Lou, Y. R., Peng, Q. Y., Wagner, G. C., & Conney, A. H. (2007). Voluntary exercise together with oral caffeine markedly stimulates UVB light-induced apoptosis and decreases tissue fat in SKH-1 mice. *Proceedings of the National Academy of Sciences of the United States of America, 104*, 12936–12941.

Mellor, H., & Parker, P. J. (1998). The extended protein kinase C superfamily. *Biochemical Journal, 332*, 281–292.

Mochly-Rosen, D., & Kauvar, L. M. (1998). Modulating protein kinase C signal transduction. *Advances in Pharmacology, 44*, 91–145.

Moore, R. J., Owens, D. M., Stamp, G., Arnott, C., Burke, F., East, N., et al. (1999). Mice deficient in tumor necrosis factor-alpha are resistant to skin carcinogenesis. *Nature Medicine, 5*, 828–831.

Morris, R. J., Tryson, K. A., & Wu, K. Q. (2000). Evidence that the epidermal targets of carcinogen action are found in the interfollicular epidermis of infundibulum as well as in the hair follicles. *Cancer Research, 60*, 226–229.

Moss, M. L., Jin, S. L., Milla, M. E., Bickett, D. M., Burkhart, W., Carter, H. L., et al. (1997). Cloning of a disintegrin metalloproteinase that processes precursor tumour-necrosis factor-alpha. *Nature, 385*, 733–736.

Muller, M., Wilder, S., Bannasch, D., Israeli, D., Lehlbach, K., Li-Weber, M., et al. (1998). p53 activates the CD95 (APO-1/Fas) gene in response to DNA damage by anticancer drugs. *The Journal of Experimental Medicine, 188*, 2033–2045.

Newton, A. C. (2001). Protein kinase C: Structural and spatial regulation by phosphorylation, cofactors, and macromolecular interactions. *Chemical Reviews, 101*, 2353–2364.

Nikitakis, N. G., Siavash, H., & Sauk, J. J. (2004). Targeting the STAT pathway in head and neck cancer: Recent advances and future prospects. *Current Cancer Drug Targets, 4*, 637–651.

Old, L. J. (1985). Tumor necrosis factor (TNF). *Science, 230*, 630–632.

Quesnelle, K. M., Boehm, A. L., & Grandis, J. R. (2007). STAT-mediated EGFR signaling in cancer. *Journal of Cellular Biochemistry, 102*, 311–319.

Reddig, P. J., Dreckschmidt, N. E., Ahrens, H., Simsiman, R., Tseng, C. P., Zou, J., et al. (1999). Transgenic mice overexpressing protein kinase Cdelta in the epidermis are resistant to skin tumor promotion by 12-O-tetradecanoylphorbol-13-acetate. *Cancer Research, 59*, 5710–5718.

Reddig, P. J., Dreckschmidt, N. E., Zou, J., Bourguignon, S. E., Oberley, T. D., & Verma, A. K. (2000). Transgenic mice overexpressing protein kinase C epsilon in their epidermis exhibit reduced papilloma burden but enhanced carcinoma formation after tumor promotion. *Cancer Research, 60*, 595–602.

Rivat, C., Rodrigues, S., Bruyneel, E., Piétu, G., Robert, A., Redeuilh, G., et al. (2005). Implication of STAT3 signaling in human colonic cancer cells during intestinal trefoil factor 3 (TFF3) – and vascular endothelial growth factor-mediated cellular invasion and tumor growth. *Cancer Research, 65*, 195–202.

Rossi, D., & Gaidano, G. (2003). Messengers of cell death: Apoptotic signaling in health and disease. *Haematologica, 88*, 212–218.

Sano, S., Chan, K. S., Kira, M., Kataoka, K., Takagi, S., Tarutani, M., et al. (2005). Signal transducer and activator of transcription 3 is a key regulator of keratinocyte survival and proliferation following UV irradiation. *Cancer Research, 65*, 5720–5729.

Sheikh, M. S., & Huang, Y. (2003a). Death receptor activation complexes: It takes two to activate TNF receptor 1. *Cell Cycle, 2*, 550–552.

Sheikh, M. S., & Huang, Y. (2003b). The FADD is going nuclear. *Cell Cycle, 2*, 346–347.
Shen, Y., Schlessinger, K., Zhu, X., Meffre, E., Quimby, F., Levy, D. E., et al. (2004). Essential role of STAT3 in postnatal survival and growth revealed by mice lacking STAT3 serine 727 phosphorylation. *Molecular and Cellular Biology, 24*, 407–419.
Starcher, B. (2000). Role for tumour necrosis factor-alpha receptors in ultraviolet-induced skin tumours. *British Journal of Dermatology, 142*, 1140–1147.
Stephanou, A., & Latchman, D. S. (2005). Opposing actions of STAT-1 and STAT-3. *Growth Factors, 23*, 177–182.
Suganuma, M., Okabe, S., Marino, M. W., Sakai, A., Sueoka, E., & Fujiki, H. (1999). Essential role of tumor necrosis factor alpha (TNF-alpha) in tumor promotion as revealed by TNF-alpha-deficient mice. *Cancer Research, 59*, 4516–4518.
Thorburn, A. (2004). Death receptor-induced cell killing. *Cellular Signalling, 16*, 139–144.
Trempus, C. S., Morris, R. J., Bortner, C. D., Cotsarelis, G., Faircloth, R. S., Reece, J. M., et al. (2003). Enrichment for living murine keratinocytes from the hair follicle bulge with the cell surface marker CD34. *The Journal of Investigative Dermatology, 120*, 501–511.
Tumbar, T., Guasch, G., Greco, V., Blanpain, C., Lowry, W. E., Rendl, M., et al. (2004). Defining the epithelial stem cell niche in skin. *Science, 303*, 359–363.
Turkson, J., Bowman, T., Garcia, R., Caldenhoven, E., De Groot, R. P., & Jove, R. (1998). Stat3 activation by Src induces specific gene regulation and is required for cell transformation. *Molecular and Cellular Biology, 18*, 2545–2552.
Varinou, L., Ramsauer, K., Karaghiosoff, M., lbe, T., Pfeffer, K., Müller, M., et al. (2003). Phosphorylation of the Stat1 transactivation domain is required for full-fledged IFN-gamma-dependent innate immunity. *Immunity, 19*, 793–802.
Vassar, R., Rosenberg, M., Ross, S., Tyner, A., & Fuchs, E. (1989). Tissue-specific and differentiation-specific expression of a human K14 keratin gene in transgenic mice. *Proceedings of the National Academy of Sciences of the United States of America, 86*, 1563–1567.
Verma, A. K., Wheeler, D. L., Aziz, M. H., & Manoharan, H. T. (2006). Protein kinase Cepsilon and development of squamous cell carcinoma, the nonmelanoma human skin cancer. *Molecular Carcinogenesis, 45*, 381–388.
Vinkemeier, U. (2004). Getting the message across, STAT! Design principles of a molecular signaling circuit. *The Journal of Cell Biology, 167*, 197–201.
Wang, R., Cherukuri, P., & Luo, J. (2005). Activation of Stat3 sequence-specific DNA binding and transcription by p300/CREB-binding protein-mediated acetylation. *The Journal of Biological Chemistry, 280*, 11528–11534.
Wang, H. Q., & Smart, R. C. (1999). Overexpression of protein kinase C-alpha in the epidermis of transgenic mice results in striking alterations in phorbol ester-induced inflammation and COX-2, MIP-2 and TNF-alpha expression but not tumor promotion. *Journal of Cell Science, 112*, 3497–3506.
Wheeler, D. L., Li, Y., & Verma, A. K. (2005). Protein kinase C epsilon signals ultraviolet light-induced cutaneous damage and development of squamous cell carcinoma possibly through induction of specific cytokines in a paracrine mechanism. *Photochemistry and Photobiology, 81*, 9–18.
Wheeler, D. L., Martin, K. E., Ness, K. J., Li, Y., Dreckschmidt, N. E., Wartman, M., et al. (2004). Protein kinase C epsilon is an endogenous photosensitizer that enhances ultraviolet radiation-induced cutaneous damage and development of squamous cell carcinomas. *Cancer Research, 64*, 7756–7765.
Wheeler, D. L., Ness, K. J., Oberley, T. D., & Verma, A. K. (2003). Protein kinase Cepsilon is linked to 12-O-tetradecanoylphorbol-13-acetate-induced tumor necrosis factor-alpha ectodomain shedding and the development of metastatic squamous cell carcinoma in protein kinase Cepsilon transgenic mice. *Cancer Research, 63*, 6547–6555.
Yuspa, S. H., Dlugosz, A. A., Denning, M. F., & Glick, A. B. (1996). Multistage carcinogenesis in the skin. *The Journal of Investigative Dermatology Symposium Proceedings, 1*, 147–150.
Ziegler, A., Jonason, A. S., Leffell, D. J., Simon, J. A., Sharma, H. W., Kimmelman, J., et al. (1994). Sunburn and p53 in the onset of skin cancer. *Nature, 372*, 773–776.

Chapter 16
PKC Isozymes and Skin Cancer

Mitchell F. Denning

Abstract Cancers of the skin are a very common type of human malignancy, encompassing basal cell carcinoma, squamous cell carcinoma, and melanoma. Of the three main types of skin cancer, the roles of PKC isozymes in squamous cell carcinoma have been most extensively studied due to the PKC agonist activity of phorbol ester tumor promoters used in the mouse skin chemical carcinogenesis model. These studies have identified PKC isozymes with oncogenic (PKCε) or tumor-suppressive (PKCδ, PKCη) activities. The activation of PKC isozymes in response to UV radiation, the main etiological agent for human skin cancers, has also implicated unique roles for PKC isozymes in human squamous cell carcinoma etiology. More recently, studies examining PKC isozyme expression and function in basal cell carcinoma and malignant melanoma have uncovered isozyme-specific changes and roles in these skin cancers as well. This chapter will summarize our current understanding of PKC isozymes in skin cancer, as well as their function in normal keratinocyte and melanocyte biology.

Keywords Protein kinase C • Skin • Squamous cell carcinoma • Basal cell carcinoma • Melanoma

16.1 Introduction to Skin Cancer

Cancers of the skin are by far the most common type of human malignancy, with approximately 1.3 million new cases diagnosed annually in the United States alone (American Cancer Society 2008). Skin cancers are broadly divided into melanoma and nonmelanoma skin cancers, with nonmelanoma cancers further divided into squamous cell carcinoma (SCC) and basal cell carcinoma (BCC). BCC is the most

M.F. Denning (✉)
Department of Pathology, Cardinal Bernardin Cancer Center, Loyola University Chicago, 2160 S. First Avenue, Maywood, IL 60153, USA
e-mail: mdennin@lumc.edu

M.G. Kazanietz (ed.), *Protein Kinase C in Cancer Signaling and Therapy*, Current Cancer Research, DOI 10.1007/978-1-60761-543-9_16,

common subtype of skin cancer and is a locally invading but indolent neoplasm of basal layer or hair follicle-derived keratinocytes (KCs) which almost never metastasizes (Fan et al. 1997). SCC is also relatively common, with ~250,000 cases diagnosed annually, and consists of transformed KCs which produce significant numbers of squamous differentiating KCs. SCC also is frequently curable with surgery and local treatments, but it can metastasize and in some cases be fatal. Due to their extraordinarily high incidence and tendency to arise in sun-exposed sites (i.e. nose, lip, ears), there is significant morbidity and cost associated with the treatment and removal of these nonmelanoma skin cancers.

Melanoma is a cancer of melanocytes, the pigment-producing cells within the epidermis. Melanomas have a much lower incidence than nonmelanoma skin cancers but are responsible for the majority of skin cancer deaths, estimated to be ~8,400 for 2008. Furthermore, unlike cancers of the lung and stomach, the incidence of melanoma has been increasing by approximately 3–6% per year since the 1970s and is now the most common cancer in women aged 25–29 in the United States (American Cancer Society 2008).

16.2 Squamous Cell Carcinoma

The study of PKC and skin cancer dates back to the early 1980s when it was reported that phorbol ester tumor promoters specifically bind to and activate PKC (Kikkawa et al. 1983; Sharkey et al. 1984; Nishizuka 1984). Phorbol esters had been used for many years as tumor-promoting agents in the mouse skin chemical carcinogenesis model (Van Duuren et al. 1973). This "2-stage" model involves treating mouse skin with a subcarcinogenic dose of a mutagen, such as 7,12-dimethylbenz[a]anthracene (DMBA) followed by repeated exposure to a tumor-promoting agent, frequently the phorbol ester 12-*O*-tetradecanolyphorbol 13-acetate (TPA). This model is very powerful as it temporally divides carcinogenesis into discrete phases of initiation, promotion, and progression. Furthermore, since the tumors arise in the epidermis, tumor latency, number, and size can be monitored in real time, and cocarcinogens, inhibitors, or chemopreventive agents are easily applied topically.

The chemically induced tumors arise on the dorsum of the mouse initially as benign papillomas, which convert to SCC at a relatively low frequency. Interestingly, cessation of TPA treatment after only 5 weeks significantly reduced papillomas incidence but had no effect on the SCC yield (Hennings et al. 1985). Thus, the early emerging papillomas carry almost all the risk for malignant conversion and have been termed "high-risk" papillomas. The implication of the conversion of these high-risk papillomas in the absence of TPA is that the chronic PKC activation by TPA is only required for the early selection and expansion of initiated cells into benign tumors.

The chemical carcinogenesis model has been extensively studied at the molecular/genetic level. The initiating mutation elicited by DMBA exposure is codon 61 of the c-Ha-Ras gene (Balmain and Pragnell 1983; Roop et al. 1986); however, the mechanism of skin tumor promotion by TPA remains an area of controversy and active investigation. Chronic inflammation, hyperplasia, and long-term PKC isozyme downregulation are all considered to be important, but the key process remains elusive (Hansen et al. 1990; Moore et al. 1999). Direct activation of PKC by TPA has profound and complex effects on epidermal KC cell proliferation. TPA treatment of normal mouse skin initially causes an inhibition of DNA synthesis, followed by several waves of increased proliferation (Raick 1973; Raick et al. 1972). After 24–48 h, the resulting epidermis is hyperplastic, with increased cell numbers in all suprabasal layers, including the stratum corneum. This complex response to PKC activation may be due to differential responses of KCs at different stages of maturation, or compensatory proliferative response of the epidermis to rapid induction of differentiation (Reiners and Slaga 1983; Yuspa et al. 1982). In addition, TPA treatment induces the activation/proliferation of hair follicle stem cells, a critical target cell population for skin carcinogenesis (Trempus et al. 2007). TPA is a potent inducer of normal KC growth arrest and terminal differentiation in culture (Tibudan et al. 2002). PKC activation by endogenous activators such as diacylglycerol, or pharmacological activators such as phorbol esters like TPA stimulate the granular layer differentiation program and cornification of normal KCs while simultaneously inhibiting the spinous layer differentiation (Denning et al. 1995a; Dlugosz and Yuspa 1993; Efimova et al. 1998). PKC also becomes activated by inducers of differentiation such as calcium (Denning et al. 1995a; Chakravarthy et al. 1995) or confluency (Lee et al. 1998; Yang et al. 2003), and inhibition of PKC activity can block the induction of differentiation gene products (Dlugosz and Yuspa 1993; Denning et al. 1995a; Lee et al. 1998; Yang et al. 2003).

Throughout this chapter, the utility of dissecting PKC isozyme function in normal cells will be demonstrated to be a very fruitful and informative approach for understanding PKC isozyme function in neoplastic cells. In normal KCs, five PKC isozymes have been described at the protein and mRNA level: PKC alpha (α), PKC delta (δ), PKC epsilon (ε), PKC eta (η), and PKC zeta (ζ) (Dlugosz et al. 1992; Denning et al. 1993; Longthorne and Williams 1997). These PKC isozymes can be classified as calcium-dependent (PKCα), calcium-independent (PKCδ, ε, η), and phorbol ester-independent (PKCζ) based upon structural and regulatory features. The diversity of PKC regulatory mechanisms and large number of isozymes has prompted investigation into distinct functions for individual PKC isozymes (Fig. 16.1).

The ability of TPA to trigger KC differentiation and growth arrest may seem at odds with its activity as a potent mouse skin tumor promoter. However, normal KCs transduced with active Ha-Ras, mouse KC cell lines derived from papillomas, and human SCC cell lines are all resistant to the differentiating effects of TPA, and in fact TPA can stimulate DNA synthesis in Ras-transformed mouse KCs (Yuspa et al. 1985). Given this divergent response between normal and neoplastic KC, it is clear that TPA would provide a strong selective advantage for clonal expansion of KCs with activating Ha-Ras mutations. Mechanisms responsible for this differential response of normal and Ras-initiated KC to TPA are not entirely understood, but selective inactivation or downregulation of the proapoptotic and prodifferentiating PKCδ isozyme in Ras transformed KCs may be part of the explanation (Denning et al. 1993; Geiges et al. 1995).

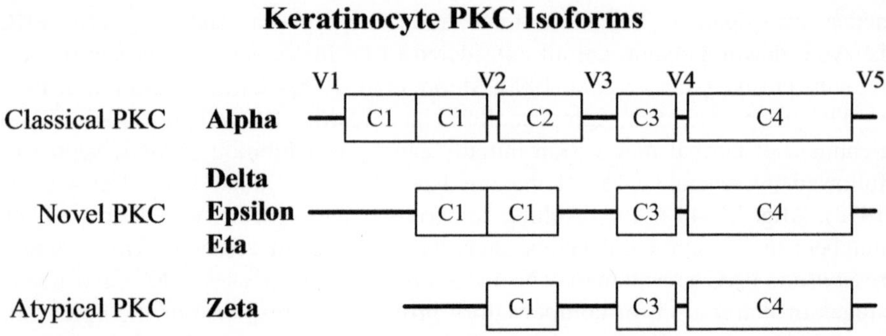

Fig. 16.1 PKC isozymes expressed in keratinocytes. Five PKC Isozymes are expressed in keratinocytes. PKCα is the only classical PKC isozyme and is responsive to both Ca^{2+} and diacylglycerol or phorbol esters. PKCδ, PKCε, and PKCη are the novel isozymes and are Ca^{2+} unresponsive. PKCζ is the only atypical PKC isozyme expressed in keratinocytes. Shown are the conserved (C1–C4) and variable (V1–V5) domains of each PKC isozyme

Studies on the chemical skin carcinogenesis model have pioneered many fundamental advances in epithelial cancers and initially attracted the attention of cancer researchers to PKC. Despite the power of the chemical carcinogenesis mouse skin model, its relevance to human SCC is limited by the different etiology of these cancers. Human SCCs are almost exclusively caused by UV radiation exposure from the sun, not exposure to genotoxic and PKC-activating chemicals (Brash et al. 1991). Early genetic changes are also different between chemical and UV-induced SCCs. Ha-Ras activation is found in >90% of benign papillomas induced by chemical carcinogenesis protocols, while Ha-Ras activation is relatively late and found in ~50% of human cutaneous SCCs (Pierceall et al. 1991). In contrast, mutations in the p53 tumor suppressor gene are common in human SCC and readily detected in actinic keratoses, a precursor lesion for human SCCs (Balmain and Pragnell 1983; Roop et al. 1986), while p53 mutations in chemically induced mouse skin tumors are a late event and found in less than 50% of SCCs (Ruggeri et al. 1991). Oncogenic Ha-Ras mutations are known to profoundly alter PKC isozyme activation status and signaling (Denning et al. 1993; Dlugosz et al. 1994; Lee et al. 1997), while the effect of p53 mutation on PKC signaling in KC is less well understood. It is clear that p53 can be directly phosphorylated by PKC in vitro; however, in vivo phosphorylation is more controversial (Milne et al. 1996; Delphin et al. 1997; Takenaka et al. 1995).

16.2.1 PKC Alpha and SCC

PKCα is the only Ca^{2+}-responsive PKC isozyme expressed in KCs, and considerable data support its role in the normal KC granular layer differentiation program (Denning et al. 1995a; Yang et al. 2003; Lee et al. 1997; Seo et al. 2004). This prodifferentiation function of PKCα is consistent with the ability of elevated extracellular Ca^{2+} to trigger a rise in intracellular Ca^{2+} and subsequent KC differentiation

(Hennings et al. 1980; Li et al. 1995). PKCα has also been linked to differentiation-induced growth arrest in normal KCs (Tibudan et al. 2002) as well as other self-renewing cell types (Saxon et al. 1994; Hizli et al. 2006; Frey et al. 2000; Nakagawa et al. 2006). PKCα has been localized to the membrane in the first suprabasal layer of epidermal KC, suggestive of activation in the epidermal compartment where cell cycle withdrawal occurs (Tibudan et al. 2002; Cataisson et al. 2006). SCC cell lines are resistant to growth arrest and terminal differentiation signals, and this may be related to altered PKCα activation. TGF-β-induced growth arrest in KC also involves PKCα in a Smad-independent signaling pathway (Sakaguchi et al. 2004).

PKCα signaling is altered in Ras-transformed KCs (Dlugosz et al. 1994; Lee et al. 1997). Ras transformation elevates cellular DAG levels, in part via increased autocrine EGFR ligand production, resulting in increased basal PKCα activity (Lee et al. 1992; Dlugosz et al. 1995, 1997). This increased PKCα activity is associated with the increased granular and decreased spinous layer differentiation marker expression observed in these transformed cell lines. Human SCC cell lines also fail to express differentiation markers in response to Ca^{2+}, and this defect can be related to the inability of Ca^{2+} to activate multiple PKC isozymes, especially PKCα, in squamous cell carcinoma lines (Yang et al. 2003). The nature of this defective PKCα activation in human SCC lines is unclear and is at odds with the increase in PKCα activity observed in Ha-Ras-transformed mouse KC cell lines. What is clear is that PKCα activation is aberrant in SCC in response to the altered growth factor production and differentiation status of these tumors.

Given the role of PKCα in normal KC differentiation and growth arrest and its defective activation in human SCC lines, it is reasonable to assume that PKCα has tumor suppressor activity for SCC. However, transgenic mice with epidermal overexpression of PKCα have been generated and characterized by several independent investigators, and these mice display no altered sensitivity to chemical carcinogenesis than wild type mice (Jansen et al. 2001; Wang and Smart 1999). In contrast, mice deficient for PKCα were significantly more susceptible to chemical carcinogenesis than control mice, consistent with a tumor suppressor activity for PKCα (Hara et al. 2005). These PKCα null mice were also resistant to PKC-induced hyperplasia, and this was associated with decreased production of EGFR ligands and TNF-α following TPA. One way to reconcile the differences between PKCα transgenic and null mice tumor susceptibility is that PKCα is already abundantly expressed in KCs and may not be rate-limiting for the development of chemically induced skin tumors.

Inflammation is critically important to carcinogenesis, and several studies link PKCα to proinflammatory cytokine release and cutaneous inflammation. TPA is a potent inducer of epidermal inflammation, and transgenic mice overexpressing PKCα in the basal epidermal layer exhibited dramatically enhanced TPA-induced inflammation, neutrophil infiltration, and proinflammatory cytokine production (Wang and Smart 1999; Cataisson et al. 2003). Significant KC apoptosis and microabscess formation were observed in TPA-treated PKCα transgenic mice, but the skin of these animals appeared to recover quickly. The cytokine induction and inflammatory response was dependent on NF-κB activation, and involved GM-CSF and CXCR2 ligands (MIP-2, cytokine-induced neutrophil chemoattractant), but was independent of AP-1 and TNF-α (Cataisson et al. 2003, 2005, 2006). Despite the expected enhanced regenerative KC hyperplasia, PKCα transgenic mice have not been reported to have altered sensitivity to chemical carcinogenesis. However, PKCα null mice have reduced TPA-induced hyperplasia and cytokine production, but are more sensitive to two-stage chemical carcinogenesis (Hara et al. 2005). Thus, the role of PKCα-mediated inflammation in skin carcinogenesis remains unclear, possibly due to the tumor-suppressive effects of PKCα on KC growth arrest and differentiation, which may partially compensate for the procarcinogenic increase in epidermal hyperplasia.

16.2.2 PKC Delta and SCC

PKCδ is an abundantly expressed, Ca^{2+}-independent PKC isozyme associated with both KC differentiation and DNA damage-induced apoptosis. While the ability of PKCδ to induced KC differentiation marker expression in TPA-treated KC is well established, the role of PKCδ in normal KC differentiation is unclear (Deucher et al. 2002; Efimova and Eckert 2000; Ohba et al. 1998). PKCδ becomes activated (membrane translocation) during Ca^{2+}-induced KC differentiation, but its localization in the epidermis is diffuse and levels actually slightly decrease in more differentiation layers (Denning et al. 1995a; D'Costa et al. 2006).

PKCδ has a firmly established role in apoptosis, including UV apoptosis (Denning et al. 1998, 2002; D'Costa and Denning 2005; Sitailo et al. 2006; Li et al. 1999; Sitailo et al. 2004). PKCδ is proteolytically activated in a caspase-dependent manner in KCs exposed to UV, and this activation is responsible for ~50% of UV apoptosis (D'Costa and Denning 2005; Denning et al. 2002). The proapoptotic function of PKCδ is not restricted to KCs or UV, but PKCδ appears to be a major apoptotic effector kinase in a wide variety of cell systems (Reyland 2007). Ectopic expression of the constitutively active PKCδ catalytic fragment induced apoptosis in KCs, in addition to other cell types (Denning et al. 2002; Sitailo et al. 2004, 2006). The mechanism of PKCδ-mediated apoptosis is controversial, as multiple PKCδ substrates implicated in apoptosis have been described. Examples include a destabilizing phosphorylation of the antiapoptotic Bcl-2 family member Mcl-1 (Sitailo et al. 2006), a stimulatory phosphorylation of phospholipid scramblase-3 (Liu et al. 2003; He et al. 2007), and an inhibitory phosphorylation of DNA-dependent protein kinase (Bharti et al. 1998). PKCδ activation can also induce a G2/M growth arrest, consistent with the DNA-damage cell cycle checkpoint, but the mechanism or critical substrate responsible for this cell cycle effect is unclear (Watanabe et al. 1992; Ishino et al. 1998). Thus, a main function of PKCδ in normal epidermis appears to respond to genotoxic stress, such as UV, to induce KC apoptosis.

The induction of apoptosis is a major tumor suppressive mechanism for carcinogenesis. Based on multiple criteria, PKCδ has tumor suppressive function for cutaneous SCC development. First, PKCδ becomes activated in response to genotoxic agents, such as UV, to limit the growth and survival of damaged KC by inducing apoptosis. Second, PKCδ expression is reduced in both chemically induced benign papillomas and UV-induced SCCs of mouse and human origin (Reddig et al. 1999; D'Costa et al. 2006; Aziz et al. 2006). Third, transgenic mice overexpressing PKCδ in their epidermises are highly resistant to two-stage chemical carcinogenesis (Reddig et al. 1999). Fourth, reexpression of full length PKCδ in SCC cell lines with reduced PKCδ expression induces these cells to undergo apoptosis and have significantly reduced tumorigenicity (D'Costa et al. 2006). One notable exception to the skin tumor suppressive function of PKCδ is that PKCδ transgenic mice were not resistant to UV carcinogenesis (Aziz et al. 2006). However, the tumors elicited by UV exposure in the PKCδ transgenic mice had reduced PKCδ expression, despite the PKCδ transgene being driven by a Keratin 14 promoter. Thus, PKCδ protein levels were decreased by some mechanism during the UV carcinogenesis protocol, and the overall results are consistent with a tumor suppressive function for PKCδ.

The mechanism(s) of PKCδ loss during either chemical or UV carcinogenesis are unclear, but multiple mechanisms are possible. The human PKCδ gene is located at 3p21.31, the most frequent site for deletions in human SCCs (Dobler et al. 1999; Ashton et al. 2003; Sikkink et al. 1997). In addition, Ha-Ras oncogene activation can suppress PKCδ activity, via tyrosine phosphorylation (Denning et al. 1993; Joseloff et al. 2002), and expression, via autocrine EGFR ligand production (D'Costa et al. 2006; Geiges et al. 1995). The inhibition by tyrosine phosphorylation mechanism has only been observed in cultured cell lines, but tyrosine 64 or 565 are required for inactivation of PKCδ function in Ha-Ras-transduced KCs (Joseloff et al. 2002). The tyrosine kinase responsible for PKCδ tyrosine phosphorylation in Ha-Ras-expressing keratinocytes is likely a member of the Src family (Src, Fyn, Yes). The gene deletion and reduced expression mechanism are both consistent with the reduced protein levels observed in both mouse and human SCCs. The PKCδ gene is upregulated by NF-κB subunits, and NF-κB activity is induced by UV and altered in SCC (Suh et al. 2003; Liu et al. 2006). Note that Ha-Ras activation is much more common in chemically induced skin tumors than in human SCC, and thus different mechanisms of PKCδ inhibition/loss may occur in mice and human tumors.

By integrating available data regarding PKCδ's tumor suppressor function, it is possible to construct a model for UV induced SCC development (Fig. 16.2). PKCδ becomes downregulated or lost via Ha-ras-induced EGFR ligand production or gene deletion, respectively. This results in KCs with enhanced survival following subsequent UV exposure. Since these damage KCs were not eliminated by PKCδ-mediated apoptosis, they acquire additional oncogenic changes (genetic or epigenetic) and survive to eventually produce a benign precursor lesion, the actinic keratoses. Over time, the actinic keratoses will undergo premalignant progression to SCC. An interesting prediction of this model is that it may be possible to pharmacologically induce the reexpression of PKCδ by applying inhibitors of Ha-Ras/EGFR signaling. It has been demonstrated that reexpression of even full length PKCδ is sufficient to dramatically reduce the tumorigenicity of SCC cells (D'Costa et al. 2006).

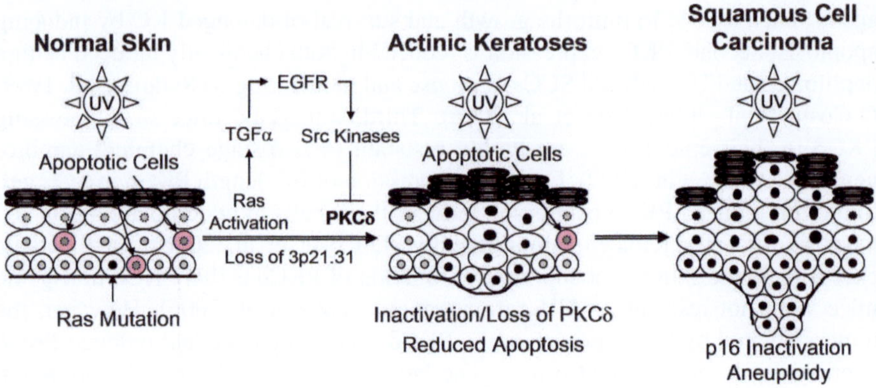

Fig. 16.2 Model of PKCδ tumor suppression in SCC. Exposure of normal skin to UV radiation induces DNA damage resulting in the formation of apoptotic cells, as well as mutations in the Ras oncogene. Activation of Ras inhibits PKCδ expression/activity via a TGFα/EGFR autocrine loop. The PKCδ gene at chromosome 3p21.31 can also become deleted. Upon subsequent UV exposure of precursor actinic keratoses lesions, KC with reduced PKCδ are relatively resistant to UV apoptosis and survive to acquire additional genetic defects resulting in the formation of SCC

16.2.3 PKC Epsilon and SCC

PKCε is strongly associated with enhanced KC proliferation. Overexpression of PKCε in mouse KC enhances TPA-induced proliferation, and transgenic mice expressing epidermal PKCε have increased basal and TPA-induced hyperplasia (Papp et al. 2004; Jansen et al. 2001). Consistent with these observations, PKCε transgenic mice are highly susceptible to both chemical and UV skin carcinogenesis protocols (Reddig et al. 2000; Wheeler et al. 2003a, b, 2004). The effects of PKCε overexpression on chemical carcinogenesis are remarkable, eliciting reduced papilloma formation but dramatically increased SCC incidence. Furthermore, the SCCs that developed were metastatic and arose even in the absence of TPA exposure (DMBA alone).

Several mechanisms have been proposed to explain the oncogenic activity of PKCε for skin cancer. TPA treatment of PKCε transgenic mice induces massive release of the proinflammatory cytokine TNF-α, epidermal inflammation, and loss of KCs from the epidermis. This is followed by regenerative hyperplasia to repopulate the epidermis with KCs. The release of TNF-α involved ROS-mediated activation of TNF-α converting enzyme, TACE, and TACE inhibition completely blocked SCC development in PKCε transgenic mice (Wheeler et al. 2003b). These results are consistent with the known role of TNF-α in skin carcinogenesis (Moore et al. 1999; Suganuma et al. 1999) and suggest that the regenerative hyperplasia following the inflammatory response and KC loss may be responsible for the SCC formation in the absence of additional TPA tumor promotion.

PKCε transgenic mice are also more sensitive to UV carcinogenesis, and the ability of PKCε to bind, phosphorylate, and activate STAT3 is an important additional oncogenic mechanism (Wheeler et al. 2004; Aziz et al. 2007). STAT3 is an oncogenic transcription factor important for EGFR signaling and constitutively activated in skin tumors (Chan et al. 2004a; Sano et al. 2005). In addition, STAT3 activation is necessary and sufficient for skin carcinogenesis (Chan et al. 2004b, 2008). PKCε transgenic mice also had increased proliferation and reduced apoptosis following UV exposure, similar to what is observed in STAT3 transgenic mice (Sano et al. 2005). Thus, multiple mechanisms are likely responsible for the profound oncogenic activity of PKCε in skin cancer.

16.2.4 PKC Eta and SCC

The localization of PKCη in the epidermis is the best characterized of all PKC isozymes. PKCη is expressed exclusively in the granular layer as determined by in situ hybridization and immunohistochemistry (Koizumi et al. 1993; Osada et al. 1993). PKCη activation has been associated with both the activation of terminal differentiation genes such as transglutaminase and involucrin (Cabodi et al. 2000; Ohba et al. 1998; Takahashi et al. 1998; Ueda et al. 1996; Efimova and Eckert 2000), as well as cell cycle withdrawal (Ishino et al. 1998; Cabodi et al. 2000; Ohba et al. 1998). The mechanism of growth arrest involves associated with cyclin e/cdk2/p21 and inhibition of cdk2 activity (Kashiwagi et al. 2000). PKCη is an upstream activator of the Src family tyrosine kinase Fyn, and Fyn is required for KC differentiation and PKCη-induced growth arrest (Cabodi et al. 2000; Calautti et al. 1995). Based upon its localization in the granular layer, the function of PKCη in normal epidermis may be to regulate the induction of KC terminal differentiation markers or the maintenance of growth arrest, but it is not expressed in the proper cells (basal and/or early spinous) to be involved in initiating growth arrest. PKCη deficient mice did have prolonged TPA-induced hyperplasia, indicating that PKCη is involved in the cessation of KC proliferation in response to phorbol ester treatment (Chida et al. 2003).

Consistent with its role in inducing terminal differentiation and growth inhibition, PKCη has been demonstrated to be a tumor suppressor for SCC (Chida et al. 2003). Mice with targeted deletion of PKCη were more susceptible to chemical carcinogenesis, although PKCη heterozygous mice had tumor yield and latency similar to wild type mice. PKCη deficient mice also had impaired wound healing. There was no alteration in normal epidermal architecture in untreated PKCη deficient mice, although compensation by other PKC isozymes may have occurred during development. Cholesterol sulfate, a lipid second messenger generated by KCs during differentiation, is able to directly active PKCη and inhibits mouse skin tumor promotion by TPA, consistent with a tumor suppressive function for PKCη (Ikuta et al. 1994; Denning et al. 1995b; Chida et al. 1995).

16.3 Basal Cell Carcinoma

BCC is the most common and most curable type of skin cancer. BCCs are composed of homogeneous nests of KCs, which morphologically resemble basal layer KC, and can invade locally (Fan et al. 1997). The vast majority of BCCs have activated hedgehog, Hh, signaling, and mutations in several components of the Hh pathway have been identified in BCCs (Gailani et al. 1996; Epstein 2008). In addition, engineered activation of Hh signaling in several transgenic models is sufficient to elicit BCCs or BCC-like tumors in mice (Fan et al. 1997; Hutchin et al. 2005; Grachtchouk et al. 2003). Thus activation Hh signaling is considered to be sufficient for BCC development.

The reliance of BCCs on Hh pathway activation makes understanding Hh signaling critically important to the pathogenesis of BCCs. Hh signaling is initiated by binding of the Hh ligand (Sonic Hh, Indian Hh, or Desert Hh) to its receptor Patched, which relieves the inhibition of Smoothened. Activation of Smoothened leads to the activation of the Gli family of transcription factors which mediate Hh effects by inducing target genes involved in survival and cell cycle progression. Several kinases have been implicated in Hh signaling, most notably the negative regulation by protein kinase A involving inhibition of Gli nuclear localization (Sheng et al. 2006). Any role of PKC in Hh signaling and BCC formation has only recently been appreciated. Note that Hh activation is associated with a large number of other common human malignancies (i.e., lung, breast, prostate), and thus PKC effects on Hh signaling may have broad applications to cancers other than cutaneous BCC.

16.3.1 PKC Alpha and BCC

Hh signaling is a major developmental morphogenic pathway, and thus evidence for PKC involvement in Hh signaling has naturally come from developmental model systems. Studies in mouse embryonic stem cells found that sonic Hh triggered elevation in intracellular Ca^{2+} and translocation of multiple PKC isozymes (α, δ, ζ) to the membrane fraction, indicative of activation (Heo et al. 2007). Induction of DNA synthesis and phosphorylation of the p65 NF-κB subunit by sonic Hh could be blocked by chelation of intracellular Ca^{2+} or the general PKC inhibitor Bisindoylmaleimide I. In the chick limb bud development model, the general PKC inhibitor chelerythrine chloride inhibited PKCα/PKCβII activation and interfered with limb bud development, as well as expression of sonic Hh (Lu et al. 2001). These studies implicate classical PKC isozyme (i.e., PKCα) activation in promoting Hh signaling.

However, PKCα expression in BCC is reported to be very low, and expression of an active PKCα suppressed Gli reporter activity in 293T cells (Neill et al. 2003). The reduced Gli activity was not due to sequestration of Gli in the cytoplasmic or alteration in MAPK activation. Reduced PKCα in BCC may reflect the low level of PKCα detected in basal layer of the epidermis and its association with terminal differentiation (Tibudan et al. 2002; Cataisson et al. 2006). Thus, despite evidence for PKCα promoting Hh signaling (Heo et al. 2007; Lu et al. 2001), the role of PKCα in Hh signaling and BCC formation is unresolved.

16.3.2 PKC Delta and BCC

Several studies have implicated PKCδ as a positive regulator of Hh signaling (Riobo et al. 2006; Neill et al. 2003). Expression of a constitutively active PKCδ stimulated Gli activity in 293T cells, and TPA induced Gli activity required active PKCδ functioning via MEK-1 (Riobo et al. 2006). Sonic Hh induced Gli or Hh target gene expression was blocked by the PKCδ inhibitor Rottlerin, but not the classical PKC inhibitor Gö6976, further implicating PKCδ as a component of the Hh signaling pathway. However, expression of PKCδ in BCC is reported to be very low, making the significance of these findings unclear (Neill et al. 2003).

16.4 Melanoma

Malignant melanoma is the most deadly type of skin cancer. Approximately, 62,000 new cases are expected in the United States in 2008 resulting in approximately 8,400 deaths (American Cancer Society 2008). In addition, there is a disquieting 3% per year increase in the incidence of melanoma. Melanoma is notoriously resistant to apoptosis induction by a variety of agents, including cancer chemotherapy drugs, and there are currently no effective treatments for metastatic disease. Melanomas arise from melanocytes, the pigment-producing cells found within the basal layer of the epidermis and hair follicle. Melanocytes normally produce pigment in the form of melanin and deliver it to keratinocytes via specialized transport vesicles called melanosomes. Since skin pigment (eumelanin) can be highly protective against UV-induced keratinocytes skin cancers (SCC and BCC), a better understanding of pigment cell biology is crucial to many fundamental aspects of skin cancer. Recently, our understanding of the molecular etiology of melanoma have advanced significantly, and drugs that target common genetic alterations (B-Raf, PI3 kinase, Ras, Notch) are under development or in clinical trials with results eagerly anticipated (Chudnovsky et al. 2005).

PKC has emerged as an ideal therapeutic target for melanoma due to the striking difference between normal melanocytes and malignant melanoma cells in their response to activators of PKC (Oka and Kikkawa 2005). Normal human melanocytes require chronic PKC activation for growth in culture, and in fact melanocyte culture medium often contains a direct activator of PKC, the phorbol ester TPA, to stimulate melanocyte proliferation (Arita et al. 1992; Halaban et al. 1986).

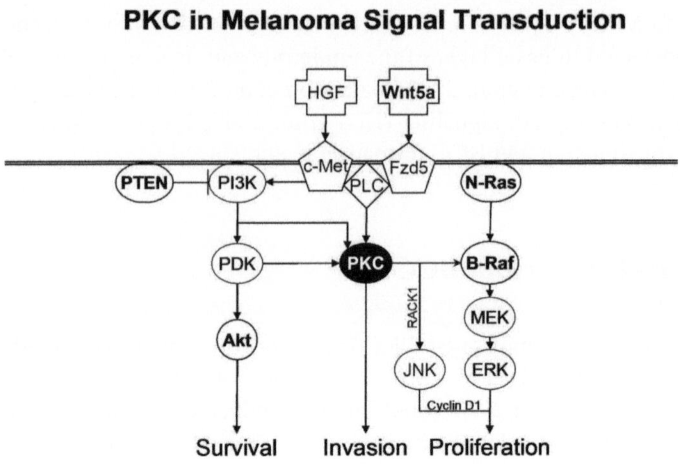

Fig. 16.3 PKC in melanoma signal transduction. PKC occupies a central point in the signal transduction pathways altered in melanoma. Molecules shown in *bold* are either activated (N-Ras, B-Raf) or inactivated (PTEN) by mutations, upregulated (Wnt5a), or enzymatically activated (Akt) in melanoma. Note that several growth factors (HGF, Wnt5a) important for melanoma growth can trigger PKC activation due to their receptors being coupled to phospholipase C. Inactivation of the PTEN phosphatase leads to activation of the PDK/Akt pathway, which promotes cell survival and can activate PKC isozymes. PKC can also directly phosphorylate and activate members of the Raf MAP kinase pathway, promoting cell proliferation

Alternatively, melanocyte culture media contains growth factors, such as endothelin-1, which activate PKC via the phospholipase C-coupled endothelin receptor (Berking et al. 2004; Swope et al. 1995). In sharp contrast, melanoma cells are growth inhibited or in some cases killed when cultured in the presence of TPA (Halaban et al. 1986; Becker et al. 1990; Brooks et al. 1990). This selective suppression of melanoma cell growth by the PKC activator TPA suggests that PKC may be a useful therapeutic target to treat melanoma.

Many of the signaling pathways altered in melanoma, including B-Raf, PTEN, PI3 Kinase, and Wnt-5a, influence PKC signaling and thus it is likely that these genetic alterations reprogram the PKC signaling in malignant melanoma. In fact, expression of active B-Raf^{V600E} can substitute for the mitogenic effects of TPA in normal melanocytes (Wellbrock et al. 2004). The roles of most PKC isozymes in normal melanocyte biology are not well understood, but the current information will be summarized here (Fig. 16.3).

16.4.1 PKC Alpha and Melanoma

PKCα is expressed in both normal melanocytes and melanoma but has been linked to increased invasion and metastasis in melanoma (Selzer et al. 2002;

Dennis et al. 1998; Jiang et al. 2005). In addition, overexpression of PKCα enhanced melanoma cell proliferation while inhibition of PKCα with antisense oligonucleotides or a chemical inhibitor inhibited cell proliferation (Krasagakis et al. 2004). The effects of PKCα on cell proliferation may be mediated by activation of Jun N-terminal kinase, JNK, phosphorylation of c-Jun, and induction of growth promoting genes such as cyclin D1 (Lopez-Bergami et al. 2005, 2007). The PKC adaptor protein RACK1 is essential for activation of JNK by PKC.

One interesting candidate for altering PKCα signaling in melanoma is Wnt-5A, a cell-associated and secreted glycoprotein that interacts with the Frizzled-5 receptor and activates the noncanonical Wnt pathway. Wnt-5A was identified from gene microarray experiments to be the single best predictor of melanoma progression out of a panel of ~7,000 genes (Weeraratna et al. 2002). Wnt-5A expression correlated with melanoma invasiveness and tumor grade, and was inversely correlated with patient survival (Weeraratna et al. 2002; Carr et al. 2003). Furthermore, it was demonstrated that overexpression of Wnt-5A triggered the phosphorylation of multiple PKC isozymes in low-grade melanoma cell lines and enhanced their invasive phenotype. These studies concluded that Wnt-5A was activating PKC in the melanoma cells. Additional studies found that classical PKC isozymes played a role in Wnt-5a mediated migration and epithelial to mesenchymal transition (Dissanayake et al. 2007). Note that the studies on Wnt-5a did not directly distinguish among which classical PKC isozymes (PKCα, PKCβ, PKCγ) was involved in melanoma invasion and metastasis; however, since both PKCβ and PKCγ expression is low in melanoma, these effects are likely mediated by PKCα.

16.4.2 PKC Beta and Melanoma

Several unique functions of PKCβ relevant to pigment cell biology have been identified. PKCβ can directly phosphorylate and activate tyrosinase, and it is recruited to melanosomes by RACK1 upon activation (Park et al. 1999, 1993, 2004a, b). PKCβ can also be activated by reactive oxygen species, leading to the phosphorylation of Shc and mitochondrial apoptosis (DelCarlo and Loeser 2006; Pinton et al. 2007). PKCβ has also been found to interfere with hepatocyte growth factor invasion by blocking phosphatidylinosital 3-kinase signaling (Oka et al. 2008). Thus, PKCβ has distinctive regulatory and functional characteristics that make it a significant molecule for melanocyte biology.

Reduced expression of PKCβ in clinical melanoma specimens and melanoma cell lines is very common and has been reported by many investigators (Oka et al. 1996, 2006; Gilhooly et al. 2001; Shields et al. 2007; Ryu et al. 2007). Although PKCβ is strongly linked with melanogenesis and is reduced in ~90% of melanomas examined, the role of PKCβ loss in melanoma has been a long-standing question in the field. This controversy exists because PKCβ is lost in some benign nevi and thus its loss also correlates well with defects in melanocyte differentiation (Gilhooly et al. 2001). The PKCβ downregulation occurs at the mRNA level and is associated with aggressive melanoma but is independent of ERK activation status (Ryu et al. 2007; Shields et al. 2007). PKCβ gene transcription is upregulated by the microphthalmia-associated transcription factor, MITF, as part of the cAMP-mediated pigmentation pathway in melanocytes; however, MITF is amplified in melanomas with poor prognosis and thus would be predicted to induce PKCβ expression in this subset of melanomas (Park et al. 2006; Garraway et al. 2005). PKCβ exists as two major spice forms, I and II, which differ in their C-terminus due to alternative exon usage (Ono et al. 1986). Although differences in subcellular localization have been reported for PKCβI and PKCβII, expression of both splice variants are reduced in melanoma (Becker and Hannun 2004; Fridberg et al. 2007; Gilhooly et al. 2001). The mechanism of PKCβ down-regulation in melanoma is likely transcriptional but requires further study.

16.4.3 PKC Zeta and Melanoma

While the activation mechanisms of the atypical PKC isozyme PKCζ are less well-defined than classical and novel PKC isoforms, PKCζ is overexpressed in melanoma cell lines at an average of ~25-fold, and can respond to a variety of lipid molecules (Hoek et al. 2004). In normal human melanocytes, the phospholipase A_2 type X product lysophosphatidylcholine simulated dendricity via activation of PKCζ (Scott et al. 2007). The increased dendricity was associated with activation of Rac and Rho small GTPases involved in the global regulation of actin remodeling and cell migration. In fact, overexpression of PKCζ in melanoma cells inhibited migration, although the clinical significance of PKCζ overexpression in melanoma remains unclear (Sanz-Navares et al. 2001).

16.4.4 Concluding Remarks

Significant progress has been made in the last 25 years of investigations focused on the roles of PKC isozymes in skin cancer. This progress lays the foundation for significant opportunities which exist to target PKC isozymes in the treatment of skin cancers. An especially exciting new development is the translation of decades of PKC research expertise to the relatively new fields of BCC and melanoma molecular carcinogenesis, and the new opportunities that these developments bring to the millions of skin cancer patients worldwide.

References

American Cancer Society. (2008). *Cancer facts and figures 2008*. Atlanta: American Cancer Society.

Arita, Y., O'Driscoll, K. R., & Weinstein, I. B. (1992). Growth of human melanocyte cultures supported by 12-O-tetradecanoylphorbol-13-acetate is mediated through protein kinase C activation. *Cancer Research, 52*, 4514–4521.

Ashton, K. J., Weinstein, S. R., Maguire, D. J., & Griffiths, L. R. (2003). Chromosomal aberrations in squamous cell carcinoma and solar keratoses revealed by comparative genomic hybridization. *Archives of Dermatology, 139*, 876–882.

Aziz, M. H., Manoharan, H. T., & Verma, A. K. (2007). Protein kinase C ε, which sensitizes skin to sun's UV radiation-induced cutaneous damage and development of squamous cell carcinomas, associates with Stat3. *Cancer Research, 67*, 1385–1394.

Aziz, M. H., Wheeler, D. L., Bhamb, B., & Verma, A. K. (2006). Protein kinase C δ overexpressing transgenic mice are resistant to chemically but not to UV radiation-induced development of squamous cell carcinomas: A possible link to specific cytokines and cyclooxygenase-2. *Cancer Research, 66*, 713–722.

Balmain, A., & Pragnell, I. B. (1983). Mouse skin carcinomas induced in vivo by chemical carcinogens have a transforming Harvey-ras oncogene. *Nature, 303*, 72–74.

Becker, D., Beebe, S. J., & Herlyn, M. (1990). Differential expression of protein kinase C and cAMP-dependent protein kinase in normal human melanocytes and malignant melanomas. *Oncogene, 5*, 1133–1139.

Becker, K. P., & Hannun, Y. A. (2004). Isoenzyme-specific translocation of protein kinase C (PKC)βII and not PKCβI to a juxtanuclear subset of recycling endosomes: Involvement of phospholipase D. *The Journal of Biological Chemistry, 279*, 28251–28256.

Berking, C., Takemoto, R., Satyamoorthy, K., Shirakawa, T., Eskandarpour, M., Hansson, J., et al. (2004). Induction of melanoma phenotypes in human skin by growth factors and ultraviolet B. *Cancer Research, 64*, 807–811.

Bharti, A., Kraeft, S. K., Gounder, M., Pandey, P., Jin, S., Yuan, Z. M., et al. (1998). Inactivation of DNA-dependent protein kinase by protein kinase Cδ: Implications for apoptosis. *Molecular and Cellular Biology, 18*, 6719–6728.

Brash, D. E., Rudolph, J. A., Simon, J. A., Lin, A., McKenna, G. J., Baden, H. P., et al. (1991). A role for sunlight in skin cancer: UV-induced p53 mutations in squamous cell carcinoma. *Proceedings of the National Academy of Sciences of the United States of America, 88*, 10124–10128.

Brooks, G., Birch, M., & Hart, I. R. (1990). Effects of biologically active tumour-promoting and non-promoting phorbol esters on in vitro growth of melanocytic cells. *Pigment Cell Research, 3*, 98–100.

Cabodi, S., Calautti, E., Talora, C., Kuroki, T., Stein, P. L., & Dotto, G. P. (2000). A PKC-eta/Fyn-dependent pathway leading to keratinocyte growth arrest and differentiation. *Molecular Cell, 6*, 1121–1129.

Calautti, E., Missero, C., Stein, P. L., Ezzell, R. M., & Dotto, G. P. (1995). Fyn tyrosine kinase is involved in keratinocyte differentiation control. *Genes and Development, 9*, 2279–2291.

Carr, K. M., Bittner, M., & Trent, J. M. (2003). Gene-expression profiling in human cutaneous melanoma. *Oncogene, 22*, 3076–3080.

Cataisson, C., Joseloff, E., Murillas, R., Wang, A., Atwell, C., Torgerson, S., et al. (2003). Activation of cutaneous protein kinase Cα induces keratinocyte apoptosis and intraepidermal inflammation by independent signaling pathways. *Journal of Immunology, 171*, 2703–2713.

Cataisson, C., Pearson, A. J., Torgerson, S., Nedospasov, S. A., & Yuspa, S. H. (2005). Protein kinase Cα-mediated chemotaxis of neutrophils requires NF-κB activity but is independent of TNFα signaling in mouse skin in vivo. *Journal of Immunology, 174*, 1686–1692.

Cataisson, C., Pearson, A. J., Tsien, M. Z., Mascia, F., Gao, J. L., Pastore, S., et al. (2006). CXCR2 ligands and G-CSF mediate PKCα-induced intraepidermal inflammation. *The Journal of Clinical Investigation, 116*, 2757–2766.

Chakravarthy, B. R., Isaacs, R. J., Morley, P., Durkin, J. P., & Whitfield, J. F. (1995). Stimulation of protein kinase C during Ca^{2+}-induced keratinocyte differentiation. Selective blockade of MARCKS phosphorylation by calmodulin. *The Journal of Biological Chemistry, 270*, 1362–1368.

Chan, K. S., Carbajal, S., Kiguchi, K., Clifford, J., Sano, S., & DiGiovanni, J. (2004). Epidermal growth factor receptor-mediated activation of Stat3 during multistage skin carcinogenesis. *Cancer Research, 64*, 2382–2389.

Chan, K. S., Sano, S., Kataoka, K., Abel, E., Carbajal, S., Beltran, L., et al. (2008). Forced expression of a constitutively active form of Stat3 in mouse epidermis enhances malignant progression of skin tumors induced by two-stage carcinogenesis. *Oncogene, 27*, 1087–1094.

Chan, K. S., Sano, S., Kiguchi, K., Anders, J., Komazawa, N., Takeda, J., et al. (2004). Disruption of Stat3 reveals a critical role in both the initiation and the promotion stages of epithelial carcinogenesis. *The Journal of Clinical Investigation, 114*, 720–728.

Chida, K., Hara, T., Hirai, T., Konishi, C., Nakamura, K., Nakao, K., et al. (2003). Disruption of protein kinase Cη results in impairment of wound healing and enhancement of tumor formation in mouse skin carcinogenesis. *Cancer Research, 63*, 2404–2408.

Chida, K., Murakami, A., Tagawa, T., Ikuta, T., & Kuroki, T. (1995). Cholesterol sulfate, a second messenger for the η isoform of protein kinase C, inhibits promotional phase in mouse skin carcinogenesis. *Cancer Research, 55*, 4865–4869.

Chudnovsky, Y., Khavari, P. A., & Adams, A. E. (2005). Melanoma genetics and the development of rational therapeutics. *The Journal of Clinical Investigation, 115*, 813–824.

D'Costa, A. M., & Denning, M. F. (2005). A caspase-resistant mutant of PKC-δ protects keratinocytes from UV-induced apoptosis. *Cell Death and Differentiation, 12*, 224–232.

D'Costa, A. M., Robinson, J. K., Maududi, T., Chaturvedi, V., Nickoloff, B. J., & Denning, M. F. (2006). The proapoptotic tumor suppressor protein kinase C-δ is lost in human squamous cell carcinomas. *Oncogene, 25*, 378–386.

DelCarlo, M., & Loeser, R. F. (2006). Chondrocyte cell death mediated by reactive oxygen species-dependent activation of PKC-βI. *American Journal of Physiology. Cell Physiology, 290*, C802–C811.

Delphin, C., Huang, K. P., Scotto, C., Chapel, A., Vincon, M., Chambaz, E., et al. (1997). The in vitro phosphorylation of p53 by calcium-dependent protein kinase C–characterization of a protein-kinase-C-binding site on p53. *European Journal of Biochemistry, 245*, 684–692.

Denning, M. F., Dlugosz, A. A., Howett, M. K., & Yuspa, S. H. (1993). Expression of an oncogenic rasHa gene in murine keratinocytes induces tyrosine phosphorylation and reduced activity of protein kinase C δ. *The Journal of Biological Chemistry, 268*, 26079–26081.

Denning, M. F., Dlugosz, A. A., Williams, E. K., Szallasi, Z., Blumberg, P. M., & Yuspa, S. H. (1995). Specific protein kinase C isozymes mediate the induction of keratinocyte differentiation markers by calcium. *Cell Growth and Differentiation, 6*, 149–157.

Denning, M. F., Kazanietz, M. G., Blumberg, P. M., & Yuspa, S. H. (1995). Cholesterol sulfate activates multiple protein kinase C isoenzymes and induces granular cell differentiation in cultured murine keratinocytes. *Cell Growth and Differentiation, 6*, 1619–1626.

Denning, M. F., Wang, Y., Nickoloff, B. J., & Wrone-Smith, T. (1998). Protein kinase Cδ is activated by caspase-dependent proteolysis during ultraviolet radiation-induced apoptosis of human keratinocytes. *The Journal of Biological Chemistry, 273*, 29995–30002.

Denning, M. F., Wang, Y., Tibudan, S., Nickoloff, B. J., & Qin, J. Z. (2002). Caspase activation and disruption of mitochondrial membrane potential during UV radiation-induced apoptosis of human keratinocyte requires activation of protein kinase C. *Cell Death and Differentiation, 9*, 40–52.

Dennis, J. U., Dean, N. M., Bennett, C. F., Griffith, J. W., Lang, C. M., & Welch, D. R. (1998). Human melanoma metastasis is inhibited following ex vivo treatment with an antisense oligonucleotide to protein kinase C-alpha. *Cancer Letters, 128*, 65–70.

Deucher, A., Efimova, T., & Eckert, R. L. (2002). Calcium-dependent involucrin expression is inversely regulated by protein kinase C (PKC)alpha and PKCdelta. *The Journal of Biological Chemistry, 277*, 17032–17040.
Dissanayake, S. K., Wade, M., Johnson, C. E., O'Connell, M. P., Leotlela, P. D., French, A. D., et al. (2007). The Wnt5A/protein kinase C pathway mediates motility in melanoma cells via the inhibition of metastasis suppressors and initiation of an epithelial to mesenchymal transition. *The Journal of Biological Chemistry, 282*, 17259–17271.
Dlugosz, A. A., Cheng, C., Williams, E. K., Darwiche, N., Dempsey, P. J., Mann, B., et al. (1995). Autocrine transforming growth factor alpha is dispensible for v-rasHa-induced epidermal neoplasia: Potential involvement of alternate epidermal growth factor receptor ligands. *Cancer Research, 55*, 1883–1893.
Dlugosz, A. A., Cheng, C., Williams, E. K., Dharia, A. G., Denning, M. F., & Yuspa, S. H. (1994). Alterations in murine keratinocyte differentiation induced by activated rasHa genes are mediated by protein kinase C-α. *Cancer Research, 54*, 6413–6420.
Dlugosz, A. A., Hansen, L., Cheng, C., Alexander, N., Denning, M. F., Threadgill, D. W., et al. (1997). Targeted disruption of the epidermal growth factor receptor impairs growth of squamous papillomas expressing the v-rasHa oncogene but does not block in vitro keratinocyte responses to oncogenic ras. *Cancer Research, 57*, 3180–3188.
Dlugosz, A. A., Mischak, H., Mushinski, J. F., & Yuspa, S. H. (1992). Transcripts encoding protein kinase C-α, -δ, -ε, -ζ, and -η are expressed in basal and differentiating mouse keratinocytes in vitro and exhibit quantitative changes in neoplastic cells. *Molecular Carcinogenesis, 5*, 286–292.
Dlugosz, A. A., & Yuspa, S. H. (1993). Coordinate changes in gene expression which mark the spinous to granular cell transition in epidermis are regulated by protein kinase C. *The Journal of Cell Biology, 120*, 217–225.
Dobler, M., Schuh, J., Kiesewetter, F., Schell, H., Liehr, T., & Gebhart, E. (1999). Deletion monitoring in skin tumors by interphase-FISH using band-specific DNA probes. *International Journal of Oncology, 14*, 571–576.
Efimova, T., & Eckert, R. L. (2000). Regulation of human involucrin promoter activity by novel protein kinase C isoforms. *The Journal of Biological Chemistry, 275*, 1601–1607.
Efimova, T., LaCelle, P., Welter, J. F., & Eckert, R. L. (1998). Regulation of human involucrin promoter activity by a protein kinase C, Ras, MEKK1, MEK3, p38/RK, AP1 signal transduction pathway. *The Journal of Biological Chemistry, 273*, 24387–24395.
Epstein, E. H. (2008). Basal cell carcinomas: Attack of the hedgehog. *Nature Reviews. Cancer, 8*, 743–754.
Fan, H., Oro, A. E., Scott, M. P., & Khavari, P. A. (1997). Induction of basal cell carcinoma features in transgenic human skin expressing Sonic Hedgehog. *Nature Medicine, 3*, 788–792.
Frey, M. R., Clark, J. A., Leontieva, O., Uronis, J. M., Black, A. R., & Black, J. D. (2000). Protein kinase C signaling mediates a program of cell cycle withdrawal in the intestinal epithelium. *The Journal of Cell Biology, 151*, 763–778.
Fridberg, M., Servin, A., Anagnostaki, L., Linderoth, J., Berglund, M., Soderberg, O., et al. (2007). Protein expression and cellular localization in two prognostic subgroups of diffuse large B-cell lymphoma: Higher expression of ZAP70 and PKC-β II in the non-germinal center group and poor survival in patients deficient in nuclear PTEN. *Leukaemia and Lymphoma, 48*, 2221–2232.
Gailani, M. R., Stahle-Backdahl, M., Leffell, D. J., Glynn, M., Zaphiropoulos, P. G., Pressman, C., et al. (1996). The role of the human homologue of *Drosophila* patched in sporadic basal cell carcinomas. *Nature Genetics, 14*, 78–81.
Garraway, L. A., Widlund, H. R., Rubin, M. A., Getz, G., Berger, A. J., Ramaswamy, S., et al. (2005). Integrative genomic analyses identify MITF as a lineage survival oncogene amplified in malignant melanoma. *Nature, 436*, 117–122.
Geiges, D., Marks, F., & Gschwendt, M. (1995). Loss of protein kinase C delta from human HaCaT keratinocytes upon ras transfection is mediated by TGF alpha. *Experimental Cell Research, 219*, 299–303.

Gilhooly, E. M., Morse-Gaudio, M., Bianchi, L., Reinhart, L., Rose, D. P., Connolly, J. M., et al. (2001). Loss of expression of protein kinase C β is a common phenomenon in human malignant melanoma: A result of transformation or differentiation? *Melanoma Research, 11*, 355–369.

Grachtchouk, V., Grachtchouk, M., Lowe, L., Johnson, T., Wei, L., Wang, A., et al. (2003). The magnitude of hedgehog signaling activity defines skin tumor phenotype. *EMBO Journal, 22*, 2741–2751.

Halaban, R., Ghosh, S., Duray, P., Kirkwood, J. M., & Lerner, A. B. (1986). Human melanocytes cultured from nevi and melanomas. *The Journal of Investigative Dermatology, 87*, 95–101.

Hansen, L. A., Monteiro-Riviere, N. A., & Smart, R. C. (1990). Differential down-regulation of epidermal protein kinase C by 12-O-tetradecanoylphorbol-13-acetate and diacylglycerol: Association with epidermal hyperplasia and tumor promotion. *Cancer Research, 50*, 5740–5745.

Hara, T., Saito, Y., Hirai, T., Nakamura, K., Nakao, K., Katsuki, M., et al. (2005). Deficiency of protein kinase Cα in mice results in impairment of epidermal hyperplasia and enhancement of tumor formation in two-stage skin carcinogenesis. *Cancer Research, 65*, 7356–7362.

He, Y., Liu, J., Grossman, D., Durrant, D., Sweatman, T., Lothstein, L., et al. (2007). Phosphorylation of mitochondrial phospholipid scramblase 3 by protein kinase C-δ induces its activation and facilitates mitochondrial targeting of tBid. *Journal of Cellular Biochemistry, 101*, 1210–1221.

Hennings, H., Michael, D., Cheng, C., Steinert, P., Holbrook, K., & Yuspa, S. H. (1980). Calcium regulation of growth and differentiation of mouse epidermal cells in culture. *Cell, 19*, 245–254.

Hennings, H., Shores, R., Mitchell, P., Spangler, E. F., & Yuspa, S. H. (1985). Induction of papillomas with a high probability of conversion to malignancy. *Carcinogenesis, 6*, 1607–1610.

Heo, J. S., Lee, M. Y., & Han, H. J. (2007). Sonic hedgehog stimulates mouse embryonic stem cell proliferation by cooperation of Ca^{2+}/protein kinase C and epidermal growth factor receptor as well as Gli1 activation. *Stem Cells, 25*, 3069–3080.

Hizli, A. A., Black, A. R., Pysz, M. A., & Black, J. D. (2006). Protein kinase C α signaling inhibits cyclin D1 translation in intestinal epithelial cells. *The Journal of Biological Chemistry, 281*, 14596–14603.

Hoek, K., Rimm, D. L., Williams, K. R., Zhao, H., Ariyan, S., Lin, A., et al. (2004). Expression profiling reveals novel pathways in the transformation of melanocytes to melanomas. *Cancer Research, 64*, 5270–5282.

Hutchin, M. E., Kariapper, M. S., Grachtchouk, M., Wang, A., Wei, L., Cummings, D., et al. (2005). Sustained Hedgehog signaling is required for basal cell carcinoma proliferation and survival: Conditional skin tumorigenesis recapitulates the hair growth cycle. *Genes and Development, 19*, 214–223.

Ikuta, T., Chida, K., Tajima, O., Matsuura, Y., Iwamori, M., Ueda, Y., et al. (1994). Cholesterol sulfate, a novel activator for the eta isoform of protein kinase C. *Cell Growth and Differentiation, 5*, 943–947.

Ishino, K., Ohba, M., Kashiwagi, M., Kawabe, S., Chida, K., & Kuroki, T. (1998). Phorbol ester-induced G1 arrest in BALB/MK-2 mouse keratinocytes is mediated by δ and η isoforms of protein kinase C. *Japanese Journal of Cancer Research, 89*, 1126–1133.

Jansen, A. P., Dreckschmidt, N. E., Verwiebe, E. G., Wheeler, D. L., Oberley, T. D., & Verma, A. K. (2001). Relation of the induction of epidermal ornithine decarboxylase and hyperplasia to the different skin tumor-promotion susceptibilities of protein kinase C alpha, -delta and -epsilon transgenic mice. *International Journal of Cancer, 93*, 635–643.

Jiang, Y., Berk, M., Singh, L. S., Tan, H., Yin, L., Powell, C. T., et al. (2005). KiSS1 suppresses metastasis in human ovarian cancer via inhibition of protein kinase C alpha. *Clinical and Experimental Metastasis, 22*, 369–376.

Joseloff, E., Cataisson, C., Aamodt, H., Ocheni, H., Blumberg, P., Kraker, A. J., et al. (2002). Src family kinases phosphorylate protein kinase C delta on tyrosine residues and modify the neoplastic phenotype of skin keratinocytes. *The Journal of Biological Chemistry, 277*, 12318–12323.

Kashiwagi, M., Ohba, M., Watanabe, H., Ishino, K., Kasahara, K., Sanai, Y., et al. (2000). PKCeta associates with cyclin E/cdk2/p21 complex, phosphorylates p21 and inhibits cdk2 kinase in keratinocytes. *Oncogene, 19*, 6334–6341.

Kikkawa, U., Takai, Y., Tanaka, Y., Miyake, R., & Nishizuka, Y. (1983). Protein kinase C as a possible receptor protein of tumor-promoting phorbol esters. *The Journal of Biological Chemistry, 258*, 11442–11445.

Koizumi, H., Kohno, Y., Osada, S., Ohno, S., Ohkawara, A., & Kuroki, T. (1993). Differentiation-associated localization of nPKC η, a Ca^{2+}-independent protein kinase C, in normal human skin and skin diseases. *The Journal of Investigative Dermatology, 101*, 858–863.

Krasagakis, K., Lindschau, C., Fimmel, S., Eberle, J., Quass, P., Haller, H., et al. (2004). Proliferation of human melanoma cells is under tight control of protein kinase C alpha. *Journal of Cellular Physiology, 199*, 381–387.

Lee, E., Punnonen, K., Cheng, C., Glick, A., Dlugosz, A., & Yuspa, S. H. (1992). Analysis of phospholipid metabolism in murine keratinocytes transformed by the v-ras oncogene: Relationship of phosphatidylinositol turnover and cytokine stimulation to the transformed phenotype. *Carcinogenesis, 13*, 2367–2373.

Lee, Y. S., Dlugosz, A. A., McKay, R., Dean, N. M., & Yuspa, S. H. (1997). Definition by specific antisense oligonucleotides of a role for protein kinase C α in expression of differentiation markers in normal and neoplastic mouse epidermal keratinocytes. *Molecular Carcinogenesis, 18*, 44–53.

Lee, Y. S., Yuspa, S. H., & Dlugosz, A. A. (1998). Differentiation of cultured human epidermal keratinocytes at high cell densities is mediated by endogenous activation of the protein kinase C signaling pathway. *The Journal of Investigative Dermatology, 111*, 762–766.

Li, L., Lorenzo, P. S., Bogi, K., Blumberg, P. M., & Yuspa, S. H. (1999). Protein kinase C δ targets mitochondria, alters mitochondrial membrane potential, and induces apoptosis in normal and neoplastic keratinocytes when overexpressed by an adenoviral vector. *Molecular and Cellular Biology, 19*, 8547–8558.

Li, L., Tucker, R. W., Hennings, H., & Yuspa, S. H. (1995). Chelation of intracellular Ca^{2+} inhibits murine keratinocyte differentiation in vitro. *Journal of Cellular Physiology, 163*, 105–114.

Liu, J., Chen, J., Dai, Q., & Lee, R. M. (2003). Phospholipid scramblase 3 is the mitochondrial target of protein kinase C δ-induced apoptosis. *Cancer Research, 63*, 1153–1156.

Liu, J., Yang, D., Minemoto, Y., Leitges, M., Rosner, M. R., & Lin, A. (2006). NF-κB is required for UV-induced JNK activation via induction of PKCδ. *Molecular Cell, 21*, 467–480.

Longthorne, V. L., & Williams, G. T. (1997). Caspase activity is required for commitment to Fas-mediated apoptosis. *EMBO Journal, 16*, 3805–3812.

Lopez-Bergami, P., Habelhah, H., Bhoumik, A., Zhang, W., Wang, L. H., & Ronai, Z. (2005). Receptor for RACK1 mediates activation of JNK by protein kinase C. *Molecular Cell, 19*, 309–320.

Lopez-Bergami, P., Huang, C., Goydos, J. S., Yip, D., Bar-Eli, M., Herlyn, M., et al. (2007). Rewired ERK-JNK signaling pathways in melanoma. *Cancer Cell, 11*, 447–460.

Lu, H. C., Swindell, E. C., Sierralta, W. D., Eichele, G., & Thaller, C. (2001). Evidence for a role of protein kinase C in FGF signal transduction in the developing chick limb bud. *Development, 128*, 2451–2460.

Milne, D. M., McKendrick, L., Jardine, L. J., Deacon, E., Lord, J. M., & Meek, D. W. (1996). Murine p53 is phosphorylated within the PAb421 epitope by protein kinase C in vitro, but not in vivo, even after stimulation with the phorbol ester o-tetradecanoylphorbol 13-acetate. *Oncogene, 13*, 205–211.

Moore, R. J., Owens, D. M., Stamp, G., Arnott, C., Burke, F., East, N., et al. (1999). Mice deficient in tumor necrosis factor-alpha are resistant to skin carcinogenesis. *Nature Medicine, 5*, 828–831.

Nakagawa, R., Soh, J. W., & Michie, A. M. (2006). Subversion of protein kinase Cα signaling in hematopoietic progenitor cells results in the generation of a B-cell chronic lymphocytic leukemia-like population in vivo. *Cancer Research, 66*, 527–534.

Neill, G. W., Ghali, L. R., Green, J. L., Ikram, M. S., Philpott, M. P., & Quinn, A. G. (2003). Loss of protein kinase C alpha expression may enhance the tumorigenic potential of gli1 in basal cell carcinoma. *Cancer Research, 63*, 4692–4697.

Nishizuka, Y. (1984). The role of protein kinase C in cell surface signal transduction and tumour promotion. *Nature, 308*, 693–698.

Ohba, M., Ishino, K., Kashiwagi, M., Kawabe, S., Chida, K., Huh, N. H., et al. (1998). Induction of differentiation in normal human keratinocytes by adenovirus-mediated introduction of the eta and delta isoforms of protein kinase C. *Molecular and Cellular Biology, 18*, 5199–5207.

Oka, M., & Kikkawa, U. (2005). Protein kinase C in melanoma. *Cancer and Metastasis Reviews, 24*, 287–300.

Oka, M., Kikkawa, U., & Nishigori, C. (2008). Protein kinase C-βII represses hepatocyte growth factor-induced invasion by preventing the association of adapter protein Gab1 and phosphatidylinositol 3-kinase in melanoma cells. *The Journal of Investigative Dermatology, 128*, 188–195.

Oka, M., Nishigori, C., Kageshita, T., Hsu, M. Y., Penmatcha, S., & Herlyn, M. (2006). Expression of PKC isoforms in human melanocytic cells in situ. *Journal of Dermatological Science, 41*, 157–161.

Oka, M., Ogita, K., Ando, H., Horikawa, T., Hayashibe, K., Saito, N., et al. (1996). Deletion of specific protein kinase C subspecies in human melanoma cells. *Journal of Cellular Physiology, 167*, 406–412.

Ono, Y., Kurokawa, T., Fujii, T., Kawahara, K., Igarashi, K., Kikkawa, U., et al. (1986). Two types of complementary DNAs of rat brain protein kinase C. Heterogeneity determined by alternative splicing. *FEBS Letters, 206*, 347–352.

Osada, S., Hashimoto, Y., Nomura, S., Kohno, Y., Chida, K., Tajima, O., et al. (1993). Predominant expression of the nPKCη, a Ca^{2+}-independent isoform of protein kinase C in epithelial tissues, in association with epithelial differentiation. *Cell Growth and Differentiation, 4*, 167–175.

Papp, H., Czifra, G., Bodo, E., Lazar, J., Kovacs, I., Aleksza, M., et al. (2004). Opposite roles of protein kinase C isoforms in proliferation, differentiation, apoptosis, and tumorigenicity of human HaCaT keratinocytes. *Cellular and Molecular Life Sciences, 61*, 1095–1105.

Park, H. Y., Lee, J., Gonzalez, S., Middelkamp-Hup, M. A., Kapasi, S., Peterson, S., et al. (2004). Topical application of a protein kinase C inhibitor reduces skin and hair pigmentation. *The Journal of Investigative Dermatology, 122*, 159–166.

Park, H. Y., Perez, J. M., Laursen, R., Hara, M., & Gilchrest, B. A. (1999). Protein kinase C-β activates tyrosinase by phosphorylating serine residues in its cytoplasmic domain. *The Journal of Biological Chemistry, 274*, 16470–16478.

Park, H. Y., Russakovsky, V., Ohno, S., & Gilchrest, B. A. (1993). The β isoform of protein kinase C stimulates human melanogenesis by activating tyrosinase in pigment cells. *The Journal of Biological Chemistry, 268*, 11742–11749.

Park, H. Y., Wu, C., Yonemoto, L., Murphy-Smith, M., Wu, H., Stachur, C. M., et al. (2006). MITF mediates cAMP-induced protein kinase C-β expression in human melanocytes. *Biochemical Journal, 395*, 571–578.

Park, H. Y., Wu, H., Killoran, C. E., & Gilchrest, B. A. (2004). The receptor for activated C-kinase-I (RACK-I) anchors activated PKC-β on melanosomes. *Journal of Cell Science, 117*, 3659–3668.

Pierceall, W. E., Goldberg, L. H., Tainsky, M. A., Mukhopadhyay, T., & Ananthaswamy, H. N. (1991). Ras gene mutation and amplification in human nonmelanoma skin cancers. *Molecular Carcinogenesis, 4*, 196–202.

Pinton, P., Rimessi, A., Marchi, S., Orsini, F., Migliaccio, E., Giorgio, M., et al. (2007). Protein kinase C β and prolyl isomerase 1 regulate mitochondrial effects of the life-span determinant p66Shc. *Science, 315*, 659–663.

Raick, A. N. (1973). Ultrastructural, histological, and biochemical alterations produced by 12-*O*-tetradecanoyl-phorbol-13-acetate on mouse epidermis and their relevance to skin tumor promotion. *Cancer Research, 33*, 269–286.

Raick, A. N., Thumm, K., & Chivers, B. R. (1972). Early effects of 12-O-tetradecanoyl-phorbol-13-acetate on the incorporation of tritiated precursor into DNA and the thickness of the interfollicular epidermis, and their relation to tumor promotion in mouse skin. *Cancer Research, 32*, 1562–1568.

Reddig, P. J., Dreckschimdt, N. E., Ahrens, H., Simsiman, R., Tseng, C. P., Zou, J., et al. (1999). Transgenic mice overexpressing protein kinase C δ in the epidermis are resistant to skin tumor promotion by 12-O-tetradecanoylphorbol-13-acetate. *Cancer Research, 59*, 5710–5718.

Reddig, P. J., Dreckschmidt, N. E., Zou, J., Bourguignon, S. E., Oberley, T. D., & Verma, A. K. (2000). Transgenic mice overexpressing protein kinase C ε in their epidermis exhibit reduced papilloma burden but enhanced carcinoma formation after tumor promotion. *Cancer Research, 60*, 595–602.

Reiners, J. J., Jr., & Slaga, T. J. (1983). Effects of tumor promoters on the rate and commitment to terminal differentiation of subpopulations of murine keratinocytes. *Cell, 32*, 247–255.

Reyland, M. E. (2007). Protein kinase Cδ and apoptosis. *Biochemical Society Transactions, 35*, 1001 1004.

Riobo, N. A., Haines, G. M., & Emerson, C. P., Jr. (2006). Protein kinase C-δ and mitogen-activated protein/extracellular signal-regulated kinase-1 control GLI activation in hedgehog signaling. *Cancer Research, 66*, 839–845.

Roop, D. R., Lowy, D. R., Tambourin, P. E., Strickland, J., Harper, J. R., Balaschak, M., et al. (1986). An activated Harvey ras oncogene produces benign tumours on mouse epidermal tissue. *Nature, 323*, 822–824.

Ruggeri, B., Caamano, J., Goodrow, T., DiRado, M., Bianchi, A., Trono, D., et al. (1991). Alterations of the p53 tumor suppressor gene during mouse skin tumor progression. *Cancer Research, 51*, 6615–6621.

Ryu, B., Kim, D. S., Deluca, A. M., & Alani, R. M. (2007). Comprehensive expression profiling of tumor cell lines identifies molecular signatures of melanoma progression. *PloS one, 2*, e594.

Sakaguchi, M., Miyazaki, M., Sonegawa, H., Kashiwagi, M., Ohba, M., Kuroki, T., et al. (2004). PKCα mediates TGFβ-induced growth inhibition of human keratinocytes via phosphorylation of S100/A11. *The Journal of Cell Biology, 164*, 979–984.

Sano, S., Chan, K. S., Kira, M., Kataoka, K., Takagi, S., Tarutani, M., et al. (2005). Signal transducer and activator of transcription 3 is a key regulator of keratinocyte survival and proliferation following UV irradiation. *Cancer Research, 65*, 5720–5729.

Sanz-Navares, E., Fernandez, N., Kazanietz, M. G., & Rotenberg, S. A. (2001). Atypical protein kinase Cζ suppresses migration of mouse melanoma cells. *Cell Growth and Differentiation, 12*, 517–524.

Saxon, M. L., Zhao, X., & Black, J. D. (1994). Activation of protein kinase C isozymes is associated with post- mitotic events in intestinal epithelial cells in situ. *The Journal of Cell Biology, 126*, 747–763.

Scott, G. A., Arioka, M., & Jacobs, S. E. (2007). Lysophosphatidylcholine mediates melanocyte dendricity through PKCζ activation. *The Journal of Investigative Dermatology, 127*, 668–675.

Selzer, E., Okamoto, I., Lucas, T., Kodym, R., Pehamberger, H., & Jansen, B. (2002). Protein kinase C isoforms in normal and transformed cells of the melanocytic lineage. *Melanoma Research, 12*, 201–209.

Seo, H. R., Kwan, Y. W., Cho, C. K., Bae, S., Lee, S. J., Soh, J. W., et al. (2004). PKCα induces differentiation through ERK1/2 phosphorylation in mouse keratinocytes. *Experimental and Molecular Medicine, 36*, 292–299.

Sharkey, N. A., Leach, K. L., & Blumberg, P. M. (1984). Competitive inhibition by diacylglycerol of specific phorbol ester binding. *Proceedings of the National Academy of Sciences of the United States of America, 81*, 607–610.

Sheng, T., Chi, S., Zhang, X., & Xie, J. (2006). Regulation of Gli1 localization by the cAMP/protein kinase A signaling axis through a site near the nuclear localization signal. *The Journal of Biological Chemistry, 281*, 9–12.

Shields, J. M., Thomas, N. E., Cregger, M., Berger, A. J., Leslie, M., Torrice, C., et al. (2007). Lack of extracellular signal-regulated kinase mitogen-activated protein kinase signaling shows a new type of melanoma. *Cancer Research, 67*, 1502–1512.

Sikkink, S. K., Rehman, I., & Rees, J. L. (1997). Deletion mapping of chromosome 3p and 13q and preliminary analysis of the FHIT gene in human nonmelanoma skin cancer. *The Journal of Investigative Dermatology, 109*, 801–805.

Sitailo, L. A., Tibudan, S. S., & Denning, M. F. (2004). Bax activation and induction of apoptosis in human keratinocytes by the protein kinase C δ catalytic domain. *The Journal of Investigative Dermatology, 123*, 434–443.

Sitailo, L. A., Tibudan, S. S., & Denning, M. F. (2006). The protein kinase Cδ catalytic fragment targets Mcl-1 for degradation to trigger apoptosis. *The Journal of Biological Chemistry, 28*, 29703–29710.

Suganuma, M., Okabe, S., Marino, M. W., Sakai, A., Sueoka, E., & Fujiki, H. (1999). Essential role of tumor necrosis factor alpha (TNF-α) in tumor promotion as revealed by TNF-alpha-deficient mice. *Cancer Research, 59*, 4516–4518.

Suh, K. S., Tatunchak, T. T., Crutchley, J. M., Edwards, L. E., Marin, K. G., & Yuspa, S. H. (2003). Genomic structure and promoter analysis of PKC-δ. *Genomics, 82*, 57–67.

Swope, V. B., Medrano, E. E., Smalara, D., & Abdel-Malek, Z. A. (1995). Long-term proliferation of human melanocytes is supported by the physiologic mitogens alpha-melanotropin, endothelin-1, and basic fibroblast growth factor. *Experimental Cell Research, 217*, 453–459.

Takahashi, H., Asano, K., Manabe, A., Kinouchi, M., Ishida-Yamamoto, A., & Iizuka, H. (1998). The α and η isoforms of prtoein kinase C stimulate transcription of human involucrin gene. *The Journal of Investigative Dermatology, 110*, 218–223.

Takenaka, I., Morin, F., Seizinger, B. R., & Kley, N. (1995). Regulation of the sequence-specific DNA binding function of p53 by protein kinase C and protein phosphatases. *The Journal of Biological Chemistry, 270*, 5405–5411.

Tibudan, S. S., Wang, Y., & Denning, M. F. (2002). Activation of protein kinase C triggers irreversible cell cycle withdrawal in human keratinocytes. *The Journal of Investigative Dermatology, 119*, 1282–1289.

Trempus, C. S., Morris, R. J., Ehinger, M., Elmore, A., Bortner, C. D., Ito, M., et al. (2007). CD34 expression by hair follicle stem cells is required for skin tumor development in mice. *Cancer Research, 67*, 4173–4181.

Ueda, E., Ohno, S., Kuroki, T., Livneh, E., Yamada, K., Yamanishi, K., et al. (1996). The eta isoform of protein kinase C mediates transcriptional activation of the human transglutaminase 1 gene. *The Journal of Biological Chemistry, 271*, 9790–9794.

Van Duuren, B. L., Sivak, A., Segal, A., Seidman, I., & Katz, C. (1973). Dose-response studies with a pure tumor-promoting agent, phorbol myristate acetate. *Cancer Research, 33*, 2166–2172.

Wang, H. Q., & Smart, R. C. (1999). Overexpression of protein kinase C-α in the epidermis of transgenic mice results in striking alterations in phorbol ester-induced inflammation and COX-2, MIP-2 and TNF-alpha expression but not tumor promotion. *Journal of Cell Science, 112*, 3497–3506.

Watanabe, T., Ono, Y., Taniyama, Y., Hazama, K., Igarashi, K., Ogita, K., et al. (1992). Cell division arrest induced by phorbol ester in CHO cells overexpressing protein kinase C-δ subspecies. *Proceedings of the National Academy of Sciences of the United States of America, 89*, 10159–10163.

Weeraratna, A. T., Jiang, Y., Hostetter, G., Rosenblatt, K., Duray, P., Bittner, M., et al. (2002). Wnt5a signaling directly affects cell motility and invasion of metastatic melanoma. *Cancer Cell, 1*, 279–288.

Wellbrock, C., Ogilvie, L., Hedley, D., Karasarides, M., Martin, J., Niculescu-Duvaz, D., et al. (2004). ^{V599E}B-RAF is an oncogene in melanocytes. *Cancer Research, 64*, 2338–2342.

Wheeler, D. L., Martin, K. E., Ness, K. J., Li, Y., Dreckschmidt, N. E., Wartman, M., et al. (2004). Protein kinase C ε is an endogenous photosensitizer that enhances ultraviolet radiation-induced cutaneous damage and development of squamous cell carcinomas. *Cancer Research, 64*, 7756–7765.

Wheeler, D. L., Ness, K. J., Oberley, T. D., & Verma, A. K. (2003a). Inhibition of the development of metastatic squamous cell carcinoma in protein kinase C epsilon transgenic mice by alpha-difluoromethylornithine accompanied by marked hair follicle degeneration and hair loss. *Cancer Research, 63*, 3037–3042.

Wheeler, D. L., Ness, K. J., Oberley, T. D., & Verma, A. K. (2003b). Protein kinase C ε is linked to 12-O-tetradecanoylphorbol-13-acetate-induced tumor necrosis factor-alpha ectodomain shedding and the development of metastatic squamous cell carcinoma in protein kinase C epsilon transgenic mice. *Cancer Research, 63*, 6547–6555.

Yang, L. C., Ng, D. C., & Bikle, D. D. (2003). Role of protein kinase C alpha in calcium induced keratinocyte differentiation: Defective regulation in squamous cell carcinoma. *Journal of Cellular Physiology, 195*, 249–259.

Yuspa, S. H., Ben, T., Hennings, H., & Lichti, U. (1982). Divergent responses in epidermal basal cells exposed to the tumor promoter 12-O-tetradecanoylphorbol-13-acetate. *Cancer Research, 42*, 2344–2349.

Yuspa, S. H., Kilkenny, A. E., Stanley, J., & Lichti, U. (1985). Keratinocytes blocked in phorbol ester-responsive early stage of terminal differentiation by sarcoma viruses. *Nature, 314*, 459–462.

Chapter 17
PKC and Breast Cancer*

Sofia D. Merajver, Devin T. Rosenthal, and Lauren Van Wassenhove

Abstract PKC expression is intimately associated with breast cancer initiation, progression, and therapy responsiveness, and these effects are highly isozyme-specific. PKC isozymes play key roles in proliferation and apoptosis of breast cancer cells and exert important modulatory roles in cell cycle progression. A close relationship exists between specific PKC isozymes and estrogen signaling.

Keywords Protein kinase C • Breast cancer progression • Estrogen receptor signaling • Drug resistance

17.1 Introduction

As discussed earlier in this book, PKC is an important component of numerous key cell signaling pathways: it serves as a crucial hub for translating a variety of extracellular stimuli into cellular responses. In this chapter, we illustrate the role of PKC isozymes in breast cancer development and progression, with a focus on how the tumorigenic properties of PKC reflect its role in normal mammary gland development. There are some general attributes shared by most PKC isoforms, yet each isozyme retains its own identity through distinct roles in breast cancer progression. An enhanced understanding of the involvement of specific isozymes is essential to the development of effective targeted therapies with minimal side effects, as is already evidenced through the early successes of small molecule inhibitors of PKC such as enzastaurin.

---

*Work supported by NIH RO1CA-61722 (SDM), the Burroughs Wellcome Fund (SDM), the Department of Defense Breast Cancer Program (predoctoral grant to DTR), the UM Cellular Biotechnology Training Grant (LVW) and the Breast cancer Research Foundation (SDM).

S.D. Merajver (✉), D.T. Rosenthal, and L.V. Wassenhove
Department of Internal Medicine, Division of Hematology and Oncology,
University of Michigan, Ann Arbor, MI, USA
e-mail: smerajve@med.umich.edu

M.G. Kazanietz (ed.), *Protein Kinase C in Cancer Signaling and Therapy*,
Current Cancer Research, DOI 10.1007/978-1-60761-543-9_17,

17.2 Mammary Gland Development Overview

The mammary gland develops through several defined stages in the course of a mammal's life; each stage utilizes distinct developmental processes which, if misregulated, can be viewed as hallmarks of cancer (Hanahan and Weinberg 2000). In this chapter, we focus on mouse mammary gland development, as this is the best-characterized model of mammary gland development and, with a few important exceptions, largely resembles the human mammary developmental process. Our description of mammary gland development is necessarily brief; for more detailed information, we recommend several comprehensive recent reviews (Richert et al. 2000; Hovey et al. 2002; Lanigan et al. 2007).

At birth, the mammary gland consists of a small epithelial anlage contained within the mammary fat pad. During puberty, these epithelial cells proliferate and invade into the surrounding fat pad, generating the rudimentary ductal structure of the gland. These cells remain generally quiescent until pregnancy, at which point the gland again undergoes extensive remodeling as the epithelial cells rapidly proliferate to form alveolar structures in preparation for lactation. After lactation and weaning, the mammary gland undergoes widespread apoptosis of the alveoli in a process known as involution, thus returning the gland to a quiescent state.

17.3 Roles of PKC Isozymes in Breast Cancer

PKC plays a variety of roles in breast cancer through involvement in apoptosis, cell cycle regulation, metastasis, growth regulation, hormonal regulation, and drug resistance. In the following sections, we elaborate on the roles specific isozymes play in these processes relevant to tumorigenesis, as well as highlight some of the conflicts within the literature regarding isozyme function.

17.4 Apoptosis

Apoptosis, or programmed cell death, is a crucial developmental tool to shape tissues and organs in space and time in living organisms, as well as a means of combating expansion of transformed cells. The involuting mammary gland is a classic example of developmentally regulated apoptosis. In contrast, cancer cells may develop ways to evade these processes. The PKC isozymes play a varied role in apoptosis in cancer; however, the present state of knowledge reveals some conflicting conclusions for the roles of individual isoforms in apoptosis.

The classical α and β isoforms are typically considered to be antiapoptotic proteins. Inhibition of PKCα, β, or ε in MCF-7 breast cancer cells restores their sensitivity to radiation-induced apoptosis (Jasinski et al. 2008). This effect can be

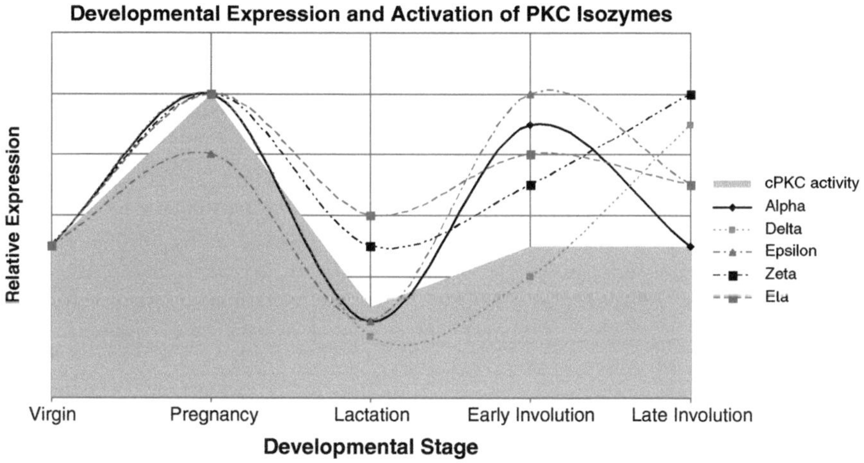

Fig. 17.1 Variation of expression as a function of developmental stages in murine mammary development (Adapted from Foncea et al. 1995; Masso-Welch et al. 1998; Masso-Welch et al. 1999)

at least in part attributed to the activation of p21^{CIP1} and bcl-2 by α (Soh et al. 2003) and bcl-2 by ε (Gubina et al. 1998).

A conflicting report by de Vente et al. (1995) demonstrated that treating PKCα-transfected MCF-7 cells with phorbol esters (TPA), an activator of PKC activity, resulted in increased apoptosis. This discrepancy may be reasoned by observing the developmental expression and activation of PKCα. PKCα expression peaks during both pregnancy and early involution – highly proliferative and highly apoptotic time points, respectively (see Fig. 17.1). These expression changes also mirror the activation peaks of Ca^{2+}-dependent PKC isoforms. It is therefore plausible that PKCα plays roles in both proliferation and apoptosis, and that this decision may be driven by differential regulation from upstream stimuli. Additionally, the use of a broad PKC activator like TPA generates nonspecific responses from other PKC isoforms, thus making it difficult to ascribe the functional consequences to any one isoform.

PKCδ also plays an ambiguous role in apoptosis. It has been shown to have proapoptotic activity in MCF-7 cells in response to UV damage by an activating cleavage at the hinge region through a caspase-dependent mechanism (Denning et al. 1998). Activation by this mechanism leads to phosphorylation of ASMase by PKCδ, which results in ceramide production and potentiation of the apoptotic signal (Zeidan et al. 2008). In contrast, PKCδ is antiapoptotic in both MCF-7 and MDA-MB-231 cells in response to ionizing radiation (McCracken et al. 2003), and in response to TNF-related apoptosis-inducing ligand- (TRAIL)-mediated apoptosis (Zhang et al. 2005). Additionally, Grossini et al. (2007) demonstrated that PKCδ enhances proliferation and survival of murine mammary cells.

Because of evidence supporting both a pro- and antiapoptotic role of PKCδ, it is difficult to classify the role of this isozyme in breast cancer. What is clear, however, is that the role of PKCδ in apoptosis is primarily dependent on the specific upstream stimulus, as UV damage and ionizing radiation elicit opposite effects in MCF-7 cells. Whatever the reason, it is apparent that the regulation of PKCδ warrants further investigation, before it can be validated as a drug target in breast cancer therapy.

As previously mentioned, PKCε can inhibit apoptosis by regulating the function of bcl-2 directly, but it also regulates several bcl-2 family members. PKCε has been shown to prevent the activation of Bax and its translocation to the mitochondria, as well as regulate the function of Bcl-2 and Bid (Lu et al. 2007; Sivaprasad et al. 2007). In addition, PKCε protects MCF-7 breast cancer cells from Tumor Necrosis Factor α (TNFα)-induced cell death by phosphorylating Akt, a potent prosurvival protein, via DNA Protein Kinase (DNA-PK) (Lu et al. 2006).

PKCι and ζ promote cell survival by interacting with IKKβ, one of two inhibitors of kappa B (IκB) kinases. This interaction prevents the inhibition of NF-κB signaling and thus activates NF-κB, leading to increased cellular proliferation (Sanz et al. 1999) and chemokine synthesis. PKCζ has also been shown to protect against UV-C-induced apoptosis (Charruyer et al. 2007). In summary, even though related, the isoforms of PKC play distinct and important roles in apoptosis, depending on the integration of upstream stimuli, a fact that may be in the future modeled mathematically to better discern how to selectively modulate PKC actions to aid in cancer therapy. The notion of one-target one-inhibitor is likely to be ineffective as an anticancer approach in this set of proteins.

17.5 Cell Cycle Regulation

PKC has been classically thought of as a promoter of proliferation, and thus, indirectly, an inhibitor of differentiation. This function is evidenced by PKC expression throughout mammary gland development, as PKC activity and expression of most isoforms are highest during pregnancy, a period of intense proliferation. Tight regulation of proliferation – in an organism by circulating hormone and cytokine signals and intracellularly by cell cycle-regulating proteins – is crucial for preventing malignant transformation of normal cells. It is therefore not surprising that expression of several proliferation-promoting PKC isoforms is altered in breast cancer.

Evidence for the role of PKCα in promoting proliferation primarily derives from experiments utilizing the MCF-7 breast cancer cell line. Transfection of MCF-7 cells with PKCα results in increased proliferation is due at least in part to increased levels of Erk2 (Ways et al. 1995; Gupta et al. 1996). Interestingly, PKCα-transfected MCF-7 cells also show increased levels of PKCβ (Ways et al. 1995) and γ (Morse-Gaudio et al. 1998), and decreased PKCδ and η expression (Ways et al. 1995). Activation of PKCβ increases proliferation in MCF-7, MDA-MB-231, and BT474 breast cancer cell lines through the activation of cyclin D1 and *c-fos* transcription (Li and Weinstein 2006), thus likely contributing to the observed PKCα-mediated increase in proliferation.

The array of changes in isozyme expression upon transfection with PKCα once again makes it difficult to assign a direct role to PKCα in regulating cell proliferation. Broad effects such as this do, however, open the door for future research into the role of cross talk between isozymes (Ventura et al. 2009). Furthering the support of cross talk between PKCα and β, small hairpin RNA (shRNA) knockdown of PKCα in T47D breast cancer cells results in reduced PKCβ levels (Lin et al. 2006). Of note, however, the reciprocal experiment results in upregulation of both PKCβ and δ (Tonetti et al. 2000). Once again, context is crucial when analyzing the effects and expression of PKC isozymes, as both characteristics vary between cell lines (as shown above) and between in vitro and in vivo settings (Lin et al. 2006).

Both PKCδ and η are involved in cell cycle regulation through influencing the G_1/S phase transition. Both isozymes exert their regulatory effects through the cyclin E/Cdk2 complex. PKCδ mediates G_1 arrest through a p21-dependent pathway in SKBR-3 breast cancer cells and a $p27^{Kip1}$-dependent pathway in MCF-7 (Yokoyama et al. 2005; Shanmugam et al. 2001; Vucenik et al. 2005). Both p21 and $p27^{Kip1}$ can bind to the cyclin E/Cdk2 complex, and thereby inhibit its activity. In contrast, PKCη directly associates with the cyclin E/Cdk2 complex, leading to G_1 arrest in MCF-7 and NIH-3T3 cells (Shtutman et al. 2003). PKCδ can also mediate G_1 arrest through a phosphorylated retinoblastoma protein (pRb)-dependent mechanism in MCF-7 cells (Vucenik et al. 2005). Because of these tumor suppressive effects, it is not surprising that PKC δ is downregulated in the previously referenced more aggressive PKCα-transfected MCF-7 cells, and that PKCη is downregulated in invasive breast cancer (Masso-Welch et al. 2001). In order to fully comprehend the role of PKC in tumorigenesis, it will be required that cell cycle regulation be temporally integrated into their dynamic modulation of signal transduction.

17.6 Metastasis

Primary tumors constitute severe disease in their own right, but the vast majority of cancer-related deaths are due to metastatic spread to vital organs. Metastasis is a complex process involving extensive interaction with, and remodeling of, the tumor microenvironment by invading cancer cells, and subsequent colonization of a new microenvironment – most notably lungs, bone, brain, pleura, and liver in breast cancer (Saenz and Phillips 1998). With the array of processes required of cancer cells in order to metastasize – motility, invasion, and (lymph)angiogenesis, to name a few – it is not surprising that PKC, one of the major intracellular signaling hubs, plays a broad and important role.

In order for cancer cells to invade both locally and into secondary sites, they need to be able to move. PKCδ has been shown to suppress migration in MCF-7 cells (Jackson et al. 2005). This downregulation may be due in part to regulation by PKCα. As previously described, MCF-7 cells transfected with PKCα showed decreased levels of PKCδ. Injecting these cells into nude mice resulted in an increased number of metastases compared to vector controls (Ways et al. 1995), perhaps due in part to lower PKCδ expression, as well as decreased cell–cell adhesion (Williams and Noti 2001).

Once again, however, the weakness of these findings is the almost exclusive reliance on cell lines, directly or indirectly through the use of xenografts. Contrasting with the aforementioned studies, work by Kiley et al. (1999) demonstrated that PKCδ is actually required for metastatic spread of the highly metastatic MTLn3 cell line. Further studies using patient tissue samples should prove highly informative in resolving these discrepancies, as well as developmental studies focusing on the invasive epithelial growth during puberty.

Chemotaxis, or the movement of cells in response to stimuli such as growth factors, is presumed to be an important event in the development of breast cancer metastasis. PKCζ has been shown to be required for epidermal growth factor (EGF)-induced chemotaxis in the aggressive breast cancer cell line MDA-MB-231 (Sun et al. 2005). The same group later showed that PKCζ acts downstream of Akt and is activated by Akt directly, which is significant given that Akt is required for chemotaxis in MDA-MB-231, T47D, and MCF-7 breast cell lines (Wang et al. 2008).

In order for cells from the primary tumor to metastasize, they need to penetrate the protective basement membrane layer, which is part of the tissue boundary that helps prevent disorganized epithelial cell outgrowth. To accomplish this, cancer cells secrete matrix metalloproteinases (MMPs), which somewhat easily degrade an assortment of basement membrane proteins, depending on the particular MMP; this process, however, is biosynthetically expensive to the cell, and thus it is coordinated carefully or the cell undergoes apoptosis when it is unable to sustain the biosynthetic demands. PKCδ has been shown to play a role in greatly activating MMP-9 in breast cancer cells (Lin et al. 2008, Alonso-Escolano et al. 2006), again adding to the ambiguity in the role of PKCδ in breast cancer progression.

After escaping the primary tumor and intravasating into the lymphatic system or blood stream, survival of cancer cells in this environment becomes paramount. Although very harsh on cells, the survivors of the blood or lymphatic vessel invasion have made powerful adaptations in specific mechanisms. One example is the assembly of a protective layer made up of fibronectin deposited on the surface of cancer cells, thus preventing damage to the cell during intravasation into the blood stream. This peculiar survival mechanism was shown to be mediated by PKCε in rat breast cancer cells (MTF7L), and helps to promote metastases in the lung (Huang et al. 2008).

Once a cancer cell has removed itself from the primary tumor and entered the circulation, extravasation and adhesion to a secondary site becomes the next important step in metastasis. This process has been shown to be mediated by PKCε and μ (Palmantier et al. 2001). In these studies, *cis*-polyunsaturated fatty acids activated PKCε and μ, which then stimulated adhesion of MDA-MB 435 breast carcinoma cells to type IV collagen, a major component of many basement membranes. Once attached, MMPs and invasive mechanisms similar to those that freed the cancer cell from its original microenvironment can be utilized to colonize a new organ.

17.7 Growth Regulation

Another characteristic of cancer cells is growth factor independent proliferation. While normal cells respond to signals from their environment dictating proliferation, quiescence, or apoptosis, cancer cells develop the ability to ignore or bypass these environmental stimuli, and even generate their own growth signals (termed autocrine signaling). This regulation or insensitivity to growth regulation plays an especially important role in breast cancer, as the breast regularly undergoes cyclical modification as a result of menstrual cycles and is capable of extensive remodeling and expansion during pregnancy.

One of the major pathways involved in growth regulation is the extracellular signal-related kinases 1 and 2 pathway, or ERK1/2. PKC isozymes δ, ε, ζ, η, and μ all play a role in this pathway. PKCα and β were also shown to be upregulated by bradykinin stimulation of primary breast cancer and adjacent normal breast tissue culture, although the functional relevance was not determined (Greco et al. 2005). In a follow-up study, however, PKCδ and ε have been shown to mediate the phosphorylation of ERK1/2 in MCF-7 cells through Akt in response to bradykinin stimulation (Greco et al. 2006). In addition, the activation of PKCμ by PKCη regulates ERK and JNK signaling, which are both important signaling pathways for growth regulation (Brändlin et al. 2002). PKCζ was shown to be required for the Angiotensin II-induced activation of ERK and synthesis of C-FOS in MCF-7 cells (Muscella et al. 2003).

Insulin signaling is another major pathway involved in growth regulation. Insulin growth factor-I (IGF-I) signaling enhances both proliferation and survival of tumor and normal cells. Interactions between PKCδ and mTOR were shown to regulate stress and IGF-I induced signaling target of the insulin receptor (IRS-1) Ser^{312} phosphorylation in MCF-7 breast cancer cells (Mingo-Sion et al. 2005). In addition, PKCδ was also shown to be involved in the degradation of IRS-1 in breast cancer cells after exposure to retinoic acid (del Rincón et al. 2004).

17.8 Hormonal Regulation

Breast cancer, unlike most other cancers, is driven by hormonal regulation – particularly estrogen. The PKC isozymes play important roles in regulating the hormone dependent signaling pathways, and in promoting hormone independence.

Estrogen upregulates PKCη expression in estrogen-responsive breast cancer cells (MCF-7 and T47D), but not in estrogen insensitive cells (MDA-MB 231) (Karp et al. 2007). This suggests a role of PKCη in cell proliferation, as the treatment of MCF-7 cells with estradiol resulted in an increased rate of proliferation and increased G_1 to S phase transition, along with increased expression of PKCη.

Inhibition of PKCδ was shown to block most of the 17-β-estriadiol-induced Erk1/2 activation in MCF-7 breast cancer cells, but not block TNFα-induced Erk activation (Keshamouni et al. 2002). Thus, PKCδ expression may be responsible for some of the ERK1/2 activation in breast cancer cells, which, as previously mentioned, can lead to increased proliferation. In somewhat of a converse experiment, Shanmugam et al. (1999) demonstrated that estrogen upregulates PKCδ in MCF-7, thereby affirming the estrogen-PKC δ-ERK1/2 connection.

PKCε was shown to be an important regulator of parathyroid hormone-related protein expression in MCF-7 cells (Lindemann et al. 2003). In addition, the synthetic antiestrogen tamoxifen induced selective membrane association of PKCε in MCF-7 cells. Because PKCε is a proposed oncogene (Basu and Sivaprasad 2007), tamoxifen's effect of changing the localization of PKCε may help to explain the role of the drug on cancer cells (Lavie et al. 1998).

In contrast to the function of the majority of PKC isozymes, PKCα does not appear to be involved in ER signaling. Numerous studies have found that PKCα is increased in ER- and low ER cell lines (Lahn et al. 2004; Assender et al. 2007) and can actually promote ER independence (Assender et al. 2007), possibly through decreased ER mRNA expression (Ways et al. 1995). These findings are in line with the idea that PKCα is oncogenic, and are strongly supported by the increased overall tumorigenicity and metastatic potential reported in PKC α-transfected ER/PR+ MCF-7 cells (Ways et al. 1995; Jasinski et al. 2008).

17.9 Drug Resistance

The resistance of a tumor to chemotherapeutic drugs is an important consideration when pursuing cancer treatment. Many commonly used breast cancer therapeutics, such as tamoxifen, are only efficacious for a short period of time before some tumors become resistant to their effects. One mechanism for cancer cells to become resistant to drugs is through the use of transmembrane pumps that remove cytotoxic drugs from the cell. These pumps are encoded for by the MDR (MultiDrug Resistance) gene. The PKC isozymes play a part in several aspects of these pathways.

Gill et al. (2001) discovered that transcription of MDR1 is strongly activated by PKCα and weakly activated by PKCθ in MCF-7 cells, in both cases by mediating binding to, and activation of, the MDR1 promoter through an undefined intermediate. This finding provides mechanistic evidence for previous observations that PKCα can confer a MDR phenotype (Yu et al. 1991; Osborn et al. 1999) and is overexpressed, along with δ and epsilon, in MDR MCF-7 cells (Ratnasinghe et al. 1998).

PKCα also plays a role in tamoxifen resistance (TAM-R). Assender et al. (2007) found that PKCα is increased in ER+, TAM-R cell lines, and that high PKC α in clinical samples correlated with decreased endocrine responsiveness (and thus tamoxifen resistance), findings which were corroborated by an independent group (Frankel et al. 2007). Assender et al. (2007) also determined that there is an inverse relationship between PKCα and δ in regard to both ER status and endocrine responsiveness, with PKCδ being associated with endocrine responsive, ER+ cell lines and clinical samples, and PKCα being associated with nonresponsive, ER-cell lines and clinical samples. Not surprisingly, these findings dispute previous work which showed that TAM-R breast cancer cells have high levels of total and activated PKCδ, and that overexpression of PKCδ led to TAM-R in MCF-7 breast cancer cells (Nabha et al. 2005) – and so the plot continues to thicken around PKCδ.

Radiation treatment affects PKC isozyme expression, and these changes in expression can confer resistance to radiotherapy. PKCα, β II, and epsilon are all increased by radiation treatment in MCF-7 cells. This change in expression is functionally significant, as inhibition of the isozymes restored radiosensitivity of MCF-7 cells xenografted in nude mice (Jasinski et al. 2008). Another group observed that doxorubicin-resistant MCF-7/Adr breast carcinoma cells cultured in media lacking doxorubicin lost their resistance and became like wild-type cells. In the same time frame, the expression of PKC isozymes α, ε, and θ was also lost, implicating these isozymes in doxorubicin resistance (Budworth et al. 1997).

17.10 PKC Isozymes as Drug Targets or Therapeutics in Breast Cancer

Because of the diverse role PKC isozymes play in breast cancer, they make excellent drug targets. However, because certain isozymes can function as oncogenes and others as tumor suppressors (or, in the case of PKC δ, either depending on context) it will be important to develop very specific inhibitors that only target one or several of the desired isozymes.

The only current PKCα-specific inhibitor is Aprinocarsen, or LY900003, a 20-mer oligonucleotide that targets the 3′-UTR of human PKCα (Dean et al. 1994). Unfortunately, due to difficulties in administering antisense oligos and lack of clinical response in a trial of Aprinocarsen in combination with gemcitabine and cisplatin in nonsmall-cell lung cancer (Paz-Ares et al. 2006), the authors were unable to find any current studies using Aprinocarsen.

Enzastaurin is a promising PKCβ inhibitor. It is an acyclic bisindolylmaleimide that inhibits PKC substrate phosphorylation by competitively binding the ATP binding pocket (Sledge and Gökmen-Polar 2006). Although it is most potent against PKCβ, it is capable of inhibiting other PKC isozymes at higher concentrations. Though very little has been described regarding PKCβ involvement in breast cancer, in several other cancer types it has been implicated in transmitting angiogenic signals via the vascular endothelial growth factor (VEGF) signaling pathway (Xia et al. 1996; Takahashi et al. 1999), and working either downstream or synergistically with PKCα (Ways et al. 1995). In vitro studies have already demonstrated that enzastaurin can confer radiosensitivity on breast cancer cells (Jasinski et al. 2008), so further studies on other aspects of breast cancer management should prove illuminating regarding the potential for enzastaurin as an antibreast cancer drug.

One group showed that a synthetic heptapeptide made from seven amino acid residues of the PKCδ isozyme necessary for heat shock protein 27 (HSP27) binding inhibited heat HSP27, thus decreasing the resistance of cancer cells to DNA damaging agents in NCI-H1299 cells when the cells were treated with the peptide (Kim et al. 2007). Because this is only a small fragment of the isozyme, the ambiguous role of PKCδ may not be applicable here. However, further testing and validation will need to be done to show that this heptapeptide will only increase tumor suppression not promotion as well.

17.11 Implications and Future Directions

As shown in the sampling of work presented in this chapter, PKC expression is intimately associated with breast cancer initiation, progression, and therapy responsiveness, and these effects are highly isozyme-specific. Our current understanding of the roles of PKC in the developing breast greatly aids our interpretation of cancer-based findings, although most of the developmental research conducted to this point is observational and correlative. Detailed mechanistic studies probing isozyme-specific contributions to mammary gland development going back as far as puberty, which provides an excellent model for epithelial cell invasion, will have far-reaching and immediate impact on the cancer field.

Though an impressive amount of research has been done on PKC in breast cancer, there is still much more to be explored regarding the disease stage and context-dependent contributions of each isozyme. This information is absolutely necessary for focusing isozyme-specific PKC drug development – an avenue that is bursting with potential. Future studies should expand beyond cell line models and into spontaneously occurring mouse tumor models and clinical samples, so that relevance and context can be probed more accurately.

References

Alonso-Escolano, D., Medina, C., Cieslik, K., Radomski, A., Jurasz, P., Santos-Martínez, M. J., et al. (2006). Protein kinase Cδ mediates platelet induced breast cancer cell invasion. *The Journal of Pharmacology and Experimental Therapeutics*, *318*, 373–380.

Assender, J., Gee, J., Lewis, I., Ellis, I. O., Robertson, J. F., & Nicholson, R. I. (2007). Protein kinase C isoform expression as a predictor of disease outcome on endocrine therapy in breast cancer. *Journal of Clinical Pathology*, *60*, 1216–1221.

Basu, A., & Sivaprasad, U. (2007). Protein kinase Cε makes the life and death decision. *Cell Signal*, *19*, 1633–1642.

Brändlin, I., Hübner, S., Eiseler, T., Martinez-Moya, M., Horschinek, A., Hausser, A., et al. (2002). Protein kinase C (PKC)η-mediated PKCμ activation modulates ERK and JNK signal pathways. *The Journal of Biological Chemistry*, *8*, 6490–6496.

Budworth, J., Gant, T., & Gescher, A. (1997). Co-ordinate loss of protein kinase C and multidrug resistance gene expression in revertant MCF-7/Adr breast carcinoma cells. *British Journal of Cancer*, *75*, 1330–1335.

Charruyer, A., Jean, C., Columba, A., Jaffrézou, J. P., Quillet-Mary, A., Laurent, G., et al. (2007). PKCzeta protects against UV-C induced apoptosis by inhibiting acid sphingomyelinase-dependent ceramide production. *Biochemical Journal*, *405*, 77–83.

de Vente, J. E., Kukoly, C. A., Bryant, W. O., Posekany, K. J., Chen, J., Fletcher, D. J., et al. (1995). Phorbol esters induce death in MCF-7 breast cancer cells with altered expression of protein kinase C isoforms. *The Journal of Clinical Investigation*, *96*, 1874–1886.

Dean, N. M., McKay, R., Condon, T. P., & Bennett, C. F. (1994). Inhibition of protein kinase C-alpha expression in human A549 cells by antisense oligonucleotides inhibits induction of intercellular adhesion molecule 1 (ICAM-1) mRNA by phorbol esters. *Journal of Biological Chemistry*, *269*, 16416–16424.

del Rincón, S., Guo, Q., Morelli, C., Shiu, H. Y., Surmacz, E., & Miller, W. H. (2004). Retinoic acid mediates degradation of IRS-1 by the ubiquitin-proteosome pathway, via a PKC-dependent mechanism. *Oncogene*, *23*, 9269–9279.

Denning, M. F., Wang, Y., Nickoloff, B. J., & Wrone-Smith, T. (1998). Protein kinase C delta is activated by caspase-dependent proteolysis during ultraviolet radiation-induced apoptosis of human keratinocytes. *Journal of Biological Chemistry*, *273*, 29995–30002.

Foncea, R., Varela, S., Sapag-Hagar, M., & Lavandero, S. (1995). Changes in protein kinase C activity, subcellular distribution and protein phosphorylation during the lactogenic cycle in the rat mammary tissue. *Research Communications in Molecular Pathology and Pharmacology*, *87*, 253–268.

Frankel, L. B., Lykkesfeldt, A. E., Hansen, J. B., & Stenvang, J. (2007). Protein kinase C alpha is a marker for antiestrogen resistance and is involved in the growth of tamoxifen resistant human breast cancer cells. *Breast Cancer Research and Treatment*, *104*, 165–179.

Gill, P., Gescher, A., & Gant, T. (2001). Regulation of MDR1 promoter activity in human breast carcinoma cells by protein kinase C isozymes α and θ. *European Journal of Biochemistry*, *268*, 4151–4157.

Greco, S., Elia, M. G., Muscella, A., Romano, S., Storelli, C., & Marsigliante, S. (2005). Bradykinin stimulates cell proliferation through an extracellular-regulated kinase 1 and 2-dependent mechanism in breast cancer cells in primary culture. *The Journal of Endocrinology*, *186*, 291–301.

Greco, S., Storelli, C., & Marsigliante, S. (2006). Protein kinase C (PKC)-δ/-ε mediate the PKC/Akt-dependent phosphorylation of extracellular signal-regulated kinases 1 and 2 in MCF-7 cells stimulated by bradykinin. *The Journal of Endocrinology*, *188*, 79–89.

Grossini, V., Falbo, K., Kazanietz, M., de Kier Joffé, E. D., & Urtreger, A. J. (2007). Protein kinase Cδ enhances proliferation and survival of murine mammary cells. *Molecular Carcinogenesis*, *46*, 381–390.

Gubina, E., Rinaudo, M. S., Szallasi, Z., Blumberg, P. M., & Mufson, R. A. (1998). Overexpression of protein kinase C isoform epsilon but not delta in human interleukin-3-dependent cells suppresses apoptosis and induces bcl-2 expression. *Blood*, *91*, 823–829.

Gupta, A. K., Galoforo, S. S., Berns, C. M., Martinez, A. A., Corry, P. M., Guan, K. L., et al. (1996). Elevated levels of ERK2 in human breast carcinoma MCF-7 cells transfected with protein kinase C alpha. *Cell Proliferation*, *29*, 655–663.

Hanahan, D., & Weinberg, R. A. (2000). The hallmarks of cancer. *Cell*, *100*, 57–70.

Hovey, R. C., Trott, J. F., & Vonderhaar, B. K. (2002). Establishing a framework for the functional mammary gland: From endocrinology to morphology. *Journal of Mammary Gland Biology and Neoplasia*, *7*, 17–38.

Huang, L., Cheng, H., Isom, R., Chen, C. S., Levine, R. A., & Pauli, B. U. (2008). Protein kinase Cε mediates polymeric fibronectin assembly on the surface of blood-borne rat breast cancer cells to promote pulmonary metastasis. *Journal of Biological Chemistry*, *283*, 7616–7627.

Jackson, D., Zheng, Y., Lyo, D., Shen, Y., Nakayama, K., Nakayama, K. I., et al. (2005). Suppression of cell migration by protein kinase Cδ. *Oncogene*, *24*, 3067–3072.

Jasinski, P., Terai, K., Zwolak, P., & Dudek, A. Z. (2008). Enzastaurin renders MCF-7 breast cancer cells sensitive to radiation through reversal of radiation-induced activation of protein kinase C. *European Journal of Cancer*, *44*, 1315–1322.

Karp, G., Maissel, A., & Livneh, E. (2007). Hormonal regulation of PKC: Estrogen up-regulates PKCη in estrogen-responsive breast cancer cells. *Cancer Letters*, *246*, 173–181.

Keshamouni, V., Mattingly, R., & Reddy, K. (2002). Mechanism of 17-β-Estradiol-induced Erk1/2 activation in breast cancer cells. *Journal of Biological Chemistry*, *277*, 22558–22565.

Kiley, S. C., Clark, K. J., Goodnough, M., Welch, D. R., & Jaken, S. (1999). Protein kinase C delta involvement in mammary tumor cell metastasis. *Cancer Research*, *59*, 3230–3238.

Kim, E., Lee, H., Lee, D., Bae, S., Soh, J. W., Jeoung, D., et al. (2007). Inhibition of heat shock protein 27-mediated resistance to DNA damaging agents by a novel PKCδ-V5 heptapeptide. *Cancer Research*, *67*, 6333–6341.

Lahn, M., Köhler, G., Sundell, K., Su, C., Li, S., Paterson, B. M., et al. (2004). Protein kinase C alpha expression in breast and ovarian cancer. *Oncology*, *67*, 1–10.

Lanigan, F., O'Conner, D., Martin, F., & Gallagher, W. M. (2007). Molecular links between mammary gland development and breast cancer. *Cellular and Molecular Life Sciences*, *64*, 3161–3184.

Lavie, Y., Zhang, Z., Cao, H., Han, T. Y., Jones, R. C., Liu, Y. Y., et al. (1998). Tamoxifen induces selective membrane association of protein kinase C epsilon in MCF-7 human breast cancer cells. *International Journal of Cancer*, *77*, 928–932.

Li, H., & Weinstein, I. B. (2006). Protein kinase C beta enhances growth and expression of cyclin D1 in human breast cancer cells. *Cancer Research*, *66*, 11399–11408.

Lin, C., Hou, W., Shen, S., Juan, S. H., Ko, C. H., Wang, L. M., et al. (2008). Quercetin inhibition of tumor invasion via suppressing PKCδ/ERK/AP-1-dependent matrix metalloproteinase-9 activation in breast carcinoma cells. *Carcinogenesis*, *29*, 1807–1815.

Lin, X., Yu, Y., Zhao, H., Zhang, Y., Manela, J., & Tonetti, D. A. (2006). Overexpression of PKC alpha is required to impart estradiol inhibition and tamoxifen-resistance in a T47D human breast cancer tumor model. *Carcinogenesis*, *27*, 1538–1546.

Lindemann, R., Braig, M., Hauser, C., Nordheim, A., & Dittmer, J. (2003). Ets2 and protein kinase Cε are important regulators of parathyroid hormone-related protein expression in MCF-7 breast cancer cells. *The Biochemical Journal*, *372*, 787–797.

Lu, D., Huang, J., & Basu, A. (2006). Protein kinase Cε activates protein kinase B/Akt via DNA-PK to protect against tumor necrosis factor-α-induced cell death. *Journal of Biological Chemistry*, *281*, 22799–22807.

Lu, D., Sivaprasad, U., Huang, J., Shankar, E., Morrow, S., & Basu, A. (2007). Protein kinase C-ε protects MCF-7 cells from TNF-mediated cell death by inhibiting Bax translocation. *Apoptosis*, *12*, 1893–1900.

Masso-Welch, P. A., Verstovsek, G., Darcy, K., Tagliarino, C., & Ip, M. M. (1998). Protein kinase C eta upregulation and secretion during postnatal rat mammary gland differentiation. *European Journal of Cell Biology*, *77*, 48–59.

Masso-Welch, P. A., Verstovsek, G., & Ip, M. M. (1999). Alterations in the expression and localization of protein kinase C isoforms during mammary gland differentiation. *European Journal of Cell Biology*, *78*, 497–510.

Masso-Welch, P., Winston, J., & Edge, S. (2001). Altered expression and localization of PKC eta in human breast tumors. *Breast Cancer Research and Treatment*, *68*, 211–223.

McCracken, M., Miraglia, L., McKay, R., & Strobl, J. S. (2003). Protein kinase Cδ is a prosurvival factor in human breast tumor cell lines. *Molecular Cancer Therapeutics*, *2*, 273–281.

Mingo-Sion, A., Ferguson, H., Koller, E., Reyland, M. E., & Van Den Berg, C. L. (2005). PKCδ and mTOR interact to regulate stress and IGF-I induced IRS-1 Ser^{312} phosphorylation in breast cancer cells. *Breast Cancer Research and Treatment*, *91*, 259–269.

Morse-Gaudio, M., Connolly, J. M., & Rose, D. P. (1998). Protein kinase C and its isoforms in human breast cancer cells: Relationship to the invasive phenotype. *International Journal of Oncology*, *12*, 1349–1354.

Muscella, A., Greco, S., Elia, M., Storelli, C., & Marsigliante, S. (2003). PKC-ζ is required for angiotensin II-induced activation of ERK and synthesis of C-FOS in MCF-7 cells. *Journal of Cellular Physiology*, *197*, 61–68.

Nabha, S., Glaros, S., Hong, M., Lykkesfeldt, A. E., Schiff, R., Osborne, K., et al. (2005). Upregulation of PKC-δ contributes to antiestrogen resistance in mammary tumor cells. *Oncogene*, *24*, 3166–3176.

Osborn, M. T., Berry, A., Ruberu, M. S., Ning, B., Bell, L. M., & Chambers, T. C. (1999). Phorbol ester induced MDR1 expression in K562 cells occurs independently of mitogen-activated protein kinase signaling pathways. *Oncogene*, *18*, 5756–5764.

Palmantier, R., George, M., Akiyama, S., Wolber, F. M., Olden, K., & Roberts, J. D. (2001). *cis*-Polyunsaturated fatty acids stimulate β_1 integrin-mediated adhesion of human breast carcinoma cells to type IV collagen by activating protein kinases C-ε and -μ. *Cancer Research*, *61*, 2445–2452.

Paz-Ares, L., Douillard, J. Y., Koralewski, P., Manegold, C., Smit, E. F., Reyes, J. M., et al. (2006). Phase III study of Gemcitabine and Cisplatin with or without Aprinocarsen, a protein kinase C-alpha antisense oligonucleotide, in patients with advanced-stage non-small-cell lung cancer. *Journal of Clinical Oncology*, *24*, 1428–1434.

Ratnasinghe, D., Phang, J. M., & Yeh, G. C. (1998). Differential expression and activity of phosphatases and protein kinases in adriamycin sensitive and resistant human breast cancer MCF-7 cells. *International Journal of Oncology*, *13*, 79–84.

Richert, M. M., Schwertfeger, K. L., Ryder, J. W., & Anderson, S. M. (2000). An atlas of mouse mammary gland development. *Journal of Mammary Gland Biology and Neoplasia*, *5*, 227–241.

Saenz, R., & Phillips, D. (1998). Breast Cancer. *Primary Care*, *25*, 309–321.

Sanz, L., Sanchez, P., Lallena, M., Diaz-Meco, M. T., & Mosat, J. (1999). The interaction of p62 with RIP links the atypical PKCs to NF-κB activation. *The EMBO Journal*, *18*, 3044–3053.

Shanmugam, M., Krett, N., Maizels, E., Cutler, R. E. Jr, Peters, C. A., Smith, L. M., et al. (1999). Regulation of protein kinase Cδ by estrogen in the MCF-7 human breast cancer cell line. *Molecular and Cellular Endocrinology*, *148*, 109–118.

Shanmugam, M., Krett, N. T., Maizels E. T., Murad, F. M., Rosen, S. T., & Hunzicker-Dunn, M. (2001). A role for protein kinase Cδ in the differential sensitivity of MCF-7 and MDA-MB 231 human breast cancer cells to phorbol esterinduced growth arrest and $p21^{WAF1/CIP1}$ induction. *Cancer Letters*, *172*, 43–53.

Shtutman, M., Hershko, T., Maissel, A., Fima, E., & Livneh, E. (2003). PKCη associates with cyclin E/Cdk2 complex in serum-starved MCF-7 and NIH-3T3 cells. *Experimental Cell Research*, *286*, 22–29.

Sivaprasad, U., Shankar, E., & Basu, A. (2007). Downregulation of bid is associated with PKCε-mediated TRAIL resistance. *Cell Death and Differentiation*, *14*, 851–860.

Sledge, G. W., Jr., & Gökmen-Polar, Y. (2006). Protein kinase C-beta as a therapeutic target in breast cancer. *Seminars in Oncology*, *33*(3 Suppl 9), S15–S18.

Soh, J. W., Lee, Y. S., & Weinstein, I. B. (2003). Effects of regulatory domains of specific isoforms of protein kinase C on growth control and apoptosis in MCF-7 breast cancer cells. *Journal of Experimental Therapeutics & Oncology*, *3*, 115–126.

Sun, R., Gao, P., Chen, L., Ma, D., Wang, J., Oppenheim, J. J., et al. (2005). Protein kinase Cζ is required for epidermal growth factor induced chemotaxis of human breast cancer cells. *Cancer Research*, *65*, 1433–1441.

Takahashi, T., Ueno, H., & Shibuya, M. (1999). VEGF activates protein kinase C-dependent, but Ras-independent Raf-MEK-MAP kinase pathway for DNA synthesis in primary endothelial cells. *Oncogene*, *18*, 2221–2230.

Tonetti, D. A., Chisamore, M. J., Grdina, W., Schurz, H., & Jordan, V. C. (2000). Stable transfection of protein kinase C alpha cDNA in hormone-dependent breast cancer cell lines. *British Journal of Cancer*, *83*, 782–791.

Ventura, A. C., Jackson, T. L., & Merajver, S. D. (2009). On the role of cell signaling models in cancer research. *Cancer Research*, *69*, 400–402.

Vucenik, I., Ramakrishna, G., Tantivejkul, K., Anderson, L. M., & Ramljak, D. (2005). Inositol hexaphosphate (IP_6) blocks proliferation human breast cancer cells through a PKCδ-dependent increase in $p27^{Kip1}$ and decrease in retinoblastoma protein (pRb) phosphorylation. *Breast Cancer Research and Treatment*, *91*, 35–45.

Wang, J., Wan, W., Sun, R., Liu, Y., Sun, X., Ma, D., et al. (2008). Reduction of Akt2 expression inhibits chemotaxis signal transduction in human breast cancer cells. *Cell Signalling*, *20*, 1025–1034.

Ways, D. K., Kukoly, C. A., deVente, J., Hooker, J. L., Bryant, W. O., Posekany, K. J., et al. (1995). MCF-7 breast cancer cells transfected with protein kinase C alpha exhibit altered expression of other protein kinase C isoforms and display a more aggressive neoplastic phenotype. *Journal of Clinical Investigation*, *95*, 1906–1915.

Williams, C. L., & Noti, J. D. (2001). Reduced expression of Wnt-1 and E-cadherin, and diminished beta-catenin stability in MCF-7 breast cancer cells that overexpress protein kinase C-alpha. *International Journal of Oncology, 19*, 1227–1233.

Xia, P., Aiello, L. P., Ishii, H., Jiang, Z. Y., Park, D. J., Robinson, G. S., et al. (1996). Characterization of vascular endothelial growth factor's effect on the activation of protein kinase C, its isoforms, and endothelial cell growth. *J Clin Invest, 98*, 2018–2026.

Yokoyama, G., Fujii, T., Tayama, K., Yamana H, Kuwano, M., & Shirouzu, K. (2005). PKCδ and MAPK mediate G_1 arrest induced by PMA in SKBR-3 breast cancer cells. *Biochemical and Biophysical Research Communications, 327*, 720–726.

Yu, G., Ahmad, S., Aquino, A., Fairchild, C. R., Trepel, J. B., Ohno, S., et al. (1991). Transfection with protein kinase C alpha confers increased multidrug resistance to MCF-7 cells expressing P-glycoprotein. *Cancer Communications, 3*, 181–189.

Zeidan, Y., Wu, B., Jenkins, R., Obeid, L. M., & Hannun, Y. A. (2008). A novel role for protein kinase Cδ-mediated phosphorylation of acid sphingomyelinase in UV light induced mitochondrial injury. *The FASEB Journal, 22*, 183–193.

Zhang, J., Liu, N., Zhang, J., Liu, S., Liu, Y., & Zheng, D. (2005). PKCδ protects human breast tumor MCF-7 cells against tumor necrosis factor-related apoptosis-inducing ligand-mediated apoptosis. *Journal of Cellular Biochemistry, 96*, 522–532.

Chapter 18
PKC and Prostate Cancer

Jeewon Kim and Marcelo G. Kazanietz

Abstract PKC isozymes regulate multiple aspects of tumorigenesis, including cell proliferation, apoptosis, angiogenesis, and metastasis in a variety of experimental models, making it a major regulator in the transformation to malignant phenotype. Various PKC isozymes play key roles during the progression of prostate cancer in humans and in rodent models. Interestingly, PKC isozymes have often been found to mediate different and sometimes opposing roles in prostate cancer growth and metastasis. Furthermore, expression levels of PKCs are altered when compared to normal prostatic tissue or benign prostatic hyperplasia, and some of these changes correlate with poor prognosis. This review focuses on the current understanding of PKC-mediated regulation of cell proliferation, apoptosis, angiogenesis, and metastasis in prostate cancer. We also discuss the relevance of signaling events modulated by PKC isozymes in prostate cancer models as well as the potential of modulating PKC activity as a means for the treatments of this disease.

Keywords Angiogenesis • Apoptosis • Proliferation • Prostate cancer • Protein kinase C

18.1 Introduction

Prostate cancer is the second leading cause of cancer-related deaths among men in the US. According to the American Cancer Society, there will be more than about 186,000 new cases of prostate cancer in the United States in 2008, and more than

J. Kim (✉)
Stanford Comprehensive Cancer Center, Stanford University, School of Medicine, Stanford, CA 94305, USA
e-mail: jwonkim@standford.edu

M.G. Kazanietz
Department of Pharmacology, University of Pennsylvania School of Medicine, 1256 Biomedical Research Building II/III, 421 Curie Blvd, Philadelphia, PA 19104-6160, USA
e-mail: marcelog@upenn.edu

M.G. Kazanietz (ed.), *Protein Kinase C in Cancer Signaling and Therapy*,
Current Cancer Research, DOI 10.1007/978-1-60761-543-9_18,

28,000 men will die of this disease in 2008. It is predicted that one man in six will get prostate cancer during his lifetime (Nelson 2007; Jemal et al. 2008).

Growth of primary prostate cancer cells is dependent on androgens. Androgen ablation therapy is the standard clinical procedure used to inhibit prostate tumor growth. However, in most cases, cancer recurs and progresses to a terminal stage. The androgenic hormones exert their cellular effects by means of interactions with the androgen receptor (AR). Ligand-activated AR translocates to the nucleus where it forms complexes with co-activators and other nuclear factors that recognize cis-acting DNA sequences defined as androgen response elements (AREs). Numerous genes involved in prostate proliferation and differentiation are regulated by androgens (Rigas et al. 2003). Deregulation of autocrine/paracrine mechanisms that involve factors secreted by either neoplastic epithelial cells or by prostatic stromal cells also plays important roles in the progression to androgen independence (Charlesworth and Harris 2006; Sakamoto et al. 2008; Augsten et al. 2009).

Genetic and epigenetic changes that lead to deregulation of mitogenic and survival signals are dominant events in prostate cancer (Nelson et al. 2007). Hyperactivation of ERK and functional inactivation of PTEN (a phosphatase for the PI3K lipid products) are among the most common signaling pathways alterations in prostate cancer, as well as in several other cancer types (Majumder and Sellers 2005). Many reports described highly relevant roles for PKC isozymes in prostate cancer cell survival, angiogenesis, apoptosis, cell proliferation, and the acquisition of an androgen-independent state (Henttu and Vihko 1998; Wu et al. 2002; Gavrielides et al. 2006; Kim et al. 2008; Xiao et al. 2008; Xiao et al. 2009).

18.2 Expression Patterns of PKC Isozymes in Prostate Cancer

As in many other cancer types, the balance in PKC isozyme expression is markedly altered in human prostate tumors, potentially reflecting their involvement in the etiology and progression of the disease. There have been several reports studying mRNA and protein levels of PKC during different stages of prostate cancer progression in rodents and humans (Hofmann 2004). In the early stage of human prostate adenocarcinomas, levels of PKC α, ε, and ζ are elevated whereas PKCβ levels are decreased compared to normal or benign hyperplastic prostate tissues (Cornford et al. 1999), as shown by immunohistochemistry. Although normal and both androgen-sensitive and -insensitive cells express PKCα, δ, ε, and η, interestingly only DU145 cells and normal prostate showed expression of PKCθ mRNA (Powell et al. 1996a). Also, the levels of PKCα mRNA and protein were 6- to 38-fold less in androgen-sensitive cells than in the androgen-insensitive cells (Powell et al. 1996a), suggesting a role for PKCα in the development of resistance to androgen ablation therapy.

In Dunning R-3327 rat prostatic tumors, PKCα, β, γ, δ, ε, η, θ and ξ mRNAs were found (Powell et al. 1996b). The mRNA levels of PKCβ, γ, and η were decreased in the aggressive subline of Dunning R-3327 rat prostatic tumor cells, MAT-Lu, compared to H or G sublines that grow much slower. Also, a spliced form of PKCζ was found in the G subline of Dunning R-3327 rat prostate cancer cells, and in a subsequent study it was shown that cells overexpressing PKCζ showed decreased metastatic potential, as revealed by *in vitro* invasion assays using MAT-LyLu cells (Powell et al. 1996b). In adult Wistar rats, treatment with flutamide, a nonsteroidal antiandrogen, increased PKCα kinase activity, membrane translocation of PKCε, and total protein levels of PKCα, βI, ε, and ζ compared to nontreated controls, suggesting roles for these kinases in the resistance to androgen ablation therapy and supporting previous results on the roles of active PKCα and ε in inducing aggressiveness of prostate cancer cells (Montalvo et al. 2002; Wu et al. 2002). Also, it has been recently shown that the levels of two splice variants PKCβI and PKCβII are quite different in PC-3 cells as measured by Western blots and immunohistochemistry (Kim et al. 2008).

More recently, PKC protein levels were compared between benign prostatic hyperplasia (BPH) and prostate cancer tissues from patients (Koren et al. 2004) by immunohistochemistry and Western blots. Protein levels of PKCα, β, ε, and η were higher in cancers compared to BPH, especially those of PKCε. Notably, deregulation of PKCε gene (PRKCE) has been reported in other cancer types, such as in lung, breast, and thyroid cancer (Knauf et al. 1999; Ding et al. 2002; Lindemann et al. 2003). Also, McJilton et al. presented immunohistochemical evidence of an outgrowth/clonal selection of PKCε-positive cells in recurrent prostate cancer (McJilton et al. 2003). A more recent analysis by Verma's lab found that high-grade human prostate tumors express very high PKCε levels. PKCε expression is markedly upregulated in prostate tumors from TRAMP mice and correlates with high phospho-Akt (active) levels (Aziz et al. 2007a). A positive correlation was also found between PKCε and Stat3 expression in prostate cancer cells (Aziz et al. 2007a). It is conceivable that PKCε overexpression in human prostate tumors has a causal relationship with the initiation and progression of prostate cancer, but this has not been formally demonstrated yet. Notably, many reports suggest the involvement of PKCε in prostate cancer cell survival, resistance to apoptosis, and increased invasiveness (Wu et al. 2002; Aziz et al. 2007a; Xiao et al. 2008).

Furthermore, the protein levels of c-Jun and c-Fos in patient samples before and after development of androgen independency have been studied to determine whether transcription factor activated protein (AP-1) complex formed by these two proteins is functionally relevant to the progression of human prostate cancers in relation to PKC (Edwards et al. 2004). Correlation and survival analyses in clinical samples revealed that increase in the active level of phosphorylated c-Jun and total levels of PKC independently correlated with decreased survival from relapse in androgen-independent prostate cancer patients, suggesting a role for the AP-1 complex in prostate cancer progression. Table 18.1 summarizes PKC expression in human and rat prostate cancer tissues.

Table 18.1 Expression of PKC isozymes in prostate cancer

Normal human prostate (IHC) (Cornford et al. 1999)	PKCα, β, ι, λ, μ, ξ, and RACK1
Human prostate cancer (IHC, organ confined) (Cornford et al. 1999)	PKCα ↑, β ↓, ε ↑, ι, λ, μ, ξ ↑, and RACK1
BPH (IHC and WB) (Koren et al. 2004)	PKCα, β, ε, η
Human prostatic carcinoma (IHC and WB) (Koren et al. 2004) vs. BPH	PKCα ↑, β ↑, ε ↑, η ↑
Dunning R-3327 rat prostatic tumor subline H (mRNA) (Powell et al. 1996b)	PKCα, β, γ, δ, ε, η, θ, ξ
Dunning R-3327 rat prostatic tumor subline MAT-Lu (more aggressive cell line, mRNA) (Powell et al. 1996b)	PKCα ↑, β ↓, γ ↓, η ↓
Human prostate cancer cells (LNCaP, PC-3, DU145 (mRNA)) (Powell et al. 1996a)	PKCα δ, ε, η, and μ

BPH benign prostatic hyperplasia; *IHC* immunohistochemistry; *WB* western blot

18.3 PKC and Cell Proliferation in Prostate Cancer

Accumulating evidence suggests that PKC family members are critical in regulating cell proliferation (Levin et al. 1990; Kiley and Parker 1995; Takahashi et al. 2000). Several studies have reported that PKC is required for the mitogenic activity of growth factors. For example, bradikynin-induced mitogenesis and ERK activation in PC-3 cells is blocked by the PKC inhibitor bisindolylmaleimide or PKC down-regulation (Barki-Harrington and Daaka 2001). Cyclin D1 induction by epidermal growth factor (EGF) is dependent on PKC (Perry et al. 1998).

More recent studies have identified specific roles for PKC isozymes in mitogenesis. Interestingly, an important role has been shown for the interaction of PKCβII and pericentrin, a centrosomal protein, in microtubule organization, spindle assembly, chromosome segregation, and proliferative activity in prostate cancer and kidney cells (Chen et al. 2004; Kim et al. 2008). This is in support of the proposed use of centrosomal protein pericentrin as a biomarker for detecting and grading prostate cancer (Pihan et al. 2001). PKCβII is activated during the growth of prostate cancer in PC-3 xenografts, and inhibition of its activity decreases tumor endothelial cell proliferation and increases apoptosis in the tumors, thereby reducing tumor growth. Cell proliferation was directly measured by the deuterium labeling method with animals being administered 4% deuterated water to label newly synthesized DNA in proliferating cells through *de novo* synthesis of the ribose moiety (Kim et al. 2004; Kim et al. 2008). It was also found that conditioned medium from PC-3 cells contained factors that induced dysregulated formation of microtubules and cell proliferation in tumor endothelial cells (Kim et al. 2008) suggesting that secreted factors from PC-3 cells may be regulating cell proliferation directly or indirectly by regulating PKCβII activity. Furthermore, the specific PKCβII inhibitor peptide βIIV5-3 facilitated protein–protein interaction between PKCβII and pericentrin, which correlated with reduced angiogenesis, cell proliferation, and tumor growth (Kim et al. 2008). These data suggest that PKCβII phosphorylation of pericentrin may be critical for regulating normal cytokinesis. Direct phosphorylation substrates of PKCβII were not studied here. Further research is needed to clarify the role of PKCβII-pericentrin interaction in prostate cancer cell proliferation and to determine if the phenotype seen here involves PKCβII substrates. Another possibility is that receptor for active C kinases (RACK1) may be positively regulating prostate cancer cell proliferation when PKCβII is bound to RACK1 (i.e., in activated PKCβII) by interacting with AR (Rigas et al. 2003).

Another PKC isozyme that plays important roles in mitogenic signaling in prostate cancer cells is PKCε. PKCε plays a critical role in the transition of androgen-dependent LNCaP cells into androgen-independent cells and also increases cell proliferation (Wu et al. 2002). Overexpression of PKCε in LNCaP cells enabled LNCaP cells to proliferate in the absence of androgens or serum. Cell cycle analysis revealed that PKCε overexpression increases the number of cells in S phase and accelerates G1/S transition (Wu et al. 2002). Elevated levels of Raf-1 and ERK 1/2 phosphorylation are observed in PKCε-overexpressing cells, as well as high levels of phosphorylated RB, cyclins D1, cyclin D3, and cyclin E. These cells also present increased levels of c-myc (Wu et al. 2002). Taken together, these studies suggest that PKCε is an active regulator of the ERK and RB signaling in LNCaP cells to promote proliferation.

18.4 PKC Isozymes in Apoptosis and Cell Survival

A remarkable feature of androgen-sensitive prostate cancer cell lines, such as LNCaP, C4-2, and CWR22-Rv1 cells, is that they undergo apoptosis in response to phorbol esters. This was initially shown by Day et al. and subsequently by several other groups (Day et al. 1994; Xiao et al. 2009). Analysis of the PKC isozymes involved in this response revealed an important role for PKCδ. Indeed cell death induced by phorbol 12-myristate 13-acetate (PMA) in LNCaP cells can be inhibited by expression of a dominant-negative PKCδ mutant as well as by PKCδ depletion using RNAi (Gavrielides et al. 2006). Moreover, overexpression of PKCδ in LNCaP cells markedly enhances the apoptotic response of PMA. Induction of LNCaP cell apoptosis by PKCδ does not involve its proteolytic cleavage, as described in many other cell types, suggesting that it depends on allosteric activation of the enzyme upon translocation to the plasma membrane (Fujii et al. 2000). It is interesting that androgen-independent prostate cancer cells such as PC-3 or DU-145 do not undergo apoptosis in response to phorbol esters, although growth inhibition is observed (Sugibayashi et al. 2002). Studies have also revealed that PKCδ also mediates prostate cancer cell death by chemotherapeutic agents. Etoposide and paclitaxel induced apoptosis through ceramide formation in LNCaP and DU145 cells, and inhibition of PKCδ significantly blocks ceramide formation and apoptosis in LNCaP cells (Sumitomo et al. 2002). Ceramide leads to the translocation of PKCδ to mitochondria, causing cytochrome c release and caspase-9 activation (Sumitomo et al. 2002). Therefore, PKCδ is a crucial mediator of apoptosis induced by phorbol esters or anticancer drugs.

Signaling analysis revealed that PMA induces the activation of mitogen-activated protein kinase (MAPK) cascades in prostate cancer cells. Phorbol esters strongly activate p38 MAPK and JNK in LNCaP cells, and inhibition of these pathways impairs PMA-induced apoptosis. On the other hand, inhibition of the ERK cascade with a MEK inhibitor enhances PMA-induced apoptosis. This suggests opposite roles for MAPK cascades in apoptosis of androgen-dependent prostate cancer cells (Tanaka et al. 2003).

PTEN loss occurs in a large percentage of human prostate tumors, and Akt is a dominant survival pathway in prostate cancer. The PI3K-Akt pathway is hyperactivated in LNCaP cells due to loss of PTEN function (Majumder and Sellers 2005). Remarkably, PKC activators promote a rapid dephosphorylation of Akt. The reduction in Akt activity does not involve an inhibition of upstream inputs, as phosphorylation of the PI3K effector PDK1 is not altered by PMA treatment (Tanaka et al. 2003). On the other hand, okadaic acid prevents Akt dephosphorylation, suggesting that PKC activation leads to the dephosphorylation of Akt by activation of the phosphatase PP2A (Li et al. 2003). Interestingly, Akt dephosphorylation seems to be dependent on PKCα rather than PKCδ, suggesting the contribution of at least two PKC isozymes to apoptotic cell death by phorbol esters (Tanaka et al. 2003).

PKC activation is known to promote the release of factors from cells and to trigger autocrine and paracrine loops (Gonzalez-Guerrico and Kazanietz 2005). It is interesting that phorbol ester-induced apoptosis is mediated by the autocrine release of death factors. Indeed, conditioned medium from PMA-treated LNCaP, DU145, or PC-3 cells has apoptotic activity when added to LNCaP cells (Gonzalez-Guerrico and Kazanietz 2005; Xiao et al. 2009). PKCδ RNAi depletion in LNCaP cells treated by PMA impairs the secretion of death factors, and therefore conditioned medium from these cells lost its apoptotic activity. It became clear that the autocrine effect is mediated primarily by tumor-necrosis factor α (TNFα) and TRAIL (TNF-related apoptosis-inducing ligand) but not by Fas. This has been demonstrated through numerous approaches (Gonzalez-Guerrico and Kazanietz 2005). For example, the apoptotic effect of PMA in LNCaP cells was lost in the presence of TAPI-2, an inhibitor of TNFα converting enzyme (TACE), the enzyme involved in TNFα shedding, or after TACE RNAi depletion. TNFα or TRAIL neutralizing antibodies, as well as blockade or RNAi depletion of death receptors, inhibited the apoptotic effect of PMA. PMA caused a marked TNFα mRNA induction as well as TNFα release from prostate cancer cells, effects that were prevented by PKCδ or TACE RNAi (Gonzalez-Guerrico and Kazanietz 2005). Signaling analysis revealed that conditioned medium from PMA-treated LNCaP cells promotes the activation of p38, JNK, and caspase-8, suggesting that PKCδ-mediated apoptosis involves the activation of the extrinsic apoptotic cascade (Xiao et al. 2009). Moreover, RNAi depletion of caspase-8 or a dominant-negative mutant of the death receptor adaptor Fas-associated death domain (FADD), impaired of the apoptotic effect of PMA in LNCaP cells (Gonzalez-Guerrico and Kazanietz 2005; Gavrielides et al. 2006).

As expected from the opposite effects of PKCδ and PKCε in many cell types, PKCε is a prosurvival kinase in prostate cancer cells. Overexpression of PKCε conferred resistance to phorbol ester-induced apoptosis, an effect associated with an inhibition of Bax oligomerization, which is required for its mitochondrial integration and cytochrome c release (McJilton et al. 2003). A functional interplay among integrin receptors, PKCε, and Akt has been also described in CWR-R1 prostate cancer cells. Coimmunoprecipitation analysis revealed the presence of signaling complexes containing PKCε, β1-integrin, Src, and Akt in prostate cancer cells (Wu et al. 2004). Because PKCε can interact with several other binding partners involved in cell survival (Budas et al. 2007a; Budas et al. 2007b), and its ratio with proapoptotic PKCδ is probably important in the cell fate between survival and death, identification of downstream targets of these PKC isozymes that regulate their balance may reveal important PKC effectors that could be potential therapeutic targets for prostate cancer.

Caveolin-1 is a secreted protein known to increase survival of prostate cancer cells (Tahir et al. 2001). LNCaP cells transfected with PKCε showed higher level of caveolin-1 and conditioned medium from these cells promoted LNCaP cell growth and survival. Accordingly, an anticaveolin antibody abrogated the effect on cell viability. These results suggest that PKCε regulates the expression or secretion of caveolin-1, important in the survival of prostate cancer cells. The importance of PKCε was confirmed by a reduction in viable cells with knockdown of PKCε using anti-sense oligonucleotides (Wu et al. 2002). The importance of PKCε in cell survival has been also highlighted in recent studies that revealed an interaction of this PKC with signal transducers and activators of transcription-3 (Stat3). In this study by Aziz et al. (2007a), PKCε was shown to interact with Stat3 in various human prostate cancer cells and in transgenic adenocarcinoma of the mouse prostate model (TRAMP), suggesting its importance in prostate cancer regardless of androgen sensitivity. Phosphorylation of Ser727 of Stat3 by PKCε was essential for Stat3 DNA binding and transcriptional activity of downstream target genes in PC-3 cells important in cell proliferation and survival (Aziz et al. 2007a). PKCε-mediated phosphorylation of Ser727 in Stat3 was reversed by siRNA of PKCε. With siRNA of PKCε, invasiveness of DU145 cells decreased suggesting the role of PKCε not only in cell survival but also in cell invasion. These data suggest that PKCε is important in the regulation of prostate cancer cell proliferation and invasion by regulating Stas3 Ser727 phosphorylation. Increased expression of PKC and Stat3 in prostate cancer from TRAMP mice is accompanied by decreased expression of the cell cycle inhibitors p21 and p27 and increased expression of Bcl-xL, Bcl-2, survivin, Akt (total and phosphorylated) and COX-2. Considering the fact that Stat3 was also shown to interact with PKCε and was phosphorylated at Ser727 in UV-induced squamous cell carcinoma (Aziz et al. 2007b), this interaction may have a role in other types of cancer as well.

18.5 PKC and Angiogenesis in Prostate Cancer

Increases in microvessel density and expression of proangiogenic factors are correlated with negative outcomes in patients with prostate cancer (Charlesworth and Harris 2006; Sakamoto et al. 2008). PKCs have been shown to play important roles in angiogenesis both *in vitro* and *in vivo* (Griner and Kazanietz 2007; Kim et al. 2008). Briefly, indication of involvement of PKC in tumor-induced angiogenesis was shown by increased growth and tube formation of bovine microvascular endothelial cells on a collagen layer when treated with PMA (Montesano and Orci 1985). Subsequently, PKCα, βII, δ, and ζ have been shown to regulate angiogenesis in different cell types including corneal, vascular, and tumor endothelial cells (reviewed in (Griner and Kazanietz 2007)). Among these isozymes, PKCβII has been most implicated in tumor-induced angiogenesis as shown in many different experimental models including brain, breast, colon, lung, and ovarian cancer (Teicher et al. 2001; Teicher et al. 2002a; Teicher et al. 2002b; Graff et al. 2005).

A recent study has shown that there is an alternating pattern of endothelial and epithelial cell proliferation during the early phase of PC-3 tumor growth subcutaneously implanted in nude mice (Kim et al. 2008). During this period, tumors were collected every week for up to 6 weeks and tumor endothelial cells and tumor epithelial cells were isolated using flow cytometry. As mentioned previously, by using the deuterium-labeling method, *in vivo* cell proliferation rates of tumor endothelial cells and tumor epithelial cells were measured separately (Kim et al. 2004; Kim et al. 2005; Kim et al. 2008). Endothelial cell proliferation rate (i.e., angiogenesis) was found to be upregulated before the tumor cell proliferation rate increased, and this continued until 4 weeks posttumor implantation but not later. This suggests that "an angiogenic switch" occurs during the early growth stage of prostate tumor cells and that this may be an optimal time window for antiangiogenic treatment in prostate cancer. This is somewhat different from a previous study, which showed that there is an early and also a late molecular switch for tumor angiogenesis in prostate cancer using a TRAMP model when they quantified CD31 staining, VEGF, and HIF-1α levels (Huss et al. 2001). The difference may be due to different model systems or use of a different method for measuring angiogenesis, but both studies suggest that there is a period of active angiogenesis in the early phase of prostate tumor growth.

Administration of a PKCβII isozyme specific inhibitor peptide (developed by the Mochly-Rosen laboratory) during this angiogenically active period decreased tumor angiogenesis and tumor growth of PC-3 xenografts (Kim et al. 2008). UCN-1, which inhibits PKC, was also shown to inhibit hypoxia-induced angiogenesis (Kruger et al. 1998). Currently, the PKCβ inhibitor enzastaurin is being tested in clinical trials for the anticancer effects in phase II studies of high-grade relapsed or refractory diffuse large B-cell lymphoma (Robertson et al. 2007).

The Par proteins ("Par" derived from "partitioning defective") were first identified in *Caenorhabditis elegans* (*C. elegans*) in screening for mutants with dysregulated partitioning of proteins in the early embryo (Kemphues et al. 1988). Then, the six Par proteins have been found in organisms from *C. elegans* to mammals (Goldstein and Macara 2007). The Par complex includes two of these Par proteins, Par3 and Par6, the serine/threonine kinase (atypical PKC), and small GTPases, such as Cdc42 or Rac1 (Aranda et al. 2008). Recent studies have shown that some members of this complex display prooncogenic activities (Aranda et al. 2008). The use of an atypical PKC inhibitor that selectively targets Par6–atypical PKC protein–protein interaction shows promising results in ovarian and lung carcinoma (Stallings-Mann et al. 2006; Fields et al. 2007; Regala et al. 2008). Par1 gene expression was also shown to induce tumor angiogenesis by increasing transcription of VEGF and production of functional VEGF in a melanoma cell line transfected with Par1. Also, Dunning rat prostatic carcinoma cells AT2.1 transfected with inducible Par1 expression vectors injected subcutaneously in rats grew significantly bigger with more angiogenesis by induction with a tetracycline analog, doxycycline, compared to vector-only transfected cells (Yin et al. 2003). Par1 expression increased angiogenic activity as measured by tube formation assay and *in vitro* cell proliferation assay. Also, Par1-regulated angiogenesis mediated by VEGF was confirmed when neutralizing antibodies for VEGF blocked angiogenesis induced by Par1 expression in melanoma cells. These data suggest that Par1 is important in mediating VEGF-induced tumor angiogenesis in prostate and other tumors. Because Par complex regulates cell polarity, it will be interesting to elucidate whether Par1 and PKC activity are involved in the regulation of cell migration and metastasis of prostate cancer.

18.6 PKC in Invasion and Metastasis of Prostate Cancer Cells

PKC activation correlates with cell migration and invasion in many types of tumors and can regulate metastatic potential of tumor cells. However, there are only a few studies implicating PKC isozymes in the control of metastasis (Powell et al. 1996b; Zeng et al. 2006; Dashevsky et al. 2009; Herman et al. 2009).

Previously, the Cartwright laboratory showed that the SH2 domain of Src directly binds RACK1, a protein with multiple WD-40 units (Chang et al. 1998; Chang et al. 2001; Schechtman and Mochly-Rosen 2001). The interactions of c-Src, PKCβII, and RACK1 negatively regulate Src activity and this interaction reduces G1/S cell entry (Chang et al. 2002; Mamidipudi et al. 2004). Recently, PKC was shown to mediate PC-3 cell proliferation and invasion by interacting with a scaffolding protein, an actin filament-associated protein (AFAP-110) (Zhang et al. 2007). Downregulation of AFAP-110 using stable clones transfected with AFAP-110 siRNA showed decreased colony formation in soft agar, decreased orthotopic growth of tumor cells in the prostates of nude mice and reduced adhesion to extracellular matrix proteins as shown by growth in laminin or collagen type IV coated plates, and by *in vitro* invasion assays. Invasiveness of PC-3 cells overexpressing AFAP-110 increased with elevated levels of β1-integrin and focal adhesion contacts as stained with vinculin. The role of PKC in the AFAP-110-mediated growth and invasion was confirmed by using cells with siRNA of AFAP-110 that are stably transfected to express wild-type or mutant-type AFAP-110 either lacking a domain to bind Src or to bind PKC. The inability to form focal contacts in cells transfected with siRNA of AFAP-110 was restored by ectopic expression of AFP-110 or AFAP-110 lacking an Src-binding sequence but not by introduction of

AFAP-110 lacking a domain that interacts with PKC. This suggests that AFAP-110-induced invasiveness is mediated by interaction with PKC. PKC contains sequences that participate in protein–protein interactions with actin (Prekeris et al. 1996). Identifying a sequence in PKC that interacts with AFAP-110 may allow for the design of a peptide or a small molecule interfering with these protein–protein interactions that could potentially inhibit growth and invasiveness of prostate cancer. These findings indicate that PKC is a critical factor in AFAP-110-mediated cell proliferation and invasion. It will be worth identifying which PKC isozyme is involved in this regulation. Also, AFAP-110-independent effects on cell proliferation, for example, PKC interaction with RACK cannot be ruled out.

With relevance to PKC and metastatic potential of prostate cancer cells, PMA was shown to increase the levels of KAI1 (CD82), a metastatic suppressor protein generally downregulated in advanced human cancers (Rowe et al. 2008). Binding of the Tip60/Pontin complex to the promoter region of KAI1 is critical for KAI1 transcription. Previously, sumoylation/desumoylation was reported to be important during the progression of prostate cancer (Baek 2006; Cheng et al. 2006). Metastatic cells express higher levels of an enzyme called small ubiquitin like modifier (SUMO)-conjugating enzyme Ubc9, which attaches SUMO to Reptin. SUMOylated Reptin represses KAI1 transcription by forming a repressive complex with beta-catenin, which binds to the KAI1 promoter and inhibits binding of Tip60 (Kim et al. 2006). Accordingly, the repressive SUMOylated form of Reptin is upregulated in metastatic tumors compared to nonmetastatic cells (Kim et al. 2006). PMA increased the recruitment of Tip60/Pontin complex to the binding site of the promoter region of KAI1 as shown by two-step chromatin immunoprecipitation, which was inhibited by treatment with a PKC inhibitor. These results suggest that PMA-activated PKC upregulates the metastasis suppressor protein KAI1 by recruiting transcriptional activator protein complex to its promoter region. The possible role of non-PKC phorbol ester-sensitive proteins like chimaerins (α1, β1) and Ras guanylyl nucleotide releasing proteins (RasGRPs) (Griner and Kazanietz 2007) in the regulation of Tip60/Pontin-induced KAI1 transcription and in metastasis of prostate cancer cannot be excluded.

Another relevant protein that interacts with PKC signaling in metastasis is PCPH/ENTPD5 (ectonucleoside triphosphate diphosphohydrolase 5). The oncogenic form of the protooncogene PCPH, mt-PCPH, is a truncated form of PCPH (Villar et al. 2007). Recently, it has been shown that levels of cytoplasmic PCPH correlate with prostate cancer progression (Villar et al. 2007). Overexpression of PCPH or mt-PCPH in PC-3 cells increased invasiveness of the cells, as shown by increased mRNA levels of collagen 1A1 and 1A2 genes, which are known to be highly expressed in metastatic prostate tumors (Ramaswamy et al. 2003; Stanbrough et al. 2006). When PCPH was stably knocked down by shRNA, the increase in collagenase gene expression was reduced to control levels. Interestingly, the PKCδ levels also decreased in PCPH-depleted cells. In PKCδ knockdown cells, levels of collagen 1A1 and 1A2 mRNA decreased, suggesting PKCδ mediates invasion by PCPH in human prostate cancer cells. Reconstitution of PKCδ levels in PCPH-depleted LNCaP cells restored the expression of collagen I. Interestingly, expression of PCPH in PC-3 cells decreased colony formation, suggesting that PCPH may affect cell growth and invasion by distinct mechanisms. These studies strongly support the notion that PKCδ is a key mediator of PCPH functions related to cell morphology, growth, and invasiveness in human prostate cancer cells. More recently, stable knockdown of PCPH or mt-PCPH was shown to increase cisplatin-sensitivity by inhibiting stabilization of the antiapoptotic protein Bcl-2 (Villar et al. 2009). The increased resistance to chemotherapy in PC-3 cells overexpressing PCPH was mediated by increased phosphorylation of PKCα in Thr638, an auto-phosphorylation site, suggestive of the role of PKCα autophosphorylation. On the other hand, RNAi depletion of PCPH in LNCaP cells resulted in reduced PKCα phosphorylation. PKCα knockdown sensitized prostate cancer cells to cisplatin-induced apoptosis by enhancing Bcl-2 downregulation. From these data, it is evident that distinct PKC isozymes play separate roles in prostate cancer cells.

PKCδ was also shown to be a critical downstream molecular player in the signaling pathway of fibronectin peptide (PHSRN)-α5β1 interaction leading to invasion of DU145 cells (Zeng et al. 2006). There is also evidence that atypical PKCs play a role in metastatic dissemination of prostate cancer cells. PKCζ has been implicated in metastasis of Dunning R3327 model of rat prostate cancer cells (Powell et al. 1996b). The Dunning R3327 subline MAT-LyLu, which overexpresses PKCζ, showed decreased metastatic potential as determined by *in vitro* invasion

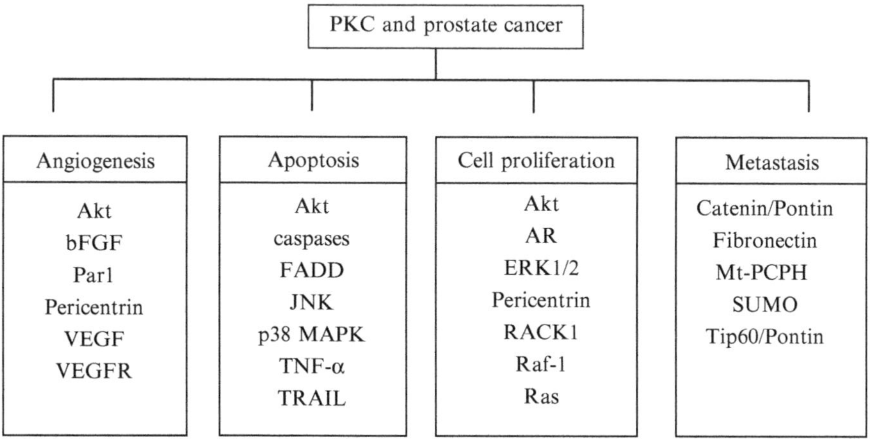

Fig. 18.1 List of molecules involved in the PKC-mediated regulation of prostate carcinogenesis. *AR* androgen receptor, *bFGF* basic fibroblast growth factor, *ERK* extracellular signal-regulated kinase, *FADD* Fas-associated death domain, *JNK* c-Jun N-terminal kinase, *MAPK* mitogen-activated protein kinase, *mTOR* mammalian target of rapamycin, *Par 1* Partitioning-defective 1, *VEGF* vascular endothelial growth factor, *VEGFR* vascular endothelial growth factor receptor, *RACK1* receptor for active kinases, *SUMO* small ubiquitin like modifier, *TNF-α* tumor-necrosis factor- α, *TRAIL* TNF-related apoptosis-inducing ligand

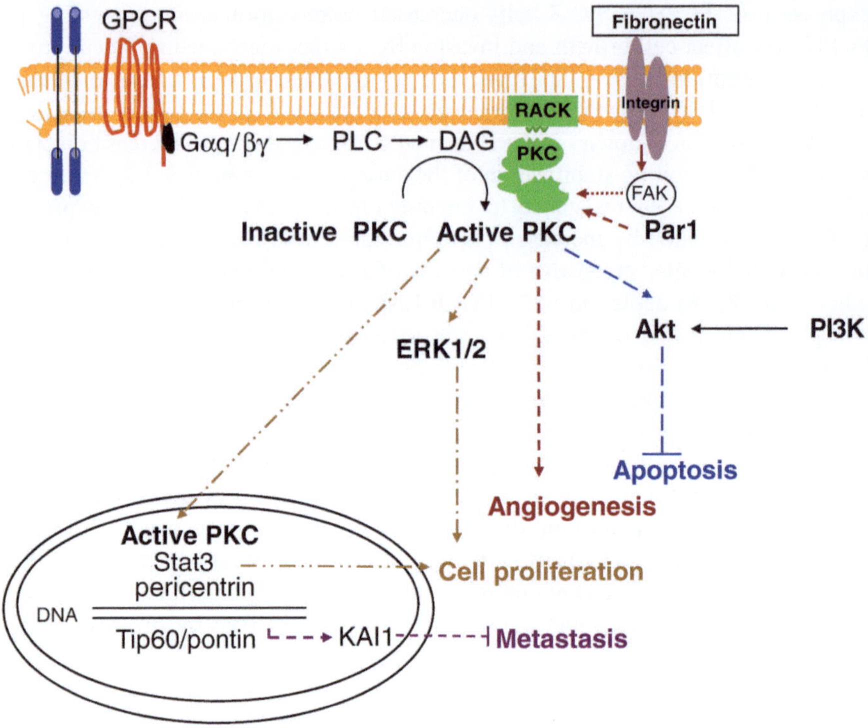

Fig. 18.2 PKC regulates cell proliferation, angiogenesis, apoptosis and metastasis of prostate cancer cells. PKC regulates cell proliferation by ERK1/2, pericentrin and Stat3 (shown in brown dashed lines). PKC induces angiogenesis by way of FAK, PAR1 and pericentrin (shown in red dashed line). PKC also regulates apoptosis by way of Akt signaling (shown in blue dashed line). Furthermore, PKC regulates metastasis by regulating transcription of KAI1 (shown in purple lines). More details on the signaling molecules involved are shown in Fig. 18.1

assays and also *in vivo*, contrary to known PKCζ's role in tumor angiogenesis in other types of cancer (Arbiser 2004; Neid et al. 2004; Xu et al. 2008). Because of these opposing roles of PKCζ in angiogenesis and invasion, two important interrelated aspects of carcinogenesis, inhibition of PKCζ in prostate carcinogenesis has to be approached with caution. Figures 18.1 and 18.2 summarize molecules involved in PKC-mediated regulation of prostate cancer progression.

18.7 Conclusions

In this chapter, we summarized the relevance of PKC isozymes in cell proliferation, apoptosis, angiogenesis, and invasion in prostate cancer. In general, increased protein levels of PKCα, ε, η, and ζ, and decreased levels of PKCβ, are found in human prostate cancer specimens compared to normal or hyperplastic prostate. Increases in

PKCβII activity were found to augment cell proliferation mediated by abnormal pericentrin localization and levels, which was inhibited by an isozyme-selective inhibitor of PKCβII, βIIV5-3. Upregulation of PKCε increased cell survival and proliferation by increasing the level of the scaffolding protein, caveolin-1, and Stat3 activation. On the other hand, activation of PKCδ confers an apoptotic response in androgen-dependent prostate cancer cells, an effect mediated by the autocrine secretion of death factors and the activation of the extrinsic apoptotic pathway. PKCβII was also found to regulate tumor-induced angiogenesis in prostate cancer as shown by an isozyme-selective inhibitor peptide of PKCβII, enzastaurin, and siRNA. Also, Par1 gene was shown to increase angiogenic activities in prostate tumorigenesis, which was mediated by PKC activity. Interestingly, AFAP-110 increased PC-3 cell proliferation and invasion by interaction with PKC, but PKC activation by PMA decreased metastatic potential of PC-3 cells by inducing transcription of KAI1, a tumor suppressor protein. PKCξ was found to decrease metastasis of Dunning R3327 rat adenocarcinoma cells whereas PCPH, an oncogenic protein, increased metastatic activity and resistance to chemotherapy in human prostate cancer cells mediated by PKCδ and α, respectively. These data suggest that each PKC isozyme plays different and sometimes opposing roles in prostate cancer progression. Peptide inhibitors can be used as effective pharmacological tools to identify and correct dysregulation of critical PKC isozymes in different stages of prostate tumorigenesis. These peptides were also found to be safe even when given for prolonged periods (Kim et al. 2008). Considering that most prostate cancer patients are elderly who are in need of the least toxic adjuvant therapies, a combination of isozyme-specific regulators of different PKC isozymes in the appropriate stages of the disease may aid in the development of improved therapeutic approaches.

Acknowledgments We thank Dr. Adrienne Gordon for critical reading of the manuscript. Work is supported by grants CA09151 (J.K.) and CA89202 (M.G.K.) from NIH, and PC061328 (M.G.K.) from Department of Defense.

References

Aranda, V., Nolan, M. E., & Muthuswamy, S. K. (2008). Par complex in cancer: a regulator of normal cell polarity joins the dark side. *Oncogene, 27*(55), 6878–6887.

Arbiser, J. L. (2004). Molecular regulation of angiogenesis and tumorigenesis by signal transduction pathways: evidence of predictable and reproducible patterns of synergy in diverse neoplasms. *Seminars in Cancer Biology, 14*(2), 81–91.

Augsten, M., Hagglof, C., Olsson, E., Stolz, C., Tsagozis, P., Levchenko, T., et al. (2009). CXCL14 is an autocrine growth factor for fibroblasts and acts as a multi-modal stimulator of prostate tumor growth. *Proceedings of the National Academy of Sciences of the United States of America, 106*(9), 3414–3419.

Aziz, M. H., Manoharan, H. T., Church, D. R., Dreckschmidt, N. E., Zhong, W., Oberley, T. D., et al. (2007). Protein kinase Cepsilon interacts with signal transducers and activators of transcription 3 (Stat3), phosphorylates Stat3Ser727, and regulates its constitutive activation in prostate cancer. *Cancer Research, 67*(18), 8828–8838.

Aziz, M. H., Manoharan, H. T., Sand, J. M., & Verma, A. K. (2007). Protein kinase Cepsilon interacts with Stat3 and regulates its activation that is essential for the development of skin cancer. *Molecular Carcinogenesis, 46*(8), 646–653.

Baek, S. H. (2006). A novel link between SUMO modification and cancer metastasis. *Cell Cycle, 5*(14), 1492–1495.

Barki-Harrington, L., & Daaka, Y. (2001). Bradykinin induced mitogenesis of androgen independent prostate cancer cells. *Journal of Urology, 165*(6 Pt 1), 2121–2125.

Budas, G. R., Churchill, E. N., & Mochly-Rosen, D. (2007). Cardioprotective mechanisms of PKC isozyme-selective activators and inhibitors in the treatment of ischemia-reperfusion injury. *Pharmacological Research, 55*(6), 523–536.

Budas, G. R., Koyanagi, T., Churchill, E. N., & Mochly-Rosen, D. (2007). Competitive inhibitors and allosteric activators of protein kinase C isoenzymes: a personal account and progress report on transferring academic discoveries to the clinic. *Biochemical Society Transactions, 35*(Pt 5), 1021–1026.

Chang, B. Y., Chiang, M., & Cartwright, C. A. (2001). The interaction of Src and RACK1 is enhanced by activation of protein kinase C and tyrosine phosphorylation of RACK1. *Journal of Biological Chemistry, 276*(23), 20346–20356.

Chang, B. Y., Conroy, K. B., Machleder, E. M., & Cartwright, C. A. (1998). RACK1, a receptor for activated C kinase and a homolog of the beta subunit of G proteins, inhibits activity of src tyrosine kinases and growth of NIH 3T3 cells. *Molecular and Cellular Biology, 18*(6), 3245–3256.

Chang, B. Y., Harte, R. A., & Cartwright, C. A. (2002). RACK1: a novel substrate for the Src protein-tyrosine kinase. *Oncogene, 21*(50), 7619–7629.

Charlesworth, P. J., & Harris, A. L. (2006). Mechanisms of disease: angiogenesis in urologic malignancies. *Nature Clinical Practice Urology, 3*(3), 157–169.

Chen, D., Purohit, A., Halilovic, E., Doxsey, S. J., & Newton, A. C. (2004). Centrosomal anchoring of protein kinase C betaII by pericentrin controls microtubule organization, spindle function, and cytokinesis. *Journal of Biological Chemistry, 279*(6), 4829–4839.

Cheng, J., Bawa, T., Lee, P., Gong, L., & Yeh, E. T. (2006). Role of desumoylation in the development of prostate cancer. *Neoplasia, 8*(8), 667–676.

Cornford, P., Evans, J., Dodson, A., Parsons, K., Woolfenden, A., Neoptolemos, J., et al. (1999). Protein kinase C isoenzyme patterns characteristically modulated in early prostate cancer. *American Journal of Pathology, 154*(1), 137–144.

Dashevsky, O., Varon, D., & Brill, A. (2009). Platelet-derived microparticles promote invasiveness of prostate cancer cells via upregulation of MMP-2 production. *International Journal of Cancer, 124*(8), 1773–1777.

Day, M. L., Zhao, X., Wu, S., Swanson, P. E., & Humphrey, P. A. (1994). Phorbol ester-induced apoptosis is accompanied by NGFI-A and c-fos activation in androgen-sensitive prostate cancer cells. *Cell Growth & Differentiation, 5*(7), 735–741.

Ding, L., Wang, H., Lang, W., & Xiao, L. (2002). Protein kinase C-epsilon promotes survival of lung cancer cells by suppressing apoptosis through dysregulation of the mitochondrial caspase pathway. *Journal of Biological Chemistry, 277*(38), 35305–35313.

Edwards, J., Krishna, N. S., Mukherjee, R., & Bartlett, J. M. (2004). The role of c-Jun and c-Fos expression in androgen-independent prostate cancer. *Journal of Pathology, 204*(2), 153–158.

Fields, A. P., Frederick, L. A., & Regala, R. P. (2007). Targeting the oncogenic protein kinase Ciota signalling pathway for the treatment of cancer. *Biochemical Society Transactions, 35*(Pt 5), 996–1000.

Fujii, T., Garcia-Bermejo, M. L., Bernabo, J. L., Caamano, J., Ohba, M., Kuroki, T., et al. (2000). Involvement of protein kinase C delta (PKCdelta) in phorbol ester-induced apoptosis in LNCaP prostate cancer cells. Lack of proteolytic cleavage of PKCdelta. *Journal of Biological Chemistry, 275*(11), 7574–7582.

Gavrielides, M. V., Gonzalez-Guerrico, A. M., Riobo, N. A., & Kazanietz, M. G. (2006). Androgens regulate protein kinase Cdelta transcription and modulate its apoptotic function in prostate cancer cells. *Cancer Research, 66*(24), 11792–11801.

Goldstein, B., & Macara, I. G. (2007). The PAR proteins: fundamental players in animal cell polarization. *Developmental Cell, 13*(5), 609–622.

Gonzalez-Guerrico, A. M., & Kazanietz, M. G. (2005). Phorbol ester-induced apoptosis in prostate cancer cells via autocrine activation of the extrinsic apoptotic cascade: a key role for protein kinase C delta. *Journal of Biological Chemistry, 280*(47), 38982–38991.

Graff, J. R., McNulty, A. M., Hanna, K. R., Konicek, B. W., Lynch, R. L., Bailey, S. N., et al. (2005). The protein kinase Cbeta-selective inhibitor, Enzastaurin (LY317615.HCl), suppresses signaling through the AKT pathway, induces apoptosis, and suppresses growth of human colon cancer and glioblastoma xenografts. *Cancer Research, 65*(16), 7462–7469.

Griner, E. M., & Kazanietz, M. G. (2007). Protein kinase C and other diacylglycerol effectors in cancer. *Nature Reviews Cancer, 7*(4), 281–294.

Henttu, P., & Vihko, P. (1998). The protein kinase C activator, phorbol ester, elicits disparate functional responses in androgen-sensitive and androgen-independent human prostatic cancer cells. *Biochemical and Biophysical Research Communications, 244*(1), 167–171.

Herman, J. G., Stadelman, H. L., & Roselli, C. E. (2009). Curcumin blocks CCL2-induced adhesion, motility and invasion, in part, through down-regulation of CCL2 expression and proteolytic activity. *International Journal of Oncology, 34*(5), 1319–1327.

Hofmann, J. (2004). Protein kinase C isozymes as potential targets for anticancer therapy. *Current Cancer Drug Targets, 4*(2), 125–146.

Huss, W. J., Maddison, L. A., & Greenberg, N. M. (2001). Autochthonous mouse models for prostate cancer: past, present and future. *Seminars in Cancer Biology, 11*(3), 245–260.

Jemal, A., Siegel, R., Ward, E., Hao, Y., Xu, J., Murray, T., et al. (2008). Cancer statistics, 2008. *CA: A Cancer Journal for Clinicians, 58*(2), 71–96.

Kemphues, K. J., Priess, J. R., Morton, D. G., & Cheng, N. S. (1988). Identification of genes required for cytoplasmic localization in early *C. elegans* embryos. *Cell, 52*(3), 311–320.

Kiley, S. C., & Parker, P. J. (1995). Differential localization of protein kinase C isozymes in U937 cells: evidence for distinct isozyme functions during monocyte differentiation. *Journal of Cell Science, 108*(Pt 3), 1003–1016.

Kim, J., Choi, Y. L., Vallentin, A., Hunrichs, B. S., Hellerstein, M. K., Peehl, D. M., et al. (2008). Centrosomal PKCbetaII and pericentrin are critical for human prostate cancer growth and angiogenesis. *Cancer Research, 68*(16), 6831–6839.

Kim, J. H., Choi, H. J., Kim, B., Kim, M. H., Lee, J. M., Kim, I. S., et al. (2006). Roles of sumoylation of a reptin chromatin-remodelling complex in cancer metastasis. *Nature Cell Biology, 8*(6), 631–639.

Kim, S. J., Cheung, S., & Hellerstein, M. K. (2004). Isolation of nuclei from label-retaining cells and measurement of their turnover rates in rat colon. *American Journal of Physiology Cell Physiology, 286*(6), C1464–1473.

Kim, S. J., Turner, S., Killion, S., & Hellerstein, M. K. (2005). In vivo measurement of DNA synthesis rates of colon epithelial cells in carcinogenesis. *Biochemical and Biophysical Research Communications, 331*(1), 203–209.

Knauf, J. A., Elisei, R., Mochly-Rosen, D., Liron, T., Chen, X. N., Gonsky, R., et al. (1999). Involvement of protein kinase Cepsilon (PKCepsilon) in thyroid cell death. A truncated chimeric PKCepsilon cloned from a thyroid cancer cell line protects thyroid cells from apoptosis. *Journal of Biological Chemistry, 274*(33), 23414–23425.

Koren, R., Ben Meir, D., Langzam, L., Dekel, Y., Konichezky, M., Baniel, J., et al. (2004). Expression of protein kinase C isoenzymes in benign hyperplasia and carcinoma of prostate. *Oncology Reports, 11*(2), 321–326.

Kruger, E. A., Blagosklonny, M. V., Dixon, S. C., & Figg, W. D. (1998). UCN-01, a protein kinase C inhibitor, inhibits endothelial cell proliferation and angiogenic hypoxic response. *Invasion and Metastasis, 18*(4), 209–218.

Levin, D. E., Fields, F. O., Kunisawa, R., Bishop, J. M., & Thorner, J. (1990). A candidate protein kinase C gene, PKC1, is required for the *S. cerevisiae* cell cycle. *Cell, 62*(2), 213–224.

Li, L., Ren, C. H., Tahir, S. A., Ren, C., & Thompson, T. C. (2003). Caveolin-1 maintains activated Akt in prostate cancer cells through scaffolding domain binding site interactions with and inhibition of serine/threonine protein phosphatases PP1 and PP2A. *Molecular and Cellular Biology, 23*(24), 9389–9404.

Lindemann, R. K., Braig, M., Hauser, C. A., Nordheim, A., & Dittmer, J. (2003). Ets2 and protein kinase C epsilon are important regulators of parathyroid hormone-related protein expression in MCF-7 breast cancer cells. *Biochemical Journal, 372*(Pt 3), 787–797.

Majumder, P. K., & Sellers, W. R. (2005). Akt-regulated pathways in prostate cancer. *Oncogene, 24*(50), 7465–7474.

Mamidipudi, V., Zhang, J., Lee, K. C., & Cartwright, C. A. (2004). RACK1 regulates G1/S progression by suppressing Src kinase activity. *Molecular and Cellular Biology, 24*(15), 6788–6798.

McJilton, M. A., Van Sikes, C., Wescott, G. G., Wu, D., Foreman, T. L., Gregory, C. W., et al. (2003). Protein kinase Cepsilon interacts with Bax and promotes survival of human prostate cancer cells. *Oncogene, 22*(39), 7958–7968.

Montalvo, L., Sanchez-Chapado, M., Prieto, J. C., & Carmena, M. J. (2002). Regulation of the expression of protein kinase C isoenzymes in rat ventral prostate: effects of age, castration and flutamide treatment. *Life Sciences, 71*(19), 2257–2266.

Montesano, R., & Orci, L. (1985). Tumor-promoting phorbol esters induce angiogenesis in vitro. *Cell, 42*(2), 469–477.

Neid, M., Datta, K., Stephan, S., Khanna, I., Pal, S., Shaw, L., et al. (2004). Role of insulin receptor substrates and protein kinase C-zeta in vascular permeability factor/vascular endothelial growth factor expression in pancreatic cancer cells. *The Journal of Biological Chemistry, 279*(6), 3941–3948.

Nelson, W. G. (2007). Prostate cancer prevention. *Current Opinion in Urology, 17*(3), 157–167.

Nelson, W. G., Yegnasubramanian, S., Agoston, A. T., Bastian, P. J., Lee, B. H., Nakayama, M., et al. (2007). Abnormal DNA methylation, epigenetics, and prostate cancer. *Frontiers in Bioscience, 12*, 4254–4266.

Perry, J. E., Grossmann, M. E., & Tindall, D. J. (1998). Epidermal growth factor induces cyclin D1 in a human prostate cancer cell line. *Prostate, 35*(2), 117–124.

Pihan, G. A., Purohit, A., Wallace, J., Malhotra, R., Liotta, L., & Doxsey, S. J. (2001). Centrosome defects can account for cellular and genetic changes that characterize prostate cancer progression. *Cancer Research, 61*(5), 2212–2219.

Powell, C. T., Brittis, N. J., Stec, D., Hug, H., Heston, W. D., & Fair, W. R. (1996a). Persistent membrane translocation of protein kinase C alpha during 12-O-tetradecanoylphorbol-13-acetate-induced apoptosis of LNCaP human prostate cancer cells. *Cell Growth and Differentiation, 7*(4), 419–428.

Powell, C. T., Gschwend, J. E., Fair, W. R., Brittis, N. J., Stec, D., & Huryk, R. (1996b). Overexpression of protein kinase C-zeta (PKC-zeta) inhibits invasive and metastatic abilities of Dunning R-3327 MAT-LyLu rat prostate cancer cells. *Cancer Research, 56*(18), 4137–4141.

Prekeris, R., Mayhew, M. W., Cooper, J. B., & Terrian, D. M. (1996). Identification and localization of an actin-binding motif that is unique to the epsilon isoform of protein kinase C and participates in the regulation of synaptic function. *The Journal of Cell Biology, 132*(1–2), 77–90.

Ramaswamy, S., Ross, K. N., Lander, E. S., & Golub, T. R. (2003). A molecular signature of metastasis in primary solid tumors. *Nature Genetics, 33*(1), 49–54.

Regala, R. P., Thompson, E. A., & Fields, A. P. (2008). Atypical protein kinase C iota expression and aurothiomalate sensitivity in human lung cancer cells. *Cancer Research, 68*(14), 5888–5895.

Rigas, A. C., Ozanne, D. M., Neal, D. E., & Robson, C. N. (2003). The scaffolding protein RACK1 interacts with androgen receptor and promotes cross-talk through a protein kinase C signaling pathway. *The Journal of Biological Chemistry, 278*(46), 46087–46093.

Robertson, M. J., Kahl, B. S., Vose, J. M., de Vos, S., Laughlin, M., Flynn, P. J., et al. (2007). Phase II study of enzastaurin, a protein kinase C beta inhibitor, in patients with relapsed or refractory diffuse large B-cell lymphoma. *Journal of Clinical Oncology, 25*(13), 1741–1746.

Rowe, A., Weiske, J., Kramer, T. S., Huber, O., & Jackson, P. (2008). Phorbol ester enhances KAI1 transcription by recruiting Tip60/Pontin complexes. *Neoplasia, 10*(12), 1421–1432. following 1432.

Sakamoto, S., Ryan, A. J., & Kyprianou, N. (2008). Targeting vasculature in urologic tumors: mechanistic and therapeutic significance. *Journal of Cellular Biochemistry, 103*(3), 691–708.

Schechtman, D., & Mochly-Rosen, D. (2001). Adaptor proteins in protein kinase C-mediated signal transduction. *Oncogene, 20*(44), 6339–6347.

Stallings-Mann, M., Jamieson, L., Regala, R. P., Weems, C., Murray, N. R., & Fields, A. P. (2006). A novel small-molecule inhibitor of protein kinase Ciota blocks transformed growth of non-small-cell lung cancer cells. *Cancer Research, 66*(3), 1767–1774.

Stanbrough, M., Bubley, G. J., Ross, K., Golub, T. R., Rubin, M. A., Penning, T. M., et al. (2006). Increased expression of genes converting adrenal androgens to testosterone in androgen-independent prostate cancer. *Cancer Research, 66*(5), 2815–2825.

Sugibayashi, R., Kiguchi, Y., Shimizu, T., Suzuki, T., Hamada, H., & Takeda, K. (2002). Up-regulation of p21(WAF1/CIP1) levels leads to growth suppression of prostate cancer cell lines. *Anticancer Research, 22*(2A), 713–719.

Sumitomo, M., Ohba, M., Asakuma, J., Asano, T., Kuroki, T., & Hayakawa, M. (2002). Protein kinase Cdelta amplifies ceramide formation via mitochondrial signaling in prostate cancer cells. *The Journal of Clinical Investigation, 109*(6), 827–836.

Tahir, S. A., Yang, G., Ebara, S., Timme, T. L., Satoh, T., Li, L., et al. (2001). Secreted caveolin-1 stimulates cell survival/clonal growth and contributes to metastasis in androgen-insensitive prostate cancer. *Cancer Research, 61*(10), 3882–3885.

Takahashi, M., Mukai, H., Oishi, K., Isagawa, T., & Ono, Y. (2000). Association of immature hypophosphorylated protein kinase cepsilon with an anchoring protein CG-NAP. *The Journal of Biological Chemistry, 275*(44), 34592–34596.

Tanaka, Y., Gavrielides, M. V., Mitsuuchi, Y., Fujii, T., & Kazanietz, M. G. (2003). Protein kinase C promotes apoptosis in LNCaP prostate cancer cells through activation of p38 MAPK and inhibition of the Akt survival pathway. *The Journal of Biological Chemistry, 278*(36), 33753–33762.

Teicher, B. A., Alvarez, E., Menon, K., Esterman, M. A., Considine, E., Shih, C., et al. (2002). Antiangiogenic effects of a protein kinase Cbeta-selective small molecule. *Cancer Chemotherapy and Pharmacology, 49*(1), 69–77.

Teicher, B. A., Menon, K., Alvarez, E., Galbreath, E., Shih, C., & Faul, M. M. (2001). Antiangiogenic and antitumor effects of a protein kinase Cbeta inhibitor in murine lewis lung carcinoma and human Calu-6 non-small-cell lung carcinoma xenografts. *Cancer Chemotherapy and Pharmacology, 48*(6), 473–480.

Teicher, B. A., Menon, K., Alvarez, E., Shih, C., & Faul, M. M. (2002). Antiangiogenic and antitumor effects of a protein kinase Cbeta inhibitor in human breast cancer and ovarian cancer xenografts. *Investigational New Drugs, 20*(3), 241–251.

Villar, J., Arenas, M. I., MacCarthy, C. M., Blanquez, M. J., Tirado, O. M., & Notario, V. (2007). PCPH/ENTPD5 expression enhances the invasiveness of human prostate cancer cells by a protein kinase C delta-dependent mechanism. *Cancer Research, 67*(22), 10859–10868.

Villar, J., Quadri, H. S., Song, I., Tomita, Y., Tirado, O. M., & Notario, V. (2009). PCPH/ENTPD5 expression confers to prostate cancer cells resistance against cisplatin-induced apoptosis through protein kinase Calpha-mediated Bcl-2 stabilization. *Cancer Research, 69*(1), 102–110.

Wu, D., Foreman, T. L., Gregory, C. W., McJilton, M. A., Wescott, G. G., Ford, O. H., et al. (2002). Protein kinase cepsilon has the potential to advance the recurrence of human prostate cancer. *Cancer Research, 62*(8), 2423–2429.

Wu, D., Thakore, C. U., Wescott, G. G., McCubrey, J. A., & Terrian, D. M. (2004). Integrin signaling links protein kinase Cepsilon to the protein kinase B/Akt survival pathway in recurrent prostate cancer cells. *Oncogene, 23*(53), 8659–8672.

Xiao, L., Caino, M. C., von Burstin, V. A., Oliva, J. L., & Kazanietz, M. G. (2008). Phorbol ester-induced apoptosis and senescence in cancer cell models. *Methods in Enzymology, 446*, 123–139.

Xiao, L., Gonzalez-Guerrico, A., & Kazanietz, M. G. (2009). PKC-mediated secretion of death factors in LNCaP prostate cancer cells is regulated by androgens. *Molecular Carcinogenesis, 48*(3), 187–195.

Xu, H., Czerwinski, P., Hortmann, M., Sohn, H. Y., Forstermann, U., & Li, H. (2008). Protein kinase C alpha promotes angiogenic activity of human endothelial cells via induction of vascular endothelial growth factor. *Cardiovascular Research, 78*(2), 349–355.

Yin, Y. J., Salah, Z., Maoz, M., Ram, S. C., Ochayon, S., Neufeld, G., et al. (2003). Oncogenic transformation induces tumor angiogenesis: a role for PAR1 activation. *FASEB Journal, 17*(2), 163–174.

Zeng, Z. Z., Jia, Y., Hahn, N. J., Markwart, S. M., Rockwood, K. F., & Livant, D. L. (2006). Role of focal adhesion kinase and phosphatidylinositol 3'-kinase in integrin fibronectin receptor-mediated, matrix metalloproteinase-1-dependent invasion by metastatic prostate cancer cells. *Cancer Research, 66*(16), 8091–8099.

Zhang, J., Park, S. I., Artime, M. C., Summy, J. M., Shah, A. N., Bomser, J. A., et al. (2007). AFAP-110 is overexpressed in prostate cancer and contributes to tumorigenic growth by regulating focal contacts. *The Journal of Clinical Investigation, 117*(10), 2962–2973.

Chapter 19
Protein Kinase C and Lung Cancer

Lei Xiao

Abstract The protein kinase C (PKC) family of serine/threonine kinases has been linked to the carcinogenic process of many types of human cancers including lung cancer. Lung carcinogenesis is a multistep process involving both genetic and epigenetic alterations in oncogenes and tumor suppressor genes, and changes in activation of signal transduction pathways, resulting in progressive deregulation of cell proliferation and survival mechanisms. Alterations in PKC isoform expression and/or activity have been observed in human lung cancer, and functional studies have suggested that individual PKC isoforms play distinct, sometimes opposite, effects in transformation, proliferation, and survival of human lung cancer cells. This chapter provides a brief review of current knowledge regarding PKC isoform-specific roles in the pathogenesis of human lung cancer and therapeutic potential of targeting specific PKC isoforms.

Keywords Apoptosis • Cell cycle • Chemoresistance • Nonsmall cell lung cancer • Protein kinase C isoform • Small cell lung cancer • Therapeutics • Tumorigenicity

19.1 Introduction

19.1.1 Lung Cancer

Lung cancer is the leading cause of cancer-related deaths worldwide today. It was projected that in 2008 there would be 215,020 new cases of lung cancer and 161,840 disease-related deaths in the United States alone (Jemal et al. 2008). Human lung cancer is a disease of heterogeneous histology, which can be divided into two major categories based on their clinical presentation: small cell lung cancer

L. Xiao (✉)
McGuire Center for Lepidoptera and Biodiversity, Florida Museum of Natural History, University of Florida, Gainesville, FL 32610, USA
e-mail: xiaotyler@gmail.com

M.G. Kazanietz (ed.), *Protein Kinase C in Cancer Signaling and Therapy*,
Current Cancer Research, DOI 10.1007/978-1-60761-543-9_19,

(SCLC) and nonsmall cell lung cancer (NSCLC). SCLC represents ~20% of all lung cancer worldwide. The remaining 80% of lung cancers fall into one of three major subtypes of NSCLC carcinomas: adenocarcinoma, squamous cell carcinoma (SCC), and large cell carcinoma (LCC). Tobacco smoking is the most important cause of lung cancer with 80–90% of the disease arising in cigarette smokers. Early epidemiologic studies of the smoking-caused lung cancer indicated that squamous cell carcinoma was the most frequently diagnosed type of lung cancer, followed by small cell carcinoma. Adenocarcinoma of the lung is the most common histologic type of lung cancer in the world today, and is the most frequent type of lung cancer in women, nonsmokers, and in young people (Josen et al. 2002; Minna et al. 2002).

SCLC is distinct from NSCLC in biology and clinicopathology. SCLCs are neuroendocrine (NE) tumors that are strongly smoking-associated and are characterized by early metastasis and initial marked responsiveness to chemotherapy and radiation. However, nearly all patients with SCLC relapse and develop resistance to cytotoxic therapies. The overall 5-year survival rate is only 3–8% (Facchini and Spiro 1999; Rathore and Weitberg 2002). NSCLCs, at large, are lacking of neuroendocrine features. They respond poorly to chemotherapy as compared to SCLCs. The treatment strategies of NSCLCs are based on the stage of the disease at the time of diagnosis, which include surgery, chemotherapy, radiotherapy, or combined therapy (Weitberg 2002). Despite significant efforts to improve patient survival, the overall treatment results have been disappointing. Over the past 30 years, the 5-year lung cancer survival rate remains between 8 and 14%.

In the past decade, an increasing understanding of the pathogenesis of lung cancer at the cellular and molecular levels has provided significant insights into the molecular process underlying lung carcinogenesis and the progression of lung cancer. Lung cancer arises as the result of multiple genetic lesions due to exposure to cigarette smoke or other environmental carcinogens as well as inherited predisposition(s) (Hecht 1999; Alberg and Samet 2003). It is becoming clear that genetic changes acquired by lung cancer are complex and heterogeneous. NSCLC and SCLC exhibit distinct but overlapping patterns of genetic and epigenetic alterations. These abnormalities include chromosomal deletion and/or amplification, epigenetic changes in DNA methylation, the activation of protooncogenes and other growth-promoting genes, and the inactivation of tumor suppressor genes (Osada and Takahashi 2002; Sekido et al. 2003; Xiao 2006). The knowledge of the molecular characteristics of lung cancers has started a novel era in the development of new compounds that target cancer-specific genetic and molecular alterations and their associated signal transduction pathways (Auberger et al. 2006).

19.1.2 Protein Kinase C

Protein kinase C (PKC) is a family of ubiquitously expressed, structurally related, phospholipid-dependent serine/threonine protein kinases which play crucial roles in transducing signals that regulate diverse biological functions, including proliferation, transformation, differentiation, and apoptosis (Nishizuka 1995; Dempsey et al. 2000).

All of PKC isoforms consist of an NH_2-terminal regulatory region and a COOH-terminal catalytic region. The PKC isoforms are classified into three subgroups based on the differences in their domain structures and biochemical properties: the classical isoforms (cPKC: α, βI, βII, and γ), which are Ca^{2+} and phorbol ester (e.g., PMA)/diacylglycerol (DAG)-dependent; the novel isoforms (nPKC: δ, ε, η, and θ), which are PMA/DAG-dependent but Ca^{2+}-independent; and the atypical isoforms (aPKC: ζ and ι/λ), which are Ca^{2+} and PMA/DAG-independent. Under physiological conditions, PKC is activated in response to various stimuli. The generation of diacylglycerol, a natural lipid produced in receptor-coupled hydrolysis of inositol phospholipids upon cell stimulation, plays a central role in the activation of cPKCs and nPKCs. Importantly, tumor-promoting phorbol esters such as 12-O-tetradecanoylphorbol-13-acetate (TPA) were found to activate PKC directly in a manner similar to diacylglycerol (Castagna et al. 1982). Unlike diacylglycerol which results in transient PKC activation, phorbol esters lead to prolonged activation of PKCs (Blumberg et al. 1984; Nishizuka 1995). The discovery that PKC is a major cellular target for the tumor-promoting phorbol esters has led to a surge of research interest in elucidating the roles of PKC in carcinogenesis.

The physiological function of PKC is defined by its phosphorylation state, conformation, and subcellular localization (Newton 2001). PKC action can be localized to multiple subcellular compartments including the plasma membrane, mitochondria, cytoskeleton, Golgi, and the nucleus, where PKC can directly or indirectly (via scaffolding proteins) interact with its downstream substrates. This kinase-substrate interaction contributes to the specificity/efficiency of PKC action (Jaken and Parker 2000; Newton 2001, 2003). Individual PKC isoforms display unique but sometimes overlapping expression patterns in many tissues. The differences in tissue expression, subcellular localization, and activator/substrate specificity indicate that individual PKC isoforms have distinct cellular functions (Dempsey et al. 2000; Jaken and Parker 2000; Way et al. 2000). Additionally, there is an extensive cross-talk among different PKC isoforms. Therefore, the overall response to PKC activation seems to depend on the presence and activity of the other isoforms in tissue and/or particular cell types studied. There is growing evidence that the role of PKC in tumorigenesis is cell context-dependent and/or isoform-specific (Griner and Kazanietz 2007). This chapter provides a brief overview of recent advances in understanding the function and mechanisms of the PKC family of proteins in the pathogenesis of human lung cancer, with emphasis on isoform-specific actions in transformation, proliferation, and apoptosis.

19.2 PKC Isoform Expression in Human Lung Cancer

19.2.1 Expression Profiling in NSCLC Specimens (Table 19.1)

Expression of multiple PKC isoforms including PKCα, PKCβII, PKCε, and PKCζ were reported at the protein and mRNA levels in normal human lung and airway smooth muscle (Webb et al. 1997). Limited investigations into the expression of

Table 19.1 Expression and function of PKC isoforms in lung cancer

PKC isoform	Expression in lung cancer specimens/cells	References	Function	References
α	≤20% NSCLC (high level)	Lahn et al. (2004)	Proproliferation and tumorigenicity;PMA-induced G_2/M phase arrest and cellular senescence; Promoting nicotine-mediated SCLC survival; Up-regulation of the activity of doxorubicin transporters	Wang et al. (1999); Oliva et al. (2008); Mai et al. (2003); Singhal et al. (2006)
β	71% NSCLC	Lahn et al. (2006)		
δ	~7% NSCLC tumor cells (frequently expressed in normal and tumor-associated stroma); Ubiquitously expressed in NSCLC cell lines	Chen et al. (2008); Ding et al. (2002); Clark et al. (2003); Kim et al. (2007)	PMA-mediated growth inhibition and G_1 phase arrest; Pro- and antiapoptosis in cell-context and/or stimuli-specific manners	Nakagawa et al. (2005); Clark et al. (2003); Hurbin et al. (2005); Lee et al. (2005); Persaud et al. (2005); Kim et al. (2007)
ε	>90% NSCLC	Bae et al. (2007)	Promoting G_1/S transition and transformed growth; Antiapoptosis and chemoresistance	Bae et al. (2007); Ding et al. (2002); Pardo et al. (2006); Felber et al. (2007)
ι	~70% NSCLC (gene amplification in ~70% SCC)	Regala et al. (2005a); Fields and Regala (2007)	Transformed growth and tumorigenicity; Promoting NNK-mediated cell survival; Promoting migration and invasion	Regala et al. (2005a); Regala et al. (2005b); Jin et al. (2005); Xu and Deng (2006); Frederick et al. (2008)
ζ	Low or undetectable in NSCLC	Regala et al. (2005a); Galvez et al. (2009)	Tumor suppressor and antiproliferation; Promoting nicotine-mediated survival; Promoting EGF-induced chemotataxis	Galvez et al. (2009); Xin et al. (2007); Liu et al. (2009)

individual PKC isoforms in human nonsmall cell lung cancer (NSCLC) have revealed a distinct expression pattern. However, there is currently no report regarding the expression of PKC isoforms in human small cell lung cancer (SCLC). Immunohistochemical and quantitive PCR studies have shown that PKCα is highly expressed in ≤20% of NSCLC specimens, and expression of PKCα appears more common in adenocarcinoma than in squamous cell carcinoma (SCC) (Lahn et al. 2004). The preferential expression of PKCα in adenocarcinoma was further confirmed by the gene expression array data. Similar to PKCα expression, high levels of PKCβ were detected in ~16% of NSCLC specimens while 71% of all NSCLC specimens examined showed positive staining of PKCβ by immunohistochemistry (Lahn et al. 2006). However, these studies did not evaluate the expression of these two PKC isoforms in normal lung tissues; it is unclear whether the observed expression of PKCα or PKCβ is tumor-specific.

Expression of PKCε in NSCLC was recently assessed by immunohistochemical analysis (Bae et al. 2007). This study demonstrated a significant increase in PKCε expression in >90% of primary human NSCLC when compared to normal lung epithelium. By evaluating PKCε expression in relationship with clinicopathologic variables, it was found that PKCε expression is significantly higher in adenocarcinoma than in SCC, and in patients with T_1 tumors than those with T_2 to T_4 tumors. No significant correlations were observed between PKCε expression and age, pathologic stages, and lymph node involvement (Bae et al. 2007). In contrast to PKCε, expression of PKCδ is negative in 93% of NSCLC tumor cells (Chen et al. 2008). Interestingly, PKCδ expression is present in the tumor stroma, particularly in smooth muscle cells, but consistently negative in the majority of tumor cells. This stromal-associated expression of PKCδ appears not to be tumor-specific, as it was also observed in the stromal compartment of the normal lung tissues (Chen and Xiao, unpublished observations). The distinct expression pattern between PKCε and PKCδ in NSCLC appears to correlate well with their in vitro biological functions. PKCε that is overexpressed in the majority of NSCLC is known to play a positive role in proliferation, transformation, and survival; whereas carcinoma-negative PKCδ is believed to be antiproliferation and antisurvival in general.

Expression of aPKC isoforms PKCι and PKCζ was studied with details by Fields and colleagues (Fields and Regala 2007). The distinct expression patterns of PKCι and PKCζ in NSCLC seem directly associated with their opposite effects on lung tumorigenesis (Regala et al. 2005b; Galvez et al. 2009) (see Sect. 19.3.1). PKCι mRNA and protein is overexpressed in ~70% of primary NSCLC whereas PKCζ mRNA and protein is extremely low or undetectable in both normal and cancerous lung tissues (Regala et al. 2005a). Overexpression of PKCι was predominantly confined to lung tumor cells, with little or no expression in tumor-associated stroma. Importantly, PKCι expression is a prognostic marker for predicting poor clinical outcome independent of tumor stage; NSCLC patients with elevated PKCι are 2.6 times more likely to die from the disease than patients without elevated PKCι. Although expression of PKCι does not correlate with tumor stage in NSCLC, its expression profiling can be used to identify patients with early stage lung cancer who may have elevated risk of relapse (Regala et al. 2005a).

The molecular mechanism underlying elevated PKCι expression in NSCLC is in part mediated through PKCι gene amplification. PKCι gene amplification was tumor-specific and occurred more frequently in SCC (~70%) but rarely in adenocarcinoma. Furthermore, sequence analysis of all 18 exons of the PKCι gene in adenocarcinoma and SCC without PKCι gene amplification have failed to detect any mutations (Fields and Regala 2007), suggesting that somatic mutation of the PKCι gene is unlikely to account for its oncogenic activation in lung cancer.

19.2.2 Expression Profiling in Lung Cancer Cell Lines (Table 19.1)

Expression of multiple PKC isoforms has been reported in various human lung cancer cell lines. The expression profiling of individual PKC isoforms in lung cancer cell lines appears less distinct than that in NSCLC specimens. In general, the α, βII, δ, ε, and ι isoforms are expressed ubiquitously at the protein level in many cell lines (Ding et al. 2002; Clark et al. 2003; Regala et al. 2005b; Kim et al. 2007). The expression profile of PKCι and PKCζ in NSCLC cells is consistent with that in NSCLC specimens (Regala et al. 2005b), whereas PKCδ appears ubiquitously expressed in NSCLC cells (Ding et al. 2002; Clark et al. 2003; Kim et al. 2007) but is rarely detected in tumor cells in NSCLC specimens (Chen et al. 2008). A differential expression pattern for PKCε was observed between NSCLC and SCLC phenotypes: its expression was detected in all of NSCLC lines but in none of the SCLC lines examined. A lack of PKCε expression in these SCLC lines appears due to transcriptional inactivation of the gene (Ding et al. 2002). However, overexpression of PKCε or its constitutively active catalytic fragment has also been reported in a subset of SCLC lines that exhibited increased chemoresistance or rapid growth compared to other SCLC line (Baxter et al. 1992; Pardo et al. 2006). Importantly, compared to primary or nontransformed human lung epithelial cells, expression of PKCα, PKCβII, PKCε, and PKCι was significantly increased (Clark et al. 2003; Regala et al. 2005b), suggesting that transformation of lung epithelial cells may be accompanied with changes in the expression of these PKC isoforms.

19.3 Roles of PKC in the Pathogenesis of Lung Cancer

19.3.1 Transformed Growth and Tumorigenicity

The role of PKC in carcinogenesis has been recognized for decades. However, the relative contribution of individual PKC isoforms to this process remains elusive. Several recent studies have started to reveal PKC isoforms-specific functions in lung tumorigenesis and in the control of transformed growth of human NSCLC cells in vitro and in vivo (Regala et al. 2005b; Bae et al. 2007; Galvez et al. 2009). A series of studies carried out by Fields and colleagues have identified PKCι as a critical cancer gene for human NSCLC (Regala et al. 2005a, b; Fields and Regala 2007). PKCι is overexpressed in the majority of NSCLC cell lines and in human NSCLC tumors. Distribution of PKCι signaling through expression of a dominant negative kinase deficient PKCι ιmutant (kdPKCι) in NSCLC cells results in significant inhibition of anchorage-independent growth in soft agar. However, kdPKCι has little effects on anchorage-dependent growth and survival (Regala et al. 2005a, b). Furthermore, expression of kdPKCι also inhibits the tumorigenicity of A549 NSCLC cells in vivo: athymic nude mice inoculated with kdPKCι-expressing A549 cells displayed a significant reduction in tumor growth, which was accompanied by the reduction in the rate of tumor cell proliferation without apparent effects on tumor cell survival and angiogenesis (Regala et al. 2005b). The transforming effect of PKCι is mediated by a PKCι-Rac1-Pak-MEK1/2-ERK1/2 signaling pathway. The Phox Bem 1 (PB1) domain within the regulatory region of PKCι is important for PKCι-dependent activation of Rac1 and transformed growth (Regala et al. 2005b). Because Rac1 is a critical effector of PKCι-mediated transformation, it is postulated that the so called polarity complex including PKCι, scaffold protein Par6, and small GTPases Rac1 or Cdc42 (Lin et al. 2000; Joberty et al. 2000) may be important for PKCι oncogenic signaling in lung cancer. Of great interest is the recent development of selective PKCι inhibitors that target the PB1 domain-mediated interaction between PKCι and Par6. These selective inhibitors display dose dependent inhibition of PKCι-Par6 interaction and block PKCι-mediated signaling to Rac1 and transformed growth of NSCLC cells in vitro and tumorigenicity in vivo (Stallings-Mann et al. 2006).

Despite a high homology (72%) in their amino acid sequences, two members of aPKCs, PKCζ and PKCι, are functionally distinct in normal physiology, embryonic development, and transformation (Akimoto et al. 1994; Kovac et al. 2007; Soloff et al. 2004). The role of PKCζ in carcinogenesis appears to be antiproliferative and proapoptotic; forced expression of PKCζ causes decreased anchorage-independent growth, increased differentiation, and enhanced apoptosis (Way et al. 1994; Mao et al. 2000; Mustafi et al. 2006). Consistent with its antitransforming function, expression of PKCζ is undetectable in the majority of NSCLC tumors and cell lines (Regala et al. 2005a; Galvez et al. 2009). A very recent study by Diaz-Meco, Moscat, and colleagues has demonstrated a potential tumor suppressor role for PKCζ in lung tumorigenesis: PKCζ-deficient mice display increased Ras-induced lung carcinogenesis in vivo (Galvez et al. 2009). The loss of PKCζ accelerated the progression of Ras-initiated lung tumors, leading to a more severe tumor phenotype. The increased tumor burden in the Ras-expressing PKCζ–/– lungs is associated with a significant increase in cyclin D1 expression, a higher percentage of Ki-67-positive cells, and the induction of intratumoral vessels, indicative of increased proliferation and neoangiogenesis. The tumor suppressing function of PKCζ is mediated through a mechanism involving the increased expression of IL-6, which promotes tumorigenesis by increased angiogenesis and enhanced proliferation of Ras-transformed cells under nutrient-deprived condition (Galvez et al. 2009). Interestingly, the mechanism by which Ras and PKCζ control of IL-6 expression is somewhat unexpected. Unlike IL-6 production in other cellular systems that requires PKCζ and NF-κB (Duran et al. 2003), increased expression of IL-6 in the Ras-expressing PKCζ–/– cells is NF-κB-independent but requires derepression of histone acetylation at the C/EBPβ element in the IL-6 promoter (Galvez et al. 2009). The negative impact of PKCζ in Ras-induced lung tumorigenesis may be functionally significant to human lung cancer as ~30% of NSCLC, particularly adenocarcinoma, harbor activating mutations in the Ki-*Ras* gene (Slebos et al. 1990; Mills et al. 1995), and the presence of Ki-*Ras* mutations is significantly associated with a shortened survival in surgically treated patients (Slebos et al. 1990).

Among all of the PKC isoforms, PKCε is unique in its oncogenic potential. Upon overexpression, PKCε acts as an oncogene that induces transformation in fibroblast and colonic epithelial cells (Cacace et al. 1993; Perletti et al. 1996; Mischak et al. 1999). The ability of PKCε to induce oncogenic transformation appears to depend on the cellular context. Significant increases in PKCε levels are present in the vast majority of primary human NSCLCs and PKCε is ubiquitously expressed in NSCLC cell lines (Ding et al. 2002; Clark et al. 2003; Bae et al. 2007). Disruption of the PKCε signaling using a kinase-inactive, dominant-negative PKCε mutant leads to a significant inhibition of anchorage-dependent and -independent growth of human NSCLC cells in a p53-independent manner (Bae et al. 2007), indicating that PKCε function is important for maintaining the transformed phenotype in NSCLC cells. The transforming function of PKCε is mediated by the suppression of p21/Cip1 in a Myc-dependent mechanism, leading to an accelerated G_1-S progression (see Sect. 19.3.2) (Bae et al. 2007). The significance of p21/Cip1 as a negative effector of the PKCε oncogenic action is underscored by the existence of an in vivo inverse correlation between the levels of PKCε and expression/location of p21/Cip1 in primary human NSCLC (Bae et al. 2007). Given its positive role in proliferation, transformation, and tumor cell invasion, it is postulated that aberrant activation of the PKCε signaling in vivo may predispose the normal lung epithelium to excessive proliferation, increased survival, and enhanced metastatic potential that may facilitate malignant transformation.

There is also evidence that PKCα may be involved in lung tumorigenesis. Expression of antisense PKCα mRNA in LTEPa-2 lung cancer cells significantly inhibits cell proliferation, anchorage-independent growth, and tumorigenicity in nude mice (Wang et al. 1999). One potential mechanism by which PKCα exerts its function on cell growth and transformation appears through up-regulation of the activity of the AP-1 transcription factor. Accumulating data indicate that multiple PKC isoforms are likely involved in the initiation and progression of lung cancer, and individual PKC isoforms may employ distinct mechanisms that contribute positively or negatively to the transformed growth and tumorigenicity of lung cancer cells in vitro and in vivo.

19.3.2 Proliferation and Cell-Cycle Regulation

It is well documented that PKC plays an important role in modulating cell proliferation and is implicated in both positive and negative regulation of cell cycle progression at two critical sites: the G_1/S and the G_2/M transitions. However, the specific role of individual PKC isoforms in controlling cell-cycle machinery remains controversial. Numerous studies indicate that the role of PKC in cell cycle regulation is highly dependent on the timing of PKC activation during the cell cycle, the specific PKC isoforms involved, and/or the cell types being examined (Black 2000). The complexity of PKC signaling in cell cycle control may be attributed to the fact that multiple PKC isoforms are present in a given cell type, and commonly used PKC activators (e.g., phorbol esters) and PKC inhibitors can simultaneously affect several, if not all, PKC isoforms.

Phorbol esters, such as phorbol 12-myristate 13-acetate (PMA), have profound antiproliferative effects in human NSCLC cells. This antiproliferative action of PMA is coupled with the regulation of cell cycle machinery in a PKC isoform-dependent manner. Recent studies by Kazanietz and colleagues have demonstrated that PKCα and PKCδ play a critical role in mediating the growth inhibitory effect of PMA in human NSCLC cells in a cell cycle phase-specific manner (Nakagawa et al. 2005; Oliva et al. 2008). Treatment of human NSCLC cells with PMA causes cell cycle arrest in different phases of the cell cycle: transient activation of PKC by PMA in early G_1 impairs the progression of NSCLC cells into S phase leading to the G_1 phase arrest, which requires the function of PKCδ (Nakagawa et al. 2005); whereas transient activation of PKC by PMA in late G_1 or early S phase leads to an irreversible cell cycle arrest in G_2/M and induction of cellular senescence, and this later effect of PMA is mediated by PKCα (Oliva et al. 2008). A common mechanism underlying phorbol esters induced cell cycle arrest in NSCLC cells is the induction of the cyclin-dependent kinase inhibitor p21/Cip1. The PMA-induced G_1 arrest is also accompanied with decreased Rb hyperphosphorylation and cyclin A expression, and PKCδ is required for the PMA-mediated, E2F-dependent repression of the cyclin A promoter (Nakagawa et al. 2005). This suggests that by controlling the expression of cyclin A, PKCδ may also impair the function of the cdk2-cyclin A complex that is required for G_1-S transition. Consistently, suppression of PKCδ expression significantly increases bryostatin 1-induced cell proliferation; whereas overexpression of PKCδ is associated with a lower rate of cell proliferation and is insensitive to proliferative stimulation in the HOP-92 NSCLC cells (Choi et al. 2006).

PKCε has proliferative effects in various cell types (Basu and Sivaprasad 2007). Increasing evidence has pinpointed to a positive role for PKCε in control of G_1/S transition of the cell cycle (Yan and Wenner 2001; Soh and Weinstein 2003; Tapinos and Rambukkana 2005). In human NSCLC cells, persistent inhibition of PKCε by expressing a kinase-inactive, dominant negative PKCε mutant, PKCε(KR), leads to a marked inhibition of cell proliferation accompanied by a significant delay in the S phase entry during the cell cycle (Bae et al. 2007). The antiproliferative effect of PKCε(KR) requires p21/Cip1 but is independent of p53. The PKCε(KR)-induced elevation of p21/Cip1 and the sustained association of p21/Cip1 with cdk2 are responsible for the inactivation of cdk2 complexes, thereby leading to a prolonged G_1/S transition. Furthermore, the antiproliferative action of PKCε(KR) is mediated at least in part through the downregulation of c-Myc, an oncogenic transcription factor that is overexpressed in ~50% of lung cancers (Osada and Takahashi 2002), which in turn negatively regulates the expression of p21/Cip1 in NSCLC cells (Bae et al. 2007). Significantly, this in vitro relationship among PKCε, c-Myc, and p21/Cip1 is also observed in vivo in human NSCLC specimens.

The fact that coexistence of multiple PKC isoforms in a given cell type suggests that individual PKC isoforms may function in a coordinated manner to regulate cellular function. Deeds et al. (2003) reported that the concurrent inhibition, rather than separate inhibition, of PKCα and PKCθ induces cell cycle arrest in the G_1 phase. Interestingly, this PKC cosuppression-mediated G_1 cell cycle arrest also requires a p53-independent induction of p21/Cip1. Both transcriptional and post-transcriptional mechanisms are involved in the up-regulation of p21/Cip1 in response to coinhibition of PKCα and PKCθ. It appears that p53-independent induction of p21/Cip1 serves as a common mechanism underlying negative regulation of cell cycle progression via either activation or inhibition of specific PKC isoforms in human NSCLC cells.

There is evidence that PKC plays an important role in the growth regulation of SCLC cells in response to neuropeptide growth factors. The mitogenic effects of neuropeptides such as galanin and neurotensin are mediated predominantly through the activation of the motigen-activated protein kinase (MAPK) or ERK pathway, and PKC activation is required for the ERK activation in this event (Seufferlein and Rozengurt 1996). However, the precise signaling mechanisms and the identity of PKC isoforms involved have not been fully defined.

19.3.3 Apoptosis and Chemoresistance

Apoptosis, or programmed cell death, is a highly specific and regulated process that plays a very critical role in development and in the maintenance of tissue homeostasis of multicellular organisms as well as tumorigenesis (Vaux and Korsmeyer 1999; Evan and Vousden 2001). Evading apoptosis is a hallmark of malignant transformation leading to carcinogenesis (Hanahan and Weinberg 2000). It has become evident that resistance to apoptosis is one potential mechanism whereby tumor cells escape from chemotherapy induced cytotoxicity, leading to cell survival and chemoresistance (Hannun 1997).

The PKC family has been shown to regulate survival and apoptosis in various cell types. The role of individual PKC isoforms in this process may be either antiapoptotic or proapoptotic, which is likely dependent on the nature of the apoptotic stimuli and specific cell types involved. In human lung cancer cells, PKC isoforms including

PKCα, PKCε, PKCι and PKCζ function as antiapoptotic kinases. In contrast, the function of PKCδ can be proapoptotic or antiapoptotic.

PKCε has been shown to promote cell survival and contribute to chemoresistance in human lung cancer cells (Ding et al. 2002; Pardo et al. 2006). Increased PKCε expression is specifically linked to chemotherapy resistance. PKCε is highly expressed in NSCLC cell lines that are resistant to chemotherapy, but not expressed in chemo-sensitive SCLC cell lines (Ding et al. 2002). Induction of PKCε in the PKCε-deficient, chemo-sensitive SCLC cells significantly increases the survival of SCLC cells against the chemotherapeutic drugs, etoposide and doxorubicin; whereas down-regulation of PKCε using antisense cDNA sensitizes NSCLC cells to these drugs. The chemo-protective effect of PKCε is mediated primarily by suppression of drug-induced apoptosis through a mechanism involving the inhibition of the mitochondrial-dependent caspase-3 activation and cytochrome c release (Ding et al. 2002). Furthermore, induction of PKCε expression in SCLC cells also enhances the anchorage-independent growth without affecting cell proliferation and cell cycle progression, indicating that PKCε could raise the apoptotic threshold of the cells, thereby promoting survival. Increased expression PKCε is also linked to the fibroblast growth factor-2 (FGF-2)-mediated chemoresistance of SCLC cells (Pardo et al. 2006). Elevated serum concentration of FGF-2 is an independent prognostic factor for adverse outcome in SCLC (Ruotsalainen et al. 2002). FGF-2 induces the activation of the extracellular-regulated kinase pathway (the MEK/ERK pathway), which enhances the expression of the antiapoptotic molecules Bcl-X_L and X-linked inhibitor of apoptosis (XIAP), thereby triggering chemoresistance in SCLC (Pardo et al. 2002, 2003). The prosurvival effect of FGF-2 requires PKCε, which forms a signaling complex specific with B-Raf and ribosomal S6 kinase-2 (S6K2). The direct interaction of PKCε with B-Raf and S6K2 is necessary and sufficient for the activation of ERK and translational up-regulation of Bcl-X_L and XIAP in response to FGF-2 treatment (Pardo et al. 2006). Interestingly, PKCε overexpression alone is capable to induce up-regulation of Bcl-X_L and XIPA and confer resistance to etoposide in SCLC cells (Pardo et al. 2006). Together, current studies suggest that the expression level of PKCε is an important determinant of cellular susceptibility to etoposide in lung cancer cells. Besides its role in chemoresistance, PKCε also protects NSCLC cells from apoptosis induced by the tumor necrosis factor (TNF)-related apoptosis-inducing ligand (TRAIL) (Felber et al. 2007). Inhibition of PKCε with a selective peptide inhibitor, myr-PKC-epsilon V1-2, significantly amplifies TRAIL-induced cytotoxic activity in NSCLC cells (Felber et al. 2007).

Tobacco-related carcinogens such as nicotine and the tobacco-specific carcinogen 4-(methylnitrosamino)-1-(3-pyridyl)-1-butanone (NNK) have been shown to cause PKC activation and to promote proliferation and survival of normal and neoplastic lung cells through a PKC-dependent mechanism (Schuller 1994; Heusch and Maneckjee 1998; Mai et al. 2003; Jin et al. 2004). Nicotine and NNK protect lung cancer cells from chemotherapeutic drug-induced apoptosis through regulating the function of the Bcl-2 family of proteins in a PKC isoform-dependent mechanism. Nicotine-mediated survival of H82 SCLC cells requires the phosphorylation of the antiapoptotic Bcl-2 exclusively at the Ser-70, and PKCα appears to be an upstream kinase responsible for the nicotine induced phosphorylation of Bcl-2 (Mai et al. 2003). Nicotine also exerts its prosurvival action through the activation of PKCζ, which in turn directly phosphorylates the proapoptotic Bax at Ser-184 (Xin et al. 2007). The activity of PKCζ is required for its interaction with Bax, and increased expression of wild-type or activated PKCζ leads to sequestration of Bax in the cytoplasm and prevents Bax from undergoing a conformational change thereby inhibiting its proapoptotic function (Xin et al. 2007). Unlike nicotine, NNK promotes survival through a PKCι-dependent mechanism, resulting in phosphorylation and inactivation of the proapoptotic Bad in NSCLC cells (Jin et al. 2005). NNK activates PKCι through a Src-dependent mechanism, and activated PKCι directly phosphorylates Bad at multiple serine sites and promotes the disassociation of the Bad/Bcl-X_L complex thereby leading to cell survival. The existence of multiple nicotine/NNK-PKC survival pathways suggests that the activation of specific pathways may be controlled in a cell context-dependent manner.

Heat shock protein 27 (HSP27) has been implicated in protecting cells from apoptosis triggered by a variety of stimuli including radiation and chemotherapeutic drugs (Bruey et al. 2000; Rane et al. 2003). HSP27 is overexpressed in human NSCLC tumor tissues compared to the corresponding normal lung tissues, and down-regulation of highly expressed HSP27 in NSCLC cell lines results in enhanced apoptosis in response to radiation or chemotherapeutic drug treatment (Kim et al. 2007). Inhibition of PKCδ kinase activity through a direct interaction between HSP27 and PKCδ contributes to HSP27-mediated chemo- and radiation-resistance in lung cancer cells (Lee et al. 2005; Kim et al. 2007). The amino acid residues 668–674 of the V5 region of PKCδ is necessary for HSP27 binding, and treatment of NSCLC cells with the PKCδ-V5 heptapeptide containing the corresponding amino acid sequences required for HSP27 binding restores the PKCδ activity and significantly increases cisplatin or radiation-induced cell death in vitro and in vivo (Kim et al. 2007), suggesting that PKCδ activity plays an essential role in mediating DNA damage-induced cell death in NSCLC cells. In contrast to this finding, Clark et al. (2003) reported an antiapoptotic function of PKCδ in NSCLC cells. It was shown that rottlerin, a selective PKCδ inhibitor, effectively induces apoptosis in NSCLC cells and enhanced chemotherapy-induced apoptosis in a cell line- and drug-specific manner. Although rottlerin has nonspecific effects toward other kinases (Davies et al. 2000) and displays a widespread cytotoxicity in many cancer cell lines, its antiapoptotic effect observed in NSCLC cells appears PKCδ-dependent as expression of a kinase-dead mutant of PKCδ in NSCLC cells induces apoptosis accompanied with decreased PKCδ phosphorylation (Clark et al. 2003). PKCδ also functions as an antiapoptotic kinase in mediating the prosurvival effect of the combination of two growth factors, amphiregulin (AR) and insulin-like growth factor-1 (IGF1), in human NSCLC cells (Hurbin et al. 2005). AR/IGF1 protects NSCLC cells from serum deprivation-induced apoptosis associated with increased phosphorylation of PKCδ at Thr-505, and rottlerin or siRNA mediated gene silencing of PKCδ abrogates this effect. Additionally, expression of constitutively activated PKCδ alone is able to inhibit serum deprivation-induced apoptosis (Hurbin et al. 2005), suggesting the involvement of PKCδ activation in this event. In contrast, proteolytic activation of PKCδ plays a small role in apoptosis triggered by DNA damage agent cisplatin in human H69 SCLC cells (Persaud et al. 2005). Perhaps, the difference in cellular contexts and the mode of PKCδ activation in response to different apoptotic signals may ultimately determine the specificity of PKCδ function.

Increased expression/activity of PKC has been linked to the intrinsic doxorubicin (DOX)-resistance in human NSCLC cells and in NSCLC patients (Ahmad et al. 1992; Volm and Pommerenke 1995). One potential mechanism that contributes to DOX-resistance in lung cancer involves PKCα-mediated modulation of the transport activity of the Ral-interacting protein (RLIP76) (Singhal et al. 2006). RLIP76 is a novel nonABC-transporter of DOX, which contributes to about two third of total DOX-transport activity (Awasthi et al. 2003a, b). PKCα has been shown to phosphorylate and stimulate the drug transport activity of RLIP76. Expression of a PKCα-phosphorylation deficient mutant of RLIP76 or siRNA-mediated gene silencing of PKCα in NSCLC cells reduces the DOX-transport activity of RLIP76 and enhances DOX-cytotoxicity to the level comparable to or greater than that in DOX-sensitive SCLC cells (Singhal et al. 2006), indicating that PKCα-mediated up-regulation of RLIP76 activity is a primary determinant of DOX-resistance in NSCLC cells.

19.3.4 Invasion and Metastasis

Accumulating evidence indicates that PKCs are involved in invasion and metastasis of human cancer. However, the underlying mechanisms, particularly the role of individual PKC isoforms in regulating these processes, remain largely undefined. Several recent studies indicate that atypical PKC isoforms may promote lung tumor invasion and metastasis through regulation of matrix metalloproteinase (MMP) activity and modulation of growth factor-mediated chemotaxis and integrin-mediated adhesion. In addition to its role in NSCLC survival, PKCι also functions in invasion. Overexpression of PKCι enhances, and inhibition of PKCι expression inhibits, migration and invasion of NSCLC cells in response to nicotine (Xu and Deng 2006). PKCι can directly phosphorylate μ- and m-calpains and is required for nicotine-mediated phosphorylation and suppression of calpain activity, which is associated with increased wound healing, migration, and invasion (Xu and Deng 2006). Additionally, PKCι can promote transformed growth and invasion of NSCLC cells through up-regulation of matrix metalloproteinase-10 (MMP-10) expression (Frederick et al. 2008). In NSCLC cells, the formation of a PKCι-Par6α-Rac1 complex is important for MMP-10 expression and invasion. Knockdown of MMP-10 expression blocks, and addition of recombinant MMP-10 restores, the PKCι-mediated anchorage-independent growth and invasion, indicating that MMP-10 is a critical effector of the PKCι oncogenic signaling pathway. Importantly, MMP-10 and PKCι are coordinately overexpressed in primary human NSCLC tumor cells, and MMP-10 expression is predictive of poor survival of NSCLC patients (Frederick et al. 2008).

Contrary to its tumor suppressor function demonstrated in oncogenic Ras-induced lung tumorigenesis (Galvez et al. 2009), PKCζ appears important for the chemotaxis of NSCLC cells in vitro. Liu et al. (2009) recently showed that PKCζ is involved in the regulation of epidermal growth factor (EGF)-induced chemotaxis of NSCLC cells. PKCζ is activated in response to EGF stimulation in a phosphatidylinositol 3 kinase (PI3K)/Akt-dependent manner. Specific inhibition of PKCζ, but not other PKC isoforms, blocks EGF-induced chemotataxis and cell adhesion to fibronectin accompanied with a reduction in actin polymerization. It is possible that the function of PKCζ may be cell-context dependent. Some early studies indicated that PMA-sensitive PKCs (cPKCs and nPKCs) may also contribute to invasion and metastasis through regulating integrin-mediated adhesion and production of proteinase in human lung cancer cells (Jakowlew et al. 1997; Quigley et al. 1998).

19.4 Therapeutics: Targeting PKC Isoforms

19.4.1 PKCα Inhibitor: Aprinocarsen (LY900003; ISIS3521)

Aprinocarsen is a 20-mer antisense oligonucleotide that specifically inhibits the transcription of PKCα (Dean et al. 1994). In preclinical studies, aprinocarsen has shown antitumor effects in a range of tumor cell lines and xenograft models with selective inhibition of PKCα mRNA and protein expression (Dean et al. 1996). Aprinocarsen was the first PKC isoform-specific agent that has gone on phase I–III studies in patients with advanced NSCLC (Lynch et al. 2003; Villalona-Calero et al. 2004; Ritch et al. 2006). However, in two randomized phase III studies, aprinocarsen failed to show additional survival benefit when applied in conjunction with chemotherapy regiments (Lynch et al. 2003; Paz-Ares et al. 2006). These disappointing results led to the early termination of the studies. A couple of issues may be related to the unsuccessful clinical development of aprinocarsen: there is no validated biomarker(s) for evaluating the effectiveness of aprinocarsen (e.g., it is unclear whether aprinocarsen is able to accumulate in tumor tissues); and patients were not screened for PKCα expression. PKCα expression is not significantly altered in NSCLC and <20% of NSCLC patients show high levels of PKCα (Lahn et al. 2004). It remains possible that a subset of patients with high levels of PKCα may be benefited from aprinocarsen therapy. This underscores the importance of selection of patients for the targeted therapy in clinical studies.

19.4.2 PKCβ Inhibitor: Enzastaurin (LY317615)

Enzastaurin is a selective PKCβ small molecular inhibitor that disrupts the intrinsic phosphotransferase activity of PKCβ. Enzastaurin was developed primarily as an antiangiogenic agent. At low concentrations it is relatively specific for PKCβ, but at higher concentrations it also inhibits other PKC isoforms (Teicher et al. 2002). Enzastaurin exhibits proapoptotic and antiproliferative activities in various cancer cell lines through the Akt pathway, suppressing the phosphorylation of glycogen synthase kinase-3β (GSK3β) (Graff et al 2005; Hanauske et al. 2007). Enzastaurin inhibits the growth of NSCLC and SCLC cell lines accompanied with the reduced phosphorylation of GSK3β (Nakajima et al. 2006). In animal models, enzastaurin shows antitumor and antiangiogenic activities in murine Lewis lung carcinoma and human Calu-6 NSCLC xenografts (Teicher et al. 2001). The combination of enzastaurin with chemotherapeutic drugs currently used in NSCLC therapy shows synergistic antiproliferative and proapoptotic effects in NSCLC cells (Morgillo et al. 2008; Tekle et al. 2008). However, the synergistic effect was only observed when chemotherapy was followed by treatment with enzastaurin (Morgillo et al. 2008), suggesting the sequence of administration of enzastaurin in a combination therapy is critical. Clinically, phase I studies of enzastaurin in NSCLC have shown encouraging results, and additional studies of a combination of enzastaurin with other anticancer agents for NSCLC are being planned (Herbst et al. 2007).

19.4.3 PKCι Inhibitors: Aurothiomalate (ATM)

The gold compound aurothioglucose (ATG) and the related compound aurothiomalate (ATM) were identified by a high throughput screen of a small molecule library as a potent inhibitor of the PB1 domain-mediated interaction between PKCι and Par6 (Stallings-Mann et al. 2006). By targeting the unique cysteine residue 69 (Cys69) of PKCι located at the binding interface between PKCι and Par6, ATM specifically inhibits the PB1 domain interaction involving PKCι but not other PB1-PB1 domain interaction, thereby leading to inhibition of PKCι-dependent oncogenic function (Erdogan et al. 2006; Fields and Regala 2007). Both ATG and ATM block the PKCι-mediated signaling to Rac1 and inhibit the transformed growth of NSCLC cells in vitro and tumor growth in nude mice (Stallings-Mann et al. 2006). In NSCLC cells, ATM sensitivity is not associated with general sensitivity of chemotherapeutic drugs including cisplatin, placitaxel, and gemcitabine, but correlates positively with expression of PKCι and Par6 (Regala et al. 2008). The antitumor effect of ATM in vivo is associated with inhibition of the MEK/ERK signaling and decreased cell proliferation without affecting tumor apoptosis and vascularization similar to that observed in A549/kdPKCι NSCLC xenografts (Regala et al. 2005b). A phase I clinical study of ATM in patients with NSCLC is currently underway.

19.5 Conclusion

Current studies have demonstrated that alterations in expression and function of specific PKC isoforms are associated with the development of human lung cancer, and the abnormal activation of PKC isoform-dependent signaling pathways can lead to transformed cell growth, dysregulation of cell cycle control machinery, and enhanced therapeutic resistance of lung cancer cells. The differential, sometimes overlapping, expression profiles of individual PKC isoforms in human NSCLC highlight the importance for understanding isoform-specific function and signaling as well as coordinated effects among different isoforms in the carcinogenic process of the lung. The modulation of tumor-specific PKC isoform function may be an attractive strategy for developing novel mechanism-based therapeutics against human lung cancer.

References

Ahmad, S., Trepel, J. B., Ohno, S., et al. (1992). Role of protein kinase C in the modulation of multidrug resistance: Expression of the atypical gamma isoform of protein kinase C does not confer increased resistance to doxorubicin. *Molecular Pharmacology, 42*, 1004–1009.

Akimoto, K., Mizuno, K., Osada, S., et al. (1994). A new member of the third class in the protein kinase C family, PKC lambda, expressed dominantly in an undifferentiated mouse embryonal carcinoma cell line and also in many tissues and cells. *The Journal of Biological Chemistry, 269*, 12677–12683.

Alberg, A. J., & Samet, J. M. (2003). Epidemiology of lung cancer. *Chest, 123*, 21S–49S.

Auberger, J., Loeffler-Ragg, J., Wurzer, W., & Hilbe, W. (2006). Targeted therapeutics in non-small cell lung cancer: Proven concepts and unfulfilled promises. *Current Cancer Drug Targets, 6*, 271–294.

Awasthi, S., Singhal, S. S., Sharma, R., et al. (2003). Transport of glutathione-conjugates and chemotherapeutic drugs by RLIP76 (RLIP76): A novel link between G-protein and tyrosine kinase signaling and drug resistance. *International Journal of Cancer, 106*, 635–646.

Awasthi, S., Singhal, S. S., Singhal, J., et al. (2003). Role of RLIP76 in lung cancer doxorubicin resistance II. Doxorubicin transport in lung cancer by RLIP76. *International Journal of Oncology, 22*, 713–720.

Bae, K.-M., Wang, H., Jiang, G., et al. (2007). Protein kinase Cε is overexpressed in primary human non-small cell lung cancers and functionally required for proliferation of non-small cell lung cancer cells in a p21/Cip1-dependent manner. *Cancer Research, 67*, 6053–6063.

Basu, A., & Sivaprasad, U. (2007). Protein kinase Cε makes the life and death decision. *Cellular Signalling, 19*, 1633–1642.

Baxter, G., Oto, E., Daniel-Issakani, S., & Strulovici, B. (1992). Constitutive presence of a catalytic fragment of protein kinase Cε in a small cell lung carcinoma cell line. *The Journal of Biological Chemistry, 267*, 1910–1917.

Black, J. D. (2000). Protein kinase C-mediated regulation of the cell cycle. *Frontiers Bioscience, 5*, d406–d423.

Blumberg, P. M., Jaken, S., Konig, B., et al. (1984). Mechanism of action of the phorbol ester tumor promoters: Specific receptors for lipophilic ligands. *Biochemical Pharmacology, 33*, 933–940.

Bruey, J. M., Ducasse, C., Bonniaud, P., et al. (2000). Hsp27 negatively regulates cell death by interacting with cytochrome c. *Nature Cell Biology, 2*, 645–652.

Cacace, A. M., Guadagno, S. N., Krauss, R. S., et al. (1993). The epsilon isoform of protein kinase C is an oncogene when overexpressed in rat fibroblasts. *Oncogene, 8*, 2095–2104.

Castagna, M., Takai, Y., Kaibuchi, K., et al. (1982). Direct activation of calcium-activated, phospholipids-dependent protein kinase by tumor-promoting phorbol esters. *The Journal of Biological Chemistry, 257*, 7847–7851.

Chen, M. G., Chen, M., Wang, H., et al. (2008). Alterations in the expression of protein kinase C isoforms in primary human non-small cell lung cancers [abstract #5223]. In *Abstracts of the Annual Meeting of American Association for Cancer Research, San Diego, 12–16, April 2008.*

Choi, S. H., Hyman, T., & Blumberg, P. M. (2006). Differential effect of bryostatin 1 and phorbol 12-myristate 13-acetate on HOP-92 cell proliferation is mediated by down-regulation of protein kinase Cdelta. *Cancer Research, 66*, 7261–7269.

Clark, A. S., West, K. A., Blumberg, P. M., & Dennis, P. A. (2003). Altered protein kinase C (PKC) isoforms in non-small cell lung cancer cells: PKCdelta promotes cellular survival and chemotherapeutic resistance. *Cancer Research, 63*, 780–786.

Davies, S. P., Reddy, H., Caivano, M., & Cohen, P. (2000). Specificity and mechanism of action of some commonly used protein kinase inhibitors. *Biochemical Journal, 351*, 95–105.

Dean, N. M., McKay, R., Condon, T. P., et al. (1994). Inhibition of protein kinase C-α expression in humans A549 cells by antisense oligonucleotides inhibits induction of intracellular adhesion molecule 1 (ICAM-1) mRNA by phorbol esters. *The Journal of Biological Chemistry, 269*, 16416 16424.

Dean, N. M., McKay, R., Miraglia, L., et al. (1996). Inhibition of growth of human tumor cell lines in nude mice by an antisense oligonucleotide inhibitor of protein kinase C α expression. *Cancer Research, 56*, 3499–3507.

Deeds, L., Teodorescu, S., Chu, M., et al. (2003). A p53-independent G1 cell cycle checkpoint induced by the suppression of protein kinase C alpha and theta isoforms. *The Journal of Biological Chemistry, 278*, 39782–39793.

Dempsey, E. C., Newton, A. C., Mochly-Rosen, D., et al. (2000). Protein kinase C isozymes and the regulation of diverse cell responses. *American Journal of Physiology. Lung Cellular and Molecular Physiology, 279*, L429–L438.

Ding, L., Wang, H., Lang, W., & Xiao, L. (2002). Protein kinase C-ε promotes survival of lung cancer cells by suppressing apoptosis through dysregulation of the mitochondrial caspase pathway. *The Journal of Biological Chemistry, 277*, 35305–35313.

Duran, A., Diaz-Meco, M. T., & Moscat, J. (2003). Essential role of RelA Ser311 phosphorylation by zetaPKC in NF-kappaB transcriptional activation. *EMBO Journal, 22*, 3910–3918.

Erdogan, E., Lamark, T., Stallings-Mann, M., et al. (2006). Aurothiomalate inhibits transformed growth by targeting the PB1 domain of protein kinase Cι. *The Journal of Biological Chemistry, 281*, 28450–28459.

Evan, G. I., & Vousden, K. H. (2001). Proliferation, cell cycle and apoptosis in cancer. *Nature, 411*, 342–348.

Facchini, F. M., & Spiro, S. G. (1999). Chemotherapy in small-cell lung cancer. In C. Brambilla & E. Brambilla (Eds.), *Lung Biology in Health and Disease: Lung Tumors* (Vol. 124, pp. 611–630). New York: Marcel Dekker.

Felber, M., Sonnemann, J., & Beck, J. F. (2007). Inhibition of novel protein kinase C-epsilon augments TRAIL-induced cell death in A549 lung cancer cells. *Pathology Oncology Research, 13*, 295–301.

Fields, A. P., & Regala, R. P. (2007). Protein kinase Cι: Human oncogenes, prognostic marker and therapeutic target. *Pharmacological Research, 55*, 487–497.

Frederick, L. A., Matthews, J. A., Jamieson, L., et al. (2008). Matrix metalloproteinase-10 is a critical effector of protein kinase Cι-Par6α-mediated lung cancer. *Oncogene, 27*, 4841–4853.

Galvez, A. S., Duran, A., Linares, J. F., Pathrose, P., Castilla, E. A., Abu-Baker, S., et al. (2009). PKCzeta represses the IL-6 promoter and impairs tumorigenesis in vivo. *Molecular and Cellular Biology, 29*, 104–115.

Graff, J. R., McNulty, A. M., Hanna, K. R., et al. (2005). The protein kinase Cβ-selective inhibitor, enzastaurin (LY317615.HCl), suppresses signaling through the AKT pathway, induces apoptosis, and suppresses growth of human colon cancer and glioblastoma xenografts. *Cancer Research, 65*, 7462–7469.

Griner, E. M., & Kazanietz, M. G. (2007). Protein kinase C and other diacylglycerol effectors in cancer. *Nature Reviews. Cancer, 7*, 281–294.

Hanahan, D., & Weinberg, R. A. (2000). The hallmarks of cancer. *Cell, 100*, 57–70.

Hanauske, A., Oberscchmidt, O., Hanauske-Abel, H., et al. (2007). Antitumor activity of enzastaurin (LY317615.HCl) against human cancer cell lines and freshly explanted tumors investigated in vitro soft-agar cloning experiments. *Investigational New Drugs, 25*, 205–210.

Hannun, Y. A. (1997). Apoptosis and the dilemma of cancer chemotherapy. *Blood, 89*, 1845–1853.

Hecht, S. S. (1999). Tobacco smoke carcinogens and lung cancer. *Journal of the National Cancer Institute, 91*, 1194–1210.

Herbst, R. S., Oh, Y., Wagle, A., et al. (2007). Enzastaurin, a protein kinase Cβ-selective inhibitor, and its potential application as an anticancer agent in lung cancer. *Clinical Cancer Research, 13*, 4641s–4646s.

Heusch, W. L., & Maneckjee, R. (1998). Signalling pathways involved in nicotine regulation of apoptosis of human lung cancer cells. *Carcinogenesis, 19*, 551–556.

Hurbin, A., Coll, J. L., Dubrez-Daloz, L., et al. (2005). Cooperation of amphiregulin and insulin-like growth factor-1 inhibits Bax- and Bad-mediated apoptosis via a protein kinase C-dependent pathway in non-small cell lung cancer cells. *The Journal of Biological Chemistry, 280*, 19757–19767.

Jaken, S., & Parker, P. J. (2000). Protein kinase C binding partners. *BioEssays, 22*, 245–254.

Jakowlew, S. B., Mariano, J. M., & You, L. (1997). Effects of transforming growth factor-β 1 and phorbol ester on PAI-1 and PA gene in human lung cancer cell. *Growth Factors, 14*, 177–197.

Jemal, A., Siegel, R., Ward, E., et al. (2008). Cancer Statistics, 2008. *CA: A Cancer Journal for Clinicians, 58*, 71–96.

Jin, Z., Gao, F., Flagg, T., & Deng, X. (2004). Tobacco-specific nitrosamine NNK promotes functional cooperation of Bcl2 and c-Myc through phosphorylation in regulating cell survival and proliferation. *The Journal of Biological Chemistry, 279*, 40209–40219.

Jin, Z., Xin, M., & Deng, X. (2005). Survival function of PKCι as a novel nitrosamine 4-(methylnitrosamino)-1-(3-pyridyl)-1-butanone-activated Bad kinase. *The Journal of Biological Chemistry, 280*, 16045–16052.

Joberty, G., Petersen, C., Gao, L., & Macara, I. G. (2000). The cell-polarity protein Par6 links Par3 and atypical protein kinase C to Cdc42. *Nature Cell Biology, 2*, 531–539.

Josen, K., Siegel, R., & Kamp, D. (2002). Incidence and etiology. In A. B. Weitberg (Ed.), *Cancer of the Lung* (pp. 3–34). Totowa: Humana Press.

Kim, E.-H., Lee, H.-J., Lee, D.-H., et al. (2007). Inhibition of heat shock protein 27-mediated resistance to DNA damaging agents by a novel PKCδ-V5 heptapeptide. *Cancer Research, 67*, 6333–6341.

Kovac, J., Oster, H., & Leitges, M. (2007). Expression of the atypical protein kinase C (aPKC) isoforms iota/lambda and zeta during mouse embryogenesis. *Gene Expression Patterns, 7*, 187–196.

Lahn, M., McClelland, P., Ballard, D., et al. (2006). Immunohistochemical detection of protein kinase C-beta (PKC-β) in tumour specimens of patients with non-small cell lung cancer. *Histopathology, 49*, 426–445.

Lahn, M., Su, C., Li, S., et al. (2004). Expression levels of protein kinase C-alpha in non-small-cell lung cancer. *Clinical Lung Cancer, 6*, 184–189.

Lee, Y.-J., Lee, D.-H., Cho, C.-K., et al. (2005). HSP25 inhibits protein kinase Cδ-mediated cell death through direct interaction. *The Journal of Biological Chemistry, 280*, 18108–18119.

Lin, D., Edwards, A. S., Fawcett, J. P., et al. (2000). A mammalian PAR-3-PAR-6 complex implicated in Cdc42/Rac1 and aPKC signaling and cell polarity. *Nature Cell Biology, 2*, 540–547.

Liu, Y., Wang, B., Wang, J., et al. (2009). Down-regulation of PKCζ expression inhibits chemotaxis signal transduction in human lung cancer cells. *Lung Cancer, 63*, 210–218.

Lynch, T. J., Raju, R., Lind, M., et al. (2003). Randomized phase III trial of chemotherapy and antisense oligonucleotide LY900003 (ISIS 3521) in patients with advanced NSCLC: Initial report. *Proceedings of the American Society of Clinical Oncology, 22*, 623 [abstract 2504].

Mai, H., May, W. S., Gao, F., et al. (2003). A functional role for nicotine in Bcl2 phosphorylation and suppression of apoptosis. *The Journal of Biological Chemistry, 278*, 1886–1891.

Mao, M., Fang, X., Lu, Y., et al. (2000). Inhibition of growth-factor-induced phosphorylation and activation of protein kinase B/Akt by atypical protein kinase C in breast cancer cells. *Biochemical Journal, 352*, 475–482.

Mills, N. E., Fishman, C. L., Scholes, J., et al. (1995). Detection of K-ras oncogene mutations in bronchoalveolar lavage fluid for lung cancer diagnosis. *Journal of the National Cancer Institute, 87*, 1056–1060.

Minna, J. D., Roth, J. A., & Gazdar, A. F. (2002). Focus on lung cancer. *Cancer Cell, 1*, 49–52.

Mischak, H., Goodnight, J., Kolch, W., et al. (1999). Overexpression of protein kinase C-delta and -epsilon in NIH 3T3 cells induces opposite effects on growth, morphology, anchorage dependence, and tumorigenicity. *The Journal of Biological Chemistry, 268*, 6090–6096.

Morgillo, F., Martinelli, E., Troiani, T., et al. (2008). Sequence-dependent, synergistic antiproliferative and proapoptotic effects of the combination of cytotoxic drugs and enzastaurin, a protein kinase Cβ inhibitor in non-small cell lung cancer cells. *Molecular Cancer Therapeutics, 7*, 1698–1707.

Mustafi, R., Cerda, S., Chumsangsri, A., et al. (2006). Protein kinase-zeta inhibits collagen I-dependent and anchorage-independent growth and enhances apoptosis of human Caco-2 cells. *Molecular Cancer Research, 4*, 683–694.

Nakagawa, M., Oliva, J. L., Kothapalli, D., et al. (2005). Phorbol ester-induced G1 phase arrest selectively mediated by protein kinase Cdelta-dependent induction of p21. *The Journal of Biological Chemistry, 280*, 33926–33934.

Nakajima, E., Helfrich, B., Chan, D., et al. (2006). Enzastaurin, a protein kinase Cβ-selective inhibitor, inhibits the growth of SCLC and NSCLC cell lines [abstract 13138]. ASCO Annual Meeting Proceedings Part I. *Journal of Clinical Oncology, 24*, 13138.

Newton, A. C. (2001). Protein kinase C: Structural and spatial regulation by phosphorylation, cofactors, and macromolecular interactions. *Chemical Reviews, 101*, 2353–2364.

Newton, A. C. (2003). Regulation of the ABC kinases by phosphorylation: Protein kinase C as a paradigm. *Biochemical Journal, 370*, 361–371.

Nishizuka, Y. (1995). Protein kinase C and lipid signaling for sustained cellular responses. *FASEB Journal, 9*, 484–496.

Oliva, J. L., Caino, M. C., Senderowicz, A. M., & Kazanietz, M. G. (2008). S-Phase-specific activation of PKC alpha induces senescence in non-small cell lung cancer cells. *The Journal of Biological Chemistry, 283*, 5466–5476.

Osada, H., & Takahashi, T. (2002). Genetic alterations of multiple tumor suppressors and oncogenes in the carcinogenesis and progression of lung cancer. *Oncogene, 21*, 7421–7434.

Pardo, O. E., Arcaro, A., Salerno, G., et al. (2002). Fibroblast growth factor-2 induces translational regulation Bcl-XL and Bcl-2 via a MEK-dependent pathway: Correlation with resistance to etoposide-induced apoptosis. *The Journal of Biological Chemistry, 277*, 12040–12046.

Pardo, O. E., Lesay, A., Arcaro, A., et al. (2003). Fibroblast growth factor 2-mediated translational control of IAPs blocks mitochondrial release of Smac/DIABLO and apoptosis in small cell lung cancer cells. *Molecular and Cellular Biology, 23*, 7600–7610.

Pardo, O. E., Wellbrock, C., Khanzada, U. K., et al. (2006). EGF-2 protects small cell lung cancer cells from apoptosis through a complex involving PKCε B-Raf and S6K2. *EMBO Journal, 25*, 3078–3088.

Paz-Ares, L., Douillard, J. Y., Koralewski, P., et al. (2006). Phase III study of gemcitabine and cisplatin with or without aprinocarsen, a protein kinase C-α antisense oligonucleotide, in patients with advanced stage non-small cell lung cancer. *Journal of Clinical Oncology, 24*, 1428–1434.

Perletti, G., Folini, M., Lin, H. C., et al. (1996). Overexpression of protein kinase C epsilon is oncogenic in rat colonic epithelial cells. *Oncogene, 12*, 847–854.

Persaud, S. D., Hoang, V., Huang, J., & Basu, A. (2005). Involvement of proteolytic activation of PKCδ in cisplatin-induced apoptosis in human small cell lung cancer H69 cells. *International Journal of Oncology, 27*, 149–154.

Quigley, R. L., Shafer, S. H., & Williams, C. L. (1998). Regulation of integrin-mediated adhesion by muscarinic acetylcholine receptors and protein kinase C in small cell lung carcinoma. *Chest, 114*, 839–846.

Rane, M. J., Pan, Y., Singh, S., et al. (2003). Heat shock protein 27 controls apoptosis by regulating Akt activation. *The Journal of Biological Chemistry, 278*, 27828–27835.

Rathore, R., & Weitberg, A. B. (2002). Small-cell lung cancer. In A. B. Weitberg (Ed.), *Cancer of the Lung* (pp. 3–34). Totowa: Humana Press.

Regala, R. P., Thompson, E. S., & Fields, A. P. (2008). Atypical protein kinase C iota expression and aurothiomalate sensitivity in human lung cancer cells. *Cancer Research, 68*, 5888–5895.

Regala, R. P., Weems, C., Jamieson, L., et al. (2005a). Atypical protein kinase C iota is an oncogene in human non-small cell lung cancer. *Cancer Research, 65*, 8905–8911.

Regala, R. P., Weems, C., Jamieson, L., et al. (2005b). Atypical protein kinase Ciota plays a critical role in human lung cancer cell growth and tumorigenicity. *The Journal of Biological Chemistry, 280*, 31109–31115.

Ritch, P., Rudin, C. M., Bitran, J. D., et al. (2006). Phase II study of PKC-alpha antisense oligonucleotide aprinocarsen in combination with gemcitabine and carboplatin in patients with advanced non-small cell lung cancer. *Lung Cancer, 52*, 173–180.

Ruotsalainen, T., Joensuu, H., Mattson, K., & Salven, P. (2002). High pretreatment serum concentration of basic fibroblast growth factor is a predictor of poor prognosis in small cell lung cancer. *Cancer Epidemiology, Biomarkers and Prevention, 11*, 1492–1495.

Schuller, H. M. (1994). Carbon dioxide potentiates the mitogenic effects of nicotine and its carcinogenic derivative, NNK, in normal and neoplastic neuroendocrine lung cells via stimulation of autocrine and protein kinase C-dependent mitogenic pathways. *Neurotoxicology, 15*, 877–886.

Sekido, Y., Fong, K. M., & Minna, J. D. (2003). Molecular genetics of lung cancer. *Annual Review of Medicine, 54*, 73–87.

Seufferlein, T., & Rozengurt, E. (1996). Galanin, neurotensin, and phorbol esters rapidly stimulate activation of mitogen-activated protein kinase in small cell lung cancer cells. *Cancer Research, 56*, 5758–5764.

Singhal, S. S., Wickramarachchi, D., Singhal, J., et al. (2006). Determinants of differential doxorubicin sensitivity between SCLC and NSCLC. *FEBS Letters, 580*, 2258–2264.

Slebos, R. J., Kibbelaar, R. E., Dalesio, O., et al. (1990). K-ras oncogene activation as a prognostic marker in adenocarcinoma of the lung. *The New England Journal of Medicine, 323*, 561–565.

Soh, J. W., & Weinstein, I. B. (2003). Roles of specific isoforms of protein kinase C in the transcriptional control of cyclin D1 and related genes. *The Journal of Biological Chemistry, 278*, 34709–34716.

Soloff, R. S., Katayama, C., Lin, M. Y., et al. (2004). Targeted deletion of protein kinase C lambda reveals a distribution of functions between the two atypical protein kinase C isoforms. *Journal of Immunology, 173*, 3250–3260.

Stallings-Mann, M., Jamieson, L., Regala, R. P., et al. (2006). A novel small-molecular inhibitor of protein kinase Cι blocks transformed growth of non-small cell lung cancer cells. *Cancer Research, 66*, 1767–1774.

Tapinos, N., & Rambukkana, A. (2005). Insights into regulation of human Schwann cell proliferation by Erk1/2 via a MEK-independent and p56Lck-dependent pathway from leprosy bacilli. *Proceedings of the National Academy of Sciences of the United States of America, 102*, 9188–9193.

Teicher, B. A., Alvarez, E., Menon, K., et al. (2002). Antiangiogenic effects of a protein kinase Cβ-selective small molecule. *Cancer Chemotherapy and Pharmacology, 49*, 69–77.

Teicher, B. A., Menon, K., Alvarez, E., et al. (2001). Antiangiogenic and antitumor effects of a protein kinase Cbeta inhibitor in murine lewis lung carcinoma and human Calu-6 non-small-cell lung carcinoma xenografts. *Cancer Chemotherapy and Pharmacology, 48*, 473–480.

Tekle, C., Giovannetti, E., Sigmond, J., et al. (2008). Molecular pathways involved in the synergistic interaction of the PKC beta inhibitor enzastaurin with the antifolate pemetrexed in non-small cell lung cancer cells. *British Journal of Cancer, 99*, 750–759.

Vaux, D. L., & Korsmeyer, S. J. (1999). Cell death in development. *Cell, 96*, 245–254.

Villalona-Calero, M. A., Ritch, P., Figueroa, J. A., et al. (2004). A phase I/II study of LY900003, an antisense inhibitor of protein kinase C-alpha, in combination with cisplatin and gemcitabine in patients with advanced non-small cell lung cancer. *Clinical Cancer Research, 10*, 6086–6093.

Volm, M., & Pommerenke, E. W. (1995). Associated expression of protein kinase C with resistance to doxorubicin in human lung cancer. *Anticancer Research, 15*, 463–466.

Wang, X. Y., Repasky, E., & Liu, H. T. (1999). Antisense inhibition of protein kinase Calpha reverses the transformed phenotype in human lung carcinoma cells. *Experimental Cell Research, 250*, 253–263.

Way, K. J., Chou, E., & King, G. L. (2000). Identification of PKC-isoform-specific biological actions using pharmacological approaches. *Trends Pharmacol, 21*, 181–187.

Way, D. K., Posekany, K., deVente, J., et al. (1994). Overexpression of protein kinase C-zeta stimulates leukemic cell differentiation. *Cell Growth and Differentiation, 5*, 1195–1203.

Webb, B. L., Lindsay, M. A., Seybold, J., et al. (1997). Identification of the protein kinase C isoenzymes in human lung and airways smooth muscle at the protein and mRNA level. *Biochemical Pharmacology, 54*, 199–205.

Weitberg, A. B. (2002). *Cancer of the lung*. Totowa: Humana Press.

Xiao, L. (2006). Lung cancer. In M. S. Runge & C. Patterson (Eds.), *Principles in Molecular Medicine* (2nd ed., pp. 736–742). Totowa: Humana Press.

Xin, M., Gao, F., May, W. S., Flagg, T., & Deng, X. (2007). Protein kinase Czeta abrogates the proapoptotic function of Bax through phorylation. *The Journal of Biological Chemistry, 282*, 21268–21277.

Xu, L., & Deng, X. (2006). Protein kinase Cι promotes nicotine-induced migration and invasion of cancer cells via phosphorylation of μ- and m-calpains. *The Journal of Biological Chemistry, 281*, 4457–4466.

Yan, S., & Wenner, C. E. (2001). Modulation of cyclin D1 and its signaling components by the phorbol ester TPA and the tyrosine phosphatase inhibitor vanadate. *Journal of Cellular Physiology, 186*, 338–349.

Part IV
PKC Isozymes as Targets for Cancer Therapy

Chapter 20
Introduction

Patricia S. Lorenzo

Keywords Protein kinase C inhibitors • Cancer therapy • Clinical trials

The first association of PKC with cancer was its identification in the 1980s as the receptor protein for a group of phorbol-12, 13-diesters isolated from croton oil, a seed-derived oil from the plant *Croton tiglium*, which was shown to have potent tumor-promoting effects on mouse skin carcinogenesis models (Hecker 1968). It was later revealed that the phorbol esters act as ultrapotent structural analogs of the second messenger diacylglycerol (DAG), the physiological activator of classical and novel PKC isozymes (Castagna et al. 1982; Leach et al. 1983). In subsequent years, the discovery that phorbol esters targeted PKC resulted in multiples studies to investigate what aspects of tumorigenesis in skin were modulated by the different PKC isozymes, and also led to the analysis of PKC in other malignancies, as discussed in the previous chapter. The initial findings about phorbol esters also promoted the exploration for novel compounds with the ability to modulate PKC activity, not only as biological tools but also as potential pharmacological agents in cancer treatment. This chapter focuses on those aspects related to PKC as a target for cancer therapy.

Intervention approaches to modulate PKC in cancer envision not only the control of cell growth and metastasis but also the reversion of the resistance of malignant cells to die when exposed to standard chemotherapeutic treatment. The role of PKC in neoplastic drug resistance has been known for quite some time, and the two main mechanisms involved are inhibition of apoptosis (Cheng et al. 1998; Clark et al. 2003; Ruvolo et al. 1998) and regulation of multidrug-resistance associated transporters like P-glycoprotein (Gollapudi et al. 1992; Yu et al. 1991). One should note that some PKC isozymes have also been reported to confer sensitivity to chemotherapy (Masanek et al. 2002). Given the complexity of the PKC family, it is not surprising to find opposite roles for PKC isozymes depending on the tissue and malignancy.

P.S. Lorenzo (✉)
Natural Products and Cancer Biology Program, Cancer Research Center of Hawaii, University of Hawaii at Manoa, Honolulu, HI 96813, USA
e-mail: plorenzo@crch.hawaii.edu

M.G. Kazanietz (ed.), *Protein Kinase C in Cancer Signaling and Therapy*, Current Cancer Research, DOI 10.1007/978-1-60761-543-9_20,

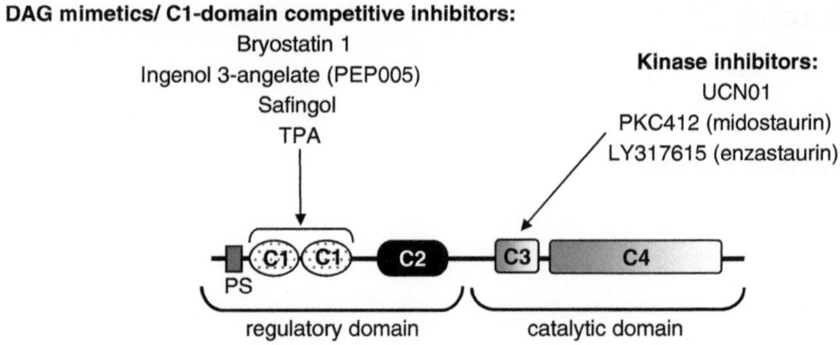

Fig. 20.1 PKC domains targeted for cancer therapy. Linear representation of a classical PKC isoform, indicating the two main domains and the subdomains targeted by some of the PKC inhibitors tested in cancer clinical trials. *PS* pseudosubtrate site, *C1* diacylglycerol (DAG)-binding domain, *C2* calcium-binding domain, *C3* ATP-binding domain, *C4* PKC-substrate binding domain

The two major targeting areas for modulation of PKC activity are the regulatory domain and the catalytic domain (Fig. 20.1). If a compound with structural similarities to DAG binds to the regulatory domain of PKC (classical and novel isozymes), it can lead to PKC inhibition or, alternatively, induce unusual PKC activation with a final biological effect very different from that triggered by DAG. On the other hand, if the catalytic site responsible for the kinase activity of the enzyme is blocked, PKC will not be able to execute its function. A different approach has been developed for modulation of the atypical PKC iota – a putative oncogene in lung cancer. It is based on blocking the ability of this isozyme to bind to effector molecules via its PB1 domain. At present, the gold compound aurothiomalate has shown promising effects by interfering with PKC iota-Par6 interaction involved in the Rac1-Pak-Mek 1/2-Erk 1/2 signaling pathway in nonsmall cell lung cancer cells (Stallings-Mann et al. 2006).

Natural products have been one of the major resources for lead compounds in the development of PKC modulators. A classic example is bryostatin 1, a macrocyclic lactone derived from the marine sponge *Bugula neritina* (Pettit et al. 1982). Bryostatin 1 binds to the regulatory domain responsible for DAG recognition – the C1 domain – with very strong affinity, and induces PKC activation and subcellular redistribution. However, the biological effects of bryostatin 1 can be very different from those of DAG and of phorbol esters, including antitumor effects in mouse models of skin carcinogenesis as well as antiproliferative actions on leukemia-derived cells (Hennings et al. 1987; Szallasi et al. 1994; Kraft et al. 1989). While bryostatin 1 has failed in most of the clinical trials as a single anticancer agent, it still shows promising activity as an adjuvant of various chemotherapeutic drugs. One major barrier in the use of bryostatin 1 in cancer treatment is its production; synthesis is very challenging and the main resource of bryostatin 1 at the present time is aquaculture. More recently, bryostatin analogs that are easier to synthesize than bryostatin 1 have been developed (Wender et al. 1998) and are currently explored in preclinical studies.

The first compound described as a PKC inhibitor targeting the catalytic domain was also a natural product, staurosporine, isolated from the microorganism *Streptomyces staurosporeus* (Tamaoki and Nakano 1990). It acts as a competitive inhibitor of ATP on the C3 domain of PKC, although it is now known to have no specificity for PKC as it is able to bind to the ATP-site of many other kinases. The staurosporine analogs 7-hydroxystaurosporine (UCN-01) and 4′-*N*-benzoyl staurosporine (PKC412, also known as CGP 41251 and midostaurin) are also potent PKC inhibitors, albeit nonspecific (Senderowicz 2000; Sato et al. 2002; Fabbro et al. 2000). Nevertheless, these compounds are currently being tested in clinical trials against cancer since they inhibit kinases such as CDKs, PDK1, and KIT, which are also involved in proliferative and survival pathways in cancer cells.

Pharmacological approaches other than small-molecule compounds to target PKC have also been explored in recent years. One of them is an antisense oligonucleotide against PKC alpha (ISIS 3521). Unfortunately, clinical trials performed so far on different solid tumors and hematological malignancies have shown no therapeutic benefits of this approach.

One of the challenges in the development of novel PKC-anticancer therapies is to obtain PKC inhibitors that do not cross-react with other kinases and that display isozyme selectivity. The acyclic bisindolylmaleimide enzastaurin (LY317615.HCl) is an example of that class of inhibitors. It blocks the activity of PKC at the ATP-binding site, with selectivity towards the PKC beta isozyme (Faul et al. 2003). One of the first activities reported for enzastaurin was antiangiogenesis, which is not surprising given the role of PKC in VEGF signaling (Keyes et al. 2004; Graff et al. 2005). In addition, it exerts antitumor activity in vitro on various cancer cell lines (Hanauske et al. 2007). Ongoing trials are evaluating the potential therapeutic use of this drug against several solid and hematological malignancies, such as glioblastoma, lung cancer, and lymphomas.

It is important to consider that many of the observations about the role of PKC in cancer and, therefore, its potential value as chemotherapeutic target have come from in vitro studies on cell lines and from animal models. In the consideration of the use of some of the PKC inhibitors in clinical trials, it is imperative that patients enrolled are evaluated in terms of PKC isozyme expression and/or activity (Lorenzo and Dennis 2003). That way, the usefulness of a given PKC modulator in a particular cancer will be better judged, for the benefit of the trial and, primarily, of the cancer patient.

References

Castagna, M., Takai, Y., Kaibuchi, K., Sano, K., Kikkawa, U., & Nishizuka, Y. (1982). Direct activation of calcium-activated, phospholipid-dependent protein kinase by tumor-promoting phorbol esters. *Journal of Biological Chemistry*, *257*, 7847–7851.

Cheng, A. L., Chuang, S. E., Fine, R. L., Yeh, K. H., Liao, C. M., Lay, J. D., et al. (1998). Inhibition of the membrane translocation and activation of protein kinase C, and potentiation of doxorubicin-induced apoptosis of hepatocellular carcinoma cells by tamoxifen. *Biochemical Pharmacology*, *55*, 523–531.

Clark, A. S., West, K. A., Blumberg, P. M., & Dennis, P. A. (2003). Altered protein kinase C (PKC) isoforms in non-small cell lung cancer cells: PKCdelta promotes cellular survival and chemotherapeutic resistance. *Cancer Research*, *63*, 780–786.

Fabbro, D., Ruetz, S., Bodis, S., Pruschy, M., Csermak, K., Man, A., et al. (2000). PKC412–a protein kinase inhibitor with a broad therapeutic potential. *Anti-cancer Drug Design*, *15*, 17–28.

Faul, M. M., Gillig, J. R., Jirousek, M. R., Ballas, L. M., Schotten, T., Kahl, A., et al. (2003). Acyclic N-(azacycloalkyl)bisindolylmaleimides: isozyme selective inhibitors of PKCbeta. *Bioorganic and Medicinal Chemistry Letters*, *13*, 1857–1859.

Gollapudi, S., Patel, K., Jain, V., & Gupta, S. (1992). Protein kinase C isoforms in multidrug resistant P388/ADR cells: a possible role in daunorubicin transport. *Cancer Letters*, *62*, 69–75.

Graff, J. R., McNulty, A. M., Hanna, K. R., Konicek, B. W., Lynch, R. L., Bailey, S. N., et al. (2005). The Protein Kinase C{beta}-Selective Inhibitor, Enzastaurin (LY317615.HCl), Suppresses Signaling through the AKT Pathway, Induces Apoptosis, and Suppresses Growth of Human Colon Cancer and Glioblastoma Xenografts. *Cancer Research*, *65*, 7462–7469.

Hanauske, A. R., Oberschmidt, O., Hanauske-Abel, H., Lahn, M. M., & Eismann, U. (2007). Antitumor activity of enzastaurin (LY317615.HCl) against human cancer cell lines and freshly explanted tumors investigated in in-vitro [corrected] soft-agar cloning experiments. *Investigational New Drugs*, *25*, 205–210.

Hecker, E. (1968). Cocarcinogenic principles from the seed oil of Croton tiglium and from other Euphorbiaceae. *Cancer Research*, *28*, 2338–2349.

Hennings, H., Blumberg, P. M., Pettit, G. R., Herald, C. L., Shores, R., & Yuspa, S. H. (1987). Bryostatin 1, an activator of protein kinase C, inhibits tumor promotion by phorbol esters in SENCAR mouse skin. *Carcinogenesis*, *8*, 1343–1346.

Keyes, K. A., Mann, L., Sherman, M., Galbreath, E., Schirtzinger, L., Ballard, D., et al. (2004). LY317615 decreases plasma VEGF levels in human tumor xenograft-bearing mice. *Cancer Chemotherapy and Pharmacology*, *53*, 133–140.

Kraft, A. S., William, F., Pettit, G. R., & Lilly, M. B. (1989). Varied differentiation responses of human leukemias to bryostatin 1. *Cancer Research*, *49*, 1287–1293.

Leach, K. L., James, M. L., & Blumberg, P. M. (1983). Characterization of a specific phorbol ester aporeceptor in mouse brain cytosol. *Proceedings of the National Academy of Sciences USA*, *80*, 4208–4212.

Lorenzo, P. S., & Dennis, P. A. (2003). Modulating protein kinase C (PKC) to increase the efficacy of chemotherapy: stepping into darkness. *Drug Resistance Updates*, *6*, 329–339.

Masanek, U., Stammler, G., & Volm, M. (2002). Modulation of multidrug resistance in human ovarian cancer cell lines by inhibition of P-glycoprotein 170 and PKC isoenzymes with antisense oligonucleotides. *Jounal of Experimental Therapeutics and Oncology*, *2*, 37–41.

Pettit, G., Herald, C., Doubek, D., Herald, D., Arnold, E., & Clardy, J. (1982). Isolation and structure of bryostatin 1. *Journal of the American Chemical Society*, *104*, 6846–6848.

Ruvolo, P. P., Deng, X., Carr, B. K., & May, W. S. (1998). A functional role for mitochondrial protein kinase Calpha in Bcl2 phosphorylation and suppression of apoptosis. *Journal of Biological Chemistry*, *273*, 25436–25442.

Sato, S., Fujita, N., & Tsuruo, T. (2002). Interference with PDK1-Akt survival signaling pathway by UCN-01 (7-hydroxystaurosporine) *Oncogene*, *21*, 1727–1738.

Senderowicz, A. M. (2000) Small molecule modulators of cyclin-dependent kinases for cancer therapy. *Oncogene*, *19*, 6600–6606.

Stallings-Mann, M., Jamieson, L., Regala, R. P., Weems, C., Murray, N. R., & Fields, A. P. (2006). A novel small-molecule inhibitor of protein kinase Ciota blocks transformed growth of non-small-cell lung cancer cells. *Cancer Research*, *66*, 1767–1774.

Szallasi, Z., Smith, C. B., Pettit, G. R., & Blumberg, P. M. (1994). Differential regulation of protein kinase C isozymes by bryostatin 1 and phorbol 12-myristate 13-acetate in NIH 3T3 fibroblasts. *Journal of Biological Chemistry*, *269*, 2118–2124.

Tamaoki, T., & Nakano, H. (1990). Potent and specific inhibitors of protein kinase C of microbial origin. *Biotechnology (NY)*, *8*, 732–735.

Wender, P. A., DeBrabander, J., Harran, P. G., Jimenez, J. M., Koehler, M. F., Lippa, B., et al. (1998). The design, computer modeling, solution structure, and biological evaluation of synthetic analogs of bryostatin 1. *Proceedings of the National Academy of Sciences USA*, *95*, 6624–6629.

Yu, G., Ahmad, S., Aquino, A., Fairchild, C. R., Trepel, J. B., Ohno, S., et al. (1991). Transfection with protein kinase C alpha confers increased multidrug resistance to MCF-7 cells expressing P-glycoprotein. *Cancer Communications*, *3*, 181–189.

Chapter 21
PKC and Resistance to Chemotherapeutic Agents

Alakananda Basu

Abstract Despite considerable efforts that have been invested in identifying novel therapeutic targets for the treatment of cancer, conventional chemotherapeutic drugs continue to be the major treatment option for cancer patients. However, intrinsic and acquired resistance to these antineoplastic drugs is the major cause of therapy failure. Understanding the molecular basis of chemoresistance is critical to manage this disease successfully. The mechanism(s) of chemoresistance is(are) often multifactorial. The protein kinase C (PKC) family plays an important role in regulating cell proliferation and cell death. Numerous studies have implicated members of the PKC family as contributors of chemoresistance. Thus, the PKC signaling pathway could be exploited to overcome chemoresistance. The objective of this chapter is to provide a comprehensive review of literature on the involvement of PKC in classical multiple drug resistance (MDR), cisplatin resistance, and resistance to apoptosis, which may affect the sensitivity of tumor cells to numerous anticancer drugs.

Keywords Protein kinase C • Multiple drug resistance • Chemotherapeutic agents • Cisplatin • Cell Survival

21.1 Introduction

The ultimate goal in cancer chemotherapy is the selective eradication of malignant cells. Despite significant advancement in our understanding of the molecular basis of cancer, the primary treatment option continues to be conventional chemotherapeutic agents. However, the single major cause of therapy failure is resistance to these anticancer drugs. Most chemotherapeutic drugs affect cell division and target

A. Basu (✉)
Department of Molecular Biology and Immunology, University of North Texas Health Science Center, 3500 Camp Bowie Boulevard, Fort Worth, TX 76107, USA
e-mail: abasu@hsc.unt.edu

M.G. Kazanietz (ed.), *Protein Kinase C in Cancer Signaling and Therapy*, Current Cancer Research, DOI 10.1007/978-1-60761-543-9_21,

DNA. Because chemotherapeutic agents often kill actively proliferating cells, several slow-growing tumors do not respond to these drugs effectively. In addition, some tumors are inherently resistant to anticancer treatments. Although the majority of tumors initially respond to chemotherapy, they often become refractory to subsequent treatments. Both intrinsic and acquired resistance to these chemotherapeutic drugs poses a significant problem in cancer chemotherapy, and there have been concerted efforts to understand the bases of chemoresistance. There are two broad mechanisms of resistance to anticancer drugs: (1) a decreased availability of drugs to interact with its target DNA, and (2) a failure to recognize and respond to DNA damage.

Protein kinase C (PKC) is a potential target for cancer therapy because of its important role in carcinogenesis. It has also been shown to regulate cellular sensitivity to anticancer agents. In 1984, Joe Bartino and colleagues (Schornagel et al. 1984) demonstrated that PKC inhibitors were active against cells with both intrinsic and acquired resistance to methotrexate (MTX). The association between PKC and drug resistance was heralded by the seminal observation of Fine et al. that the activation of PKC could induce multiple drug resistance (MDR) (Fine et al. 1988). Since then, there has been a veritable deluge of scientific literature in the late 1980s and 1990s linking PKC with the multiple drug-resistant phenotype. This excitement, however, subsided when it was demonstrated that PKC-mediated phosphorylation of the major drug efflux pump P-glycoprotein that contributes to MDR had little effect on drug resistance. Alternate mechanisms by which PKC could contribute to MDR have been explored. Furthermore, an attempt was made to associate a particular PKC isozyme with a drug-resistant phenotype. The involvement of PKC was also extended to resistance to other anticancer drugs such as cisplatin that do not belong to the group of drugs that contribute to MDR.

Although it was originally believed that the inhibition of cell proliferation was the major cause of anticancer activity of the conventional cytotoxic chemotherapeutic drugs, it was later realized that these anticancer agents could kill cancer cells by inducing apoptosis (Fisher 1994). Thus, a failure to undergo apoptosis due to deregulation in apoptotic signaling pathways could also contribute to chemoresistance. Several members of the PKC family, including PKCδ, -θ, -ε, and -ζ have been shown to be substrates for caspases. While some members of the PKC family are needed for cell death by apoptosis, others could in fact inhibit cell death and contribute to chemoresistance. There are two major pathways of cell death by apoptosis: intrinsic and extrinsic. DNA damaging agents primarily affect the intrinsic or mitochondrial cell death pathway, and PKC isozymes have been shown to regulate members of the Bcl-2 family proteins that regulate the mitochondrial cell death pathway. An increase in antiapoptotic and a decrease in proapoptotic Bcl-2 family members can also contribute to chemoresistance. The purpose of this book chapter is to assimilate recent evidence on how the PKC signaling pathway contributes to chemoresistance. Some earlier studies will be discussed to provide a historical perspective. The focus of this chapter is in three main areas: (1) MDR, which includes a majority of conventional chemotherapeutic drugs; (2) resistance to cisplatin, which is highly effective for the treatment of solid tumors; and (3) a defect in apoptosis, which also contributes to resistance to multiple chemotherapeutic drugs.

21.2 PKC and MDR

21.2.1 What Is MDR?

When cancer cells become simultaneously resistant to several structurally unrelated natural product cytotoxic drugs, this phenotype is called multiple drug resistance or MDR. Several anticancer drugs, including anthracyclines (e.g., doxorubicin and daunorubicin), vinca alkaloids (e.g., vincristine and vinblastine), epipodophyllotoxins (e.g., etoposide or VP-16), antibiotics (e.g., actinomycin D and gramicidine), and taxanes (e.g., taxol and paclitaxel), belong to this group of drugs. The major mechanism contributing to MDR is the extrusion of drugs by an energy-dependent drug efflux pump, resulting in a decrease in intracellular drug accumulation (Higgins 1993). The 170-kDa plasma membrane glycoprotein (P-glycoprotein or P-gp) encoded by *MDR1* gene is the major drug efflux pump contributing to MDR (Dickson and Gottesman 1990). It contains several transmembrane domains and two ATP binding sites. A few other drug transporters, including the multidrug resistance-associated protein (MRP) and the lung resistance-associated protein (LRP), have also been shown to contribute to MDR (Rumsby et al. 1998).

21.2.2 Correlation Between PKC Activity and MDR

Fine et al. first demonstrated that PKC activity was elevated in breast cancer MCF-7 cells selected for resistance to doxorubicin compared to the parental drug-sensitive counterpart (Fine et al. 1988). Activation of PKC by the phorbol ester phorbol 12,13-dibutyrate (PDBu) caused transient induction of MDR in drug-sensitive cells and this was accompanied by a decreased intracellular accumulation of doxorubicin and vincristine. This observation led to numerous reports that tried to establish a link between PKC and MDR. First, an increase in PKC level/activity was correlated with both the inherent and acquired resistance to drugs involved in MDR. Several cell lines that developed MDR by drug selection, including MCF-7 breast cancer, human promyelocytic leukemia HL-60, human lymphoblastic leukemia MOLT3, murine fibrosarcoma UV-2237M, human epidermoid carcinoma KB-V1, and Sarcoma S180 cells, displayed an increase in PKC activity (Aquino et al. 1988; Chambers et al. 1990; Dolci et al. 1993; O'Brian et al. 1989; Palayoor et al. 1987; Posada et al. 1989; Schwartz et al. 1991). In most cases, cells were selected for resistance to adriamycin or doxorubicin except for KB-V1 cells, which were resistant to vinblastine. PKC activity, however, decreased in murine leukemia P388 cells selected with VP-16 or in MOLT leukemia selected with trimetrexate (Ido et al. 1987; Schwartz et al. 1991) compared to parental cells. An increased expression of PKC in 18 primary cultures of human renal cell carcinomas correlated with the level of P-gp and resistance to doxorubicin (Efferth and Volm 1992). In addition, the inherent resistance of untreated human nonsmall cell lung carcinomas to doxorubicin was associated with an increase in PKC and P-gp expression (Volm and Pommerenke 1995). These results suggest the importance of PKC in intrinsic resistance. Second, the activation of PKC by phorbol esters induced MDR and reduced drug accumulation (Bates et al. 1993; Chambers et al. 1990; Chambers et al. 1992; Dong et al. 1991; Ferguson and Cheng 1987; Ido et al. 1986; Wielinga et al. 1997). Third, numerous studies attempted to reverse MDR by using pharmacological inhibitors of PKC (Bates et al. 1993; Beltran et al. 1997; Budworth et al. 1996; Ganeshaguru et al. 2002; Killion et al. 1995; Merritt et al. 1999; Miyamoto et al. 1993; Sachs et al. 1995, 1996; Sato et al. 1990). While most PKC inhibitors enhanced drug accumulation and reversed MDR, there was little correlation between the ability of these inhibitors to inhibit PKC activity and to increase drug accumulation (Budworth et al. 1996). A major problem with these earlier studies was that most of the available PKC inhibitors, such as staurosporine or its derivatives (e.g., CGP-41251 and bisindolylmaleimides), lack specificity. In addition, these inhibitors affect multiple PKC isozymes that may have a distinct or even opposite effect on MDR.

21.2.3 Involvement of PKC Isozymes in MDR

Since PKC is a family of isozymes, it was realized that proper targeting of PKC to reverse MDR would require identification of a specific subtype of PKC in mediating drug resistance. Most of the studies point to the involvement of PKCα in contributing to the MDR phenotype. First, PKCα is frequently elevated in MDR cell lines (Blobe et al. 1993; Budworth et al. 1997; Cloud-Heflin et al. 1996; Matsumoto et al. 1995; O'Brian et al. 1991; Scala et al. 1995). PKCα was also associated with intrinsic resistance of human colon cancer (O'Brian et al. 1995) and nonsmall cell lung carcinoma cells to doxorubicin (Singhal et al. 2006). Second, altered regulation of PKCα was noted in several cell lines. For example, an elevated level of a slightly altered form of PKCα was present in the nucleus of MCF-7/ADR cells but not in MCF-7/WT cells (Lee et al. 1992). Third, the stable transfection of PKCα in MCF-7 cells induced MDR (Ahmad et al. 1994; Gill et al. 2001). Conversely, antisense oligonucleotides to PKCα increased sensitivity of colon carcinomas and reversed taxol resistance in ovarian cancer cells (Masanek et al. 2002). Finally, the reversion of MDR phenotype of MCF-7/Adr (Budworth et al. 1997) and KB-V1 cells (Cloud-Heflin et al. 1996) by culturing in drug-free media was associated with decrease in expression of PKCα and loss of P-gp.

How is PKCα upregulated in MDR cells? Upregulation of PKCα in MDR cells could be at the transcriptional or posttranscriptional level. PKCα was increased at the message level in MCF-7-MDR cells (Blobe et al. 1993), whereas TPA-induced downregulation of PKCα was attenuated in MDR UV-2237M cells (Ward and O'Brian 1991) and KB-V1 cells (Cloud-Heflin et al. 1996). Thus, a reduced rate of PKCα degradation may also contribute to an elevated PKCα level. Although most of the studies suggest activation or overexpression of PKCα to be associated with

MDR, a recent report demonstrated that PKCα expression is elevated in parental murine leukemia L1210 cells compared to its drug-resistant counterpart L1210/R, and an increase in PKCα protects parental cells but not drug-resistant cells against histone deacetylase inhibitors (Castro-Galache et al. 2007).

A few reports also implicated other PKC isozymes besides PKCα in MDR. Blobe et al. demonstrated that while conventional PKCs, such as PKCα and PKCβ, were increased both at the mRNA and protein levels, novel PKCs, such as PKCδ and PKCε, were decreased when cells acquired an MDR phenotype (Blobe et al. 1993). The stable expression of PKCβ1 in rat embryo fibroblasts induced resistance to adriamycin, actinomycin D, vincristine, and vinblastine (Fan et al. 1992). Drug resistance of P388/ADR cells was associated with an increase in PKCβ isozyme, and the introduction of anti-PKCβ, but not anti-PKCα, antibody to these cells reversed resistance to daunorubicin and partially corrected a drug accumulation defect (Gollapudi et al. 1995). PKCβ inhibitor Ly379196 increased sensitivity of neuroblastoma cells to doxorubicin, etoposide, paclitaxel, and vincristine but had no effect on the sensitivity to carboplatin (Svensson and Larsson 2003). Both PKCα and PKCβ belong to conventional group of PKCs, and it is conceivable that depending on the cell type, either PKCα or PKCβ may play a role in MDR.

Among the novel PKCs, PKCη appears to be important in conferring drug resistance. There was a significant correlation between the expression of PKCη with MDR-related gene products, such as *MDR1*, *MRP*, and *LRP*, in patients with acute myelogenous leukemia (Beck et al. 1996), in specimens from patients with primary breast cancer (Beck et al. 1998a), and in ascites aspirates from ovarian cancer patients (Beck et al. 1998b). Downregulation of PKCη by antisense oligonucleotides enhanced sensitivity of nonsmall cell lung cancer A549 cells to vincristine and paclitaxel (Sonnemann et al. 2004). PKCη was preferentially expressed in Hodgkin's lymphoma cells that are resistant to camptothecin and doxorubicin, and knockdown of PKCη in the resistant L428 cells rendered them more sensitive to doxorubicin and camptothecin (Abu-Ghanem et al. 2007).

An increase in atypical PKCζ was noted in MDR glioma cells (Matsumoto et al. 1995). In addition, treatment of ovarian cancer A2780 cells with adriamycin, camptothecin, etoposide, and vincristine increased the levels of MDR1, LRP, as well as PKCζ (Brugger et al. 2002). While these results demonstrate that chemotherapeutic agents involved in MDR can induce PKCζ, they do not provide direct evidence linking PKCζ with MDR.

21.2.4 Targets of PKC

Since P-gp is the major drug efflux pump contributing to MDR, it was considered a logical target for PKC. Several investigators embarked on the identification of P-gp as a substrate for PKC. Chambers et al. first demonstrated that membrane vesicles from KB-V1 cells were phosphorylated by an endogenous kinase similar to PKC, and P-gp could be phosphorylated by a highly purified rat brain

PKC in vitro (Chambers et al. 1990). The same group demonstrated that P-gp phosphorylation was stimulated by PKC activator TPA and inhibited by PKC inhibitors, and a decrease in P-gp phosphorylation was associated with an increase in drug accumulation (Chambers et al. 1992). It was proposed that PKC is primarily responsible for P-gp phosphorylation and the phosphorylation of P-gp regulates its drug pumping activity (Chambers et al. 1992). The phosphorylation sites in P-gp were identified in the linker region located between two homologous halves of P-gp at Ser 661, 671, and perhaps serine 667, 675, and 683 (Chambers et al. 1993). PKCα was identified as the kinase responsible for P-gp phosphorylation since the introduction of PKCα in BC-19 cells overexpressing P-gp increased P-gp phosphorylation and decreased drug accumulation (Ahmad and Glazer 1993). P-gp was phosphorylated when coexpressed with PKCα in a baculovirus expression system and coimmunoprecipitated with PKCα, and this phosphorylation was inhibited by the PKC inhibitor Ro 31-8220 (Ahmad et al. 1994). Furthermore, ATPase activity of P-gp was abolished by the mutation of Ser-671 site to Asn in the linker region of P-gp. These studies fit nicely with the concept that phosphorylation of P-gp by PKC increased drug efflux, causing a decrease in intracellular drug and conferring resistance to chemotherapeutic drugs that exhibit MDR phenotype.

Several studies, however, contradicted this simple concept. Although PKCα was overexpressed in MDR MCF-7TH cells (generated by intermittent exposure to doxorubicin), and bryostatin 1, a partial agonist of PKC, decreased P-gp phosphorylation, it did not affect drug transport or reverse MDR (Scala et al. 1995). To directly demonstrate the importance of P-gp phosphorylation on its drug efflux activity, the phosphorylation sites were mutated to nonphosphorylatable Ala or phosphomimicking Asp residues (Germann et al. 1996; Goodfellow et al. 1996). Mutation in PKC phosphorylation sites in P-gp, however, had no effect on drug transport or the MDR phenotype. Thus, mutational analysis of PKC phosphorylation sites in P-gp argued against the involvement of PKC-mediated P-gp phosphorylation in regulating drug efflux activity of P-gp.

Several investigators came up with an alternate explanation based on the observation that PKC inhibitors could directly bind to P-gp and thus compete with anticancer drugs for binding to P-gp (Bates et al. 1993; Budworth et al. 1996; Castro et al. 1999; Conseil et al. 2001; Merritt et al. 1999; Sato et al. 1990; Sha et al. 1996; Smith and Zilfou 1995; Wakusawa et al. 1993). This could provide an explanation of how PKC inhibitors could enhance drug accumulation and reverse MDR via P-gp phosphorylation-independent mechanism. Most of the inhibitors that were transported by P-gp were staurosporine derivatives. Safingol and natural isomers of sphingosine that inhibited PKC by binding to the regulatory domain of PKC also inhibited basal and phorbol ester-induced P-gp phosphorylation and increased drug accumulation (Sachs et al. 1995, 1996). These inhibitors had no effect on the binding of anticancer drugs to P-gp, suggesting that binding of PKC inhibitors to P-gp was not adequate to explain PKC-mediated drug resistance.

An increase in P-gp was often accompanied by an increase in PKC or vice versa. TPA, as well as diacylglycerol, a physiological stimulator of PKC, increased *MDR1* gene expression both at the protein and mRNA level in several cell lines derived from different types of leukemias and solid tumors (Chaudhary and Roninson 1992). This induction of *MDR1* expression by PKC activators was suppressed by staurosporine (Chaudhary and Roninson 1992). It was proposed that an increase in MDR1 expression by PKC activators was responsible for the emergence of the MDR phenotype. This group later found that chemotherapeutic drugs that are not transported by P-gp could also cause MDR1 induction, which was inhibited by PKC inhibitors (Chaudhary and Roninson 1993). To directly demonstrate the involvement of PKC in *MDR1* gene expression, Gill *et al.* generated MCF-7 cells in which PKCα was expressed under an inducible promoter, and transfected these cells with *MDR1* promoter or deletion mutants linked to a chloramphenicol acetyl transferase (CAT) reporter to rule out the possibility of PKC inhibitors binding to P-gp (Gill et al. 2001). Treatment of cells with TPA caused an induction of *MDR1* promoter activity that could be inhibited by PKC-specific inhibitor GF 109203X. Overexpression of PKCα increased TPA-inducible *MDR1* promoter transcription. These results suggest that one mechanism by which PKC regulates MDR is by upregulating *MDR1* gene expression. A recent report demonstrated that a specific peptide inhibitor of PKCε suppressed the induction of P-gp in LNCaP prostate cancer cells (Flescher and Rotem 2002). The transcription factor c-Jun (Ratnasinghe et al. 2001) as well as NF-κB (Kameyama et al. 2008) have been implicated in PKC-mediated gene transcription of *MDR1*. While these studies are consistent with the notion that PKC mediates *MDR1* transcription through phosphorylation of a transcription factor, PKC could also regulate the MDR phenotype in the absence of P-gp overexpression. Overexpression of PKCβ1 in rat embryo fibroblasts induced the MDR phenotype without altering P-gp expression (Fan et al. 1992). In addition, TPA reduced uptake of both adriamycin and vincristine in human colon cancer cells lacking P-gp and in the absence of any induction of MDR1 expression (Bergman et al. 1997). Therefore, the quest for a potential PKC target that can contribute to MDR continued.

There are several problems with the earlier studies. First, most of the PKC activators and inhibitors used in these studies were nonspecific and clearly had additional targets. Second, the focus was primarily on P-gp. TPA decreased intracellular drug accumulation even in cells that lack P-gp, suggesting that TPA may also influence drug uptake in a P-gp-independent mechanism (Bergman et al. 1997; Wielinga et al. 1997). Although P-gp is the major drug efflux pump associated with MDR, there are other drug efflux pumps that could also contribute to MDR. There are a few reports that implicated other transporters such as MRP (Beck et al. 1998b; Gekeler et al. 1995), LRP (Brugger et al. 2002), or Ral-interacting protein (RLIP76) (Singhal et al. 2006) in PKC-mediated induction of MDR. TPA has been shown to cause transcriptional upregulation of *MRP2* and *MRP3* promoters (Pulaski et al. 2005). Interestingly, nonspecific PKC inhibitors staurosporine and H7 stimulated expression from *MRP2* promoter and rendered cells more resistant to etoposide.

Third, drug resistance could be due to mechanisms distinct from drug efflux. An aberration in apoptotic signaling could also contribute to MDR, and PKC plays an important role in apoptosis (see Section 21.4). It has been reported that antisense oligonucleotides to PKCη sensitized A549 cells to vincristine and paclitaxel by decreasing the levels of antiapoptotic Bcl-xL mRNA and protein (Sonnemann et al. 2004). PKCα, the major isozyme implicated in MDR, has been shown to phosphorylate Bcl-2 (May et al. 1994). Interestingly, Bcl-2 overexpressing clones displayed an increased rather than decreased sensitivity to adriamycin, vincristine, vinblastine, and actinomycin D (Del Bufalo et al. 2002). Perhaps, the phosphorylation status of Bcl-2 plays a role. Alternatively, the presence of other antiapoptotic proteins, such as Bcl-xL or Mcl-1, may be important. Bcl-2 overexpression had no effect on P-gp expression but decreased ATP levels and basal PKC activity, which could influence MDR phenotype.

Fourth, although an altered drug efflux is the common mechanism contributing to drug resistance of the drugs associated with MDR, the other targets of these drugs can also be regulated by PKC. For example, both topoisomerase I (Cardellini and Durban 1993; Samuels et al. 1989) and topoisomerase II (Mouchel and Jenkins 2006), which are targets for camptothecins, anthracyclines, and etoposide, are regulated by PKC-mediated phosphorylation. Increased phosphorylation of topoisomerase II was observed in etoposide-resistant mutants of human glioma cell line (Matsumoto et al. 1999). Topoisomerase II is a substrate for PKCζ, and overexpression of PKCζ reduced topoisomerase II catalytic activity and etoposide-induced cytotoxicity in monocytic U937 leukemic cells (Plo et al. 2002).

In summary, there is convincing accumulating evidence that suggest the involvement of PKC in regulating MDR. In most cases, PKCα appears to be the isozyme responsible for the MDR phenotype, although depending on the cell type other PKC isozymes may also play a role. There is still considerable debate with regard to the mechanism by which PKC regulates MDR. Although phosphorylation of P-gp at the known Ser/Thr sites may not be important, it is conceivable that there are other unidentified PKC phosphorylation sites but the functional significance of those sites is yet to be elucidated. In addition, both P-gp-dependent and P-gp-independent mechanisms may operate, and they may not be mutually exclusive. Since most of the chemotherapeutic agents involved in MDR also induce apoptosis, PKC isozymes can influence cell death by altering the function of apoptotic regulators. Thus, depending on drug-target interaction, the presence of PKC isozymes and their intracellular localization, the levels and phosphorylation status of P-gp and other drug efflux pumps, and the status of pro- and antiapoptotic proteins, the regulation of drug resistant phenotype by PKC could significantly differ from cell type to cell type. Thus, it will be difficult to apply a one size fits all strategy to explain the involvement of PKC in MDR.

21.3 PKC and Cisplatin Resistance

21.3.1 The Mechanism of Action of Cisplatin

Cisplatin or *cis*-diamminedichloroplatinum(II) (cDDP) is one of the most important anticancer agents used in the treatment of solid tumors. It forms adducts with DNA and resembles bifunctional alkylating agents. The interaction of cisplatin with two adjacent guanine residues is believed to be responsible for its cytotoxic action. Cisplatin is very effective for the treatment of solid tumors, especially ovarian, testicular, cervical, and small cell lung cancers. However, many patients experience a relapse and its success is often compromised due to innate and acquired resistance by tumor cells to cisplatin. The mechanism(s) of cisplatin resistance is often multi-factorial and could be due to decreased drug accumulation, increased drug detoxification by cellular thiols glutathione and metallothionein, increased DNA repair and tolerance, and a defect in apoptosis (Kelland 2007). Unlike MDR, drug efflux does not appear to be the major cause of cisplatin resistance. Many cell lines with acquired resistance to cisplatin often exhibit reduced drug accumulation, but this is due to decrease in uptake of cisplatin rather than an increase in drug efflux (Gately and Howell 1993; Kelland 2007). Cisplatin is believed to enter cells by passive or facilitated diffusion (Hall et al. 2008), although recent reports suggest that cisplatin may also be transported via an active transport. A plasma membrane copper transporter-1 has been shown to play a role in cisplatin uptake (Ishida et al. 2002), whereas copper transporter ATP7A and ATP7B have been implicated in cisplatin export (Safaei et al. 2004).

Although the antitumor activity of cisplatin is believed to be due to its interaction with chromosomal DNA (Sherman et al. 1985), only a small fraction of cisplatin actually interacts with DNA, and the inhibition of DNA replication cannot solely account for its biologic activity (Eastman 1990). Following DNA damage, cells may either repair the damage and start progressing through the cell cycle or if they cannot repair the damage, cells are destined to die (Eastman 1990). The efficacy of chemotherapeutic drugs, including cisplatin, not only depends on their ability to induce DNA damage but also on the cell's ability to detect and respond to DNA damage (Kerr et al. 1994; Siddik 2003). Cisplatin, like many other anticancer agents, causes activation of caspases, and a defect in apoptosis may also contribute to cisplatin resistance.

21.3.2 PKC and Cisplatin

In the course of our study to understand the mechanism of cisplatin resistance by metallothionein, we inadvertently found that PKC activator TPA sensitized cells to cisplatin. In the meantime, Hofmann et al. reported that cellular sensitivity to cisplatin could be enhanced by inhibition or downregulation of PKC (Hofmann et al. 1988). Steve Howell and coworkers (Isonishi et al. 1990) and our laboratory (Basu et al. 1990) simultaneously reported that PKC activators, such as phorbol esters, enhanced sensitivity of human ovarian cancer 2008 and human cervical cancer HeLa cells to cisplatin. TPA caused sensitization of both drug-sensitive 2008 cells and its cisplatin-resistant variant 2008/C13*5.25 cells to cisplatin (Isonishi et al. 1990, 1994) and its analogues, carboplatin and (glycolato-O,O′) diammineplatinum(II) (254-S) (Isonishi et al. 1994). TPA also enhanced sensitivity of cisplatin-sensitive (KF) and -resistant (KK and MH) ovarian cancer cell lines derived from ascites of patients with clear-cell carcinoma and serous cystadenocarcinoma of the ovary who showed clinical resistance to cisplatin (Hirata et al. 1993). Decrease in cisplatin sensitivity in the resistant cells was associated with an increase in PKC activity.

While there is little controversy that PKC is an important regulator of cisplatin sensitivity, the most contested issue is whether activation or downregulation of PKC confers cisplatin resistance. Brief exposure to TPA was necessary to increase sensitivity of cisplatin-sensitive and -resistant ovarian carcinoma cells to cisplatin and continuous treatment with TPA that caused downregulation of PKC reduced cisplatin sensitivity (Hirata et al. 1993; Isonishi et al. 1990). These results suggest that activation of PKC is necessary for cisplatin sensitization. In contrast, downregulation of PKC by persistent treatment with TPA was associated with sensitization of human osteosarcoma U2-OS cell line and its cisplatin-resistant variant U2-OS/Pt cells (Perego et al. 1993), but short-term exposure to TPA had no effect. These results suggest that downregulation rather than activation of PKC was necessary to enhance sensitivity of these osteosarcoma cells to cisplatin. Prolonged treatment with phorbol esters was also associated with sensitization of HeLa cells to cisplatin (Basu et al. 1990). However, based on the comparison of a series of novel structural analogues of the PKC activator lynbyatoxin A, we found that PKC activators that failed to induce substantial downregulation of PKC could also sensitize HeLa cells to cisplatin (Basu et al. 1991). One of the problems with these earlier studies was that PKC activation and downregulation was monitored based on PKC activity assay (Basu et al. 1991). Since PKC is a family of isozymes and the ability of PKC activators to downregulate individual PKC isozyme may vary, it is difficult to conclude from these studies whether activation or downregulation of a particular PKC isozyme is associated with cisplatin resistance.

21.3.3 Involvement of PKC Isozymes in Cisplatin Resistance

In contrast to MDR, where an increase in conventional PKCs and a decrease in novel PKCs were associated with the MDR (Blobe et al. 1993), cisplatin resistance was accompanied by a decrease in conventional PKCs and an increase in novel PKCs. For example, PKCα level was decreased in human ovarian cancer 2008 cells (Basu and Weixel 1995), and both PKCα and PKCβ were decreased in human small cell lung cancer H69 cells (Basu et al. 1996) that acquired resistance to cisplatin. The levels of novel PKCδ and PKCε were increased in cisplatin-resistant 2008/C13*5.25 and H69/CP cells, respectively (Basu et al. 1996; Basu and Weixel 1995). Overexpression of PKCε in rat embryo fibroblasts protected cells against cisplatin cytotoxicity (Basu and Cline 1995), suggesting the importance of this isozyme in conferring cisplatin resistance. PKCδ level was also elevated in HeLa cells that acquired resistance to cisplatin, but there was little change in the levels of other PKC isozymes (Huang et al. 2004). Interestingly, bryostatin 1 exhibited a biphasic response on cisplatin sensitivity as well as PKCδ downregulation (Basu and Akkaraju 1999; Basu and Lazo 1992). Bryostatin 1 caused a parallel increase in cisplatin sensitivity and PKCδ downregulation with up to 1 nM, then the ability of bryostatin 1 to cause cisplatin sensitization and PKCδ downregulation gradually reversed such that 1 μM bryostatin 1 had little effect. Downregulation of PKCδ by bryostatin 1 correlated with increase in cisplatin sensitivity in both HeLa and HeLa/CP cells (Mohanty et al. 2005). In addition, we have observed that downregulation of PKCδ by PKC activators, such as PDBu, TPA, and indolactam V was compromised in HeLa cells that acquired resistance to cisplatin, whereas downregulation of other PKC isozymes, such as PKCα or PKCε, was not affected (Huang et al. 2004; Mohanty et al. 2005). Taken together, these results implicate PKCδ in cisplatin resistance. TPA also failed to induce downregulation of membrane associated PKC in cisplatin-resistant ovarian cancer KK and MH cells (Hirata et al. 1993), although the identity of the membrane-associated PKC isozyme is not known.

There are two observations that led us to believe that overexpression of PKCδ is not sufficient to explain cisplatin resistance. First, we have found that siRNA against PKCδ caused a modest decrease rather than an increase in cisplatin sensitivity (Mohanty et al. 2005). PKCδ has been implicated in both cell survival and cell death (Basu 2003; Brodie and Blumberg 2003; Clark et al. 2003). We believe that while full-length PKCδ functions as an antiapoptotic protein, its cleavage products function as proapoptotic proteins. Depletion of PKCδ by siRNA not only removes the full-length antiapoptotic PKCδ but it also prevents generation of cleaved fragments of PKCδ that act as proapoptotic proteins. Depending on the experimental conditions and cell types, the ratio of full-length versus cleaved PKCδ will vary and this may decide if PKCδ will functions as an anti- or proapoptotic protein during DNA damage-induced apoptosis. Second, even though phorbol esters failed to downregulate PKCδ in cisplatin-resistant HeLa cells, prolonged treatment with PDBu sensitized these cells to cisplatin (Mohanty et al. 2005). Inhibition of PKCα by Gö 6976 and depletion of PKCα by siRNA enhanced sensitivity of both parental and cisplatin-resistant HeLa cells to cisplatin (Mohanty et al. 2005), suggesting that PKCα is a prosurvival protein and downregulation of PKCα by PDBu was associated with cisplatin sensitization. This notion is corroborated by the observation that antisense oligonucleotide against PKCα (CGP 64128A or ISIS 3521) in combination with cisplatin demonstrated superadditive antitumor activities against MCF-7 breast cancer and PC3 prostate cancer cells with complete response (Geiger et al. 1998). It is not clear, however, why PKCα level is decreased rather than increased in ovarian cancer 2008 cells and small cell lung cancer H69 cells that acquired resistance to cisplatin (Basu et al. 1996; Basu and Weixel 1995).

The atypical PKCι has been associated with chemoresistance of glioblastoma multiforme, an aggressive form of brain cancer (Baldwin et al. 2006). Depletion of PKCι in glioblastoma cells increased cisplatin sensitivity. The mechanism of PKCι-mediated chemoresistance involved negative regulation of GMFβ, an enhancer or p38 mitogen-activated protein kinase. Therefore, depending on the cellular context, PKCα, -δ, -ε, or -ι can contribute to cisplatin resistance.

21.3.4 Mechanism of PKC-Mediated Cisplatin Resistance

In contrast to MDR, there is no clear target for PKC that could explain PKC-mediated cisplatin resistance. Sensitization of cisplatin-sensitive and -resistant ovarian cancer 2008 cells to cisplatin by TPA was not associated with decrease in intracellular accumulation of cisplatin or an increase in cellular thiol, such as glutathione or metallothionein (Isonishi et al. 1990). In HeLa cells, TPA increased cellular accumulation of cisplatin modestly, but the increase in cisplatin accumulation was not sufficient to explain the magnitude of cisplatin sensitization by PKC activators.

Since cisplatin-induced DNA damage triggers apoptosis, it is conceivable that PKC alters the function of pro- and antiapoptotic proteins and thereby regulates cisplatin sensitivity. It has been reported that treatment of cisplatin-resistant human squamous cell carcinoma SCC-25 (SCC25/CP) cells with cisplatin failed to induce caspase-3 activation and cleavage of PKCδ due to an increase in antiapoptotic Bcl-xL (Segal-Bendirdjian and Jacquemin-Sablon 1995).

Overexpression of heat shock protein 27 (HSP27) has been associated with cisplatin resistance. HSP27 was upregulated in cisplatin-resistant ovarian cancer cells (Yamamoto et al. 2001), and introduction of HSP27 in drug-sensitive ovarian (Yamamoto et al. 2001) and testicular (Richards et al. 1996) cancer cells conferred cisplatin resistance. In addition, the levels of HSP27 were shown to be high in various lung and breast tissues that display chemoresistance (Kim et al. 2007). Recently, it has been shown that HSP27 directly interacts with the amino acid 668–674 in the V5 region of PKCδ and inhibits both its catalytic activity and proapoptotic activity. Introduction of a heptapeptide targeted to the 668–674 region in NCI-H1299 lung cancer cells restored PKCδ activity and dramatically enhanced cisplatin sensitivity (Kim et al. 2007).

21.4 PKC and Apoptosis Resistance

Most of the chemotherapeutic drugs induce apoptosis, and a defect in apoptosis could also confer resistance to multiple drugs via a mechanism that does not involve increase in drug extrusion and hence is distinct from classical MDR. DNA damage induced by chemotherapeutic drugs causes release of cytochrome c resulting in activation of apical caspase-9 followed by activation of effector caspase-3 and -7 resulting in cell death. Bcl-2 family proteins regulate DNA damage-induced apoptosis by regulating the release of mitochondrial cytochrome c in response to DNA damage. The antiapoptotic Bcl-2 family members, such as Bcl-2 and Bcl-xL, prevent release of mitochondrial cytochrome c whereas proapoptotic Bcl-2 family members, such as Bid, Bax, and Bak facilitate cytochrome c release. Thus, the status of these Bcl-2 family proteins could impact on chemoresistance.

Several members of the PKC family, including PKCα, PKCε, PKCη, and PKCζ have been shown to function as antiapoptotic proteins. May and coworkers first demonstrated that Bcl-2 is a substrate for PKC and phosphorylation of Bcl-2 causes suppression of apoptosis (May et al. 1994). It has been shown that PKCα could phosphorylate Bcl-2 at Ser70 site (Ruvolo et al. 1998). Bcl-2 phosphorylation was important for its antiapoptotic action since HL-60 cells, which contain highly phosphorylated Bcl-2, were resistant to etoposide, cytosine arabinoside, and adriamycin compared to human pre-B REH cells, which express high levels of unphosphorylated Bcl-2 (Ruvolo et al. 1998). Treatment with PKC activators or ectopic expression of PKCα in REH cells induced translocation of PKCα to the mitochondrial membrane, phosphorylation of Bcl-2, and increased resistance to chemotherapeutic drugs (Ruvolo ct al. 1998). PKCα increased Bcl-2 phosphorylation not only by acting as a Bcl-2 kinase, but it also inhibited mitochondrial Ser/Thr phosphatase PP2A that acts as a Bcl-2 phosphatase (Jiffar et al. 2004). The interaction of PKCα with PICK1 (protein that interacts with C-kinase) has been shown to be necessary for anchoring PKCα to the mitochondria (Wang et al. 2003). Recently, it has been shown that overexpression of PICK1 in REH cells confers resistance to etoposide-induced apoptosis (Wang et al. 2007). Interaction of PKCα by PICK1 at the mitochondria facilitates phosphorylation of Bcl-2 at Ser70 site (Wang et al. 2007). Bcl-2 phosphorylation and active PKCα was also associated with poor survival in patients with acute lymphoblastic leukemia (AML) (Kurinna et al. 2006).

We have shown that overexpression of PKCε contributes to cisplatin resistance by inhibiting cisplatin-induced apoptosis (Basu and Cline 1995). The expression of PKCε but not other PKC isoforms was associated with chemoresistance of nonsmall cell lung carcinomas (Ding et al. 2002). Introduction of PKCε in NCI-H82 cells conferred resistance to etoposide and doxorubicin, whereas downregulation of PKCε by antisense cDNA in NSCLC enhanced sensitivity to etoposide (Ding et al. 2002). Overexpression of PKCε inhibited release of cytochrome c from the mitochondria and activation of caspase-9 and -3 (Ding et al. 2002). Recently, it has been reported that the mechanism by which PKCε contributes to chemoresistance in human small cell lung cancer cells is by upregulation of X-linked inhibitor of apoptosis protein (XIAP) and Bcl-xL via activation of ribosomal S6 kinase-2 (Pardo et al. 2006).

Downregulation of PKCη sensitized A549 lung cancer cells to vincristine and paclitaxel by inducing mitochondrial depolarization, demonstrating that PKCη also acts upstream of mitochondrial cell death pathway (Sonnemann et al. 2004). Antiapoptotic function of PKCη has been implicated in conferring chemoresistance to Hodgkin's lymphoma-derived cell lines L428 and KMH2 (Abu-Ghanem et al. 2007). The drug resistant L428 cells overexpressed PKCη and knockdown of PKCη sensitized these cells to camptothecin and doxorubicin by increasing cytochrome c release, caspase-7 activation, and PARP cleavage.

Chemotherapeutic agents, such as etoposide, have been shown to cause activation of PKCζ in U937 cells (Filomenko et al. 2002). Sensitization of U937 cells to etoposide by inhibition of PKCζ was associated with decrease in Bcl-2, increase in Bax, inhibition of nuclear translocation of NF-κB, and inhibition of XIAP accumulation (Filomenko et al. 2002). On the other hand, Bax was identified as a substrate for PKCζ (Xin et al. 2007). Thus, PKCζ could exert its antiapoptotic function by phosphorylating Bax and sequestering it in the cytoplasm (Xin et al. 2007).

In contrast, PKCι but not PKCζ has been shown to protect human K562 leukemia cells against taxol-induced apoptosis (Murray and Fields 1997). It has been reported that PKCι acts downstream of Bcr-Abl and it induces taxol resistance via activation of NF-κB (Lu et al. 2001). Expression of constitutively active PKCι in K562 cells increased NF-κB transactivation. Conversely, downregulation or inhibition of PKCι sensitized cells to taxol-induced apoptosis but overexpression of NF-κB rescued these cells from apoptosis.

21.5 Conclusion

It has been twenty years since PKC was first identified as an important regulator of chemoresistance. Significant advancements have been made to our understanding of the mechanisms of drug resistance. Earlier studies relied on pharmacological PKC activators and inhibitors that are now considered to be less specific. Also, we had limited tools available to study the involvement of individual PKC isozymes. With the advent of siRNA technology, it has become easier to elucidate the function of a particular PKC isozyme. It is clear that the PKC signaling pathway may act at various stages starting from the entry of the drugs to the execution of cell death. We now know that failure to undergo apoptosis can also contribute to chemoresistance and PKC plays a significant role in this process.

Although this chapter focuses on the involvement of PKC in drug resistance, there are other signaling pathways, such as Akt and mitogen-activated protein kinase, which play important roles in drug resistance. In addition, there may be cooperation among these pathways. Future studies should consider how these signaling networks contribute to chemoresistance. These pathways may vary from one cell type to another and within patient populations. Consequently, the mechanism(s) of chemoresistance will differ from one patient to another. It is now clear that cancer therapy must be tailored towards the individual patients. The status of PKC isozymes and their potential targets may provide insight into whether a patient will respond to a particular therapy. Even though targeting the PKC signaling pathway alone may not be very successful in the clinic, the PKC signaling pathway could be intervened in combination with conventional chemotherapeutic agents that are already in the clinic to combat chemoresistance, the most significant problem in cancer therapy.

Acknowledgment The author wishes to thank Shalini Persaud and Anindita Basu Sempere for critical reading of this chapter. We apologize if we inadvertently left out any major contribution in this field. The author is supported by a grant CA071727 from the National Cancer Institute.

References

Abu-Ghanem, S., Oberkovitz, G., Benharroch, D., Gopas, J., & Livneh, E. (2007). PKCeta expression contributes to the resistance of Hodgkin's lymphoma cell lines to apoptosis. *Cancer Biology & Therapy, 6*, 1375–1380.

Ahmad, S., & Glazer, R. I. (1993). Expression of the antisense cDNA for protein kinase C alpha attenuates resistance in doxorubicin-resistant MCF-7 breast carcinoma cells. *Molecular Pharmacology, 43*, 858–862.

Ahmad, S., Safa, A. R., & Glazer, R. I. (1994). Modulation of P-glycoprotein by protein kinase C alpha in a baculovirus expression system. *Biochemistry, 33*, 10313–10318.

Aquino, A., Hartman, K. D., Knode, M. C., Grant, S., Huang, K. P., Niu, C. H., et al. (1988). Role of protein kinase C in phosphorylation of vinculin in adriamycin-resistant HL-60 leukemia cells. *Cancer Research, 48*, 3324–3329.

Baldwin, R. M., Garratt-Lalonde, M., Parolin, D. A., Krzyzanowski, P. M., Andrade, M. A., & Lorimer, I. A. (2006). Protection of glioblastoma cells from cisplatin cytotoxicity via protein kinase Ciota-mediated attenuation of p38 MAP kinase signaling. *Oncogene, 25*, 2909–2919.

Basu, A. (2003). Involvement of protein kinase C-delta in DNA damage-induced apoptosis. *Journal of Cellular and Molecular Medicine, 7*, 341–350.

Basu, A., & Akkaraju, G. R. (1999). Regulation of caspase activation and cis-diamminedichloroplatinum(II)-induced cell death by protein kinase C. *Biochemistry, 38*, 4245–4251.

Basu, A., & Cline, J. S. (1995). Oncogenic transformation alters cisplatin-induced apoptosis in rat embryo fibroblasts. *International Journal of Cancer, 63*, 597–603.

Basu, A., Kozikowski, A. P., Sato, K., & Lazo, J. S. (1991). Cellular sensitization to cis-diamm inedichloroplatinum(II) by novel analogues of the protein kinase C activator lyngbyatoxin A. *Cancer Research, 51*, 2511–2514.

Basu, A., & Lazo, J. S. (1992). Sensitization of human cervical carcinoma cells to cis-diamm inedichloroplatinum(II) by bryostatin 1. *Cancer Research, 52*, 3119–3124.

Basu, A., Teicher, B. A., & Lazo, J. S. (1990). Involvement of protein kinase C in phorbol ester-induced sensitization of HeLa cells to cis-diamminedichloroplatinum(II). *The Journal of Biological Chemistry, 265*, 8451–8457.

Basu, A., & Weixel, K. M. (1995). Comparison of protein kinase C activity and isoform expression in cisplatin-sensitive and -resistant ovarian carcinoma cells. *International Journal of Cancer, 62*, 457–460.

Basu, A., Weixel, K., & Saijo, N. (1996). Characterization of the protein kinase C signal transduction pathway in cisplatin-sensitive and -resistant human small cell lung carcinoma cells. *Cell Growth and Differentiation, 7*, 1507–1512.

Bates, S. E., Lee, J. S., Dickstein, B., Spolyar, M., & Fojo, A. T. (1993). Differential modulation of P-glycoprotein transport by protein kinase inhibition. *Biochemistry, 32*, 9156–9164.

Beck, J., Bohnet, B., Brugger, D., Bader, P., Dietl, J., Scheper, R. J., et al. (1998). Multiple gene expression analysis reveals distinct differences between G2 and G3 stage breast cancers, and correlations of PKC eta with MDR1, MRP and LRP gene expression. *British Journal of Cancer, 77*, 87–91.

Beck, J. F., Bohnet, B., Brugger, D., Dietl, J., Scheper, R. J., Bader, P., et al. (1998). Expression analysis of protein kinase C isozymes and multidrug resistance associated genes in ovarian cancer cells. *Anticancer Research, 18*, 701–705.

Beck, J., Handgretinger, R., Klingebiel, T., Dopfer, R., Schaich, M., Ehninger, G., et al. (1996). Expression of PKC isozyme and MDR-associated genes in primary and relapsed state AML. *Leukemia, 10*, 426–433.

Beltran, P. J., Fan, D., Fidler, I. J., & O'Brian, C. A. (1997). Chemosensitization of cancer cells by the staurosporine derivative CGP 41251 in association with decreased P-glycoprotein phosphorylation. *Biochemical Pharmacology, 53*, 245–247.

Bergman, P. J., Gravitt, K. R., Ward, N. E., Beltran, P., Gupta, K. P., & O'Brian, C. A. (1997). Potent induction of human colon cancer cell uptake of chemotherapeutic drugs by N-myristoylated protein kinase C-alpha (PKC-alpha) pseudosubstrate peptides through a P-glycoprotein-independent mechanism. *Investigational New Drugs, 15*, 311–318.

Blobe, G. C., Sachs, C. W., Khan, W. A., Fabbro, D., Stabel, S., Wetsel, W. C., et al. (1993). Selective regulation of expression of protein kinase C (PKC) isoenzymes in multidrug-resistant MCF-7 cells. Functional significance of enhanced expression of PKC alpha. *The Journal of Biological Chemistry, 268*, 658–664.

Brodie, C., & Blumberg, P. M. (2003). Regulation of cell apoptosis by protein kinase c delta. *Apoptosis, 8*, 19–27.

Brugger, D., Brischwein, K., Liu, C., Bader, P., Niethammer, D., Gekeler, V., et al. (2002). Induction of drug resistance and protein kinase C genes in A2780 ovarian cancer cells after incubation with antineoplastic agents at sublethal concentrations. *Anticancer Research, 22*, 4229–4232.

Budworth, J., Davies, R., Malkhandi, J., Gant, T. W., Ferry, D. R., & Gescher, A. (1996). Comparison of staurosporine and four analogues: their effects on growth, rhodamine 123 retention and binding to P-glycoprotein in multidrug-resistant MCF-7/Adr cells. *British Journal of Cancer, 73*, 1063–1068.

Budworth, J., Gant, T. W., & Gescher, A. (1997). Co-ordinate loss of protein kinase C and multidrug resistance gene expression in revertant MCF-7/Adr breast carcinoma cells. *British Journal of Cancer, 75*, 1330–1335.

Cardellini, E., & Durban, E. (1993). Phosphorylation of human topoisomerase I by protein kinase C in vitro and in phorbol 12-myristate 13-acetate-activated HL-60 promyelocytic leukaemia cells. *Biochemical Journal, 291*(Pt 1), 303–307.

Castro, A. F., Horton, J. K., Vanoye, C. G., & Altenberg, G. A. (1999). Mechanism of inhibition of P-glycoprotein-mediated drug transport by protein kinase C blockers. *Biochemical Pharmacology, 58*, 1723–1733.

Castro-Galache, M. D., Menendez-Gutierrez, M. P., Carrasco Garcia, E., Garcia-Morales, P., Martinez-Lacaci, I., Saceda, M., et al. (2007). Protein kinase C-alpha antagonizes apoptosis induction by histone deacetylase inhibitors in multidrug resistant leukaemia cells. *International Journal of Biochemistry and Cell Biology, 39*, 1877–1885.

Chambers, T. C., McAvoy, E. M., Jacobs, J. W., & Eilon, G. (1990). Protein kinase C phosphorylates P-glycoprotein in multidrug resistant human KB carcinoma cells. *The Journal of Biological Chemistry, 265*, 7679–7686.

Chambers, T. C., Pohl, J., Raynor, R. L., & Kuo, J. F. (1993). Identification of specific sites in human P-glycoprotein phosphorylated by protein kinase C. *The Journal of Biological Chemistry, 268*, 4592–4595.

Chambers, T. C., Zheng, B., & Kuo, J. F. (1992). Regulation by phorbol ester and protein kinase C inhibitors, and by a protein phosphatase inhibitor (okadaic acid), of P-glycoprotein phosphorylation and relationship to drug accumulation in multidrug-resistant human KB cells. *Molecular Pharmacology, 41*, 1008–1015.

Chaudhary, P. M., & Roninson, I. B. (1992). Activation of MDR1 (P-glycoprotein) gene expression in human cells by protein kinase C agonists. *Oncology Research, 4*, 281–290.

Chaudhary, P. M., & Roninson, I. B. (1993). Induction of multidrug resistance in human cells by transient exposure to different chemotherapeutic drugs. *Journal of the National Cancer Institute, 85*, 632–639.

Clark, A. S., West, K. A., Blumberg, P. M., & Dennis, P. A. (2003). Altered protein kinase C (PKC) isoforms in non-small cell lung cancer cells: PKCdelta promotes cellular survival and chemotherapeutic resistance. *Cancer Research, 63*, 780–786.

Cloud-Heflin, B. A., McMasters, R. A., Osborn, M. T., & Chambers, T. C. (1996). Expression, subcellular distribution and response to phorbol esters of protein kinase C (PKC) isozymes in drug-sensitive and multidrug-resistant KB cells evidence for altered regulation of PKC-alpha. *European Journal of Biochemistry, 239*, 796–804.

Conseil, G., Perez-Victoria, J. M., Jault, J. M., Gamarro, F., Goffeau, A., Hofmann, J., et al. (2001). Protein kinase C effectors bind to multidrug ABC transporters and inhibit their activity. *Biochemistry, 40*, 2564–2571.

Del Bufalo, D., Biroccio, A., Trisciuoglio, D., Bruno, T., Floridi, A., Aquino, A., et al. (2002). Bcl-2 has differing effects on the sensitivity of breast cancer cells depending on the antineoplastic drug used. *European Journal of Cancer, 38*, 2455–2462.

Dickson, R. B., & Gottesman, M. M. (1990). Understanding of the molecular basis of drug resistance in cancer reveals new targets for chemotherapy. *Trends in Pharmacological Sciences, 11*, 305–307.

Ding, L., Wang, H., Lang, W., & Xiao, L. (2002). Protein kinase C-epsilon promotes survival of lung cancer cells by suppressing apoptosis through dysregulation of the mitochondrial caspase pathway. *The Journal of Biological Chemistry, 277*, 35305–35313.

Dolci, E. D., Abramson, R., Xuan, Y., Siegfried, J., Yuenger, K. A., Yassa, D. S., et al. (1993). Anomalous expression of P-glycoprotein in highly drug-resistant human KB cells. *International Journal of Cancer, 54*, 302–308.

Dong, Z. Y., Ward, N. E., Fan, D., Gupta, K. P., & O'Brian, C. A. (1991). In vitro model for intrinsic drug resistance: Effects of protein kinase C activators on the chemosensitivity of cultured human colon cancer cells. *Molecular Pharmacology, 39*, 563–569.

Eastman, A. (1990). Activation of programmed cell death by anticancer agents: Cisplatin as a model system. *Cancer Cells, 2*, 275–280.

Efferth, T., & Volm, M. (1992). Expression of protein kinase C in human renal cell carcinoma cells with inherent resistance to doxorubicin. *Anticancer Research, 12*, 2209–2211.

Fan, D., Fidler, I. J., Ward, N. E., Seid, C., Earnest, L. E., Housey, G. M., et al. (1992). Stable expression of a cDNA encoding rat brain protein kinase C-beta I confers a multidrug-resistant phenotype on rat fibroblasts. *Anticancer Research, 12*, 661–667.

Ferguson, P. J., & Cheng, Y. C. (1987). Transient protection of cultured human cells against antitumor agents by 12-O-tetradecanoylphorbol-13-acetate. *Cancer Research, 47*, 433–441.

Filomenko, R., Poirson-Bichat, F., Billerey, C., Belon, J. P., Garrido, C., Solary, E., et al. (2002). Atypical protein kinase C zeta as a target for chemosensitization of tumor cells. *Cancer Research, 62*, 1815–1821.

Fine, R. L., Patel, J., & Chabner, B. A. (1988). Phorbol esters induce multidrug resistance in human breast cancer cells. *Proceedings of the National Academy of Sciences of the United States of America, 85*, 582–586.

Fisher, D. E. (1994). Apoptosis in cancer therapy: Crossing the threshold. *Cell, 78*, 539–542.

Flescher, E., & Rotem, R. (2002). Protein kinase C epsilon mediates the induction of P-glycoprotein in LNCaP prostate carcinoma cells. *Cellular Signalling, 14*, 37–43.

Ganeshaguru, K., Wickremasinghe, R. G., Jones, D. T., Gordon, M., Hart, S. M., Virchis, A. E., et al. (2002). Actions of the selective protein kinase C inhibitor PKC412 on B-chronic lymphocytic leukemia cells in vitro. *Haematologica, 87*, 167–176.

Gately, D. P., & Howell, S. B. (1993). Cellular accumulation of the anticancer agent cisplatin: A review. *British Journal of Cancer, 67*, 1171–1176.

Geiger, T., Muller, M., Dean, N. M., & Fabbro, D. (1998). Antitumor activity of a PKC-alpha antisense oligonucleotide in combination with standard chemotherapeutic agents against various human tumors transplanted into nude mice. *Anti-Cancer Drug Design, 13*, 35–45.

Gekeler, V., Boer, R., Ise, W., Sanders, K. H., Schachtele, C., & Beck, J. (1995). The specific bis-indolylmaleimide PKC-inhibitor GF 109203X efficiently modulates MRP-associated multiple drug resistance. *Biochemical and Biophysical Research Communications, 206*, 119–126.

Germann, U. A., Chambers, T. C., Ambudkar, S. V., Licht, T., Cardarelli, C. O., Pastan, I., et al. (1996). Characterization of phosphorylation-defective mutants of human P-glycoprotein expressed in mammalian cells. *The Journal of Biological Chemistry, 271*, 1708–1716.

Gill, P. K., Gescher, A., & Gant, T. W. (2001). Regulation of MDR1 promoter activity in human breast carcinoma cells by protein kinase C isozymes alpha and theta. *European Journal of Biochemistry, 268*, 4151–4157.

Gollapudi, S., Soni, V., Thadepalli, H., & Gupta, S. (1995). Role of protein kinase beta isozyme in multidrug resistance in murine leukemia P388/ADR cells. *Journal of Chemotherapy, 7*, 157–159.

Goodfellow, H. R., Sardini, A., Ruetz, S., Callaghan, R., Gros, P., McNaughton, P. A., et al. (1996). Protein kinase C-mediated phosphorylation does not regulate drug transport by the human multidrug resistance P-glycoprotein. *The Journal of Biological Chemistry, 271*, 13668–13674.

Hall, M. D., Okabe, M., Shen, D. W., Liang, X. J., & Gottesman, M. M. (2008). The role of cellular accumulation in determining sensitivity to platinum-based chemotherapy. *Annual Review of Pharmacology and Toxicology, 48*, 495–535.

Higgins, C. F. (1993). The multidrug resistance P-glycoprotein. *Current Opinion in Cell Biology, 5*, 684–687.

Hirata, J., Kikuchi, Y., Kita, T., Imaizumi, E., Tode, T., Ishii, K., et al. (1993). Modulation of sensitivity of human ovarian cancer cells to cis-diamminedichloroplatinum(II) by 12-O-tetradecanoylphorbol-13-acetate and D, L-buthionine-S, R-sulphoximine. *International Journal of Cancer, 55*, 521–527.

Hofmann, J., Doppler, W., Jakob, A., Maly, K., Posch, L., Uberall, F., et al. (1988). Enhancement of the antiproliferative effect of cis-diamminedichloroplatinum(II) and nitrogen mustard by inhibitors of protein kinase C. *International Journal of Cancer, 42*, 382–388.

Huang, J., Mohanty, S., & Basu, A. (2004). Cisplatin resistance is associated with deregulation in protein kinase C-delta. *Biochemical and Biophysical Research Communications, 316*, 1002–1008.

Ido, M., Asao, T., Sakurai, M., Inagaki, M., Saito, M., & Hidaka, H. (1986). An inhibitor of protein kinase C, 1-(5-isoquinolinylsulfonyl)-2-methylpiperazine(H-7) inhibits TPA-induced reduction of vincristine uptake from P388 murine leukemic cell. *Leukemia Research, 10*, 1063–1069.

Ido, M., Sato, K., Sakurai, M., Inagaki, M., Saitoh, M., Watanabe, M., et al. (1987). Decreased phorbol ester receptor and protein kinase C in P388 murine leukemic cells resistant to etoposide. *Cancer Research, 47*, 3460–3463.

Ishida, S., Lee, J., Thiele, D. J., & Herskowitz, I. (2002). Uptake of the anticancer drug cisplatin mediated by the copper transporter Ctr1 in yeast and mammals. *Proceedings of the National Academy of Sciences of the United States of America, 99*, 14298–14302.

Isonishi, S., Andrews, P. A., & Howell, S. B. (1990). Increased sensitivity to cis-diamm inedichloroplatinum(II) in human ovarian carcinoma cells in response to treatment with 12-O-tetradecanoylphorbol 13-acetate. *The Journal of Biological Chemistry, 265*, 3623–3627.

Isonishi, S., Hom, D. K., Eastman, A., & Howell, S. B. (1994). Enhancement of sensitivity to platinum(II)-containing drugs by 12-O-tetradecanoyl-phorbol-13-acetate in a human ovarian carcinoma cell line. *British Journal of Cancer, 69*, 217–221.

Jiffar, T., Kurinna, S., Suck, G., Carlson-Bremer, D., Ricciardi, M. R., Konopleva, M., et al. (2004). PKC alpha mediates chemoresistance in acute lymphoblastic leukemia through effects on Bcl2 phosphorylation. *Leukemia, 18*, 505–512.

Kameyama, N., Arisawa, S., Ueyama, J., Kagota, S., Shinozuka, K., Hattori, A., et al. (2008). Increase in P-glycoprotein accompanied by activation of protein kinase Calpha and NF-kappaB p65 in the livers of rats with streptozotocin-induced diabetes. *Biochimica et Biophysica Acta, 1782*, 355–360.

Kelland, L. (2007). The resurgence of platinum-based cancer chemotherapy. *Nature Reviews. Cancer, 7*, 573–584.

Kerr, J. F., Winterford, C. M., & Harmon, B. V. (1994). Apoptosis. Its significance in cancer and cancer therapy. *Cancer, 73*, 2013–2026.

Killion, J. J., Beltran, P., O'Brian, C. A., Yoon, S. S., Fan, D., Wilson, M. R., et al. (1995). The antitumor activity of doxorubicin against drug-resistant murine carcinoma is enhanced by oral administration of a synthetic staurosporine analogue, CGP 41251. *Oncology Research, 7*, 453–459.

Kim, E. H., Lee, H. J., Lee, D. H., Bae, S., Soh, J. W., Jeoung, D., et al. (2007). Inhibition of heat shock protein 27-mediated resistance to DNA damaging agents by a novel PKC delta-V5 heptapeptide. *Cancer Research, 67*, 6333–6341.

Kurinna, S., Konopleva, M., Palla, S. L., Chen, W., Kornblau, S., Contractor, R., et al. (2006). Bcl2 phosphorylation and active PKC alpha are associated with poor survival in AML. *Leukemia, 20*, 1316–1319.

Lee, S. A., Karaszkiewicz, J. W., & Anderson, W. B. (1992). Elevated level of nuclear protein kinase C in multidrug-resistant MCF-7 human breast carcinoma cells. *Cancer Research, 52*, 3750–3759.

Lu, Y., Jamieson, L., Brasier, A. R., & Fields, A. P. (2001). NF-kappaB/RelA transactivation is required for atypical protein kinase C iota-mediated cell survival. *Oncogene, 20*, 4777–4792.

Lu, D., Sivaprasad, U., Huang, J., Shankar, E., Morrow, S., & Basu, A. (2007). Protein kinase C-epsilon protects MCF-7 cells from TNF-mediated cell death by inhibiting Bax translocation. *Apoptosis, 12*, 1893–1900.

Masanek, U., Stammler, G., & Volm, M. (2002). Modulation of multidrug resistance in human ovarian cancer cell lines by inhibition of P-glycoprotein 170 and PKC isoenzymes with antisense oligonucleotides. *Journal of Experimental Therapeutics & Oncology, 2*, 37–41.

Matsumoto, Y., Kunishio, K., & Nagao, S. (1999). Increased phosphorylation of DNA topoisomerase II in etoposide resistant mutants of human glioma cell line. *Journal of Neuro-Oncology, 45*, 37–46.

Matsumoto, T., Tani, E., Yamaura, I., Miyaji, K., & Kaba, K. (1995). Effects of protein kinase C modulators on multidrug resistance in human glioma cells. *Neurosurgery, 36*, 565–571. discussion 572.

May, W. S., Tyler, P. G., Ito, T., Armstrong, D. K., Qatsha, K. A., & Davidson, N. E. (1994). Interleukin-3 and bryostatin-1 mediate hyperphosphorylation of BCL2 alpha in association with suppression of apoptosis. *The Journal of Biological Chemistry, 269*, 26865–26870.

Merritt, J. E., Sullivan, J. A., Drew, L., Khan, A., Wilson, K., Mulqueen, M., et al. (1999). The bisindolylmaleimide protein kinase C inhibitor, Ro 32-2241, reverses multidrug resistance in KB tumour cells. *Cancer Chemotherapy and Pharmacology, 43*, 371–378.

Miyamoto, K., Inoko, K., Wakusawa, S., Kajita, S., Hasegawa, T., Takagi, K., et al. (1993). Inhibition of multidrug resistance by a new staurosporine derivative, NA-382, in vitro and in vivo. *Cancer Research, 53*, 1555–1559.

Mohanty, S., Huang, J., & Basu, A. (2005). Enhancement of cisplatin sensitivity of cisplatin-resistant human cervical carcinoma cells by bryostatin 1. *Clinical Cancer Research, 11*, 6730–6737.

Mouchel, N. A., & Jenkins, J. R. (2006). The identification of a functional interaction between PKC and topoisomerase II. *FEBS Letters, 580*, 51–57.

Murray, N. R., & Fields, A. P. (1997). Atypical protein kinase C iota protects human leukemia cells against drug-induced apoptosis. *The Journal of Biological Chemistry, 272*, 27521–27524.

O'Brian, C. A., Fan, D., Ward, N. E., Dong, Z., Iwamoto, L., Gupta, K. P., et al. (1991). Transient enhancement of multidrug resistance by the bile acid deoxycholate in murine fibrosarcoma cells in vitro. *Biochemical Pharmacology, 41*, 797–806.

O'Brian, C. A., Fan, D., Ward, N. E., Seid, C., & Fidler, I. J. (1989). Level of protein kinase C activity correlates directly with resistance to adriamycin in murine fibrosarcoma cells. *FEBS Letters, 246*, 78–82.

O'Brian, C. A., Ward, N. E., Gravitt, K. R., & Gupta, K. P. (1995). The tumor promoter receptor protein kinase C: A novel target for chemoprevention and therapy of human colon cancer. *Progress in Clinical and Biological Research, 391*, 117–120.

Palayoor, S. T., Stein, J. M., & Hait, W. N. (1987). Inhibition of protein kinase C by antineoplastic agents: Implications for drug resistance. *Biochemical and Biophysical Research Communications, 148*, 718–725.

Pardo, O. E., Wellbrock, C., Khanzada, U. K., Aubert, M., Arozarena, I., Davidson, S., et al. (2006). FGF-2 protects small cell lung cancer cells from apoptosis through a complex involving PKCepsilon, B-Raf and S6K2. *The EMBO Journal, 25*, 3078–3088.

Perego, P., Casati, G., Gambetta, R. A., Soranzo, C., & Zunino, F. (1993). Effect of modulation of protein kinase C activity on cisplatin cytotoxicity in cisplatin-resistant and cisplatin-sensitive human osteosarcoma cells. *Cancer Letters, 72*, 53–58.

Plo, I., Hernandez, H., Kohlhagen, G., Lautier, D., Pommier, Y., & Laurent, G. (2002). Overexpression of the atypical protein kinase C zeta reduces topoisomerase II catalytic activity, cleavable complexes formation, and drug-induced cytotoxicity in monocytic U937 leukemia cells. *The Journal of Biological Chemistry, 277*, 31407–31415.

Posada, J. A., McKeegan, E. M., Worthington, K. F., Morin, M. J., Jaken, S., & Tritton, T. R. (1989). Human multidrug resistant KB cells overexpress protein kinase C: Involvement in drug resistance. *Cancer Communications, 1*, 285–292.

Pulaski, L., Szemraj, J., Uchiumi, T., Kuwano, M., & Bartosz, G. (2005). Transcriptional upregulation of the human MRP2 gene expression by serine/threonine protein kinase inhibitors. *Journal of Biological Regulators and Homeostatic Agents, 19*, 113–119.

Ratnasinghe, D., Daschner, P. J., Anver, M. R., Kasprzak, B. H., Taylor, P. R., Yeh, G. C., et al. (2001). Cyclooxygenase-2, P-glycoprotein-170 and drug resistance; is chemoprevention against multidrug resistance possible? *Anticancer Research, 21*, 2141–2147.

Richards, E. H., Hickey, E., Weber, L., & Master, J. R. (1996). Effect of overexpression of the small heat shock protein HSP27 on the heat and drug sensitivities of human testis tumor cells. *Cancer Research, 56*, 2446–2451.

Rumsby, M. G., Drew, L., & Warr, J. R. (1998). Protein kinases and multidrug resistance. *Cytotechnology, 27*, 203–224.

Ruvolo, P. P., Deng, X., Carr, B. K., & May, W. S. (1998). A functional role for mitochondrial protein kinase Calpha in Bcl2 phosphorylation and suppression of apoptosis. *The Journal of Biological Chemistry, 273*, 25436–25442.

Sachs, C. W., Ballas, L. M., Mascarella, S. W., Safa, A. R., Lewin, A. H., Loomis, C., et al. (1996). Effects of sphingosine stereoisomers on P-glycoprotein phosphorylation and vinblastine accumulation in multidrug-resistant MCF-7 cells. *Biochemical Pharmacology, 52*, 603–612.

Sachs, C. W., Safa, A. R., Harrison, S. D., & Fine, R. L. (1995). Partial inhibition of multidrug resistance by safingol is independent of modulation of P-glycoprotein substrate activities and correlated with inhibition of protein kinase C. *The Journal of Biological Chemistry, 270*, 26639–26648.

Safaei, R., Holzer, A. K., Katano, K., Samimi, G., & Howell, S. B. (2004). The role of copper transporters in the development of resistance to Pt drugs. *Journal of Inorganic Biochemistry, 98*, 1607–1613.

Samuels, D. S., Shimizu, Y., & Shimizu, N. (1989). Protein kinase C phosphorylates DNA topoisomerase I. *FEBS Letters, 259*, 57–60.

Sato, W., Yusa, K., Naito, M., & Tsuruo, T. (1990). Staurosporine, a potent inhibitor of C-kinase, enhances drug accumulation in multidrug-resistant cells. *Biochemical and Biophysical Research Communications, 173*, 1252–1257.

Scala, S., Dickstein, B., Regis, J., Szallasi, Z., Blumberg, P. M., & Bates, S. E. (1995). Bryostatin 1 affects P-glycoprotein phosphorylation but not function in multidrug-resistant human breast cancer cells. *Clinical Cancer Research, 1*, 1581–1587.

Schornagel, J. H., Chang, P. K., Sciarini, L. J., Moroson, B. A., Mini, E., Cashmore, A. R., et al. (1984). Synthesis and evaluation of 2, 4-diaminoquinazoline antifolates with activity against methotrexate-resistant human tumor cells. *Biochemical Pharmacology, 33*, 3251–3255.

Schwartz, G. K., Arkin, H., Holland, J. F., & Ohnuma, T. (1991). Protein kinase C activity and multidrug resistance in MOLT-3 human lymphoblastic leukemia cells resistant to trimetrexate. *Cancer Research, 51*, 55–61.

Segal-Bendirdjian, E., & Jacquemin-Sablon, A. (1995). Cisplatin resistance in a murine leukemia cell line is associated with a defective apoptotic process. *Experimental Cell Research, 218*, 201–212.

Sha, E. C., Sha, M. C., & Kaufmann, S. H. (1996). Evaluation of 2, 6-diamino-N-([1-(1-oxotridecyl)-2-piperidinyl]methyl)- hexanamide (NPC 15437), a protein kinase C inhibitor, as a modulator of P-glycoprotein-mediated resistance in vitro. *Investigational New Drugs, 13*, 285–294.

Sherman, S. E., Gibson, D., Wang, A. H., & Lippard, S. J. (1985). X-ray structure of the major adduct of the anticancer drug cisplatin with DNA: cis-[Pt(NH3)2(d(pGpG))]. *Science, 230*, 412–417.

Siddik, Z. H. (2003). Cisplatin: mode of cytotoxic action and molecular basis of resistance. *Oncogene, 22*, 7265–7279.

Singhal, S. S., Wickramarachchi, D., Singhal, J., Yadav, S., Awasthi, Y. C., & Awasthi, S. (2006). Determinants of differential doxorubicin sensitivity between SCLC and NSCLC. *FEBS Letters, 580*, 2258–2264.

Smith, C. D., & Zilfou, J. T. (1995). Circumvention of P-glycoprotein-mediated multiple drug resistance by phosphorylation modulators is independent of protein kinases. *The Journal of Biological Chemistry, 270*, 28145–28152.

Sonnemann, J., Gekeler, V., Ahlbrecht, K., Brischwein, K., Liu, C., Bader, P., et al. (2004). Down-regulation of protein kinase Ceta by antisense oligonucleotides sensitises A549 lung cancer cells to vincristine and paclitaxel. *Cancer Letters, 209*, 177–185.

Svensson, K., & Larsson, C. (2003). A protein kinase Cbeta inhibitor attenuates multidrug resistance of neuroblastoma cells. *BMC Cancer, 3*, 10.

Volm, M., & Pommerenke, E. W. (1995). Associated expression of protein kinase C with resistance to doxorubicin in human lung cancer. *Anticancer Research, 15*, 463–466.

Wakusawa, S., Inoko, K., Miyamoto, K., Kajita, S., Hasegawa, T., Harimaya, K., et al. (1993). Staurosporine derivatives reverse multidrug resistance without correlation with their protein kinase inhibitory activities. *The Journal of Antibiotics, 46*, 353–355.

Wang, W. L., Yeh, S. F., Chang, Y. I., Hsiao, S. F., Lian, W. N., Lin, C. H., et al. (2003). PICK1, an anchoring protein that specifically targets protein kinase Calpha to mitochondria selectively upon serum stimulation in NIH 3T3 cells. *The Journal of Biological Chemistry, 278*, 37705–37712.

Wang, W. L., Yeh, S. F., Huang, E. Y., Lu, Y. L., Wang, C. F., Huang, C. Y., et al. (2007). Mitochondrial anchoring of PKCalpha by PICK1 confers resistance to etoposide-induced apoptosis. *Apoptosis, 12*, 1857–1871.

Ward, N. E., & O'Brian, C. A. (1991). Distinct patterns of phorbol ester-induced downregulation of protein kinase C activity in adriamycin-selected multidrug resistant and parental murine fibrosarcoma cells. *Cancer Letters, 58*, 189–193.

Wielinga, P. R., Heijn, M., Broxterman, H. J., & Lankelma, J. (1997). P-glycoprotein-independent decrease in drug accumulation by phorbol ester treatment of tumor cells. *Biochemical Pharmacology, 54*, 791–799.

Xin, M., Gao, F., May, W. S., Flagg, T., & Deng, X. (2007). Protein kinase Czeta abrogates the proapoptotic function of Bax through phosphorylation. *The Journal of Biological Chemistry, 282*, 21268–21277.

Yamamoto, K., Okamoto, A., Isonishi, S., Ochiai, K., & Ohtake, Y. (2001). Heat shock protein 27 was up-regulated in cisplatin resistant human ovarian tumor cell line and associated with the cisplatin resistance. *Cancer Letters, 168*, 173–181.

Chapter 22
PKCδ as a Target for Chemotherapeutic Drugs

Chaya Brodie and Stephanie L. Lomonaco

Abstract PKCδ, a member of the novel PKC family, has been widely implicated as mediator of apoptosis in response to phorbol esters and chemotherapeutic agents, and it is differentially expressed in various human cancer types. A characteristic of PKCδ is that it is regulated by tyrosine phosphorylation, and in some cases this post-translational modification seems to be essential for the response to stimuli. PKCδ may be an attractive target for cancer therapeutics.

Keywords PKCδ • Chemotherapeutic drugs • Tyrosine phosphorylation • Apoptosis

22.1 Introduction

PKCδ is a ubiquitously expressed isoform of the novel PKC subfamily which plays major roles in various cell signaling and in a large array of cellular processes in a tissue- and cell-specific manner. PKCδ has been shown to exert both inhibitory and promoting effects on cell proliferation and tumor progression in most tumor cells; however, it exerts opposite effects in other cancer cells. Similarly, PKCδ can regulate both cell apoptosis and survival. Thus, analyzing the role of PKCδ in tumorigenesis and its consideration as a target for chemotherapeutic drugs are complicated by the ability of PKCδ to exert both pro- and antitumor effects in various tumor cells.

This chapter summarizes data indicating differential expression of PKCδ in various human tumors and the diverse effects of PKCδ on cell proliferation, cell cycle control, migration, invasion, angiogenesis, regulation of cell apoptosis, and

C. Brodie (✉) and S.L. Lomonaco
William and Karen Davidson Laboratory of Cell Signaling and Tumorigenesis, Hermelin Brain Tumor Center, Department of Neurosurgery, Henry Ford Hospital, Detroit, MI, USA
e-mail: nscha@neuro.hfh.edu

C. Brodie
Mina and Everard Goodman Faculty of Life-Sciences, Bar-Ilan University, Ramat-Gan, Israel

M.G. Kazanietz (ed.), *Protein Kinase C in Cancer Signaling and Therapy*, Current Cancer Research, DOI 10.1007/978-1-60761-543-9_22,

response to different antitumor therapies. Different substrates of PKCδ and downstream signaling pathways in cancer cells will be discussed as well.

22.2 Protein Kinase C and Cancer

PKC is a family of serine threonine kinases that play a major role in cell signaling and a variety of cellular processes including proliferation, differentiation, cell motility, and apoptosis (Dempsey et al. 2000). The PKC family is comprised of at least 12 isoforms with distinct cellular functions that are divided into three subgroups: the classical PKCs (PKCα, β1, PKCβ2, and PKCγ) that are activated by Ca^{2+} and DAG, the novel PKCs (PKCδ, PKCε, PKCθ and PKCη) that are only activated by DAG, and the atypical PKCs (PKCζ and PKCι) that do not respond to either Ca^{2+} or DAG (Mellor and Parker 1998; Nakamura and Nishizuka 1994; Nishizuka 1995; Toker 1998). The findings that PKC acts as a high-affinity intracellular receptor for the tumor promoter phorbol esters indicated an important role for PKC in carcinogenesis and positioned it as an important molecular therapeutic target in cancer (Castagna et al. 1982; Kikkawa et al. 1983; Leach et al. 1983). Indeed, the role of PKC in carcinogenesis and as a potential target in cancer therapy has been studied in various cellular systems, and the results of these studies are summarized in a large number of reviews (da Rocha et al. 2002; Fields et al. 2007; Griner and Kazanietz 2007; Koivunen et al. 2006; Lorenzo and Dennis 2003; Mackay and Twelves 2007; Martiny-Baron and Fabbro 2007; O'Brian et al. 2001; Podar et al. 2007; Serova et al. 2006). The role of PKC in cancer is not due to mutations in PKC genes and other than rare mutations in PKCα (Alvaro et al. 1993; Prevostel et al. 1997), there have been no reports of mutations in other PKC isoforms. Therefore, it seems that the contribution of a specific PKC isoform to carcinogenesis may result from aberrant expression, enhanced activation downstream to growth factors receptors, and changes in subcellular localization or depletion as a result of prolonged activation. In addition, the interaction of PKC isoforms with different oncogenes or tumor suppressors may also impact their contribution to carcinogenesis.

22.2.1 *PKCδ*

PKCδ is a ubiquitously expressed isoform that belongs to the novel PKC family (Gschwendt 1995). In addition to its classical mode of activation, PKCδ activity is mediated by a series of phosphorylation on serine, threonine, and tyrosine residues (Denning et al. 1993; Durgan et al. 2007; Kronfeld et al. 2000). Numerous studies of PKC expression in normal and tumor cells as well as functional studies of tumor progression, cell proliferation, motility, and apoptosis confirmed the involvement of PKCδ in various processes related to malignant transformation.

The human PKCδ gene is located on chromosome 3p (Huppi et al. 1994) in a region that is characterized by loss of heterozygosity (LOH) in a large number of tumors; however, a role of LOH in the decreased expression of PKCδ in various tumors has not yet been confirmed. Studies in PKCδ null mice indicated that these mice developed normally but had a significantly higher number of smooth muscle cells (Leitges et al. 2001) and B cell (Miyamoto et al. 2002), suggesting that PKCδ is not required for the proliferation of normal cells.

The expression of PKCδ in different tumors and its diverse functions in cancer cells are addressed below.

22.3 Expression and Functions of PKCδ in Cancer Cells

Altered PKCδ expression has been reported in various tumors, indicating a possible role of this isoform in tumorigenesis. In addition to its differential expression, PKCδ has been shown to play major roles in the regulation of various tumor cell functions such as proliferation, migration and invasion, angiogenesis and cell apoptosis, and survival.

22.3.1 Expression of PKCδ in Different Tumors

PKCδ has been shown to display variable expression in different types of tissues and cancers and in different stages of cancer progression. PKCδ expression is related to the degree of malignancy in many cancers. Normal tissue and low-grade bladder tumors express high levels of PKCδ, whereas a lower level is observed in high-grade bladder tumors (Koren et al. 2000; Langzam et al. 2001; Varga et al. 2004). PKCδ expression is increased in highly metastatic mammary tumors compared to less metastatic parental cell lines (Kiley et al. 1999a, b). Over-expression of PKCδ in low and moderate metastatic cells showed no change in cell proliferation, but significantly increased anchorage-independent growth. In addition, antiestrogen-resistant breast cancer cells expressed significantly high levels of both total and activated PKCδ levels compared to sensitive cells (Nabha et al. 2005). Inhibition of PKCδ by rottlerin, a specific PKCδ inhibitor, or by siRNA significantly inhibited estrogen and tamoxifen-induced growth in estrogen-resistant cells. Thus, PKCδ plays a major role in antiestrogen resistance in breast cancer tumor cells and may provide a new target for treatment. There is also an increased PKCδ expression and activity in pancreatic cancer cells and these cells become sensitive to apoptosis by inhibition of PKCδ (El-Rayes et al. 2008). Colorectal cancer has varied levels of PKCδ, whereas adenocarcinomas express lower levels of PKCδ compared to the surrounding mucosa (Craven and DeRubertis 1994) and highly invasive colorectal cancer has increased levels of PKCδ as well as other PKC isoforms (Li et al. 2004; Pongracz et al. 1995). The expression of PKCδ was recently examined in endometrioid carcinomas of increasing grade. PKCδ exhibited abundant nuclear and cytoplasmic staining in normal endometrium, whereas endometrial tumors showed decreased PKCδ expression and increasing tumor grade, with PKCδ being preferentially lost from the nucleus. Similarly, reduced PKCδ levels were observed in endometrial cancer cell lines derived from poorly differentiated tumors compared to well-differentiated lines, suggesting that loss of PKCδ is an indicator of endometrial malignancy and that PKCδ may function as a tumor suppressor in endometrial cancer (Reno et al. 2008).

The expression of specific PKC isoforms was also determined in grade II astrocytomas and glioblastomas (grade IV). It was found that the expression of PKCα and ε was increased in GBM, whereas that of PKCδ was decreased (Mandil et al. 2001). However, a larger study indicated that a subpopulation of GBM and anaplastic astrocytomas express higher levels of PKCδ than low-grade astrocytomas.

In addition to changes in PKCδ expression, other changes in its activation, phosphorylation, and subcellular localization should also be considered. For example, activation of PLC downstream of growth factor receptors, which are overexpressed or are constitutively active in various tumors, such as EGFR and PDGFR, could lead to phosphorylation and prolonged activation of PKCδ, which may eventually lead to its down regulation.

22.3.2 Tumor Suppression and Progression

The first indications of the tumor-suppressing effect of PKCδ came from studies that examined the role of specific PKC isoforms in the effect of the tumor promoters, phorbol esters. These compounds can initially activate and translocate various PKC isoforms, whereas prolonged activation results in proteolytic degradation. Using c-Src transformed fibroblast, it has been shown that depletion of PKCδ by phorbol esters contributed to the transformed phenotype of these cells (Lu et al. 1997). These results were further validated by the use of a PKCδ-KD mutant, suggesting that in these cells PKCδ acts as a tumor suppressor (Lu et al. 1997). Similarly, inactivation of PKCδ causes the progression of keratinocytes and the malignant transformation to squamous cell carcinoma (Yuspa 1998). Various experiments using the PKCδ inhibitor, rottlerin (Gschwendt et al. 1994) supported a role of PKCδ as a tumor suppressor in human keratinocytes and rat fibroblasts overexpressing the EGF receptor or Src; however, these studies should be cautiously interpreted since rottlerin has been shown to exert multiple effects in a PKCδ-independent manner (Basu et al. 2008a; Song et al. 2008). Expression of a PKCδ-KD mutant induced transformed phenotypes in various cell types further supporting a role of PKCδ as a tumor suppressor. A recent study (Cai et al. 2009) reported that PKCδ acted as a tumor suppressor by negatively regulating hedgehog signaling in human hepatoma cells. Constitutive activation of the hedgehog pathway plays a role in the tumorigenesis of various tumors. PKCδ negatively regulates the expression of the hedgehog target genes downstream of the hedgehog antagonist Smoothened. Collectively, these studies suggest that PKCδ acts as a tumor suppressor in most human cellular systems.

22.3.3 Role of PKCδ in Cell Proliferation and Cell Cycle Regulation

PKCδ has been shown to regulate cell proliferation in a cell-specific manner (Jackson and Foster 2004). Although most studies report that PKCδ negatively regulates cell proliferation and cell cycle progression, there are some studies that demonstrated that PKCδ can also promote cell proliferation. A negative effect of PKCδ on CHO cell proliferation was reported (Watanabe et al. 1992). Similarly, Mischak et al. and Acs et al. reported that PKCδ inhibited the proliferation of NIH3T3 cells (Acs et al. 2000; Mischak et al. 1993). Interestingly, the inhibitory effect of PKCδ was dependent on its phosphorylation of tyrosine 155 since mutation in this tyrosine residue increased cell proliferation.

PKCδ effects have also been examined in glioma cells. In these cells, PKCδ and PKCα played opposite effects, where PKCα promoted cell proliferation and PKCδ decreased it. Using PKC chimeras it was found that the regulatory domain of PKCδ mediated the inhibitory effect of PKCδ on cell proliferation.

The effects of PKCδ on the proliferation of breast cancer cells are controversial. On the one hand, PKCδ has been shown to mediate the antiproliferative effects of inositol hexaphosphate in MCF-7 cells via inhibition of Erk1/2 and pRb (Vucenik et al. 2005). Similarly, a PKCδ-KD mutant abolished the G1 arrest of SKRB-3 breast cancer cells induced by phorbol esters (Yokoyama et al. 2005). In contrast, other studies implicated PKCδ as a positive regulator of breast cancer cells via the activation of the Ras/Erk1/2 pathway (Keshamouni et al. 2002). In addition, overexpression of PKCδ in immortalized mammary cells induced anchorage-independent growth and enhanced the survival of the cells, supporting a role of PKCδ in promoting cell proliferation and tumor progression in these cells (Kiley et al. 1999a). A positive effect of PKCδ was also demonstrated for the proliferation signals of the insulin-like growth factor-1 (IGF-1) receptor in transformed cells via the association and tyrosine phosphorylation of this isoform (Li et al. 1998).

In addition to its effect on cell proliferation, there have been numerous studies demonstrating the effect of PKCδ on cell cycle progression in both normal and cancer cells, and PKCδ has been shown to arrest cells in G1/S and G2/M phases of the cell cycle. In lung adenocarcinoma cells, PKCδ induced a G1 arrest via upregulation of the cell cycle inhibitor p21 (Nakagawa et al. 2005). PKCδ blocked vascular smooth muscle cells in G0/G1 phase and prevented progression to S phase of capillary endothelial cells (Ashton et al. 1999). G1 block was also demonstrated in A431 cells treated with an antitumor somatostatin analog (Stetak et al. 2001). PKCδ mediated the induction of the G1 cyclin-dependent kinase inhibitor $p21^{cip1}$ in various cell types and inhibit the expression of cyclin D1 (Cerda et al. 2006; Fukumoto et al. 1997). A recent study showed that PKCδ mediated the effect of PMA on the inhibition of proliferation and cell cycle progression at the G1/S phase of thyroid cancer cells via an increase in $p21^{cip1}$ and p27kip1 (Afrasiabi et al. 2008). A similar inhibitory effect of PKCδ on cell proliferation and G1 arrest in thyroid cancer cells was also observed (Hung et al. 2008; Koike et al. 2006). In lung cancer cells, PKCδ was essential for the G1 arrest induced by phorbol esters and the up-regulation of p21 and inhibition of cyclin A which was required for the activation of cdk2 in the S phase (Nakagawa et al. 2005). The inhibitory effect of PKCδ on the cyclin A promoter in these cells was secondary to the pocket protein inactivation and E2F release. PKCδ has also been shown to mediate the antiproliferative effects and cell cycle arrest of anticancer agents. Apicidin, a histone deacetylase inhibitor, increased the transcriptional activity of cyclin D3, which was regulated by PKCδ and was mediated through Sp1 site (Kim et al. 2007b).

In addition to inhibition progression into the S-phase, PKCδ can also arrest CHO cells (Watanabe et al. 1992) and human melanoma cells (Watters et al. 1998) at the G2/M phase. These results implicate PKCδ as a "gatekeeper" that can prevent cell cycle progression through the G1/S and G2/M checkpoints of the cell cycle.

22.3.4 Role of PKCδ in Tumor Cell Migration and Invasion

Cell migration is a complex process that involves receptor-mediated adhesion, membrane protrusions, and the formation of defined cell-matrix adhesion sites linked to a reorganization of the actin cytoskeleton that is required for tumor invasion and metastasis (Ridley et al. 2003).

PKCδ has been shown to play a role in cell migration in various cellular systems. Using PKCδ null mice, Li et al., has demonstrated that PKCδ plays a role in the migration of smooth muscle cells following mechanical stress and the phosphorylation of vinculin, FAK, and paxillin (Li et al. 2003). PKCδ has been shown to play an important role in the sustained phase of cell migration of EGFR overexpressing breast cancer cells. PKCδ signaling in the EGFR overexpressing invasive cells circumvented the loss of MAP-mediated signaling, which plays a role in the early stages of cell migration (Kruger and Reddy 2003). The proposed mechanisms for the effect of PKCδ on cell migration included the activation of Src and FAK and the establishment of the Cas–Crk complex for sustained cell migration. Similarly, PKCδ played a role in EGF-induced fibroblast motility via the phosphorylation of myosin light chain (MLC) (Iwabu et al. 2004). PKCδ mediated the inhibitory effect of phospho-protein enriched in astrocytes-15 kDa (PEA-15) on astrocyte migration (Renault-Mihara et al. 2006). In renal cell carcinoma, PKCδ regulated the migration of these cells by affecting the expression and activity of β1 integrins and FAK (Brenner et al. 2008).

PKCδ has also been shown to play a major role in the metastasis of mammary tumors cells. The expression of PKCδ was significantly increased in highly metastatic mammary tumor cells as compared to less metastatic cells (Kiley et al. 1999b). Overexpression of PKCδ increased the anchorage-independent growth of the cells and inhibition of PKCδ with the regulatory domain of this isoform inhibited the phosphorylation of the PKC cytoskeletal substrate adducin (Kiley et al. 1999a) and reduced the number of lung metastases without affecting cell growth, suggesting that inhibition of PKCδ selectively interfered with the formation of metastases (Kiley et al. 1999b). Increased PKCδ expression is also correlated with hepatic metastasis in colorectal carcinomas (Li et al. 2004).

In addition to its effects on cell migration and metastasis, PKCδ also affects the invasion of cancer cells. PKCδ has been shown to play a crucial role in mediating the effect of platelets on the invasion of breast cancer cells via secretion of MMP9 (Alonso-Escolano et al. 2006). In human mammary epithelial cells overexpressing erbB2, PKCδ mediated the invasion of these cells downstream of AKT (Woods Ignatoski et al. 2003). The role of PKCδ was also studied in the invasion of prostate cancer cells stimulated with a peptide composed of the PHSRN sequence of fibronectin, which associates with the α5β1 integrin. Stimulation of prostate cancer cells with this peptide induced cell invasion and induction of MMP-1, which was abolished on silencing of PKCδ. In these cells, the activation of PKCδ was attributed to the activation of PI3-Kinase (Zeng et al. 2006). PKCδ has been shown to be activated in melanoma cells following interaction of CD44 and epidermal growth factor receptor. Activation of PKCδ in these cells was downstream of AKT and associated with activation of MMP-2 (Kim et al. 2008). The expression of another member of the metalloproteinase family, MMP-9 has been associated with the activation of PKCδ in the invasion of human pituitary adenoma cells (Hussaini et al. 2007). A role of PKCδ in MMP-9 induction was also demonstrated in cells treated with PMA (Woo et al. 2004). Resveratrol inhibited PMA-mediated MMP-9 induction by inhibiting the activation of PKCδ and JNK (Woo et al. 2004). PCPH, a known oncogene, which is highly expressed in prostate carcinoma, regulates invasiveness and collagen-1 expression by a mechanism involving PKCδ (Villar et al. 2007). In summary, PKCδ appears to positively regulate the invasion of various cancer cells downstream of AKT and PI3-kinase and upstream of the induction of various members of the MMP family.

22.3.5 PKCδ and Angiogenesis

Few studies have demonstrated the role of PKCδ in the regulation of tumor angiogenesis.

Pal et al. demonstrated the role of PKCδ in the angiogenesis of renal cell carcinoma (Pal et al. 1997). The von-Hippel-Lindau (VHL) suppressor gene downregulates VEGF levels at both the transcriptional and posttranscriptional levels. Loss of VHL leads to increased VEGF expression and well vascularized tumors. Overexpression of VHL was directly associated with PKCδ and prevented its translocation to the plasma membrane, the activation of the MAPK pathway, and the decreasing aberrant VEGF expression and angiogenesis.

The role of PKCδ in angiogenesis has also been demonstrated in mediating the effects of VEGF on endothelial cell survival and eNOS activation. Thus, PKCδ has been shown to activate the AKT pathway downstream of VEGF (Gliki et al. 2002).

Recent studies also indicate that PKCδ is a novel regulator of hypoxia-induced angiogenesis. Thus, hypoxia activates PKCδ which in turn increases the protein stability and transcriptional activity of HIF-1α in human cervical adenocarcinoma cells (Lee et al. 2007). Knockdown of PKCδ under hypoxic conditions inhibited VEGF expression and angiogenic activity (Basu et al. 2008b). Collectively, these studies suggest that PKCδ plays a positive role in the regulation of VEGF expression, its signaling pathway, and angiogenesis.

22.4 Role of PKCδ in the Apoptosis and Survival of Cancer Cells

Many of the studies regarding the role PKCδ in the function of cancer cells have been focused on the response of cancer cells to different antitumor therapeutics. Indeed, PKCδ has been reported to play a critical role in the control of cell apoptosis of both normal and cancer cells in various cellular systems. PKCδ is involved in cell apoptosis induced by a variety of apoptotic stimuli in different cellular systems: in neutrophils undergoing spontaneous apoptosis (Khwaja and Tatton 1999), in various cell types in response to H_2O_2 (Konishi et al. 2001), and in response to ceramide (Kajimoto et al. 2001), TNF-α (Emoto et al. 1995) and Fas ligation (Scheel-Toellner et al. 1999). PKCδ also plays a role in the apoptosis induced by UV radiation (Denning et al. 1998) and by DNA-damaging treatments such as ionizing radiation (Yuan et al. 1998), etoposide (Blass et al. 2002; Reyland et al. 1999) and by cytosine arabinoside (Datta et al. 1996). Although the majority of the studies indicate that PKCδ is involved in the induction of cell apoptosis, there are other studies pointing to a role of PKCδ in cell survival and in antiapoptotic responses. Thus, PKCδ plays a role in the antiapoptotic effect of TNF-α in human neutrophils (Kilpatrick et al. 2002), in the antiapoptotic effect of basic FGF (Peluso et al. 2001), and in serum-deprived PC-12 cells (Wert and Palfrey 2000). PKCδ also protects the RAW 264.7 macrophages from nitric oxide-induced apoptosis (Jun et al. 1999), lung cancer cells from chemotherapy-induced apoptosis (Clark et al., Cancer Research, 2003) and is required for the survival of Ras overexpressing cells (Xia et al. 2007). In addition, it was recently reported that PKCδ protected glioma cells from cell apoptosis induced by infection with Sindbis virus (Zrachia et al. 2002) and by TRAIL (Okhrimenko et al. 2005).

22.4.1 Factors That Modulate PKCδ Effects

The diverse effects of PKCδ on cell apoptosis and survival are usually associated with phosphorylation on tyrosine residues, translocation to distinct subcellular sites, and caspase-dependent cleavage. These factors affect the ability of PKCδ to interact and activate different downstream signaling pathways that mediate its apoptotic effects.

Therefore, understanding the different parameters that affect PKCδ function is essential for the design of PKCδ inhibitors that can selectively target specific PKC functions.

22.4.1.1 Phosphorylation of PKCδ on Tyrosine Residues

One of the earliest events that are common to many apoptotic stimuli that affect PKCδ is its phosphorylation of specific tyrosine residues (Brodie and Blumberg 2003). Indeed, PKCδ undergoes tyrosine phosphorylation in response to diverse stimuli (Brodie et al. 1998; Denning et al. 1993; Kronfeld et al. 2000; Li et al. 1996; Lu et al. 2007a; Steinberg, 2004). It was recently found that etoposide induced phosphorylation of PKCδ on tyrosines 64 and 187 in the regulatory domain, and this phosphorylation was essential for the cleavage of PKCδ by caspase 3 and for the apoptotic effect of PKCδ (Blass et al. 2002; Lomonaco et al. 2008). In addition, H_2O_2 induced phosphorylation of PKCδ on multiple phosphorylation sites (Konishi et al. 2001; Lu et al. 2007a), cisplatin induced phosphorylation of tyrosine 332 and this phosphorylation was essential for the cleavage of PKCδ (Lu et al. 2007b). In contrast, infection of the cells with SV induced phosphorylation of PKCδ on tyrosines 52, 64, and 155 and this phosphorylation was essential for the antiapoptotic effect of PKCδ (Zrachia et al. 2002). Similarly, TRAIL induced phosphorylation of PKCδ on tyrosine 155 that was essential for the translocation of PKCδ to the ER (Okhrimenko et al. 2005).

Although tyrosine phosphorylation of PKCδ is common to many apoptotic stimuli, its role in the antiapoptotic function of this kinase is still not completely understood. A number of tyrosine kinases have been implicated in the phosphorylation of PKCδ. c-Abl is one of the important tyrosine kinases that is activated by apoptotic stimuli and phosphorylates PKCδ in response to DNA damage and oxidative stress (Steinberg 2004; Sun et al. 2000). c-Abl is a ubiquitously expressed kinase that can localize to the nucleus, ER, and mitochondria and plays a role in cell apoptosis in a p53- and p73-dependent manner (Deng et al. 2004). c-Abl and PKCδ interact in response to oxidative and genotoxic stresses (Sun et al. 2000). It was proposed that c-Abl associates with PKCδ via the SH3 domain and that following radiation it phosphorylates PKCδ on tyrosine residues and mediates its nuclear translocation (Yuan et al. 1998). c-Abl also phosphorylates PKCδ in response to oxidative stress on tyrosine 311 in glioma cells (Lu et al. 2007a). The association of PKCδ and c-Abl is important for the activation of the latter since the inhibition of PKCδ by overexpression of its regulatory domain decreased the activity of c-Abl (Sun et al. 2000). Similarly, cells treated with cisplatin and methylglyoxal increased the association of these two kinases and the activity of c-Abl (Godbout et al. 2002).

Src kinases also play an important role in the phosphorylation of PKCδ (Joseloff et al. 2002; Rybin et al. 2007; Song et al. 1998). Src phosphorylates PKCδ in glioma cells on tyrosine 332 in response to PMA and cisplatin (Lu et al. 2007b) and

Fyn on tyrosine 187 in response to PDGF (Kronfeld et al. 2000). Identification of additional kinases that phosphorylate PKCδ in response to apoptotic stimuli is essential for understanding the contribution of PKCδ to the apoptotic signaling. Tyrosine phosphorylation of PKCδ acts as a molecular switch in PKCδ function. Thus, a promising approach for the development of specific PKCδ therapeutics is by targeting specific tyrosine residues. Indeed, expression of different PKCδ mutants in which specific tyrosine residues were mutated to phenylalanine were able to either antagonize or sensitize cancer cells to different chemotherapeutic drugs (Okhrimenko et al. 2005; Li et al. 2007).

22.4.1.2 Translocation and Subcellular Localization

Another important factor that contributes to the distinct effects of PKCδ on cell apoptosis is the differential patterns of PKC translocation in response to apoptotic stimuli. Translocation of PKCδ to specific cellular compartments may lead to different cellular effects due to the phosphorylation of specific substrates and to the association of PKCδ with distinct proteins present in these locations. Apoptotic stimuli induce distinct patterns of cellular localization of PKCδ. Thus, translocation of PKCδ to the membrane was observed in response to UV radiation (Denning et al. 2002) and to the mitochondria in response to oxidative stress (Li et al. 1999). PKCδ also translocates to the nucleus in cytosine arabinoside-treated cells (Datta et al. 1996) and in cells treated with etoposide (Reyland et al. 1999). Ceramide induces translocation of PKCδ to the golgi (Kajimoto et al. 2001). Recently, we reported that PKCδ translocates to the endoplasmic reticulum in response to Sindbis virus (SV) infection (Zrachia et al. 2002) and TRAIL (Okhrimenko et al. 2005). Targeting of PKCδ to distinct subcellular sites using pShooter vectors demonstrated that the expression of PKCδ in the nucleus, mitochondria, and cytosol resulted in cell apoptosis, whereas its translocation in the ER exerted antiapoptotic effects (Gomel et al. 2007).

Since translocation of PKCδ is one of the hallmarks of its activation and the localization of PKCδ to distinct subcellular sites determines it diverse effect, one approach to selectively affect the effects of PKCδ on cell apoptosis is by using translocation inhibitors. Indeed, peptides targeting the translocation of PKCδ have been recently described as potential inhibitors of PKCδ (Budas et al. 2007; Tanaka et al. 2004); however, it is unclear whether these inhibitors can selectively inhibit translocation to a specific subcellular site in tumor cells.

22.4.1.3 Caspase-3-Dependent Cleavage

PKCδ is proteolytically cleaved in response to apoptotic stimuli (Emoto et al. 1995). The caspase-3 cleavage site in PKCδ was mapped to the V3 domain adjacent to the aspartic acid at the DMQD^{330}N site (Ghayur et al. 1996). In various systems, PKCδ inhibitors and a PKCδKD mutant inhibited both the activation of caspase 3 and the cleavage of PKCδ suggesting that PKCδ may also act upstream of caspase 3 and pointing to the existence of a positive regulatory loop (Blass et al. 2002). PKCδ undergoes cleavage in response to various antitumor drugs such as etoposide (Blass et al. 2002; Reyland et al. 1999), cisplatin (Li et al. 2007), cytosine arabinoside (Koriyama et al. 1999), and mitomycin (Emoto et al. 1996). The cleavage of PKCδ generates a catalytic fragment that is constitutively active and which has been shown to promote cell apoptosis when localized to the nucleus or mitochondria. Although the cleavage of PKCδ has been mainly associated with the apoptotic function of PKCδ, it can also provide antiapoptotic signals particularly when it is not localized to the nucleus (Okhrimenko et al. 2005). The cleavage of PKCδ has been shown to be modulated by its tyrosine phosphorylation, especially tyrosines 311 and 332 that flank the caspase cleavage site of this isoform. Indeed, tyrosine 311 was essential for the cleavage of PKCδ by oxidative stress (Kaul et al. 2005), whereas tyrosine 332 was important for the cleavage of PKCδ by cisplatin in glioma cells (Lu et al. 2007b).

22.4.2 Role of PKCδ in the Response of Tumor Cells to Chemotherapeutic Drugs

Etoposide: The topoisomerase II inhibitor, etoposide, is an important chemotherapeutic agent that is used to treat a wide spectrum of tumors (Meresse et al. 2004). Various studies demonstrated the role of PKCδ in mediating the apoptotic effect of etoposide in various cell types. Etoposide induces tyrosine phosphorylation of PKCδ, caspase-3-dependent cleavage, and its translocation to the nucleus, which are essential for its apoptotic effect (Blass et al. 2002; DeVries-Seimon et al. 2007). In a recent study, the proapoptotic effect of PKCδ in etoposide-treated glioma cells was shown to be mediated by Erk1/2. The phosphorylation of PKCδ on tyrosines 64 and 187 induced the ubiquitination and proteosomal degradation of MKP-1, which resulted in prolonged Erk1/2 activation and cell apoptosis (Lomonaco et al. 2008). Shin et al. (2004) demonstrated that etoposide induced transcriptional and posttranscriptional regulation of the PKCδ gene in murine leukemic cells that were dependent on the activation of PKCδ, suggesting a mechanism of autoregulation of PKCδ.

Cisplatin: PKCδ has been shown to mediate the apoptotic effect of cisplatin and related compounds in various tumor cells. The sensitivity of small cell lung cancer cells and ovarian carcinoma cells was associated with an increase in the expression of PKCδ and a decrease in the expression of classical PKC isoforms (Basu et al. 1996). *cis*-Diamminedichloroplatinum (II) (cDDP) induced translocation of PKCδ to the cytosol and heavy membrane and its cleavage and inhibition of PKCδ blocked cDDP-induced cell apoptosis at an early stage that preceded caspase activation (Basu et al. 2001). In gastric cancer cells, PKCδ increased the sensitivity of the cells to cisplatin when it was overexpressed with p53 (Iioka et al. 2005). The importance of PKCδ cleavage in the apoptotic response of cancer cells to cisplatin is cell dependent.

Thus, the catalytic domain of PKCδ does not play a critical role in cisplatin-induced apoptosis of small-cell lung cancer cells H69 (Persaud et al. 2005) but did affect the apoptotic response to glioma cells. The activation of PKCδ by cisplatin involved phosphorylation on tyrosine 332 by Src (Lu et al. 2007b) and PKCδ-mediated cisplatin apoptotic effect occurred via activation of Erk1/2 (Basu and Tu 2005).

Doxorubicin and related compounds: Doxorubicin is an anthracyclin that exerts major antitumor effects and is used in the treatment of various human cancers. The intracellular effect of this drug includes the formation of free radicals, inhibition of topoisomerase II, and DNA intercalation (Muller et al. 1998). These effects lead to the inhibition of DNA replication and DNA strand break damage (Gewirtz 1999). The apoptotic effect of doxorubicin appears to mediate its main antiproliferative effect, and it has been shown to involve the activation of PKCδ. Panaretakis et al. (2005) identified PKCδ as a novel substrate of caspase 2 in doxorubicin-treated leukemic cells and showed that the apoptotic effect of PKCδ was mediated by JNK. In a recent study, PKCδ triggered TP53-dependent cell apoptosis in doxorubicin-treated leukemia and osteosarcoma cells (Liu et al. 2007a). In this cellular system, PKCδ increased the transcription of TP53 via the TP53 core promoter element (CPE-TP53). PKCδ interacted and activated the death promoting transcription factor Btf to co-occupy CPE-TP53.

The C1b domain of PKCδ has been recently identified as the molecular target of a new extranuclear-targeted anthracycline derivative, *N*-benzyladtiamycin-14-valerate (AD198). Ad198 promoted a rapid translocation of PKCδ to the mitochondria and the phosphorylation of phospholipids scramblase 3 (PLS3) on Thr21, which mediated the apoptotic role of PKCδ in AD198 effect (He et al. 2005).

Docetaxel: The role of PKCδ in the effect of docetaxel was studied in melanoma cells, where it was mediated by c-jun-NH2-terminal kinase (JNK) and inhibited by the Erk1/2 pathway (Mhaidat et al. 2007). PKCδ and PKCε have been found to mediate the sensitivity and resistance of melanoma cells to docetaxel, respectively. The pro-apoptotic effect of PKCδ was upstream of the JNK pathway in these cells (Mhaidat et al. 2007).

IFN-α: The cytokine IFN-α, which is routinely used in the treatment of chronic myelogenous leukemia (CML), induces antileukemic effect in BCR-ABL expressing cells by the activation of PKCδ, which phosphorylates Stat1 and activates IFN-α inducible gene transcription (Kaur et al. 2005).

In a recent study, PKCδ together with Erk and JNK, downstream of PI3-kinase and mTOR have been shown to mediate the effect of IFN-α on cell apoptosis in multiple myeloma cells in the absence of de-novo transcription (Panaretakis et al. 2008).

Aplidin: Aplidin is a new antitumor drug currently in phase-II clinical trials with both in vitro and in vivo activity against cancer cells. Aplidin induces oxidative stress that results in the phosphorylation of JNK, p38, and Erk and the activation of the mitochondrial death pathway (Garcia-Fernandez et al. 2002). In addition, Aplidin induces cleavage of PKCδ late in the apoptotic pathway, which acts as an important component in the activation of the caspase cascade and in the execution of the apoptotic pathway.

TRAIL: Tumor necrosis factor-related apoptosis inducing ligand (TRAIL; Apo2 ligand) belongs to the tumor necrosis factor superfamily (Kelley and Ashkenazi 2004) and induces apoptosis in many transformed cells including some glioma cell lines and primary cultures (Song et al. 2003; Wiley et al. 1995). TRAIL acts by formation of the death-inducing signaling complex that is also common to other members of the death receptors.

In contrast to its proapoptotic effect in most antitumor agents, PKCδ appears to mediate the resistance of cancer cells to TRAIL or antiapoptotic effects of TRAIL in various cellular systems. PKCδ induced an increase in FLIP expression, which acts as a negative regulator of FAS and TRAIL-mediated apoptosis via transactivation of NF-kB, suggesting that the PKCδ/NF-kB pathway plays a major role in the sensitivity of colon cancer cells to both FAS and TRAIL (Wang, International journal of cancer, 2007).

In prostate cancer cells, the apoptotic effect of PMA was mediated by a novel autocrine proapoptotic loop that was triggered by PKCδ and involved the release of death receptor ligand and activation of the extrinsic apoptotic pathway (Gonzalez-Guerrico and Kazanietz 2005).

In glioma cells, TRAIL induced the phosphorylation of PKCδ on tyrosine 155, its translocation to the ER, and cleavage. Silencing of PKCδ increased the sensitivity of glioma cells to TRAIL, whereas overexpression of PKCδ exerted an opposite effect, suggesting that in these cells, PKCδ mediates the antiapoptotic effects of TRAIL. Interestingly, in this cellular system, the cleavage of PKCδ was essential for its protective effect (Okhrimenko et al. 2005). In a recent study, Ndebele et al. (2008) demonstrated that exogenous expression of phospholipids scramblase 3 (PLS3) enhances the apoptotic effect of PKCδ and that activation of PKCδ by TRAIL mediates changes in cardiolipin induced by PLS3.

22.4.3 PKCδ Interacting Proteins and Downstream Signaling Pathways

Various interacting proteins of PKCδ have been identified that mediate the apoptotic function of PKCδ in response to various apoptotic stimuli and chemotherapeutic drugs. In the mitochondria, PKCδ interacts with and phosphorylates scramblase 3 (Liu et al. 2003). Other PKCδ targets/substrates are nuclear proteins that function in cell apoptosis such as DNA-dependent protein kinase (DNA-PK), which plays an essential role in the repair of DNA double strand breaks and interacts with c-Abl during genotoxic stress (Bharti et al. 1998). Lamin proteins represent another important potential target for the apoptotic function of PKCδ, which acts as an apoptotic lamin kinase (Cross et al. 2000). P73 is a p53 homolog, which is also phosphorylated by PKCδ, and this phosphorylation is associated with accumulation of p73β and the induction of p73β transactivation and apoptotic functions (Ren et al. 2002).

PKCδ activates multiple signaling pathways in response to apoptotic stimuli and chemotherapeutic drugs. In addition to the signaling molecules that were discussed previously, PKCδ activates various members of the MAP kinase family and the activation and inactivation of Erk1/2, p38, and JNK have been associated with PKCδ signaling during cell apoptosis. Activation of JNK by PKCδ has been associated with the apoptotic function of this isoform in cells treated with PMA, etoposide, and γ-radiation (Humphries et al. 2006; Tanaka et al. 2003). Similarly, the apoptotic effect of PKCδ in IFN-α and docetaxel cells was also mediated by JNK (Mhaidat et al. 2007; Panaretakis et al. 2008). Both JNK and p38 have been shown to mediate the apoptotic effect of Aplidin via the mitochondria pathway (Garcia-Fernandez et al. 2002). The activation of JNK by PKCδ was downstream of MKK7 and was associated with the localization of PKCδ in the nucleus (Gomel et al. 2007).

Similarly to JNK, activation of Erk1/2 was mainly associated with the proapoptotic effect of PKCδ in cisplatin (Basu and Tu 2005), γ-irradiated (Lee et al. 2002), PEP005 (Ingenol-3-angelate) (Serova et al. 2008), and etoposide-treated cells (Lomonaco et al. 2008).

In contrast, activation of the AKT pathway by PKCδ was associated with the antiapoptotic effect of this isoform. PKCδ was required for the survival of cells expressing active p21Ras, and this effect was mediated by activation of the PI3-kinase/AKT pathway (Xia et al. 2007). PKCδ also protected cells against cell apoptosis by mediating the association of Erb3-binding protein to nuclear AKT and preventing its apoptotic degradation (Liu et al. 2007b).

22.4.3.1 Additional PKCδ Targets in Cancer Cells

One of the important targets of PKCδ in cancer cells is heat shock protein 27 (HSP27). This protein is highly expressed in various tumors and confers resistance against various antitumor agents. Tumor progression of hepatocellular carcinoma has been correlated with the attenuated phosphorylation of HSP27. PKCδ has been shown to regulate the phosphorylation of HSP27 via p38 in these cells (Takai et al. 2007). In a recent study, Kim et al. (2007a) identified a specific sequence in the V5 region of PKCδ that mediates the interaction between PKCδ and HSP27. This interaction is implicated in the sequestering of HSP27, which leads to cell sensitization to chemotherapy.

Another target of PKCδ in cancer cells is hTERT. PKCδ, similar to the PKC isoforms α, β, ε, and ζ regulates telomerase activity in head and neck cancer cells by phosphorylating hTERT, which is essential for telomerase holoenzyme assembly, telomerase activation, and oncogenesis (Chang et al. 2006).

22.5 Summary

PKCδ is implicated in many aspects of carcinogenesis. It is differentially expressed in various tumors compared to the normal surrounding tissues and plays a major role in a variety of cellular functions in tumor cells including proliferation, cell cycle control, migration, invasion, angiogenesis and cell apoptosis, and survival.

Moreover, PKCδ plays a role in determining the response to tumor cells to a variety of chemotherapeutic drugs. Therefore, PKCδ may be an attractive therapeutic target in many types of cancers. Limitations to the use of PKCδ as a therapeutic target include its ubiquitous expression and its diverse effects in a given cellular system. Thus, targeting PKCδ may decrease cell invasion and sensitize tumor cells to chemotherapy along with concomitantly increasing cell proliferation. Thus, an alternative approach for inhibiting PKCδ activity or expression may be the inhibition of specific functions of PKCδ by either interfering with a specific substrate, blocking translocation to a specific subcellular site, or inhibiting the phosphorylation of PKCδ on a distinct tyrosine residue that specifically mediates the distinct effect of this isoform. Additional studies are required to identify molecular switches that finely control the ability of PKCδ to exert distinct effects in tumor cells and inhibitors that can selectively target these effects.

References

Acs, P., Beheshti, M., Szállási, Z., Li, L, Yuspa, S.H., & Blumberg, P. M. (2000). Effect of a tyrosine 155 to phenylalanine mutation of protein kinase cdelta on the proliferative and tumorigenic properties of NIH 3T3 fibroblasts. *Carcinogenesis*, *21*, 887–891.

Afrasiabi, E., Ahlgren, J., Bergelin, N., & Törnquist K. (2008). Phorbol 12-myristate 13-acetate inhibits FRO anaplastic human thyroid cancer cell proliferation by inducing cell cycle arrest in G1/S phase: Evidence for an effect mediated by PKCdelta. *Molecular and Cellular Endocrinology*, *292*, 26–35.

Alonso-Escolano, D., Medina, C., Cieslik, K., Radomski, A., Jurasz, P., Santos-Martínez, M. J., et al. (2006). Protein kinase C delta mediates platelet-induced breast cancer cell invasion. *The Journal of Pharmacology and Experimental Therapeutics*, *318*, 373–380.

Alvaro, V., Lévy, L., Dubray, C., Roche, A., Peillon, F., Quérat, B., et al. (1993). Invasive human pituitary tumors express a pointmutated alpha-protein kinase-C. *The Journal of Clinical Endocrinology and Metabolism*, *77*, 1125–1129.

Ashton, A. W., Watanabe, G., Albanese, C., Harrington, E. O., Ware, J. A., & Pestell, R. G. (1999). Protein kinase Cdelta inhibition of S-phase transition in capillary endothelial cells involves the cyclin-dependent kinase inhibitor p27(Kip1). *The Journal of Biological Chemistry*, *274*, 20805–20811.

Basu, A., Adkins, B., & Basu, C. (2008). Down-regulation of caspase-2 by rottlerin via protein kinase C-delta-independent pathway. *Cancer Research*, *68*, 2795–2802.

Basu, A., Contreras, A. G. Datta, D., Flynn, E., Zeng, L., Cohen, H. T., et al. (2008). Overexpression of vascular endothelial growth factor and the development of post-transplantation cancer. *Cancer Research*, *68*, 5689–5698.

Basu, A., Mohanty, S., & Sun, B. (2001). Differential sensitivity of breast cancer cells to tumor necrosis factor-alpha: Involvement of protein kinase C. *Biochemical and Biophysical Research Communications*, *280*, 883–891.

Basu, A., & Tu, H. (2005). Activation of ERK during DNA damage-induced apoptosis involves protein kinase Cdelta. *Biochemical and Biophysical Research Communications*, *334*, 1068–1073.

Basu, A., Weixel, K., & Saijo, N. (1996). Characterization of the protein kinase C signal transduction pathway in cisplatin-sensitive and -resistant human small cell lung carcinoma cells. *Cell Growth and Differentiation*, *7*, 1507–1512.

Bharti, A., Kraeft, S. K., Gounder, M., Pandey, P., Jin, S., Yuan, Z.M., et al. (1998). Inactivation of DNA-dependent protein kinase by protein kinase Cdelta: Implications for apoptosis. *Molecular and Cellular Biology*, *18*, 6719–6728.

Blass, M., Kronfeld, I., Kazimirsky, G., Blumberg, P. M., & Brodie, C. (2002). Tyrosine phosphorylation of protein kinase Cdelta is essential for its apoptotic effect in response to etoposide. *Molecular and Cellular Biology*, *22*, 182–195.

Brenner, W., Greber, I., Gudejko-Thiel, J., Beitz, S., Schneider, E., Walenta, S., et al. (2008). Migration of renal carcinoma cells is dependent on protein kinase Cdelta via beta1 integrin and focal adhesion kinase. *International Journal of Oncology*, *32*, 1125–1131.

Brodie, C., & Blumberg, P. M. (2003). Regulation of cell apoptosis by protein kinase c delta. *Apoptosis*, *8*, 19–27.

Brodie, C., Bogi, K., Acs, P., Lorenzo, P. S., Baskin, L., & Blumberg, P. M. (1998). Protein kinase C delta (PKCdelta) inhibits the expression of glutamine synthetase in glial cells via the PKCdelta regulatory domain and its tyrosine phosphorylation. *The Journal of Biological Chemistry*, *273*, 30713–30718.

Budas, G. R., Churchill, E. N., & Mochly-Rosen, D. (2007). Cardioprotective mechanisms of PKC isozyme-selective activators and inhibitors in the treatment of ischemia-reperfusion injury. *Pharmacological Research*, *55*, 523–536.

Cai, J., Donaldson, A., Yang, M., German, M. S., Enikolopov, G., & Iacovitti, L. (2009). The role of Lmx1a in the differentiation of human embryonic stem cells into midbrain dopamine neurons in culture and after transplantation into a Parkinson's disease model. *Stem Cells*, *27*, 220–229.

Castagna, M., Takai, Y., Kaibuchi, K., Sano, K., Kikkawa, U., & Nishizuka, Y. (1982). Direct activation of calcium-activated, phospholipid-dependent protein kinase by tumor-promoting phorbol esters. *The Journal of Biological Chemistry*, *257*, 7847–7851.

Cerda, S. R., Mustafi, R., Little, H., Cohen, G., Khare, S., Moore, C., et al. (2006). Protein kinase C delta inhibits Caco-2 cell proliferation by selective changes in cell cycle and cell death regulators. *Oncogene*, *25*, 3123–3138.

Chang, J. T., Lu, Y. C., Chen, Y. J., Tseng, C. P., Chen, Y. L., Fang, C. W., et al. (2006). hTERT phosphorylation by PKC is essential for telomerase holoprotein integrity and enzyme activity in head neck cancer cells. *British Journal of Cancer*, *94*, 870–878.

Craven, P. A., & DeRubertis, F. R. (1994). Loss of protein kinase C delta isozyme immunoreactivity in human adenocarcinomas. *Digestive Diseases and Sciences*, *39*, 481–489.

Cross, T., Griffiths, G., Deacon, E., Sallis, R., Gough, M., Watters, D., et al. (2000). PKC-delta is an apoptotic lamin kinase. *Oncogene*, *19*, 2331–2337.

da Rocha, A. B., Mans, D. R., Regner, A., & Schwartsmann, G. (2002). Targeting protein kinase C: New therapeutic opportunities against high-grade malignant gliomas? *The Oncologist*, *7*, 17–33.

Datta, R., Banach, D., Kojima, H., Talanian, R. V., Alnemri, E. S., Wong, W. W., et al. (1996). Activation of the CPP32 protease in apoptosis induced by 1-beta- d -arabinofuranosylcytosine and other DNA-damaging agents. *Blood*, *88*, 1936–1943.

Dempsey, E. C., Newton, A. C., Mochly-Rosen, D., Fields, A. P., Reyland, M. E., Insel, P. A., et al. (2000). Protein kinase C isozymes and the regulation of diverse cell responses. *American Journal of Physiology. Lung Cellular and Molecular Physiology*, *279*, L429–L438.

Deng, X., Hofmann, E. R., Villanueva, A., Hobert, O., Capodieci, P., Veach, D. R., et al. (2004). Caenorhabditis elegans ABL-1 antagonizes p53-mediated germline apoptosis after ionizing irradiation. *Nature Genetics*, *36*, 906–912.

Denning, M. F., Dlugosz, A. A., Howett, M. K., & Yuspa, S. H. (1993). Expression of an oncogenic rasHa gene in murine keratinocytes induces tyrosine phosphorylation and reduced activity of protein kinase C delta. *The Journal of Biological Chemistry*, *268*, 26079–26081.

Denning, M. F., Wang, Y., Nickoloff, B. J., & Wrone-Smith, T. (1998). Protein kinase Cdelta is activated by caspase-dependent proteolysis during ultraviolet radiation-induced apoptosis of human keratinocytes. *The Journal of Biological Chemistry*, *273*, 29995–30002.

Denning, M. F., Wang, Y., Tibudan, S., Alkan, S., Nickoloff, B. J., & Qin, J. Z. (2002). Caspase activation and disruption of mitochondrial membrane potential during UV radiation-induced apoptosis of human keratinocytes requires activation of protein kinase C. *Cell Death and Differentiation*, *9*, 40–52.

DeVries-Seimon, T. A., Ohm, A. M., Humphries, M. J., & Reyland M. E. (2007). Induction of apoptosis is driven by nuclear retention of protein kinase C delta. *The Journal of Biological Chemistry*, *282*, 22307–22314.

Durgan, J., Michael, N., Totty, N., & Parker, P. J. (2007). Novel phosphorylation sites of protein kinase C delta activation. *FEBS Letters*, *581*, 3377–3381.

El-Rayes, B. F., Ali, S., Philip, P. A., & Sarkar, F. H. (2008). Protein kinase C: A target for therapy in pancreatic cancer. *Pancreas*, *36*, 346–352.

Emoto, Y., Kisaki, H., Manome, Y., Kharbanda, S., & Kufe, D. (1996). Activation of protein kinase Cdelta in human myeloid leukemia cells treated with 1-beta-D-arabinofuranosylcytosine. *Blood*, *87*, 1990–1996.

Emoto, Y., Manome, Y., Meinhardt, G., Kisaki, H., Kharbanda, S., Robertson, M., et al. (1995). Proteolytic activation of protein kinase C delta by an ICE-like protease in apoptotic cells. *EMBO Journal*, *14*, 6148–6156.

Fields, A. P., Frederick, L. A., & Regala, R. P. (2007). Targeting the oncogenic protein kinase Ciota signalling pathway for the treatment of cancer. *Biochemical Society Transactions*, *35*, 996–1000.

Fukumoto, S., Nishizawa, Y., Hosoi, M., Koyama, H., Yamakawa, K., Ohno, S., et al. (1997). Protein kinase C delta inhibits the proliferation of vascular smooth muscle cells by suppressing G1 cyclin expression. *The Journal of Biological Chemistry*, *272*, 13816–13822.

García-Fernández, L. F., Losada, A., Alcaide, V., Alvarez, A. M., Cuadrado, A., et al. (2002). Aplidin induces the mitochondrial apoptotic pathway via oxidative stress-mediated JNK and p38 activation and protein kinase C delta. *Oncogene*, *21*, 7533–7544.

Gewirtz, D. A. (1999). A critical evaluation of the mechanisms of action proposed for the antitumor effects of the anthracycline antibiotics adriamycin and daunorubicin. *Biochemical Pharmacology*, *57*, 727–741.

Ghayur, T., Hugunin, M., Talanian, R. V., Ratnofsky, S., Quinlan, C., Emoto, Y., et al. (1996). Proteolytic activation of protein kinase C delta by an ICE/CED 3-like protease induces characteristics of apoptosis. *The Journal of Experimental Medicine*, *184*, 2399–2404.

Gliki, G., Wheeler-Jones, C., & Zachary, I. (2002). Vascular endothelial growth factor induces protein kinase C (PKC)-dependent Akt/PKB activation and phosphatidylinositol 3′-kinasemediates PKC delta phosphorylation: Role of PKC in angiogenesis. *Cell Biology International*, *26*, 751–759.

Godbout, J. P., Pesavento, J., Hartman, M. E., Manson, S. R., & Freund, G. G. (2002). Methylglyoxal enhances cisplatininduced cytotoxicity by activating protein kinase Cdelta. *The Journal of Biological Chemistry*,*277*, 2554–2561.

Gomel, R., Xiang, C., Finniss, S., Lee, H. K., Lu, W., Okhrimenko, H., et al. (2007). The localization of protein kinase Cdelta in different subcellular sites affects its proapoptotic and antiapoptotic functions and the activation of distinct downstream signaling pathways. *Molecular Cancer Research*, *5*, 627–639.

Gonzalez-Guerrico, A. M., & Kazanietz, M. G. (2005). Phorbol ester-induced apoptosis in prostate cancer cells via autocrine activation of the extrinsic apoptotic cascade: A key role for protein kinase C delta. *The Journal of Biological Chemistry*, *280*, 38982–38991.

Griner, E., & Kazanietz, M. G. (2007). Protein kinase C and other effectors in cancer. *Nature Reviews. Cancer*, *7*, 281–294.

Gschwendt, M. (1995). Protein kinase C delta. *European Journal of Biochemistry*, *259*, 555–564.

Gschwendt, M., Müller, H. J., Kielbassa, K., Zang, R., Kittstein, W., Rincke, G., et al. (1994). Rottlerin, a novel protein kinase inhibitor. *Biochemical and Biophysical Research Communications*, *199*, 93–98.

He, Y., Liu, J., Durrant, D., Yang, H. S., Sweatman, T., Lothstein, L., et al. (2005). N-benzy ladriamycin-14-valerate (AD198) induces apoptosis through protein kinase C-delta-induced phosphorylation of phospholipid scramblase 3. *Cancer Research*, *65*, 10016–10023.

Humphries, M. J., Limesand, K. H., Schneider, J. C., Nakayama, K. I., Anderson, S. M., Reyland, M. E. (2006). Suppression of apoptosis in the protein kinase C delta null mice. *The Journal of Biological Chemistry, 281*, 9728–9737.

Hung, J. H., Lu, Y. S., Wang, Y. C., Ma, Y. H., Wang, D. S., Kulp, S. K., et al. (2008). FTY720 induces apoptosis in hepatocellular carcinoma cells through activation of protein kinase C delta signaling. *Cancer Research, 68*, 1204–1212.

Huppi, K., Siwarski, D., Goodnight, J., & Mischak, H. (1994). Assignment of the protein kinase C delta polypeptide gene (PRKCD) to human chromosome 3 and mouse chromosome 14. *Genomics, 19*, 161–162.

Hussaini, I. M., Trotter, C., Zhao, Y., Abdel-Fattah, R., Amos, S., Xiao, A., Agi, C. U., et al. (2007). Matrix metalloproteinase-9 is differentially expressed in nonfunctioning invasive and noninvasive pituitary adenomas and increases invasion in human pituitary adenoma cell line. *American Journal of Pathology, 170*, 356-365.

Iioka, Y., Mishima, K., Azuma, N., Tsuchida, A., Takagi, Y., Aoki, T., et al. (2005). Overexpression of protein kinase Cdelta enhances cisplatin-induced cytotoxicity correlated with p53 in gastric cancer cell line. *Pathobiology, 72*, 152–159.

Iwabu, A., Smith, K., Allen, F. D., Lauffenburger, D. A., & Wells, A. (2004). Epidermal growth factor induces fibroblast contractility and motility via a protein kinase C delta-dependent pathway. *The Journal of Biological Chemistry, 279*, 14551–14560.

Jackson, D. N., & Foster, D. A. (2004). The enigmatic protein kinase Cdelta: Complex roles in cell proliferation and survival. *FASEB Journal, 18*, 627–636.

Joseloff, E., Cataisson, C., Aamodt, H., Ocheni, H., Blumberg, P., Kraker, A. J., et al. (2002). Src family kinases phosphorylate protein kinase C delta on tyrosine residues and modify the neoplastic phenotype of skin keratinocytes. *The Journal of Biological Chemistry, 277*, 12318–12323.

Jun, C. D., Oh, C. D., Kwak, H. J., Pae, H. O., Yoo, J. C., Choi, B. M., et al. (1999). Overexpression of protein kinase C isoforms protects RAW 264.7 macrophages from nitric oxide-induced apoptosis: Involvement of c-Jun N-terminal kinase/stress-activated protein kinase, p38 kinase, and CPP-32 protease pathways. *Journal of Immunology, 162*, 3395–3401.

Kajimoto, T., Ohmori, S., Shirai, Y., Sakai, N., & Saito, N. (2001). Subtype-specific translocation of the delta subtype of protein kinase C and its activation by tyrosine phosphorylation induced by ceramide in HeLa cells. *Molecular and Cellular Biology, 21*, 1769–1783.

Kaul, S., Anantharam, V., Yang, Y., Choi, C. J., Kanthasamy, A., & Kanthasamy, A. G. (2005). Tyrosine phosphorylation regulates the proteolytic activation of protein kinase Cdelta in dopaminergic neuronal cells. *The Journal of Biological Chemistry, 280*, 28721–28730.

Kaur, S., Parmar, S., Smith, J., Katsoulidis, E., Li, Y., Sassano, A., et al. (2005). Role of protein kinase C-delta (PKC-delta) in the generation of the effects of IFN-alpha in chronic myelogenous leukemia cells. *Experimental Hematology, 33*, 550–557.

Kelley, S. K., & Ashkenazi, A. (2004). Targeting death receptors in cancer with Apo2L/TRAIL. *Current Opinion in Pharmacology, 4*, 333–339.

Keshamouni, V. G., Mattingly, R. R., & Reddy, K. B. (2002). Mechanism of 17-beta-estradiolinduced Erk1/2 activation in breast cancer cells. A role for HER2 AND PKC-delta. *The Journal of Biological Chemistry, 277*, 22558–22565.

Khwaja, A., & Tatton, L. (1999). Caspase-mediated proteolysis and activation of protein kinase Cdelta plays a central role in neutrophil apoptosis. *Blood, 94*, 291–301.

Kikkawa, U., Takai, Y., Tanaka, Y., Miyake, R., & Nishizuka, Y. (1983). Protein kinase C as a possible receptor protein of tumor-promoting phorbol esters. *The Journal of Biological Chemistry, 258*, 11442–11445.

Kiley, S. C., Clark, K. J., Duddy, S. K., Welch, D. R., & Jaken, S. (1999). Increased protein kinase C delta in mammary tumor cells: Relationship to transformtion and metastatic progression. *Oncogene, 18*, 6748–6757.

Kiley, S. C., Clark, K. J., Goodnough, M., Welch, D. R., Jaken, S. (1999). Protein kinase C delta involvement in mammary tumor cell metastasis. *Cancer Research, 59*, 3230–3238.

Kilpatrick, L. E., Lee, J. Y., Haines, K. M., Campbell, D. E., Sullivan, K. E., & Korchak, H. M. (2002). A role for PKC-delta and PI 3-kinase in TNF-alpha-mediated antiapoptotic signaling in the human neutrophil. *American Journal of Physiology. Cell Physiology, 283*, C48–C57.

Kim, Y., Lee, Y. S., Choe, J., Lee, H., Kim, Y. M., & Jeoung, D. (2008). CD44-epidermal growth factor receptor interaction mediates hyaluronic acid-promoted cell motility by activating protein kinase C signalling involving Akt, Rac1, Phox, reactive oxygen species, focal adhesion kinase, and MMP-2. *The Journal of Biological Chemistry, 283*, 22513–22528.

Kim, E. H., Lee, H. J., Lee, D. H., Bae, S., Soh, J. W., Jeoung, D., et al. (2007). Inhibition of heat shock protein 27-mediated resistance to DNA damaging agents by a novel PKC delta-V5 heptapeptide. *Cancer Research,67*, 6333–6341.

Kim, Y. H., Lim, J. H., Lee, T. J., Park, J. W., & Kwon, T. K. (2007). Expression of cyclin D3 through Sp1 sites by histone deacetylase inhibitors is mediated with protein kinase C-delta (PKC-delta) signal pathway. *Journal of Cellular Biochemistry, 101*, 987–995.

Koike, K., Fujii, T., Nakamura, A. M., Yokoyama, G., Yamana, H., Kuwano, M., et al. (2006). Activation of protein kinase C delta induces growth arrest in NPA thyroid cancer cells through extracellular signal-regulated kinase mitogen-activated protein kinase. *Thyroid, 16*, 333–341.

Koivunen, J., Aaltonen, V., & Peltonen, J. (2006). Protein kinase C (PKC) family in cancer progression. *Cancer Letters, 235*, 1–10.

Konishi, H., Yamauchi, E., Taniguchi, H., Yamamoto, T., Matsuzaki, H., Takemura, Y., et al. (2001). Phosphorylation sites of protein kinase C delta in H2O2-treated cells and its activation by tyrosine kinase in vitro. *Proceedings of the National Academy of Sciences of the United States of America, 98*, 6587–6592.

Koren, R., Langzam, L., Paz, A., Livne, P. M., Gal, R., & Sampson, S.R. (2000). Protein kinase C (PKC) isoenzymes immunohistochemistry in lymph node revealing solution-fixed, paraffin-embedded bladder tumors. *Applied Immunohistochemistry & Molecular Morphology, 8*, 166–171.

Koriyama, H., Kouchi, Z., Umeda, T., Saido, T. C., Momoi, T., Ishiura, S., et al. (1999). Proteolytic activation of protein kinase C delta and epsilon by caspase-3 in U937 cells during chemotherapeutic agent-induced apoptosis. *Cellular Signalling, 11*, 831–838.

Kronfeld, I., Kazimirsky, G., Lorenzo, P. S., Garfield, S. H., Blumberg, P. M., & Brodie, C. (2000). Phosphorylation of protein kinase Cdelta on distinct tyrosine residues regulates specific cellular functions. *The Journal of Biological Chemistry, 275*, 35491–35498.

Kruger, J. S., & Reddy, K. B. (2003). Distinct mechanisms mediate the initial and sustained phases of cell migration in epidermal growth factor receptor-overexpressing cells. *Molecular Cancer Research, 1*, 801–809.

Langzam, L., Koren, R., Gal, R., Kugel, V., Paz, A., Farkas, A., et al. (2001). Patterns of protein kinase C isoenzyme expression in transitional cell carcinoma of bladder. Relation to degree of malignancy. *American Journal of Clinical Pathology, 116*, 377–385.

Leach, K. L., James, M. L., & Blumberg, P. M. (1983). Characterization of a specific phorbol ester aporeceptor in mouse brain cytosol. *Proceedings of the National Academy of Sciences of the United States of America, 80*, 4208–4212.

Lee, J. W., Park, J. A., Kim, S. H., Seo, J. H., Lim, K. J., Jeong, J. W., et al. (2007). Protein kinase C-delta regulates the stability of hypoxia-inducible factor-1 alpha under hypoxia. *Cancer Science, 98*, 1476–1481.

Lee, Y. J., Soh, J. W., Dean, N. M., Cho, C. K., Kim, T. H., Lee, S. J., et al. (2002). Protein kinase Cdelta overexpression enhances radiation sensitivity via extracellular regulated protein kinase 1/2 activation, abolishing the radiation-induced G(2)-M arrest. *Cell Growth and Differentiation, 13*, 237–246.

Leitges, M., Mayr, M., Braun, U., Mayr, U., Li, C., Pfister, G., et al. (2001). Exacerbated vein graft arteriosclerosis in protein kinase Cdelta-null mice. *The Journal of Clinical Investigation, 108*, 1505–1512.

Li, W., Chen, X. H., Kelley, C. A., Alimandi, M., Zhang, J., Chen, Q., et al. (1996). Identification of tyrosine 187 as a protein kinase C-delta phosphorylation site. *The Journal of Biological Chemistry, 271*, 26404–26409.

Li, W., Jiang, Y. X., Zhang, J., Soon, L., Flechner, L., Kapoor, V., et al. (1998). Protein kinase C-delta is an important signaling molecule in insulin-like growth factor I receptor-mediated cell transformation. *Molecular and Cellular Biology, 18*, 5888–5898.

Li, L., Lorenzo, P. S., Bogi, K., Blumberg, P. M., & Yuspa, S. H. (1999). Protein kinase Cdelta targets mitochondria, alters mitochondrial membrane potential, and induces apoptosis in normal and neoplastic keratinocytes when overexpressed by an adenoviral vector. *Molecular and Cellular Biology*, *19*, 8547–8558.

Li, J., Vesey, D. A., Johnson, D. W., & Gobe, G. (2007). Erythropoietin reduces cisplatin-induced apoptosis in renal carcinoma cells via a PKC dependent pathway. *Cancer Biology & Therapy*, *6*, 1944–1950.

Li, C., Wernig, F., Leitges, M., Hu, Y., & Xu, Q. (2003). Mechanical stress-activated PKCdelta regulates smooth muscle cell migration. *FASEB Journal*, *17*, 2106–2108.

Li, S.Y., Yu, B., An, P., Liang, Z. J., Yuan, S. J., & Cai, H. Y. (2004). Effects of cell membrane phospholipid level and protein kinase C isoenzyme expression on hepatic metastasis of colorectal carcinoma. *Hepatobiliary & Pancreatic Diseases International*, *3*, 411–416.

Liu, J., Chen, J., Dai, Q., & Lee, R. M. (2003). Phospholipid scramblase 3 is the mitochondrial target of protein kinase C delta-induced apoptosis. *Cancer Research*, *63*, 1153–1156.

Liu, Z., Liu, X., Nakayama, K. I., Nakayama, K., & Ye, K. (2007). Protein kinase C-delta phosphorylates Ebp1 and prevents its proteolytic degradation, enhancing cell survival. *Journal of Neurochemistry*, *100*, 1278–1288.

Liu, H., Lu, Z. G., Miki, Y., & Yoshida, K. (2007). Protein kinase C delta induces transcription of the TP53 tumor suppressor gene by controlling death-promoting factor Btf in the apoptotic response to DNA damage. *Molecular and Cellular Biology*, *27*, 8480–8491.

Lomonaco, S. L., Kahana, S., Blass, M., Brody, Y., Okhrimenko, H., Xiang, C., et al. (2008). Phosphorylation of protein kinase Cdelta on distinct tyrosine residues induces sustained activation of Erk1/2 via down-regulation of MKP-1: Role in the apoptotic effect of etoposide. *The Journal of Biological Chemistry*, *283*, 17731–17739.

Lorenzo, P. S., & Dennis, P. A. (2003). Modulating protein kinase C (PKC) to increase the efficacy of chemotherapy: Stepping into darkness. *Drug Resistance Updates*, *6*, 329–339.

Lu,W., Finnis, S., Xiang, C., Lee, H. K., Markowitz, Y., Okhrimenko, H., et al. (2007). Tyrosine 311 is phosphorylated by c-Abl and promotes the apoptotic effect of PKCdelta in glioma cells. *Biochemical and Biophysical Research Communications*, *352*, 431–436.

Lu, Z., Hornia, A., Jiang, Y. W., Zang, Q., Ohno, S., & Foster, D. A. (1997). Tumor promotion by depleting cells of protein kinase C delta. *Molecular and Cellular Biology*, *17*, 3418–3428.

Lu, W., Lee, H. K., Xiang, C., Finniss, S., & Brodie, C. (2007). The phosphorylation of tyrosine 332 is necessary for the caspase 3-dependent cleavage of PKCdelta and the regulation of cell apoptosis. *Cellular Signalling*, *19*, 2165–2173.

Mackay, H. J., & Twelves, C. J. (2007). Targeting the protein kinase C family: Are we there yet? *Nature Reviews. Cancer*, *7*, 554–562.

Mandil, R., Ashkenazi, E., Blass, M., Kronfeld, I., Kazimirsky, G., Rosenthal, G., et al. (2001). Protein kinase Calpha and protein kinase Cdelta play opposite roles in the proliferation and apoptosis of glioma cells. *Cancer Research*,*61*, 4612–4619.

Martiny-Baron, G., & Fabbro, D. (2007). Classical PKC isoforms in cancer. *Pharmacological Research*, *55*, 477–486.

Mellor, H., & Parker, P. J. (1998). The extended protein kinase C superfamily. *Biochemical Journal*, *332* (Pt 2), 281–292.

Meresse, P., Dechaux, E., Monneret, C., & Bertounesque, E. (2004). Etoposide: Discovery and medicinal chemistry. *Current Medicinal Chemistry*, *11*, 2443–2466.

Mhaidat, N. M., Thorne, R. F., Zhang, X. D., & Hersey, P. (2007). Regulation of docetaxel-induced apoptosis of human melanoma cells by different isoforms of protein kinase C. *Molecular Cancer Research*, *5*, 1073–1081.

Mischak, H., Goodnight, J. A., Kolch, W., Martiny-Baron, G., Schaechtle, C., Kazanietz, M. G., et al. (1993). Overexpression of protein kinase C-delta and -epsilon in NIH 3T3 cells induces opposite effects on growth, morphology, anchorage dependence, and tumorigenicity. *The Journal of Biological Chemistry*, *268*, 6090–6096.

Miyamoto, A., Nakayama, K., Imaki, H., Hirose, S., Jiang, Y., Abe, M., et al. (2002). Increased proliferation of B cells and autoimmunity in mice lacking protein kinase Cdelta. *Nature*, *416*, 865–869.

Muller, I., Niethammer, D., & Bruchelt, G. (1998). Anthracycline-derived chemotherapeutics in apoptosis and free radical cytotoxicity (Review). *International Journal of Molecular Medicine*, *1*, 491–494.

Nabha, S. M., Glaros, S., Hong, M., Lykkesfeldt, A. E., Schiff, R., Osborne, K., et al. (2005). Upregulation of PKC-delta contributes to antiestrogen resistance in mammary tumor cells. *Oncogene*, *24*, 3166–3176.

Nakagawa, M., Oliva, J. L., Kothapalli, D., Fournier, A., Assoian, R. K, & Kazanietz, M. G. (2005). Phorbol ester-induced G1 phase arrest selectively mediated by protein kinase Cdelta-dependent induction of p21. *The Journal of Biological Chemistry*, *280*, 33926–33934.

Nakamura, S., & Nishizuka, Y. (1994). Lipid mediators and protein kinase C activation for the intracellular signaling network. *Journal of Biochemistry*, *115*, 1029–1034.

Ndebele, K., Gona, P., Jin, T. G., Benhaga, N., Chalah, A., Degli-Esposti, M., et al. (2008). Tumor necrosis factor (TNF)-related apoptosisinducing ligand (TRAIL) induced mitochondrial pathway to apoptosis and caspase activation is potentiated by phospholipid scramblase-3. *Apoptosis*, *13*, 845–856.

Nishizuka, Y. (1995). Protein kinase C and lipid signaling for sustained cellular responses. *FASEB Journal*, *9*, 484–496.

O'Brian, C. A., Ward, N. E., Stewart, J. R., & Chu, F. (2001). Prospects for targeting protein kinase C isozymes in the therapy of drug-resistant cancer – an evolving story. *Cancer and Metastasis Reviews*, *20*, 95–100.

Okhrimenko, H., Lu, W., Xiang, C., Ju, D., Blumberg, P. M., Gomel, R., et al. (2005). Roles of tyrosine phosphorylation and cleavage of protein kinase Cdelta in its protective effect against tumor necrosis factor-related apoptosis inducing ligand-induced apoptosis. *The Journal of Biological Chemistry*, *280*, 23643–23652.

Pal, S., Claffey, K. P., Dvorak, H. F., & Mukhopadhyay, D. (1997). The von Hippel-Lindau gene product inhibits vascular permeability factor/vascular endothelial growth factor expression in renal cell carcinoma by blocking protein kinase C pathways. *The Journal of Biological Chemistry*, *272*, 27509–27512.

Panaretakis, T., Hjortsberg, L., Tamm, K. P., Björklund, A. C., Joseph, B., & Grandér, D. (2008). Interferon alpha induces nucleusindependent apoptosis by activating extracellular signal-regulated kinase 1/2 and c-Jun NH2- terminal kinase downstream of phosphatidylinositol 3-kinase and mammalian target of rapamycin. *Molecular Biology of the Cell*, *19*, 41–50.

Panaretakis, T., Laane, E., Pokrovskaja, K., Björklund, A. C., Moustakas, A., Zhivotovsky, B., et al. (2005). Doxorubicin requires the sequential activation of caspase-2, protein kinase Cdelta, and c-Jun NH2-terminal kinase to induce apoptosis. *Molecular Biology of the Cell*, *16*, 3821–3831.

Peluso, J. J., Pappalardo, A., & Fernandez, G. (2001). Basic fibroblast growth factor maintains calcium homeostasis and granulosa cell viability by stimulating calcium efflux via a PKC delta-dependent pathway. *Endocrinology*, *142*, 4203–4211.

Persaud, S. D., Hoang, V., Huang, J., & Basu, A. (2005). Involvement of proteolytic activation of PKCdelta in cisplatin-induced apoptosis in human small cell lung cancer H69 cells. *International Journal of Oncology*, *27*, 149–154.

Podar, K., Raab, M. S., Chauhan, D., & Anderson, K. C. (2007). The therapeutic role of targeting protein kinase C in solid and hematologic malignancies. *Expert Opinion on Investigational Drugs*, *16*, 1693–1707.

Pongracz, J., Clark, P., Neoptolemos, J. P., & Lord, J. M. (1995). Expression of protein kinase C isoenzymes in colorectal cancer tissue and their differential activation by different bile acids. *International Journal of Cancer*, *61*, 35–39.

Prévostel, C., Martin, A., Alvaro, V., Jaffiol, C., & Joubert, D. (1997). Protein kinase C alpha and tumorigenesis of the endocrine gland. *Hormone Research*, *47*, 140–144.

Ren, J., Datta, R., Shioya, H., Li, Y., Oki, E., Biedermann, V., et al. (2002). p73beta is regulated by protein kinase Cdelta catalytic fragment generated in the apoptotic response to DNA damage. *The Journal of Biological Chemistry*, *277*, 33758–33765.

Renault-Mihara, F., Beuvon, F., Iturrioz, X., Canton, B., De Bouard, S., Léonard, N., et al. (2006). Phosphoprotein enriched in astrocytes-15 kDa expression inhibits astrocyte migration by a protein kinase C delta-dependent mechanism. *Molecular Biology of the Cell*, *17*, 5141–5152.

Reno, E. M., Haughian, J. M., Dimitrova, I. K., Jackson, T. A., Shroyer, K. R., & Bradford, A. P. (2008). Analysis of protein kinase C delta (PKC delta) expression in endometrial tumors. *Human Pathology*, *39*, 21–29.

Reyland, M. E., Anderson, S. M., Matassa, A. A., Barzen, K. A., & Quissell, D. O. (1999). Protein kinase C delta is essential for etoposide-induced apoptosis in salivary gland acinar cells. *The Journal of Biological Chemistry*, *274*, 19115–19123.

Ridley, A. J., Schwartz, M. A., Burridge, K., Firtel, R. A., Ginsberg, M. H., Borisy, G., et al. (2003). Cell migration: Integrating signals from front to back. *Science*, *302*, 1704–1709.

Rybin, V. O., Guo, J., Gertsberg, Z., Elouardighi, H., & Steinberg, S. F. (2007). Protein kinase Cepsilon (PKCepsilon) and Src control PKCdelta activation loop phosphorylation in cardiomyocytes. *The Journal of Biological Chemistry*, *282*, 23631–23638.

Scheel-Toellner, D., Pilling, D., Akbar, A. N., Hardie, D., Lombardi, G., Salmon, M., et al. (1999). Inhibition of T cell apoptosis by IFNbeta rapidly reverses nuclear translocation of protein kinase C-delta. *European Journal of Immunology*, *29*, 2603–2612.

Serova, M., Ghoul, A., Benhadji, K. A., Cvitkovic, E., Faivre, S., Calvo, F., et al. (2006). Preclinical and clinical development of novel agents that target the protein kinase C family. *Seminars in Oncology*, *33*, 466–478.

Serova, M., Ghoul, A., Benhadji, K. A., Faivre, S., Le Tourneau, C., Cvitkovic, E., et al. (2008). Effects of protein kinase C modulation by PEP005, a novel ingenol angelate, on mitogen-activated protein kinase and phosphatidylinositol 3-kinase signaling in cancer cells. *Molecular Cancer Therapeutics*, *7*, 915–922.

Shin, S. Y., Kim, C. G., Ko, J., Min, D. S., Chang, J. S., Ohba, M., et al. (2004). Transcriptional and post-transcriptional regulation of the PKC delta gene by etoposide in L1210 murine leukemia cells: Implication of PKC delta autoregulation. *Journal of Molecular Biology*, *340*, 681–693.

Song, K. S., Kim, J. S., Yun, E. J., Kim, Y. R., Seo, K. S., Park, J. H., et al. (2008). Rottlerin induces autophagy and apoptotic cell death through a PKC-delta independent pathway in HT1080 human fibrosarcoma cells: The protective use of autophagy in apoptosis. *Autophagy*, *4*, 650–658.

Song, J. H., Song, D. K., Pyrzynska, B., Petruk, K. C., Van Meir, E. G., & Hao, C. (2003). TRAIL triggers apoptosis in human malignant glioma cells through extrinsic and intrinsic pathways. *Brain Pathology*, *13*, 539–553.

Song, J. S., Swann, P. G., Szallasi, Z., Blank, U., Blumberg, P. M., & Rivera, J. (1998). Tyrosine phosphorylation-dependent and -independent associations of protein kinase C-delta with Src family kinases in the RBL-2H3 mast cell line: Regulation of Src family kinase activity by protein kinase C-delta. *Oncogene*, *16*, 3357–3368.

Steinberg, S. F. (2004). Distinctive activation mechanisms and functions for protein kinase Cdelta. *Biochemical Journal*, *384*, 449–459.

Steták, A., Lankenau, A., Vántus, T., Csermely, P., Ullrich, A., & Kéri, G. (2001). The antitumor somatostatin analogue TT-232 induces cell cycle arrest through PKCdelta and c-Src. *Biochemical and Biophysical Research Communications*, *285*, 483–488.

Sun, X., Wu, F., Datta, R., Kharbanda, S., & Kufe, D. (2000). Interaction between protein kinase C delta and the c-Abl tyrosine kinase in the cellular response to oxidative stress. *The Journal of Biological Chemistry*, *275*, 7470–7473.

Takai, S., Matsushima-Nishiwaki, R., Tokuda, H., Yasuda, E., Toyoda, H., Kaneoka, Y., et al. (2007). Protein kinase C delta regulates the phosphorylation of heat shock protein 27 in human hepatocellular carcinoma. *Life Sciences*, *81*, 585–591.

Tanaka, Y., Gavrielides, M. V., Mitsuuchi, Y., Fujii, T., & Kazanietz, M. G. (2003). Protein kinase C promotes apoptosis in LnCAP prostate cancer cells through activation of p38MAP kinase and inhibitin of the Akt survival pathway. *The Journal of Biological Chemistry*, *278*, 33753–33762.

Tanaka, M., Terry, R. D., Mokhtari, G. K., Inagaki, K., Koyanagi, T., Kofidis, T., et al. (2004). Suppression of graft coronary artery disease by a brief treatment with a selective epsilonPKC activator and a deltaPKC inhibitor in murine cardiac allografts. *Circulation, 110*, II194–II199.

Toker, A. (1998). Signaling through protein kinase C. *Frontiers in Bioscience, 3*, D1134–D1147.

Varga, A., Czifra, G., Tallai, B., et al. (2004). Tumor grade-dependent alterations in the protein kinase C isoform pattern in urinary bladder carcinomas. *European Urology, 46*, 462–465.

Villar, J., Arenas, M. I., MacCarthy, C. M., Blánquez, M. J., Tirado, O. M., & Notario, V. (2007). PCPH/ENTPD5 expression enhances the invasiveness of human prostate cancer cells by a protein kinase C delta-dependent mechanism. *Cancer Research, 67*, 10859–10868.

Vucenik, I., Ramakrishna, G., Tantivejkul, K., Anderson, L. M., & Ramljak, D. (2005). Inositol hexaphosphate (IP6) blocks proliferation of human breast cancer cells through a PKCdelta-dependent increase in p27Kip1 and decrease in retinoblastoma protein (pRb) phosphorylation. *Breast Cancer Research and Treatment, 91*, 35–45.

Watanabe, T., Ono, Y., Taniyama, Y., Hazama, K., Igarashi, K., Ogita, K., et al. (1992). Cell division arrest induced by phorbol ester in CHO cells overexpressing protein kinase C-delta subspecies. *Proceedings of the National Academy of Sciences of the United States of America, 89*, 10159–10163.

Watters, D., Garrone, B., Coomer, J., Johnson, W. E., Brown, G., & Parsons, P. (1998). Stimulation of melanogenesis in a human melanoma cell line by bistratene A. *Biochemical Pharmacology, 55*, 1691–1699.

Wert, M. M., & Palfrey, H. C. (2000). Divergence in the anti-apoptotic signalling pathways used by nerve growth factor and basic fibroblast growth factor (bFGF) in PC12 cells: Rescue by bFGF involves protein kinase C delta. *Biochemical Journal, 352* (Pt 1), 175–182.

Wiley, S. R., Schooley, K., Smolak, P. J., Din, W. S., Huang, C. P., Nicholl, J. K., et al. (1995). Identification and characterization of a new member of the TNF family that induces apoptosis. *Immunity, 3*, 673–682.

Woo, J. H., Lim, J. H., Kim, Y. H., Suh, S. I., Min, D. S., Chang, J. S., et al. (2004). Resveratrol inhibits phorbol myristate acetateinduced matrix metalloproteinase-9 expression by inhibiting JNK and PKC delta signal transduction. *Oncogene, 23*, 1845–1853.

Woods Ignatoski, K. M., Livant, D. L., Markwart, S., Grewal, N. K., & Ethier, S. P. (2003). The role of phosphatidylinositol 3′-kinase and its downstream signals in erbB-2-mediated transformation. *Molecular Cancer Research, 1*, 551–560.

Xia, S., Forman, L. W., & Faller, D. V. (2007). Protein kinase C delta is required for survival of cells expressing activated p21RAS. *The Journal of Biological Chemistry, 282*, 13199–13210.

Yokoyama, G., Fujii, T., Tayama, K., Yamana, H., Kuwano, M., & Shirouzu, K. (2005). PKCdelta and MAPK mediate G(1) arrest induced by PMA in SKBR-3 breast cancer cells. *Biochemical and Biophysical Research Communications, 327*, 720–726.

Yuan, Z. M., Utsugisawa, T., Ishiko, T., Nakada, S., Huang, Y., Kharbanda, S., et al. (1998). Activation of protein kinase C delta by the c-Abl tyrosine kinase in response to ionizing radiation. *Oncogene, 16*, 1643–1648.

Yuspa, S. H. (1998). The pathogenesis of squamous cell cancer: Lessons learned from studies of skin carcinogenesis. *Journal of Dermatological Science, 17*, 1–7.

Zeng, Z. Z., Jia, Y., Hahn, N. J., Markwart, S. M., Rockwood, K. F., & Livant, D. L. (2006). Role of focal adhesion kinase and phosphatidylinositol 3′-kinase in integrin fibronectin receptor-mediated, matrix metalloproteinase-1-dependent invasion by metastatic prostate cancer cells. *Cancer Research, 66*, 8091–8099.

Zrachia, A., Dobroslav, M., Blass, M., Kazimirsky, G., Kronfeld, I., Blumberg, P. M., et al. (2002). Infection of glioma cells with Sindbis virus induces selective activation and tyrosine phosphorylation of protein kinase C delta. Implications for Sindbis virus-induced apoptosis. *The Journal of Biological Chemistry, 277*, 23693–23701.

Chapter 23
Atypical PKCs as Targets for Cancer Therapy

Verline Justilien and Alan P. Fields

Abstract The Protein Kinase Cs (PKCs) were identified nearly thirty years ago as major cellular receptors for the tumor promoting phorbol esters; a finding that has prompt an intense search for the role of the individual PKC isozymes in cancer. The PKCs, including the two atypical isozymes PKCι/l and PKCζ, have been linked to multiple aspects of transformation in different tumor types. Although PKCι and PKCζ show high sequence homology, studies have shown that these two PKC isozymes have distinct and non-redundant roles in normal and oncogenic cellular signaling. In fact, to date, PKCι is the only member of the PKC family that has been shown to be a bonafide human oncogene. In this chapter, we review pertinent aspects of atypical PKC structure, function and regulation that relate to their role in human tumor biology. We discuss the evidence that PKCι is a human oncogene and describe the molecular pathways involved in PKCι-mediated oncogenic signaling. Finally, we will discuss a novel mechanism-based therapeutic drug that targets oncogenic PKCι signaling and is currently in clinical trials for treatment of human lung cancer.

Keywords Atypical Protein Kinase C • PKCι/l • PKCζ • Cancer • Hyper-proliferation • Invasion • Metastasis • K-ras • Phox-Bem1 (PB1) Domain • Par6 • Rac1 • MMP10 • Mechanism-based therapeutics • Aurothiomalate

23.1 Introduction

Protein kinase C (PKC) was identified more than 30 years ago as a novel serine-threonine protein kinase that is activated by phosphatidylserine (PS) and diacylglycerol (DAG) in a calcium-dependent manner (Kishimoto et al. 1980; Takai et al. 1979; Takai et al. 1977). PKC was initially thought to be a single protein, but subsequent biochemical and molecular cloning studies revealed that PKC is a

V. Justilien and A.P. Fields (✉)
Department of Cancer Biology, Mayo Clinic College of Medicine, Jacksonville, FL, USA
e-mail: fields.alan@mayo.edu

M.G. Kazanietz (ed.), *Protein Kinase C in Cancer Signaling and Therapy*,
Current Cancer Research, DOI 10.1007/978-1-60761-543-9_23,

family of related lipid-dependent kinases. The PKCs consist of at least 11 distinct but structurally related isozymes grouped into three separate subclasses based on their structure and regulatory properties. The conventional PKCs (cPKCs) were the first to be identified and consist of α, βI, βII, and γ isozymes. The second class, the novel PKCs (nPKCs), consists of δ, ε, η, θ, and μ isozymes. Finally, the atypical PKCs (aPKCs) consist of ζ and ι (also known as λ in rodents). The PKC isozymes are the products of independent genes with the exception of PKCβI and PKCβII, which are splice variants of the PKCβ gene.

The PKCs play essential roles in numerous signal transduction pathways that control cell proliferation, cell cycle, differentiation, survival, cell migration, and polarity (reviewed in (Griner and Kazanietz 2007), (Joberty et al. 2000; Larsson 2006). Participation of the PKCs in these complex intracellular signaling pathways is regulated by different extracellular stimuli (reviewed in (Liu and Heckman 1998)), intracellular localization (Disatnik et al. 1995; Ron and Mochly-Rosen 1995), tissue distribution (Hug and Sarre 1993; Nishizuka 1995; Wetsel et al. 1992), phosphorylation status (Chou et al. 1998; Dutil et al. 1998; Le Good et al. 1998), and intermolecular interactions (Jaken and Parker 2000; Mochly-Rosen 1995; Poole et al. 2004). In the early 1980s, PKCs were identified as major cellular receptors for the tumor-promoting phorbol esters (Castagna et al. 1982) leading to an intense effort to define the roles of individual PKC isozymes in the development of cancer. Extensive investigation has demonstrated the participation of numerous PKC isozymes in multiple aspects of transformation, including hyperproliferation, migration, invasion, metastasis, angiogenesis, and resistance to apoptosis (reviewed in (Fields and Gustafson 2003). Alterations in PKC activity, localization, phosphorylation, and/or expression have been found in virtually all tumor types examined. Furthermore, accumulating evidence shows that distinct aspects of the transformed phenotype are mediated by individual PKC isozymes (reviewed in (Fields and Gustafson 2003)). However, despite the close link between PKC isozymes and cancer, to date only one PKC isozyme, atypical PKCι has been shown to function as a bonafide human oncogene (Regala et al. 2005b). This chapter will focus on the structure, function, biochemistry, and biology of the two aPKC isozymes, PKCζ and PKCι, in the context of human cancer. We will specifically discuss the importance of these enzymes as viable therapeutic targets for cancer therapy. Particular emphasis will be placed on the discovery of a mechanism-based therapeutic agent that targets oncogenic PKCι signaling.

23.2 Structure and Function of the aPKCs

23.2.1 The aPKCs Are Structurally Divergent from the Other PKC Isozymes

The aPKCs (PKCζ and PKCι) are structurally and functionally distinct from the other PKCs. Unlike the cPKCs and nPKCs, the catalytic activity of the aPKCs does not require DAG, PS, or calcium, and the aPKCs do not serve as cellular receptors for phorbol esters (Nishizuka 1995; Ono et al. 1989). Instead, aPKC activity can be regulated by 3-phosphoinositides (Nakanishi et al. 1993) phosphorylation by phosphoinositide-dependent kinase-1 (PDK-1) (Chou et al. 1998; Dong et al. 1999; Le Good et al. 1998) and through specific protein–protein interactions (Erdogan et al. 2006; Frederick et al. 2008; Moscat and Diaz-Meco 2000; Puls et al. 1997; Qiu et al. 2000; Sanchez et al. 1998; Suzuki et al. 2001). Important protein–protein interactions between aPKC and effector molecules are mediated through a Phox Bem1 (PB1) domain within the N-terminal regulatory domain of the aPKCs. The PB1 domain is a structurally conserved motif found on a family of signaling molecules (reviewed in (Moscat et al. 2006)) that mediates their homo- and heterotypic interactions through specific interaction codes (Lamark et al. 2003). The PB1 domain interactions formed between aPKCs and other PB1 domain containing proteins such as ZIP/p62 (Hirano et al. 2004; Puls et al. 1997), Par-6 (partitioning-defective 6) (Frederick et al. 2008; Joberty et al. 2000; Lin et al. 2000; Noda et al. 2001; Qiu et al. 2000), and MEK5 [MAPK(mitogen-activated protein kinase)/ERK(extracellular-signal-regulated kinase)kinase 5] (Diaz-Meco and Moscat 2001; Hirano et al. 2004) are critical for aPKC activation and localization in several contexts, including cell polarity, cell proliferation, cell survival, and more recently cell transformation (Frederick et al. 2008; Grunicke et al. 2003; Jamieson et al. 1999; Moscat and Diaz-Meco 2000; Murray and Fields 1997; Regala et al. 2005a).

23.2.2 PKCζ and PKCι are Structurally Related but Functionally Distinct

PKCζ and PKCι/λ are highly related, exhibiting 72% overall amino acid sequence homology and 86% identity within the kinase domain. Due to this high level of homology, PKCζ and PKCι/λ share many in vitro enzymatic characteristics which have made the identification of physiologically relevant isozyme-specific biological functions difficult. Many early studies of aPKC function used dominant negative mutants or pseudo-substrate peptide inhibitors to infer a specific function to one or the other of these isozymes. However, due to the high similarity between the kinase domains of the aPKCs, these reagents inhibit both of these aPKC isozymes, making interpretation of results from these studies problematic. In addition, some early studies characterizing aPKC expression were performed using PKCζ antibodies that cross-react with PKCι/λ. As a consequence, many of these studies concluded that the aPKCs perform overlapping or redundant functions in most cells. With the advent of more recent isozyme-specific aPKC reagents and the application of specific genetic disruption techniques, conclusive evidence of distinct and often dramatic functional differences between PKCζ and PKCι/λ has been obtained in many cell types and tissues. Northern blot analysis of adult mouse tissues reveal that PKCι/λ is nearly ubiquitously expressed in mouse tissues whereas PKCζ expression is largely limited to the brain, kidney, intestine, and testes (Akimoto et al. 1994). Consistent with these results, isozyme-specific riboprobes demonstrated that PKCζ and PKCι/λ exhibit different temporal–spatial expression patterns in cells and tissues of the developing mouse embryo (Kovac et al. 2007).

By embryonic day 16.5 (E16.5) PKCι/λ is the more abundant aPKC isoform and is broadly expressed in most tissues. In contrast, PKCζ expression is largely restricted to specific regions including the bladder, intestine, kidney, testes, and lung. Given the divergent expression pattern between the aPKCs, it is not surprising that the consequences of genetic disruption of the PKCζ and PKCι/λ genes in mice are strikingly different. Complete loss of PKCι/λ expression is embryonically lethal (Bandyopadhyay et al. 2004; Soloff et al. 2004), demonstrating that PKCι/λ is required for embryogenesis. In contrast, PKCζ knockout mice are viable and develop essentially normally, with the exception of subtle inefficiencies in the immune response (Duran et al. 2004; Leitges et al. 2001). NFκB signaling is a well-characterized downstream effector pathway of the aPKCs, and PKCζ knockout mice show impaired activation of NFκB and IL-4 signaling pathways (Duran et al. 2004; Leitges et al. 2001). Since homozygous PKCι/λ knockout mice do not survive past E9.5, Soloff et al. generated mouse embryonic fibroblast (MEF) cells from chimeric PKCλ -deficient embryonic stem cells and C57BL/6 or *Rag2*-deficient blastocysts to assess the function of PKCλ in mouse cells (Soloff et al. 2004). MEF cells lacking PKCλ expression exhibited normal activation of NFκB in response to TNFα, serum, epidermal growth factor, IL-1, and lipopolysacchride (LPS) (Soloff et al. 2004), demonstrating that PKCλ is not required for NFκB activation through these signaling pathways. Rather, MEFs deficient in PKCλ expression showed an increase in stress fibers compared to normal MEFs, suggesting that PKCλ may play a role in normal cellular morphology through control of cytoskeletal functions. These differences in the abundance, tissue distribution and function of the aPKCs during embryonic development predict that these proteins may also have distinct roles in cellular transformation. This prediction has been borne out by the majority of the studies of aPKCs in human cancer and mouse cancer models.

23.3 Atypical PKCs in Human Cancer

23.3.1 PKCζ Exhibits Tumor-suppressor Functions in Human Cancer

In the majority of transformed systems evaluated to date, PKCζ exhibits either tumor-suppressor activity, or no discernible role in tumorigenesis (Table 23.1). Genetic manipulation of PKCζ has shed some light on its roles in human leukemia, colon, breast, and lung cancers. Human myeloid U927 leukemia cells engineered to overexpress PKCζ show a longer doubling time, lower saturation density at confluency, and increased adherence to plastic in vitro than control cells (de Vente et al. 1995; Ways et al. 1994). These cells also exhibit changes in morphology, surface antigen expression, and lysosomal enzyme activities characteristic of a more differentiated phenotype (de Vente et al. 1995; Ways et al. 1994). Finally, treatment of U927 cells overexpressing PKCζ with phorbol esters induces an

Table 23.1 Summary of atypical PKC function in human cancer. Although the aPKCs (PKCζ and PKCι/λ) share high sequence homology, they possess distinct and often dramatic functional differences in both normal and tumor tissues. In the majority of transformed systems evaluated to date, PKCζ exhibits either tumor-suppressor activity, or little to no discernible role in tumorigenesis. PKCι, in contrast, plays a critical promotive role in transformed growth, invasion, migration, survival, chemoresistance, and tumor proliferation in numerous tumor model systems *in vitro* and *in vivo*. In addition, PKCι is the first and only PKC isozyme to be identified as an oncogene in human cancer

	PKCι/λ		PKCζ	
Tumor type	Primary tumors	In vitro and in vivo animal models	Primary tumors	In vitro and in vivo animal models
Nonsmall cell lung cancer	Overexpressed, amplified, oncogene, prognostic indicator (Regala et al. 2005a; b)	Overexpressed and amplified in NSCLC cell lines; promotes chemoresistance transformed growth, invasion, migration and tumor proliferation (Regala et al. 2005a; b; Frederick et al. 2008)	Low levels of expression in normal and tumor tissues; no change in expression between normal and tumor tissues	May act as tumor suppressor (Galvez et al. 2009)
Leukemia		Overexpressed (Gustafson et al. 2004), mediates chemoresistance (Jamieson et al. 1999; Lu et al. 2001; Murray et al. 1999)		
Colon cancer	Increased expression in human colon carcinoma (Murray et al. 2004)	Overexpressed and promotes tumor development / progression in AOM, $APC^{Min/+}$ and $K\text{-}ras^{LA}$ colon carcinogenesis models (Murray et al. 2004; Gokmen-Polar et al. 2001; Murray et al. 2009)		May act as tumor suppressor; inhibits transformed growth of colon cells in vitro and colon tumor formation in vivo (Mustafi et al. 2006; Oster and Leitges 2006)

(continued)

Table 23.1 (continued)

	PKCι/λ		PKCζ	
Tumor type	Primary tumors	In vitro and in vivo animal models	Primary tumors	In vitro and in vivo animal models
Pancreatic cancer	Increased expression in pancreatic ductal and ampullary carcinomas (Evans et al. 2003; Scotti et al. 2010). Increased expression correlates with poor patient survival (Scotti et al. 2010)	Required for transformed growth, invasion, tumor associated angiogenesis and metastasis (Scotti et al. 2010)	Increased expression in pancreatic ductal ampullary carcinomas (Evans et al. 2003)	Stimulates directed motility (Laudanna 2003)
Glioma/Glioblastoma	Overexpressed in gliomas, benign and malignant meningiomas (Patel et al. 2008)	Overexpressed (Patel et al. 2008) mediates chemoresistance (Baldwin et al. 2006); directs cell motility and invasion (Baldwin 2006)		
Ovarian	Overexpressed, amplified oncogene, prognostic indicator (Eder et al. 2005; Weichert et al. 2003; Zhang et al. 2006)	Promotes transformed growth, (Zhang et al. 2006)	Decreased expression (Zhang et al. 2006)	
Esophageal squamous cell carcinoma	Overexpressed, amplified, prognostic indicator (Yang et al. 2008)	Overexpressed and amplified in ESCC cell lines (Yang et al. 2008)		

Prostate	Expressed by both benign and malignant epithelia ; trend of increased expression of PKCλ (Conford et al. 1999)	Activates NF-κB pathways in tumorigenic prostate cancer cells (Win and Acevedo-Duncan 2008)	Increased expression in malignant epithelia (Conford et al. 1999)	Overexpression inhibits invasion in vitro and metastasis in vivo. (Powell 1996). Activate NF-κB pathways in non-tumorigenic prostate cancer cells (Win and Acevedo-Duncan 2008)
Breast cancer	Overexpressed and mislocalized in invasive ductal carcinomas (Kojima et al. 2008)			Overexpression inhibits growth (Mao et al. 2000); Stimulates directed motility (Sun et al. 2005)
Cholangiocarcinomas	Increased expression in malignant tissues; prognostic marker (Li et al. 2008)			

enhanced apoptotic response (de Vente et al. 1995). These cellular changes suggest a role for PKCι as a tumor suppressor in these cells.

PKCζ has also been reported to show tumor suppressor activity in colon cancer cells in vitro. Forced expression of PKCζ in $CaCo_2$ human colon cancer cells causes inhibition of anchorage-independent growth, increased differentiation, and cell death (Mustafi et al. 2006). In contrast, expression of a kinase-deficient dominant negative mutant of PKCζ (kdPKCζ) in $CaCo_2$ cells stimulates growth in soft agar (Mustafi et al. 2006). Interestingly, azoxymethane (AOM)-induced colon tumors in rats show decreased PKCζ expression (Mustafi et al. 2006), and PKCζ expression is also downregulated in intestinal tumors developed by $Apc^{Min/+}$ (multiple intestinal neoplasia) mice suggesting a tumor suppressor function in these colon cancer models in vivo. However, genetic deletion of PKCζ does not affect intestinal carcinogenesis in $APC^{Min/+}$ mice directly demonstrating that PKCζ does not serve as a tumor suppressor in the context of this model (Oster and Leitges 2006). Thus, whereas PKCζ may function to inhibit the transformed growth of colon cells in vitro, it may not do so in vivo. These studies reinforce the importance of in vivo genetic models in the study of aPKC isozymes in cancer.

Forced overexpression of PKCζ also inhibits the growth of human MDA-MB-468 breast cancer cells in vitro (Mao et al. 2000). In contrast, PKCζ reportedly stimulates directed motility of human MDA-MB-231 breast cancer cells (Sun et al. 2005) and pancreatic cancer cells in vitro (Laudanna et al. 2003). The results of these experiments must be interpreted cautiously since they rely solely on the use of pseudosubstrate peptide inhibitors of aPKC and dominant negative mutants to assign isozyme-specific function. As discussed above, these reagents are not specific due to the high homology between the kinase domains of PKCζ and PKCι. In this regard, several commonly used antibodies recognize sequences that are identical in the aPKCs (Bareggi et al. 1995; Diaz-Meco et al. 1994) and peptide pseudosubstrate inhibitors used to inhibit PKCζ also impair the functions of other PKCs including PKCι (Dominguez et al. 1992; Dominguez et al. 1993). Thus, some of the cellular effects observed in these studies may be attributable to other PKC isozymes, including PKCι.

In a recent study, PKCζ-deficient mice show an increase in oncogenic Ras-mediated tumor growth in the lung (Galvez et al. 2009) suggesting a role for PKCζ as a tumor suppressor in vivo. In line with its role in the inflammatory response, PKCζ suppressed IL-6 expression, which was found to be required for Ras-transformed cells to grow under nutrient-deprived conditions in vitro and in vivo. Taken together, the majority of in vitro and in vivo studies on PKCζ indicate a role either in tumor suppression or only a limited role in tumorigenesis.

23.3.2 PKCι Is an Oncogene and Prognostic Marker in Human Cancer

In contrast to PKCζ, the literature overwhelmingly demonstrates that PKCι plays a critical promotive role in tumorigenesis in numerous tumor model systems in vitro and in vivo (Table 23.1). In addition, PKCι, but not PKCζ, is frequently targeted for tumor-specific genetic alteration in primary human cancers. PKCι is overexpressed in many tumor types, including gliomas (Patel et al. 2008), myelogenous leukemias (Gustafson et al. 2004), and cancers of the lung (Regala et al. 2005b), colon (Murray et al. 2004), pancreas (Evans et al. 2003; Scotti et al. 2010), ovary (Eder et al. 2005; Weichert et al. 2003; Zhang et al. 2006), and breast (Kojima et al. 2008). In fact, PKCι is the first and, to date, only PKC isozyme to be identified as an oncogene in human cancer; it was first demonstrated in nonsmall cell lung cancer (NSCLC) (Regala et al. 2005b), and subsequently in ovarian cancer (Zhang et al. 2006).

PKCι is overexpressed at the mRNA and protein levels in established NSCLC cell lines as well as primary NSCLC tumors (Regala et al. 2005b). Immunohistochemical staining for PKCι in primary NSCLC tumors shows PKCι expression predominantly in tumor cells with little or no staining in tumor associated stroma and low but detectable staining in normal lung epithelial cells (Regala et al. 2005b). Interestingly, PKCζ mRNA and protein levels are low in both normal lung and NSCLC tumors (Regala et al. 2005b) indicating that PKCι is selectively targeted for overexpression in human NSCLC tumors. PKCι expression in tumors from NSCLC patients is associated with poor prognosis and lower overall survival. Thus, PKCι expression is of clinical relevance since it predicts clinical outcome in NSCLC patients (Regala et al. 2005b). The correlation between PKCι expression and patient survival is striking; patients diagnosed with early stage (Stage I or II) NSCLC whose tumors express high PKCι are 11 times more likely to die than stage-matched patients whose tumors express low PKCι (Regala et al. 2005b). The prognostic significance of PKCι expression is similar to tumor stage, the standard prognostic indicator in NSCLC patients. Since PKCι overexpression is similar in patients with early and late stages of NSCLC, PKCι expression profiling may be useful in identifying patients with early stage NSCLC that are likely to relapse. These patients are attractive candidates for more aggressive adjuvant chemotherapeutic treatment, perhaps with a newly developed mechanism-based therapy specifically targeting oncogenic PKCι signaling (Erdogan et al. 2006; Regala et al. 2008; Stallings-Mann et al. 2006).

PKCι is overexpressed in primary ovarian cancers (Eder et al. 2005; Weichert et al. 2003; Zhang et al. 2006). PKCι expression in ovarian cancers correlates with tumor stage suggesting the involvement of PKCι in tumor progression and aggressiveness (Eder et al. 2005; Weichert et al. 2003; Zhang et al. 2006). Though PKCι expression is not an independent prognostic marker in ovarian cancer, when combined in a multivariate analysis with tumor cyclin E expression, it is a strong predictor of survival (Eder et al. 2005). The mechanistic significance of the association between PKCι and cyclin E is currently unknown since no functional link between PKCι and cyclin E has been established in these tumors.

PKCι expression is also a potential prognostic marker in cholangiocarcinomas (Li et al. 2008). Cholangiocarcinoma patients whose tumors express high levels of PKCι showed significantly shorter survival time than patients with low PKCι expressing tumors. PKCι expression correlated with differentiation, infiltration, and invasion of these tumors into adjacent lymph nodes. In this same study, loss of E-cadherin expression also correlated with these clinicopathologic features. Whether a mechanistic link exists between PKCι and E-cadherin in cholangiocarcinoma is currently unknown but merits further investigation.

A major mechanism by which PKCι is overexpressed in human tumor cells is tumor-specific amplification of the PKCι gene (*PRKCI*) on chromosome 3q26 (Regala et al. 2005b; Zhang et al. 2006), a region of DNA frequently amplified in human cancers (Han et al. 2002). An increase in PKCι mRNA and protein expression correlates with gene copy number gains of the *PRKCI* gene (Regala et al. 2005b). *PRKCI* gene amplification is observed in 70% of lung squamous cell carcinomas (LSCC), but rarely in lung adenocarcinoma (LAC) (Regala et al. 2005b). These findings are in agreement with the prevalence of chromosome 3q26 amplification in LSCCs and the rare occurrence of 3q26 amplification in LACs (Balsara et al. 1997; Brass et al. 1997). PKCι is an important functional target of 3q26 amplification since LSCC cells harboring *PRKCI* amplification require PKCι for both anchorage-independent growth and invasive behavior (Frederick et al. 2008; Regala et al. 2005b). *PRKCI* gene amplification also correlates with PKCι overexpression in primary ovarian cancer, implicating gene amplification as an important mechanism driving PKCι expression in these tumors (Eder et al. 2005; Zhang et al. 2006). Decreased PKCι expression inhibits anchorage-independent growth of ovarian cancer cells, whereas overexpression of PKCι promoted murine ovarian surface epithelium transformation (Zhang et al. 2006) demonstrating that PKCι is also a relevant gene target of 3q26 amplification in ovarian cancer. A recent study demonstrated amplification of *PRKCI* in 53% of esophageal squamous cell carcinomas (ESCC) and PKCι protein expression correlated with *PRKCI* gene amplification in these tumors (Yang et al. 2008). Examination of clinicopathologic features of ESCC tumors revealed a significant correlation between *PRKCI* expression and tumor size, stage, and lymph node metastasis suggesting that *PRKCI* overexpression is associated with tumor progression and metastasis in ESCC (Yang et al. 2008). Chromosomal gains at 3q26 are also frequently observed in SCC of the head and neck (Snaddon et al. 2001) and cervix (Sugita et al. 2000). It will be of interest to determine if an increase in *PRKCI* copy number drives PKCι overexpression and whether PKCι expression profiling may be of use as a prognostic indicator in these tumors as well.

Though *PRKCI* gene amplification is a major mechanism driving PKCι overexpression in many human tumor types, alternative mechanisms must also exist in the many tumor types that express elevated PKCι levels but do not harbor frequent 3q26 amplification. In this regard, lung adenocarcinomas (LACs) exhibit high PKCι expression, but lack alterations of *PRKCI* gene copy number (Regala et al. 2005b). Likewise, PKCι is frequently overexpressed in colon cancer (Murray et al. 2004), pancreatic cancer (Evans et al. 2003; Scotti et al. 2010), breast cancer (Kojima et al. 2008), and chronic myelogenous leukemia (CML) (Gustafson et al. 2004) despite the fact that chromosome 3q26 amplification is rarely observed in these tumors. An alternative mechanism for oncogenic PKCι expression involving tumor cell-specific transcriptional activation has been elucidated in CML cells. Analysis of the human PKCι promoter identified an Elk1 element within the proximal 5′ region of the PKCι gene that mediates transcriptional activation of PKCι expression through a Bcr-Abl/Mek/Erk signaling mechanism in CML cells (Gustafson et al. 2004). Whether this mechanism is operative in other tumor types requires further investigation. Other possible mechanisms for oncogenic activation of PKCι such as posttranslational modification and/or somatic mutations have not been extensively analyzed and also warrant further investigation, particularly in light of the prognostic significance of PKCι expression in multiple human tumor types. Sequence analysis of all 18 exons of *PRKCI* in primary human LAC and SCC tumors that do not harbor *PRKCI* amplification did not reveal any mutations, suggesting that somatic mutations of *PRKCI* either do not occur, are restricted to noncoding sequences, or are relatively rare in NSCLC (Regala and Fields, unpublished observation). However, given the functional importance of PKCι in human tumor biology, a more extensive analysis of PKCι mutational status in human tumors is warranted. In conclusion, PKCι is the first and thus far the only PKC isozyme shown to possess the characteristics of a human oncogene. PKCι is overexpressed as a result of tumor-specific amplification of *PRKCI* in NSCLCs, ovarian cancers, and ESCCs. In addition, PKCι expression and function is required for the transformed phenotype of NSCLC and ovarian cancer cells. In contrast, PKCζ appears to play a less significant role in most tumor types; in some systems, PKCζ plays no apparent role in tumorigenesis, while in others it serves a tumor suppressive function.

23.4 PKCι Is a Critical Effector of Oncogenic Ras

aPKCs function at the crossroads of a number of signaling pathways that contribute to the transformed phenotype (Fig. 23.1). Therefore, it is not surprising that aPKCs appear to participate in several aspects of transformation. PKCι is itself an oncogene, which is activated through tumor-specific overexpression. In addition, PKCι is activated downstream of other oncogenes including oncogenic Ras, Bcr-Abl, and Src. Of these activation mechanisms, the critical role of PKCι in oncogenic Ras is the best characterized.

23.4.1 Ras-Mediated Activation of aPKCs

Oncogenic mutations of Ras are one of the most frequent molecular alterations in human cancers occurring in ~30% of all human tumors (Adjei 2001). Early studies indicate that Ras can bind and activate the aPKCs (Diaz-Meco et al. 1994; Mwanjewe et al. 2001), and the aPKCs have been implicated in oncogenic Ras signaling in vitro (Bjorkoy et al. 1997; Hellbert et al. 2000). Whereas PKCζ appears to be a negative regulator of Ras-mediated transformation in the lung in vivo (Galvez et al. 2009), PKCι is required for oncogenic Ras-mediated transformation in the colon, lung and pancreas (Murray et al. 2004; Regala et al. 2005a; Scotti et al. 2010).

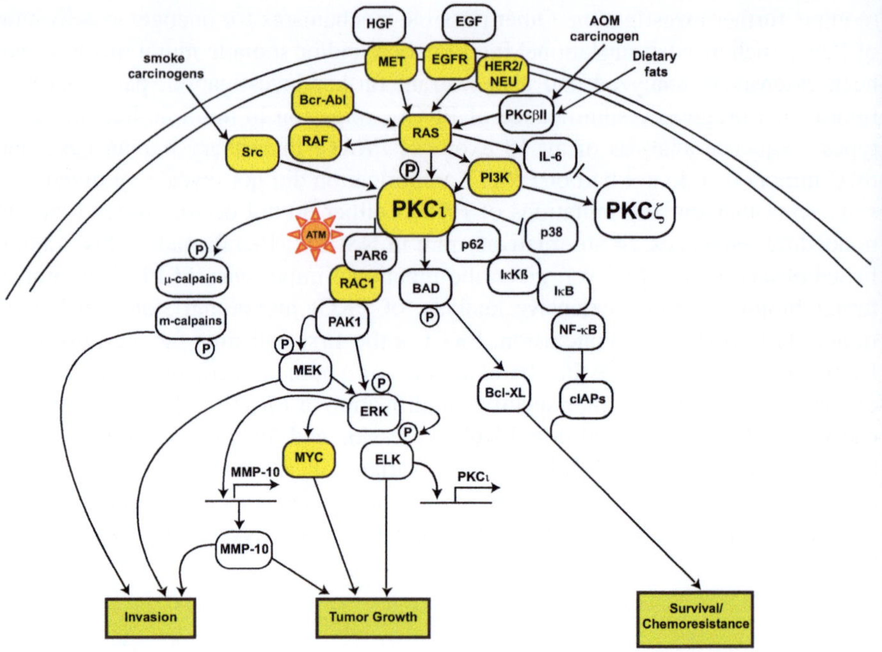

Fig. 23.1 Schematic representation of oncogenic atypical PKC signaling. The atypical PKCs reside in several major signaling pathways implicated in human cancer. The aPKCs can be activated by known oncogenes such as Ras, Bcr-Abl, Src, and PI3K or cytokines such as TNF and IL-1 or growth factors such NGF and EGF. The aPKCs signal to downstream effectors such as Rac1 and NFkB, which are important for different aspects of transformed phenotypes. Many components in aPKC signal pathways are mutated, often by multiple mechanisms (i.e., gene amplification and somatic mutation), in human tumors (indicated by *yellow boxes*). Arrows indicate flow through signaling pathways; touching boxes indicate directly binding of signaling components. Phosphorylation events are indicated by circled Ps

23.4.1.1 Colon Cancer

Oncogenic K-ras mutations are frequently found in colon cancers, and Ras signaling plays an important role in colon carcinogenesis (Slattery et al. 2001; Takayama et al. 2001). The role of aPKCs in colon carcinogenesis has been studied in three complementary in vivo mouse models, the AOM carcinogen model, the spontaneous K-ras mutation model, and the APC$^{Min/+}$ mouse model.

In the AOM carcinogen model, colon tumors are induced in mice by exposure to the colon-selective carcinogen azoxymethane (AOM). AOM induces dose- and time-dependent formation of colonic lesions termed aberrant crypt foci (ACF) and subsequently colon adenomas. ACF are the likely precursors to colon cancer in both mice and humans (McLellan et al. 1991; Takayama et al. 1998). The number and multiplicity of ACF (number of aberrant crypts/lesion) are predictive of the development of colon tumors in rodents (Magnuson et al. 1993) and the presence of ACF correlate with increased risk of colon cancer in humans (Takayama et al. 1998). ACF harbor many of the same genetic and biochemical alterations found in colon tumors, including increased expression of PKCβII (Gokmen-Polar et al. 2001) and activating K-ras mutations (Zhang et al. 2004).

An analysis of PKC isozyme expression in AOM-induced ACF and colon tumors demonstrated a loss of PKCα and increased expression of PKCβII and PKCι (Gokmen-Polar et al. 2001). Both PKCβII and PKCι play critical promotive roles in AOM-induced colon carcinogenesis (Murray et al. 1999; Murray et al. 2004; Murray et al. 2009). Overexpression of PKCβII in the colonic epithelium of transgenic mice (Tg PKCβII mice) leads to increased susceptibility to AOM-induced ACF and colonic tumors (Murray et al. 1999). Conversely, nullizygous PKCβ mice (PKCβ$^{-/-}$ mice) are highly resistant to AOM-mediated tumorigenesis (Murray et al. 2002). However, bi-transgenic Tg-PKCβII/PKCβ$^{-/-}$ mice, which express PKCβII only in the colon, exhibit AOM-induced colon tumor formation indistinguishable from wild-type mice, demonstrating that PKCβII expression in colonic epithelial cells is required for colon carcinogenesis (Murray et al. 2002).

PKCι is overexpressed in both AOM-induced colon tumors and human colon carcinomas, and like PKCβII plays a promotive role in AOM-induced colon carcinogenesis (Murray et al. 2004). Transgenic mice that express a constitutively active PKCι allele (caPKCι) in the colonic epithelium exhibit a higher incidence of colon tumors in response to AOM than nontransgenic control mice (Murray et al. 2004). In addition, AOM treatment causes mostly malignant carcinomas in transgenic caPKCι mice whereas nontransgenic mice develop mainly benign tubular adenomas (Murray et al. 2004). Thus, PKCι facilitates formation of colonic tumors and promotes colon tumor progression from benign adenoma to malignant carcinoma in vivo (Murray et al. 2004).

Given the similar tumor-promoting phenotype of transgenic PKCβII and caPKCι mice, it was possible that PKCβII and PKCι serve redundant functions in colonic carcinogenesis. Alternatively, these two PKC isozymes could collaborate during colon carcinogenesis. To assess this question, and to determine the relative contribution of these two PKC isozymes to colon carcinogenesis, compound transgenic mice were generated and analyzed (Murray et al. 2009). Mice expressing PKCβII in the presence of either caPKCι (Tg-PKCβII/caPKCι mice) or a dominant negative kinase deficient PKCι allele (Tg-PKCβII/kdPKCι mice) developed similar numbers of colonic tumors in response to AOM. In contrast, PKCβ$^{-/-}$ mice develop no tumors after AOM exposure even when these mice are engineered to also express caPKCι (PKCβ$^{-/-}$/caPKCι mice) (Murray et al. 2009). Therefore, early colon tumor promotion is driven predominantly by PKCβII, whereas PKCι plays a relatively minor role in the early stages of colon carcinogenesis. Interestingly however, whereas tumors in Tg-PKCβII/kdPKCι mice are predominantly colonic adenomas, the majority of tumors in Tg-PKCβII/caPKCι mice are intramucosal carcinomas. Thus, PKCι plays a prominent role in tumor progression from benign adenoma to malignant carcinoma. These results indicate that PKCβII and PKCι play distinct roles in colon cancer, collaborating to promote initiation of colon carcinogenesis and progression to carcinoma, respectively.

PKCι is also elevated in intestinal tumors formed in $APC^{Min/+}$ mice compared to normal intestinal epithelium (Murray et al. 2009; Oster and Leitges 2006). To determine if PKCι plays a role in tumor development in $APC^{Min/+}$ mice, the mouse homolog of PKCι (PKCλ) was inactivated in the intestinal epithelium of triple transgenic $APC^{Min/+}$/floxPKCλ/villin-Cre mice by Cre-mediated recombination. $APC^{Min/+}$/floxPKCλ/villin-Cre mice exhibit loss of intestinal epithelial PKCλ expression and a decrease in the number of tumors formed in the small intestine compared to $APC^{Min/+}$/flox PKCλ mice that harbor intact alleles of the PKCι gene (Murray et al. 2009). Thus, PKCι is important for tumor progression in $APC^{Min/+}$-induced intestinal epithelial tumorigenesis.

PKCι is also required for oncogenic K-Ras-mediated colon carcinogenesis in vivo (Murray et al. 2004). K-ras^{LA} mice containing a latent oncogenic K-ras allele (G12D) that is activated by spontaneous recombination develop K-Ras-dependent ACF in the colonic epithelium. K-ras^{LA}/kdPKCι mice, which express kdPKCι in the colonic epithelium, develop significantly fewer ACF than K-ras^{LA} mice (Murray et al. 2004). Thus, PKCι is required for oncogenic Ras-mediated transformation of intestinal epithelial cells in vivo. PKCι activity is elevated in Ras transformed intestinal epithelial cells and is required for invasion and anchorage-independent growth of Ras-transformed intestinal epithelial cells in vitro (Murray et al. 2004). The Rho family GTPase, Rac1 is important for Ras-mediated invasion of intestinal epithelial cells (Murray et al. 2004; Qiu et al. 1995). Interestingly, expression of kdPKCι in Ras transformed intestinal epithelial cells decreases Rac1 activity and inhibits cellular invasion (Murray et al. 2004) whereas expression of a constitutively active Rac1 allele, RacV12, overcomes kdPKCι-mediated inhibition of invasion (Murray et al. 2004). Thus, Ras activates a PKCι/Rac1 signaling axis that is necessary for Ras-mediated colon carcinogenesis. We recently elucidated a similar oncogenic PKCι signaling mechanism in pancreatic cancer cells harboring mutant K-Ras (Scotti et al. 2010).

PKCι is a downstream effector of PKCβII in intestinal epithelial cells in vitro. Increased expression of PKCβII induces an invasive phenotype in rat epithelial intestinal cells that is dependent on PKCι activity (Zhang et al. 2004). K-ras is important in PKCβII induced cellular invasion in RIE cells since expression of PKCβII in RIE cells increases K-ras activity and invasion in RIE/PKCβII can be blocked by repressing the K-ras effectors PKCι, Rac1, or Mek (Zhang et al. 2004). Expression of K-ras induces PKCι activity in RIE cells, and PKCι is required for Ras activation of Rac1 in Ras transformed RIE cells (Murray et al. 2004). In addition, expression of kdPKCι in RIE/PKCβII cells inhibits cellular invasion. Thus, PKCβII induces invasion of intestinal epithelial cells in vitro through activation of a PKCβII→Ras→PKCι/Rac1→Mek signaling pathway.

23.4.1.2 PKCι and Oncogenic K-ras Signaling in NSCLC

In addition to its promotive role in colon cancer, PKCι is also a critical downstream effector of oncogenic K-ras in lung cancer (Regala et al. 2005a). Oncogenic mutation of K-ras is one of the most frequent genetic changes in NSCLC, occurring in some

25% of NSCLC tumors. Expression analysis demonstrates that PKCι is the predominant atypical PKC isozyme expressed in both normal lung epithelium and NSCLC cells whereas PKCζ is expressed at much lower levels than PKCι in both tissues (Regala et al. 2005b). In addition, PKCι is overexpressed in NSCLC tumor cells compared to matched nontransformed lung epithelial cells, whereas no increase in PKCζ expression is observed in NSCLC tumor cells (Regala et al. 2005a). RNAi-mediated inhibition of PKCι, or expression of kdPKCι, in NSCLC cells harboring an oncogenic mutation in K-ras blocks Rac1 activation and transformed growth and invasion in vitro and tumorigenicity in vivo (Frederick et al. 2008; Regala et al. 2005a; Regala et al. 2005b; Stallings-Mann et al. 2006). NSCLC cells expressing kdPKCι showed decreased anchorage independent growth *in vitro* and tumorigenicity in vivo without affecting tumor apoptotic cell death or tumor associated vascularity (Regala et al. 2005a). Thus, just as in the colon, a primary function of PKCι in NSCLC is to promote transformed growth. RNAi-mediated depletion of PKCζ, in contrast has no effect on the transformed phenotype of NSCLC cells (Frederick et al. 2008). Interestingly, inhibition of PKCι has no significant effects on the growth rate, saturation density, or survival of NSCLC cells grown as adherent cultures. This finding is consistent with observations made in human leukemia cells, intestinal epithelial cells, ovarian tumor cells and pancreatic cancer cells, which also require PKCι for anchorage-independent transformed growth, but is dispensable for adherent cell growth and survival (Jamieson et al. 1999; Murray et al. 2004; Zhang et al. 2004; Scotti et al. 2010).

PKCι has also been implicated in other aspects of the transformed phenotype of NSCLC cells including invasiveness, survival, and resistance to chemotherapy. RNAi-mediated depletion of PKCι expression in NSCLC cells inhibits cellular invasion (Frederick et al. 2008). NSCLC cells grown in three-dimensional Matrigel cultures exhibit long cellular projections and spikes protruding into the surrounding matrix characteristic of morphologically transformed, highly invasive cells (Kleinman and Martin 2005). In contrast, cells made deficient in PKCι grow in clusters of rounded cells, indicating that PKCι is important for cellular invasion of NSCLC cells. Indeed, PKCι-deficient NSCLC cells exhibit decreased invasion through Matrigel-coated chambers. Consistent with these findings, overexpression of PKCι enhances, and inhibition of PKCι expression blocks, migration and invasion of NSCLC cells in response to nicotine (Xu and Deng 2006). In NSCLC cells, PKCι can directly phosphorylate μ- and m-calpains, which are associated with increased wound healing, migration, and invasion (Xu and Deng 2006). Inhibition of PKCι expression causes A549 NSCLC cells to undergo apoptosis in response to the smoke carcinogen Nitrosamine 4-(methylnitrosamino)-1-(3-pyridyl)-1-butanone (NNK) as well as the chemotherapeutic agents taxol and cisplatin (Jin et al. 2005). These results are consistent with those observed in CML cells, in which PKCι inhibition causes increased sensitivity to taxol-mediated apoptosis (Jamieson et al. 1999; Murray and Fields 1997). Thus, PKCι plays a critical role in the control of anchorage-independent growth, cellular motility, invasion, and resistance to chemotherapeutic agent- and carcinogen-induced apoptosis in multiple human cancer cell types.

23.4.2 Bcr-Abl and Src Activation of aPKCs

PKCι functions as a survival gene downstream of oncogenic Brc-Abl signaling in chronic myelogenous leukemia (CML). The chimeric tyrosine kinase oncogene Brc-Abl oncogene mediates cell survival and resistance to chemotherapeutic drugs such as taxol in K562 CML cells (Bedi et al. 1995). Brc-Abl activates a Ras/Mek/Erk signaling pathway that stimulates PKCι expression through an Elk1 transcription factor site in the proximal promoter of PKCι that is important for the resistance of K562 cells to chemotherapeutic agents (Gustafson et al. 2004). PKCι is highly expressed in human K562 leukemia cells whereas PKCζ is undetectable in these cells (Murray and Fields 1997). Overexpression of PKCι, but not PKCζ, leads to enhanced resistance of K562 cells to taxol-induced apoptosis (Murray and Fields 1997). Conversely, inhibition of PKCι expression enhances taxol mediated apoptotic cell death of K562 cells indicating that the antiapoptotic effect of PKCι in K562 cells is isozyme-specific. Treatment of K562 cells with taxol leads to sustained activation of PKCι, whereas taxol treatment of the Bcr-Abl negative myeloid leukemia cell line, HL60, showed only transient and weak activation of PKCι (Jamieson et al. 1999). Treatment of K562 cells with Bcr-Abl inhibitor results in decreased PKCι activation and sensitization to apoptosis after chemotherapeutic treatment (Jamieson et al. 1999). In contrast, expression of caPKCι protects K562 cells from chemotherapeutic drug-mediated apoptosis. Thus, activation of PKCι downstream of Bcr-Abl is necessary and sufficient to mediate apoptotic resistance to chemotherapy in K562 cells.

Nitrosamine 4-(methylnitrosamino)-1-(3-pyridyl)-1-butanone (NNK) has been identified as one of the most potent carcinogens in cigarette smoke (Schuller 2002). PKCι is important for NNK-mediated survival of lung cancer cells. Src can directly bind aPKCs and promote their activation (Wooten et al. 2001). In NSCLC cells, NNK increases c-Src-activated PKCι, which in turn phosphorylates Bad resulting in reduced Bad/Bcl-XL interaction. PKCι-mediated phosphorylation abrogates the proapoptotic function of Bad and enhances cell survival and decreased sensitivity of NNK treated NSCLC cells to VP-16 and cisplatin (Jin et al. 2005).

23.5 The Importance of Protein–Protein interactions in oncogenic aPKC Signaling

The aPKCs are unique among the PKC isozymes in having a Phox/Bem1 (PB1) protein–protein interaction domain at their N-terminus. The PB1 motif serves a regulatory function by coupling aPKCs to various signaling pathways through formation of PB1–PB1 domain interactions with other PB1 domain containing proteins. The aPKCs form PB1–PB1 domain interactions with at least three different signaling molecules; ZIP/p62 (Hirano et al. 2004; Puls et al. 1997), Par6 (Joberty et al. 2000; Lin et al. 2000; Noda et al. 2001; Qiu et al. 2000), and Mek5 (Diaz-

Meco and Moscat 2001; Hirano et al. 2004). ZIP/p62 is a scaffolding protein that links the aPKCs to NFκB activation downstream of extracellular signals from tumor necrosis factorα (TNFα), interlukin-1 (IL-1), and nerve growth factor (NGF) (Moscat and Diaz-Meco 2000). MEK5 and aPKCs form a complex upon EGF stimulation that has been implicated in proliferative and prosurvival signaling, again through NFκB (Diaz-Meco and Moscat 2001).

Perhaps the best characterized PB1 domain interaction involving the aPKCs is the formation of a polarity complex consisting of Par6, aPKCs, and a GTPase (Rac1 or Cdc42) that is evolutionary conserved from nematodes to mammals. The aPKC/Par6 polarity complex is involved in various polarity related processes including asymmetric cell division, directed cell migration, tight junction formation, cell adhesion, cytoskeletal reorganization, axon specification, and establishment of apical-basolateral polarity. Cellular polarity is necessary for organization and normal function of epithelial cells, and loss of epithelial organization is an early event in carcinoma development. Defects or loss of cell polarity may directly contribute to carcinogenesis through disregulation of normal proliferation and differentiation processes in cells (Addeo et al. 2007; Aranda et al. 2006; Bilder 2004; Cochand-Priollet et al. 1998; Curto et al. 2007). In addition, there is mounting evidence that loss of polarity participates in tumorigenesis by facilitating an epithelial to mesenchymal transition (EMT) that is a critical step for tumor cells to acquire motility and invasiveness (Thiery 2002).

23.5.1 aPKC-Mediated Cell Survival Through Activation of NFκB

ZIP/p62 links the aPKCs to TNFα, IL-1, and NGF activation of NFκB and cell survival. Increased PKCι expression and/or activity have been widely implicated in promoting pro-survival signals in a variety of cancer cell types. PKCι associates with IKKαβ and IκBα in TNFα-treated DU-145 prostate carcinoma cells (Win and Acevedo-Duncan 2008). PKCι phosphorylates IKKαβ, which subsequently phosphorylates IκBα resulting in NFκB/p65 translocation into the nucleus and potential transcriptional activation. Interestingly, PKCι does not associate with IKKαβ and IκBα in transformed nontumorigenic RWPE-1 prostate cells. Rather, TNFα treatment of RWPE-1 induces association of PKCζ with both IKKαβ and IκBα indicating that PKCι may be preferentially used to activate NFκB pathways in tumorigenic prostate cancer cells (Win and Acevedo-Duncan 2008). It will be interesting to determine whether this preference is due to the selective increase in PKCι expression in prostate cancer cells (and other tumor cells), or from some other selection mechanism. In this regard, CML cells expressing the Bcr-Abl oncogene, which expresses PKCι but not PKCζ, are resistant to chemotherapy-induced apoptosis as a result of increased PKCι activation (Jamieson et al. 1999; Lu et al. 2001; Murray and Fields 1997). Treatment of K562 CML cells with taxol induces IκB phosphorylation and NFκB nuclear translocation and transcriptional activation (Lu et al. 2001). Disruption of PKCι function sensitizes K562 cells to taxol-induced apoptosis and inhibits RelA transcriptional activity. Overexpression of NFκB in K562 cells with disrupted PKCι function, rescues taxol-induced apoptosis. In addition, overexpression of constitutively active PKCι further upregulates NFκB transcriptional activity. Thus, PKCι induction of NFκB transactivation is important for Bcr-Abl–dependent resistance to taxol-induced apoptosis (Lu et al. 2001; Murray and Fields 1997). Glioblastoma multiforme are highly resistant to most standard cancer chemotherapeutics (Baldwin et al. 2006). RNAi-mediated depletion of PKCι results in sensitization of U87MG glioblastoma cells to cisplatin (Baldwin et al. 2006). However, unlike in CML and prostate cancer cells, activation of the NFκB pathway does not appear to be a major mechanism driving PKCι-dependent chemoresistance in glioblastoma cells. Rather, PKCι-mediated survival in glioblastoma cells appears to result from PKCι-mediated attenuation of p38 mitogen-activated protein kinase signaling that protects these cells against cytotoxicity to chemotherapeutic agents (Baldwin et al. 2006). Thus, PKCι appears to function in several signaling pathways that promote cell survival in different tumor cell types.

23.5.2 Oncogenic PKCι-Mediated Activation of Rac1 Signaling

Interestingly, several components of the polarity complex, particularly PKCι, Par6, and Rac1 have been directly implicated in oncogenesis. PKCι activates Rac1 by forming an oncogenic complex with Par6 through PB1-PB1 domain interactions. Expression of the PB1 domain of PKCι in NSCLC cells uncouples PKCι and Par6 from Rac1 activation and inhibits transformed growth (Regala et al. 2005a). Likewise, RNAi-mediated knock down of PKCι, Par6, or Rac1 inhibits transformed growth and cellular invasion in NSCLC cancer cells (Frederick et al. 2008). Expression of wild-type PKCι in PKCι KD cells restores transformation, whereas expression of a PB1 domain mutant of PKCι, PKCι-D63A, that cannot bind Par6, does not (Frederick et al. 2008). Similarly, expression of wild type Par6 in Par6 KD cells restores transformation whereas expression of Par6 mutants that either cannot bind PKCι (Par6-K19A) or couple to Rac1 (Par6-ΔCRIB) does not (Frederick et al. 2008). Expression of a constitutively active Rac1 allele (RacV12) in PKCι or Par6 depleted NSCLC cells restores transformed growth and cellular invasion (Frederick et al. 2008). These data demonstrate that Rac1 is a key effector of PKCι-mediated transformation in the lung. Interestingly, RNAi-mediated knockdown of Mek5 does not affect the transformed phenotype of NSCLC cells (L.A. Frederick and A.P. Fields, unpublished observations) suggesting that the PKCι-Par6 interaction is the key mediator of PKCι signaling in NSCLC cells. These findings also suggest that the PB1–PB1 domain interaction formed between PKCι and Par6 may be a promising therapeutic target for disruption of PKCι signaling in cancer cells.

23.6 Transcriptional Targets of Oncogenic aPKC Signaling

The expression of matrix metalloproteinase-9 (MMP9) correlates with the progression in gliomas (Rao et al. 1993) and inhibition of MMP9 significantly reduces the invasiveness of glioma cells in vitro and in vivo (Kondraganti et al. 2000). In glioma cells, IL-6 and TNFα induce activation of PKCζ leading to an NFκB-mediated increase in MMP9 expression (Esteve et al. 2002). Our recent work has identified another member of the MMP family, MMP10 or stromelysin 2, as a major transcriptional target of oncogenic PKCι/Par6 signaling in NSCLC cells and primary NSCLC tumors (Frederick et al. 2008). MMP10 was identified as a potential transcriptional target of PKCι through genome-wide gene expression analysis of NSCLC cells expressing either a control lentiviral RNAi or an RNAi that specifically knocks down PKCι expression (Frederick et al. 2008). Analysis of primary NSCLC samples revealed that MMP10 is overexpressed in NSCLC and that MMP10 expression correlates positively with PKCι expression (Frederick et al. 2008). Depletion of PKCι, Par6, or Rac1 by RNAi inhibits MMP10 expression in NSCLC cells. Futhermore, expression of exogenous wild-type Par6 in Par6 KD cells restored MMP10 expression whereas expression of Par6 mutants that either cannot bind PKCι or Rac1 did not. Similar to depletion of PKCι and Par6, RNAi mediated knockdown of MMP10 blocks anchorage-independent growth and cell invasion in NSCLC cells. In addition, loss of transformed growth and invasion in PKCι KD or Par6 KD NSCLC cells is rescued by the addition of catalytically active MMP10. Thus, expression of MMP10 is regulated through the PKCι-Par6-Rac1 signaling axis and MMP10 represents a key downstream effector in PKCι-mediated transformation in lung cancer cells. The molecular mechanisms by which PKCι-mediated overexpression of MMP10 leads to transformation is unclear and merits further investigation.

In order to identify other potential downstream targets of oncogenic PKCι, we recently conducted a meta-analysis of gene expression in primary lung adenocarcinomas (LAC) from three independent public domain databases (Erdogan et al. 2009). Using these data, we identified genes whose expression correlates with that of PKCι in primary LAC tumors. Seven genes were identified whose expression was coordinately induced with PKCι expression in all three databases (Erdogan et al. 2009). QPCR analysis of a panel of 60 primary LAC samples showed that four of these seven genes were highly overexpressed in tumors compared to matched normal control lung tissue, and that expression of each of these four genes exhibited a strong positive correlation with PKCι expression (Erdogan et al. 2009). RNAi-mediated knock down of PKCι in three LAC cell lines led to significant reduction in expression of each of the four target genes, indicating that PKCι regulates the expression of these four genes in LAC cells (Erdogan et al. 2009). RNAi-mediated knock down of each of these genes led to significant inhibition of anchorage-independent growth and cellular invasion demonstrating that each of them are important for transformation in LAC cells (Erdogan et al. 2009). Furthermore, several of these PKCι-regulated genes are coordinately overexpressed with PKCι in other major tumor types including lung squamous cell carcinoma, breast, colon, prostate, pancreatic, and glioblastoma cancers. This analysis revealed novel signaling mechanisms that participate in PKCι-mediated transformation. These PKCι-regulated genes may serve as useful biomarkers in determining the effectiveness of PKCι-directed therapies and may themselves serve as targets for the development of novel prognostic markers and/or therapeutic agents.

23.7 aPKCs as Therapeutic Targets

Given the role of aPKCs in human cancer, it is perhaps not surprising that changes in aPKC activity and function have been implicated in the mechanism of action of chemotherapeutic agents. In this regard, the action of rituximab in follicular lymphoma cells (Leseux et al. 2008), spisulosine [a compound isolated from the marine organism, Spisula polynyma (Cuadros et al. 2000)] in PC-3 and LNCaP prostate tumor cells (Sanchez et al. 2008) and ursolic acid in rat C6 glioma cells (Huang et al. 2008) have all been linked to changes in aPKC activity, implicating aPKC in their mechanism of action. Oncrasin-1 (oncogenic Ras tumor inhibiting compound 1) is a small molecule identified from a chemical library as an effective inducer of apoptosis in lung cancer cell lines harboring oncogenic K-ras mutation (Guo et al. 2008). Treatment of mutant K-ras lung cancer cells with Oncrasin-1 leads to changes in the subcellular localization of PKCι, whereas inhibition of PKCι expression by RNAi inhibits the antiproliferative effects of Oncrasin-1 (Guo et al. 2008). It is currently unclear what causes the changes in the subcellular localization of PKCι in oncrasin-1 treated cells and whether the change of subcellular localization of PKCι plays a critical role in oncrasin-1-mediated effects. Although the aforementioned antitumor agents appear to modulate aPKC activity, they do not specifically target aPKCs; rather the aPKCs may serve as indirect targets of their cellular effects. These studies underscore the need to develop specific, mechanism-based chemotherapeutic agents that directly target oncogenic aPKC signaling.

23.8 Targeting the oncogenic PKCι-Par6 signaling complex for treatment of NSCLC

The catalytic domain of PKCι is highly related to the other PKC isozymes. On the other hand, the PB1 domain of PKCι is uniquely present in atypical PKCs but not other isozymes. The PB1–PB1 domain interaction formed between PKCι and Par6 is required for the oncogenic PKCι-Par6-Rac1-MMP10 signaling axis that mediates anchorage-independent growth and invasion of human NSCLC cells in vitro and tumorigenicity in vivo. Thus, we reasoned that the PB1–PB1 domain interaction between PKCι and Par6 is an attractive target for development of novel mechanism-based therapeutics for treatment of PKCι- dependent tumors such as NSCLC.

We therefore developed and implemented a novel fluorescence resonance energy transfer (FRET)-based assay and performed high throughput screening for small molecular weight compounds that can disrupt the PB1–PB1 domain interaction between PKCι and Par6 (Stallings-Mann et al. 2006). Among the most potent inhibitors identified in our screen were the gold-containing compounds aurothioglucose (ATG) and aurothiomalate (ATM). These compounds have been used for many years in the treatment of rheumatoid arthritis (Messori and Marcon 2004). ATG and ATM exhibited dose dependent inhibition of PKCι-Par6 binding with IC50s of ~1μM (Stallings-Mann et al. 2006). Treatment of NSCLC cells with these compounds inhibits PKCι-mediated Rac1 activation and blocks anchorage-independent growth of NSCLC cells in vitro and tumorigenicity in vivo (Stallings-Mann et al. 2006). Expression of Rac1V12 rescues ATM-mediated inhibition of transformed growth in NSCLC cells, confirming that ATM targets the interaction between PKCι and Par6, which in turn couples PKCι to Rac1 (Stallings-Mann et al. 2006).

Although ATG and ATM have been used extensively in the treatment of rheumatoid arthritis (RA) (Messori and Marcon 2004), their precise mechanism of action against RA is still unknown. A proposed mechanism of action for ATG/ATM is the formation of gold-cysteine adducts with target cellular proteins. ATM can inhibit the activity of thioredoxin reductases through formation of a gold adduct with a critical cysteine residue within the active site of the enzymes, and this mechanism has been suggested to play a role in the antioxidant effects of ATM (Pia Rigobello et al. 2004). In a similar report, ATM was suggested to exhibit antiinflammatory properties through a mechanism involving a critical cysteine residue in IκK that may participate in the inhibition of NF-κB signaling (Bratt et al. 2000; Jeon et al. 2000; Yamashita et al. 2003). Since cysteine residues have been implicated as targets for ATM action, we assessed whether the PB1 domains of PKCι and/or Par6 contain cysteine residues that could serve as potential targets for ATM binding (Erdogan et al. 2006). The PB1 domain of the aPKCs contains a cysteine residue, (Cys69) that is not found in other PB1 domain-containing proteins. The crystal structure of the PKCι-Par6 complex reveals that Cys69 is located within the conserved OPR, PC, and AID (OPCA) motif of PKCι at the binding interface between PKCι and Par6 in the complex (Erdogan et al. 2006). Cys69 interacts with Arg28, a residue within the basic cluster of Par6 involved in PKCι binding (Hirano et al. 2004; Lamark et al. 2003). Molecular modeling of ATM within the crystal structure of the PKCι PB1 domain, predicts formation of an adduct between Cys69 and ATM that protrudes into the binding cleft between PKCι and Par6 causing steric hinderance to Par6 binding (Erdogan et al. 2006). Mutation of Cys69 of PKCι to isoleucine (C69I) or valine (C69V), amino acids that frequently exist at this position in other PB1 domains, does not alter Par6 binding. However, the C69I and C69V mutations make PKCι resistant to the inhibitory effects of ATM on Par6 binding *in vitro* (Erdogan et al. 2006). Furthermore, expression of the C69I PKCι mutant in NSCLC cells supports transformed growth, but renders these cells resistant to the inhibitory effects of ATM on transformed growth (Erdogan et al. 2006). Thus, Cys69 is a critical target for the inhibitory effects of ATM on transformed growth. Since Cys69 is unique to the PB1 domain of aPKCs, it was predicted that ATM selectively inhibits PB1–PB1 domain interactions involving PKCι, but not other PB1–PB1 domain interactions. Consistent with this prediction, ATM inhibits the interaction between PKCι and the two scaffolding proteins, Par6 and p62 but has no appreciable inhibitory effects on p62–p62, p62-NBR1, or MEK5-MEKK3 PB1–PB1 domain interactions in vitro (Erdogan et al. 2006). Thus, ATM selectively inhibits PKCι-Par6 interactions in vitro and in vivo and blocks NSCLC cell transformation by targeting Cys69 within the PB1 domain of PKCι.

23.9 PKCι Expression Correlates with ATM Sensitivity in Human Lung Cancer Cells

Given the clinical potential of ATM as a therapeutic agent we assessed the inhibitory efficacy of ATM on the transformed growth of cell lines representing the major subtypes of lung cancer including lung adenocarcinoma (LAC), lung squamous cell carcinoma (LSCC), large cell carcinoma (LCC), and small cell lung carcinoma (SCLC) (Regala et al. 2008). ATM potently inhibited anchorage-independent growth in all lines tested with IC50s ranging from ~300 nM – 100 μM. The lung cancer cell lines clustered into two categories based on ATM sensitivity; those that are highly sensitive to ATM (IC50 < 5 μM) and those that are relatively insensitive to ATM (IC50 > 40 μM). Interestingly, ATM sensitivity among this group of cell lines did not correlate with tumor sub-type, K-ras status, or sensitivity to a panel of standard chemotherapeutic agents frequently used to treat lung cancer patients, including cis-platin, placitaxel, and gemcitabine (Regala et al. 2008). Rather, PKCι expression was the major molecular characteristic of lung cancer cells that correlates with ATM responsiveness (Regala et al. 2008). Specifically, ATM-sensitive lung cancer cell lines express significantly higher PKCι mRNA and protein than ATM-insensitive cell lines. Interestingly, ATM sensitivity also correlated positively with expression of Par6. In contrast, there was no correlation between ATM sensitivity and expression of p62 or the proposed molecular targets of ATM in rheumatoid arthritis (RA), thioredoxin reductase 1 or 2 (TrxR1 and TrxR2) (Regala et al. 2008). PKCι expression profiling revealed that a significant subset of primary NSCLC tumors express PKCι at or above the level associated with ATM sensitivity *in vitro* (Regala et al. 2008). Consistent with our in vitro observations, ATM inhibits tumorigenicity of both sensitive and insensitive lung cell tumors in vivo at plasma drug concentrations consistent with the IC50 of the cell lines to ATM in vitro. Furthermore, measurements of plasma drug concentrations demonstrated that both sensitive as well as insensitive cell lines exhibit an antitumor response to ATM at plasma levels routinely achieved in RA patients undergoing ATM therapy (Regala et al. 2008). Biochemical analysis demonstrated that ATM exhibits its antitumor effects in vivo through direct inhibition of PKCι-mediated activation of the Mek/Erk proliferative signaling axis (Regala et al. 2008). Thus, ATM exhibits antitumor activity against major lung cancer subtypes, particularly tumor cells that express high levels of PKCι. Therefore, PKCι expression profiling in lung tumor samples may be useful in identifying lung cancer patients most likely to respond to ATM therapy. A phase I clinical trial is currently accruing at Mayo Clinic to determine an appropriate dosing regimen for ATM, and to assess antitumor activity.

23.10 Conclusions

The aPKCs (PKCι and PKCζ) have been implicated in several aspects of transformation, and the roles of the aPKCs in human cancer appear to be nonredundant. In most of the carcinogenesis models investigated, PKCζ either plays no demonstrable role or inhibits transformation; PKCι on the other hand, promotes transformed growth, invasion, resistance to chemotherapeutics, and tumor cell survival in a growing number of tumor types. PKCι is the first PKC isozyme to be identified as a human oncogene. Specifically, PKCι is frequently overexpressed and is a target of tumor-specific gene amplification in multiple human tumor types. PKCι overexpression is prognostic of poor clinical outcome in several human cancers and also shows promise as a means of identifying patients with early stage lung cancer at elevated risk of relapse. The PB1-PB1 domain interaction formed between PKCι and Par6 is important for oncogenic PKCι signaling and has been successfully targeted using ATM, a novel mechanism-based therapy that disrupts this interaction. ATM is actively being evaluated in the clinic for the treatment of NSCLC and other tumor types. *PRKCI* amplification and PKCι overexpression is a frequent event in squamous carcinomas of the head and neck (Snaddon et al. 2001), esophagus (Imoto et al. 2001; Pimkhaokham et al. 2000), ovary (Eder et al. 2005; Weichert et al. 2003; Zhang et al. 2006), and cervix (Sugita et al. 2000). These findings suggest that therapies designed to target PKCι signaling in NSCLC may also be effective therapeutic approaches in these tumors. ATM holds promise as a novel, mechanism-based agent for the effective treatment of multiple human cancers.

Acknowledgments The authors wish to thank the members of the Fields laboratory for their key contributions to the data described in this chapter. The authors also wish to apologize to any of our colleagues whose important contributions to this area have been inadvertently omitted in our citations. Though we attempted to cite as much relevant literature as possible, space limitations made comprehensive citation impossible. Work from the Fields laboratory discussed in this article was supported by grants to A.P.F. from the National Institutes of Health, the American Lung Association, The V Foundation for Cancer Research, The James and Esther King Biomedical Research Program, and the Mayo Foundation.

References

Addeo, R., Crisci, S., D'Angelo, V., Vincenzi, B., Casale, F., Pettinato, G., et al. (2007). Bax mutation and overexpression inversely correlate with immature phenotype and prognosis of childhood germ cell tumors. *Oncology Reports, 17*, 1155–61.

Adjei, A. A. (2001). Blocking oncogenic Ras signaling for cancer therapy. *Journal of the National Cancer Institute, 93*, 1062–74.

Akimoto, K., Mizuno, K., Osada, S., Hirai, S., Tanuma, S., Suzuki, K., et al. (1994). A new member of the third class in the protein kinase C family, PKC lambda, expressed dominantly in an undifferentiated mouse embryonal carcinoma cell line and also in many tissues and cells. *Journal of Biological Chemistry, 269*, 12677–83.

Aranda, V., Haire, T., Nolan, M. E., Calarco, J. P., Rosenberg, A. Z., Fawcett, J. P., et al. (2006). Par6-aPKC uncouples ErbB2 induced disruption of polarized epithelial organization from proliferation control. *Nature Cell Biology, 8*, 1235–45.

Baldwin, R. M., Garratt-Lalonde, M., Parolin, D. A., Krzyzanowski, P. M., Andrade, M. A., & Lorimer, I. A. (2006). Protection of glioblastoma cells from cisplatin cytotoxicity via protein kinase Ciota-mediated attenuation of p38 MAP kinase signaling. *Oncogene, 25*, 2909–19.

Balsara, B. R., Sonoda, G., du Manoir, S., Siegfried, J. M., Gabrielson, E., & Testa, J. R. (1997). Comparative genomic hybridization analysis detects frequent, often high-level, overrepresentation of DNA sequences at 3q, 5p, 7p, and 8q in human non-small cell lung carcinomas. *Cancer Research, 57*, 2116–20.

Bandyopadhyay, G., Standaert, M. L., Sajan, M. P., Kanoh, Y., Miura, A., Braun, U., et al. (2004). Protein kinase C-lambda knockout in embryonic stem cells and adipocytes impairs insulin-stimulated glucose transport. *Molecular Endocrinology, 18*, 373–83.

Bareggi, R., Grill, V., Zweyer, M., Narducci, P., & Martelli, A. M. (1995). Distribution of the extended family of protein kinase C isoenzymes in fetal organs of mice: an immunohistochemical study. *Cell Tissue Research, 280*, 617–25.

Bedi, A., Barber, J. P., Bedi, G. C., el-Deiry, W. S., Sidransky, D., Vala, M. S., et al. (1995). BCR-ABL-mediated inhibition of apoptosis with delay of G2/M transition after DNA damage: a mechanism of resistance to multiple anticancer agents. *Blood, 86*, 1148–58.

Bilder, D. (2004). Epithelial polarity and proliferation control: links from the Drosophila neoplastic tumor suppressors. *Genes & development, 18*, 1909–25.

Bjorkoy, G., Perander, M., Overvatn, A., & Johansen, T. (1997). Reversion of Ras- and phosphatidylcholine-hydrolyzing phospholipase C-mediated transformation of NIH 3T3 cells by a dominant interfering mutant of protein kinase C lambda is accompanied by the loss of constitutive nuclear mitogen-activated protein kinase/extracellular signal-regulated kinase activity. *Journal of Biological Chemistry, 272*, 11557–65.

Brass, N., Racz, A., Heckel, D., Remberger, K., Sybrecht, G. W., & Meese, E. U. (1997). Amplification of the genes BCHE and SLC2A2 in 40% of squamous cell carcinoma of the lung. *Cancer Research, 57*, 2290–4.

Bratt, J., Belcher, J., Vercellotti, G. M., & Palmblad, J. (2000). Effects of anti-rheumatic gold salts on NF-kappa B mobilization and tumour necrosis factor-alpha (TNF-alpha)-induced neutrophil-dependent cytotoxicity for human endothelial cells. *Clinical and Experimental Immunology, 120*, 79–84.

Castagna, M., Takai, Y., Kaibuchi, K., Sano, K., Kikkawa, U., & Nishizuka, Y. (1982). Direct activation of calcium-activated, phospholipid-dependent protein kinase by tumor-promoting phorbol esters. *Journal of Biological Chemistry, 257*, 7847–51.

Chou, M. M., Hou, W., Johnson, J., Graham, L. K., Lee, M. H., Chen, C. S., et al. (1998). Regulation of protein kinase C zeta by PI 3-kinase and PDK-1. *Current Biology, 8*, 1069–77.

Cochand-Priollet, B., Raison, D., Molinie, V., Guillausseau, P. J., Wassef, M., & Bouchaud, C. (1998). Altered gap and tight junctions in human thyroid oncocytic tumors: a study of 8 cases by freeze-fracture. *Ultrastructural Pathology, 22*, 413–20.

Cornford, P., Evans, J., Dodson, A., Parsons, K., Woolfenden, A., Neoptolemos, J., et al. (1999). Protein kinase C isoenzyme patterns characteristically modulated in early prostate cancer. *American Journal of Pathology, 154*, 137–144.

Cuadros, R., Montejo de Garcini, E., Wandosell, F., Faircloth, G., Fernandez-Sousa, J. M., & Avila, J. (2000). The marine compound spisulosine, an inhibitor of cell proliferation, promotes the disassembly of actin stress fibers. *Cancer Letters, 152*, 23–9.

Curto, M., Cole, B. K., Lallemand, D., Liu, C. H., & McClatchey, A. I. (2007). Contact-dependent inhibition of EGFR signaling by Nf2/Merlin. *Journal of Cell Biology, 177*, 893–903.

de Vente, J., Kiley, S., Garris, T., Bryant, W., Hooker, J., Posekany, K., et al. (1995). Phorbol ester treatment of U937 cells with altered protein kinase C content and distribution induces cell death rather than differentiation. *Cell Growth & Differentiation, 6*, 371–82.

Diaz-Meco, M. T., Lozano, J., Municio, M. M., Berra, E., Frutos, S., Sanz, L., et al. (1994). Evidence for the in vitro and in vivo interaction of Ras with protein kinase C zeta. *Journal of Biological Chemistry, 269*, 31706–10.

Diaz-Meco, M. T., & Moscat, J. (2001). MEK5, a new target of the atypical protein kinase C isoforms in mitogenic signaling. *Molecular and Cellular Biology, 21*, 1218–27.

Disatnik, M. H., Jones, S. N., & Mochly-Rosen, D. (1995). Stimulus-dependent subcellular localization of activated protein kinase C; a study with acidic fibroblast growth factor and transforming growth factor-beta 1 in cardiac myocytes. *Journal of Molecular and Cellular Cardiology, 27*, 2473–81.

Dominguez, I., Diaz-Meco, M. T., Municio, M. M., Berra, E., Garcia de Herreros, A., Cornet, M. E., et al. (1992). Evidence for a role of protein kinase C zeta subspecies in maturation of *Xenopus laevis* oocytes. *Molecular and Cellular Biology, 12*, 3776–83.

Dominguez, I., Sanz, L., Arenzana-Seisdedos, F., Diaz-Meco, M. T., Virelizier, J. L., & Moscat, J. (1993). Inhibition of protein kinase C zeta subspecies blocks the activation of an NF-kappa B-like activity in *Xenopus laevis* oocytes. *Molecular and Cellular Biology, 13*, 1290–5.

Dong, L. Q., Zhang, R. B., Langlais, P., He, H., Clark, M., Zhu, L., et al. (1999). Primary structure, tissue distribution, and expression of mouse phosphoinositide-dependent protein kinase-1, a protein kinase that phosphorylates and activates protein kinase Czeta. *The Journal of Biological Chemistry, 274*, 8117–22.

Duran, A., Rodriguez, A., Martin, P., Serrano, M., Flores, J. M., Leitges, M., et al. (2004). Crosstalk between PKCzeta and the IL4/Stat6 pathway during T-cell-mediated hepatitis. *EMBO Journal, 23*, 4595–605.

Dutil, E. M., Toker, A., & Newton, A. C. (1998). Regulation of conventional protein kinase C isozymes by phosphoinositide-dependent kinase 1 (PDK-1). *Current Biology, 8*, 1366–75.

Eder, A. M., Sui, X., Rosen, D. G., Nolden, L. K., Cheng, K. W., Lahad, J. P., et al. (2005). Atypical PKCiota contributes to poor prognosis through loss of apical-basal polarity and cyclin E overexpression in ovarian cancer. *Proceedings of the National Academy of Sciences of the United States of America, 102*, 12519–24.

Erdogan, E., Lamark, T., Stallings-Mann, M., Lee, J., Pellecchia, M., Thompson, E. A., et al. (2006). Aurothiomalate inhibits transformed growth by targeting the PB1 domain of protein kinase Ciota. *The Journal of Biological Chemistry, 281*, 28450–9.

Erdogan, E., Klee, E. W., Thompson, E. A., & Fields, A. P. (2009). Meta-analysis of oncogenic protein kinase ciota signaling in lung adenocarcinoma. *Clinical Cancer Research, 15*(5), 1527–33.

Esteve, P. O., Chicoine, E., Robledo, O., Aoudjit, F., Descoteaux, A., Potworowski, E. F., et al. (2002). Protein kinase C-zeta regulates transcription of the matrix metalloproteinase-9 gene induced by IL-1 and TNF-alpha in glioma cells via NF-kappa B. *The Journal of Biological Chemistry, 277*, 35150–5.

Evans, J. D., Cornford, P. A., Dodson, A., Neoptolemos, J. P., & Foster, C. S. (2003). Expression patterns of protein kinase C isoenzymes are characteristically modulated in chronic pancreatitis and pancreatic cancer. *American Journal of Clinical Pathology, 119*, 392–402.

Fields, A. P., & Gustafson, W. C. (2003). Protein kinase C in disease: cancer. *Methods in Molecular Biology, 233*, 519–37.

Frederick, L. A., Matthews, J. A., Jamieson, L., Justilien, V., Thompson, E. A., Radisky, D. C., et al. (2008). Matrix metalloproteinase-10 is a critical effector of protein kinase Ciota-Par6alpha-mediated lung cancer. *Oncogene, 27*, 4841–53.

Galvez, A. S., Duran, A., Linares, J. F., Pathrose, P., Castilla, E. A., Abu-Baker, S., et al. (2009). PKC{zeta} represses the IL-6 promoter and impairs tumorigenesis in vivo. *Molecular and Cellular Biology, 29*, 104–115.

Gokmen-Polar, Y., Murray, N. R., Velasco, M. A., Gatalica, Z., & Fields, A. P. (2001). Elevated protein kinase C betaII is an early promotive event in colon carcinogenesis. *Cancer Research, 61*, 1375–81.

Griner, E. M., & Kazanietz, M. G. (2007). Protein kinase C and other diacylglycerol effectors in cancer. *Nature Reviews Cancer, 7*, 281–94.

Grunicke, H. H., Spitaler, M., Mwanjewe, J., Schwaiger, W., Jenny, M., & Ueberall, F. (2003). Regulation of cell survival by atypical protein kinase C isozymes. *Advances in Enzyme Regulation, 43*, 213–28.

Guo, W., Wu, S., Liu, J., & Fang, B. (2008). Identification of a small molecule with synthetic lethality for K-ras and protein kinase C iota. *Cancer Research, 68*, 7403–8.

Gustafson, W. C., Ray, S., Jamieson, L., Thompson, E. A., Brasier, A. R., & Fields, A. P. (2004). Bcr-Abl regulates protein kinase Ciota (PKCiota) transcription via an Elk1 site in the PKCiota promoter. *The Journal of Biological Chemistry, 279*, 9400–8.

Han, Y., Wei, F., Xu, X., Cai, Y., Chen, B., Wang, J., et al. (2002). Establishment and comparative genomic hybridization analysis of human esophageal carcinomas cell line EC9706. *Zhonghua Yi Xue Yi Chuan Xue Za Zhi, 19*, 455–7.

Hellbert, K., Kampfer, S., Maly, K., Hochholdinger, F., Mwanjewe, J., Baier, G., et al. (2000). Implication of atypical protein kinase C isozymes lambda and zeta in Ras mediated reorganization of the actin cytoskeleton and cyclin D1-induction. *Advances in Enzyme Regulation, 40*, 49–62.

Hirano, Y., Yoshinaga, S., Ogura, K., Yokochi, M., Noda, Y., Sumimoto, H., et al. (2004). Solution structure of atypical protein kinase C PB1 domain and its mode of interaction with ZIP/p62 and MEK5. *The Journal of Biological Chemistry, 279*, 31883–90.

Huang, H. C., Huang, C. Y., Lin-Shiau, S. Y., & Lin, J. K. (2008). Ursolic acid inhibits IL-1beta or TNF-alpha-induced C6 glioma invasion through suppressing the association ZIP/p62 with PKC-zeta and downregulating the MMP-9 expression. *Molecular Carcinogenesis, 48*, 517–31.

Hug, H., & Sarre, T. F. (1993). Protein kinase C isoenzymes: divergence in signal transduction? *Biochemical Journal, 291*(Pt 2), 329–43.

Imoto, I., Pimkhaokham, A., Fukuda, Y., Yang, Z. Q., Shimada, Y., Nomura, N., et al. (2001). SNO is a probable target for gene amplification at 3q26 in squamous-cell carcinomas of the esophagus. *Biochemical and Biophysical Research Communications, 286*, 559–65.

Jaken, S., & Parker, P. J. (2000). Protein kinase C binding partners. *Bioessays, 22*, 245–54.

Jamieson, L., Carpenter, L., Biden, T. J., & Fields, A. P. (1999). Protein kinase Ciota activity is necessary for Bcr-Abl-mediated resistance to drug-induced apoptosis. *The Journal of Biological Chemistry, 274*, 3927–30.

Jeon, K. I., Jeong, J. Y., & Jue, D. M. (2000). Thiol-reactive metal compounds inhibit NF-kappa B activation by blocking I kappa B kinase. *Journal of Immunology, 164*, 5981–9.

Jin, Z., Xin, M., & Deng, X. (2005). Survival function of protein kinase C{iota} as a novel nitrosamine 4-(methylnitrosamino)-1-(3-pyridyl)-1-butanone-activated bad kinase. *The Journal of Biological Chemistry, 280*, 16045–52.

Joberty, G., Petersen, C., Gao, L., & Macara, I. G. (2000). The cell-polarity protein Par6 links Par3 and atypical protein kinase C to Cdc42. *Nature Cell Biology, 2*, 531–9.

Kishimoto, A., Takai, Y., Mori, T., Kikkawa, U., & Nishizuka, Y. (1980). Activation of calcium and phospholipid-dependent protein kinase by diacylglycerol, its possible relation to phosphatidylinositol turnover. *The Journal of Biological Chemistry, 255*, 2273–6.

Kleinman, H. K., & Martin, G. R. (2005). Matrigel: basement membrane matrix with biological activity. *Seminars in Cancer Biology, 15*, 378–86.

Kojima, Y., Akimoto, K., Nagashima, Y., Ishiguro, H., Shirai, S., Chishima, T., et al. (2008). The overexpression and altered localization of the atypical protein kinase C lambda/iota in breast cancer correlates with the pathologic type of these tumors. *Human Pathology, 39*, 824–31.

Kondraganti, S., Mohanam, S., Chintala, S. K., Kin, Y., Jasti, S. L., Nirmala, C., et al. (2000). Selective suppression of matrix metalloproteinase-9 in human glioblastoma cells by antisense gene transfer impairs glioblastoma cell invasion. *Cancer Research, 60*, 6851–5.

Kovac, J., Oster, H., & Leitges, M. (2007). Expression of the atypical protein kinase C (aPKC) isoforms iota/lambda and zeta during mouse embryogenesis. *Gene Expression Patterns, 7*, 187–96.

Lamark, T., Perander, M., Outzen, H., Kristiansen, K., Overvatn, A., Michaelsen, E., et al. (2003). Interaction codes within the family of mammalian Phox and Bem1p domain-containing proteins. *The Journal of Biological Chemistry, 278*, 34568–81.

Larsson, C. (2006). Protein kinase C and the regulation of the actin cytoskeleton. *Cellular Signalling, 18*, 276–84.

Laudanna, C., Sorio, C., Tecchio, C., Butcher, E. C., Bonora, A., Bassi, C., et al. (2003). Motility analysis of pancreatic adenocarcinoma cells reveals a role for the atypical zeta isoform of protein kinase C in cancer cell movement. *Laboratory Investigation, 83*, 1155–63.

Le Good, J. A., Ziegler, W. H., Parekh, D. B., Alessi, D. R., Cohen, P., & Parker, P. J. (1998). Protein kinase C isotypes controlled by phosphoinositide 3-kinase through the protein kinase PDK1. *Science, 281*, 2042–5.

Leitges, M., Sanz, L., Martin, P., Duran, A., Braun, U., Garcia, J. F., et al. (2001). Targeted disruption of the zetaPKC gene results in the impairment of the NF-kappaB pathway. *Molecular Cell, 8*, 771–80.

Leseux, L., Laurent, G., Laurent, C., Rigo, M., Blanc, A., Olive, D., et al. (2008). PKC zeta mTOR pathway: a new target for rituximab therapy in follicular lymphoma. *Blood, 111*, 285–91.

Li, Q., Wang, J. M., Liu, C., Xiao, B. L., Lu, J. X., & Zou, S. Q. (2008). Correlation of aPKC-iota and E-cadherin expression with invasion and prognosis of cholangiocarcinoma. *Hepatobiliary & Pancreatic Diseases International, 7*, 70–5.

Lin, D., Edwards, A. S., Fawcett, J. P., Mbamalu, G., Scott, J. D., & Pawson, T. (2000). A mammalian PAR-3-PAR-6 complex implicated in Cdc42/Rac1 and aPKC signalling and cell polarity. *Nature Cell Biology, 2*, 540–7.

Liu, W. S., & Heckman, C. A. (1998). The sevenfold way of PKC regulation. *Cellular Signalling, 10*, 529–42.

Lu, Y., Jamieson, L., Brasier, A. R., & Fields, A. P. (2001). NF-kappaB/RelA transactivation is required for atypical protein kinase C iota-mediated cell survival. *Oncogene, 20*, 4777–92.

Magnuson, B. A., Carr, I., & Bird, R. P. (1993). Ability of aberrant crypt foci characteristics to predict colonic tumor incidence in rats fed cholic acid. *Cancer Research, 53*, 4499–504.

Mao, M., Fang, X., Lu, Y., Lapushin, R., Bast, R. C., Jr., & Mills, G. B. (2000). Inhibition of growth-factor-induced phosphorylation and activation of protein kinase B/Akt by atypical protein kinase C in breast cancer cells. *Biochemical Journal, 352*(Pt 2), 475–82.

McLellan, E. A., Medline, A., & Bird, R. P. (1991). Sequential analyses of the growth and morphological characteristics of aberrant crypt foci: putative preneoplastic lesions. *Cancer Research, 51*, 5270–4.

Messori, L., & Marcon, G. (2004). Gold complexes in the treatment of rheumatoid arthritis. *Metal Ions in Biological Systems, 41*, 279–304.

Mochly-Rosen, D. (1995). Localization of protein kinases by anchoring proteins: a theme in signal transduction. *Science, 268*, 247–51.

Moscat, J., & Diaz-Meco, M. T. (2000). The atypical protein kinase Cs. Functional specificity mediated by specific protein adapters. *EMBO Reports, 1*, 399–403.

Moscat, J., Diaz-Meco, M. T., Albert, A., & Campuzano, S. (2006). Cell signaling and function organized by PB1 domain interactions. *Molecular Cell, 23*, 631–40.

Murray, N. R., Davidson, L. A., Chapkin, R. S., Clay Gustafson, W., Schattenberg, D. G., & Fields, A. P. (1999). Overexpression of protein kinase C betaII induces colonic hyperproliferation and increased sensitivity to colon carcinogenesis. *The Journal of Cell Biology, 145*, 699–711.

Murray, N. R., & Fields, A. P. (1997). Atypical protein kinase C iota protects human leukemia cells against drug-induced apoptosis. *The Journal of Biological Chemistry, 272*, 27521–4.

Murray, N. R., Jamieson, L., Yu, W., Zhang, J., Gokmen-Polar, Y., Sier, D., et al. (2004). Protein kinase Ciota is required for Ras transformation and colon carcinogenesis in vivo. *The Journal of Cell Biology, 164*, 797–802.

Murray, N. R., Weems, J., Braun, U., Leitges, M., & Fields, A. P. (2009). Protein kinase C betaII and PKCiota/lambda: collaborating partners in colon cancer promotion and progression. *Cancer Research, 69*, 656–62.

Murray, N. R., Weems, C., Chen, L., Leon, J., Yu, W., Davidson, L. A., et al. (2002). Protein kinase C betaII and TGFbetaRII in omega-3 fatty acid-mediated inhibition of colon carcinogenesis. *The Journal of Cell Biology, 157*, 915–20.

Mustafi, R., Cerda, S., Chumsangsri, A., Fichera, A., & Bissonnette, M. (2006). Protein Kinase-zeta inhibits collagen I-dependent and anchorage-independent growth and enhances apoptosis of human Caco-2 cells. *Molecular Cancer Research, 4*, 683–94.

Mwanjewe, J., Spitaler, M., Ebner, M., Windegger, M., Geiger, M., Kampfer, S., et al. (2001). Regulation of phospholipase D isoenzymes by transforming Ras and atypical protein kinase C-iota. *Biochemical Journal, 359*, 211–7.

Nakanishi, H., Brewer, K. A., & Exton, J. H. (1993). Activation of the zeta isozyme of protein kinase C by phosphatidylinositol 3, 4, 5-trisphosphate. *The Journal of Biological Chemistry, 268*, 13–6.

Nishizuka, Y. (1995). Protein kinase C and lipid signaling for sustained cellular responses. *FASEB Journal, 9*, 484–96.

Noda, Y., Takeya, R., Ohno, S., Naito, S., Ito, T., & Sumimoto, H. (2001). Human homologues of the Caenorhabditis elegans cell polarity protein PAR6 as an adaptor that links the small GTPases Rac and Cdc42 to atypical protein kinase C. *Genes Cells, 6*, 107–19.

Ono, Y., Fujii, T., Ogita, K., Kikkawa, U., Igarashi, K., & Nishizuka, Y. (1989). Protein kinase C zeta subspecies from rat brain: its structure, expression, and properties. *Proceedings of the National Academy of Sciences of the United States of America, 86*, 3099–103.

Oster, H., & Leitges, M. (2006). Protein kinase C alpha but not PKCzeta suppresses intestinal tumor formation in ApcMin/+ mice. *Cancer Research, 66*, 6955–63.

Patel, R., Win, H., Desai, S., Patel, K., Matthews, J. A., & Acevedo-Duncan, M. (2008). Involvement of PKC-iota in glioma proliferation. *Cell Proliferation, 41*, 122–35.

Pia Rigobello, M., Messori, L., Marcon, G., Agostina Cinellu, M., Bragadin, M., Folda, A., et al. (2004). Gold complexes inhibit mitochondrial thioredoxin reductase: consequences on mitochondrial functions. *Journal of Inorganic Biochemistry, 98*, 1634–41.

Pimkhaokham, A., Shimada, Y., Fukuda, Y., Kurihara, N., Imoto, I., Yang, Z. Q., et al. (2000). Nonrandom chromosomal imbalances in esophageal squamous cell carcinoma cell lines: possible involvement of the ATF3 and CENPF genes in the 1q32 amplicon. *Japanese Journal of Cancer Research, 91*, 1126–33.

Poole, A. W., Pula, G., Hers, I., Crosby, D., & Jones, M. L. (2004). PKC-interacting proteins: from function to pharmacology. *Trends in Pharmacological Sciences, 25*, 528–35.

Powell, C. T., Gschwend, J. E., Fair, W. R., Brittis, N. J., Stec, D., & Huryk, R. (1996). Overexpression of protein kinase C-zeta (PKC-zeta) inhibits invasive and metastatic abilities of Dunning R-3327 MAT-LyLu rat prostate cancer cells. *Cancer Research, 56*, 4137–4141.

Puls, A., Schmidt, S., Grawe, F., & Stabel, S. (1997). Interaction of protein kinase C zeta with ZIP, a novel protein kinase C-binding protein. *Proc Natl Acad Sci U S A., 94*, 6191–6.

Qiu, R. G., Abo, A., & Steven Martin, G. (2000). A human homolog of the *C. elegans* polarity determinant Par-6 links Rac and Cdc42 to PKCzeta signaling and cell transformation. *Current Biology, 10*, 697–707.

Qiu, R. G., Chen, J., Kirn, D., McCormick, F., & Symons, M. (1995). An essential role for Rac in Ras transformation. *Nature, 374*, 457–9.

Rao, J. S., Steck, P. A., Mohanam, S., Stetler-Stevenson, W. G., Liotta, L. A., & Sawaya, R. (1993). Elevated levels of M(r) 92, 000 type IV collagenase in human brain tumors. *Cancer Research, 53*, 2208–11.

Regala, R. P., Thompson, E. A., & Fields, A. P. (2008). Atypical protein kinase C iota expression and aurothiomalate sensitivity in human lung cancer cells. *Cancer Research, 68*, 5888–95.

Regala, R. P., Weems, C., Jamieson, L., Copland, J. A., Thompson, E. A., & Fields, A. P. (2005). Atypical protein kinase Ciota plays a critical role in human lung cancer cell growth and tumorigenicity. *The Journal of Biological Chemistry, 280*, 31109–15.

Regala, R. P., Weems, C., Jamieson, L., Khoor, A., Edell, E. S., Lohse, C. M., et al. (2005). Atypical protein kinase C iota is an oncogene in human non-small cell lung cancer. *Cancer Research, 65*, 8905–11.

Ron, D., & Mochly-Rosen, D. (1995). An autoregulatory region in protein kinase C: the pseudo-anchoring site. *Proceedings of the National Academy of Sciences of the United States of America, 92*, 492–6.

Sanchez, P., De Carcer, G., Sandoval, I. V., Moscat, J., & Diaz-Meco, M. T. (1998). Localization of atypical protein kinase C isoforms into lysosome-targeted endosomes through interaction with p62. *Molecular and Cellular Biology, 18*, 3069–80.

Sanchez, A. M., Malagarie-Cazenave, S., Olea, N., Vara, D., Cuevas, C., & Diaz-Laviada, I. (2008). Spisulosine (ES-285) induces prostate tumor PC-3 and LNCaP cell death by de novo synthesis of ceramide and PKCzeta activation. *European Journal of Pharmacology, 584*, 237–45.

Schuller, H. M. (2002). Mechanisms of smoking-related lung and pancreatic adenocarcinoma development. *Nature Reviews Cancer, 2*, 455–63.

Scotti, M. L., Bamlet, W. R., Smyrk, T. C., Fields, A. P., & Murray, N. R. (2010). Protein Kinase C iota is required for pancreatic cancer cell transformed growth and tumorigenesis. *Cancer Research*. Epub ahead of print.

Slattery, M. L., Anderson, K., Curtin, K., Ma, K., Schaffer, D., Edwards, S., et al. (2001). Lifestyle factors and Ki-ras mutations in colon cancer tumors. *Mutation Research, 483*, 73–81.

Snaddon, J., Parkinson, E. K., Craft, J. A., Bartholomew, C., & Fulton, R. (2001). Detection of functional PTEN lipid phosphatase protein and enzyme activity in squamous cell carcinomas of the head and neck, despite loss of heterozygosity at this locus. *British Journal of Cancer, 84*, 1630–4.

Soloff, R. S., Katayama, C., Lin, M. Y., Feramisco, J. R., & Hedrick, S. M. (2004). Targeted deletion of protein kinase C lambda reveals a distribution of functions between the two atypical protein kinase C isoforms. *Journal of Immunology, 173*, 3250–60.

Stallings-Mann, M., Jamieson, L., Regala, R. P., Weems, C., Murray, N. R., & Fields, A. P. (2006). A novel small-molecule inhibitor of protein kinase Ciota blocks transformed growth of non-small-cell lung cancer cells. *Cancer Research, 66*, 1767–74.

Sugita, M., Tanaka, N., Davidson, S., Sekiya, S., Varella-Garcia, M., West, J., et al. (2000). Molecular definition of a small amplification domain within 3q26 in tumors of cervix, ovary, and lung. *Cancer Genetics and Cytogenetics, 117*, 9–18.

Sun, R., Gao, P., Chen, L., Ma, D., Wang, J., Oppenheim, J. J., et al. (2005). Protein kinase C zeta is required for epidermal growth factor-induced chemotaxis of human breast cancer cells. *Cancer Research, 65*, 1433–41.

Suzuki, A., Yamanaka, T., Hirose, T., Manabe, N., Mizuno, K., Shimizu, M., et al. (2001). Atypical protein kinase C is involved in the evolutionarily conserved par protein complex and plays a critical role in establishing epithelia-specific junctional structures. *The Journal of Cell Biology, 152*, 1183–96.

Takai, Y., Kishimoto, A., Kikkawa, U., Mori, T., & Nishizuka, Y. (1979). Unsaturated diacylglycerol as a possible messenger for the activation of calcium-activated, phospholipid-dependent protein kinase system. *Biochemical and Biophysical Research Communications, 91*, 1218–24.

Takai, Y., Yamamoto, M., Inoue, M., Kishimoto, A., & Nishizuka, Y. (1977). A proenzyme of cyclic nucleotide-independent protein kinase and its activation by calcium-dependent neutral protease from rat liver. *Biochemical and Biophysical Research Communications, 77*, 542–50.

Takayama, T., Katsuki, S., Takahashi, Y., Ohi, M., Nojiri, S., Sakamaki, S., et al. (1998). Aberrant crypt foci of the colon as precursors of adenoma and cancer. *The New England Journal of Medicine, 339*, 1277–84.

Takayama, T., Ohi, M., Hayashi, T., Miyanishi, K., Nobuoka, A., Nakajima, T., et al. (2001). Analysis of K-ras, APC, and beta-catenin in aberrant crypt foci in sporadic adenoma, cancer, and familial adenomatous polyposis. *Gastroenterology, 121*, 599–611.

Thiery, J. P. (2002). Epithelial-mesenchymal transitions in tumour progression. *Nature Reviews Cancer, 2*, 442–54.

Ways, D. K., Posekany, K., deVente, J., Garris, T., Chen, J., Hooker, J., et al. (1994). Overexpression of protein kinase C-zeta stimulates leukemic cell differentiation. *Cell Growth and Differentiation, 5*, 1195–203.

Weichert, W., Gekeler, V., Denkert, C., Dietel, M., & Hauptmann, S. (2003). Protein kinase C isoform expression in ovarian carcinoma correlates with indicators of poor prognosis. *International Journal of Oncology, 23*, 633–9.

Wetsel, W. C., Khan, W. A., Merchenthaler, I., Rivera, H., Halpern, A. E., Phung, H. M., et al. (1992). Tissue and cellular distribution of the extended family of protein kinase C isoenzymes. *The Journal of Cell Biology, 117*, 121–33.

Win, H. Y., & Acevedo-Duncan, M. (2008). Atypical protein kinase C phosphorylates IKKalphabeta in transformed non-malignant and malignant prostate cell survival. *Cancer Letters, 270*, 302–11.

Wooten, M. W., Vandenplas, M. L., Seibenhener, M. L., Geetha, T., & Diaz-Meco, M. T. (2001). Nerve growth factor stimulates multisite tyrosine phosphorylation and activation of the atypical protein kinase C's via a src kinase pathway. *Molecular and Cellular Biology, 21*, 8414–27.

Xu, L., & Deng, X. (2006). Protein kinase Ciota promotes nicotine-induced migration and invasion of cancer cells via phosphorylation of micro- and m-calpains. *The Journal of Biological Chemistry, 281*, 4457–66.

Yamashita, M., Ashino, S., Oshima, Y., Kawamura, S., Ohuchi, K., & Takayanagi, M. (2003). Inhibition of TPA-induced NF-kappaB nuclear translocation and production of NO and PGE2 by the anti-rheumatic gold compounds. *The Journal of Pharmacy and Pharmacology, 55*, 245–51.

Yang, Y. L., Chu, J. Y., Luo, M. L., Wu, Y. P., Zhang, Y., Feng, Y. B., et al. (2008). Amplification of PRKCI, located in 3q26, is associated with lymph node metastasis in esophageal squamous cell carcinoma. *Genes Chromosomes Cancer, 47*, 127–36.

Zhang, J., Anastasiadis, P. Z., Liu, Y., Thompson, E. A., & Fields, A. P. (2004). Protein kinase C (PKC) betaII induces cell invasion through a Ras/Mek-, PKC iota/Rac 1-dependent signaling pathway. *The Journal of Biological Chemistry, 279*, 22118–23.

Zhang, L., Huang, J., Yang, N., Liang, S., Barchetti, A., Giannakakis, A., et al. (2006). Integrative genomic analysis of protein kinase C (PKC) family identifies PKCiota as a biomarker and potential oncogene in ovarian carcinoma. *Cancer Research, 66*, 4627–35.

Index

F

G

H

I

K

L

R

S

MIX
Papier aus verantwortungsvollen Quellen
Paper from responsible sources
FSC® C105338

FSC
www.fsc.org

If you have any concerns about our products,
you can contact us on
ProductSafety@springernature.com

In case Publisher is established outside the EU,
the EU authorized representative is:
Springer Nature Customer Service Center GmbH
Europaplatz 3, 69115 Heidelberg, Germany

Printed by Libri Plureos GmbH
in Hamburg, Germany